CONTENTS

CRUSTAL MOTIONS

THE GLOBAL VIEW

ORE DEPOSITS

D.W. Strangway, J.T. Wilson and Sir Edward Bullard
on the occasion of the Wilson Symposium.

PREFACE

In May of 1979 a group of friends and colleagues gathered to honour J. Tuzo Wilson, a long time member of the faculty at the University of Toronto and now the Director of the Ontario Science Centre. The papers in the symposium on "The Continental Crust and its Mineral Deposits" were presented under the following groupings: i) The Early Earth; ii) Evolution of the Precambrian Crust; iii) Vertical Geometry of the Crust; iv) Crustal Motions; v) The Global View; vi) Ore Deposits. From a study of the Table of Contents, one can see that the conference covered a wide range of topics on the current view of the Earth and its processes ranging from its formation to the movements of fluids leading to the concentration of useful mineral deposits.

It is particularly fitting that this conference should have taken place in Toronto with this broad range of topics, since Tuzo's views of the Earth have had a profound influence on models of terrestrial processes and since there is a great deal of mineral exploration work based in Toronto. Tuzo's early work involved field mapping in the Canadian shield and in Montana. He was instrumental in compiling the first glacial map of Canada and was among the first to recognize how isotopic dating could be used to divide the shield into various age provinces. He used this as the base for his models of stable, fixed continental nucleii. He compiled information on island arcs and championed the contracting Earth theories. This was converted later to support of the expanding Earth hypothesis, but eventually after a thorough synthesis of all known information on ocean islands, he became one of the staunch supporters of the new plate tectonics. He predicted transform faults and championed the concept of hot spots.

There is no doubt that Tuzo's thinking on Earth processes has contributed in a major way to our present models of the Earth. This conference was about this topic and how these models have affected geologic thinking. Each speaker was asked to give his view of current thinking in his sector of the discipline. They were not asked to give a history. In a sense then, the conference was really about the second generation of the plate tectonics revolution. Forty-three papers were presented at the conference with adequate time for discussion. There were no multiple sessions, so that workers from all of the Earth Science disciplines, listened to papers from other disciplines and participated in animated discussion. We have been able to publish in this volume thirty-nine of these papers. This volume represents a snapshot of the 1979 view of the Earth and we hope it will stand as a useful reference work.

Over 300 people attended the conference representing industry, government and university and they came from Canada, the United States, Britain, India, Belgium, Venezuela, Japan and Australia. During the conference, a banquet was held in Tuzo's honour. The featured speaker was Sir Edward Bullard. It was particularly appropriate that Sir Edward was able to come to this conference. Sir Edward had been head of the Physics Department at Toronto when Tuzo was a struggling young professor. He had been at Cambridge when the transform fault theory was generated and sea floor lineations were recognized. Sir Edward himself was a pioneer, who had been responsible for much of the data on which Tuzo's models were built. In fact, one even supposes that Sir Edward may have helped in the conversion of Wilson, when he jumped on the moving continents. Sir Edward passed on the Albatross award to Tuzo at this meeting. His "unusual contribution to oceanography" was "making the faults run backwards". It is therefore with much regret that we record that early in 1980, Sir Edward passed away after a struggle with cancer.

D.W. Strangway
June 1980

THE EARLY EARTH

The Continental Crust and Its Mineral Deposits, edited by D.W. Strangway,
Geological Association of Canada Special Paper 20

NUMERICAL CALCULATIONS
RELEVANT TO THE
ACCUMULATION OF THE TERRESTRIAL PLANETS

George W. Wetherill
Department of Terrestrial Magnetism,
Carnegie Institution of Washington,
Washington, D.C. 20015

ABSTRACT

A well-defined problem relevant to planetary accumulation is the velocity distribution of a swarm of bodies in heliocentric orbits undergoing mutual collisions and gravitational perturbations. The extent to which the gravitational perturbations counteract the tendency of collisions to circularize the orbits determines whether the swarm will accumulate into four terrestrial planets, or into a larger number of smaller bodies in concentric orbits.

The most extensive treatment of this problem has been given by Safronov. In this analytical theory Keplerian motion is not explicitly introduced, but its effects are calculated by use of an analogy with the dynamics of a rotating fluid. In order to test the validity of this analogy, numerical calculations have been carried out for a problem which can also be treated by Safronov's approach: the evolution of the velocity distribution of a swarm of non-accreting bodies of equal mass. It is found that steady-state relative velocities are achieved which vary with the mass of the bodies in the manner predicted by Safronov, i.e., the mean relative velocity is always comparable to the escape velocity. In addition to the increase in eccentricity and inclination associated with the velocity increase, a radial "diffusion" in semi-major axes accompanies it as well. The velocity distribution is somewhat non-maxwellian.

When accretion is permitted, four or five large planets result, independent of the initial eccentricity. In earlier two-dimensional calculations, Cox found that a small number of large planets were formed only when the initial eccentricity was as high as 0.15. In the present work, Cox's result is reproduced for the two-dimensional case. Thus, although it is possible that the difference between the present and the earlier work arises from the approximations made here, it is more likely that Cox's result is peculiar to the two-dimensional case. Further work is required to learn the conditions under which a more realistic swarm of bodies with varying mass and hence greater collisional damping can evolve into a system resembling that of the present planets.

RÉSUMÉ

La répartition des vitesses d'un ensemble de corps célestes en orbite héliocentrique, subissant des collisions et des perturbations gravitationnelles mutuelles est un problème classique relatif à l'accumulation planétaire. L'accumulation de cet ensemble en quatre planètes terrestres ou en un grand nombre de plus petits corps en orbites concentriques est determinée par la capacité des perturbations gravitationnelles à contrebalancer la circularisation des orbites sous l'effet des collisions.

Le traitement le plus complet de ce probleme a été donné par Safronov. Dans son analyse théorique le mouvement Keplerien n'est pas introduit explicitement mais ses effets sont calculés par analogie avec la dynamique d'un fluide en rotation. Afin d'évaluer la validité de cette analogie on effectue numériquement la résolution d'un problème qui peut être abordé par la methode de Safronov: évolution de la distribution des vitesses dans un ensemble de corps célestes de même masse ne subissant pas l'accrétion. On trouve que les vitesses relatives, à l'état stationnaire, varient avec la masse des corps selon les prédictions de Safronov, c'est à dire que la vitesse relative moyenne est toujours du même ordre que la vitesse de libération. En plus de l'accroissement de l'excentricité et de l'inclinaison associées à l'augmentation de la vitesse, une "diffusion" radiale des demis grands axes intervient aussi. La distribution des vitesses n'est pas tout à fait maxwellienne.

Lorsqu'on introduit la possibilité d'accrétion, il en résulte quatre on cinq grandes planètes independamment de l'excentricité initiale. Précédemment sur la base de calculs bidimensionnels, Cox a trouvé qu'un petit nombre de planètes se forme seulement si l'excentricité initiale est au moins 0,15. Dans le présent travail nous avons pu reproduire le résultat de Cox pour un modèle bidimensionnel. En conséquence, bien que la divergence entre ces deux travaux puisse provenir des approximations que nous avons faites, il est plus probable que le resultat de Cox reflète une particularité du modèle à deux dimensions. D'autres études sont nécessaires pour savoir dans quelles conditions, le cas plus réaliste d'un groupe de corps de masses variées, ayant donc un amortissement des collisions plus important, peut evoluer en un système ressemblant à celui des planètes actuelles.

INTRODUCTION

In the past several years there has been increasing interest in "comparative planetology" – the relationships between the larger and smaller bodies of the solar system and what they have to teach us about one another. The initial boundary conditions for these bodies, in other words, knowledge of the origin of the solar system, are now coming to occupy a more central place in the thinking of the working scientist. In the past such theories were principally used to satisfy our need for a creation myth and to serve as introductory material for textbooks with no substantial consequences to the further development of the subject. Unfortunately, we are far from an adequate understanding of these initial conditions, and there is much work which must be done if this situation is to improve.

For the most part theories of the origin of the solar system have attempted to describe a single mechanism responsible for the formation of at least the major and terrestrial planets. Once past the initial stages, which are properly part of the problem of star formation, these mechanisms fall into two general categories: 1) those in which massive gravitational instabilities lead to the formation of large gaseous protoplanets (e.g., Kuiper, 1951; McCrea, 1960; Cameron, 1978), and 2) those in which gravitational instability plays only a minor role, if any, and the planets grow by the continuing sweeping up of smaller bodies by larger ones (e.g., Chamberlin, 1904; Safronov, 1969; Weidenschilling, 1974, 1976; Hayashi *et al.*, 1977; Greenberg *et al.*, 1978; Cox, 1978). Qualitative "scenarios" for the formation of all of the planets by one or the other of these mechanisms have been described. The problem with such scenarios is that experience shows that only a small fraction of one's qualitatively appealing ideas survive more detailed quantitative discussion. Progress toward the goal of understanding how the solar system formed will therefore require "rolling up our sleeves" and making detailed quantitative studies of each of the stages of planetary growth described in less quantitative theories. It seems quite possible that this will show that fundamentally different processes were involved in the formation of the major and terrestrial planets.

An insufficient quantity of detailed work of this kind exists at present. The few detailed studies have concentrated on a quantitative treatment of the gravitational accumulation of planets by mutual collisions of planetesimals in the 1 to 1000 km range, principally assuming gas-free accumulation. Gas-free accumulation is not an attractive way for major planets to have formed. However, the volatile-poor composition of the terrestrial planets together with the difficulty of obtaining planet-sized gravitational instabilities at small heliocentric distances have led a number of workers to explore more thoroughly the possibility that at least these planets formed by the "sweeping-up" process under conditions in which gas drag was not important.

Major contributions to the solution of this limited problem have been made by Safronov and his co-workers (Safronov, 1969; Zvyagina *et al.*, 1973; Safronov, 1978). These investigators applied the stellar dynamical techniques of Chandrasekhar (1942) to the problem of the dynamical evolution of a growing swarm of planetesimals: relative velocities increase by mutual gravitational perturbations, and decrease by dissipative collisions and the averaging of velocities accompanying the cohesive collisions responsible for planetary growth. Subject to certain assumptions, these workers calculated that such a swarm can achieve quasi-steady states for both the velocity and the size distribution of the planetesimals.

The result for the velocity distribution is particularly significant. Most other workers have introduced the relative velocity of the planetesimals as a free parameter. If the relative velocity is too high, e.g., several times the gravitational escape velocity of the growing planet, fragmentation rather than accumulation will be dominant, and planets won't grow. On the other hand, if relative velocities are too low, the system of planetesimals will be in nearly circular concentric orbits, and the collisions required for growth will not take place. Building upon the earlier work of Gurevich and Lebedinskii (1950), Safronov (1962) showed that it was not necessary to avoid this problem by the ad hoc introduction of a suitable relative velocity. Rather, he showed that for plausible assumptions regarding dissipation of energy in collisions, and size distribution of the bodies, the mutual gravitational perturbations of the bodies caused their mean relative velocity to be only somewhat less than the escape velocity of the larger bodies. Thus throughout the entire course of planetary growth, from 1-km planetesimals to Earth-sized planets, the system regulated itself in such a way that the larger bodies could always grow, whereas smaller objects would fragment, establishing a spectrum of sizes in the smaller planetesimal mass range.

Despite this success, as well as other important achievements, this theoretical work has its limitations, and even its major conclusions have been questioned (Greenberg *et al.*, 1978; Levin, 1978). One limitation is the difficulty of treating analytically the coupled evolution of the mass and velocity distribution. While the existence of consistent and acceptable approximate steady-state solutions for both of these quantities can be demonstrated, it is difficult to show that the system will actually evolve into these steady-states. For example, Greenberg *et al.* (1978) present numerical simulations of the earliest stages of this growth which suggest that the early formation of a few larger (~1000 km diameter) bodies while most of the mass of the system remains in about 1 km bodies may preclude the development of the higher velocities inferred by Safronov, which may be necessary if planetary growth is to proceed (e.g., discussed in Wetherill, 1978).

Another possible limitation results from the fact that the dynamical theory of Chandrasekhar, upon which Safronov's work is based, does not explicitly consider the planetesimals to be constrained to move in heliocentric Keplerian orbits. Instead, the swarm is treated as analogous to molecules in the kinetic theory of gases. Safronov introduced the effects of Keplerian motion by an ingenious analogy with Prandtl's semi-empirical theory of turbulence in a rotating viscous fluid. In this analogy, the shearing forces resulting from the differential circular velocities of bodies at different heliocentric distances give rise to a partition between turbulent random motion and organized circular motion. More recently Kaula (1980) has used Chandrasekhar's work in a somewhat different way, but still within the "kinetic theory of gases" framework.

In the present work numerical calculations are reported which explicitly involve Keplerian motion. They address two problems relevant to terrestrial planet accumulation:

1) *The orbital evolution in a gas-free medium of a swarm of equal-sized non-accreting bodies that are subject to mutual gravitational perturbations and collisional damping.* This calculation, although far from being a simulation of real terrestrial planet formation, includes the essential physical basis of the Safronov steady-state velocity distribution, and permits a comparison of analytical theory with numerical calculations.

2) *The three-dimensional accumulation of a swarm of large bodies, taking into account mutual gravitational perturbations, collisional damping and cohesion (accretion).* This is a three-dimensional version of a similar problem investigated in two dimensions by Cox (1978). Cox's work, involving more rigorous dynamical procedures than those of the present work, led to the result that unless initial velocities were implausibly large, the swarm would evolve into a system of 8 to 10 small terrestrial planets, rather than into the observed two large bodies and two smaller bodies. In the present work, the two-dimensional results of Cox are reproduced and extended to the three-dimensional case.

STEADY-STATE VELOCITY DISTRIBUTION
OF A NON-ACCRETING SWARM

One hundred bodies of equal mass are assumed to have their initial semi-major axes distributed at random over the narrow range 0.96 to 1.04 A.U., with random initial eccentricities less than 0.01 and initial inclinations less than 0.01 radians. Because of their non-zero eccentricities, many of these bodies will be in crossing orbits, permitting both collisions and gravitational perturbations by close approaches. The "Öpik (1951) collision formula is used to calculate the probability of each pair undergoing collision or close encounter during successive time steps. Use of the "Öpik collision formula introduces the assumption that the arguments of perihelion, longitude of nodes, and position of the body in its orbit are random. The pairs which actually encounter one another during each time step are chosen at random in accordance with these Öpik probabilities. The procedure of Arnold (1965) for calculating orbital changes following a close encounter is used as modified previously. In addition to some earlier modifications (Wetherill and Williams, 1968), further changes are made to treat the encounter as a two-body problem, rather than assuming one of the bodies to be much more massive than the other. The calculation also differs in that the Tisserand (1896) definition of the sphere of influence is used ($R_s \propto m^{2/5}$) rather than $R_s \propto m^{1/3}$, which was used by Öpik (1951) and Arnold (1965). In order to reduce computing time, gravitational perturbations are usually calculated only for encounters within four gravitational radii; the inclusion of longer-range encounters out to the edge of the sphere of influence is accomplished by the statistical procedure used in Arnold's (1965) equation 11. As a check, a few calculations include perturbations out to ten gravitational radii, and the difference in the results is not significant.

Encounters within one gravitational radius constitute collisions. In this case the relative velocity of the two colliding bodies is set equal to zero, but the two bodies retain their identity (assumption of no accretion). In the case of both gravitational perturbation and collision, new orbital elements for both bodies are calculated. Both angular momentum and energy (including dissipation as heat during collision) are accurately conserved in these calculations. This point is quite important, since the mutual gravitational perturbations of small bodies are quite small, and their effects can easily be obscured by approximations or rounding-off errors which result in small but finite non-conservation of angular momentum and energy.

Following the first time step most of the bodies will have slightly different orbital elements (a, e, and i), which constitute the initial conditions for the next time step. The length of the time steps are chosen so that the probability of the same pair making two encounters or collisions during the interval is $< 0.2\%$. The time steps therefore vary with the masses of the bodies chosen, and for a given mass they change as the relative velocity distribution of the bodies evolves. For the smallest bodies investigated (1.0×10^{24}g) the time steps varied from 6×10^5 years to 2.5×10^6 years, while for the largest bodies (1.0×10^{26}g) the time steps ranged from 4×10^4 years to 6×10^5 years.

The evolution of the swarm can be followed in this manner through successive time steps for as long as is deisred. After every time step various quantities are calculated, e.g., the root mean square (r.m.s.) encounter velocities averaged over all pairs of crossing bodies, the r.m.s. encounter velocities of bodies which actually interacted during that time step, as well as the mean velocities of the swarm with respect to circular reference orbits. In addition, several quantities are computed for use in subsequent work in which the dimensionless parameters in the Safronov (1969, Chapter 7) theory will be compared with those found by these numerical experiments. After a regular number of time steps, complete distributions of the various types of velocities and the complete current orbital elements of the swarm are printed.

The evolution of the r.m.s. velocity at which a swarm of 9×10^{24}g (830-km radius) bodies encounter one another is shown in Figure 1. In three of the four cases shown, the initial r.m.s. relative velocity is very low, about 0.3 km/sec (open circles, open and closed squares). As time increases, at first the effect of mutual perturbations dominate over collisions, and the r.m.s. velocity of the swarm increases rapidly, approaching a velocity of about 1.4 km/sec. At these higher velocities, a steady state is gradually established, wherein the increases in relative velocity caused by close encounters are on the average cancelled by decreases in relative velocity following collisions. On the time scale shown in Figure 1, this steady state is reached after about four million years. The time scale shown in the figure differs from the one actually calculated in that it has been compressed by a factor of 2 to 5 which varies throughout the calculation so as to take into consideration the opposing effects of the total mass of the swarm being less than the mass of the actual terrestrial planets, and the radial spread of the swarm covering a narrower range of heliocentric distance than do the actual planets. This permits a more convenient comparison between the time required to achieve a steady state and the time required to actually accrete the real planets. Thus the "four million year" point of Figure 1 corresponds to an elapsed time of 16.3 million years in the actual computation.

It is apparent from Figure 1 that the steady-state velocity achieved after about four

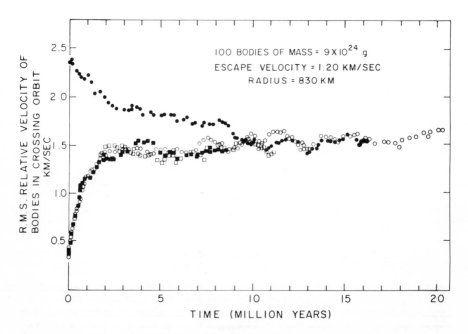

Figure 1. Approach to the steady-state distribution of r.m.s. velocity, averaged over all bodies which can encounter one another. The solid squares, open squares, and open circles represent three calculations starting from initial relative velocities of 0.3 km/sec, but with different random number sequences. The solid circles correspond to an initial velocity of 2.4 km/sec. In both cases an approximately constant velocity of 1.55 km/sec is reached after a few million years.

million years is not absolutely constant but exhibits short period (about 1 Ma) fluctuations as well as a long-range small but monotonic increase in velocity. The short-term effects are associated with stochastic fluctations resulting from the small number of bodies. Whether or not this effect would have a counterpart in a real solar system swarm evolving in this way depends upon the size distribution of the actual swarm. The long-term trend is a consequence of never actually being in a final equilibrium state. Because collisions cause dissipation of energy while conserving angular momentum, the swarm will continue to spread indefinitely. However the time scale for this effect to cause a significant increase in velocity is comparable to the full accretion times of $\sim 10^8$ years, and will not be important in a real solar system.

The approach to the steady state from the high velocity side is shown by the solid circles in Figure 1. The initial velocities of about 2.4 km/sec decrease to about 1.5 km/sec on a time scale somewhat longer than that required for the growth from initial low velocities. This is a consequence of the lower encounter probabilities of a high velocity swarm. However the fact that the final velocities are very similar, regardless of whether the approach is from the high velocity or the low velocity side, demonstrates that there is an equilibrium r.m.s. velocity toward which the swarm evolves.

For bodies of this size the equilibrium velocity was found to be about 1.55 km/sec, slightly above the 1.20 km/sec escape velocity of the bodies. This steady-state velocity is obtained by averaging over all pairs of bodies with orbits which cross one another. In many cases the crossing is "partial", i.e., the perihelion of one body is less than the perihelion of the other, but the aphelion of the first body is not beyond the aphelion of the second. Under these circumstances in calculating the average, the squared velocities are weighted in accordance with the fraction of the time the line of nodes of one body in the plane of the other is oriented so as to permit encounters as the argument of perihelion of the first body precesses.

Another "r.m.s. velocity" of interest is obtained by averaging the velocities of bodies actually selected to encounter. This quantity approaches a steady state on the same time scale as the r.m.s. velocity plotted on Figure 1, but is systematically smaller in value, because of the increase in encounter probability with decrease in relative velocity. The r.m.s. value of this velocity was found to be 1.23 ± .05 km/sec.

Safronov (1969) described the velocity of the swarm in a third way, by referring the motion of each body to that of a hypothetical body in a circular reference orbit. This has been done in the present work by calculating the heliocentric velocity components of each body at a random orientation of its argument of perihelion and at a random value of its true anomaly. From these quantities one subtracts the velocity components of a reference body with the same heliocentric longitude and zero inclination moving in a circular orbit of radius equal to the heliocentric distance of the body belonging to the swarm. These differences are squared and summed to attain squared velocities with respect to circular motion. Safronov does not explicitly define "velocity with respect to the circular orbit", but as far as I can tell, this calculation corresponds to his definition. The present calculations give an r.m.s. velocity defined in this way of 0.84 ± .02 km/sec. Use of Safronov's equation 7.32, corresponding to a swarm of bodies of equal mass, gives a result of 1.21 km/sec. Thus the numerical and analytical theories are in fair agreement, despite the differences in physical modelling introduced by Safronov's hydrodynamical analogue. In an earlier abstract (Wetherill, 1979) different

conclusions were stated regarding the difference between the numerical value and
Safronov's value. This resulted from my misunderstanding of Safronov's use of the
term "relative velocity", and the present conclusion therefore supersedes the earlier
one.

In more recent work (Kaula, 1980) it appears that a slightly different definition of
velocity with respect to a circular orbit is given (Kaula, pers. commun. 1979). In this
case the radius of the reference body's circular orbit is taken to be the semi-major axis
of the swarm body. Where r.m.s. velocities are calculated using this definition, a value
of 0.89 ± 0.03 km/sec is found, similar to but slightly higher than that obtained using
Safronov's definition.

In addition to swarms consisting of 9×10^{24}g bodies, the same calculations were
carried out for bodies of 1×10^{24}g, 3×10^{24}g, 3×10^{25}g, and 1×10^{26}g. In every case
a steadystate was achieved in a manner very similar to that shown in Figure 1 for
9×10^{24}g bodies. When adjusted to take into consideration differences in density of the
swarm resulting from the different total mass and radial spread, the time scales
required to reach the steady state remained at about 5 Ma.

The calculated steady-state velocities for all five assumed masses are shown in
Figure 2. For each of the three "relative velocities" discussed above, the steady-state
velocity is found to be a linear function of the radius and escape velocities of the body.
This can be expressed in terms of the dimensionless Safronov number $\Theta = \dfrac{Ve^2}{2v^2}$ where

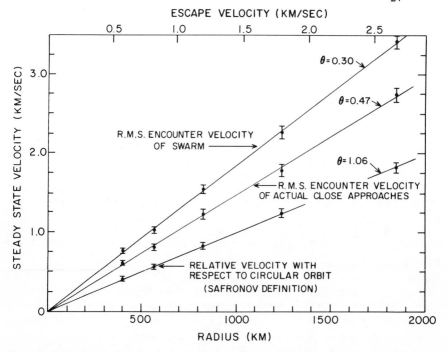

Figure 2. Dependence of steady-state velocities as a function of radius (lower scale) and escape
velocity (upper scale).

Ve is the escape velocity and v is a measure of the relative velocity of the swarm. For the r.m.s. encounter velocity of the pairs, $\Theta_e = 0.30$; for the r.m.s. encounter velocity of the pairs selected, $\Theta_a = 0.47$, and for Safronov's relative velocity with respect to the circular orbit, $\Theta_s = 1.06$. This last value can be compared with Safronov's value of 0.49, when the same assumption is made regarding the effect of non-central collisions on the collision cross-section (ξ = in Safronov's equation 7.32).

Within the accuracy of the calculations, the value of Θ is independent of the mass. This is in agreement with the analytical theory in which Θ remains of order unity during the course of accretion.

In a real swarm, the bodies will not all have the same mass. It is planned to evaluate the effect of varying mass in subsequent work. However the essential physics of Safronov's approach is contained in the present idealized problem, and it is reasonable that its validity will not be greatly affected by the introduction of a different mass distribution.

A typical calculated steady-state distribution of eccentricity and inclination is shown in Figure 3 for a swarm of 9 x 10^{24}g bodies with initial eccentricities and inclinations confined to the region e less than .01, sin i less than .01 radians. Both the range and the average value of the inclination are found to be about twice that of the eccentricity.

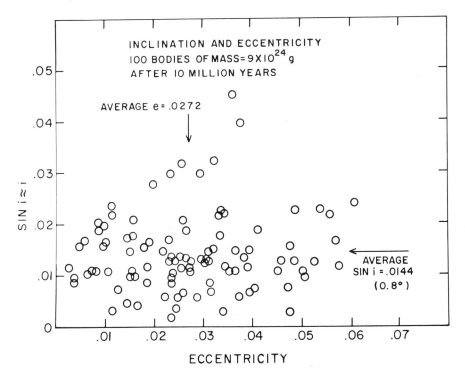

Figure 3. Typical steady-state distribution of eccentricity and inclination.

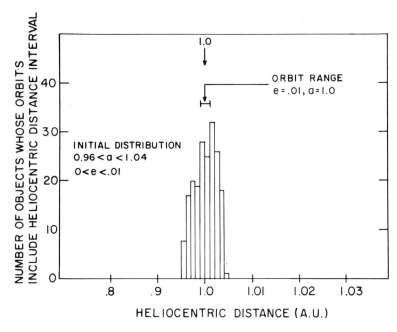

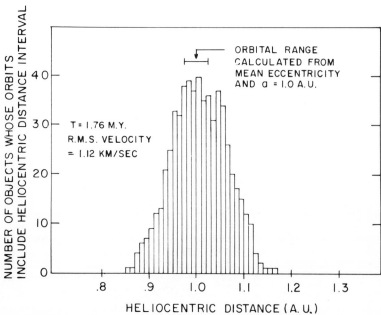

Figure 4. Increase in radial spread of swarm as a function of time, as a consequence of collisions and gravitational perturbations. (a) Initial state; (b) after 1.8 Ma, approach to steady-state; (c) 6.7

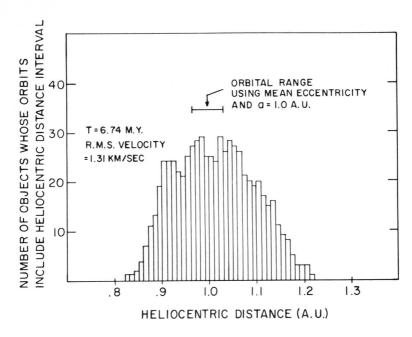

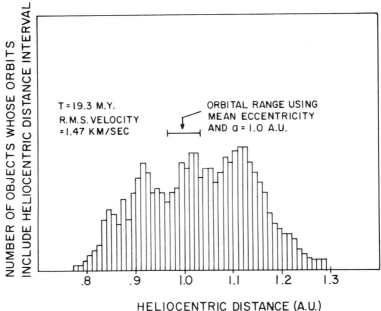

Ma, steady-state approximately reached; (d) 19.3 Ma, swarm continues to spread even after reaching steady-state velocity.

The radial distribution of the bodies is shown in Figure 4 for various elapsed times (time-scaled as in Figure 1). For the case of $T = 6.74$ Ma (Fig. 4c) the steady state has been reached, and is accompanied by a spread in the radial distribution of the bodies, as well as by the spread in inclination and eccentricity shown in Figure 3. This result has no counterpart in the analytical theory, since differences in heliocentric distance were not explicitly introduced in the earlier work. The range of heliocentric distance traversed by the bodies is considerably greater than that which would be inferred simply from the growth in eccentricity of a swarm of bodies with the same initial semi-major axis. This spread in heliocentric distance is relevant to the question of maintaining "orbital linkage" during the course of accretion (Levin, 1978; Wetherill, 1978; Greenberg et al., 1978).

More details of the calculated steady-state distribution of velocities are shown in Figure 5. In the analytical treatment of this problem (Chandrasekhar, 1942; Safronov, 1969; Kaula, 1980) it is assumed that the individual velocities of the bodies form a Maxwellian distribution. In Figure 5 calculated distributions are compared with Maxwellian distributions fitted to the peak corresponding to the most probable velocity. It may be seen that the calculated distributions resemble Maxwellian distributions, but that some excess of high velocity bodies is found.

THREE-DIMENSIONAL ACCUMULATION OF A SWARM OF BODIES OF EQUAL MASS

As illustrated in the preceding section, I feel that at the present time we should concentrate on the quantitative treatment of problems relevant to more detailed aspects of the various accretion theories. This is preferable to prematurely attempting to carry out very complex simulations of the entire accretional process, and certainly it is necessary before taking strong adversary positions on which theory of solar system formation will ultimately turn out to be correct.

In addition to the problem discussed in the previous section, another consequence of this kind is that of the consequences of treating accretion in a two-dimensional approximation. The most accurate numerical calculations involving simultaneous collision, gravitational perturbations and planetesimal accumulation have been carried out by Cox (1978). The precision of the algorithms employed limited at least his first investigation to a two-dimensional treatment of the problem (Cox et al., 1978). Cox calculated the orbital evolution of a swarm of 100 1.2×10^{26}g bodies initially distributed over the terrestrial planet region. For initial maximum eccentricities < 0.10, he found that the final state consisted of 8 to 10 planetary bodies, rather than the four observed terrestrial planets. A result more similar to the present system of planets was found by using a maximum initial eccentricity of 0.15. This eccentricity is only slightly higher than that associated with the steady-state value consistent with the Safronov numbers shown in Figure 2.

However, for a real swarm the eccentricities would certainly be lower, because the effects of a large fraction of the mass being in small bodies as well as the averaging of velocities accompanying accretion would decrease the eccentricities and relative velocities. Therefore Cox's result, as discussed elsewhere (Wetherill, 1978; Greenberg et al., 1978), emphasizes the fact that there is a serious question as to whether or not a

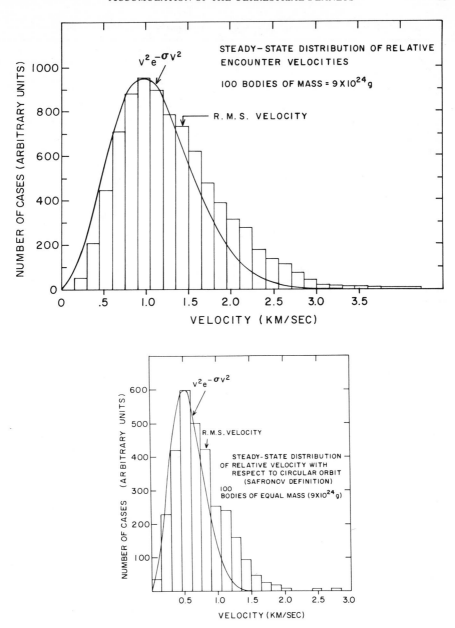

Figure 5. Distribution of velocities after the steady-state is reached. (a) relative velocities of encountering pairs (corresponds to velocities of Fig. 1 and uppermost set of points in Fig. 2). (b) relative velocity of bodies with respect to circular orbit (corresponds to lowermost set of points of Fig. 2.)

self-contained swarm of this kind can avoid premature orbital isolation of embryos and consequent production of too many small planets. It is also possible, as discussed in these references, that Jupiter perturbations may be important in determining whether or not this premature isolation will occur. If Jupiter is necessary, serious implications arise regarding the time needed to form that planet; when time scale considerations are introduced, massive gravitational instabilities in the outer solar system may be required at an early stage.

Therefore it was thought worthwhile to investigate the extent to which Cox's result depends on his calculation being two-dimensional.

The problem with treating the accretion in three dimensions is that it is necessary to forego the more exact numerical procedures used by Cox. In the present work, perturbations following close encounters are calculated in the manner developed by Öpik and Arnold, which was in the preceding discussion of the steady-state velocity distribution of a non-accreting swarm. In Cox's calculation the encounter is treated as a two-body interaction in the centre of mass frame of reference, and the centre of mass of the two encountering bodies is assumed to move in an elliptical orbit throughout the encounter. The duration of the encounter is determined by the time required for the bodies to tranverse their sphere of mutual influence. Cox has shown this to be a good approximation to the results of the exact numerical integration. In the present work, the relative motion of the higher eccentricity body with respect to the body with the lower eccentricity is calculated as if the latter were fixed and its mass increased to be equal to the sum of the masses of the two bodies, as in the standard gravitational two-body problem.

The change in relative velocity of a body of mass m_1 calculated in this way is the same as that which would be found in a reference frame in which the center of mass is moving at constant velocity, and in which the gravitational attraction is calculated as due to a mass $\dfrac{m_2^3}{(m_1+m_2)^2}$, where m_2 is the mass of the body encountered. The reference frame moves as well as rotates slightly during the encounter, as the x-direction is always taken to be opposite to the direction of the sun, the z-direction perpendicular to the plane of zero inclination, and the y-direction perpendicular to both these directions, positive in the direction of heliocentric motion. This reference frame is assumed to be moving at a constant velocity with respect to a body in a circular orbit at the heliocentric distance of closest encounter, the components of this constant velocity being given by the x, y and z coordinates of the velocity the "fixed" body would have at the point of encounter if no perturbation occurs. The calculated changes in relative velocity caused by the gravitational perturbations are then referred to the unperturbed frame, and apportioned between the two bodies in accordance with their masses and with the requirement that momentum be conserved during the encounter. The new velocities of both bodies are finally transformed back into the heliocentric reference frame, and from the velocities the perturbed orbital elements are calculated.

This calculation differs from that of Cox insofar as the elliptical motion of the center of mass deviates from the approximation that the mass center is moving at a constant velocity relative to circular velocity during the encounter. It also differs in that in Cox's calculation the hyperbolic two-body encountered is truncated at the boundary of the mutual sphere of influence of the two bodies, whereas in the present calculations the hyperbolic deflection is taken to its asymptotic limits. The extent to

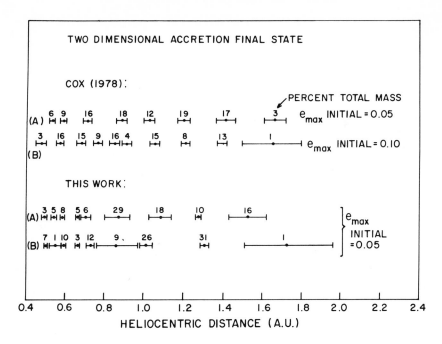

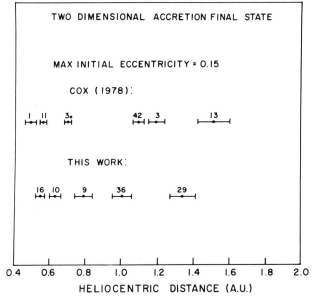

Figure 6. Final distribution of mass found in two-dimensional calculations. With the possible exception of the highest maximum eccentricity (0.15), an excessive number of planets is formed. (a) e_{max} = 0.05 and 0.10. (b) e_{max} = 0.15)

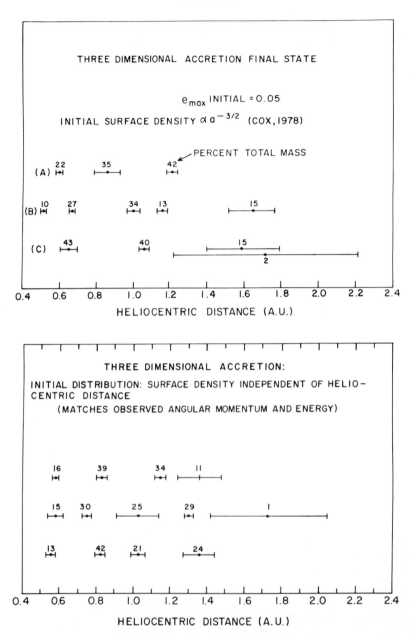

Figure 7. Final distribution of mass found in three-dimensional calculations. A reasonable number of planets is formed even for low initial eccentricities.
(a) a $^{3/2}$ initial surface density distribution;
(b) uniform initial surface density distribution.

which these approximations differ from those used by Cox and the extent to which both of these differ from the results of detailed numerical integration of the motion throughout the encounter are currently under investigation.

In order to compare my results with those of Cox, the two-dimensional case was calculated first. This was done by specifying that all inclinations were initially zero and that all encounters took place in the common orbital plane, so that no changes in inclination could occur. The probability of encounters was calculated by replacing the Öpik (1951) collision formula with an equivalent two-dimensional expression.

The final results of the two-dimensional calculations for a random distribution of initial eccentricities ranging up to .05 are shown in Figure 6a. Following Cox (1978), the initial semi-major axes were randomly distributed between 0.5 and 1.5 A.U. in accordance with the mean $R^{3/2}$ dependence of surface density on heliocentric distance. In agreement with Cox's result for $e_{max} = 0.05$ and 0.10, in the final state 8 to 10 bodies remain, and these tend to be concentrated at the smaller heliocentric distances. When the maximum initial eccentricity is increased to 0.15, the final number of bodies decreases to 5 or 6 (Fig.6b).

When the same calculation is carried out in three dimensions, again with $e = 0.05$, the final number of bodies is reduced to 3 to 5 (Fig. 7a). As Figure 7b shows, this result is not strongly dependent upon the initial distribution of semi-major axes. The latter results were obtained with an initial semi-major axis distribution having a mean surface density independent of heliocentric distance. As shown earlier (Wetherill, 1978), this agrees more closely with the actual total energy and angular momentum of the present terrestrial planets than does the $R^{3/2}$ dependence.

The difference between the three-dimensional and two-dimensional results is a consequence of the lesser importance of collisions relative to gravitational perturbations in three dimensions. In two dimensions the ratio of collisions to close encounters within a sphere of influence corresponding to a given number (N) of planetesimal radii varies inversely with N, while in three dimensions it varies inversely with N^2. Therefore in three dimensions, close encounters are sufficiently frequent to "pump up" the relative velocities of the bodies and maintain linkage of their orbits much more effectively.

This effect is seen in Figure 8. The initial eccentricities are all low (Fig. 8a). As accumulation proceeds, many of the bodies are accelerated to more eccentric orbits, and the range of semi-major axes increases as well (Fig. 8b). In the final state (Fig. 8c), the eccentricities are again low, as a results of the embryos' failing to cross and therefore perturb one another, combined with the averaging of velocities of the accumulating planetesimals and the resulting circularization of the combined orbits. The time scale for accretion is about 10^8 years, which is characteristic of this type of accretion theory (Fig. 9).

This model should not be interpreted as a serious simulation of actual planetary accumulation. It simply shows that a three-dimensional treatment may permit accumulation to proceed further prior to the separation of embryos into isolated orbits. One particularly serious defect in the model that precludes interpretation of these results as actual simulations is the fact that all the bodies are assumed to be quite large initially, 1.2×10^{26}g. In a real solar system, these large bodies would be accompanied by a larger number of smaller bodies, and the effects of fragmentation would maintain a population of small bodies throughout the course of accumulation.

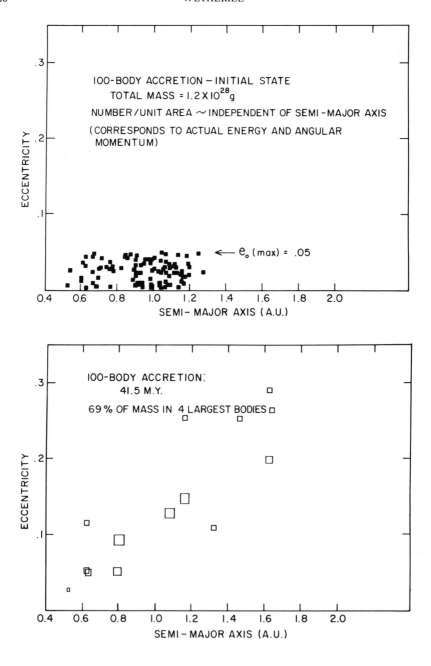

Figure 8. Evolution of eccentricity and semi-major axis distribution, three dimensional accumulation: (a) initial state; (b) after accretion 69% complete; (c) final state.

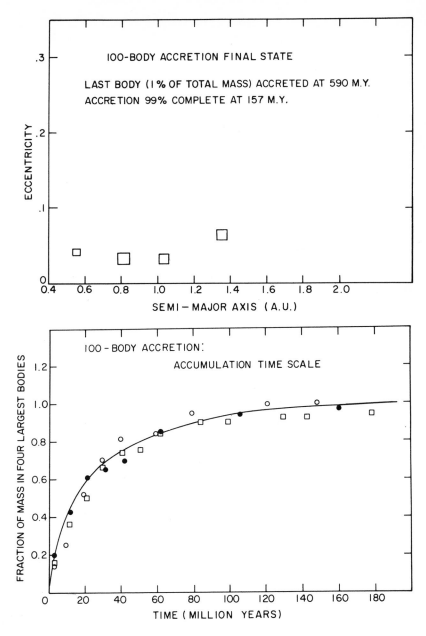

Figure 9. Accumulation time scale. Open circles, closed circles, squares, correspond to separate calculations using different random numbers. Most of the accretion occurs during the first about 20 Ma, and is 99% complete after 150 Ma. A last "straggler" may remain for several hundred million years longer, in accordance with earlier work (Wetherill 1977).

The effect of increasing the importance of damping by small factors is shown in Figure 10. A three-fold increase in damping does not change the number of final embryos, but a four-fold increase causes the result to tend back toward that found in the two-dimensional case, i.e., too many small isolated embryos. Thus to some extent an increase in damping can be accommodated without prematurely destroying the orbital linkage of the swarm. However it is not yet clear whether this is sufficient, or whether the introduction of long range perturbations, possibly involving resonances with the major planets (Greenberg *et al.*, 1978; Wetherill, 1978), should be invoked to achieve the observed number of terrestrial planets by the gas-free accumulation process.

In comparison with the large bodies, these smaller bodies would be ineffective perturbers, but nevertheless collisions with the large numbers of small bodies would diminish relative velocities, leading to early isolation of planetary embryos. A realistic quantitative calculation of this problem is premature at this time. However the magnitude of these effects can be appreciated if we assume that the primary factor determining the number of isolated embryos is simply the relative importance of gravitational perturbations in comparison with collisional damping. In the numerical calculations it is possible to artificially change the effects of perturbation relative to the damping by an arbitrary factor, without changing the accretion rate.

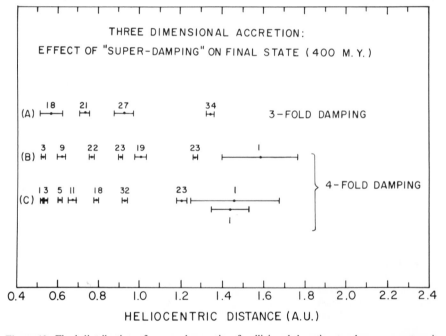

Figure 10. Final distribution of mass when ratio of collisional damping to close encounters is artificially increased, as a rough simulation of the effect of the presence of smaller bodies in the swarm. When this ratio is increased by a factor of about four, an excessive number of planets is found.

ACKNOWLEDGEMENTS

I wish to thank Larry W. Finger and Felix Chayes of the Geophysical Laboratory, Carnegie Institution of Washington, for much help in the use of their computing facility; Larry P. Cox and William M. Kaula for valuable discussions; and Mary Coder, Richard Carlson, Benjamin McCall, and Robert Hurwitz for assistance in the preparation of the manuscript. Albert Jambon contributed his linguistic skills to the preparation of the abstract. This work was supported by a grant from the NASA office of Planetary Programs.

REFERENCES

Arnold, J.R., 1965, The origin of meteorites as small bodies: Astrophys. Jour., v. 141, p. 1536-1556.

Cameron, A.G.W., 1978, Physics of the primitive solar accretion disk: The Moon and the Planets, v. 18, p. 5-40.

Chamberlin, T.C., 1904, Fundamental problems of geology: Carnegie Inst. Washington Year Book 3, p. 195-258.

Chandrasekhar, S., 1942, Principles of Stellar Dynamics: Univ. of Chicago Press, Chicago, 251 p.

Cox, L.P., Lewis, J.S., and Lecar, M., 1978, A model for close encounters in the planetary problems: Icarus, v. 34, p. 415-427.

Cox, L.P., 1978, Numerical simulation of the final stages of terrestrial planet formation: Ph.D. Thesis, Massachusetts Institute of Technology.

Greenberg, R., Wacker, J., Chapman, C.R., and Hartmann, W.K., 1978, Planetesimals to planets: a simulation of collisional evolution: Icarus, v. 35, p. 1-26.

Gurevich, L.E., and Lebedinskii, A.I., 1950, Formation of the planets: Izvestia Akad. Nauk USSR, Seriya Fizich, v. 14, p. 765-799.

Hayashi, C., Nakazawa, K., and Adachi, I., 1977, Longterm behavior of planetesimals and the formation of the planets: Publ. Astron. Soc. Japan, v. 29, p. 163-196.

Kaula, W.M., 1980, Equilibrium velocities of a planetesimal population: Submitted to Icarus.

Kuiper, G.P., 1951, On the origin of the Solar System, in Hynek, J.A., ed., Astrophysics: McGraw-Hill, N.Y.

Levin, B.J., 1978, Relative velocities of planetesimals and the early accumulation of the planets: The Moon and Planets. v. 19, p. 289-296.

McCrea, W.H., 1960, The origin of the Solar System: Proc. Royal Sci. London, v. 256A, p. 245-266.

Öpik, E.J., 1951, Collision probabilities with the planets and the distribution of interplanetary matter: Proc. Royal Irish Acad., v. 54A, p. 165-199.

Safronov, V.S., 1962, Velocity dispersion in rotating systems of gravitating bodies with inelastic collisions: Voprosy Kosmogonii, v. 8, p. 168.

Safronov, V.S., 1969, Evolution of the Protoplanetary Cloud and Formation of the Earth and Planets: Moscow, Nauka; translated for NASA and NSF by Israel Program for Scientific Translations (1972) NASA TT F-677.

Safronov, V.S., 1978, The heating of the earth during its formation, Icarus, v. 33, p. 3-12.

Tisserand, F., 1896, Traite de Mecanique Céleste: v. 4, p. 200, Paris, Gauthier-Villiers.

Weidenschilling, S.J., 1974, A model for accretion of the terrestrial planets, Icarus, V. 22, p. 426-435.

Weidenschilling, S.J., 1976, Accretion of the terrestrial planets II, Icarus, V. 27, p. 161-170.

Wetherill, G.W., and Williams, J.G., 1968, Evaluation of the Apollo asteroids as sources of stone meteorites: Jour. Geophys. Res., v. 73, p. 635-648.

Wetherill, G.W., 1977, Evolution of the earth's planetesimal swarm subsequent to the formation of the earth and moon: Proc. 8th Lunar Sci. Conf., v. 1, p. 1-16.

Wetherill, G.W., 1978, Accumulation of the Terrestrial Planets: In Gehrels, T., ed., Protostars and Planets: University of Arizona Press, p. 565-598.

Wetherill, G.W., 1979, Steady-state velocity distribution of a proto-planetary swarm: Lunar Science, v. X (Abstracts of papers presented at the 10th Lunar Science Conference), p. 1335-1337.

Zvyagina, Y.V., Pechernikova, G.V., and Safronov, V.S., 1973, A qualitative solution of the coagulation equation taking into account the fragmentation of bodies: Astron. Zh., v. 50, p. 1261-1273.

The Continental Crust and Its Mineral Deposits, edited by D.W. Strangway,
Geological Association of Canada Special Paper 20

THE BEGINNING OF THE EARTH'S THERMAL EVOLUTION

W.M. Kaula
Department of Earth and Space Sciences,
University of California, Los Angeles,
Los Angeles, California 90024

ABSTRACT

Formation of a homogeneous planet from a plausible population of planetesimals results in internal temperatures T(r) proportionate to r^2 up to a radius r where temperatures near melting occur. Above this radius, the viscosity would be low enough for convection to remove heat as fast as it is generated by impacts. The initial radial temperature gradient for convection to occur would be established by impact stirring, and convection would be abetted by horizontal temperature gradients caused by infall heterogeneities.

Temperatures well above melting, found in recent studies, resulted from either artificial modelling constraints or computational imperfections.

The portion of impact energy retained as heat significantly increases with the size of impact, to probably more than 50 per cent.

Core formation probably started before the Earth was 10 per cent formed, since melting temperatures would have been attained by that time. The peak energy release was near the core-mantle interface, and about 60 per cent went to the core. The resulting temperature gradient in the mantle was strongly superadiabatic. Energy released by differentiation of subsequently infalling material would have gone mainly into the mantle, and would have further contributed to vigorous convection in the mantle, but not in the core, where the temperature gradients were adverse (unless there was strong heterogeneity of infalls).

RÉSUMÉ

La formation d'une planète homogène à partir d'un ensemble typique de planétésimales donnerait des températures internes proportionnelles à r^2 jusqu'à une valeur de r où la température peut provoquer la fusion. Au-delà de ce rayon, la viscosité serait assez faible pour que la convection dissipe la chaleur aussi rapidement qu'elle se produit par les impacts. Le gradient radial de température qui initie la convection serait établi par l'agitation causée par les impacts, et la convection serait facilitée par les gradients horizontaux de température à cause de l'hétérogénéité des impacts.

Les températures beaucoup plus élevées que la température de fusion obtenues par des études récentes résultent surtout des contraintes artificielles des modèles ou d'imperfections dans les calculs.

La portion d'énergie d'impacts retenue sous forme de chaleur augmente significativement avec l'amplitude de l'impact pour atteindre probablement plus de 50 pour cent.

La formation du noyau a commencé probablement avant que la Terre n'ait atteint 10% de son volume, parce que les températures de fusion auraient été atteintes à ce moment. La plus grande libération d'énergie se situait à la limite entre le noyau et le manteau et à peu près 60% de l'énergie allait vers le noyau. Le gradient de température qui en résultait dans le manteau était fortement superadiabatique. L'énergie libérée par la différentiation de la matière qui tombait ensuite sur la planète serait allée surtout vers le manteau. Cette énergie aurait contribué à la convection vigoureuse dans le manteau, mais pas dans le noyau où les gradients des températures étaient contraires (sauf s'il y avait une forte hétérogénéité dans les impacts).

INTRODUCTION

As discussed by Wetherill (1980) in these proceedings, it is probable that the Earth's formation was from sizeable planetesimals. Such bodies would be very effective in heating the forming planet because much of the impact energy transformed into heat would be buried at depth. The loss of this heat would thus be governed by heat transfer within the body, rather than by radiation from its surface. While there would be some enhancement of transfer within the body through the stirring of material by subsequent impacts, the dominant uncertainty in the process is the portion of the internal energy lost by radiation from small fragments or drops thrown up at impacts.

A fairly elaborate computer algorithm for the thermal evolution of the Earth and the Moon growing by impacts has been published by Kaula (1979). This algorithm was applied to highly simplified models: compositionally homogeneous; incompressible; spherically symmetric; and smoothly growing along a sigmoid. The present paper will re-examine the simplified model, attempt to infer the leading effects for the Earth entirely independently without resort to computer, and then discuss the principal inadequacies: 1) impact energy partitioning; 2) heterogeneity, both temporal and spatial; and 3) core differentiation.

HOMOGENEOUS MODEL

The kinetic energy/unit mass at impact of a planetesimal on a planet of mass M and radius r can be expressed as:

$$\frac{1}{2}v^2 = \frac{GM}{r} \cdot (1 + 1/2\theta), \tag{1}$$

where G is the gravitational energy and θ is the ratio of $GM/2r$ to the approach kinetic energy/unit mass (Safronov, 1972). If (1) a portion h of this energy was retained as heat, (2) this heat was distributed in a uniform layer at the surface and (3) none of the energy was lost, then the temperature at radius r within the planet would be

$$T(r) = T_e + hGM(r) (1 + 1/2\theta)/rC, \tag{2}$$

where T_e is the temperature of the surroundings, $M(r)$ is the mass within radius r and C is the specific heat. Comparing the values from eq. 2 with those given for the earth at two-thirds completion by Kaula (1979), assuming the ratio $h = 0.1333$, a uniform density $\rho = 4.0$ gm/cm³, and the ratio $\theta = 4$, we see agreement within 20% except for the outermost ~500 km:

Radius km	Temperatures, °K From eq. (2)	From Fig. 1, Kaula (1979)
1000	400	400
2000	820	850
3000	1520	1500
4000	2500	2010
5000	3750	3450
5500	4220	3670
6000	5020	3230
6142	—	260

The discrepancies above 4000 km radius are apparently due to the upward transfer of heat by convection, while those above 5500 km radius are probably also affected by heat transfer due to impact stirring. To estimate the latter, we use the steady state plane layer model of Safronov (1978) for a growth rate $\dot{R}$ in which the energy deposition and convective stirring are uniformly distributed over a layer of thickness L_I. This model obtains for the temperature at depth z:

$$T(z) = T_o + \frac{H}{\rho C \dot{R}} \left[z - \frac{e^{-aL_I}}{a} (e^{az} - 1) \right], \tag{3}$$

where H is the energy/unit volume deposition rate,

$$H = h\rho \dot{R} G M(R) (1 + 1/2\theta)/L_I R , \tag{4}$$

and a is the ratio $\dot{R}/\kappa_I$, κ_I is the effective thermal diffusivity arising from impact stirring. Thence

$$T(z) = T_o + h \frac{GM(R)}{L_I RC} \left[1 + 1/2\theta \right] \left[z - \frac{e^{-aL_I}}{a} (e^{az} - 1) \right]. \tag{5}$$

An upper limit on the effective thermal diffusivity κ_I for an energy deposition rate H/h can be prescribed through a "lifting distance" l and an "event rate" $\dot{n}$:

$$\kappa_I = \frac{1}{2} l^2 \dot{n} \tag{6}$$

The lifting distance l is obtained by translating kinetic energy of ejecta into potential energy:

$$l = kmv^2/2AL_1 g\rho. \tag{7}$$

The event rate $\dot{n}$ is obtained by dividing the mass growth rate by the mass per planetesimal m:

$$\dot{n} = \rho \dot{R} A/m. \tag{8}$$

In eqs. 7 and 8, k is the portion of infall energy partitioned to kinetic energy of ejecta (O'Keefe and Ahrens, 1977), and A is the horizontal cross section of the volume affected by each impact. Substituting eqs. 7 and 8 into 6 we obtain:

$$\kappa_I = k^2 m v^4 \dot{R}/8AL_I^2 g^2 \rho \tag{9}$$

If we assume (as in Kaula, 1979) the area A has a ratio πb^2 to $(mv^2/2)^{2/3}$ and the depth L_I a ratio νb to $(mv^2/2)^{1/3}$, then (using eq. 1):

$$\kappa_I = k^2 [GM(R)(1+1/2\theta)/R]^{2/3} \dot{R}/2\pi m^{1/3} \; \nu^2 b^4 g^2 \rho \tag{10}$$

Approximate numerical values are $k = 0.10$ (O'Keefe and Ahrens, 1977), $b = 7 \times 10^{-4} cm/erg^{1/3}$ (Pike, 1974) and $\nu = 0.1$ (Kaula, 1979). For the case appearing in the table above, M(R) is $\sim 4 \times 10^{27}g$, $\theta \sim 4$, $R \sim 6140$ km, $\dot{R} \sim 2.5 \times 10^{-7} cm/sec$, $g \sim 700$ cm/sec^2, and $\rho \sim 4g/cm^3$. Instead of a uniform planetesimal mass m there is a distribution of masses from a maximum $m_1 \sim 10^{25}g$ with a number density $n(m) \propto m^{-q}$, $q \sim 1.8$. Half the infalling matter is thus in planetesimals of mass m greater than $\sim 3 \times 10^{23}g$. Values corresponding to this mass are 370 km for L_I and $\sim 0.3 cm^2/sec$ for κ_I. Hence the effective depth $1/a = \kappa_I/R$ for impact stirring is only 13 km. A positive dT/dz (necessary for convection to operate) comes about because the turbulent diffusivity defined by eq. 6 removes heat more slowly than it is deposited, except at very shallow depths of a few kilometres.

The temperatures in the table above are well above melting, ~ 2000 K, and sufficient for vapourization. Hence there must be vigorous convection, and the rate of energy/unit mass loss must approach the rate energy/unit mass gain. Therefore the temperature gradient in the outer shell can be approximated by

$$\frac{dT}{dz} \approx HL_I/\rho C\kappa_I \approx h\dot{R}GM(R)(1+1/2\theta)/RC\kappa_I \tag{11}$$

using eq. 4. For the numerical values given after eq. 10, dT/dz is about 420°/km. Hence temperatures high enough to give viscosities low enough for convection, $\sim 1100°K$, will be reached at a depth of about 2 km. This result depends on using the impact diffusivity κ_I averaged over the outer 370 km, 0.3 cm^2/sec; allowing for an exponential drop off to 0.01 cm^2/sec at 370 km would give a surface value of 2.3 cm^2/sec for κ_I and a decay depth of 70 km. Hence a better dT/dz for the outermost layers is 60°/km, leading to the attainment of convective viscosities at a depth $L_c \approx 14km$. The somewhat higher value given by Kaula (1979), 40 km, is probably affected by the integration interval of the numerical calculation, 25 km.

The peak temperature and the temperature drop off to the surface are thus dominated by the interaction between energy deposition H, eq. 4, and convective transport. For temperatures above $\sim 2000°K$, the kinematic viscosity ν will be that of molten silicate, $\sim 10^2 cm^2/sec$. In the range $1100°K - 2000°K$ the solid state viscosity is strongly temperature – dependent:

$$\nu \approx a \exp(bT_m/T), \tag{12}$$

in which a is $\sim 1.4 \times 10^9 \; cm^2/s$ and b is ~ 25 (Carter, 1976).

In Kaula (1979) convective transport was approximated by a pseudo-diffusivity $\kappa_V \propto Ra^{1/3}$, where Ra is the Rayleigh number:

$$\kappa_V = 2\left[\frac{g\alpha\Delta T_V}{\nu Ra_c}\right]^{1/3}L_V \kappa_o^{2/3} \approx V(T) (\Delta T_V)^{1/3}L_V \kappa_o^{2/3} \tag{13}$$

where (using eq. 12) V(T) ranges from 10^{-2} cgs at T>>2000°K to $\sim 10^{-11}$ cgs at T = 1100°K; Ra_c is the critical Rayleigh number; and ΔT_V is the superadiabatic temperature differential across a convecting layer of thickness L_V. (For an explanation of this convective pseudo-diffusivity, see Stevenson and Turner, 1979). The reference diffusivity κ_o used in Kaula (1979) is $\kappa_I + \kappa_c$, where κ_c is the ordinary conductive diffusivity, ~0.01cm^2/sec. However, this use of the impact stirring diffusivity κ_I entails the dubious assumption that the rate of significant impacts (i.e., penetrating to depths $\sim L_V$) is at least as great as the rate of convective overturn.

The pseudo-diffusivity κ_V at any instant is dependent on the convective regime created by previous evolution through ΔT_V and L_V. Hence a full corroboration of the numerical integration by hand calculation would be difficult. What can be done is to compare the energy transfer rate across the convecting zone (dT/dr<0, T⩾1100°K) to the energy transfer rate across the impact dominated zone (T⩽1100°K) and the energy deposition rate:

$$\frac{h\dot{R}GM(R) (1+1/2\theta)}{RC} \approx \frac{\kappa_I\Delta T_I}{L_c} \approx \frac{\kappa_{V'}\Delta T_V}{L_V}, \tag{14}$$

where $\Delta T_I \approx 800°$ and $\Delta T_V \approx 2300°$ (allowing a $\approx200°$ adiabatic contribution); $L_C = 14$ km has already been inferred; and $L_V \approx 800$ km. Substituting from eq. 13 eliminates L_V. Using the same numerical values as before on the left:

$$1.6\times10^{-3} \approx V(T) (\Delta T_V)^{4/3}\kappa_o^{2/3} \tag{15}$$

In the tabulated computed values, κ_o was of order 0.1. Using this with $\Delta T_V \approx 2300°$ gives V(T) $\approx 2\times10^{-7}$ cgs, corresponding to a viscosity of 1.5×10^{16}cm^2/sec from eq. 12-13. This value is surprisingly high, since in the averaging of $1/\nu$ only a minor fraction at $\nu = 10^2$cm^2/sec would be required to make the reciprocal appreciably lower. Hence either the situation is far from steady state, or the computer modelling did not correctly simulate the convective transport. The method used in Kaula (1979) overestimated the effectiveness of convection by (1) using $\kappa_I + \kappa_c$ for κ_o, and (2) not allowing for reduction of ΔT within a macro-time step by using matrix multiplication with constant κ_V to economize on computer time. On the other hand, the method underestimated the rate at which the convecting layer would widen on the bottom (by reducing the temperature so that the peak is shifted downward) and the top (by raising the temperature so that the viscosity is lowered enough for convection) by holding the layer thickness constant for a macro-time step, rather than recalculating it at every time step required by the normal stability criterion.

Whether the temperature curve would approach steady state can be estimated by comparing the heat removed in a convective overturn with the heat added. The time for a convective overturn is approximated by the diffusive decay time:

$$\tau = (2L_V/\pi)^2/\kappa_V, \tag{16}$$

about 40 years for the numbers suggested above. The energy removed would be that associated with driving the superadiabatic temperatures down to $1/e$ of their original values if the upper boundary were at $T_o \approx 300°K$. However, a limitation of the spherically symmetrical model (as distinguished from reality) is the surface layer of T $\leqslant 1100°K$, in which heat transfer is governed by impact stirring. The effective decay time then would be $(2L_c^2/\pi)^2/\kappa_1$, about 8000 years if thinning of the layer due to temperature build-up is neglected. The energy per unit area removed is thus roughly

$$E = \int_0^{L_V} (1-1/e)\rho C\Delta T_V \sin(\pi z/2L_V)dz : \qquad (17)$$

$\sim 3 \times 10^{18}$ergs/cm^2 for the standard case. The energy input in the same duration of 8000 years, $\rho h\bar{R}GM(R) (1+1/2\theta)\tau/R$, is only $\sim 2 \times 10^{16}$ergs/cm^2.

The temperatures well in excess of melting obtained in Kaula (1979) are thus probably artifacts, resulting from an algorithm which did not provide for a realistic rate of thickening of the convective layer, and hence which artifically trapped heat at depths of adverse dT/dr. The similarly high temperatures obtained by Safronov (1978) are the consequence of neglecting convective transport, as he mentions. A more likely outcome within the imposed constraints (homogeneity, spherical symmetry, smooth growth) would be temperatures in accord with eq. 2 up to a radius at which the melting temperature was equalled or slightly exceeded. Above this level the temperatures would follow the melting curve fairly closely, since it is steeper ($\sim 2.5°C/kbar$) than the adiabat ($\sim 1°C/kbar$).

IMPACT ENERGY PARTITIONING

The partitionings of energy to kinetic, $k \sim 0.10$, and retained heat, $h \sim 0.1333$, are based on the calculations of O'Keefe and Ahrens (1977), with the additional assumption that fragments of high energy content will be of centimetre size. The O'Keefe and Ahrens (1977: Figs. 2, 3) calculations do indeed show a high correlation of kinetic and internal energy. This circumstance, plus the assumption of comminution energy density being closely correlated as well, makes it very plausible that highly heated material will be broken into small pieces, which because of their high kinetic energy will not come back into contact with other material until they have radiated much of their heat.

However, the O'Keefe and Ahrens calculations are for an impacter of only 5 cm radius and less than 2×10^{16} ergs energy, in order to model shock-induced phase transition effects accurately. The scaling to much larger impacts necessary for inferences as to early planetary heating is far from straightforward. If a large impact is viewed as many simultaneous small impacts occurring beside and behind each other, it is evident that for many material elements kinetic energy will be converted into compressive energy and thence some into heat because of the interference of different material elements with each other. Proportionately greater volumes of material will be subjected to pressures sufficient for melting. Some compression will be converted elastically back into kinetic energy, the net effect of which would be to transfer energy from more interior, hotter elements to outer, cooler elements. Another effect would be to merge together fragments which would have been ejected separately in smaller impacts.

Qualitatively, the following effects seem unavoidable in the upward scaling of impacts: 1) A decrease in partitioning of energy to kinetic energy, k; 2) an increase in partitioning of energy to heat, h; 3) a decrease in the correlation of kinetic and internal energies; and 4) an increase in the size of fragments. Photogeologic examinations of lunar impact features (e.g., Howard *et al.*, 1974; Schultz, 1976) see effect (2) in the form of a greater proportion of melt with large impacts, and effect (4) in the sizes of secondary craters associated with large primary craters.

The effects described above should increase the heat retention factor h appreciably, probably to more than 0.5 for Imbrium-sized impacts. The implications for early thermal evolution would be higher heating rates and less impact stirring, leading to a steeper central temperature curve, earlier onset of convection and earlier tiggering of core formation. Quantitative estimates of the effects would require computations for large impacts similar to those of O'Keefe and Ahrens (1977), with approximations for the phase transition phenomena. Such estimates may make a significant difference in the modelling of the early thermal evolution of smaller bodies such as the Moon or Mars. In the case of the Earth, they would only hasten the pinning of the mean temperature curve to the melting curve for most of the planet's bulk.

HETEROGENEITY

The two-thirds mass growth rate, $\sim 2.5 \times 10^{-7}$ cm/sec or 1.5×10^{20} grams/year, if partitioned amongst a planetesimal population of $m_1 \sim 10^{25}$ grams, $n(m) \propto m^{-1.8}$, would result in the following frequencies of occurrence of major impacts:

$m > 10^{24}$g: 1/50,000 years,
$m > 10^{23}$g: 1/7,000 years
$m > 10^{22}$g: 1/1,000 years, etc.

The percentage of the surface area affected by impacts greater than a size m_n per unit time can be approximated using the area $A(m)$ and event rate $\dot{n}(m)$ (see eqs. 8-10):

$$P_I = \frac{1}{4\pi R^2} \int_{m_n}^{m_1} A(m)\dot{n}(m)dm$$

$$= \frac{1}{4\pi R^2} \int_{m_n}^{m_1} \pi b^2 \left[m \frac{GM(R)}{R}(1+1/2\theta) \right]^{2/3} \cdot \frac{(2-q)4R^2\rho\dot{R}_m}{m_1^{2-q}} dm$$

$$= \pi b^2 \left[\frac{GM(R)}{(R)}(1+1/2\theta) \right]^{2/3} \cdot \frac{(2-q)\rho\dot{R}}{m_1^{2-q}} \cdot \frac{m_n^{5/3-q}}{q-5/3}, \tag{18}$$

where m_1 is the mass of the largest planetesimal and q is the slope of the frequency, $\dot{n}(m) \propto m^{-q}$. In the second row of eq.18, we have the same substitution for A as in eq.10, as well as

$$\dot{M} = \int_{m_n}^{m_1} m \, \dot{n}(m)dm \tag{19}$$

(Safronov, 1972; Kaula, 1979)

If a "significant impact" is one for which the depth affected, μR_A, is the same as the mean lithosphere, ~ 12 km, then $m_n \sim 2 \times 10^{17}$ grams and P_I is 2×10^{-4}/yr.

The significant lateral heterogeneities will thus be those associated with convective irregularities induced by major impacts on a 10^3 to 10^4 year time scale. Energy loss would be mainly by radiation from lava flows. The portion of the surface needed to be covered by flows would be, roughly (using eq.4):

$$P_L \approx \frac{h\rho \dot{R}GM(R)\ (1+1/2\theta)}{\sigma(T_m^4 - T_e^4)R} \tag{20}$$

where σ is the Stefan-Boltzman constant, T_m is melting temperature, and T_e is environmental temperature. For the standard case, P_L is $\sim 2 \times 10^{-4}$. It seems quite plausible that this amount of volcanism could be sustained by a convective system close to melting driven by lateral temperature differentials induced by impacts. A once-in-10^4 year impact delivers at least 10^{34} ergs heat energy, through a volume of $\sim 1.6 \times 10^{25}$ cm^3, or 2% of the planet, heating it an average of 13°.

CORE FORMATION

This problem has been treated in several papers, most recently by Shaw (1978), as though core formation did not occur until the earth was fully collected. However, since the melting required to trigger core formation requires a temperature of ~ 1250°K to 1500°K (dependent on the degree to which the iron is in oxide or sulphide form), it is more plausible that iron separation started when such temperature was attained. By eq. 2, a forming earth would attain these temperatures when its radius was less than 3000 km, or when it was less than 10 per cent grown.

The simplest relevant model for the effects of core formation, consistent in spirit with the simple homogeneous model discussed above, would be one in which:

1) there are two chemical components: by volume, 6/7 of component S, of uncompressed density $\rho_S \sim 3.5$g/cm^3, and 1/7 of component I, of uncompressed density $\rho_I \sim 7.0$g/cm^3;

2) these ratios apply throughout the growth of the planet;

3) when sufficient temperatures are attained, component I separates to form a core of radius $R_c = cR$, where $c = (1/7)^{1/3} = 0.52$ (assuming incompressibility), resulting in upward displacement of material of component S from radius r to radius $[c^3R^3 + (1-c^3)r^3]^{1/3}$;

4) thereafter, adding material at a rate $\dot{R}$ results in:

a) increasing the radius of the core of component I at a rate $c\dot{R}$,

b) displacing material of component S at radius $r > cR$ outward at a rate $c^3R^2\dot{R}/r^2$, and

c) adding to component S at a rate $(1-c^3)^{1/3}\dot{R}$.

Given the above-described model, the change in gravitational energy/unit volume at core formation for any material element can be calculated. However, this calculation is not very useful, because the heating of the element will not merely be the negative of its change in gravitational energy, due to the interaction of the material with its surroundings during its shift in location. It is more practical to calculate the total gravitational energy change, and then to distribute it among the material according to plausible constraints, such as: (1) the heating is proportionate to the distance of travel and (2) the temperatures of the two components are equal at the core-mantle boundary after any shift. Constraint (1) results in no heating at the centre and the surface, and a maximum heating at the core-mantle interface. Constraint (2) requires, in general, different proportionality factors for the two components. It does *not* make the amount of heating a function of the pre-existing temperature curve because the I component at the core-mantle boundary will have come from the surface while the S component will have come from the centre. Let the proportionality factors be ι and σ for components I and S, respectively. From the expression for the gravitational energy,

$$\Omega = 4\pi \int g(r)\rho(r)r^3 dr, \tag{21}$$

the total change on core formation (neglecting compression) can be calculated:

$$\Delta\Omega = \frac{(4\pi)^2}{15} GR^5 c^3 \left[\rho_I^2 c^2(1-c) + \rho_I\rho_S(1/2 - 5c^2/2 + 2c^3) \right.$$

$$\left. + \rho_S^2(-1/2 + 3c^2/2 - c^3) \right] \tag{22}$$

Then the constraints lead to the conditions

$$4\pi \left[\iota\int_0^{cR}(r/c - r)r^2 dr + \sigma\int_{cR}^R(r - r')r^2 dr \right] = \Delta\Omega, \tag{23}$$

and

$$\frac{\iota}{\rho_I}(1-c) = \frac{\sigma}{\rho_S} c, \tag{24}$$

where r' is $\left[(r^3 - c^3R^3)/(1-c^3) \right]^{1/3}$, and equality of specific heats has been assumed. The results are:

$$\iota = \frac{\Delta\Omega}{\pi R^4} \left\{ c^3(1-c) + c^2(1-c)^2 \frac{\rho_S}{\rho_I} \right\}^{-1}$$

$$\tag{25}$$

$$\sigma = \frac{\Delta\Omega}{\pi R^4} \left\{ c^4 \frac{\rho_I}{\rho_S} + c^3(1-c) \right\}^{-1}$$

For the case $R = 3000$ km, $c^3 = 1/7$, $\rho_I = 7.0$ g/cm^3 and $\rho_S = 3.5$ g/cm^3, the resulting heating at the core-mantle boundary is $\Delta T(cR) = 994°$, compared to an average of 272° over the entire volume. About 60 per cent of the heat goes to the core and 40 per cent to the mantle.

Hence a strongly superadiabatic temperature gradient would have been set up throughout the mantle early in the formation process of the earth, leading to vigorous convection throughout subsequent evolution. This effect would have been intensified by the differentiation of subsequent infalling material, which would have deposited a maximum of energy at the bottom of the mantle. Meanwhile, the core would have been stabilized against convection by a strongly adverse temperature gradient, which would also have favoured a solid inner core and a fluid outer core. This stabilization of the core may have been modified somewhat by heterogeneity of infalls, particularly in the earlier stages.

CONCLUSIONS

Re-examination of the spherically symmetrical modelling done two years ago indicates more firmly than before that formation of the Earth from a plausible population of planetesimals would have led to significant heating, more than enough to trigger core formation. All factors seem to favour greater effective heating with larger impacts: larger early-stage partitioning to internal energy, lower correlation of kinetic and internal energies, and larger fragments. The principal modification from the conclusions of Kaula (1979) is that the temperature would never have risen much above melting, since the viscosity of molten material would have been low enough to remove heat as rapidly as it was generated by impacts at any plausible rate. The effects of lateral heterogeneity are mainly to promote convection and thus heat removal. Core formation probably started rather early in the growth of the earth, and also acted to promote mantle convection because the highest energy density was near the core-mantle boundary. However, this effect would also have acted to keep the core stratified.

More effective modelling of great impacts remains the principal research that is needed.

ACKNOWLEDGEMENTS

This work was supported by NASA grant NGL 05-007-002, a Guggenheim Fellowship, and the Australian National University.

REFERENCES

Carter, N.L., 1976, Steady State flow of rocks: Reviews of Geophysics and Space Physics, v. 14, p. 301-360.

Howard, K.A., Wilhelms, D.E. and Scott, D.H., 1974, Lunar basin formation and highland stratigraphy: Reviews of Geophysics and Space Physics, v. 12, p. 309-327.

Kaula, W.M., 1979, Thermal evolution of earth and moon growing by planetesimal impacts: Jour. Geophys. Res. v. 84, p. 999-1008.

O'Keefe, J.D. and Ahrens, T.J., 1977, Impact-induced energy partitioning, melting, and vaporization on terrestrial planets: Proceedings, 8th Lunar Science Conference, p. 3357-3374.

Pike, R.J., 1974, Depth/diameter relations of fresh lunar craters: revision from spacecraft data: Geophys. Research Letters, v. 1, p. 291-294.

Safronov, V.S., 1972, Evolution of the protoplanetary cloud and formation of the earth and planets, NASA Technical Translation, TTF-677, Washington, 206 p.

Safronov, V.S., 1978, The heating of the earth during its formation: Icarus, v. 33, p. 3-12.

Schultz, P.H., 1976, Moon Morphology: Univ. Texas Press, Austin, 626 p.

Shaw, G.H., 1978, Effects of core formation: Physics of the Earth and Planetary Interiors, v. 16, p. 361-369.

Stevenson, D.J. and Turner, J.S., 1979, Fluid models of mantle convection: in McElhinny, M.ed., The Earth: Origin, Evolution, and Structure: John Wiley, New York, p. 227-263.

Wetherill, G.W., 1980, Numerical calculations relevant to the accumulation of the terrestrial planets: in Strangway, D.W., ed., The Continental Crust and Its Mineral Deposits, Geological Association of Canada Special Paper 20, in press.

The Continental Crust and Its Mineral Deposits, edited by D.W. Strangway,
Geological Association of Canada Special Paper 20

WHEN DID THE EARTH'S CORE FORM?

J.A. Jacobs
Department of Geodesy and Geophysics,
Madingley Rise, Madingley Road,
Cambridge CB3 OEZ, England

ABSTRACT

The origin of the Earth's core cannot be divorced from the much broader issue of the origin of the Earth itself. There is one constraint, however, that can be imposed upon its possible evolution. It is generally believed that the Earth's magnetic field is due to some form of electromagnetic induction in the fluid, electrically-conducting core. Rocks as old as about 3000 Ma have been found which possess remanent magnetization, indicating that the Earth must have had a molten outer core comparable to its present size at least that long ago. A major problem in the evolution of the Earth is the means by which it could have heated up sufficiently to form a (predominantly iron) molten outer core at least 3000 Ma ago.

Fairly short times (about 20 Ma after accretion) appear to be favoured for forming the core. An upper limit is about 500 Ma, so that core formation should have been essentially complete during the first 10 per cent of the Earth's history. The subsequent evolution from an approximately homogeneous, all-fluid core into the present state of a solid, (mainly iron) inner core and a fluid outer core containing some light alloying element is another problem. This subsequent evolution may have proceeded much more slowly and still may not be complete. Gravitational energy would be released by the growth of the inner core, and this, it has been argued, may provide the main source of energy for driving the geodynamo.

RÉSUMÉ

On ne peur dissocier l'origine du noyau terrestre de la question plus générale de l'origine de la Terre elle-même. Il y a toutefois une restriction qu'on peut imposer à son évolution possible. On pense généralement que le champ magnétique terrestre est causé par une forme d'induction électromagnétique dans le noyau fluide et conducteur d'électricité. On a retrouvé des roches aussi anciennes que 3000 Ma environ qui possèdent une aimantation rémanente, ce qui indique que la Terre à dû avoir un noyau externe en fusion de dimension comparable au noyau actuel au moins jusqu'à cette date. Un problème majeur dans l'évolution de la Terre est de savoir par quel moyen elle a pu s'échauffer suffisamment pour amener la fusion du noyau externe (composé surtout de fer) il y a au moins 3000 Ma.

On suppose des temps relativement courts (environ 20 Ma après l'accrétion) pour la formation du noyau. La limite supérieure serait d'environ 500 Ma, de sorte que la formation du noyau
a dû essentiellement être complète durant le premier 10% de l'histoire de la Terre. L'évolution
subséquente à partir d'un noyau entièrement fluide, à peu près homogène, jusqu'à l'état actuel
correspondant à un noyau intérieur solide (fer surtout) et d'un noyau extérieur fluide contenant
une partie d'éléments légers en alliage pose un autre problème. Cette évolution subséquente
peut avoir procédé beaucoup plus lentement et n'est peut-être pas encore complète. L'énergie
gravitationnelle serait libérée par la croissance du noyau intérieur et on peut proposer que ceci
peut fournir la source principale d'énergie pour actionner la géodynamo.

INTRODUCTION

The major source of our information about the Earth's interior comes from
seismology. Travel times can be used to construct velocity-depth curves; from them
the variation of density and other physical properties within the Earth can be calculated. More recently, travel-time curves have been supplemented by free oscillation
data, and considerable advances have been made in dealing with this inverse problem. Many more sophisticated Earth models have now been constructed and bounds
placed on the variables, although the broad division of the Earth into crust, mantle
and core has been known for some time; Oldham deduced the existence of a core as
early as 1906. It must not be forgotten, however, that seismology only gives us a
'snapshot' of the interior of the Earth as it is today and provides no information about
its structure in the past nor of its evolution.

The outer part of the core (OC) is considered to be liquid because shear waves
have never been observed in it. However, evidence for a solid inner core (IC) has
steadily increased. Dziewonski and Gilbert (1971, 1973) have shown that free oscillation data demand a solid IC in which the average shear velocity is about 3.6 km/s.
Müller (1973) came to a similar conclusion from an analysis of the amplitudes of
long-period core phases. The observation of shear waves in the IC would establish its
rigidity directly. Julian et al. (1972) claimed to have identified the phase PKJKP on a
seismogram and deduced a value of around 2.95 km/s for the shear wave velocity in
the IC. Unfortunately, this cannot be reconciled with the value of 3.6 km/s obtained
from the normal mode data by Dziewonski and Gilbert (1973). It is not clear what
phase Julian et al. (1972) identified on their seismogram; Doornbos (1974) has suggested that the phase PKJKP is too small to be observed.

The existence of a liquid OC and a solid IC raises the question of its origin. Has
the Earth always had such a structure, or has it evolved over geologic time? This
question cannot be divorced from the much broader question of the origin of the
Earth and solar system. It is usually assumed that the embryo Earth was homogeneous and later differentiated into crust, mantle and core. Inhomogeneous models have
also been suggested in which the (mainly iron) core formed first, the silicate mantle
being deposited upon it later. Although the origin of the Earth cannot be discussed in
any detail in this paper, some aspects of the early history of the solar system will be
considered in the next two sections.

HEAT SOURCES FOR THE EARLY EARTH

The dynamo theory of the geomagnetic field demands a (predominantly iron) molten OC at least 3000 Ma ago. How the Earth could have heated up to this extent is a major problem. Possible heat sources are the decay of long-lived and short-lived radioactive isotopes. The long-lived radioactive isotopes which contribute significantly to the present heat production within the Earth are U^{238}, U^{235}, Th^{232}, and K^{40}, all of which have half lives comparable to the age of the Earth. The temperature increase due to the radioactive decay of these long-lived isotopes is thus small, about 150°K after 100 Ma (MacDonald, 1959). During the first 1000 Ma, the temperature increase would only have been about 700°K, even assuming no heat escape.

Short-lived radioactive isotopes could have contributed to the initial heat of the Earth if the time between the formation of the elements and the aggregation of the Earth was short compared with the half lives of the isotopes. The most important short-lived isotopes are U^{236}, Sm^{146}, Pu^{244}, and Cm^{247}, all of which have half lives sufficiently long to have heated up the Earth for some tens of million years after its initial formation. The decay of three shorter-lived radionuclides, Al^{26}, Cl^{36}, and Fe^{60}, would have significantly heated up accreting planetary bodies for a period of about 5 to 15 Ma after the termination of nucleosynthesis in the primitive solar system. Of these, Al^{26} is the most important. It decays to Mg^{26} with a half life of 0.74 Ma and would have remained a significant source of heat for about 10 Ma. If the Earth accreted within 20 Ma of the termination of nucleosynthesis, the heat released through the decay of Al^{26} could have been the main cause of its high internal temperature. In this regard, no trace of any anomaly in the Mg^{26}/Mg^{24} ratio had, until recently, been found in meteorites, nor in any lunar and terrestrial samples. However, in 1974 Gray and Compston found an anomalous value of this ratio in a chondrule in the Allende meteorite. Their observation has since been confirmed by Typhoon Lee *et al.* (1977) and Papanastassiou (talk presented at Wilson Symposium, October 1978) providing definite evidence for the presence of Al^{26} in the early solar system. This would require either injection of freshly synthesized nucleosynthetic material into the solar system immediately before condensation and planet formation, or local production within the solar system by intense activity of the early Sun. Mg^{26} anomalies have also been reported in the Leoville and Murchison carbonaceous chondrites.

It has been suggested that the Moon melted completely 4400 Ma ago, and that between 4000 and 3200 Ma ago it had an internal magnetic field (Stephenson *et al.* 1974). For an object the size of the Moon, the gravitational energy released by accretion (see later) is insufficient to provide the heat of melting, which must have come mainly from short-lived radioelements. Runcorn *et al.* (1977) and Runcorn (1978) have suggested that super-heavy elements (SHE) with atomic numbers between 114 and 126 may be relatively stable and could have provided an early heat source in the Moon. Runcorn *et al.* (1977) have reviewed the many efforts to search for evidence of SHE in nature; Libby *et al.* (1979) suggest the possibility of their presence in iron meteorites. It is possible that they could also have provided a heat source for the Earth.

Other possible heat sources are more difficult to calculate than radioactive decay, but nevertheless may be important. Although data (particularly on the variation with pressure of the coefficient of thermal expansion) are rather uncertain, adiabatic compression could have raised the temperature of the aggregating Earth several hundred degrees. Another possible heat source is the potential energy from the mutual gravitational attraction of the particles in the dust cloud. The kinetic energy of the aggregating particles would have been either converted into internal energy or radiated away. It is extremely difficult to estimate the contribution from this source because we lack knowledge of the physical processes of accretion. The result depends quite critically on the temperature attained at the surface of the aggregating Earth, on the transparency of the surrounding atmosphere to radiation and on the duration of the accretion process; rapid accretion is necessary to produce high temperatures and melt the OC. The dissipation of the Earth's rotational energy through tidal interaction with the Moon has been suggested as another important heat source. However, calculations indicate that tidal dissipation would have raised the temperature by less than 100°K in 1500 Ma. Even if the Moon was once close to the Roche limit, it would have receded so rapidly that the effects of tidal heating would have been negligible. It has also been suggested that if the Sun passed through a T-Tauri stage, the solar wind could have driven a unipolar generator which produced heat by electric currents (Sonett and Colburn, 1970; Herbert et al., 1977). Although this effect might have been significant, it is difficult to estimate.

Safronov (1972) believes that one of the main sources of heat was the impact of falling bodies. This is also discussed by Kaula (1980). Estimates of the temperature rise from this source depend largely on the body sizes assumed. Calculations based on the assumption that these bodies were small give a low initial temperature. Safronov believes, however, that relatively large bodies were involved in the formation of the planets, the largest bodies falling on the Earth having diameters of several hundred kilometres. In a later paper, Safronov and Kozlovskaya (1977) showed that in the final stages of growth of the Earth, the fusion point would have been reached if the radius of the largest body impacting the Earth was greater than 20 km. The presence of huge impact craters on all of the terrestrial planets would seem to affirm the existence of large-scale preplanetary bodies. It is thus highly probable that the outer layers of the Earth were fused to a thickness of some hundreds of kilometres in the final stages of growth. If a significant part of the Earth's mass came from bodies with dimensions of tens of km, then its upper layers could have been molten as early as the time of its formation, in which case the final stages of the Earth's growth could also have been the beginning of the process of gravitational differentiation and core formation.

Finally, the formation of the core itself may have played a dominant role in the thermal history of the Earth. If the Earth formed by accretion from approximately homogeneous material and later differentiated into crust, mantle and core, the formation of the core would have released a large amount of gravitational energy as a result of the concentration of the high density nickel-iron in the centre of the Earth. Tozer (1965) estimated the increase of heat arising from core formation from an originally undifferentiated Earth to be 470 cal/g. About 6 percent of this would melt the Ni-Fe phase, while the rest would raise the mean temperature of the Earth about 1500°K. Flaser and Birch (1973), in a similar calculation, estimated the energy release due to core formation to be 590 cal/g, which would be sufficient to heat the Earth by about

2000°K. It does not follow, however, that the release of such heat would cause extensive or complete melting of the mantle. The process of core formation would result in strong convection throughout the Earth leading rapidly to the establishment of an adiabatic gradient. Because of the drastic effect of pressure upon the melting point gradient, the mantle below a depth of about 500 km would most probably be solid (Ringwood, 1977b).

TIME OF CORE FORMATION

It is generally believed that the Earth's magnetic field arises from dynamo action in the fluid, electrically conducting OC. Remanent magnetization has been found in rocks about 3000 Ma old (e.g., McElhinny *et al.*, 1968), so that the OC must have been molten at least that long ago. Hanks and Anderson (1969) carried out an investigation of the thermal history of the Earth with this additional constraint and came to the conclusion that the Earth must have accreted in a period less than 0.5 Ma. There would not have been time for the decay of long-lived radioactive isotopes to have had sufficient heating effect, and they discounted the importance of the decay of short-lived radionuclides. A short accretion time is thus essential for most of the gravitational potential energy of the dust cloud to be trapped and not lost to space. (Mizutani *et al.* (1972) and Wetherill (1972) have both come to the conslusion that the Moon accreted in a very short time $\simeq$ about 1000 yr). The accretion time for the Earth could be lengthened somewhat since the core is not pure iron but contains substantial amounts of some light element (see next section). Hall and Murthy (1972) have estimated that at core pressures the melting temperature of the Fe-FeS eutectic is about 1600 K lower than that of pure iron. Ringwood (1977b) believes that FeO would also have a drastic effect in reducing the melting point of iron.

Oversby and Ringwood (1971) have also produced evidence for an early formation of the Earth's core. Ringwood had argued in 1960 that iron descending during core formation would take with it substantial amounts of lead, but not uranium. The main argument for his conclusion was that during core formation the Pb/U ratio would be higher in the metal phase than in silicates. Oversby and Ringwood carried out experiments to measure the distribution coefficient of lead between a series of relevant iron alloys and a basaltic melt, and obtained a relationship between the distribution coefficient and the time Δt that would have elapsed from the accretion of the Earth to the segregation of the core. They concluded that the Earth's core formed either during accretion or very soon afterwards (probably in < 100 Ma). Similar calculations have since been made by Vollmer (1977), using small revisions to the u decay constants (Jaffey *et al.*, 1971) and the isotopic composition of primordial lead. Vollmer concluded that core formation most probably occurred over a period of about 500 Ma after accretion. However, he assumed that deviations in isotopic compositions of oceanic basaltic leads from the lead growth curve of the terrestrial crust were caused, not by minor fractionations of lead from uranium in the source region of basalts, but by the core formation process itself. Geochemical evidence casts some doubt on this assumption; even accepting Vollmer's longer time, core formation should have been essentially complete during the first 10 per cent of the Earth's history.

The discovery of Buolos and Manuel (1971) of Xe^{129} in deep mantle gases sets limits on the time of formation of the Earth relative to nucleosynthesis (Xe^{129} is produced by the decay of the short-lived extinct radioactive I^{129} ($t_{1/2} = 17$ x 10^6 yr). Their discovery precludes a sustained high temperature regime for the mantle after a few mean lifetimes (about 80 Ma) of this extinct radionuclide. If the Earth accreted rapidly (10^3 - 10^4 yr), it is doubtful whether it would have cooled to low enough temperatures to retain Xe^{129} in this short a time. Hot accretion models for the Earth, in which an efficient trapping of the gravitational potential energy occurs, would thus seem untenable in the light of this discovery. On the other hand, low temperature accretion and early formation of the core would rule out pure Fe/Ni melting and sinking to form the core; an additional component would be necessary to lower the melting temperature.

Further evidence for a short time interval between nucleo-synthesis and the formation of planetary-sized objects has been put forward by Kelly and Wasserburg (1978). They found an isotopic anomaly of Ag^{107} in the Santa Clara iron meteorite which they attributed to the insitu decay of Pd^{107} (half life 6.5 x 10^6 yr). Their data are incompatible with Δt about 10^8 yr and suggest that the time required for the differentiation of some planets into large scale metallic and silicate phases is as short as 10^6 yr.

Slattery (1978) has carried out model calculations which indicate that for a protoplanet the mass of Saturn, core formation is very rapid. De Campli and Cameron (1979) showed that the iron, present as clumped interstellar grains, would melt and remain liquid for up to 10^5 yr. The melted iron drops would coalesce to form a cloud, and a rain-out of large iron drops would follow. Slattery showed that the coalescence and rain-out of iron drops in a protoplanet are very rapid compared to evolutionary time scales, and that mass transport or iron to the center of a protoplanet is an extremely efficient mechanism for core formation.

Formation of the core soon after accretion requires that the melting point of the iron alloy be exceeded throughout substantial regions of the Earth's interior, permitting the metal to segregate into bodies sufficiently large to sink through the mantle. Thus a large source of "initial heat" is required. This was most probably supplied during accretion by the transformation into thermal energy of the gravitational potential energy of the dust cloud (see previous section). This initial heat source does not imply that the mantle ever became completely molten. Ringwood (1977b) has presented convincing evidence that the mantle was not subject to large-scale melting during the formation of the Earth.

Apart from the time taken for the evolution of the present structure of the Earth's core, if gravitational energy is invoked we must also consider where inside the Earth the maximum energy is released and what form this takes. In the formation of a completely fluid core from a homogeneous cold Earth, the gravitational energy is presumably transferred into heat by viscous dissipation. The distribution of this energy will depend on the mechanism of core formation. In Vityazev's (1973) model, a ring of iron moves continuously through the Earth to eventually form the core, while in Elsasser's (1963) model, the core forms through a series of "drops". A critical question not considered by these authors is the energy transfer between the silicate and iron phases. (At thermal equilibrium, silicate would contain twice as much energy/g as iron because of its higher heat capacity. However, once the ring or drop of iron has passed a certain point in the Earth, the energy release process ceases in the silicate, but continues in the iron.) This problem will not be considered further here.

The subsequent evolution from an approximately homogeneous, all-fluid core into the present state of a solid (mainly iron) IC and a fluid OC containing some light alloying element is another problem. This evolution may have proceeded much more slowly and still may not be complete. Gravitational energy would be released by the growth of the IC and this, it has been argued, could provide the main source of energy for driving the geodynamo (see final section).

THE COMPOSITION OF THE CORE

The composition of the core will not be discussed in any detail here; a comprehensive review has recently been written by Brett (1976). All evidence indicates that the IC consists primarily of Fe-Ni and the OC of molten Fe alloyed with from 8 to 20 per cent of a light element, giving an average atomic number of about 23.

Jeanloz (pers. commun., 1979) has re-examined the shock wave data for Fe and confirmed earlier conclusions (see e.g., Birch, 1968) that both densitites and bulk moduli in the OC are less than those of Fe under equivalent conditions, although their gradients through the OC are consistent with gross chemical homogeneity (i.e., uniform intermixing of Fe with a lighter, more compressible element or compound). Both densities and bulk moduli for the IC are compatible with those of Fe, suggesting that the inner core boundary (ICB) is likely to be a compositional as well as a phase boundary, in which case the boundary cannot be used as a fixed temperature point on the melting curve of Fe (or a related compound). Masters (1979) has come to a similar conclusion, obtaining for the OC an average value of Bullen's (1963) stratification parameter η of 1.0 ± 0.05. The error is sufficiently small to exclude the strong chemical stratification proposed by Usselman (1975). The shock wave data that Usselman used were extrapolated to core conditions from relatively low pressures, but no detailed corrections were made for temperature effects. Masters has shown that a significant temperature correction to shock wave data is necessary when considering the relative compositions of the IC and OC. (The temperature along the Hugoniot for iron rises much more rapidly with increasing pressure than it does in the Earth. Temperatures are inferred to be similar to those in the Earth at pressures close to those at the top of the core but may be a factor of two or more larger at pressures corresponding to those at the centre of the Earth.)

There is no firm evidence as to what the light element in the OC is. It must be reasonably abundant, miscible with Fe and possess chemical properties that would allow it to enter the core. Prime candidates are silicon (for a summary see e.g., Ringwood, 1966a, 1966b) and sulphur (see e.g., Murthy and Hall, 1970). In order to incorporate Si in the metal which segregated from a mantle containing FeO, Ringwood (1966a) had to postulate a very specific model for the accretion of the Earth and subsequent core formation. Difficulties associated with this model have been summarized by Brett (1976). Herndon (pers. commun., 1979) has pointed out that Ni-Fe-S alloys can exist as liquids at temperatures hundreds of degrees below the melting point of pure Fe at atmospheric pressure. He has shown experimentally that the introduction of Si into a Ni-Fe-S liquid can cause the precipitation of solid (Ni, Fe)$_2$ Si, and suggested that the Earth's IC consists, not of nickel-iron metal, but of nickel silicide.

Ahrens (1976) estimated from shock wave data that about 14 wt % of S would be necessary in the OC. This is equivalent to 4.5% of S in the bulk Earth and implies that the Earth must have accreted about 44% of the primordial abundance of S, assuming the latter to be Type 1 Carbonaceous chondrites. The hypothesis that S in the principal light element in the OC poses a major problem: Why did the Earth accrete about half of the primordial abundance of this extremely volatile element, while at the same time it became more strongly depleted in several elements much less volatile than S?

In 1971 Lewis and Hall and Murthy independently suggested that if the light element of the OC were S, the bulk of the Earth's potassium would then be present in the Earth's OC. If this were true, it would have far-reaching implications for the thermal evolution of the Earth because of the radioactive decay of K^{40}. Energy would become available for convection in the core, thus providing a driving mechanism for the Earth's magnetic field, and the thermal regime in the mantle would also be affected. The issue is still controversial. Goettel (1972) and Goettel and Lewis (1973) have given arguments in favour, and Bukowinski (1976) has calculated that K would take on the characteristics of a transition metal at high pressures. However, as Ringwood (1977) has pointed out, this does not imply that K would be selectively partitioned into the core. Oversby and Ringwood (1972, 1973), Seitz and Kushiro (1974) and Ganguly and Kennedy (1977) have all presented arguments against there being any K in the core. It is often stated that the Earth's magnetic field demands large quantities of radioactive material in the core. This argument weakens once alternative energy sources are recognized. As discussed in the final section, if the Earth is cooling, the growth of the IC could provide ample power for magnetic field generation; the gravitational energy released by the rearrangement of material in the OC can be converted very efficiently into heat by ohmic dissipation.

Recently, oxygen has been proposed as the light element in the OC. Bullen (1973) suggested that the OC consists of liquid Fe$_2$O. However, this oxide is unstable at ordinary pressures and there is as yet no experimental evidence that it can exist at high pressures. Ringwood (1977a) has suggested that FeO may be a major constituent of the OC. FeO in solution is known to reduce the melting point of Fe quite considerably, more so than does S. Ringwood pointed out that under the high pressures at great depth in the Earth, complete miscibility might occur between Fe and FeO, which could decrease the melting point of Fe enough to enable core segregation without appreciable melting of mantle material. He estimated that the core would contain about 44 wt % of FeO, equivalent of 10 wt % of O. Although he believes that O is probably the principal light element in the Earth's OC, Ringwood concedes that significant amounts of S and C may also be present, the exact amounts depending on the conditions under which the Earth accreted (Ringwood, 1977b).

DRIVING FORCE FOR THE GEODYNAMO

It is now generally believed that the Earth's magnetic field arises from some dynamo action in the fluid OC (see Gubbins, 1974 for a review). Although it has been shown that kinematic dynamos exist, such solutions are of limited interest since there is no guarantee that there exist forces in the core that can sustain them. A number of possible force fields have been proposed. Malkus (1963, 1968) suggested that precession may produce turbulent motions in the core and hence drive the dynamo. A detailed theory of a dynamo in a precessing turbulent core is difficult, and no full treatment has as yet been given. However, Rochester *et al.* (1975) have pointed out mathematical and physical errors in Malkus' arguments and shown that precession fails by at least two orders of magnitude to satisfy the power requirements to drive the dynamo. Loper (1975) has also come to a similar conclusion.

Won and Kuo (1973) suggested that the Earth's IC might be excited by an external agency (such as an earthquake) and set up motions in the fluid OC. However, Busse (1975) pointed out that the radial velocity of such motions would be far too small to maintain the field. Mullan (1973) also suggested that seismic activity could generate the field, but Gubbins (1975) showed that the amplitude of such oscillations in the core would have to be so large that they would set up observable gravity anomalies. The predicted gravity anomalies are not seen, since there are in addition unexplained low harmonies in the field. This leaves two main candidates for the generation of the field: thermal convection and the gravitational energy released by the rearrangement of matter in the formation of the IC.

The inner core boundary (ICB) is thought to be at the melting point of core material. If the Earth is cooling slowly, the IC grows at the expense of the OC. If there is a compositional difference between the OC and IC then the rejection of impurities at the advancing ICB will provide an additional energy source to drive motions in the OC. The rejected impurities will be gravitationally buoyant and rise, re-combining with OC material further away from the ICB; this rearrangement of matter will release gravitational energy and mechanically stir the OC. Braginsky (1963) first suggested that this process may have produced the geomagnetic field. The poor efficiency of dynamos driven by thermal convection has led to revived interest in the "gravitationally powered dynamo" (Gubbins, 1976, 1977; Loper, 1978).

A gravitational power source for the dyanmo depends critically on the size of the density jump Δ_e at the ICB. Estimates of Δ_e at the ICB have varied from 0.2 to 0.9 g/cm^3 in recent Earth models (e.g., Buchbinder, 1972; Cook, 1972). An upper limit of 1.8 g/cm^3 was calculated by Bolt and Qamar (1970), using a small dataset of PKiKP/PcP amplitude ratios. One Earth model, C2 of Anderson and Hart (1976), shows a small value of Δ_e at the ICB but also a decrease in bulk modulus from the OC to the IC. This is at variance with their conclusion that the IC and OC are isochemical; experimental evidence suggests that the bulk modulus will increase on solidification unless there is a large compositional difference between solid and liquid phases.

Masters (1979) has used free oscillation data to show that Δ_e at the ICB is 0.6 to 0.7 g/cm^3, with an estimated error of about 0.4 g/cm^3. The crucial question is whether this increase is due solely to solidification or to a compositional difference as well. Masters has shown that the upper limit for the density change in the solidification of ϵ-iron is 0.15 g/cm^3, which is significantly less than that inferred from the free oscillation data. The effect of impurities will further reduce this value, and it does appear that there is a compositional difference between the IC and OC.

Ruff and Anderson (pers. commun., 1979) have proposed a different mechanism for fluid motions in the OC. They begin with a modified inhomogeneous accretion model. The early condensates, Ca- and Al-rich silicates, heavy refractory metals and Fe accrete to form the protocore. In contrast to an earlier model, (Anderson and Hanks, 1972), the early thermal history of the Earth is dominated by Al^{26}, which produces enough heat to raise core temperatures by $1000°$ K and melt it. Melting of the core results in unmixing and the emplacement of refractory material (including U, Th and possibly Al^{26}) into the lowermost mantle. Ruff and Anderson then suggest that fluid motions in the core are driven by differential heating from *above*, the resulting cyclonic motions powering the geodynamo.

THE GRAVITATIONALLY POWERED DYNAMO

Gubbins *et al.* (1979) have shown that the geomagnetic field could be generated either by radioactive heating in the core or by its slow thermal evolution. At least 10^{13}W of radioactive heating is required, and since this value is only 25% of the observed surface heat flux, the possibility of a dynamo driven by radioactive heating cannot be ruled out on thermal arguments alone.

Verhoogen (1961) has suggested that the latent heat released by the solidification of the IC could power the dynamo. Gubbins *et al.* (1979) have shown this to be untenable since the power required would freeze the liquid core more rapidly than the present size of the IC would suggest. The differences between these calculations are due to the fuller use of the entropy equation by Gubbins *et al.* and to the fact that, unlike Verhoogen, they related the growth rate of the IC to the difference in the melting point and adiabatic temperature gradients rather than to the melting point gradient alone. The melting and adiabatic temperature gradients are probably quite similar in the core, and this could lead to a substantial numerical difference.

In calculating the gravitational energy release from a cooling, differentiating core, Gubbins *et al.* found that the heat sources and entropy changes depend critically on three very poorly known quantitites: the cooling rate, the temperature gradients and the density jump at the ICB. Their sample calculation yielded a heat flux of 2.5×10^{12}W, of which 49% comes from latent heat, 25% from gravitational energy, 20% from cooling, 5% from chemical internal energy and 1% from adiabatic heating. On the other hand, the entropy available for dissipation comprises 67% gravitational, 25% latent heat, 6% cooling and 2% chemical internal energy and adiabatic heating. The entropy is dissipated by ohmic heating (44%) thermal conduction (42%) and by the diffusion of material (14%), giving a magnetic field of the order of 100 Gauss. The growth rate of the IC determines the available power; the density jump at the ICB determines the gravitational energy release and the adiabatic heating.

The cooling rate of the core is actually determined by the thermal evolution of the mantle, which may have cooled $200°K$ since the early Precambrian (McKenzie and Weiss, 1975). This cooling rate of $70°K$ per 10^9 yr (seven times the value used by Gubbins *et al.*) gives a heat flux of about 10^{13}W from cooling of the mantle alone. To retain the same rate of growth of the IC using this rate of cooling, the value of the temperature gradient would have to be increased by an amount which would change the adiabatic and melting temperature gradients by a factor of two or three, which is well within the limits of error.

Loper (1978) has also considered the question of a gravitationally powered dynamo, but while Gubbins *et al.* used the entropy equation to estimate the allowed magnetic field, including the very important entropy of conduction, Loper omitted it. This explains the difference in the two estimates of the magnetic fields that can be generated. In addition, Gubbins *et al.* included chemical terms (omitted by Loper) that add to the heat flux but at the same time lose entropy.

It must be stressed that the above calculations refer only to the energy balance of the core at the present time. It is of interest to consider the energy balance in the past and in particular to inquire whether a gravitational power source could have maintained a magnetic field throughout geological time. If the IC were absent, then only cooling and radioactive heat sources would be available to power the dynamo. The cooling rate required to sustain a magnetic field is quite large, about $150^{\circ}K/10^9$ yr, and if this cooling rate were to continue after IC formation, it would imply that the IC were a very young feature. It is also interesting to speculate whether an IC is necessary for the operation of a dynamo, and whether the very small magnetic fields observed on the moon and other terrestrial planets are due to a lack of differentiation into a solid IC and fluid OC.

REFERENCES

Ahrens, T.J., 1976, Shock wave data for pyrrhotite and constraints on the composition of the outer core, U.S.-Japan seminar on high-pressure research applications: in geophysics, p. 31.

Anderson, D.L. and Hanks, T.C., 1972, Formation of the Earth's core Nature, v. 237, p. 387.

Anderson, D.L. and Hart, R.S., 1976, An Earth model based on free oscillations and body waves: Jour. Geophys. Res. v. 81, p. 1461.

Birch, F., 1968, On the possibility of large changes in the Earth's volume: Phys. Earth Planet. Int., v. 1, p. 141.

Bolt, B.A. and Qamar, A., 1970, Upper bound to the density jump at the boundary of the Earth's inner core: Nature, v. 228, p. 148.

Branginsky, S.I., 1963, Structure of the F layer and reasons for convection in the Earth's core: Dokl. Akad, Nauk, SSSR, v. 149, p. 1311.

Brett, R., 1976, The current status of speculations on the composition of the Earth: Rev. Geophys. Space Phys., v. 14, p. 375.

Buchbinder, G.G.R., 1972, An estimate of inner core density: Phys. Earth Planet. Int., v. 5, p. 123.

Bukowinski, M.S.T., 1976, The effect of pressure on the physics and chemistry of potassium: Geophys. Res. Letters, v. 3, p. 491.

Bullen, K.E., 1963, An index of degree of chemical inhomogeneity in the Earth: Geophys. Jour. v. 7, p. 584.

——————, 1973, Cores of the terrestrial planets: Nature v. 243, p. 68.

Buolos, M.S. and Manuel, O.K., 1971, The Xenon record of extinct radioactivities in the Earth: Science, v. 174, p. 1334.

Busse, F., 1975, A necessary condition for the geodynamo, Jour. Geophys, Res. v. 80, p. 278.

Cook, A.H., 1972, The inner core of the Earth: Lincei-Rend. Sci. Fis. Mat. Nat., v. 52, p. 576.

De Campli, W.M. and Cameron, A.G.W., 1979, Giant gaseous proto-planets: Submitted to Icarus.

Doornbos, D.J., 1974, The anelasticity of the inner core: Geophys. Jour., v. 38, p. 397.

Dziewonski, A.M. and Gilbert, F., 1971, Solidity of the inner core of the Earth inferred from normal mode observations: Nature, v. 234, p. 465.

Dziewonski, A.M. and Gilbert, F., 1973, Observations of normal modes from 84 recordings of the Alaskan Earthquake of 1964 March 28-II Further remarks based on new spheroidal overtone data: Geophys. Jour., v. 35, p. 401.

Elsasser, W.M., 1963, Early history of the Earth In Geiss, J. and Goldberg, E.D.,eds., Earth Science and Meteorites: North Holland Publ. Co., Amsterdam.

Flaser, F.M. and Birch, F., 1973, Energetics of core formation: a correction: Jour. Geophys. Res., v. 78, p. 6101.

Ganguly, J. and Kennedy, G.C., 1977, Solubility of K in Fe-S liquid, silicate-K-(Fe-S) liq equilibria, and their planetary implications: Earth Planet. Sci. Letters, v. 35, p. 411.

Goettel, K.A., 1972, Partitioning of potassium between silicate and sulfide melt: experiments relevant to the Earth's core: Earth Planet Sci. Letters, v. 6, p. 161.

Geottel, K.A. and Lewis, J.S., 1973, Comments on a paper by V.M. Oversby and A.E. Ringwood: Earth Planet Sci. Letters, v. 18, p. 148.

Gray, C.M. and Compston, W., 1974, Excess ^{26}Mg in the Allende Meteorite: Nature, v. 251, p. 495.

Gubbins, D., 1974, Theories of the geomagnetic and solar dynamos: Rev. Geophys. Space Phys. v. 12, p. 137.

—————————, 1975, Can the Earth's magnetic field be sustained by core oscillations?: Geophys. Res. Letters, v. 2, p. 409.

—————————, 1976, Observational constraints on the generation process of the Earth's magnetic field: Geophys. Jour., v. 47, p. 19.

—————————, 1977, Energetics of the Earth's core: Jour. Geophys., v. 43, p. 453.

Gubbins, D., Masters, T.G. and Jacobs, J.A., 1979. Thermal evolution of the Earth's core: Geophys. Jour., v. 59, p. 57.

Hall, H.T. and V.R. Murthy, 1971, The early chemical history of the Earth: some critical elemental fractionations: Earth. Planet. Sci. Letters, v. 11, p. 239.

—————————, 1972, Comments on the chemical structure of an Fe-Ni-S core of the Earth, Int. Conf. Core-Mantle interface: Trans-Amer. Geophys. Union, v. 53, p. 602.

Hanks, T.C. and Anderson, D.L., 1969, The early thermal history of the Earth, Phys. Earth Planet. Int., v. 2, p. 19.

Herbert, F., Sonett, C.P. and Wiskerchen, M.J., 1977, Model "zero age" lunar thermal profiles resulting from electrical induction: Jour. Geophys. Res., v. 82, p. 2054.

Jaffey, A.H., Flynn, K.F., Glendenan, L.E., Bently, W.C. and Essling, A.M., 1971, Precision measurement of half lifes and specific activities of ^{235}U and ^{238}U: Phys. Rev. v. C4, p. 1889.

Julian, B.R., Davies, D. and Sheppard, R.M., 1972, PKJKP, Nature, v. 235, p. 317.

Kaula, W., 1980, Thermal evolution of a planet growing by planetesimal impacts: in Strangway, D.W., ed., The Continental Crust and Its Mineral Deposits, Geol. Assoc. Canada Spec. Paper 20.

Kelly, W.R. and Wasserburg, G.J., 1978, Evidence for the existence of ^{107}Pd in the early solar system: Geophys. Res. Letters, v. 5, p. 1079.

Lewis, J.S., 1971, Consequences of the presence of sulphur in the core of the Earth: Earth Planet Sci. Letters, v. 11, p. 130.

Libby, L.M., Libby, W.F., and Runcorn, S.K., 1979, The possibility of superheavy elements in iron meteorites: Nature, v. 278, p. 613.

Loper, D.E., 1975, Torque balance and energy budget for the precessionally driven dynamo: Phys. Earth Planet. Int., v. 11, p. 43.

—————————, 1978, The gravitationally powered dynamo: Geophys. Jour., v. 54, p. 389.

MacDonald, G.J.F., 1959, Calculations on the thermal history of the Earth: Jour. Geophys. Res. v. 64, p. 1967,

Malkus, W.V.R., 1963, Precessional torques as the cause of geomagnetism: Jour. Geophys. Res., v. 68, p. 2871.

_____, 1968, Precession of the Earth as the cause of geomagnetism: Science, v. 160, p. 259.

Masters, T.G., 1979, Observational constraints on the chemical and thermal structure of the Earth's deep interior: Geophys. Jour., v. 57, p. 507.

McElhinny, M.W., Briden, J.C., Jones, D.L. and Brock, A., 1968, Geological and geophysical implications of paleomagnetic results from Africa: Rev. Geophys., v. 6, p. 201.

McKenzie, D.P. and Weiss, N.O., 1975, Speculations on the thermal and tectonic history of the Earth: Geophys. Jour., v. 42, p. 131.

Mizutani, H., Matsui, T. and Takeuchi, H., 1972, Accretion process of the Moon: The Moon, v. 4, p. 476.

Mullan, D.J., 1973, Earthquake waves and the geomagnetic dynamo: Science, v. 181, p. 553.

Müller, G., 1973, Amplitude studies of core phases: Jour. Geophys. Res., v. 78, p. 3469.

Murthy, V.R., and Hall, T.H., 1970, The chemical composition of the Earth's core: possibility of sulphur in the core: Phys. Earth Planet. Int., v. 2, p. 276.

Oversby, V.M. and Ringwood, A.E., 1971, Time of formation of the Earth's core: Nature, v. 234, p. 463.

_____, 1972, Potassium distribution between metal and silicate and its bearing on the occurrence of potassium in the Earth's core: Earth Planet. Sci. Letters, v. 14, p. 345.

_____, 1973, Reply to comments by K.A. Goettel and J.S. Lewis: Earth Planet. Sci. Letters, v. 18, p. 151.

Ringwood, A.E., 1960, Some aspects of the thermal evolution of the Earth: Geochim. Cosmochim, Acta, v. 20, p. 241.

_____, 1966a, Chemical evolution of the terrestrial planets: Geochim. Cosmochim. Acta, v. 30, p. 41.

_____, 1966b, The chemical composition and origin of the Earth: In Hurley, P.M., ed., Advances in Earth Science: M.I.T. Press, Cambridge, Mass.

_____, 1977a, Composition of the core and implications for origin of the Earth: Australian Nat. Univ. Publ. No. 1277.

_____, 1977b, Composition and origin of the Earth, Australian Nat. Univ. Publ. No. 1299.

Rochester, M.G., Jacobs, J.A., Smylie, D.E. and Chong, K.F., 1975, Can precession power the geomagnetic dynamo?, Geophys. Jour., v. 43, p. 661.

Runcorn, S.K., 1978, On the possible existence of superheavy elements in the primaevel Moon: Earth Planet. Sci. Letters, v. 39, p. 193.

Runcorn, S.K., Libby, L.M. and Libby, W.F., 1977, Primaeval melting of the Moon: Nature, v. 270, p. 676.

Safronov, V.S., 1972, Evolution of the protoplanetary cloud and formation of the Earth and planets, NAUKA, Moscow (1969): translated NASA TTF-667.

Safronov, V.S. and Kozlovskaya, S.V., 1977, Heating of the Earth by the impact of accreted bodies, Izv. Acad. Sci. USSR: Phys. Solid Earth, v. 13, p. 677.

Seitz, M.G. and Kushiro, I., 1974, Melting relations of the Allende Meteorite: Science, v. 183, p. 954.

Slattery, W.L., 1978, Protoplanetary core formation by rain-out of iron drops: Moon and Planets, v. 19, p. 443.

Sonett, C.P. and Colburn, D.S., 1970, The role of accretionary and electrically inverted thermal profiles in lunar evolution: The Moon, v. 1, p. 483.

Stephenson, A., Collinson, D.W. and Runcorn, S.K., 1974, Lunar magnetic field palaeointensity determinations on Apollo 11, 16 and 17 rocks: Geochim. Cosmochim. Acta 3 (Suppl. 5), p. 2859.

Tozer, D.C., 1965, Thermal history of the Earth1: The formation of the core: Geophys. Jour. v. 9, p. 95.

Typhoon Lee, Papanastassiou, D.A., and Wasserburg, G.J., 1977, Aluminium-26 in the early
    solar system: fossil or fuel?: Astrophys. Jour., v. 211, p. L 107.

Usselman, T.M., 1975, Experimental approach to the state of the core II Composition and
    thermal regime: Amer. Jour. Sci., v. 275, p. 291.

Verhoogen, J., 1961, Heat balance of the Earth's core: Geophys. Jour., v. 4, p. 276.

Vityazev, A.V., 1973, On gravitational differentation energy in the Earth, Izv, Phys. Solid
    Earth: v. 10, p. 676.

Vollmer, R., 1977, Terrestrial lead isotopic evolution and formation time of the Earth's core:
    Nature, v. 270, p. 144.

Wetherill, G.W., 1972, The beginning of continental evolution: Tectonophysics, v. 13, p. 31.

Won, S.J. and Kuo, J.T., 1973, Oscillations of the Earth's inner core and its relation to the
    generation of geomagnetic field: Jour. Geophys. Res., v. 78, p. 905.

The Continental Crust and Its Mineral Deposits, edited by D.W. Strangway,
Geological Association of Canada Special Paper 20

ISOTOPES AND THE EARLY EVOLUTION OF THE EARTH

R.D. Russell
Department of Geophysics and Astronomy,
University of British Columbia,
Vancouver, B.C. V6T 1W5

ABSTRACT

The observed isotopic ratios of lead, strontium, neodynium, helium, and argon contain information about the chemical abundances of selected parent and daughter elements in the outer parts of the Earth. By necessity, we observe these isotopic ratios at the Earth's surface, which is a small, highly evolved part of the Earth. The studies of such isotopic ratios permit inferences to be made about interactions between this crust and the upper mantle.

Helium has been especially valuable for demonstrating that primordial materials are still being outgassed from the earth. Models based on the observed argon isotopic ratios have lead to contradictory conclusions about the existence of an early period of extensive outgassing of the Earth.

Lead has been a particularly interesting element because the ratio of the parents, $^{235}U/^{238}U$, was very different in the Earth's early history than it is now. Therefore there is the potential for determining constraints on the early history of the Earth. A number of recently published papers have offered lead isotope interpretations that reflect on the Earth's early history. Some of these will be reviewed, with special reference to the models of Russell and Birnie that are based upon uni-directional and bi-directional exchange between a protocrust and a residual mantle. Geochemical parameters for uranium, thorium and lead can be inferred for two evolving systems, as well as rate constants for differentiation. The principal conclusions are that the differentiation process extended beyond the first quarter of the Earth's history, and that it is possible to reproduce exactly the apparent oceanic basalt isochron by a simple two-reservoir model. In particular, such a model can explain quantitatively the observed lead-207 deficiency in the oceanic basalts.

RÉSUMÉ

Les rapports isotopiques observés pour le plomb, le strontium, le néodynium, l'hélium et l'argon contiennent de l'information sur les abondances chimiques de certains éléments parents et dérivés dans les parties extérieures de la terre. Par nécessité, nous observons ces rapports isotopiques à la surface de la Terre qui est une partie petite et très évoluée de la Terre. Les études de ces rapports isotopiques permettent de faire certaines inférences sur les interactions entre cette croûte et le manteau supérieur.

L'hélium a été particulièrement utile pour démontrer que les matériaux primordiaux continuent encore de se dégazer de la Terre. Des modèles basés sur les rapports observés d'isotopes d'argon ont mené à des conclusions contradictoires quant à l'existence d'une période précoce de dégazage intensif de la Terre.

Le plomb a été un élément particulièrement intéressant parce que le rapport des parents, $^{235}U/^{238}U$, était très différent dans l'histoire primitive de la Terre de ce qu'on observe actuellement. Par conséquent, il existe une possibilité de déterminer les contraintes sur l'histoire précoce de la Terre. Un nombre d'articles publiés récemment ont présenté des interprétations des isotopes de plomb qui ont une incidence sur les débuts de l'histoire terrestre. On fera une revue de ces interprétations avec une référence spéciale aux modèles de Russell et Birnie qui sont basés sur un échange uni-directionnel et bi-directionnel entre la croûte primitive et un manteau résiduel. On peut déduire des paramètres géochimiques pour l'uranium, le thorium et le plomb pour les deux systèmes en évolution de même que des constantes de taux de différenciation. Les conclusions principales sont que les processus de différenciation se sont prolongés après le premier quart de l'histoire de la Terre, et qu'il est possible de reproduire exactement l'isochrone apparente du basalte océanique par un simple modèle à deux réservoirs. En particulier, un tel modèle pourrait expliquer quantitativement la déficience observée en plomb-207 dans les basaltes océaniques.

INTRODUCTION

The study of the early history of the Earth is difficult practically and philosophically. Not only are the very early events obscured by the remoteness of time, but there is the fundamental difficulty of all histories that there is no *a priori* reason to believe that the essential information is still available to us. Nevertheless, by putting together evidence from many sources, a picture is beginning to emerge of the first thousand million years of our planet's life.

Controversy over hot and cold origins for the Earth has begun to give way to more general agreement that the Earth must have been molten quite early in its history. The likelihood that the planet was accreted from large objects rather than dust particles, objects perhaps as large as one-tenth or more of the Earth's radius (Safronov, 1965), leads to the conclusion that a significant portion of the impact energy was dissipated throughout the body of the early planet by attenuation of seismic waves (Levin, 1972; Wetherill, 1976). If only a small fraction of the energy was distributed throughout the Earth in this way, the initial temperature distribution would have been greatly affected. Without recounting the various agruments in detail, the favoured view seems to be that the Earth accreted in about 10^8 years (e.g., Safronov, 1969) and that the accretion concluded with a molten earth. Under these circumstances, it seems inescapable that the core of the earth was an early feature. This conclusion presents some difficulties, for it has been argued that the chemical composition of the mantle is inconsistent with chemical equilibrium between the metallic iron and silicate phases (Ringwood, 1977).

Other features of the accretion period are even more speculative. Homogeneous and heterogeneous accretion models have had different appeal for different authors, and some authors have argued at different times for one or the other (e.g., Ringwood, 1975, 1977). Similarly, there seems no clear answer to the question of whether the various seismic divisions within the Earth's mantle require chemical layering or whether phase changes observable in the laboratory are sufficient to explain the seismic evidence. In any case, the chemical composition, except for the volatile elements, is suggested to be consistent with accretion of 15% of an oxidized, volatile-rich primordial component similar to Cl-carbonaceous chondrites and 85% of a highly reduced, metal-rich devolatized material formed at high temperatures (Ringwood, 1977).

The loss of the volatiles, assumed in nearly all models, presents some difficulties as well. Probably most or all of the primitive atmosphere escaped. Ringwood (1977) argues that bulk loss of a gaseous atmosphere could be accomplished at a mean temperature of about 2000°K. This estimate is questionable, because it is based on the assumption that all five degrees of freedom of a diatomic gas contribute to the escape; if one considers only the three degrees of freedom associated with translational kinetic energy, the figure must be raised to an unacceptably high value. There is reason to believe that early in the Earth's history the intensity of the solar wind might have been as much as 10^8 times its present value. If so, it is tempting to look to it for a mechanism for removing the atmosphere, but an adequate model has not yet been presented.

The remainder of this paper will be given over to some personal speculations about the evolution of a continental-like structure from such an initially molten planet for which the core has separated and the atmosphere been lost. The evidence used will be almost exclusively isotopic arguments, and even these will be selected and interpreted in a highly subjective way.

ARGON AND HELIUM

Even if it can be agreed that the atmosphere is the result of outgassing since the formation of the Earth about 4600 Ma ago, there remains the difficult question of whether the outgassing was confined to an early period or whether it has continued throughout geologic time. This controversy has continued since the publication of the original arguments of Rubey (1951) concerning the evolution of the atmosphere and oceans. It was one of the important subjects considered by a joint Japan-U.S. conference held in Hakone in the spring of 1977 (Alexander and Ozima, 1978).

The following calculation indicates that, at least for argon, material balance causes no serious problems. The amount of potassium in the crust is estimated to be 3.9×10^{20} kg (Hamano and Ozima, 1978). It is known that potassium contains 0.0117 atom per cent of ^{40}K, that the decay constant for ^{40}K is 5.53×10^{-10}/yr, and that 10% of the decays result in the production of an argon nucleus. Thus the potassium in the Earth's crust has contributed 5.4×10^{16} kg of ^{40}Ar in 4600 Ma, whereas the mass of ^{40}Ar in the atmosphere is about 6.6×10^{16} kg. This calculation neglects the ^{40}Ar trapped in the solid Earth and the total potassium content of the mantle, which is uncertain but estimated to be comparable to that in the crust. Nevertheless, it indicates that the potassium in the Earth can account for, by radioactive decay, the radiogenic argon content of the atmosphere.

Hamano and Ozima (1978) have elaborated this model to obtain estimates of the history of the outgassing process. Like many interpretations, Ozima's model represents by first-order rate constants the material transport of minor elements between the various sub-systems. Apart from being the simplest method for handling the exchange equations, this representation has the advantage of being a good model for a system in which bulk material transport is effected by a "conveyer belt", with onloading and offloading at the ends controlled by simple partition coefficients. The model is formulated in terms of many unknowns, the principal ones being the time of catastrophic outgassing, the fraction of argon that is outgassed at that time, and the first-order rate constants describing the subsequent transport of potassium from the mantle to the crust and the subsequent loss of radiogenic argon to the atmosphere. The essential geophysical conclusions are presented in Table I, taken from the paper of Hamano and Ozima (1978). Note the conclusion that an episodic loss of at least 77% of the available argon took place prior to about 3640 Ma, and also that the rate constants are of the order of 5 x 10^{-11}/yr. The corresponding half-periods for the transport process are of the order of 10^{10} yr, longer than the history of the Earth.

The difficulty in obtaining reliable results from this type of calculation can be shown by comparing these conclusions with those of Fisher (1978). Fisher formulates a rather simpler evolutionary model, also based on first-order rate constants. The principal constraint used is the ratio of $^{40}Ar/^{36}Ar$ (approximately 15,000) inferred for the mantle from analyses of mid-ocean ridge basalts. Fisher concludes that substantial episodic outgassing during the earlier part of the Earth's history is unacceptable, and deduces a rate constant for outgassing of the radiogenic argon into the atmosphere of about $(0.9 - 1.5)$ x 10^{-9}/yr. This corresponds to a half-period for the outgassing process of about 850 Ma, much less than the age of the Earth.

Both the models of Hamano and Ozima, and of Fisher, invoke unidirectional processes only.

TABLE I

Main geophysical results from the model of Hamano and Ozima (1978). k is the first-order rate constant for argon outgassing and α is the first-order rate constant for transport of potassium from mantle to crust. t_d and f are the time and fraction of argon that correspond to the early, sudden degassing.

Degassing process	4.55 Ga ago$< t_d <$3.64 Ga ago
early sudden degassing	77%$<$f ($<$100%)
subsequent continuous degassing	$5.04 \times 10^{-11}yr^{-1} < k < 3.40 \times 10^{-10}yr^{-1}$
Present ^{40}Ar flux from the mantle	$(5.6 - 14.5) \times 10^5$ atoms cm^{-2}s^{-1}
Present ^{40}Ar flux from the crust	$(5.7 - 7.5) \times 10^5$ atoms cm^{-2}s^{-1}
Transportation rate constant, α	$(4.8 - 18.5) \times 10^{-11}yr^{-1}$
Present amounts of Ar	
(^{40}Ar) in the crust	$(6.7 - 8.8) \times 10^{18}$g
(^{40}Ar) in the mantle	$(3.0 - 18.0) \times 10^{19}$g
(^{36}Ar) in the mantle	$(0 - 3.4) \times 10^{16}$g

Tolstikhin (1978) and Craig *et al.* (1978) have recently reviewed the atmospheric characteristics of terrestrial helium, and their significance. These measurements are · particularly important because the remarkable isotopic enrichment of ^{3}He in such materials as hot brines from the Red Sea provides conclusive evidence that primordial gases are still being released to the crust, ocean, and atmosphere. One interesting observation (Craig and Lupton, 1976) is the unexpected uniformity of the ratio ^{3}He/^{4}He in tholeiite glasses for which the absolute helium concentration varies by as much as a factor of 50. Craig *et al.* (1978) comment on this as follows:

Such uniformity in the ratio of a primordial component to a presumably radiogenic component implies a very high degree of homogeneity in ^{3}He, U, and Th, over long times and in large volumes of the mantle. One possible explanation of the uniform ^{3}He/^{4}He ratio, however, is that most of the mantle ^{4}He is actually primordial rather than radiogenic. This possibility requires an early and fairly complete removal of U and Th and ^{40}Ar/^{36}Ar ratios in mantle and crustal material through time. Further

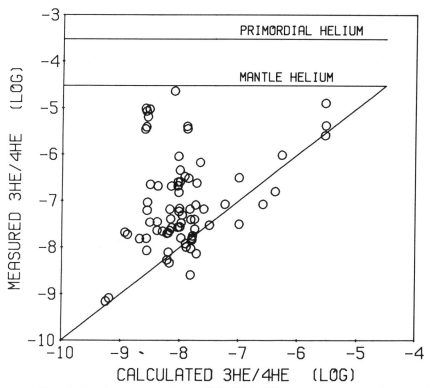

Figure 1. The calculated and measured ratio of ^{3}He/^{4}He in different rocks and minerals (adapted from Tolstikhin and Drubetskoi, 1975). The calculated values are based on the assumption that the rocks contain only radiogenic helium. The points are displaced upward due to mixing with primordial helium. The ratio inferred for the Earth's mantle is also shown. Measured points include acid, intermediate, basic and alkaline rocks of different ages; rocks and minerals rich in lithium; uranium minerals; xenoliths in basalts.

discussion must await more detailed studies of the variability of ^{3}He/^{4}He ratios in many mantle-derived samples.

Figure 1, taken from Tolstikhin (1978) shows a comparison between observed and "calculated" helium isotopic ratios. The calculated ratios are based on the assumption that the helium content is entirely radiogenic. The Figure demonstrates that a primordial component is present in substantial amounts.

Some interesting (but tentative) numerical conclusions follow from the helium isotopic studies. These are based on the hypotheses that the ^{3}He/^{4}He ratio in the Earth's deep interior is very high (perhaps 5×10^{-5} compared with about 10^{-8} for radiogenic helium), that most terrestrial helium represents mixtures of primordial and radiogenic helium, and that the ratios of the helium isotopes in young rocks of probable mantle origin are representative of mantle concentrations. Tolstikhin (1978) infers a ratio of ^{40}Ar/^{36}Ar in the mantle in the range of 500 to 1000. Other estimates range from near atmospheric (295) to more than 10^4. Craig *et al.* (1978) estimate He and Ar fluxes to the Earth's surface to be 4 atoms s^{-1} cm^{-2} (for ^{3}He), 3×10^5 atoms s^{-1} cm^{-2} (for ^{4}He), and 3×10^4 atoms s^{-1} cm^{-2} (for ^{40}Ar).

There is no question that the study of helium isotopic abundances is a very powerful tool for evaluating exchanges between the surface and the deep interior. In the present state of understanding of gaseous exchanges, it is hard to find convincing evidence to settle the question of whether major, episodic outgassing occurred during an early period of the Earth's history. Tolstikhin, at least, seems to doubt it.

STRONTIUM AND NEODYNIUM

Radiometric age determinations based on the decay of rubidium to strontium are among the most common of those used today. Such ages are normally obtained from an isochron plot, from which the initial strontium isotopic ratio, ^{87}Sr/^{86}Sr, is also obtained. Therefore, the published literature contains many determinations of that useful isotopic ratio. The ^{87}Sr/^{86}Sr ratios vary substantially, according to the sources from which the dated materials were formed. The smallest values are found in those rocks and minerals that are thought to derive from the most primitive materials.

From such initial strontium isotopic ratios, estimates have been made of the evolution of the "mantle" values during the history of the Earth. Figure 2, taken from Moorbath (1975) shows ^{87}Sr/^{86}Sr growth lines for samples from Greenland, Scotland, and Rhodesia. The line labelled "A" indicates the estimated variation with time of the strontium isotopic ratio in the mantle. It extends from about 0.699 at 4600 Ma ago to about 0.703 at the present. There is insufficient data to determine whether the line should be drawn straight, or whether it should be curved, concave upwards. Because very old materials are very rare, it is particularly difficult to determine whether the locus is steeper in the early part of the Earth's history, which would be consistent with early redistribution of the near-surface components. Moorbath uses the relationships shown in the Figure to demonstrate the unlikelihood of multiple reworking of older sialic crust, a conclusion that is still controversial.

Similar, complementary studies can be made for the ratio ^{143}Nd/^{144}Nd. Because the parent samarium and daughter neodynium are both rare earth elements with similar chemical properties, the parent/daughter ratio is much less altered by geologic processes than is the case for rubidium and strontium. The changes that do take place are generally in the opposite sense, in that processes for which rubidium is enriched often result in the depletion of samarium, each relative to their daughter elements. It has been shown that the evolution of the neodynium isotopic ratio is close to what one would expect from a reservoir in which the Sm/Nd ratio is similar to the chondritic value (Hamilton et al., 1978; McCulloch and Wasserburg, 1978). De Paolo and Wasserburg (1978), Carter et al. (1978), and others relate the strontium and neodynium isotopic patterns to uncover the effects of contamination of mantle-derived materials by the crust. The neodynium growth curve is not yet known in sufficient detail to tell whether it exhibits evolving crustal chemistry in the very early part of the Earth's history.

Russell and Ozima (1971), starting with a model of Hurley (1968a, 1978b), have

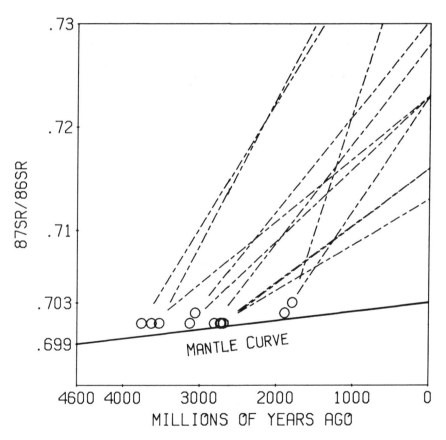

Figure 2. Average ^{87}Sr/^{86}Sr growth lines (shown as broken lines) for gneisses from Greenland, Scotland, and Rhodesia. The inferred initial ratios for each suite of samples, indicated by circles, lie close to an assumed mantle growth curve (actually drawn as a straight line). The figure is adapted from Moorbath (1975).

used the observed surface abundances of ^{40}Ar and rubidium to estimate the mean potassium/rubidium ratio for that part of the Earth that interacts with the surface. It was assumed that the rate of outgassing of argon equalled or exceeded the rate of outgassing for both potassium and rubidium. (The term 'outgassing' describes the movement of the alkalis to the surface, as well as argon). This assumption leads directly to the conclusion that the potassium/rubidium ratio cannot exceed 710. The actual value for this ratio is unknown, but the higher values observed in submarine tholeiites have frequently led to speculation that the ratio exceeds 1000, perhaps substantially. Russell and Ozima concluded that such a high value was incompatible with a continuous outgassing model, and that if the actual value could be proven to be much above 710, then the principal outgassing must have been confined to a relatively short, probably early, part of the Earth's history.

Sun and Hanson (1975) have examined trace element concentration and lead and strontium isotopic ratios with a view to contributing to the understanding of the evolution of the mantle. Among other things, they concluded that chemical heterogeneities in U/Pb and in Rb/Sr have existed in the mantle for 1000 to 3000 Ma.

LEAD

Studies of the decay of uranium into lead have proven to be particularly important for studying the age and evolution of the Earth. The two parent isotopes, ^{238}U and ^{235}U, are normally present in constant proportions; the former decays to ^{206}Pb and the latter to ^{207}Pb. There is, in addition, a third closely related decay, that of ^{232}Th to ^{208}Pb. The half-lives of the uranium isotopes, and the isotopic abundances of uranium, are such that the accumulation of ^{207}Pb was dominant early in the Earth's history, and the accumulation of ^{206}Pb is dominant now. Therefore, there is the potential of distinguishing early from late processes.

The construction of a valid historical record from the lead isotopic evidence has all of the pitfalls of any inverse problem. Perhaps the most important problems relate to uniqueness and time resolution obtainable from the data. However, much has been accomplished.

Stanton and Russell (1959) demonstrated that certain types of stratiform ore leads have isotopic ratios that define a reasonably good evolutionary sequence. The evolutionary curve defined in this way has come to be known as the *primary lead isotope growth curve for the Earth*. To a first approximation it can be represented mathematically by evolution of the daughter isotopes in a closed chemical system. It has often been corrected and revised, for example by Cumming and Richards (1975), but the features of the curve have remained remarkably consistent throughout the 20 years since its proposal. This is not to say that the curve is correct in any *absolute* sense. The data points do not fit on any single-stage curve at exactly their correct ages, and in some cases the order of the apparent ages inferred from the position of fit is the inverse of that required from proven geologic ages. The curve is somewhat diffuse, and upon close investigation the data points themselves often show internal scatter. Nevertheless, it is hard to escape the conclusion that the general concept provides a reasonable approximation to the behaviour of a major lead isotopic system.

When the isotopic composition of natural lead is compared with the growth curve, at least two other anomalies are observed that reflect on the early history of the Earth.

One is the fact that the $^{208}Pb/^{204}Pb$ versus the $^{206}Pb/^{204}Pb$ curve cannot be forced through the primordial abundances (taken to be the isotopic ratios of the troilite phase of iron meteorites) without abandoning the assumption of a closed system. In addition, lead extracted from oceanic basalts have been found to represent a population quite distinct from the above-mentioned ores. The scatter of these data (Fig. 3) extends over an elongated region much greater than would be expected for stratiform ore samples of similar age. In fact, the distribution of the basaltic lead isotope ratios is much more reminiscent of secondary isochrons, such as are exhibited by the well-known anomalous leads of Broken Hill, Australia, of the Tri-State mining district of the United States, and of many other regions.

The scatter of the basaltic leads is a characteristic that is interesting and important, for it reflects the heterogeneous character of the mantle, which is presumed to be the source of these materials. In terms of understanding the early history of the Earth, though, the most interesting characteristic is that the linear array of the basaltic leads lies distinctly *below* the primary growth curve. This is often described as the ^{207}Pb deficiency of the oceanic basalts. It has previously been mentioned that the characteristics of the uranium decay are such that major changes in the ^{207}Pb abundance must occur early in the Earth's history. Therefore, we see that the oceanic basalt leads offer direct evidence of early processes. Moreover, we see direct evidence that characteristically different isotopic regions have been preserved from early in the Earth's history to the present. Reference has already been made to a similar suggestion by Sun and Hanson (1975).

Many authors have attempted to rationalize the observed data from conformable lead ores with a growth curve starting at the assumed values of primordial lead, and fitting the data points reasonably well. Doe and Stacey (1974) obtained $t_0 = 4430$ Ma, $\mu = 9.56$ and $\kappa = 4.404$ (t_0 is the age of the Earth, μ is the $^{238}U/^{204}Pb$ ratio extrapolated to the present and κ is the $^{232}Th/^{238}U$ ratio extrapolated to the present). Cumming and Richards (1975), using quite different procedures, obtained $t_0 = 4470$ Ma, $\mu = 9.15$, and $\kappa = 3.94$. Russell (1978) obtained $t_0 = 4473$ Ma, $\mu = 9.10$ and $\kappa = 4.03$. Thus, there is fair agreement in the conclusions made by different authors.

For the conditions at the time of formation of the Earth, Stacey and Kramers (1975) used the initial lead isotopic composition and age of the meteorite system ($t_0 = 4570$) as measured by Tatsumoto et al. (1973). By imposing an episodic change in the Earth's chemistry at about 3700 Ma they obtained better agreement with the assigned geologic ages of the samples. Cumming and Richards (1975) obtained similar benefits by allowing the quantity μ to vary linearly with time. It is questionable whether our present techniques have sufficient resolution to distinguish between such episodic and continuous models. Nevertheless both of these interpretations suggest significant changes early in the Earth's history.

Of particular relevance to the evolution of the Earth is the paper of Armstrong (1968), which was the first major attempt to reconcile the isotopic abundances of lead and strontium with the concept of plate tectonics. The paper anticipated many later publications by considering material exchange between two (or more) chemical reservoirs. For lead, the model was constructed to remove anomalies in the apparent ages of lead extracted from beach sand composites. The model of Armstrong is constrained by two assumptions that may profoundly restrict the quantitative conclusions: 1) the number of ^{204}Pb atoms in the crust has not varied with time, and 2) the relevant chemical exchanges have reached a steady state. Armstrong did provide an estimate for a first-order transport constant for lead, which was 6 x 10^{-9}/yr. The corresponding half-period is much less than the age of the Earth, implying a much more rapid evolution in the early part of the Earth's history. A comment on the validity of this rate constant is given in Russell and Birnie (1974). It is well to bear in mind that such isotopic models are useful primarily in increasing our understanding of how isotopic systems behave. To relate them in detail to natural processes is a far more difficult task. The publication by Doe and Zartman (1979) represents one recent and noteworthy attempt to establish such relationships.

The present writer has published a series of papers developing open system isotopic models which are capable of reproducing the ^{207}Pb discrepancy of the oceanic basalt leads; the most recent of these papers is Russell (1978). The models are based on the concept of two interacting reservoirs, the proto-crust and the residual mantle, between which exchanges are governed by first-order rate constants. The solutions are in a closed, analytical form, for which the general properties can be evaluated. By fitting the model parameters to agree with the observed data points, both geochemical ratios and rate constants can be evaluated. The general characteristics of the resulting model are shown in Figure 3, which also shows the observed ^{207}Pb discrepancy.

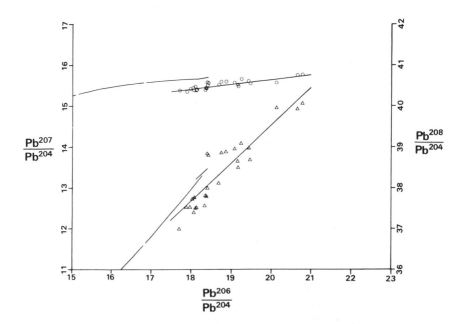

Figure 3. The relationship between the primary lead isotope growth curves, shown as curved lines broken at 1000 Ma intervals, and oceanic basalt leads, shown as circles (^{207}Pb/^{204}Pb) and triangles (^{208}Pb/^{204}Pb). The longer straight lines indicate the trends of the oceanic basalts, as inferred by Russell and Birnie (1974). The short straight lines at the right-hand end of the primary growth curves indicate the range of primary leads of zero age. The figure is from Russell (1978.)

In developing this model it was first assumed that the Earth was entirely molten at some time early in its history. Next there was assumed to be a significant period of time after the formation before the surface became sufficiently stable to support large-scale heterogeneities. After this time, it became possible to preserve large-scale systems without significant mixing of materials among them. There was then a mechanism for transport between residual mantle and proto-crust. For example, the transport process of present plate tectonics seems to operate like a "conveyer belt", and if a simple partition law governs loading and unloading of the conveyer, the overall process can be described by first-order rate constants. (This last point has also been made by M. Ozima). Current opinion is beginning to favour the idea that plate tectonics in the early Precambrian may have been rather different from what we now observe. Nevertheless, in developing this model the assumption was made that whatever forms of evolutionary process were involved, first-order rate constants are appropriate.

The primary objective of the calculations was to explain why leads from oceanic basalts generally lie along a trend distinctively below the primary growth curve from the Earth. This is the ^{207}Pb deficiency already described. The oceanic basalt leads were represented by the regression line of Russell and Birnie (1974):

$$y = 0.1158 (x + 18.322) + 15.543$$

where $x = {}^{206}Pb/{}^{204}Pb$ and $y = {}^{207}Pb/{}^{204}Pb$

The primary growth curve used was that of Russell (1978). The origin of time was taken to be the beginning of differentiation, time t_d ago, so that the present is indicated by $t = t_d$.

All the models considered evolved two quasi-isolated reservoirs (the proto-crust and the residual mantle) from a single system supposed to represent the primordial mantle, or at least the part available for interaction with the surface. As the proto-crust was formed, the elements uranium, thorium, and lead were transferred to it from the residual mantle. Before the evolution began, all of these elements were presumed to be contained within the primordial mantle.

To result in the observed ^{207}Pb deficiency in the residual mantle, uranium (and presumably thorium) must be transferred more rapidly than lead, and the evolution must result in non-zero concentrations of these elements in the residual mantle. In this way the lead remaining in the residual mantle will have experienced, in turn, a higher-μ environment, a very low-μ environment, and finally a higher-μ environment.

Two classes of models were considered: evolutionary and bi-directional. In the evolutionary model the concentration of materials into the proto-crust was strictly unidirectional. No return flow was permitted, and the isotopic composition of the lead in the residual mantle was not influenced by the isotopic changes in the lead in the proto-crust. According to this type of model, transport of the uranium, thorium, and lead-204 to the proto-crust was arbitrarily limited by the fact that some fraction of the total number of atoms of these elements was not available to the transport process. The transfer of the other isotopes of lead was determined by the isotopic composition of the lead in the residual mantle and by the rate of transfer of ^{204}Pb.

If the rate constant for lead is denoted by a, and the fraction of the ^{204}Pb not available for transport is α, then the numbers, S_M and S_C, of such atoms in the two systems are described by the pair of equations

$$-\dot{S}_M = a\{S_M - \alpha(S_M + S_C)\} \tag{1}$$

$$S_M + S_C = S_O \tag{2}$$

Similarly, for the parent uranium P, using b for the rate constant, β for the fraction unavailable for transport and λ for the decay constant, the number of atoms P_M and P_C are given by the pair of equations

$$-\dot{P}_M = \lambda P_M + b\{P_M - \beta(P_M + P_C)\} \tag{3}$$

$$P_M + P_C = P_O\, e^{-\lambda t} \tag{4}$$

For the evolutionary model, the transport of the daughter atoms, D, is controlled by the isotope ratio in the mantle, defined here by the relationship

$$X_M = D_M/S_M \tag{5}$$

The result can be expressed by the pair of equations

$$\dot{D}_M = \lambda P_M + X_M\, S_M \tag{6}$$

$$D_M + D_C = D_O + P_O\,(1 - e^{-\lambda t}) \tag{7}$$

However, these equations are very awkward because the quantity X_M is also a variable, and it is much simpler to use the equivalent equations.

$$X_M = \lambda P_M/S_M \tag{8}$$

$$S_M X_M + S_C X_C = D_O + P_O\,(1 - e^{-\lambda t}) \tag{9}$$

In Russell's (1978) paper the rate constants a and b were assumed to be independent of time. This assumption is not required by the differential equations above, and some solutions for time-dependent constants were reported by Russell and Birnie (1974). The concentrations at the beginning of differentiation were taken to be (P_O, S_O, D_O) for the number of atoms in the residual mantle, and zero for the numbers of the corresponding atoms in the proto-crust.

Because P_M is linear in β, and S_M is independent of β, it follows that the isotopic ratios in the residual mantle (and consequently in the proto-crust) will vary linearly with β. To permit β to take a range of values is the most convenient way of providing for a variable μ in the residual mantle.

For the bi-directional model, equation (8) does not apply, but it can be shown that the solution for P and for S remain valid if the rate constant describing the transport of S from residual mantle to proto-crust is taken to be $(1 - \alpha)a$ and the constant for transport in the return direction, from proto-crust to residual mantle, is taken to be αa. Of course, equation (7) remains valid provided the two interacting systems together make up a closed system.

Russell (1978) reformulated the equations as simultaneous matrix equations in the sums and differences of the isotopic populations, a form amenable to simple computer evaluation. A rather broad range of numerical solutions was found that fit exactly the stated constraints. For the evolutionary model, acceptable times for the beginning of differentiation range from 2410 Ma ago to 2900 Ma ago. For bi-directional model, the

TABLE II

Values inferred from the parameters corresponding to the earliest evolving bi-directional model (after Russell, 1978). The symbols μ and κ refer to the ratios $^{238}U/^{204}Pb$ and $^{232}Th/^{238}U$, both ratios extrapolated to the present by means of the appropriate decay constants.

Quantity	Time Related Parameters	Geochemical Parameters	
		Residual Mantle	Proto-Crust
Start of differentiation	4,000 Ma		
Rate constant for lead	1.34×10^{-9}yr^{-1}		
Mean transport interval for lead	4,000−3,250 Ma		
Rate constant for uranium and thorium	8.3×10^{-9}yr^{-1}		
Mean transport interval for uranium	4,000−3,880 Ma		
Range of μ-values		7.41−21.1	7.83−9.20
Geometric means of μ-values		14.4	8.7
Range of κ-values		1.81−3.72	3.69−4.14
Geometric means of κ-values		2.70	4.0

computation gave acceptable times ranging from 2600 Ma ago to 4000 Ma ago. Table 2 gives the essential quantitative data for the solution with the earliest start of differentiation.

Note that models such as Russell and Birnie's (1974) and Russell's (1978) correctly describe the transport of lead as well as uranium, while models in which μ is the only variable assume that the systems are only open to the movement of uranium.

CONCLUSIONS

In this paper I have brought together some information that demonstrates how isotopic measurements of terrestrial materials contribute to our understanding of the evolution of the Earth. I have selected the material subjectively and arbitrarily − to make an exhaustive review would be an enormous task! I have considered the results of some research projects that are interesting to me because they relate to the evolution of the crust and atmosphere, with particular emphasis on evolution during the first quarter of the Earth's history.

The helium isotopic ratios seem to provide the most convincing evidence that primordial materials are still being released from the mantle: the isotopic ratios of the helium presently being released are quite different from those expected in radiogenic helium. Although helium provides evidence that the mantle is incompletely outgassed, at present it is difficult to determine whether the outgassing has occurred continuously, or whether there was episodic outgassing early in the Earth's history. Helium studies provide estimates of the rates at which helium and argon isotopes are released to the surface, but different estimates disagree by about an order of magnitude.

Various authors have appealed to argon isotopic ratios for rate constants and a time profile for mantle outgassing. Certainly argon has the potential to give this kind of information, but so far the conclusions are quite contradictory. Combined argon and rubidium concentrations have been used to relate the outgassing profile to the mean K/Rb ratio of the outer part of the Earth. It has been proposed that if this ratio can be shown to be much greater than 710, then there must have been early periods of episodic outgassing.

Initial strontium isotopic analyses are sufficiently understood to derive a rough mantle or primary growth curve for radiogenic strontium in the Earth, but the resolution is not sufficient to determine whether or not this curve is linear.

Variations in the isotopic abundance of ^{207}Pb give the most convincing evidence of early processes, the effects of which have been preserved to the present. In refining the primary lead isotope growth curve, it has been necessary to postulate either an episodic or an evolutionary change in the ratio of U/Pb during the early part of the Earth's history. Quantitative interpretations of the ^{207}Pb discrepancy in oceanic basalts imply structure in the chemical evolution of the outer part of the Earth during the first 1000 Ma.

ACKNOWLEDGEMENTS

The excellent facilities of the Computing Centre of the University of British Columbia have played an important part in my isotopic research. Conversations with M. Ozima, I. Tolstikhin, and W.F. Slawson have been stimulating and helpful. The technical contributions of R.D. Meldrum are acknowledged with thanks. I would like to express my sincere appreciation to the Office of Grants and Scholarships of the Natural Sciences and Engineering Research Council of Canada for continued research support.

REFERENCES

Alexander, E.C. Jr. and Ozima, M. eds., 1978, Terrestrial Rare Gases: Centre for Academic Publications Japan, Japan Scientific Societies Press, 229, p.

Armstrong, R.L., 1968, A model for the evolution of strontium and lead isotopes in a dynamic earth: Rev. Geophys., v. 6, p. 175-199.

Carter, S.R., Evensen, N.M., Hamilton, P.J. and O'Nions, R.K., 1978, Neodynium and strontium isotope evidence for crustal contamination of continental volcanics: Nature, v. 202, p. 743-746.

Craig, H. and Lupton, J.E., 1976, Primordial neon, helium and hydrogen in oceanic basalts: Earth Planet. Sci. Letters, v. 26, p. 125-132.

Craig, H., Lupton, J.E. and Horibe, Y., 1978, A mantle component in circum-Pacific volcanic gases: Hakone, the Marianas and Mt. Lassen: in Alexander, E.C., Jr. and Ozima, M., eds., Terrestrial Rare Gases: Center for Academic Publications Japan, Japan Scientific Societies Press, p. 3-16.

Cumming, G.L. and Richards, J.R., 1975, Ore lead isotope ratios in a changing earth: Earth Planet. Sci. Letters, v. 28, p. 155-171.

De Paolo, D.J. and Wasserburg, G.J., 1978, Petrogenetic mixing models and Nd-Sr isotopic patterns: Geochim. Cosmochim. Acta, v. 43, p. 615-627.

Doe, B.R. and Stacey, J.S., 1974, The application of lead isotopes to the problems of ore geneses and ore prospect evaluation: a review: Econ. Geol., v. 69, p. 757-776.

Doe, B.R. and Zartman, R.E., 1979, Plumbo-tectonics I, the Phanerozoic: in Barnes, H., ed., Geochemistry of ore deposits: John Wiley and Sons, New York.

Fisher, D.E., 1978, Terrestrial potassium and argon abundances as limits to models of atmospheric evolution: in Alexander, E.C., Jr., and Ozima, M., eds., Terrestrial Rare Gases: Center for Academic Publications Japan, Japan Scientific Societies Press, p. 173-184.

Hamano, Y. and Ozima, M., 1978, Earth-atmosphere evolution model based on Ar isotopic data: in Alexander, E.C., Jr. and Ozima, M., eds., Terrestrial Rare Gases: Center for Academic Publications Japan, Japan Scientific Societies Press, p. 155-172.

Hamilton, P.J., O'Nions, R.K., Evensen, N.M., Bridgwater, D. and Allaart, J.H., 1978, Sm-Nd isotopic investigations of Isua supracrustals and implications for mantle evolution: Nature, v. 272, p. 41-43.

Hurley, P.M., 1968a, Absolute abundance and distribution of Rb, K and Sr in the earth: Geochim. Cosmochim. Acta, v. 32, p. 273-283.

_____, 1968b, Correction to : Absolute abundance and distribution of Rb, K and Sr in the earth: Geochim. Cosmochim. Acta, v. 32, p. 1025-1030.

Levin, B.J., 1972, Origin of the earth: Tectonophysics, v. 13, p. 7-29.

McCulloch, M.T. and Wasserburg, G.J., 1978, Sm-Nd and Rb-Sr chronology of continental crust formation: Science, v. 200, p. 1003-1010.

Moorbath, S., 1975, Evolution of Precambrian crust from strontium isotopic evidence: Nature, London, v. 254, p. 395-398.

Ringwood, A.E., 1975, Composition and Petrology of the Earth's Mantle: McGraw Hill, New York, 618 p.

_____, 1977, Composition and Origin of the Earth: Publication No. 1299, Research School of Earth Sciences, Australian National University, Canberra, 65 p.

Rubey, W.S., 1951, Geological history of sea water. An attempt to state the problem: Geol. Soc. Amer. Bull., v. 62, p. 1111-1147.

Russell, R.D., 1978, Lead isotope constraints on the early history of the earth: in Alexander, E.C., Jr. and Ozima, M., eds., Terrestrial Rare Gases: Center for Academic Publications Japan, Japan Scientific Societies Press, p. 207-224.

Russell, R.D. and Birnie, D.J., 1974, a Bi-directional mixing model for lead isotope evolution: Phys. Earth Planet. Interiors, v. 8, p. 158-166.

Russell, R.D. and Ozima, M., 1971, The potassium/rubidium ratio of the earth: Geochim. Cosmochim. Acta, v. 35, p. 679-685.

Safronov, V.S., 1965, Sizes of largest bodies fallen on planets in the process of their formation: Astron. Zh, v. 42, p. 1270-1276.

_____, 1969, Evolution of the preplanetary cloud and the formation of the earth and planets: Nauka, Moscow, 241 p.

Stacey, J.S. and Kramers, J.D., 1975, Approximation of terrestrial lead isotope evolution by a two stage model: Earth Planet. Sci. Letters, v. 26, p. 207-221.

Stanton, R.L. and Russell, R.D., 1959, Anomalous leads and the emplacement of lead sulphide ores: Economic Geology, v. 54, p. 588-607.

Sun, S.S. and Hanson, G.N., 1975, Evolution of the mantle: geochemical evidence from alkali basalt: Geology, v. 3, p. 297-302.

Tatsumoto, M., Knight, R.J. and Allegre, C.J., 1973, Time differences in the formation of meteorites as determined from the ratio of lead-207 to lead-206: Science, v. 180, p. 1279-1283.

Tolstikhin, I.N., 1978, A review: Some recent advances in isotope geochemistry of light rare gases: in Terrestrial Rare Gases (E.C. Alexander, Jr. and M. Ozima, eds) Center for Academic Publications Japan, Japan Scientific Societies Press, p. 33-62.

Tolstikhin, I.N. and Drubetskoi, E.R., 1975, The isotopic ratios of $^3He/^4He$ and $^4He/^{40}Ar$ in the rocks of the earth's crust: Geokhimiya, v. 8, p. 1125-1136.

Wetherill, G.W., 1976, The role of large bodies in the formation of the earth and moon: Proc. Lunar. Sci. Conf. 7th, v. 3, p. 3245-3257.

The Continental Crust and Its Mineral Deposits, edited by D.W. Strangway,
Geological Association of Canada Special Paper 20

EVOLUTIONARY TECTONICS OF THE EARTH IN THE LIGHT OF EARLY CRUSTAL STRUCTURE*

Denis M. Shaw
Department of Geology,
McMaster University,
Hamilton, Ontario L8S 4M1

ABSTRACT

The evolution of the primitive Earth necessitated that the first crust was thin, world-wide and submarine. The present-day Earth is dominated by plate-tectonic processes which require the existence of contrasting continental and oceanic plates.

A qualitative model of evolutionary Earth tectonics outlines the transition from the Protoarchean to the present crust, and satisfactorily explains a wide diversity of geochemical, geological and geophysical observations.

INTRODUCTION

One of the major problems in geology concerns the history of the crust of the Earth throughout geological time. Has the present division of the crust into two contrasting components, continental and oceanic, always been so? Has the material composing the continents always been at the surface? Has its mass increased or decreased, and if so, how?

RÉSUMÉ

L'évolution de la Terre primitive exigeait que la première croûte soit mince, sous-marine et à l'échelle du globe. La Terre actuelle est dominée par les processus de tectonique des plaques dépendant de l'existence de plaques continentales et océaniques contrastantes.

Un modèle qualitatif pour l'évolution de la tectonique terrestre met en relief la transition de la croûte du Protoarchéen à la croûte actuelle et explique de façon satisfaisante une grande diversité d'observations géochimiques, géologiques et géophysiques.

Questions of this kind have been before us since geological reasoning began. However, they are in sharper focus now that a coherent body of theory, plate tectonics, has developed during the last 20 years. They have been subjected to close scrutiny in recent years, particularly as a result of study of radiogenic isotope systems bearing on the evolution of the crust-mantle system.

The history of the crust is dominated by two well-established facts, namely that much continental crust is old and that all oceanic crust is young. These are axioms in any reconsideration of crustal history. The present paper will begin by inferential reasoning concerning the nature of the original crust of the Earth.

THE EARLY CRUST OF THE EARTH

Many lines of reasoning (see Shaw, 1976) suggest that the Earth, soon after its formation, passed through a high-temperature phase, when much of its material was molten. This heating was provoked by gravitational infall and by subsequent separation of a metallic core. Volatile elements and compounds, condensed earlier from a stellar nebula similar in composition to the Sun, were lost from the Earth at this stage, including H, noble gases, C, N, S, halogens. Crystallization differentiation followed the melting phase, permitting some early-condensing metals (Ca, REE, U, Th, etc.) to concentrate in lower-melting compounds in the upper part of the mantle.

Some authors have rejected this early hot Earth concept, on the grounds that *all* the rare gases and volatile elements would have been eliminated. It is evident that the Earth still contains primitive ^{3}He, and that terrestrial xenon contains components derived from short-lived ^{129}I and ^{244}Pu formed during nucleosynthesis: these and other gaseous components could however have been retained in magmatic solution during a heating phase. Another argument against a hot Earth is that Ni appears much too abundant in mantle rocks for them to have been in equilibrium with a metallic core (in comparison with meteorites) so the sources of core and mantle must have been independent. This is a stubborn problem which cannot be solved at the present state of our knowledge, but does not provide sufficient evidence to reject a hot early Earth.

If it can be accepted that the Earth was at one time largely molten, then heat conduction from the interior, aided by convection and surface radiation, would lead initially to crystallization upwards from the base of the mantle. This would be followed by crystallization downward from the surface, sometime after the rate of supply of heat from the interior was exceeded by the rate of surface radiative heat loss. At this time the molten geoid would begin to secrete its first crust.

Since there is no evidence to the contrary, it may be accepted (by application of Ockham's razor) that this first crust was world-wide. It was of course unstable, composed of polygonal segments or rafts of crystals, remelting and resolidifying in response to local heat variations, floating on residual convecting magma beneath. As time went on, the crystal rafts grew in area and thickness, and thermal contraction led to "jostling" of these ice-flow-like polygons. Eventually there was little or no magma at the surface, and from then on the release of magma to the surface took place through conduits, initiating volcanic action as we know it today.

Crystallization of a peridotitic mantle led to upward concentration of residual basaltic magma. Near-surface low-pressure crystallization of this residual magma permitted plagioclase to separate and, because of its lower density, to rise towards the surface. The initial crust was thus of dioritic or anorthositic composition, but would also contain entrapped gabbroic material as well as granitic pockets resulting from further differentiation. Between these dioritic mini-plates were areas of more rapidly chilled basaltic or gabbroic composition, representing magma on which the diorite had previously floated.

An additional factor must be mentioned. The evidence of extensive impact cratering on the Moon, Mercury, Mars and (probably) Venus, extending from perhaps 4.5 to 3.9 Ga ago makes it highly probable that Earth also experienced this high flux of large meteorites and asteroidal fragments, even though direct evidence is lacking. The effect of this impact flux was to remelt and brecciate the early crust, perhaps to the extent of destroying all the initial solid rock, leaving a pattern of craters over much of the Earth's surface.

As soon as solid rock covered most of the Earth's surface an insulating effect came into play, with two effects. First, the crust acted rather like a plug in a modern Peléean volcano, restraining heat loss for a while but then yielding to explosive volcanic action through zones of weakness: this early volcanism was on a scale difficult to imagine in its intensity. Second, the proto-atmosphere released by volcanic degassing, was able to cool rapidly and water began to condense. The acid rains which ensued initially evaporated quickly on contact with hot rock, but slowly liquid water began to accumulate on the surface. Eventually most of the water had condensed and a universal shallow warm sea covered the Earth, interspersed here and there with emergent volcanic islands. If the Moon was at that time an Earth satellite, it lay much closer to Earth than now, and the proto-ocean was subjected twice daily to tides of much greater amplitude than now. Since there was little emergent land, however, the erosional powers of these tides were restricted to rapid lowering of volcanic islands.

In summary, this Protoarchean era was characterized by a universal brecciated and cratered crust of rather subdued relief, mostly submerged below a primitive universal ocean. There was at this time no meaning to the present-day terms *continental crust* and *oceanic crust,* and the entities which we call continents and ocean basins did not exist. Plate tectonics consequently had no meaning at this time.

THE PRESENT-DAY CRUST OF THE EARTH

The present-day Earth surface is quite different from that of the Protoarchean era. Continental masses of granodioritic composition and density rise substantially above sea level, terminate downwards at depths of 30 to 50 km in a seismic discontinuity, but constitute the upper parts of structurally coherent lithospheric plates. Beneath the deep oceans and their thin sedimentary floors lies a 5 km basaltic crust, the upper layer of oceanic lithospheric plates.

Both continental and oceanic plates are in movement. New oceanic crust is generated at spreading centres, while oceanic lithosphere is consumed where it collides with continental plates. Other types of plate motion are well-documented but this is the principal one. Age measurements show that oceanic crust is young, that very little fossil oceanic crust has been incorporated into the continents, and that large areas of the continents are old. It has been concluded that ocean basins are transitory, whereas continental masses are more stable (about 0.55 of the mass of the continental crust is Precambrian).

Although some of the newly formed oceanic crust persists as fresh basalt, much is now believed to undergo profound chemical alteration, through the action of heated sea-water in convective percolation (Fyfe, 1978). In considering the fate of subducted oceanic lithosphere, this must be borne in mind.

In addition to crustal generation at spreading centres, new crust is formed by volcanic action along colliding plate margins (circum-Pacific belt, island arcs, etc.) and in areas of plateau basalt eruption, rifts and other intra-plate volcanic centres.

PROBLEMS

The salient features described in the two preceding sections lead to a number of major problems in interpreting the Earth's history, which may be summarized in a series of questions: 1) How did the present-day processes of plate tectonics begin? 2) Why is oceanic lithosphere almost totally consumed during plate collisions, whereas (some? much?) continental lithosphere survives? 3) What becomes of the surficial layers of sediments and altered basalt when oceanic lithosphere is subducted? 4) Has there been a net growth in continental crustal mass throughout geological time? 5) When was the transition achieved from the Proto-archean universal thin crust and overlying shallow ocean to the present-day dichotomy of continental land-mass and ocean basin?

The first four of these questions have been examined repeatedly in recent years. However they can only be correctly addressed in the context supplied by question 5, i.e., the great contrast between the processes and geometry of the present-day and those of the remote past. The importance of the inference that the first crust and the first ocean covered all the Earth has been stressed in a number of publications (Shaw, 1972, 1976, in press; Hargraves, 1976; Lowman, 1978; Fyfe, 1978) but not sufficiently yet to establish its central role in a coherent account of the evolution of plate tectonics.

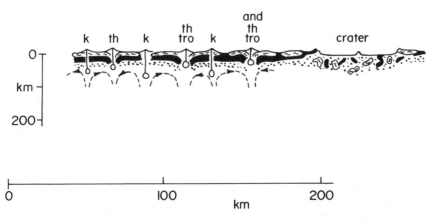

Figure 1. The Protoarchean crust. The primitive crust (solid black) is underlain by depleted mantle (dotted) and overlain by volcanic accumulations of komatiitic (k), tholeiitic (th), andesite (a) and trondjhemitic (tro) composition. A nearby crater has brecciated these materials. All is overlain by the primitive ocean (not shown) except for occasional volcanic peak islands. Convection cells in the mantle are shown in dashed lines.

DISCUSSION

An attempt will now be made to outline a theory of crust-mantle evolution which will provide at least partial answers to the questions stated above.

During the Protoarchean, the thin submarine world-wide crust of intermediate composition broke into contraction polygons (Drury, 1978) and was extensively brecciated by asteroidal impact cratering (Fig. 1). Because the generation of radioactive heat was faster then than now (Dickinson and Luth, 1971; Schilling *et al.*, 1978), volcanism was wide-spread and was fed by innumerable, small near-surface closely-spaced mantle convection cells. Deep and extensive mantle melting produced komatiitic volcanism; shallower partial melting of lherzolitic mantle, followed by minor fractional crystallization, produced tholeiitic magma (Davis and Condie, 1977; Condie and Harrison, 1976) in large amounts; more extensive fractionation of the basaltic magma formed andesitic magma; trondhjemitic-tonalitic magma was produced by melting of down-warped earlier basaltic piles at depths of 30 to 80 km (Barker *et al.*, 1976); extreme fractionation led to minor rhyolitic volcanism.

Volcanism led to upward growth of the surface, partially but not completely compensated by gravitational subsidence, and welded crustal polygons together into larger platelets. Upward growth was accompanied by submarine alteration and erosion, followed by sub-aerial erosion which produced greywackes and other sediments more enriched in quartz than the parental volcanic rocks: these sediments would later serve, by anatexis, as sources of more granitic rocks. As the surface rose, major island masses began to emerge from the hitherto world-wide ocean.

Coalescence of polygonal segments into larger island platelets was aided by the compressional effects of pairs of adjacent near-surface convection cells. Downwarping near points of convergence led to metamorphism and remelting of basal crust and upper mantle, permitting renewed plutonic and volcanic activity. Over the course of time most, if not all, of the original crust must thus have been recycled.

Complementary tensional effects between convection cells led to rifting which created low-lying basins into which the seas accumulated. Mantle material was thus brought close to the surface and, as time went on, tholeiitic magma began to dominate in the accompanying volcanism (Fig. 2). The cooling effect of the water occupying such basins tended to localize the upward currents of convection cells, thereby stabilizing spreading centres for longer duration. Plate tectonics was now beginning.

The processes described in the previous paragraphs began in the Protoarchean

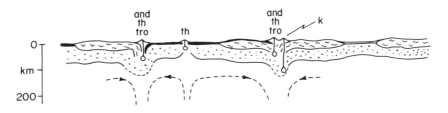

Figure 2. Mid-Precambrian crust. Crustal segments are now discrete land-masses, separated by low-lying small ocean basins where rifting and basaltic volcanism (black) is beginning to form oceanic crust. Only minor komatiitic volcanicm now occurs.

they have been formed by 25 to 30 per cent melting of peridotite: this had previously undergone an earlier partial melting, removing a substantial proportion of incompatible elements such as K, Rb, Ba, REE. The evidence supporting this interpretation is abundant and has been summarized elsewhere (Shaw, 1980). Such peridotite may be referred to as *depleted mantle* material, and the abundance of MORB suggests an extensive layer of this material underlying the major ocean basins.

Another major group of basalts comprises the extensive outpourings on numerous continental areas. These plateau basalts are tholeiitic in character, like the MORB, but show less depletion in the incompatible elements. A third and minor group are the alkali basalts and associated other alkalic rocks. These occur at intra-plate volcanic centres and are substantially richer in incompatible elements. The evidence is strong (Shaw, 1980) that these rocks represent the melting products of different peridotites, perhaps similar to the mantle lherzolite xenoliths which they often contain; their source may be referred to as undepleted or *fertile mantle material*.

Archean basalts tend to be intermediate in chemical compostion between alkali basalts and MORB (Hart *et al.*, 1970) although they are generally tholeiitic. Alkali basalts are rare or non-existent in Archean terrances.

The Protoarchean mantle was constituted by residual Mg-Fe silicates remaining after separation of the first crust. The extensive outpouring of basalts which dominated Archean times came from mantle melting and resulted in formation of residual depleted mantle. It has recently been demonstrated that at high temperature and pressure such depleted peridotite would have a density (3.30) lower to a significant degree than fertile (3.39) mantle (Boyd and McCallister, 1976). Depleted mantle will consequently tend to float above fertile mantle. The upward growth of Protoarchean island platelets was thus accompanied by accumulation of depleted mantle under the lower parts of the crust. As time went on, there developed a near-surface layer of depleted mantle, with more fertile mantle below, with two consequences: first, the near-surface melting beneath the developing ocean basins led to production of basalt of MORB character; second, the depleted mantle beneath the proto-continental island masses was in relative gravitational equilibrium and in effect developed into downward-growing roots.

This concept has been explored by Jordan (1978) who, from the standpoint of thermal behaviour, concluded that the present-day continents possess roots of this nature, extending to depths as great as 400 km and constituting the *continental tectosphere*. The material composing these roots has suffered "irreversible basaltic depletion" (Jordan, 1978, p. 546) and has lower density, lower temperature, higher viscosity and a more elevated solidus than similar material at equivalent depths below the oceans. As a consequence the present-day continental plates behave as rigid bodies to a much greater depth than the oceanic plates.

This, if substantiated, will explain both the longevity of continental plates and the ephemeral character of oceanic lithosphere. In colliding processes it is the oceanic plate which will deform more readily, although continental margins will undergo a limited degree of metamorphism and melting (Fig. 3).

At the present-day the convection cells are larger, fewer and deeper than in the remote past. Melting of fertile mantle at depths of several hundred km accompanies tensional rifting in mid-continental areas permitting alkalic magmas to reach the

At the present-day the convection cells are larger, fewer and deeper than in the remote past. Melting of fertile mantle at depths of several hundred km accompanies tensional rifting in mid-continental areas permitting alkalic magmas to reach the surface: melting of both fertile and depleted mantle in larger volumes leads to plateau basalt extrusions. Both of these processes form a more depleted mantle, which tends to rise towards the surface either below the continents or below the oceans. Sub-oceanic rising currents will thus replenish the depleted mantle surface layer as it melts beneath a spreading centre to produce MORB. The same currents, on descending and enforcing subduction, will carry altered MORB and incompatible-element-rich oceanic sediments to undergo transformation and recycling below continental margins.

It may in fact happen that some of the ocean floor rock, sediment, H_2O and CO_2 escape immediate re-melting and accumulate in fertile mantle, re-emerging later as carbonatite, kimberlite and diamond: a recent paper (Smirnov et al, 1979) demonstrates that some diamonds have very light isotopic ratios, in the range of surface carbon compounds and quite distinct from meteoritic values. If subduction replenishment of the deeper mantle contributes to the generation of some alkalic magmas then the near-absence of such magmas in Archean times would be explained, because deep subduction of oceanic materials had not yet started.

The present-day scale of plate interactions is larger than in the past, and has kept pace with the waning heat production of the Earth and the consequent accumulation of depleted mantle. It is difficult to estimate with any confidence the present proportion of depleted material, but it probably forms at least half the mantle mass (see Shaw, 1980). This demands a considerable increase in continental crustal mass since Protoarchean times, which could be measured by using estimates of the addition of oceanic and continental basalt, modified by the effects of surficial geo-chemical processes which took place before metamorphism, subduction and anatexis. The overall result will be granitic crustal growth and peridotitic return to the mantle: details of the process are strongly dependent on the quantitative extent of seawater-basalt interaction, which is under active scrutiny at the present time (e.g., Bloch and Bischoff, 1979; Fyfe, 1978).

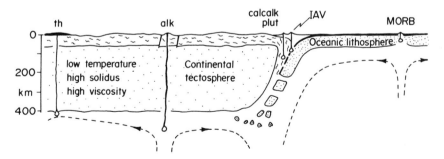

Figure 3. Present-day crust. Extensive continental masses with marginal island arc volcanics (IAV) contrast with thinner oceanic crust (black). Both continent and ocean are underlain by more abundant (than in Proto-archean times) depleted mantle (dotted), much of which has plated on to continental crust to form deep roots (Jordan, 1978).

The principal *direct* source of granitic material for the growing continental platelets (disregarding the ultimate basaltic source) was the melting of sandstones, andesitic volcanics and similar rocks (Arth and Hanson, 1975). These materials were not available to any degree until extensive emergence of island masses permitted gradation. This helps explain the lower abundance in early Archean times of K-rich granitic rocks (Burke and Kidd, 1978), and the increase of Si, K, Ca (relative to Al, Mg, Na) in sediments of younger geological times (Schwab, 1978). Such processes constitute recycling of crustal material and have been energetically advocated by many authors (e.g., Veizer and Jansen, 1979). They are nevertheless subject to severe constraints imposed by Sr and Pb isotopic systems (see Moorbath, 1977), which require granitic materials produced in this way to be reprocessed quickly enough that their parent mantle isotopic ratios do not change greatly.

CONCLUSIONS

According to the model proposed here, plate tectonic processes have grown in scale from the Protoarchean to the present and did not operate at all until the Earth developed continental nuclei and incipient ocean basins. The development was gradual, probably in stochastic response to mantle convection patterns. This does not exclude other, perhaps catastrophic causes (Moon capture?). Although there is little evidence for plate tectonic processes in the early Archean, evidence has been presented for a significant increase in continental crustal material 2700 to 2500 m.y. ago (McCullough and Wasserburg, 1978; Jacobsen and Wasserburg, 1978) and elucidation of this event is desirable.

All the questions posed in an earlier section appear to be resolved qualitatively by this model of crustal evolution, but need substantiation in quantitative terms.

ACKNOWLEDGEMENTS

Financial aasistance from NRC grant no. A0155 is acknowledged. Mrs. Lynn Falkiner and Ms. Helen Elliott quickly and efficiently typed the manuscript.

It is a pleasure and a privilege to dedicate this paper to J. Tuzo Wilson, in appreciation of the role he has played in the fostering of the earth sciences in Canada, in the development of challenging ideas in his chosen field of geotopology and in the pleasure of frequent personal association with him over many years.

REFERENCES

Arth, J.G. and Hanson, G.N., 1975, Geochemistry and origin of the early Precambrian crust of northeastern Minnesota: Geochim. Cosmochim. Acta, v. 39, p. 325-362.

Barker, F., Arth, J.G., Peterman, Z.E. and Friedman, I., 1976, The 1.7- to 1.8-b.y.-old trondhjemites of southwestern Colorado and northern New Mexico: Geochemistry and depths of genesis: Geol. Soc. Amer. Bull., v. 87, p. 189-198.

Bloch, S., and Bischoff, J.L., 1979, The effect of low-temperature alteration of basalt on the oceanic budget of potassium: Geology, v. 7, p. 193-196.

Boyd, F.R., and McCallister, R.H., 1976, Densities of fertile and sterile garnet peridotites: Geophys. Res. Letters, v. 3, p. 509-512.

Burke, K., and Kidd, W.S.F., 1978, Were Archean continental geothermal gradients much steeper than those of today?: Nature, v. 272, p. 240-241.

Condie, K.C., and Harrison, N.M., 1976, Geochemistry of the Archean Bulawayan Group, Midlands greenstone belt, Rhodesia: Precambrian Res., v. 3, p. 253-271.

Davis, P.A. Jr., and Condie, K.C., 1977, Trace element model studies of Nyanzian greenstone belts, western Kenya: Geochim. Cosmochim. Acta, v. 41, p. 271-278.

Dickinson, W.R., and Luth, W.C., 1971, A model for plate tectonic evolution of mantle layers: Science, v. 174, p. 400-404.

Drury, S.A., 1978, Basic Factors in Archaean Geotectonics: in Windley and Naqvi, eds., Archean Geochemistry: Elsevier Scientific, p. 3-23.

Fyfe, W.S., 1978, The evolution of the earth's crust: modern plate tectonics to ancient hot spot tectonics: Chemical Geology, v. 23, no. 2, p. 89-114.

Hargraves, R.B., 1976, Precambrian geologic history: Science, v. 193, p. 363-370.

Hart, S.R., Brooks, C., Krogh, T.E., Davis, G.L., Nava, D., 1970, Ancient and modern volcanic rocks: a trace element model: Earth and Planet. Sci. Letters, v. 10, p. 17-28.

Jacobsen, S.B., and Wasserburg, G.J., 1978. Interpretation of Nd, Sr and Pb istope data from Archean migmatites in Lofoten-Vesteralen, Norway: Earth and Planet. Sci. Letters, v. 41, p. 245-253.

Jordan, T.H., 1978, Composition and development of the continental tectosphere: Nature, v. 274, p. 544-548.

Lowman, P.D., 1978, Comparative planetology and the origin of continents: Comparative Planetology, Academic Press, p. 51-78.

McCulloch, M.T., and Wasserburg, G.J., 1978. Sm-Nd and Rb-Sr chronology of continental crust formation: Sciences, v. 200, p. 1003-1011.

Moorbath, S., 1977, Ages, isotopes and evolution of Precambrian continental crust: Chem. Geol., v. 20, p. 151-187.

Schilling, J.-G., Unni, C.K., and Bender, M.L., 1978. Origin of Cl and Br in the oceans: Nature, v. 273, p. 631-636.

Schwab, F.L., 1978, Secular trends in the composition of sedimentary rock assemblages – Archean through Phanerozoic time: Geology, v. 6, no. 9, p. 532-552.

Shaw, D.M., 1972, Development of the early continental crust. I. Use of trace element distribution coefficient models for the protoarchean crust: Canadian Jour. Earth Sci., v. 9, p. 1577-1575.

———————————, 1976, Development of the early continental crust, 2. Prearchean, Protoarchean and later eras: in Windley, B.F., ed., The John Wiley and Sons, p. 33-53.

———————————, 1980, Development of the Early Continental Crust. Part III. Depletion of incompatible elements in the mantle: Precambrian Geology, v. 10, p. 281-299.

Smirnov, G.I., Mofolo, M.M., Lerotholi, P.M., Kaminsky, F.V., Galimov, E.M., and Ivanovskaya, I.N., 1979, Isotopically light carbon in diamonds from some kimberlite pipes in Lesotho: Nature, v. 278, p. 630-631.

Veizer, J., and Jansen, S.L., 1979, Basement and sediment recycling and continental evolution: Jour. Geol., v. 87, p. 341-370.

EVOLUTION OF THE PRECAMBRIAN CRUST

The Continental Crust and Its Mineral Deposits, edited by D.W. Strangway,
Geological Association of Canada Special Paper 20

CRUST FORMATION AND DESTRUCTION

W. S. Fyfe
Department of Geology,
University of Western Ontario,
London, Ontario N6A 5B7

ABSTRACT

The formation of new ocean floor crust, the cooling process and chemical modification by exchange with seawater are becoming better understood. The modern subduction process is not well defined but there is increasing evidence that spilites and sediments may contribute to andesite genesis. Mechanisms of sediment trapping in descending oceanic lithosphere are being revealed by studies in trench environments. In particular it is noted that there is a lack of metamorphosed deep ocean sediments in blue schist assemblages. Given uncertainties in the subduction process we cannot be certain if continents are growing or being slowly returned to the mantle.

There is increasing evidence for a major change in the convective style of the Earth near the Archean-Proterozoic boundary. Models of this change are imperfect, but a change from viscous drag subduction to negative buoyancy subduction as proposed by Hargraves is attractive. In any event, the key point of modern debate is whether or not there can be massive subduction of low density crust.

RÉSUMÉ

On commence à mieux connaître la formation de nouvelle croûte dans les fonds océaniques, le processus de refroidissement et les modifications chimiques par échange avec l'eau de mer. Le processus moderne de subduction n'est pas bien défini mais il y a des preuves de plus en plus nombreuses montrant que les spilites et les sédiments peuvent contribuer à la génèse des andésites. Les mécanismes de piégeage des sédiments par descente de la lithosphère océanique sont mieux connus depuis les études dans les milieux de fossé. En particulier, on note qu'il y a une carence de sédiments métamorphisés provenant des profondeurs de l'océan dans les assemblages de schistes bleus. En tenant compte des incertitudes dans le processus de subduction, nous ne pouvons affirmer avec certitude que les continents sont en croissance ou encore qu'ils retournent lentement vers le manteau.

Il y a des indications de plus en plus nombreuses qu'il y a eu un changement majeur dans le style convectif de la Terre entre l'Archéen et le Protérozoïque. Les modèles pour ce changement sont imparfaits, mais le changement d'une subduction par entraînement visqueux à une subduction par flottabilité négative, comme l'a proposé Hargraves, est attrayant. A tout événement, le point clé de ce débat est de savoir si oui ou non il peut y avoir subduction massive de croûte à faible densité.

INTRODUCTION

Writing in *Nature* on the pioneer view of Venus, Hunt (1979) aptly stated that "detailed exploration of the planets always yields surprises and gives a new perspective on the Earth." At this workshop Papanastassiou provided evidence that rather large objects existed in the solar system a million years after the elements were synthesized. We also heard that there is no other planet quite like the Earth. In these notes I would like to suggest that many of the unique features of the evolution of the Earth may be linked to our unique hydrosphere more than to any other single cause.

Reflection on the works of J. Tuzo Wilson reminds us how rapidly our science has advanced in the past decades. I would like to discuss some of these recent advances resulting from study of the past two hundred million years of Earth history which must be considered when we try to reconstruct the past.

WATER-COOLED CRUST

We have discovered that to a significant degree new crust is water-cooled. Heat flow patterns near ocean ridges are a reflection of both conductive cooling and heat transport by circulating seawater in a porous, brecciated and cracked medium. The ocean floor crust is complex, more complex than often depicted in simple idealized ophiolite sections (Barrett and Spooner, 1977; Hall and Robinson, 1979). The heat loss of the Earth is of the order of 3.0×10^{20} cal/yr. Wolery and Sleep (1976) have estimated that about 4×10^{19} cal yr is removed by fluid circulation and that, depending on the rate of sea-floor spreading, 30 to 40 per cent of the energy at ridges is removed by this process.

Straus and Schubert (1977) have shown that the adiabatic gradient for fluid circulation is exceeded under all present crustal conditions. The vigour of such flow is thus largely limited by the local permeability. We have become aware that great care is needed in the interpretation of heat flow both on continents and sea floor. Problems of nuclear waste disposal have shown us how little we know about the permeability of the crust.

The ocean floor ophiolitic crust, covering some 70 per cent of the Earth's surface, is young (0 to 200 Ma). Subduction roughly balances production. As the total mass of preserved ophiolites is trivial, the largest known mass being Oman, this age spread (0 to 200 Ma) will continue as long as the present convection style continues; it is a natural consequence of the "Wilson cycle".

We have recently begun to appreciate and quantify the mechanisms of fluid flow through the new ocean floor crust. Most writers agree that the fluid penetration is deep, about 5 km (Lister, 1974, 1977; Spooner, 1976; Ribando *et al.*, 1976). It has even been suggested that penetration may proceed to the Moho interface (Lewis and Snydsman, 1977), but many question the validity of this proposal. Lister has analysed crack propagation and shown how it influences the ophiolite structure. He considers that fluid penetration may be limited by the swelling reactions involved in hydration processes particularly serpentinization.

SEAWATER-BASALT EXCHANGE

Studies of sea-floor crust are now sufficiently advanced to show that massive chemical exchange between this crust and seawater occur. What comes from the mantle at ridges is chemically different from that which is subducted. (This subject has recently been reviewed by Fyfe and Lonsdale, 1980.)

Species from the continental crust, atmosphere and hydrosphere are exchanged. The early studies of Vallance (1960) on the composition of spilites (see Hyndman, 1972) are proving to be amazingly correct. Spilites are enriched in a number of important elements or species such as K, Na, U, O, H_2O, CO_2 and S. Subduction of spilites recharges the mantle with certain volatiles and species such as Na, K and U. Bloch and Bischoff (1979) have recently reviewed the situation with regard to potassium. They show that of the 6.3 x 10^{13} g of K carried to the oceans annually, as much as half may be fixed in the upper layer of the basaltic crust. This is a profound reaction. Given the mass of continental crust (1.6 x 10^{25}g) and a K content of 2% (3 x 10^{23} g), at present rates all this K would be recycled to the mantle in 10^{10} years. If sediments are subducted the cycling rate is much faster. Recently R.W. Kay of Cornell University sent me an interesting manuscript dealing with the problem of K-Rb-Ba-Sr cycling and the origin of andesites. His conclusions is that larger proportions of these elements in arc volcanics are recycled crustal material.

Aumento (1979) has reviewed uranium fixation in ocean floor basalts. He concludes that the U content of an oceanic lithospheric plate may be doubled between formation and subduction.

Here I wish to mention only one volatile species, water. The average water content of old oceanic crust is about 5%, depending very much on the quality of serpentine considered. Given the mass of the oceans of 1.4 x 10^{24} g, and a crustal subduction rate of about 3 x 10^{16} g/yr (H_2O, 1.5 x 10^{15} g/yr), the ocean volume would be recycled in 10^9 years. In a cooling Earth, where the quantites of water held in solid crust must be about 50% of the oceanic mass, such figures caution regarding assumptions about constancy of mass of the oceans (Fyfe, 1976a, 1976b).

SUBDUCTION

In his delightful book "The New View of the Earth", Uyeda (1978) lists subduction problems among the great remaining questions of modern tectonics. Perhaps the greatest question of all is what happens to sediments (Uyeda, 1978, p. 186-188), a problem first emphasized by Gilluly (1971). To take an example from Uyeda, "since the Mesozoic era, nearly 10,000 kilometres of oceanic plate along with oceanic ridges seems to have descended beneath Japan." Are the scraped-off sediments present in sufficient volume?

Numerous recent writers on this topic (Gilluly, 1971; Sibley and Vogel, 1976; Garrels and Lerman, 1977) have suggested that massive sediment subduction must occur to balance recent sediment-forming processes. Mechanisms for the process are beginning to appear. First, one need only reflect on the topography of the surface near a ridge (e.g., Press and Siever, 1974, Fig. 5.28) to see how sediments may be locked into the descending slab. Lister's fault block structures (Lister, 1976, Fig. 4) is also ideal for this purpose.

Recently Molnar and Gray (1979) have shown that massive subduction of continental lithosphere may be possible. "If continental crust can remain intact, peninsulas and microcontinents might be subducted completely." In fact these writers show that there is sufficient negative buoyancy to balance subduction of a considerable mass of light material. The detailed mechanics of the process are still unclear. Hargraves (1978) has recently examined the mechanics of viscous drag subduction and negative buoyancy subduction.

It appears that light crustal material can be subducted if there is a mechanism for locking it on or within the slab. I have already noted ocean-floor structures that may do this. The same structures are reported in the *Geotimes'* account of the Leg 60 deep sea drilling project (*Geotimes*, v. 23, no. 10, 1978, p. 19-22). The final sentence of this report is worth quoting: "Though some rocks contained in the deepest holes in the trench may have originally been part of the Pacific Plate, it now appears that subduction is a smooth process, with little accretion of material from the Pacific Plate. Blocks of the fore-arc region seem to be down-dropping into the trench, perhaps to be subducted themselves." A continental edge may be nibbled away by the subduction process, in accord with the discovery of two billion year old basement on both sides of the South American continent (Dalmayrac *et al.*, 1977). At this workshop John Dewey showed mechanisms suggesting the same type of roll-under process.

A particularly significant paper by Jones *et al.*, 1979, is concerned with analysis of the mechanical behaviour of a descending lithosphere slab. These authors propose an elastic core but an outside zone where brittle failure occurs. This cracking is correlated with the topography of the slabs near trenches and earthquake focal mechanisms. Fault blocks with 400 to 800 m scarps are common. Here one is reminded of the fault blocks near ridges described by Lister (1976). Jones *et al.* suggest that the brittle bending structures may play an important role in trapping sediment which is subducted, and they stress that the expected sediments are not found in the walls of the Japan or Mariana trenches.

Here I would like to emphasize that not all continental debris are of low density once they are dragged into a region of blue schist or eclogite facies metamorphism. Thus Sibley and Vogel's (1976) average pelagic sediment (SiO_2 61.5%; Al_2O_3 18.1%; Fe 6.4%; MgO 3.2%; CaO 1.0%; Na_2O 2%; K_2O 3.2%), once buried 30 km, will be present as a dominantly quartz-kyanite-magnetite-muscovite rock with other minerals like glaucophane or pyroxene. In spite of a large excess of quartz, the density will exceed 3.0. The same will be true for typical low quartz andesites which form quartz-eclogites. In fact, when one considers the massive amounts of aluminous pelagic sediments that must be involved in any major spreading and subduction process, one cannot help but be impressed by the lack of kyanite rocks in blue schist regions. Are not most of the rocks observed in situations like the Franciscan of California simply a concentration of these lightest sediments plus those involved before the subduction zone ceases to operate or moves? The ultimate fate of descending sediments is not known, and perhaps our clues will eventually come from andesite geochemistry.

ANDESITES-BATHOLITHS

Over the past decade data have appeared which may finally resolve the old problem of the origin of andesites and the associated plutonic rocks of the granodiorite-granite family. The general situation, with emphasis on the strontium isotope data, is summarized by Hyndman (1972). Two recent papers in a single issue of "Contributions to Mineralogy and Petrology" nicely illustrate the situation. Whitford and Jezek (1979) in their study of the Banda Arc conclude that partial melting of the crustal component of the subducted slab may play a significant role in the island arc petrogenesis. In considering the andesites of the Northern Mariana Arc, Stern (1979) concludes that mantle, not subducted crust, is the main source of these rocks.

Combined studies of strontium and oxygen isotopes clearly show that in many cases a crust-sediment-mantle mix is required to explain the behaviour of these species (Margaritz et al., 1978; Taylor and Silver, 1978). The complexity of Sr-isotope systematics observed by Elston (1978) in New Mexico could hardly be explained on the basis of a single source. In fact, to make sense of all the varying results one must consider all the possible steps:

spilite $\pm$ sediment $\rightarrow$ degassing into hotter mantle, fluid transport of large amounts of K-Na-SiO_2 etc. $\rightarrow$ melting or mixing with overlying mantle $\rightarrow$ magma rise and crustal mixing $\rightarrow$ final product.

Basalt and basic andesitic lavas have densities in the range of 2.68 to 2.77 which are very similar to average crust. As well as extrusion of such lavas one must consider underplating and pervasive sill penetration of continental crust (Brown, 1977; Fyfe, 1979). In environments near the base of the continental crust, rising arc magmas induce partial fusion, particularly if they carry volatiles upwards to flux older granulite basement. In such regions vast mixing seem possible and is in accord with the total chemistry of the products. The recent work of Eichelberger (1978) is of particular relevance here for his data clearly show the presence of two magmas which can be mixed to varying degrees. Eichelberger's two-magma model is in accord with observations on two-magma dykes (Marzouki and Fyfe, 1977) and fits nicely with a recent idea proposed by G.P.L. Walker at the Durham (1977) IASPEI-IAVCEI assembly. Walker suggested that massive ignimbrite eruptions are triggered by the injection of hot basic magmas into acid magma chambers, an idea which has great merit. Finally it should be noted that continental underplate mechanisms are in accord with the complexities of the continental Moho stressed by Oliver (1978).

THE PRESENT BALANCE

At the present time new ocean-floor crust is being created at a rate of about 10 km^3/yr [1], while erosion removes about 4 km^3/yr [1] from the continents. Spilite subduction about balances ocean crust production. A massive chemical exchange occurs during sea floor spilitization, and sediments may be subducted. Andesite magma production is the main process for return flow.

Is the continental crust growing at the present time? I do not think we can answer this question, but the evidence points to continental removal. For example, if we compare K-fixation in spilites ($\sim 3 \times 10^{13}$ g/yr, Bloch and Bischoff, 1979) with return in andesites ($0.3 - 1.5 \times 10^{13}$ g/yr, Brown and Hennessy, 1978), K-fixation in the mantle seems probable, without even considering contributions from pelagic sediments. There is nothing illogical about such a process. If we mix a cooling earth, crustal components will become fixed in the mantle (e.g., H_2O in hydrates, K in phlogopite, SiO_2 in pyroxene from olivine, Na in omphacitic pyroxene), but the rates of these processes are complex and depend on factors such as sediment trapping (perhaps a function of spreading rate and subduction angle), erosion rate (a function of continental area and elevation), seawater chemistry, etc. Quantifying the present is difficult, but quantifying the past is even more difficult.

ANCIENT REGIMES

I do not wish to review here the flood of observations and ideas about ancient tectonics which have appeared over the past decade (see e.g., Windley, 1977), but I will mention some of the most critical observations, which must be fitted to any adequate evolutionary model.

Evidence for an early very hot Earth is accumulating. The time required for the assembly of major objects seems to be getting shorter (Papanastassiou, paper given at the Wilson Symposium; Uyeda, 1978); this is supported by intriguing astronomical considerations of "Purcell clustering" in dust clouds (Press and Vishniac, 1979). At this workshop Shaw spoke of a totally molten Earth (see also Shaw, 1976).

Recently Runcorn (1978) and Libby et al. (1979), have emphasized that the ancient lunar core dynamo requires "special" radioactive heating to cause core separation. One is reminded of the review of Birch (1965), where all possible energy sources were reviewed. I was also intrigued by a recent comment of Van Flandern (1979) that Lunar tidal friction may have been large and reached a maximum about 3 x 10^9 years ago; the volcanism of Io, so elegantly explained by Peale et al., 1979, reminds us that such tidal influences can dominate thermal regimes. In short, the early Earth may have become very hot, and extensive melting seems almost certain.

During the early heating up and degassing of our planet, the hydrosphere must have played a critical role, for this water would have controlled the nature of the early crust, its thickness and thermal regime. Gradients in a water-cooled crust will have an upper limit ultimately determined by fluid convection.

The heating up of a planet must be a gigantic prograde metamorphic event. Fluids will bleed to the surface as pressures reach hydraulic fracturing conditions. Salt water may be the first major fluid to reach the surface. At later stages the fluid will become increasingly rich in CO_2 and an early saline hydrosphere and CO_2 − rich atmosphere may form. This same hydrosphere will be subject to massive perturbations by meteoric events. One of the critical questions bearing on early processes is whether or not the Earth went through a "greenhouse" stage. Sagan (1977) and Anders and Owen (1977) suggest that early reducing greenhouses on Earth and Mars were possible. Temperatures may have exceeded freezing on Mars, and the Earth's ocean may have boiled off or become very saline.

In the absence of water-cooled crust, the molten or partially molten Earth would cool very fast. As stated by Birch (1965), "The original surface was engulfed and digested, although a thin solid skin must have existed over most of the surface most of the time".

As the interior of the Earth heated up following early degassing, the Fe-S system would melt out to form the core. This melt would probably be quite rich in hydrogen, for as sinking metal interacts with mantle still containing some hydrated materials, reactions such as

$$X(OH)_2 + Fe \rightarrow FeO + H_2O + XO$$

would generate large hydrogen partial pressures. The existence of hydrogen in the core is still being discussed (Stevenson, 1977), but it is difficult to quantify. As Stevenson points out, only one per cent hydrogen is necessary in the core to explain its density. The consequences of the presence of hydrogen could be significant. As the core crystallizes, H_2 would bleed out and reduce the overlying lower mantle which would add metal to the core and possibly cause melting of the lower mantle. This could provide another energy source for the core dynamo (Carrigan and Gubbins, 1979) and slowly add H_2O to the Earth.

But there are other intriguing possibilities. If H_2 dissolves in liquid iron, so does helium. Core formation could scavenge helium into the metal phase. The helium ion (He $^+$) is small (~ 0.3Å); its ionization potential (24.6 e.v.) is similar to Cu and much less than Fe^2. Could core crystallization be related to the slow bleed-out of ^{3}He that is observed at most mantle volcanic centres? Helium is known to be quite soluble in liquid metals, and recent studies of implantation in metals (Johnson and Mazey, 1978) show how easily it fits into metallic structures.

One thing is certain: by the time the rather "normal" rocks at Isua, Greenland formed about 3.8 billion years ago, the hydrosphere was present (Moorbath, 1978). The lack of much older rocks could be due either to meteoritic events or to the absence of earlier oceans to stabilize crust by water-cooling. This latter possibility may be important. Chemical sediments and carbonated rocks at Isua indicate activity of hydrothermal systems and the possibility that crustal fixation of CO_2 was massive. If ocean formation was significant only after about four billion years, then it is unlikely that the mantle could have remained liquid for very long (unless of course the tidal heating effects mentioned by Van Flandern (1979) were of sufficient magnitude).

When one reflects on the age and geochemical data which have appeared in the last decade for old rocks, certain generalities emerge. The Earth appears to have had two major periods of distinct tectonic style, with a break occurring near the Archaean-Proterozoic boundary. Typical Archaean styles can be seen in Greenland or Rhodesia and the Canadian Shield, where the same structures are repeated over a long period of time (Young, 1978). From the Proterozoic onwards (see Hoffman, 1980) igneous and tectonic events approach modern convective styles; certainly plate tectonic phenomena were operating about one billion years ago (Black et al., 1979). The early structures appear to be dominated by Ramberg-type vertical gravity tectonics at high crustal levels (Rhodesia) and by thrust tectonics at lower levels (e.g., Bridgewater et al., 1974; Hall and Friend, 1979). Recently Veiser and Jansen (1979) have reviewed the massive evidence for a drastic change in sediment chemistry over the period 3 to 2Ga. This period suggests the transition from volcanogenic oceans (Fryer et al., 1979) to oceans and sediments dominated by continental run-off. One of our great challenges is to explain this drastic change in convective style.

ARCHAEAN MODELS

Crustal models now commonly considered appear to me to be of two types:

In one type (e.g., Moorbath, 1978; McCulloch and Wasserburg, 1978) continental-type crust is considered to form episodically and irreversibly; this assumes a slow growth of continental crust. In the other models (e.g., Hargraves, 1976, 1978; Fyfe, 1978; Goodwin and Smith, 1979) continental crust (mostly submarine) is formed early, but is reworked and preservation is accidental; this assumes a recycling of continental crust.

It now appears that models must fit in with certain important constraints:

a) If lighter" granite family" crustal elements are reworked, the mechanism must insure that their isotopic signature will be "almost" obliterated and reset to near-mantle values;

b) the models must explain the lack of modern-type ocean floor structures (the modern ophiolite structure);

c) the models must be in accord with a hotter or more turbulent mantle as indicated by high temperature periodotite lavas (ultramafic Komatiites).

The problems I see with slow growth models are:

a) How did hotter mantle hold back for 2 Ga much of the most fusible materials?

b) If continental growth occurs by sweeping up "andesitic" island arcs, where is the evidence for large areas of the complementary basaltic ocean floor (ophiolites)? There are enough "low grade" belts for structures like sheeted dykes to still be present.

c) How does this model fit the complex komatiite-tholeiite-andesite-rhyolite volcanic piles of the Archaean (Goodwin, 1977), piles which appear to have all the rock types of a modern ridge and subduction zone combined?

The problem I see with "reworking" models is to explain how light material is mixed back into the mantle without evidence of the subduction process. From the increasing evidence from modern andesites, such mixing does occur.

Goodwin and Smith (1979) explain this return mixing process by the down-sagging of volcanic piles back into the upper mantle. Such centres must be closely spaced. Fyfe (1978) suggests a similar mechanism, but cold periodotite and eclogite are part of the return flow. Fyfe has suggested a "hot spot" model of Hawaiis on "continental crust" as an appropriate analogue. Recent work of Ewing (1979) has shown clear evidence of eclogite fractionation in some old calc-alkaline volcanics.

Hargraves (1978) has proposed an elegant explanation in which given a hotter, more vigorous early mantle, viscous drag subduction of a scum crust dominated early regimes, but with cooling and thickening of the lithosphere, modern negative buoyancy subduction commenced near the beginning of the Proterozoic. By this process continents would thicken and the modern separation of ocean floor crust and continental crust would develop. Hargrave's analysis and that of Molnar and Gray (1979) show that the dragging of lighter crust into the mantle is reasonable. If this is so, an early globe-encircling, light scum crust which is continually reworked through the mantle presents no problem. The basic change in convective style, presumably related to lithosphere thickness, could explain the great event reviewed by Veizer and Jansen (1979). It would also explain the preservation of the apparent great 2.7 to 2.5 Ga pulse of crust formation of McCulloch and Wasserburg (1978); this would represent material that formed about the time of the change in style and was not rapidly recycled through the mantle.

If one considers Lister's (1976, 1974) models of ocean-floor cooling processes reasonable, it is difficult to see how the rise of basaltic magma and ocean-floor spreading could have occurred without forming something like a modern ophiolite structure. Are these structures missing in old, mainly submarine crust? One should always be a little dubious about emphasizing the lack of something not yet observed, particularly structures formed in the upper 1 to 2 km of the crust, but many have searched and ancient pillow lavas abound.

A necessary condition of large scale ocean-floor spreading is mantle rise at widely spaced centres. Even if magma production was several times faster in the Archaean and if the penetration sites were closely spaced, spreading might have been on a small scale (Fyfe, 1978). Does the typical ridge structure form when spreading is erratic or small-scale?

A necessary condition for continuous dike penetration should be that the density of the penetrating liquid is lighter than the crust it penetrates. This would not be true for komatiitic liquids penetrating basaltic crust or for magnesian basalts (Elthon, 1979) penetrating "Rhodesia-type" crust. Even modern ocean ridges seem to avoid light low-melting crust (see Fyfe and Leonardos, 1977) Hargraves' (1978) scum crust might also prevent penetration.

WHEN DID THE MANTLE SOLIDIFY?

As this meeting was termed a workshop by the convenor Dave Strangway, I felt free to raise a question I have seldom seen addressed. If the Earth was once largely molten, as many suggest, how long would it have taken to cool and crystallize a molten layer sandwiched between the solid lower mantle and the outer quenched skin? It appears to me that the setting of the boundary conditions is the key to this problem.

If we assume that oceans were present very early, perhaps almost always present, then there would be a 10 to 20 km-thick light quench zone. As heating up occurred, rising mantle plumes would produce increasingly basic magmas and eventually something like komatiites. These heavy liquids would underplate the skin, float off anorthitic plagioclase and sink olivine-pyroxene. An anorthitic layer could be built up under the skin to a thickness limited by the transition of anorthite to garnet and kyanite (about 60 km at 1000°C, Henson, 1976). Hence the question: How fast would liquid mantle cool under a thick continuous anorthosite layer? As anorthosites have a rather low thermal conductivity, the answer is that it could take a long time and, by analogy with modern lithosphere, it is unlikely that such a layer of anorthosite would convect even with a 1000°C gradient across it.

A model of this type is interesting. It would make ultra basic liquids the most common type for the early Earth, their surface rarity being due only to their density (>3.0). It would explain the lack of ophiolites, with the crust being decoupled from the mantle, and the entire lithosphere being light. It would predict contraction of the planet with thrust-dominated tectonics in the skin; such thrust structures have been proposed for the Moon and Mars (Mutch et al., 1976). It would naturally lead to a vast change in convection style once the upper mantle was solid and low-density basalts became the common liquid product. And it would explain the long retention of continental elements in the molten region.

CONCLUSIONS

While many features about the convective process for the modern Earth are understood, there still exist great uncertainties about the subduction process. Various data concerning the genesis of andesites suggest that crustal components take part in their formation either via spilites or subducted sediments. Because of such unknowns, we cannot be sure if the mass of continental crust is increasing or decreasing.

There appears to be mounting evidence for a profound break in the tectonic style of the Earth between the Archaean and the Proterozoic periods. While various models are being considered which may explain this change, all have inadequacies. The key problem which must be resolved is the mechanism for returning light crustal materials to the mantle.

REFERENCES

Anders, E. and Owen, T., 1977, Mars and Earth: Origin and abundance of volatiles: Science, v. 198, p. 453-465.

Aumento, F., 1979, Distribution and evolution of uranium in the oceanic lithosphere: Phil. Trans. Royal Soc. London A., v. 291, p. 423-431.

Barrett, T.J. and Spooner, E.T.C., 1977, Ophiolitic breccias associated with allochthonous oceanic crustal rocks in the East Ligurian Apennines, Italy – A comparison with observations from rifted oceanic ridges: Earth Planet Sci. Letters, v. 35, p. 79-91.

Birch, F., 1965, Speculations on the Earth's thermal history: Geol. Soc. Amer. Bull., v. 76, p. 133-154.

Black, R., Caby, R., Moussine-Ponchkine, A., Bayer, R., Bertrand, J.M., Boullier, A.M., Fabre, J. and Lesgner, A., 1979, Evidence for late Precambrian plate tectonics in West Africa: Nature, v. 278, p. 223-227.

Bloch, S. and Bischoff, J.L., 1979, The effect of low-temperature alteration of basalt on the oceanic budget of potassium: Geology, v. 7, p. 193-196.

Bridgwater, D., McGregor, V.R. and Myers, J.S., 1974, A horizontal tectonic regime in the early Archaean of Greenland and its implications for early crustal thickening: Precambrian Research, v. 1, p. 179-197.

Brown, G.C., 1977, Mantle origin of Cordilleran granites: Nature, v. 265, p. 21-24.

Brown, G.C. and Hennessy, J., 1978, The initiation and thermal diversity of granite magmatism: Phil. Trans. Royal Soc. London A., v. 288, p. 631-643.

Corrigan, C.R. and Gubbins, D., 1979, The source of the Earth's magnetic field: Sci. Amer. v. 240, p. 118-133.

Dalmayrac, B., Lancelot, J.R. and Leyreloup, A., 1977, Two-billion year granulites in the late Precambrian metamorphic basement along the Southern Peruvian Coast: Nature, v. 198, p. 49-51.

Eichelberger, J.C., 1978, Andesitic volcanism and crustal evolution: Nature, v. 275, p. 21-27.

Elston, W.E., 1978, Tectonic significance of mid-tertiary volcanism in the Basin and Range Province: a critical review with special reference to New Mexico: New Mexico Geol. Soc. Special Publ. No. 5, p. 93-102.

Elthon, D., 1979, High magnesia liquids as the parental magma for ocean floor basalts: Nature, v. 278, p. 514-518.

Ewing, T.E., 1979, Two calc-alkaline trends in the Archaean: trace element evidence: Contrib. Mineral Petrol., in press.

Fryer, B.J., Fyfe, W.S. and Kerrich, R., 1979, Archaean volcanogenic oceans: Chem. Geol., v. 24, p. 25-33.

Fyfe, W.S., 1976a, Hydrosphere and continental crust: growing or shrinking? Geosci. Canada, v. 3, p. 82-83.

_____, 1976b, The water inventory for Earth: Proc. Colloquium on Water in Planetary Regoliths: U.S. Army Cold Regions Res. and Eng. Lab., Hanover, New Hampshire, No. 03755, p. 2-4.

_____, 1978, The evolution of the Earth's crust: modern plate tectonics to ancient hot spot tectonics?: Chem. Geol., v. 23, p. 89-114.

_____, 1979, The tectonic significance of granitic magmatism: Symp. on evolution and mineralization of the Arabian-Nubian Shield: Inst. Appl. Geol. Bull. Jiddah, in press.

Fyfe, W.S. and Lonsdale, P., 1980, Ocean floor hydrothermal activity: in Emiliani, C., ed., The Sea, v. 5: John Wiley, in press.

Fyfe, W.S. and Leonardos, O.H., 1977, Speculations on the causes of crustal rifting and subduction, with applications to the Atlantic margin of Brazil: Tectonophysics, v. 42, p. 29-36.

Garrels, R.M. and Lerman, A., 1977, The exogenic cycles: reservoirs, fluxes and problems: in Stumm, W., ed., Global Chemical Cycles and Their Alterations by Man: Abakon Press, Berlin, p. 23-32.

Gilluly, J., 1971, Plate tectonics and magmatic evolution: Geol. Soc. Amer. Bull., v. 82, p. 2382-2396.

Goodwin, A.M., 1977, Archaean volcanism in Superior Province, Canadina Shield: in Baragar, W.R.A., Coleman, L.C., and Hall, J.M., eds., Volcanic Regimes in Canada: Geol. Assoc. Canada Spec. Paper 16, p. 205-241.

Goodwin, A.M. and Smith, I.E.M., 1979, Chemical discontinuities in Archaean volcanic terrains and the development of Archaean crust: Precambrian Research, in press.

Hall, R.P. and Friend, C.R.L., 1979, Structural evolution of the Archaean rocks of Ivisartoq and the neighbouring inner Godthabsfjord region, Southern West Greenland: Geology, v. 7, p. 311-315.

Hall, J.M. and Robinson, P.T., 1979, Deep crustal drilling in the North Atlantic Ocean: Nature, v. 204, p. 573-586.

Hargraves, R.B., 1976, Precambrian geologic history: Science, v. 193, p. 363-371.

_____, 1978, Punctuated evolution of tectonic style: Nature, v. 276, p. 459-461.

Hensen, B.J., 1976, The stability of pyrope-grossular garnet with excess silica: Contrib. Mineral Petrol., v. 55, p. 279-292.

Hoffman, P.F., 1980, A Wilson cycle of early Proterozoic age in the Canadian Shield: in Strangway, D.W., ed., the Continental Crust and Its Mineral Deposits: Geol. Assoc. Canada Spec. Paper 20, in press.

Hunt, G.E., 1979, A pioneer's view of Venus: Nature, v. 278, p. 777-778.

Hyndman, D.W., 1972, Petrology of Igneous and Metamorphic Rocks: McGraw-Hill Book Co., 533 p.

Johnson, P.B. and Mazey, D.J., 1978, Helium gas bubble lattices in face-centred-cubic metals: Nature, v. 276, p. 595-596.

Jones, G.M., Hilde, T.W.C., Sharman, G.F. and Agnew, D.C., 1979, Fault patterns in outer trench walls and their tectonic significance: Jour. Phys. Earth, in press.

Lewis, B.R.T. and Snydsman, W.E., 1977, Evidence for a low velocity layer at the base of the oceanic crust: Science, v. 266, p. 340-344.

Libby, L.M., Libby, W.F. and Runcorn, S.K., 1979, The possibility of superheavy elements in iron meteorites: Nature, v. 278, p. 613-617.

Lister, C.R.B., 1976, On the penetration of water into hot rock: Geophys. Jour. Royal Astron. Soc., v. 39, p. 465-509.

_____, 1977, Qualitative models of spreading-centre processes, including hydrothermal penetration: Tectonophysics, v. 37, p. 203-218.

Magaritz, M., Witford, D.J. and James, D.E., 1978, Oxygen isotopes and the origin of high $^{87}Sr/^{86}Sr$ andesites: Earth Planet. Sci. Letters, v. 40, p. 220-230.

Marzouki, F. and Fyfe, W.S., 1977, Pan-African plates: additional evidence from igneous events in Saudi Arabia: Contrib. Mineral Petrol., v. 60, p. 219-224.

McCulloch, M.T. and Wasserburg, G.J., 1978, Sm-Nd and Rb-Sr chronology of continental crust formation: Science, v. 200, p. 1003-1011.

Molnar, P. and Gray, D., 1979, Subduction of continental lithosphere: some constraints and uncertainties: Geology, v. 7, p. 58-63.

Moorbath, S., 1978, Age and isotopic evidence for the evolution of continental crust: Phil. Trans. Royal Soc. London A., v. 288, p. 401-413.

Mutch, T.A., Arvidson, R.E., Head, J.W., Jones, R.L. and Saunders, R.S., 1976, The Geology of Mars: Princeton University Press, 400 p.

Oliver, J., 1978, Exploration of the continental basement by seismic reflection profiling: Nature, v. 275, p. 485-488.

Peale, S.J., Cassen, P. and Reynolds, R.T., 1979, Melting of Io by tidal dissipation: Science, v. 203, p. 892-894.

Press, F. and Siever, R., 1974, Earth: W.H. Freeman, San Francisco, 945 p.

Press, W.H. and Vishniac, E.T., 1979, Production of new cosmological perturbations during the radiation-dominated era: Nature, v. 279, p. 137-139.

Ribando, R.J., Torrance, K.E. and Turcotte, D.L., 1976, Numerical models for hydrothermal circulation of the ocean crust: Jour. Geophys. Res., v. 81, p. 3007-3012.

Runcorn, S.K., 1978, The ancient lumar core dynamo: Science, v. 199, p. 771-773.

Sagan, C., 1977, Reducing greenhouses and the temperature history of Earth and Mars: Nature, v. 269, p. 224-226.

Shaw, D.M., 1976, Development of the early continental crust part 2: Prearchean, Protarchean and later eras: in Windley, B.F., ed., Early History of the Earth.

Sibley, D.F. and Vogel, T.A., 1976, Chemical mass balance of the Earth's crust: The Calcium? dilemma and the role of pelagic sediment: Science, v. 192, p. 551-553.

Spooner, E.T.C., 1976, The strontium isotopic composition of sea water and sea water-oceanic crust interaction. Earth Planet Sci. Letters, v. 31, p. 167-174.

Stern, R.J., 1979, on the origin of andesite in the Northern Mariana Island Arc: implications from Agrigan: Contrib. Mineral. Petrol., v. 68, p. 207-220.

Stevenson, D.J., 1977, Hydrogen in the Earth's core: Nature, v. 268, p. 130.

Straus, J.M. and Schubert, G., 1977, Thermal convection of water in porous medium: effect of temperature- and pressure-dependent thermodynamic and transport properties: Jour Geophys. Res., v. 82, p. 325-333.

Taylor, H.P. and Silver, L.T., 1978, Oxygen isotope relationships in plutonic igneous rocks of the Peninsular Ranges batholith, Southern and Baja California: United States Geol. Survey Open-file report 78-701, p. 423-425.

Uyeda, S., 1978, The New View of the Earth: Freeman and Co., San Francisco, 217 p.

Vallance, T.G., 1969, Concerning spilites: Proc. Linn. Soc. New South Wales, v. 85, p. 8-52.

Van Flandern, T.C., 1979, Gravity and expansion of the Earth: Nature, v. 278, p. 821.

Veizer, J. and Jansen, S.L., 1980, Basement and sedimentary recycling and continental evolution: Jour. Geol. (in press).

Whitford, D.J. and Jezek, P.A., 1979, Origin of late-cenozoic lava from the Banda Arc, Indonesia: trace element and Sr isotope evidence: Contrib. Mineral. Petrol., v. 68, p. 107-115.

Windley, B.F., 1977, The Evolving Continents: John Wiley, New York, 385 p.

Wolery, T.J. and Sleep, N.H., 1976, Hydrothermal circulation and geochemical flux at mid-ocean ridges: Jour. Geol., v. 84, p. 249-275.

Young, G.M., 1978, Some aspects of the evolution of the Archaean crust: Geosci. Canada, v. 5, p. 140-149.

The Continental Crust and Its Mineral Deposits, edited by D. W. Strangway,
Geological Association of Canada Special Paper 20

ASPECTS OF THE CHRONOLOGY OF ANCIENT ROCKS
RELATED TO CONTINENTAL EVOLUTION

————

S. Moorbath
Department of Geology and Mineralogy,
University of Oxford, Parks Road, Oxford, England

————

ABSTRACT

The oldest known rocks in West Greenland comprise a 3800 Ma-old volcano-sedimentary
sequence, deposited on crust of unknown type, and intruded by voluminous, sima-derived
magmatic precursors of calc-alkaline orthogneisses which became fully stabilized as typical
continental crust by about 3700 Ma ago. Small, thick, stable continental nuclei of this type were
forming independently at different sites by 3700 to 3500 Ma ago. The production of juvenile
continental, sialic crust reached enormous proportions 3000 to 2600 Ma ago, and continues
today at destructive plate margins.

Radiogenic Sr, Pb and Nd isotopic studies suggest that continental growth by possibly
episodic irreversible differentiation of the upper mantle throughout geological time greatly
predominates over recycling and reworking of much older continental crust. Taken in conjunction with much other geochemical and geophysical evidence, age and isotope data provide a
simple, dynamic picture of the genesis, internal differentiation and stabilization of continental
crust, at the same time providing severe constraints for opposing hypotheses which involve
repeated quantitative recycling and regeneration of ancient continental protoliths, massive
recirculation of continental crust through the mantle, and unrestricted mixing between continental crust and mantle.

While the initial isotopic compositions of calc-alkaline orthogneisses generally suggest
extraction of their magmatic precursors from mantle or basic lithosphere followed by penecontemporaneous metamorphic and geochemical differentiation, true granites of Archaean (or any)
age frequently have initial isotopic compositions suggesting derivation by partial melting of
older continental crust.

RÉSUMÉ

Parmi les roches les plus anciennes connues dans l'ouest du Groënland, on observe une séquence volcano-sédimentaire de 3800 Ma déposée sur un type inconnu de croûte et injectée de précurseurs magmatiques dérivés du sima consistant en orthogneiss calco-alcalins qui devinrent complètement stabilisés sous la forme d'une croûte continentale typique il y a environ 3700 Ma. Des noyaux stables continentaux, petits et épais, de ce type se sont formés indépendamment en plusieurs endroits entre 3700 et 3500 Ma. La production de croûte continentale juvénile sialique a atteint d'énormes proportions entre 3000 et 2600 Ma et se poursuit actuellement aux bordures destructives des plaques.

Les études isotopiques du Sr, Pb et Nd radiogéniques suggèrent que la croissance continentale, probablement par différentiation épisodique irréversible du manteau supérieur à travers l'histoire géologique, a grandement dominé sur le recyclage et le remaniement d'une croûte continentale beaucoup plus ancienne. Utilisées en conjonction avec beaucoup d'autres évidences géochimiques et géophysiques, les données chronologiques et isotopiques fournissent une image simple et dynamique de la génèse, de la différentiation interne et de la stabilisation de la croûte continentale, imposant en même temps des limites rigides aux hypothèses contraires qui invoquent un recyclage quantitatif répété et une régénérescence d'anciens protolithes continentaux, la recirculation massive de croûte continentale à travers le manteau et le mélange sans limites de la croûte continentale et du manteau.

Alors que les compositions isotopiques initiales des orthogneiss calco-alcalins suggèrent généralement que leurs précurseurs magmatiques proviennent du manteau ou d'une lithosphère basique et ensuite qu'il y a eu différentiation métamorphique et géochimique pénécontemporaine, les vrais granites d'âge archéen (ou d'un autre âge) ont fréquemment des compositions isotopiques initiales qui suggèrent une dérivation par fusion partielle d'une croûte océanique plus ancienne.

INTRODUCTION

It is not the aim here to provide a comprehensive summary of all the latest age and isotope data on Archaean rocks, but to outline some recent work relevant to the study of the development of early continental crust and associated volcano-sedimentary sequences. The main emphasis will be on interpretation of age and isotope data, but several other relevant matters, such as the thickness of Archaean crust, will be mentioned. This paper should be regarded as a supplement to earlier reviews (Moorbath, 1976, 1977, 1978a), with particular emphasis on works published since those reviews were prepared. However, the list of cited works is admittedly incomplete and has been selected to show the way in which the subject appears to be developing. Detailed technicalities and methodology on all matters discussed here will be found in the cited papers, and it is hoped that those authors who disagree with the general conclusions outlined here, based on the recent work and ideas of many workers, will study these technical matters closely and show in equally compelling technical detail how the data can be interpreted differently.

It should be noted that the ages reviewed in this paper are based on the following decay constants:

$\lambda^{235}U = 9.85 \times 10^{-10}/\text{Yr}, \quad \lambda^{238}U = 1.55 \times 10^{-10}/\text{Yr}; \quad \lambda^{87}Rb = 1.39 \times 10^{-11}/\text{Yr};$
$\lambda^{147}Sm = 6.54 \times 10^{-12}/\text{Yr}.$

THE OLDEST SUPRACRUSTAL ROCKS

The oldest known terrestrial rocks, with an age of nearly 3800 Ma, consist of a complex and varied sequence of metamorphosed supracrustal rocks of igneous and sedimentary origin, exposed in an arc some 15 to 20 km in diameter around a dome of gneiss (Amîtsòq gneiss) near the margin of the Inland Ice at Isua, about 150 km northeast of Godthaab in West Greenland (Fig. 1). The Isua supracrustal belt has a variable width from 1 to 3 km, and can be followed as continuous outcrop over a distance of nearly 40 km. Contacts with the enclosing orthogneisses are sharp and nearly vertical, while clear intrusive relationships are preserved at several places along the contact. Geological descriptions of the area have been given by Allaart (1976) and Bridgwater et al. (1976). Intense deformation and amphibolite facies metamorphism have obliterated primary igneous and sedimentary textures as well as the original stratigraphy, but not the clues to the original character of the rocks, which exhibit many similarities to the much larger greenstone belts of late Archaean times. The depositional environment might be envisaged as one in which island volcanoes, erupting acid, basic and ultrabasic lavas, were surrounded by shallow-water platforms covered by coarse, volcaniclastic deposits, followed seawards by volcanigenic sands, shales and muds and ultimately by a rich variety of chemical sediments. The latter dominate the eastern part of the present outcrop of the Isua supracrustals, where they now survive as a bewildering profusion of banded iron formation, cherts and carbonate rocks. The Isua carbonates are reported by Schid-

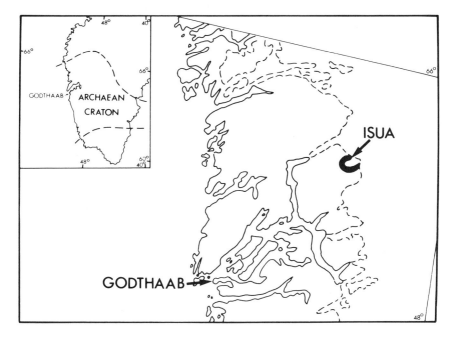

Figure 1. Sketch-map of southern West Greenland, showing the localities of Godthaab (the capital of Greenland) and the Isua supracrustal belt.

lowski *et al.* (1979) as containing organic and inorganic carbon in approximately modern proportions; they state that 'unless this early stabilization of the terrestrial carbon cycle in terms of a constant partitioning of carbon between the reduced and oxidized species is shown to have been caused by some inorganic geochemical process, a considerably earlier start of chemical evolution and spontaneous generation of life must be considered than is presently accepted.' In contrast, sulphur isotope measurements on Isua sediments have their $\delta^{34}S$ values close to zero per cent and show no evidence of bacterial sulphate reducers (Monster *et al.*, 1979).

One of the most interesting sedimentary units at Isua contains a major meta-conglomerate with pebbles, cobbles and boulders of an acid volcanic rock set in a finer grained, locally laminated matrix (Allaart, 1976; Bridgwater *et al.*, 1976). This conglomerate can be traced for nearly 30 km along strike. Over much of the outcrop the coarse clasts are highly deformed and sheared out, but in one locality there occurs a relatively little-deformed outcrop where the clasts consist of pale, fine-grained, muscovite-bearing, quartzo-feldspathic rocks with variable amounts of secondary carbonate. They contain up to 9% K_2O. The clasts are regarded as intra-formational because they are lithologically and chemically similar to muscovite-bearing, quartzo-feldspathic rocks which occur as sheets within the Isua supracrustal succession and which may represent original acid volcanic layers or volcanigenic sediments. No material has yet been found in the conglomerates which can be identified as a sample of the substratum on which the Isua supracrustal sequence was originally deposited. The substratum is not discernible in the field, because the supracrustal sequence is surrounded by, and infolded with, the younger gneisses. It may be that the degree of metamorphism and deformation is too high to permit distinction between the substratum and the supracrustal rocks themselves. At any rate, there is no evidence yet for the presence of a continental-type basement.

Published isotopic age determinations on the Isua supracrustals are relatively straightforward. A Pb/Pb isochron age of 3710 ± 70 Ma for the banded iron-formation (Moorbath *et al.*, 1973), and a Rb-Sr whole-rock age of 3740 ± 60 Ma (with an initial $^{87}Sr/^{86}Sr$ ratio of 0.7015 ± 0.0017) on combined data for boulders and matrix from the acid conglomerate unit, have been interpreted as the time of deposition and/or metamorphism of the supracrustal sequence. Zircons from a quartzite and from the matrix of the acid conglomerate have yielded $^{207}Pb/^{206}Pb$ ages of 3670 and 3770 Ma respectively (Baadsgaard, 1976), while U-Pb analyses of single zircons from con-glomerate pebbles have given an age of $3770\pm^{12}_{6}$Ma (Michard-Vitrac *et al.*, 1977). A Sm-Nd isochron age of 3770 ± 42 Ma has been reported (Hamilton *et al.*, 1978) for three pebbles and a matrix from the acid conglomerate unit, and for six samples from a basic metavolcanic unit. The initial $^{143}Nd/^{144}Nd$ ratio for the isochron is 0.507831 ± 46, which falls on the now firmly established growth line of $^{143}Nd/^{144}Nd$ ratio versus age for many Precambrian igneous and meta-igneous rocks, with charac-teristic Sm/Nd ratio of 0.308 as calculated from the slope of the growth line. (The actual Sm/Nd ratio required to generate the initial $^{143}Nd/^{144}Nd$ ratio for the analyzed Isua rocks at 3770 Ma is 0.306 ± 0.006). Hamilton *et al.* (1978) interpret the Isua Sm-Nd age (which is in perfect agreement with the U-Pb single zircon age of Michard-Vitrac *et al.*, 1977) as providing a precise estimate of the timing of the volcanism which produced the basic and acid rocks from a mantle reservoir with chondritic Sm/Nd ratio.

In addition, the Isua supracrustals contain minor sulphide mineralizations, including galena. One galena sample contains Pb with the least radiogenic ratios reported for any terrestrial material, ($^{206}Pb/^{204}Pb$ = 11.15; $^{207}Pb/^{204}Pb$ = 13.04) and a single-stage model age of 3740 Ma is obtained (Appel et al., 1978), using the model parameters of Oversby (1974). The $^{208}Pb/^{204}Pb$ ratio of 31.18 is very nearly the least thorogenic terrestrial Pb known. The isotope data for this galena fit all of the current Pb isotope mantle evolution models very closely (e.g., Cumming and Richards, 1975; Stacey and Kramers, 1975).

Most of the ages quoted above are not significantly different from published ages for the enclosing, younger (from field evidence) Amîtsoq gneisses, except for U-Pb discordia ages of zircons from the boulders and matrix of the acid conglomeratic unit ($3770 \pm ^{12}_{9}$), which are significantly different from U-Pb discordia ages of 3600 ± 50 Ma for Amîtsoq gneisses of the type area near Godthaab (Baadsgaard, 1973; Michard-Vitrac et al., 1977). Unfortunately, no zircon U-Pb ages have yet been reported for the Amîtsoq gneisses at Isua.

Even before the discovery of the Isua supracrustal belt and realization of its geological importance, McGregor (1973) had discovered widely distributed metabasaltic and metasedimentary xenoliths and enclaves in the 3600 to 3700 Ma-old Amîtsoq gneisses of the Godthaab area. They have been described in detail by McGregor and Mason (1977) and termed the 'Akilia association'. They include basic and ultrabasic rocks, some with komatiitic and others with Fe-rich tholeiitic chemical affinities, as well as banded ironstones and gneisses of detrital sedimentary origin. The range of lithologies suggests that the protoliths of the Akilia association consisted of a greenstone belt-type sequence that was intruded and disrupted by the calc-alkaline magmatic precursors of the Amîtsoq gneisses. It is probable that the Akilia association is broadly contemporaneous with the generally similar, but essentially in situ, Isua supracrustal rocks, but that the Akilia rocks are exposed at a deeper crustal level than the Isua supracrustals.

While the Isua supracrustals are the oldest rocks of their type yet discovered, slightly younger rocks of similar type are known from other continents. Some of them are less metamorphosed and deformed than the Isua rocks, preserving primary igneous and sedimentary textures. The Warrawoona Group of the eastern Pilbara Block of Western Australia, described by Barley et al. (1979), is an excellent example of an Archaean shallow-water, volcanic-sedimentary facies, and may even contain microfossils. The depositional basin was large, volcanically active and apparently shallow with subdued marginal relief. Within the basin, felsic volcanoes formed topographic highs from which sheets of volcanically derived sediments interfingered with ultramafic and mafic volcanics. U-Pb measurements on zircons from a columnar dacite in the Warrawoona succession have yielded an age of 3452 ± 16 Ma (Pidgeon, 1978), the oldest age so far reported for the Archaean of Australia. Barley et al. (1979) point out that the Onverwacht Group of the Barberton Mountain Land of South Africa (Reimer, 1975; Anhaeusser, 1978) probably represents a similar depositional environment. Acid and basic members of the Onverwacht Volcanic Sequence have yielded a Sm-Nd isochron age of 3540 ± 30 Ma, with an initial $^{143}Nd/^{144}Nd$ ratio of 0.50808 ± 8, which is interpreted as the age of mantle-derived volcanism (Hamilton et al., 1979b). A previous Rb-Sr determination on a comparatively well-preserved Onverwacht komatiite had yielded an internal isochron age of 3500 ± 200 Ma, with an initial ^{87}Sr/^{86}Sr ratio of 0.70048±5 (Jahn and Shih, 1974). In the Rhodesian shield, the Selukwe greenstone belt assigned to the Sebakwian Group is older than the cross-cutting Mont d'Or granite, which has yielded a Rb-Sr whole-rock isochron age of 3420±60 Ma, with an initial ^{87}Sr/^{86}Sr ratio of 0.711±0.001 (Moorbath et al., 1976). This high initial ratio is interpreted as indicating partial melting of older sialic crust which probably underlies the Selukwe greenstone belt and which may be equivalent to the 3600 Ma-old, high Rb/Sr gneisses from the Shabani area, some 70 km south of Selukwe (Moorbath et al., 1977; Wilson et al., 1978). In Southern India, the Sargur Supergroup of Mysore is another example of an earlier Archaean volcano-sedimentary sequence. It has not yet been dated directly, but is regarded as predating the Peninsular Gneiss (Chadwick et al., 1978), parts of which have been dated by the Rb-Sr whole-rock isochron method at 3100±70 Ma, with an initial ^{87}Sr/^{86}Sr ratio of 0.7016±0.0005 (Moorbath, Taylor and Ramakrishnan, unpublished data).

During the late Archaean, some 3100 to 2600 Ma ago, volcano-sedimentary associations of greenstone belt-type were enormous and worldwide; there are many well-described examples from several continents, particularly from Canada, the northern U.S., Rhodesia and Australia.

Field relationships of greenstone belt facies supracrustal rocks to older rocks vary greatly depending upon local geological conditions – there is no hard-and-fast rule. (Detailed discussion is outside the scope of this paper). Some greenstone belts can be demonstrated to rest unconformably on continental, sialic crust. Well-known examples from Rhodesia have been described by Bickle et al. (1975) and Wilson et al. (1978). In other cases, no continental, sialic basement is discernible from field evidence, nor is there any isotopic evidence for it in igneous plutons which so often intrude greenstone belts (see, for example, Arth and Hanson, 1975). It seems that greenstone belt assemblages can be deposited on either sialic or simatic crust. In the case of simatic crust it might be difficult or impossible to discern where the simatic substratum ends and where the greenstone belt proper begins. By the time that most Archaean greenstone belt assemblages formed, it is likely that both sialic and simatic 'basement' were sufficiently thick and mechanically strong to support them, in the same way that both modern oceanic and continental crust can support sedimentary basins.

In yet another group of cases, arguments persist as to whether a given greenstone belt rests on sialic or simatic crust. In these cases inadequate or ambiguous field evidence is sometimes balanced against a particular author's view as to whether the primordial crust of the earth was acid (granitic) or basic (basaltic). Unfortunately, no primordial terrestrial crust has yet been recognized, but it is an axiom of most modern petrogenetic hypotheses that granitoid (sensu lato) magmatic rocks always have more basic precursors. As an example of this type of controversy, a very convincing geochemical and petrological case can be made for regarding the Ancient Gneiss Complex of Swaziland as the basement to the 3500 Ma-old Barberton greenstone belt (Hunter et al., 1978), yet the field evidence does not appear to rule out the granitoid gneisses being younger than some of the supracrustal rocks (Anhaeusser, 1978).

Regardless of the nature of their basement or of their age, most Archaean green-stone belt assemblages are invaded by granitoid plutons and batholiths, predomin-antly of calc-alkaline composition, although true granites are by no means rare. The mechanical breaking-up of massive greenstone belts by voluminous calc-alkaline magmas is an awe-inspiring phenomenon, with which all Archaean geologists are familiar. The most impressive, and by far the best exposed, example I have ever seen occurs some 50 km ESE of the town of Sukkertoppen (and about 125 km north of Godthaab) in West Greenland, where the calc-alkaline magmatic precursors of the Alangua-Finnefjeld gneisses (Rb-Sr whole-rock isochron age = 2930± 60 Ma, initial $^{87}Sr/^{86}Sr$ ratio = 0.7018±0.0004, Moorbath and Taylor, unpublished data) invaded, broke up, tore apart and shredded a major supracrustal belt composed mainly of amphibolites: this belt is probably contemporaneous with the approximately 3000 Ma old Malene supracrustal belt of the Godthaab district (McGregor, 1973; Bridgwater et al., 1976). On a more modest scale the same thing can be observed in the Godthaab district, where, as already mentioned, the magmatic precursors of the 3600 to 3700 Ma-old Amîtsoq gneisses broke up the volcano-sedimentary protoliths of the Akilia association (McGregor and Mason, 1977).

THE OLDEST GNEISSES

As a logical corollary, just as most Archaean greenstone belt assemblages are invaded by voluminous calc-alkaline magmas, almost all regions of Archaean ortho-gneisses of any size show irrefutable evidence (from enclosed fragments) of having been intruded into crustal and supra-crustal sequences. Indeed, by definition, all plutonic magmatic rocks *must* have been intruded into something! The widespread recognition of many granitoid Archaean gneisses as deep-seated rocks of magmatic origin is fairly recent, but it is of major geological significance. The conventional term 'gneiss' itself becomes somewhat ambiguous under these circumstances, since gneis-sification is frequently a syn-intrusive phenomenon. Where exposures are superb and continuous, as in West Greenland, it is evident that deformation in Archaean granitoids was locally heterogeneous, so that sequences ranging from little-deformed (or undeformed) intrusive igneous rocks to strongly deformed gneisses may be clearly traced out. The development of finely layered gneisses by progressive defor-mation of a variety of undeformed rocks in West Greenland has been vividly de-scribed and illustrated by Myers (1978). Undeformed early and late Archaean gneis-ses, with or without supracrustal enclaves, show highly complex intrusive relation-ships between numerous successive phases of injection. The transition to uniformly banded gneiss may take place within a few metres, so that all magmatic features are totally obliterated. The rapid and ubiquitous alternation of deformed and undeformed orthogneisses is a highly characteristic feature of many Archaean gneisses and, as indicated above, may have originated in many cases at the magmatic/ intrusive stage.

The oldest known gneisses, in the age range 3500 to 3700 Ma, occur in West Greenland (Moorbath et al., 1972), Labrador (Barton, 1975; Hurst et al., 1975), Rhodesia (Moorbath et al., 1977), South Africa (Barton, 1977), the Minnesota River Valley (Goldich and Hedge, 1975), etc., and have been dated by Rb-Sr and/or Pb/Pb whole-rock methods, and/or by U-Pb methods on zircon. Massive worldwide pro-duction of calc-alkaline orthogneisses occurred 3100 to 2600 Ma ago. Numerous age and isotope data on such rocks are available from most continents, but are not reviewed here. The significance of these calc-alkaline orthogneisses, and similar rocks formed throughout geological time, as the fundamental, mostly mantle-derived, building blocks of continental crust is becoming more widely recognized and has been reviewed from many different aspects (Moorbath, 1976, 1977, 1978a; Windley and Smith, 1976; Brown, 1977; Brown and Hennessy, 1978; O'Nions and Pankhurst, 1978; Taylor, 1979). It should be noted that a very different picture of continental evolution is painted by Fyfe (1978, 1979), which is in accord neither with the conclusions of the quoted works, nor with the present paper.

Sr and Pb isotopic evidence, as well as age data, on calc-alkaline igneous rocks including Archaean and Proterozoic gneisses, suggests that irreversible chemical differentiation of the mantle's essentially infinite reservoir commenced at least 3800 Ma ago and produced new continental, sialic crust during several relatively short (about 100 to 300 Ma) episodes, widely separated in time. During each of these so-called 'accretion-differentiation-superevents' (Moorbath, 1977, 1978a), juvenile sial, derived from the mantle and/or basic lithosphere, underwent thorough penecontemporaneous igneous, metamorphic and geochemical differentiation. This resulted in thick, stable, compositionally layered, permanent, continental crust which exhibits close grouping of isotopic ages of rock formation, as well as mantle-type initial Sr and Pb isotopic compositions, for all major constituents. Isotopic evidence suggests that within most accretion-differentiation-superevents, and especially during the earlier ones, continental growth greatly predominated over reworking of older sialic crust.

However, it must be realized that reworking of much older sialic crust can occur in several tectonic environments and becomes more common with the progressive growth and fragmentation of continental crust throughout geological time, and with the independent migration of the separated fragments. In any given terrane reworking can be distinguished from the accretion-differentiation regime by detailed age and isotope studies. True granites are more likely to be the magmatic expression of partial melting of sialic crust than are any members of the calc-alkaline kindred. True granites of any geological age tend to have higher and more heterogeneous initial $^{87}Sr/^{86}Sr$ ratios than contemporaneous calc-alkaline rocks, which shows the greater involvement of continental crust in their genesis. The degree of isotopic divergence of a given granite from contemporaneous mantle-derived calc-alkaline or basaltic rocks is a function of the degree of crustal involvement, the geochemical nature of the crust and the magnitude of the age difference between the granite and its crustal protolith. Where the latter age difference is small, initial isotopic compositions of crustally derived granites may be very similar to values for contemporaneous mantle-derived rocks. In cases where granitic magma is formed directly by fractionation of mantle-derived basaltic or intermediate magma, initial isotope characteristics may be identical with isotopic ratios of contemporaneous mantle. Thus it is *particularly* important, in dealing with granitic rocks, that every example is considered on its own merits and on the basis of its own evidence. True granites are petrogenetically much more varied than any other igneous rocks, because they may be derived from mantle, basic lithosphere, continental sialic crust, or any mixture of these. On purely petrogenetic grounds, and indeed on the basis of much isotopic evidence, it is becoming evident that the continent-building tonalites and their calc-alkaline plutonic and volcanic brethren of any geological age have a more restricted range of precursors in the upper mantle or in petrologically and geochemically related basic lithosphere (see, for example, Ringwood, 1974; Stern *et al.*, 1975; Thorpe *et al.*, 1976; Anderson *et al.*, 1978; Hanson, 1978).

I do not wish to repeat the evidence and arguments in the review articles quoted earlier, except to focus attention in the following sections on several recently published papers, not previously reviewed, which seem to me to be of particular significance for the study of continental evolution and which fit in well with previous data and hypotheses.

ARCHAEAN CRUSTAL THICKNESS AND CONTINENTAL GEOTHERMS

Contrary to earlier views, it is now becoming more widely accepted that the thickness of Archaean continental crust and the magnitude of Archaean continental geotherms, as well as the characteristic continental distribution of the radioactive heat-producing elements (exponential decrease with depth), were much the same in late Archaean times, some 3000 to 2500 Ma ago, as they are now. Metamorphic pressure estimates on Archaean granulites and high amphibolite facies rocks range mostly between 7 and 12 kbar, corresponding to depths of about 25 to 40 km and, more rarely, up to 15 kbar (50 km), with corresponding temperatures mostly in the range of 700 to 900°C. To quote only one specific example, mineral assemblage data show that conditions of late Archaean (about 2850 Ma ago) upper amphibolite facies were about 7.0 kb at 630°C, and rose to about 10.5 kb at 810°C in adjacent granulite facies in the Buksefjorden region of southwest Greenland (Wells, 1976). More recent work by Wells (1979) in this area strongly suggests the late Archaean sialic crust at Buksefjorden was probably as much as 60 to 80 km thick. The P-T data for other late Archaean areas have been summarized by Tarney and Windley (1978), who concluded that the geothermal gradient in lower levels of the crust during the late Archaean corresponded to 18 to 35°/km.

Several workers have recently discussed the inevitable consequences of the overall similarity of continental crustal thickness as well as continental geotherms between the Archaean and today (Burke and Kidd, 1978; Bickle, 1978; Brown and Hennessy, 1978; England, 1979). In summary, although radiogenic heat production in the mantle between 3000 and 2500 Ma ago was approximately three times that of today, this extra heat was not removed by conduction through continental crust. Burke and Kidd (1978) state that 'a convective process seems necessary to dissipate this heat and plate tectonics is an efficient and familiar convective process capable of doing the job'. Bickle (1978) considers that the relation between earth heat loss, the rate of plate creation and the rate of heat transport to the base of the lithosphere suggests that a significant proportion of the heat loss in the Archaean must have taken place by the processes of plate creation and subduction. The Archaean plate processes may have involved much more rapid production of plates only slightly thinner than at present.' Similarly, England (1979) concludes that 'the lower geothermal gradients permitted by the P-T data would be consistent with the view of Archaean tectonics in which the continental thermal regime was similar to that of the present day and most of the additional heat was lost from the Earth's interior by a faster rate of creation of oceanic lithosphere.' Tarney and Windley (1978) propose a threefold mechanism for crustal thickening in late Archaean times involving tectonic underthrusting of new oceanic material into deep parts of the continental crust, nappe stacking and imbricate interthrusting of oceanic and continental material (see also Bridgwater *et al*., 1974), and tonalite magma injection with thrusting in deep crustal levels. All this explains the rapidly developing granulite metamorphism deep in the thickened crustal pile under the P-T conditions indicated earlier. Windley and Tarney state that 'to view this Archaean scheme of events as taking place in some kind of proto-Cordilleran tectonic environment is preferable to the alternative view of granulite facies conditions being widespread throughout a static crust or non-subducting lithosphere (e.g., Fyfe, 1974; Lambert, 1976) which requires a much thinner crust than at present and an unusually steep geothermal gradient; such conditions are inconsistent with the pressure, gradient and thickness estimates summarized in this paper.' Drury (1978) has modelled rare earth element distributions in high-grade Archaean gneisses of northwest Scotland, and proposes a model for the gneisses which implies generation of geochemical layering in the thick Archaean crust by fractional crystallization of parental tonalitic magmas rising under conditions of high deformation from mantle depths.

Possibly the most detailed and graphic account of the emplacement of immense volumes of mantle-derived tonalitic magmas has been given by Wells (1979). He postulates that the overthick (60 to 80 km) late Archaean sialic crust in the Buksefjorden region of West Greenland was generated by massive accumulation of sialic magma in magmatic and tectonothermal regimes similar to present-day island arcs. Wells estimates that the thickness of the pre-2800 Ma-old basic crust was between 10 and 20 km, and that this crust was thickened nearly homogeneously by the horizontally orientated intrusion of the 2900 Ma-old tonalitic magmas. Wells prefers a model of crustal over-accretion, in which successive pulses of juvenile magmatic material were added above previous intrusive phases. This novel model of crustal accretion is based closely on observed metamorphic, tectonic and geochemical features of the rocks, and can account in detail for the respective metamorphic histories of both the gneisses and of the enclosed remnants of older basic crust.

With regard to the general problem of the timing of continental growth and the emergence of land in relation to the thickness of continental crust, Windley (1977) has justly criticized the much-publicized model of Hargraves (1976), in which gradually differentiating, compositionally contrasted, primordial earth shells led to Archaean continents submerged beneath a globe-encircling sea, and where continental emergence did not take place on a large scale until after 1400 to 1000 Ma ago in the late Proterozoic. The model of Hargraves is totally at odds with the most basic geological evidence for the existence of sedimentary basins extending right back through the Proterozoic and well into the Archaean. Continental nuclei of the type described in this paper must surely always have emerged as land early in their history (see, for example, Watson, 1976).

No estimates of P-T conditions and geothermal gradients appear to have been reported yet for early Archaean (3700 to 3500 Ma) continental crust, using the same techniques as employed by previous workers for late Archaean crust. I predict that the average values for pressure, temperature, thermal gradient and crustal thickness will turn out to be quite similar for early and late Archaean crust. This may sound like a premature and subjective impression, but anyone familiar with early Archaean gneisses cannot fail to be struck by their profound geological, petrological and geochemical similarity to late Archaean and much younger gneisses. While amphibolite facies gneisses appear to predominate over granulite facies gneisses at present exposure levels, there is no evidence whatsoever for the thin, permobile, widespread granitic crust postulated by some authors (e.g., Hargraves, 1976; Lowman, 1976; Shaw, 1976; Fyfe, 1976, 1978; Collerson and Fryer, 1978). These early Archaean areas are often singular free of later major tectonic thermal disturbances. For example, at Isua in West Greenland, massive, voluminous 3700 Ma-old Amîtsoq gneisses, as well as the even older Isua supra-crustal sequence, are cut by the great Ameralik dyke swarm, which is at least 3000 Ma-old. These dykes were emplaced into relatively cold, thick, brittle, sialic crust and there is no sign of post-dyke deformation. Whole-rock Rb-Sr, Pb/Pb, Sm-Nd and zircon U-Pb ages on gneisses and supracrustals are virtually concordant, as described earlier, while only K-Ar and Rb-Sr mineral ages indicate later thermal disturbances (Pankhurst *et al.*, 1973; Baadsgaard *et al.*, 1976). Futhermore, for anyone who accepts the temporal constraints imposed by age and isotope data (for further discussion, see below), there can be no doubt that Archaean calc-alkaline orthogneisses with mantle-like initial Sr, Pb and Nd isotopic compositions were not produced by reworking or partial melting of significantly older sialic crust of the same type. Thus I find it difficult to comment in detail within the present limits of space on the recent paper by Collerson and Fryer (1978), because I do not fully understand their arguments, in particular those relating to 'primitive initial Pb and Sr isotopic compositions of early Archaean gneisses as due to volatile exchange of Pb and Sr', which they themselves recognize as contrary to most other interpretations. Rightly or wrongly, it seems to me that the mechanism for massive mantle-crust metasomatism during early crustal evolution proposed by Collerson and Fryer (1978) would lead to quite different geological, petrological, geochemical, geochronological and isotopic data than those which are actually observed in Archaean terranes. In particular, Collerson and Fryer *only* consider Sr isotopes which are admittedly subject to some interpretational ambiguities (see below), but do *not* address themselves to the Pb and Nd isotopic data, which are not subject to the same limitations.

The evidence summarized above suggests that the early sial was thick, strong and stable; it was composed of essentially indestructible calc-alkaline material derived from the mantle. Restricted in area initially, it formed the nuclei for later, voluminous continental additions derived mainly from the mantle. After continental accretion, geochemical gradients and corresponding geothermal characteristics of the sial were established very rapidly (geochronologically instantaeously). In contrast, the thickness and nature of early *basic* lithosphere are more difficult to determine: this problem still awaits enlightenment.

There is no geochronological, isotopic, or other evidence whatever that the Amîtsoq gneisses were derived from the reworking of pre-Isua supracrustal (i.e., pre-3800 Ma) basement of continental type; the same holds true for many other early or late Archaean calc-alkaline orthogneisses. Contrary to the views of Bridgwater and Fyfe (1974), Bridgwater and Collerson (1976), Collerson and Fryer (1978), Fyfe (1978, 1979), etc., I contend that the isotopic constraints on crustal history in many of these situations are very tight indeed, not least in the early Archaean of West Greenland.

In this connection I will now add some remarks to earlier papers and reviews (Moorbath, 1975a, 1975b, 1976, 1977, 1978a; O'Nions and Pankhurst, 1978), based mainly on very recent age and isotope data by several workers.

Sr ISOTOPES

It is now well-established that many calc-alkaline orthogneiss terranes yield good Rb- Sr whole-rock regression lines, and that a true isochron fit is achieved in many cases. In my experience, the smaller and more petrologically uniform the gneiss unit sampled, the better is the quality of fit to the regression line. The initial $^{87}Sr/^{86}Sr$ ratios for most gneisses either coincide with, or lie very slightly above, contemporaneous upper mantle values as deduced from plausible upper mantle evolution models (Faure and Powell, 1972; Hart, 1977; Hurst, 1978a). Despite Hurst's compelling arguments to the contrary (Hurst 1978a, 1978b), I find it difficult to accept (Moorbath, 1978b) that measured initial $^{87}Sr/^{86}Sr$ ratios of calc-alkaline orthogneisses in high-grade Archaean terrains are necessarily identical within analytical error with their mantle source region, although this may certainly hold true in individual cases.

Most workers agree that low initial $^{87}Sr/^{86}Sr$ ratios of major Archaean and Proterozoic calc-alkaline orthogneisses preclude derivation by reworking of significantly older rocks with typical crustal Rb/Sr ratios, and that whole-rock Rb- Sr isochrons broadly date crustal accretion events. However, it is an observed fact that both *within* and *between* many calc-alkaline orthogneiss terranes of a given age, initial $^{87}Sr/^{86}Sr$ ratios may exhibit significant variations, and that the actual values usually fall above the most plausible contemporaneous mantle growth line. Thus early and late Archaean gneisses commonly yield initial $^{87}Sr/^{86}Sr$ ratios mostly in the range of 0.701 to 0.703 (Moorbath, 1977; O'Nions and Pankhurst, 1978), while there is good evidence from Sr isotope measurements on late Archaean mafic volcanic rocks that by about 2700 Ma the upper mantle $^{87}Sr/^{86}Sr$ ratio was close to 0.7011 (Hart, 1977), although by then there may already have been small, but significant, isotopic heterogeneities within the upper mantle as a whole.

In my view, the observed difference between penecontemporaneous crustal and mantle $^{87}Sr/^{86}Sr$ ratios results mainly from the geochronologically short, but finite, time interval for the genesis of juvenile sial and crustal accretion, with accompanying petrological, geochemical and metamorphic differentiation. The juvenile calc-alkaline material *must* have had higher Rb/Sr ratios than its basic source region, since any plausible process of crustal accretion and internal differentiation, such as that proposed by Wells (1979), must take *at least* a few tens of millions of years. The actual date recorded by a whole-rock Rb-Sr isochron represents the time at which the rock became a closed system to migration of the geochemically unrelated elements Rb and Sr on the scale of sampling of analysed whole-rock specimens.

It is well known that Rb/Sr ratios in rocks of broadly similar composition are closely related to metamorphic grade (see, for example, Heier, 1973, 1978; Tarney and Windley, 1977). Broadly speaking, metamorphic grade is directly proportional to Rb-depletion and inversely proportional to Rb/Sr ratios. For example, the late Archaean Lewisian pyroxene – granulites of northwestern Scotland have very low Rb/Sr ratios indeed (<0.04), while the early Archaean amphibolite facies Amîtsoq gneisses of West Greenland have broadly average crustal Rb/Sr ratios (about 0.3), although there is much local variation. The data recorded by a Rb-Sr whole-rock isochron (which may present considerable analytical difficulties for very low Rb/Sr rocks) most probably refers to the time of geochemical differentiation and prograde metamorphism, and *not* to the time of genesis and emplacement of juvenile sial. In some continent-forming episodes these two groups of processes may be geochronologically instantaneous (i.e., within analytical error of the age measurements), while in other cases there may be a geochronologically significant time interval between them. In the latter case there may clearly be sufficient time for measurable enhancement of $^{87}Sr/^{86}Sr$ relative to contemporaneous mantle over a period of several tens, or even a few hundreds, of millions of years, until the characteristic geochemical gradient (progressive decrease of Rb/Sr with depth in the newly formed crust) is established and frozen in.

It follows from the above that the average $^{87}Sr/^{86}Sr$ ratio of granulite facies, low -Rb/Sr continental crust is usually close to the $^{87}Sr/^{86}Sr$ ratio of contemporaneous upper mantle. This is in line with the recognition of the lower crust as essentially a granulite facies residuum depleted in light-ion-lithophile elements. Thus, the present-day $^{87}Sr/^{86}Sr$ ratio of the late Archaean Lewisian granulites of northwest Scotland is in the range of 0.702 to 0.704. If these gneisses were partially melted today, the resulting igneous rocks would have an initial $^{87}Sr/^{86}Sr$ ratio not significantly different from present-day mantle-derived rocks. Thus, in principle, Sr isotopes *alone* cannot distinguish between an upper mantle or deep crustal origin for a given igneous rock, however improbable it may appear on general petrological and geochemical evidence to account for the production of vast amounts of juvenile calc-alkaline magma from deep, depleted, dry and geochemically barren continental crust. Fortunately, this is where Pb and Nd isotopes can help to solve the problem.

Pb ISOTOPES

In contrast to the situation for Rb and Sr, internal geochemical differentiation of sialic crust produces lower continental crust in which U/Pb ratios are *much* lower than in the upper mantle, so that radiogenic Pb isotope evolution in the lower continental crust is severely retarded relative to the mantle. Pb isotopic data, in conjunction with U and Pb analyses, show that U/Pb ratios in deep continental crust are typically 5 to 15 times lower than in the mantle. Clearly, this fundamental difference has persisted for the entire history of most continental crust. This makes it possible to use Pb isotopes to distinguish unambiguously between the accretion of mantle-derived juvenile sial and the reworking of much older sialic crust. Severe depletion of U relative to Pb is not confined to pyroxene-granulites, but extends to many amphibolite-facies gneisses. U is more mobile in dehydrating crust than Rb; hence, U is even more easily separated from Pb than Rb is from Sr.

Quoting the same two examples as in the previous section, the late Archaean Lewisian pyroxene-granulite facies gneisses of northwest Scotland as well as the early Archaean amphibolite facies Amîtsoq gneisses of West Greenland both have extremely unradiogenic present-day Pb isotopic compositions whose crustal development, judged by their close proximity to a single-stage mantle growth curve for Pb isotopic evolution, began at, or only shortly before, 2700 Ma ago for the Lewisian gneisses and 3700 Ma ago for the Amîtsoq gneisses (Moorbath *et al.*, 1969; Chapman and Moorbath, 1977; Black *et al.*, 1971; Gancarz and Wasserburg, 1977). If these two gneisses were partially melted today, the Pb isotopic ratios of the resulting granites would both be far less radiogenic than any modern mantle-derived igneous rock. (In contrast, remember that a modern Lewisian-granulite-derived granite would have a low, mantle-type initial $^{87}Sr/^{86}Sr$ ratio of about 0.703, while a modern Amîtsoq-gneiss-derived granite would have a very high initial $^{87}Sr/^{86}Sr$ ratio in the range of 0.72 to 0.74).

Exactly the same principle applies to the isotopic and genetic relationships between the late Archaean (about 2900 Ma.) Nûk gneisses of West Greenland and the early Archaean crust of Amîtsoq-type gneiss. Thus, the Sr isotope data clearly show that Nûk gneisses are not reworked, high – Rb/Sr Amîtsoq gneisses such as are widely exposed (Moorbath and Pankhurst, 1976). However, the data alone do not exclude the possibility that Nûk gneisses were produced by partial melting of deep-seated, low-Rb/Sr, early Archaean gneisses whose Rb/Sr and $^{87}Sr/^{86}Sr$ ratios were not significantly different from contemporaneous mantle values. In contrast, even the exposed amphibolite facies Amîtsoq gneisses became so severely depleted in U (relative to Pb) about 3700 Ma ago that their Pb isotopic compositions became virtually fossilised at that time (Black *et al.*, 1971; Gancarz and Wasserburg, 1977). Many samples of Amîtsoq gneiss have whole-rock Pb isotopic ratios close to a single-stage growth curve corresponding to about 3700 Ma ago. Amîtsoq gneisses at greater depths than any now exposed would be even more depleted in U (relative to Pb), if that were possible! It follows that any subsequent partial melt of such a basement would similarly contain unradiogenic Pb whose crustal development began about 3700 Ma ago. Pb/Pb whole-rock measurements on numerous Nûk-type gneisses from West Greenland demonstrate beyond any doubt that the observed Pb isotope systematics have been developing in a crustal environment only since about 2900 Ma ago and *not* since 3700 Ma ago (Black *et al.*, 1973 Taylor *et al.*, 1980).

Similarly, Pb isotope studies on the late Archaean Lewisian gneisses of northwest Scotland rule out any contribution from early Archaean Amîtsoq-type continental crust (Chapman and Moorbath, 1977). Likewise, there is no Pb isotopic evidence whatever that the 3700 Ma-old Amîtsoq gneisses have continental, sialic precursors of significantly greater age. Their initial Pb isotopic ratios provide very close temporal constraints in this connection. I do not consider that they can possibly be derived from the partial melting or reworking of continental, sialic crust older than the Isua supracrustal sequence (Black *et al.*, 1971; Baadsgaard *et al.*, 1976; Gancarz and Wasserburg, 1977; Appel *et al.*, 1978).

In contrast, the 2550 Ma-old Qôrqut granite, which is the last major rock-forming event in the Archaean of the Godthaab area of West Greenland, has an initial $^{87}Sr/^{86}Sr$ ratio of 0.709 (Moorbath and Pankhurst, 1976); most of its Pb began its isotopic development in the continental crust about 3700 – *not* 2550 – Ma ago (Moorbath, Taylor, Goodwin and McGregor, unpublished data). It is believed that this granite (*sensu stricto*) was produced by the partial melting of 3700 Ma-old Amîtsoq gneisses; there is no significant mantle component in its Pb isotopic composition. This is yet another example of the great petrogenetic contrast between calc-alkaline and granitic igneous rocks.

By analogy with Rb-Sr, Pb/Pb whole-rock isochrons represent the time when whole-rock specimens became closed systems to the migration of U and/or Pb on the scale of sampling. This is the time of metamorphic/geochemical differentiation of the gneiss complex, the duration of which can be closely constrained by initial Pb isotopic ratios. The early Archaean gneisses of West Greenland, as well as the late Archaean gneisses of West Greenland and northwest Scotland (and so many other calc-alkaline orthogneiss terrains from all continents), represent continental accretion-differentiation superevents of geochronologically short duration (Moorbath, 1977, 1978a). In many cases, the duration of such a superevent may not exceed the measured uncertainties (typically of $\pm$ 50 to 100 Ma) on Pb/Pb whole rock isochron age measurements (but see section on Sm-Nd below).

The fundamentally different ways in which Sr and Pb isotopes operate in the continental environment (because of the similarity between deep-crustal and mantle Rb/Sr, but the difference between deep-crustal and mantle U/Pb) has not been appreciated by several recent authors, for whom the isotopic data appear to be merely an inconvenience to their favoured hypotheses of continental evolution. Indeed, there is still remarkably little general understanding of Pb isotopic systematics. This is unfortunate, because in several ways it is currently the most versatile of the radiogenic isotope methods, both as regards the range of applicability to geological problems and the unambiguity of interpretation of the data. (Doe, 1970; Faure, 1977; Köppel and Grünenfelder, 1979).

The Pb/Pb isochron approach can yield spurious age results when a high-grade metamorphism affects a rock a very long time after its formation, that is to say when crustal accretion and final metamorphic/geochemical differentiation do *not* form part of the same accretion-differentiation-superevent. Taylor (1975) published a Pb/Pb whole-rock isochron age of 3410 ± 70 Ma for the migmatitic Vikan gneisses of Langoy, Vesteralen, North Norway, and originally concluded that the measured age referred to the gneiss-forming event. Some anomalous features of the Pb isotope systematics, facilitated by further Pb, Sr and Nd isotopic measurements, led to a re-interpretation, (Griffin *et al.*, 1978; Jacobsen and Wasserburg, 1978). Specifically, the high apparent $^{238}U/^{204}Pb$ value of 8.9 calculated for the source region of the Vikan gneisses (compared with typical mantle values of 7.5 to 7.8) and the lack of any particularly unradiogenic Pb isotopic compositions are unusual in comparison with other Precambrian terrains for which Pb/Pb data are available.

It is now clear that the present range of Pb isotopic compositions cannot have resulted from a single stage of Pb isotopic evolution from a uniform initial Pb isotopic composition. The quoted authors have convincingly demonstrated that the linear array of Pb isotopes reported by Taylor (1975) resulted from the formation of the protoliths of the Vikan gneisses about 2600 to 2700 Ma ago, followed by severe U (and Rb) depletion during granulite facies metamorphism about 1800 Ma ago. The linear Pb/Pb array is thus no longer regarded as a true secondary isochron, but as a 'transposed palaeoisochron' (Griffin *et al.*, 1978) in which the Pb isotope data fall on a slightly displaced line joining the points corresponding to 2700 Ma and 1800 Ma on a single-stage Pb isotope mantle evolution curve.

It is likely that many more cases of this kind will turn up and advance our understanding of long-term crustal evolution. The Pb, Sr and Nd isotope systematics in such cases are very different indeed from those observed during penecontemporaneous crustal accretion and metamorphic/geochemical differentiation, not least by the apparent discordances between different isotopic techniques. Provided, however, that all available isotopic techniques are used (i.e., Pb/Pb, Rb-Sr, Sm-Nd), the interpretations should be unambiguous. The Vikan gneisses provide a clear example of geochemical reactivation of much older continental crust and, as such, are isotopically as clearly recognizable as the contrasting case of penecontemporaneous accretion-differentiation during a single major episode of continental growth.

By analogy, I consider that the combined isotopic approach will certainly be capable, in principle, of sorting out the geochemical and geochronological complexities of any gneisses which have undergone metasomatism and/or retrogression at some time long after their primary formation. As far as I know, little detailed isotopic work has yet been reported for any gneiss terrains in which these processes are clearly recognizable. However, it is easy to predict in detail the effects of metasomatism and retrogression on radiogenic isotope systematics; these effects would, in my view, turn out to be very different indeed from those postulated by Collerson and Fryer (1978) in their unconventional model of massive mantle-crust metasomatism and crustal transformation. We shall have to wait until the relevant data appear!

Nd ISOTOPES

Pioneering advances have recently been made in the application of the ^{147}Sm - ^{143}Nd method to the study of continental evolution (De Paolo and Wasserburg, 1976a, 1976b; see also review by O'Nions *et al.*, 1979). The published results are so far in satisfying accord with conclusions regarding crustal evolution based on Sr and Pb isotope systematics, but nevertheless provide additional important constraints on the age and origin of rock units. The principal contrast between the Sm-Nd method and both the Rb-Sr and U-Pb methods lies in the comparative geochemical coherence of Sm and Nd. Unlike Rb and Sr, and U and Pb, these two rare-earth elements are not significantly fractionated by crustal processes. This has been vividly demonstrated by McCulloch and Wasserburg (1978), whose analyses of gneiss composites from the Canadian Shield, representing portions of the Superior, Slave and Churchill structural provinces, indicate that these provinces were all formed within the period 2700 to 2500 Ma ago, which is in general agreement with much published Rb-Sr isotopic work. In addition, these workers were able to determine the age of sediment provenances from many areas, since it appears that the Sm-Nd isotopic system is not significantly disturbed during sedimentation or diagenesis.

However, Nd *is* enriched relative to Sm during magmatic processes leading to the production of sialic crust from the upper mantle, and Hamilton *et al.* (1979a) have shown that mantle-crust differentiation can be precisely dated by the Sm-Nd method. They report a whole-rock Sm-Nd isochron age from Lewisian granulite facies gneisses of northwest Scotland of 2920±50 Ma, with an initial ^{143}Nd/^{144}Nd ratio of 0.508959±49. This Sm-Nd age is significantly older than reliable ages reported for the Lewisian by whole-rock Rb-Sr and Pb/Pb, and zircon U-Pb methods. These latter ages group fairly closely around 2700 Ma, while initial Sr and Pb isotopic compositions have always been interpreted as permitting a maximum time interval of 100 to 200 Ma between formation of the igneous precursors of the gneisses and their metamorphic/geochemical differentiation (Moorbath *et al.*, 1969; Chapman and Moorbath, 1977). The actual Sm-Nd age of 2920 ± 50 Ma is interpreted as the time of mantle-crust differentiation which produced the calc-alkaline magmatic precursors of the Lewisian gneisses. The time interval between the Sm-Nd age and both the Rb-Sr and Pb/Pb whole-rock ages is 240 ± 90 Ma, which is interpreted as the time between crustal accretion and metamorphic/geochemical differentiation, when whole-rock samples became closed systems to Rb, Sr, U and Pb migration. The initial $^{143}Nd/^{144}Nd$ ratio for the Lewisian magmatic precursors lies exactly on the (chondritic) growth curve for the Earth's upper mantle throughout geological time.

From simple thermal considerations based on reasonable concentrations of heat-producing elements and on conductivity and thermal diffusivity of juvenile Lewisian crust, Hamilton *et al.* (1979a) conclude that in order to avoid extensive crustal melting, heat-producing elements must have begun migrating from the lower crust less than 100 Ma after crustal thickening. While the exact mechanism of this migration is not yet clear, it does not appear to have been a transport medium which significantly affected the Sm-Nd systematics, as would a magmatic phase. Hamilton *et al.* (1979a) state that 'the combined application of Sm-Nd, U-Pb and Rb-Sr systematics to ancient crustal terrains can provide important constraints on the early phases of crustal differentiation and stabilization. This is a simple reflection of the markedly different mobilities of these elements during crustal differentiation processes. At least in the case of the Lewisian crust, the completion of metamorphic differentiation and granulite formation succeeded the primary igneous differentiation of parent materials from the mantle by 240 ± 90 Ma. The heat-producing elements were transported to higher crustal levels from where heat could be readily dissipated, and the granulite facies lower crust was maintained in a more refractory and thermally stable condition.'

The close geochemical coherence between Sm and Nd makes it possible to reliably date amphibolite sequences from greenstone belt supra-crustals, which frequently exhibit partial open system behaviour to Rb and/or Sr (Hamilton *et al.*, 1977). Initial $^{143}Nd/^{144}Nd$ ratios so far obtained from Sm-Nd isochron or model age data on Archean rocks (DePaolo and Wasserburg, 1976a, 1976b; see also review by O'Nions *et al.*, 1979) all lie on the chrondritic growth curve for the bulk Earth-mantle reservoir (Sm/Nd = 0.308). This includes the early Archaean Isua (West Greenland) and Onverwacht (Swaziland) supracrustals, the Amîtsoq gneisses (West Greenland), as well as the late Archean Bulawayan (Rhodesia) greenstone belt, the Great Dyke (Rhodesia), the Fiskenaesset (West Greenland) anorthosite, and the Lewisian gneisses (Northwest Scotland). Each of these analyzed rock units yields an age for its respective mantle-crust differentiation episode, together with the corresponding $^{143}Nd/^{144}Nd$ ratio of the mantle source region at that time. Any crustal pre-history is precluded for these rock units.

It is hoped that, in future, many more rock units of all ages will be analyzed by the Sm-Nd method, in conjunction with Rb-Sr, U-Pb and Pb/Pb methods. There is no doubt that by using these methods, precise petrogenetic and temporal constraints for the evolution of any given sector of continental crust of any age will emerge. As far as the Precambrian is concerned, particular attention should be paid to intra-cratonic mobile belts which may represent, at least in part, sites of ensialic reworking and rejuvenation, and from which ãdequate age and isotope data are not yet available (see, for example, Kröner 1977a, 1977b). In my view, reworking and rejuvenation of ancient continental crust during a later geological epoch will not, and indeed *cannot*, eradicate all previous isotopic memory. In contrast, where a given rock unit shows no isotopic memory whatsoever of older crust, then it must be concluded that the rock unit represents juvenile sial derived either from upper mantle, or from basic lithosphere with mantle-like geochemistry.

MAGMA-CRUST INTERACTION AND ISOTOPIC MIXING

Brief mention must be made of a complicating factor in the interpretation of radiogenic isotopes, which may have to be taken into account when considering mantle-derived magmatic rocks emplaced into, or onto, continental crust.

Continental igneous rocks sometimes exhibit Sr, Pb and Nd isotope systematics of greater complexity than oceanic igneous rocks. Numerous conflicting interpretations have been given for this phenomenon. While full treatment is outside the scope of this paper, certain relevant points will be discussed here. A representative introduction to this complex and important subject may be found in several recent works (Faure, 1977; Pankhurst, 1977; Brooks *et al.*, 1976b; Carter *et al.*, 1978).

In my view, systematics in isotopically 'noisy' continental igneous rocks are most often the result of crust-magma interaction which involved the mixing of incompatible elements (including those with radiogenic isotopes) derived both from mantle-derived magma and the surrounding continental crust. Bulk mixing of magma and crust is frequently ruled out by petrological and major-element data, and one is forced to postulate selective contamination and mixing processes involving only incompatible trace elements (see for example, Briqueu and Lancelot, 1979; Moorbath and Thompson, 1980). The mechanism probably involves mixing of mantle-derived calc-alkaline magmas with a fluid phase containing incompatible elements sweated out from hydrous minerals in the country rock of the magmatic zone. Provided that sufficient heat is available, the process can go further and lead to partial melting of country rock. (All this can happen only when the country rock surrounding the magma chambers is not already completely dry, geochemically depleted and refractory, such as might be expected for high-grade granulite gneisses).

The magnitude of isotopic perturbation in any given continental igneous rock is strongly dependent on the thickness, age and geochemistry of the invaded or traversed crust. Regardless of the actual mechanism of crust-magma interaction, radiogenic Sr, Pb and Nd isotopes act as very sensitive tracers for this process; Pb isotopes can often yield an approximate age for the rocks which supplied the continental Pb component, even when they are not exposed at the surface. The basic principles of this approach have been recognized for many years (see Russell and Farquhar, 1960) in ore lead studies. Recently it has become apparent that Pb isotope systematics in common igneous silicate rocks can be similarly interpreted. Mostly, radiogenic isotope studies have been carried out on Mesozoic, Tertiary and Recent continental igneous rocks which have been emplaced into, or traversed, ancient continental basement (see, for example, Moorbath and Welke, 1969a; Doe and Delevaux, 1973; Zartman, 1974; Armstrong *et al.*, 1977; Leeman and Dasch, 1978;

Lipman *et al.*, 1978). The observed combination of *radiogenic* Sr with *unradiogenic* Pb in some continental igneous rocks testifies to the interaction of magma with ancient continental crust which has maintained low U/Pb and high Rb/Sr ratios ever since its formation.

The corollary of all this is that radiogenic isotope analyses of continental igneous rocks, particularly Pb, can sometimes be used to detect and characterize ancient continental basement, even where it is not exposed at the surface. Thus, Moorbath and Welke (1969b) correctly predicted from Pb isotopic studies on the Rockall Granite that the Rockall Bank in the North Atlantic was either produced from, or contaminated with, ancient continental crust possibly similar to the Lewisian basement of northwest Scotland. Conversely, the absence of continental crust from oceanic areas is clearly demonstrated by numerous radiogenic isotope measurements. In addition, radiogenic isotope measurements have proved essential for studying geochemical heterogeneity in the upper mantle (see, for example, Sun and Hanson, 1975; Brooks *et al.*, 1976; Hanson, 1977; Hofmann and Hart, 1978; O'Nions *et al.*, 1978).

We are currently studying at Oxford the Pb isotope systematics of the widely distributed late Archaean (2850 to 3000 Ma old) gneisses of West Greenland (Taylor *et al.*, 1980). Over most of the large area of outcrop so far studied, Pb/Pb whole-rock systematics are straightforward and indicate derivation of the magmatic precursors of these calc-alkaline orthogneisses from a source region with a typical mantle U/Pb ratio at, or shortly before 2850 to 3000 Ma ago (see also Black *et al.*, 1973). However, within the Godthaab and immediately surrounding areas, where late Archaean Nûk gneisses and early Archaean Amîtsoq gneisses are both exposed (McGregor, 1973; Bridgwater *et al.*, 1976) some Nûk gneisses contain variable proportions of two quite distinct types of Pb which started their respective crustal developments about 2900 and 3700 Ma ago. Both Pb and Sr isotopic constraints clearly demonstrate that in *no* case is Nûk gneiss derived from reworking or remelting of Amîtsoq gneiss, but that the mixing of early and late Archaean Pb is due in part to selective contamination and mixing between Pb derived from the upper mantle 2850 to 3000 Ma ago with ancient, unradiogenic Pb derived from 3700 Ma-old Amîtsoq-type continental crust invaded by the Nûk magmas. Using Pb isotopes, we have not been able to detect any early Archaean Amîtsoq-type continental crust in areas more than 10 to 15 km outside the known outcrop of the Amîtsoq gneisses within the greater Godthaab region. We conclude that such crust neither exists at depth elsewhere in southern West Greenland nor, indeed, in northwest Scotland (Chapman and Moorbath, 1977). Early Archaean continental, sialic crust should be easily detectable by the isotopic perturbations it produces in much younger plutonic and volcanic rocks. This naturally applies to igneous rocks of *any* geological age invading or traversing significantly older continental crust, as outlined above.

Thus, the isotopic effects of crust-magma interaction, resulting from geochemical contamination and trace element mixing processes, appears to be a powerful tool for the detection, characterization and sub-surface mapping of ancient continental crust at depth, and for the study of the areal extent and growth of such crust. The recently developed Sm-Nd method may also be very helpful in this connection (Carter *et al.*, 1978; DePaolo and Wasserburg, 1979).

CONCLUSIONS

The study of ancient rocks provides constraints for the processes which have shaped the surface of the Earth since the beginnings of geological time. Geochronological and associated Sr, Pb and Nd isotopic constraints are particularly powerful in this respect and can be used to distinguish between juvenile, mantle-derived sial and reworked continental crust. Both growth and reworking have occurred throughout geological time and usually in totally contrasting geological environments. Today, continental growth predominates at continental margins, where subducted oceanic lithosphere and the overlying mantle wedge are partially melted and differentiated to yield calc-alkaline magmas; reworking predominates in regions of continental collision, where continental crust is thickened in mobile belts. There is no compelling reason to suppose that this state of affairs has changed fundamentally since the time when the early continental crust formed, although the proportion of growth over reworking must have decreased with time. Thus the continental growth rate has almost certainly decreased with time, due to the decaying thermal output of the Earth (Brown, 1977; Brown and Hennessy, 1978), while continental collision frequency has increased with time as more continental crust has become available. Both continental growth and some form of inter- or intra-cratonic mobility became fully established as major global tectonic processes by late Archaean times (3000 to 2500 Ma ago), when much continental crust had already come into existence. Brown (1977) calculated a contemporary global growth rate of 0.5 km^3 per year, so that it would take 10,000 Ma to produce all $5 \times 10^9 km^3$ of continental crust which now exist. This implies a higher accretion rate in the past.

The modern phase of the geological concept of continental growth by successive accretion of mantle-derived sial was pioneered by Tuzo Wilson, whom we are honouring at this symposium. It should not be forgotten, however, how the remarkable pioneering work of Patrick Hurley and his colleagues at the Massachusetts Institute of Technology in the early 1960s demonstrated the use of initial $^{87}Sr/^{86}Sr$ ratios in studying continental growth (Hurley et al., 1962). Many of these early conclusions are strongly supported by recent age and isotope work, using Sr, Pb and Nd isotopes.

At present, the evolution of the Earth's crust during the first 800 Ma of Earth history is still conjectural. Mechanical and thermal turbulence resulting from impacting planetesimals and/or intense mantle-wide convection (Elsasser et al., 1979) may have inhibited the production of continental, sialic crust. By about 3800 Ma ago, a solid crust of probably simatic character was able to support sediments and lavas, a hydrosphere, and possibly the beginnings of a biosphere. The earliest known thick, stable, permanent, geochemically/metamorphically differentiated, calc-alkaline continental crust appeared around 3700 Ma ago. During the next 200 Ma., many such small continental nuclei may have appeared for the first time in different regions.

There is little, if any, geological, geochemical, geochronological or isotopic evidence for an early Precambrian globe-encircling granitic crust, or for massive recycling of continental crust through the mantle throughout geological time, or for global mixing of continental crust and mantle on a huge scale (e.g., Fyfe, 1976, 1978; Hargraves, 1976; Lowman, 1976; Shaw, 1976, etc.). Most evidence suggests that the history of the outermost parts of the Earth is dominated by irreversible differentiation of the mantle (and particularly the *upper* mantle) throughout geological time, and by the permanence of continental crust once it is formed. I cannot agree with Fyfe (1978) that 'conservation of crustal mass is one of the basic tenets of plate tectonics.' In principle, plate tectonic processes can lead equally well to an increase in crustal mass (e.g., Ringwood, 1974).

How much continentally derived sediment is subducted and partially melted to form new igneous rocks is still a matter of considerable debate (see, for example, McKenzie, 1969; Karig and Sharman, 1975; James, 1978; Molnar and Gray, 1979). Kay *et al.* (1978) suggest that Pb and Sr isotopic data from the Aleutian and Pribilof Islands of Alaska could be explained by the mixing of about 2 per cent of continentally derived sediment with melt derived from the subducted oceanic crust and overlying mantle wedge. In my view, Fyfe (1978) has grossly exaggerated the extent of sediment subduction, continental recycling and crust-mantle mixing. This is not to deny the possibility – indeed, the probability – that a certain amount of melting of young, continentally derived sediment occurs deep within sedimentary basins marginal to a continental region, to produce the so-called 'S-type' granites described by Australian workers (Chappell and White, 1974; Flood and Shaw, 1978). As stated earlier, true granitic rocks can be produced from a considerable variety of precursors in different types of tectonic environments. However, the largely mantle-derived rocks of the predominantly intermediate calc-alkaline kindred must be regarded as the true heralds of continental growth.

ACKNOWLEDGEMENTS

I thank Dr. Paul N. Taylor for critical comments.

REFERENCES

Allaart, J.H., 1976, The pre-3760 m.y. old supracrustal rocks of the Isua area, Central West Greenland, and the associated occurrence of quartz-banded ironstone: *in* Windley, B. F., ed., The Early History to the Earth: London, Wiley, p. 177-189.

Anderson, R.N., DeLong, S.E. and Schwarz, W.M., 1978, Thermal model for subduction with dehydration in the downgoing slab: Jour. Geol., v. 86, p. 731-739.

Anhaeusser, C.R., 1978, The geological evolution of the primitive earth – evidence from the Barberton Mountain Land: *in* Tarling, D.H., ed., Evolution of the Earth's Crust: London, Academic Press, p. 71-106.

Appel, P.W.U., Moorbath, S., and Taylor, P.N., 1978, Least radiogenic terrestrial lead from Isua, West Greenland: Nature, v. 272, p. 524-526.

Armstrong, R.L., Taubeneck, W.H. and Hales, P.O., 1977, Rb-Sr and K-Ar geochronometry of Mesozoic granitic rocks and their Sr isotopic composition, Oregon, Washington, Idaho: Geol. Soc. America Bull., v. 88, p. 397-411.

Arth, J.G. and Hanson, G.N., 1975, Geochemistry and origin of the Early Precambrian crust of northeastern Minnesota: Geochim. Cosmochim. Acta, v. 39, p. 325-362.

Baadsgaard, H., 1973, U-Th-Pb dates on zircons from the early Precambrian Amîtsoq gneisses, Godthaab district, West Greenland: Earth and Planetary Sci. Letters, v. 19, p. 22-28.

――――――――, 1976, Further U-Pb dates on zircons from the early Precambrian rocks of the Godthaabsfjord area, West Greenland: Early and Planetary Sci. Letters, v. 33, p. 261-267.

Baadsgaard, H., Lambert, R.St.J. and Krupicka, J., 1976, Mineral isotopic age relationships in the polymetamorphic Amîtsoq gneisses, Godthaab District, West Greenland: Geochim. Cosmochim. Acta, v. 40, p. 513-527.

Barley, M., Dunlop, J.S.R. and Groves, D.I., 1979, Sedimentary evidence for an Archaean shallow-water volcanic-sedimentary facies, Eastern Pilbara Block, Western Australia: Earth and Planetary Sci. Letters, v. 43, p. 74-84.

Barton, J.M., 1975, Rb-Sr isotopic characteristics and chemistry of the 3.6 b.y. Hebron gneiss, Labrador: Earth and Planetary Sci. Letters, v. 27, p. 427-435.

_____, 1977, Rb-Sr ages and geological setting of ancient dykes in the Sand River area, Limpopo Mobile Belt, South Africa: Nature, v. 267, p. 487-490.

Bickle, M.J., 1978, Heat loss from the earth: a constraint on Archaean tectonics from the relation between geothermal gradients and the rate of plate production: Earth and Planetary Sci. Letters, v. 40, p. 301-315.

Bickle, M.J., Martin, A. and Nisbet, E.G., 1975, Basaltic and peridotitic komatiites and stromatolites above a basal unconformity in the Belingwe greenstone belt, Rhodesia: Earth and Planetary Sci. Letters, v. 27, p. 115-162.

Black, L.P., Gale, N.H., Moorbath, S., Pankhurst, R.J., and McGregor, V.R., 1971, Isotopic dating of very early Precambrian amphibolite facies gneisses from the Godthaab district, West Greenland: Earth and Planetary Sci. Letters, v. 12, p. 245-259.

Black, L.P., Moorbath, S., Pankhurst, R.J. and Windley, B.F., 1973, [207]Pb/[206]Pb whole rock age of the Archaean granulite facies metamorphic event in West Greenland: Nature Physical Sci., v. 244, p. 50-53.

Bridgwater, D. and Collerson, K.D., 1976, The major petrological and geochemical characters of the 3600 m.y. Uivak gneisses from Labrador: Contrib. Mineral. Petrol., v. 54, p. 43-59.

Bridgwater, D. and Fyfe, W.S., 1974, The pre-3b.y. crust: fact-fiction-fantasy: Geosci. Canada, v. 1, p. 7-11.

Bridgwater, D., Keto, L., McGregor, V.R. and Myers, J.S., 1976, Archaean Gneiss Complex of Greenland: in Escher, A., and Watt, W.S., eds., Geology of Greenland: Copenhagen, p. 19-75.

Bridgwater, D., McGregor, V.R. and Myers, J.S., 1974, A horizontal tectonic regime in the Archaean of Greenland and its implications for early crustal thickening: Precambrian Research, v. 1, p. 179-197.

Briqueu, L. and Lancelot, J.R., 1979, Rb-Sr systematics and crustal contamination models for calc-alkaline igneous rocks: Earth and Planetary Sci. Letters, v. 43, p. 385-396.

Brooks, C., Hart, S.R., Hofmann, A. and James, D.E., 1976, Rb-Sr mantle isochrons from oceanic regions: Earth and Planetary Sci. Letters, v. 32, p. 51-61.

Brooks, C., James, D.E., and Hart, S.R., 1976, Ancient lithosphere: its role in young continental volcanism: Science, v. 193, p. 1086-1094.

Brown, G.C., 1977, Mantle origin of Cordilleran granites: Nature, v. 265, p. 21-24.

Brown, G.C. and Hennessy, J., 1978, The initiation and diversity of granite magmatism: Philos. Trans. Royal Soc. London A, v. 288, p. 631-643.

Burke, K. and Kidd, W.S.F., 1978, Were Archaean continental geothermal gradients much steeper than those of today?: Nature, v. 272, p. 240-241.

Carter, S.R., Evensen, N.M., Hamilton, P.J. and O'Nions, R.K., 1978, Neodymium and strontium isotope evidence for crustal contamination of continental volcanics: Science, v. 202, p. 743-747.

Chadwick, B., Ramakrishnan, M., Viswanatha, M.N. and Murthy, V.S., 1978, Structural studies in the Archaean Sargur and Dharwar supracrustal rocks of the Karnataka Craton: Geol. Soc. India Jour., v. 19, p. 531-549.

Chapman, H. and Moorbath, S., 1977, Lead isotope measurements from the oldest recognised Lewisian gneisses of northwest Scotland: Nature, v. 268, p. 41-42.

Chappell, B.W. and White, A.J.R., 1974, Two contrasting granite types: Pacific Geol. v. 8, p. 173-174.

Cumming, G.L. and Richards, J.R., 1975, Ore lead–isotope ratios in a continuously changing earth: Earth and Planetary Sci. Letters, v. 28, p. 155-171.

Collerson, K.D. and Fryer, B.J., 1978, The role of fluids in the formation and subsequent development of early continental crust: Contrib. Mineral. Petrol., v. 67, p. 151-167.

DePaolo, D.J. and Wasserburg, G.J., 1976a, Nd-isotope variations and petrogenetic models: Geophys. Research Letters, v. 3, p. 249-252.

_____, 1976b, Inferences about magma sources and mantle structure from variations of $^{143}Nd/^{144}Nd$: Geophys. Research Letters, v. 3, p. 743-746.

_____, 1979. Petrogenetic mixing models and Nd-Sr isotopic patterns: Geochim. Cosmochim. Acta, v. 43, p. 615-627.

Doe, B.R., 1970, Lead Isotopes: Springer-Verlag, Berlin.

Doe, B.R. and Delevaux, M.H., 1973, Variations in lead-isotope compositions in Mesozoic granite rocks of California: a preliminary investigation: Geol. Soc. America Bull., v. 84, p. 3513-3526.

Drury, S.A., 1978, REE distributions in a high-grade Archaean gneiss complex in Scotland: implications for the genesis of ancient sialic crust: Precambrian Research, v. 7, p. 237-257.

Elsasser, W.M., Olson, P. and Marsh, B.D., 1979, The depth of mantle convection, Jour. Geophys. Research, v. 84, p. 147-155.

England, P.C., 1979, Continental geotherms during the Archaean: Nature, v. 277, p. 556-558.

Faure, G. 1977, Principles of Isotope Geology: New York, Wiley and Sons, 464 p.

Faure, G. and Powell, J.L., 1972, Strontium Isotope Geology: New York, Springer-Verlag, 188 p.

Flood, R.H. and Shaw, S.E., 1977, Two "S-type" granite suites with low initial $^{87}Sr/^{86}Sr$ ratios from the New England Batholith, Australia, Contrib. Mineral. Petrol., v. 61, p. 163-173.

Fyfe, W.S., 1974, Archaean tectonics: Nature v. 248, p. 338.

_____, 1976, Heat flow and magmatic activity in the Proterozoic: Philos. Trans. Royal Soc. London A, v. 280, p. 655-660.

_____, 1978, The evolution of the earth's crust: modern plate tectonics to ancient hot spot tectonics?: Chem. Geol., v. 23, p. 89-114.

_____, 1979, The geochemical cycle of uranium: Philos. Trans. Royal Soc. London A, v. 291, p. 433-445.

Gancarz, A.J. and Wasserburg, G.J., 1977, Initial Pb of the Amîtsoq Gneiss, West Greenland, and implications for the age of the earth: Geochim. Acta, v. 41, p. 1283-1301.

Goldich, S.S. and Hedge, C.E., 1975, Interpretation of apparent ages in Minnesota: Nature, v. 257, p. 722.

Griffin, W.L., Taylor, P.N. and Hakkinen, J.W., 1978, Archaean and Proterozoic crustal evolution in Lofoten-Vesteralen, Norway: Geol. Soc. London Jour., v. 135, p. 629-647.

Hamilton, P.J., O'Nions, R.K. and Evensen, N.M., 1977, Sm-Nd dating of basic and ultrabasic volcanics: Earth and Planetary Sci. Letters, v. 36, p. 263-268.

Hamilton, P.J., Evensen, N.M., O'Nions, R.K. and Tarney, J., 1979a, Sm-Nd systematics of Lewisian gneisses: implications for the origin of granulites: Nature, v. 277, p. 25-28.

Hamilton, P.J., Evensen, N.M., O'Nions, R.K., Smith, H. and Erlank, A.J. 1979b, Sm-Nd dating of Onverwacht Group volcanics, southern Africa: Nature v. 279, p. 298-300.

Hamilton, P.J., O'Nions, R.K., Evensen, N.M., Bridgwater, D. and Allaart, J.H., 1978, Sm-Nd isotopic investigations of Isua supracrustals and implications for mantle evolution: Nature, v. 272, p. 41-43.

Hanson, G.N., 1977, Geochemical evolution of the suboceanic mantle: Geol. Soc. London Jour., v. 134, p. 235-253.

_____, 1978, The application of trace elements to the petrogenesis of igneous rocks of granitic composition: Earth and Planetary Sci. Letters, v. 38, p. 26-43.

Hargraves, R.B., 1976, Precambrian geologic history: Science, v. 193, p. 363-371.

Hart, S.R., 1977, The geochemistry and evolution of early Precambrian mantle: Contrib. Mineral. Petrol., v. 61, p. 109-128.

Heier, K.S., 1973, Geochemistry of granulite facies rocks and problems of their origin: Philos. Trans. Royal Soc. London A, v. 273, p. 429-442.

_____, 1978, The distribution and redistribution of heat-producing elements in the continents: Philos. Trans. Royal Soc. London A, v. 288, p. 393-400.

Hofmann, A. and Hart, S.R., 1978, An assessment of local and regional isotopic equilibrium in the mantle: Earth and Planetary Sci. Letters, v. 38, p. 44-62.

Hunter, D.R., Barker, F. and Millard, H.T., 1978, The geochemical nature of the Archaean Ancient Gneiss Complex and Granodiorite Suite, Swaziland: a preliminary study: Precambrian Research, v. 7, p. 105-127.

Hurley, P.M., Hughes, H., Faure, G., Fairbairn, H.W. and Pinson, W.H., 1962, Radiogenic strontium-87 model of continent formation: Jour. Geol. Research, v. 67, p. 5315-5334.

Hurst, R.W., 1978a, Sr evolution in the West Greenland-Labrador craton: a model for early Rb depletion in the mantle: Geochim. Cosmochim. Acta, v. 42, p. 39-44.

_____, 1978b, Reply to critical comment on "Sr evolution in the West Greenland – Labrador craton: a model for early Rb depletion in the mantle, by R.W. Hurst", by S. Moorbath: Geochim. Cosmochim. Acta, v. 42, p. 1585-1586.

Hurst, R.W., Bridgwater, D., Collerson, K.D. and Wetherill, G.W., 1975, 3600 m.y. Rb-Sr ages from very early Archaean gneisses from Saglek Bay, Labrador: Earth and Planetary Sci. Letters, v. 27, p. 393-403.

Jacobsen, J.B. and Wasserburg, G.J., 1978, Interpretation of Nd, Sr and Pb isotope data from Archaean migmatites in Loften-Vesteralen, Norway: Earth and Planetary Sci. Letters, v. 41, p. 245-253.

Jahn, B.M. and Shih, C.Y., 1974, On the age of the Onverwacht Group, Swaziland Sequence, South Africa: Geochim. Cosmochim. Acta, v. 38, p. 873-885.

James, D.E., 1978, On the origin of the calc-alkaline volcanics of the central Andes: a revised interpretation: Carnegie Instit. Washington Year Book 77, p. 562-590.

Karig, D.E. and Sharman, G.F., 1975, Subduction and accretion in trenches: Geol. Soc. America Bull., v. 86, p. 377-389.

Kay, R.W., Sun, S.S. and Lee-Hu, C.N., 1978, Pb and Sr isotopes in volcanic rocks from the Aleutian Islands and Pribilof Islands, Alaska: Geochim. Cosmochim. Acta, v. 42, p. 263-273.

Köppel, V. and Grünenfelder, M., 1979, Isotope geochemistry of lead: in Jäger, E., and Hunziker, J.C., eds., Lectures in Isotope Geology: Springer-Verlag Berlin, p. 134-153.

Kröner, A., 1977a, Precambrian mobile belts of Southern and Eastern Africa – ancient sutures or sites of ensialic mobility, A case for crustal evolution towards plate tectonics: Tectonophysics, v. 40, p. 101-135.

_____, 1977b, The Precambrian geotectonic evolution of Africa: plate accretion versus plate destruction: Precambrian Research, v. 4, p. 163-213.

Lambert, R.St.J., 1976, Archaean thermal regimes, crustal and upper mantle temperatures, and a progressively evolutionary model for the earth: in Windley, B.F., ed., The Early History of the Earth: London, Wiley, p. 363-373.

Leeman, W.P. and Dasch, E.J., 1978, Strontium, lead and oxygen isotopic investigations of the Skaergaard intrusion, East Greenland: Earth and Planetary Sci. Letters, v. 41, p. 47-59.

Lipman, P.W., Doe, B.R., Hedge, C.A. and Steven, T.A., 1978, Petrologic evolution of the San Juan volcanic field, southwestern Colorado: Pb and Sr isotope evidence: Geol. Soc. America Bull., v. 89, p. 59-82.

Lowman, P.D., 1976, Crustal evolution in silicate planets: implications for the origin of continents: Jour. Geol., v. 84, p. 1-26.

McCulloch, M.T. and Wasserburg, G.J., 1978, Sm-Nd and Rb-Sr chronology of continental crust formation: Science, v. 200, p. 1003-1011.

McGregor, V.R., 1973, The early Precambrian gneisses of the Godthaab district, West Greenland: Philos. Trans. Royal Soc. London A, v. 273, p. 343-358.

McGregor, V.M. and Mason, B., 1977, Petrogenesis and geochemistry of metabasaltic and metasedimentary enclaves in the Amîtsoq gneisses, West Greenland: American Mineralogist, v. 62, p. 887-904.

McKenzie, D.P., 1969, Speculations of the consequences and causes of plate motions: Royal Astron. Soc. Jour., v. 18, p. 1-32.

Michard-Vitrac, A., Lancelot, J., Allègre, C.J. and Moorbath, S., 1977, U-Pb ages on single zircons from the early Precambrian rocks of West Greenland and the Minnesota River Valley: Earth and Planetary Sci. Letters, v. 35, p. 449-453.

Molnar, P. and Gray, D., 1979, Subduction of continental lithosphere: some constraints and uncertainties; Geology, v. 7, p. 58-62.

Monster, J., Appel, P.W.U., Thode, H.G., Schidlowski, M., Carmichael, C.M. and Bridgwater, D., 1979, Sulfur isotope studies in early Archaean sediments from Isua, West Greenland: implications for the antiquity of bacterial sulfate reduction: Geochim. Cosmochim. Acta, v. 43, p. 405-413.

Moorbath, S., 1975a, Evolution of Precambrian crust from strontium isotopic evidence: Nature, v. 254, p. 395-398.

_____, 1975b, Geological interpretation of whole-rock isochron dates from high grade gneiss terrains: Nature, v. 255, p. 391.

_____, 1976 Age and isotope constraints for the evolution of Archaean crust: in Windley, B.F., ed., The Early History of the Earth: London, Wiley, p. 351-360.

_____, 1977, Ages isotopes and evolution of Precambrian continental crust: Chem. Geol., v. 20, p. 151-187.

_____, 1978a, Age and isotope evidence for the evolution of continental crust: Philos. Trans. Royal Soc. London A, v. 288, p. 401-413.

_____, 1978b, Critical comment on "Sr evolution in the West Greenland – Labrador craton: a model for early Rb depletion in the mantle" by R.W. Hurst: Geochim. Cosmochim. Acta, v. 42, p. 1583-1584.

Moorbath, S. and Pankhurst, R.J., 1976, Further Rb-Sr age and isotope evidence for the nature of the late Archaean plutonic event in West Greenland: Nature, v. 262, p. 124-126.

Moorbath, S. and Thompson, R.N., 1980, Strontium isotope geochemistry and petrogenesis of the early Tertiary lava pile of the Isle of Skye, Scotland, and other basic rocks of the British Tertiary Province: an example of magma-crust interaction: Jour. Petrol., in press.

Moorbath, S. and Welke, H., 1969a, Lead isotope studies on igneous rocks from the Isle of Skye, northwest Scotland: Earth and Planetary Sci. Letters, v. 5, p. 217-230.

_____, 1969b, Isotopic evidence for the continental affinity of the Rockall Bank, North Atlantic: Earth and Planetary Sci. Letters, v. 5, p. 211-216.

Moorbath, S., O'Nions, R.K. and Pankhurst, R.J., 1973, Early Archaean age for the Isua Iron Formation, West Greenland: Nature, v. 245, p. 138-139.

Moorbath, S., Welke, H. and Gale, N.H., 1969, The significance of lead isotope studies in ancient, high grade metamorphic basement complexes as exemplified by the Lewisian rocks of northwest Scotland: Earth and Planetary Sci. Letters, v. 6, p. 245-256.

Moorbath, S., Wilson, J.F. and Cotterill, P., 1976, Early Archaean age for the Sebakwian group at Selukwe, Rhodesia: Nature, v. 264, p. 536-538.

Moorbath, S., Allaart, J.H. Bridgwater, D. and McGregor, V.R., 1977, Rb-Sr ages of early Archaean supracrustal rocks and Amîtsoq gneisses at Isua: Nature, v. 270, p. 43-45.

Moorbath, S., O'Nions, R.K., Pankhurst, R.J., Gale, N.H. and McGregor, V.R., 1972, Further rubidium-strontium age determinations on the very early Precambrian rocks of the Godthaab district, West Greenland: Nature Phys. Sci., v. 240, p. 78-82.

Moorbath, S., Wilson, J.F., Goodwin, R. and Humm, M., 1977, Further Rb-Sr and isotope data on early and late Archaean rocks from the Rhodesian craton: Precambrian Research, v. 5, p. 229-239.

Myers, J.S., 1978, Formation of banded gneisses by deformation of igneous rocks: Precambrian Research, v. 6, p. 43-64.

O'Nions, R.K. and Pankhurst, R.J., 1978, Early Archaean rocks and geochemical evolution of the earth's crust: Earth and Planetary Sci. Letters, v. 38, p. 211-236.

O'Nions, R.K., Carter, S.R., Evensen, N.M. and Hamilton, P.J., 1979. Geochemical and cosmochemical applications of Nd isotope analysis: Annual Reviews of Earth and Planetary Sciences, v. 7, p. 11-38.

O'Nions, R.K., Evensen, N.M., Hamilton, P.J. and Carter, S.R., 1978, Melting of the mantle past and present: isotope and trace element evidence: Philos. Trans. Royal Soc. London A, v. 258, p. 547-559.

Oversby, V.M., 1974, New look at the lead-isotope growth curve: Nature, v. 248, p. 132-133.

Pankhurst, R.J., 1977, Strontium isotope evidence for mantle events in the continental lithosphere: Geol. Soc. London Jour. v. 134, p. 255-268.

Pankhurst, R.J., Moorbath, S., Rex, D.C. and Turner, G., 1973, Mineral age patterns in ca. 3700 m.y.-old rocks from West Greenland: Earth and Planetary Sci. Letters, v. 20, p. 157-170.

Pidgeon, R.T., 1978, 3450 m.y.-old volcanics in the Archaean layered greenstone succession of the Pilbara block, Western Australia: Earth and Planetary Sci. Letters, v. 37, p. 421-428.

Reimer, T., 1975, Untersuchung über Abtragung, Sedimentation und Diagenese im frühen Präkambrium am Beispiel der Sheba-Formation (Südafrika): Geologisches Jahrbuch, v. 17, p. 3-108.

Ringwood, A.E., 1974, The petrological evolution of island arc systems: Geol. Soc. London Jour. v. 130, p. 183-204.

Russell, R.D. and Farquhar, R.M., 1960, Lead Isotopes in Geology: New York, Interscience Publishers, 243, p.

Schidlowski, M., Appel, P.W.U., Eichmann, R. and Junge, C.E., 1979, Carbon isotope geochemistry of the 3.7×10^9-yr-old Isua sediments, West Greenland: implications for the Archaean carbon and oxygen cycles: Geochim. Cosmochim. Acta, v. 43, p. 189-199.

Shaw, D.M., 1976, Development of the early continental crust. Part 2: Prearchaean, Protoarchaean and later eras: in Windley, B.F., ed., The Early History of the Earth: London, Wiley, p. 33-53.

Stacey, J.S., and Kramers, J.D., 1975, Approximation of terrestrial lead isotope evolution by a two-stage model: Earth and Planetary Sci. Letters, v. 26, p. 207-221.

Stern, C.R., Huang, W.L. and Wyllie, P.J., 1975, Basalt-andesite-rhyolite-H_2O: crystallisation intervals with excess H_2O and H_2O-undersaturated liquidus surfaces to 35 kilobars, with implications for magma genesis: Earth and Planetary Sci. Letters, v. 28, p. 189-196.

Sun, S.S. and Hanson, G.N., 1975, Evolution of the mantle: geochemical evidence from alkali basalt: Geology, v. 3, p. 297-302.

Tarney, J. and Windley, B.F., 1977, Chemistry, thermal gradients and evolution of the lower continental crust: Geol. Soc. London Jour. v. 134, p. 153-172.

Taylor, P.N., 1975, An early Precambrian age for migmatitic gneisses from Vikan i Bø, Vesteralen, North Norway: Earth and Planetary Sci. Letters, v. 27, p. 35-42.

Taylor, S.R., 1979, Chemical composition and evolution of the continental crust: the rare earth element evidence: in McElhinny, M.W., ed., The Earth – Its Origin, Structure and Evolution: Academic Press, Chapt. 11, p. 353-376.

Taylor, P.N., Moorbath, S., Goodwin, R. and Petrykowski, A.C., 1980, Crustal contamination as an indicator of the extent of early Archaean continental crust: Pb isotopic evidence from the late Archaean gneisses of West Greenland: Geochimica et Cosmochimica Acta, in press.

Thorpe, R.S., Potts, P.J. and Francis, P.W., 1976, Rare earth data and petrogenesis of andesite from the North Chilean Andes: Contrib. Mineral. Petrol., v. 54, p. 65-78.

Watson, J.V., 1976, Vertical movements in Proterozoic structural provinces: Philos. Trans. Royal Soc. London A, v. 280, p. 629-640.

Wells, P.R.A., 1976, Late Archaean metamorphism in the Buksefjorden region, Southwest Greenland: Contrib. Mineral. Petrol., v. 56, p. 229-242.

Wells, P.R.A., 1979, Chemical and thermal evolution of Archaean sialic crust, southern West Greenland: Jour. Petrol. v. 20, p. 187-226.

Wilson, J.F., Bickle, M.J., Hawkesworth, C.J., Martin, A, Nisbet, E.G. and Orpen, J.L., 1978, Granite-greenstone terrains of the Rhodesian Archaean craton: Nature, v. 271, p. 23-27.

Windley, B.F., 1977, Timing of continental growth and emergence: Nature, v. 270, p. 426-428.

Windley, B.F. and Smith, J.V., 1976, Archaean high grade complexes and modern continental margins: Nature, v. 260, p. 671-675.

Zartman, R.E., 1974, Lead-isotopic provinces in the Cordillera of the western United States and their geological significance: Econ. Geol., v. 69, p. 792-805.

The Continental Crust and Its Mineral Deposits, edited by D.W. Strangway,
Geological Association of Canada Special Paper 20

FORMATION OF CONTINENTAL CRUST

G.F. West
Geophysics Laboratory, Department of Physics
University of Toronto
Toronto, Ontario M5S 1A7

ABSTRACT

A geotectonic model for Archean time should be capable of explaining not only how the typical Archean volcanic-plutonic lithological assemblages were formed, but also how they were assembled into large blocks that suddenly became tectonically inactive about -2.6 Ga, for this cratonization process clearly sets the early Precambrian apart from later tectonic regimes.

It is very attractive to correlate the cratonization process with the separation of the Earth's crust into emergent continents and deep ocean basins, i.e., the onset of plate tectonics. This is compatible with current models of the planet's formation, where the planet would have melted, differentiated, and solidified (except for the core) very early, and the initial surface would have been violently disrupted by meteoritic impacts until about -4 Ga. These processes and Archean volcanism should have produced considerable sialic scum on the surface that would not easily remix into the mantle. Higher radiogenic heat in the crust and higher mantle heat flow than at present would have made it difficult for this sialic material to remain collected in thick sections of limited lateral extent, as such sections would have become overheated and very soft. Thus, it is unlikely that the early Earth had continents of anything like the present thickness.

A geotectonic model is proposed for the early Precambrian in which widespread, sporadic episodes of mafic volcanism from the upper mantle build up on a submarine protocrust containing a sialic component. A sufficiently thick layer of volcanics acts as a thermal blanket, causing softening of the substratum; the protocrust then partially ingests the volcanics by diapiric action. Irregularly, but gradually, the protocrust thickens. Meanwhile, gradual cooling of the mantle thickens the lithosphere and increases its mean density. Eventually, subduction of high-density parts of the lithosphere begins, while more sialic, hotter, softer portions are compressed laterally and thickened. Water drains into new ocean basins. Uplifted continental surfaces are deeply eroded and stabilized by the loss of a substantial part of their radioelement content into sediment.

RÉSUMÉ

Un modèle géotectonique pour l'Archéen devrait pouvoir expliquer non seulement comment les assemblages lithologiques volcano-plutoniques typiques de l'Archéen se sont formés, mais aussi comment ils se sont assemblés en de grands blocs qui sont soudainement devenus tectoniquement inactifs il y a environ 2,6 Ga, parce que ce processus de cratonisation distingue clairement le début du Précambrien des régimes tectoniques plus tardifs.

On est fortement tenté de faire une corrélation entre les processus de cratonisation et la séparation de la croûte terrestre en continents émergés et en bassins océaniques profonds, c'est-à-dire le début de la tectonique des plaques. Cette théorie est compatible avec les modèles récents de formation de la planète, dans lesquels il y aurait eu fusion, différenciation et solidification (à l'exception du noyau) de la planète très tôt; la surface initiale aurait été violemment fragmentée par des impacts météoritiques jusqu'à il y a environ 4 Ga. Ces processus et le volcanisme du début de l'Archéen auraient dû produire en surface une écume sialique qui ne se serait pas facilement réincorporée dans le manteau. De la chaleur radiogénique dans la croûte et un flux thermique dans le manteau plus élevés qu'actuellement auraient rendu difficile le maintien de ce matériel sialique dans des sections épaisses d'étendue latérale limitée puisque de telles sections seraient devenues surchauffées et très molles. Ainsi, il est peu probable que la Terre primitive ait eu des continents d'une épaisseur se comparant à ce qu'on observe actuellement.

On propose un modèle géotectonique pour le début du Précambrien dans lequel des épisodes sporadiques de volcanisme mafique provenant du manteau supérieur ont affecté une grande étendue de protocroûte sous-marine contenant une composante sialique. Une couche suffisamment épaisse de roches volcaniques a agi comme couverture thermique en provoquant l'amollissement du substratum; la croûte primitive a alors englouti les volcaniques par action diapirique. Irrégulièrement mais graduellement, la croûte primitive s'est épaissie. Pendant ce temps, le refroidissement graduel du manteau a épaissi la lithosphère et a augmenté sa densité moyenne. Eventuellement, la subduction de portions de lithosphère à densité élevée a débuté alors que les portions plus sialiques, plus chaudes et plus molles se comprimaient latéralement et épaississaient. L'eau s'est drainée dans les nouveaux bassins océaniques. Des surfaces continentales se sont soulevées et ont été profondément érodées et stabilisées par la perte d'une portion importante de leur contenu en radioéléments dans les sédiments.

INTRODUCTION

The origin of the continents has been debated since the earliest days of geology. J. T. Wilson (1949a, 1949b) set the stage for modern discussion by proposing an age and structural regionalization for the Canadian Shield, and outlining theories for continent formation. An important thesis of his papers was the lateral growth of continents both in Precambrian and later time. This concept has taken on a much more specific meaning with the advent of plate tectonics. However, Wilson noted that the Archean cratons lack the geosynclinal sediments and linear fold belts typical of later orogens, and suggested that the Archean cratons were primitive nucleii around which the continents grew by lateral accretion of marginal geosynclines and volcanic belts. He also asked the important question: What makes some parts of the continents so stable and other parts so tectonically active?

Since 1949 our understanding of Phanerozoic geotectonic processes has improved remarkably, but the debate on the origin of Precambrian cratons continues. Much of the controversy centres around whether the basement on which the Archean volcanics were laid was granitoid or mafic (see Glikson, 1978, and Baragar and McGlynn, 1978, for opposing views). Other discussion focuses on the relationship of the volcanic sequences to modern island arcs and ocean ridges. Many authors automatically equate sialic and mafic basements with continental and oceanic crusts of the modern type. Others (e.g., Hargraves, 1976; Baer, 1977) discuss the possibility that the early crust was not divided into continents and oceans as it is now, but was in some intermediate state under a universal shallow ocean.

Wilson assumed that the continental nucleii were built on primitive mafic crust and this hypothesis has been developed in different ways. Some authors (e.g., Burke *et al.*, 1976; Windley, 1976a; Condie, 1978) take a uniformitarian approach and maintain that only modern-style plate tectonics is required to account for the Archean terrane. Others, such as Anhaeusser (1973) and Glikson (1976) argue that Archean geology is explained better by processes that do not require the lateral accretion and shelf-type sedimentation implicit in the arc-trench-continental margin scenario of plate tectonics, since little evidence of these conditions has been found in the Archean cratons.

In my view, the principal obstacle to explaining the formation of the crystalline shields and platforms of the continental cratons in terms of current plate tectonics is the pattern of argon radiometric ages. From it, we see that the cratons are composed of large, roughly equant blocks, each of which arrived in approximately its present geological state via a widespread orogenic process that ended in an elevation of the crustal surface high above sea level and a consequent deep erosion. This event, cratonization, took place at -2.6 Ga or at -1.8 Ga in many of the blocks; a few areas appear to have been cratonized earlier, from -3.3 Ga onwards.

Cratonization should not be confused with the formation of the supracrustal and plutonic rocks which now make up the cratons. Radiometric age studies by penetrative methods such as the Rb-Sr whole-rock isochron and the U-Pb zircon method show that although many shield rocks were formed shortly before their cratonization, others were formed much earlier. Also, the geology suggests that the crustal surface was predominantly submarine when the rocks were formed. The formational history of the rocks in a craton may have been complicated and may have extended over a long period of time, but the final welding of those rocks into a stable, elevated, block of continental crust occurred rapidly and in most cases on a broad or even global scale.

BACKGROUND ASSUMPTIONS

In constructing a tectonic model for the first half of the Earth's history, one must consider how the Earth was formed initially, when core-mantle segregation took place, whether or not the mantle has always been solid, whether or not creeping convection always occurred there, and whether catastrophic events might have taken place.

Planetary Formation. It is generally believed that the planets formed by accretion soon after the formation of the Sun. The Moon's intense volcanic history before -4 Ga, coupled with various theoretical arguments (e.g., Kaula, 1979), makes a completely cold accretion very unlikely. Thus, the outer part of the protoplanet must have formed at or near the melting temperature and if the core was not directly formed by heterogeneous accretion, the planet melted and the core separated out very early. Even a molten planet would still have maintained a frozen surface boundary layer, however, as the planetary heat loss from a molten surface layer would have been too enormous. The solid surface was undoubtedly in constant turmoil from meteoritic bombardment and the sinking of solid mafic silicate fragments into molten silicates.

The Earth's molten stage would have been brief; heat transfer models demonstrate that the mantle would have frozen very rapidly (in about 0.1 Ga). Thus we begin with an Earth differentiated into solid silicate mantle and liquid iron core. Many authors also assume that the radioactive elements were concentrated in the uppermost levels. It also seems virtually certain that a hydrosphere and primitive atmosphere were already in place when the meteoritic pounding of the surface abated towards -4.0 Ga.

Mantle Convection. The fundamental idea of plate tectonics is that the solid mantle (*asthenosphere*) undergoes continuous, quasi-fluid, creeping convective flow, except for its cold surface layers (*lithosphere*).

Sharpe and Peltier (1978) have modelled the history of a convecting Earth and come to two important conclusions: 1) The lithosphere plays a key role in regulating mantle temperature and heat flow, and 2) the primordial heat associated with planetary formation and differentiation is nearly sufficient to explain present mantle heat flow and viscosity. Allowing for a reasonable amount of radioactivity in the mantle, we can certainly assume that the early mantle was hotter and less viscous than at present, and was convecting more rapidly.

Although the matter is debatable (e.g., McKenzie and Weiss, 1975), the present-day lithosphere seems to operate more as the asthenosphere's thermal boundary layer than as a rigid outer shell (Peltier, 1980). A transition to eclogite in the oceanic lithosphere may provide the greater density necessary to overcome the light, differentiated material in the top of the crust, allowing old, cold lithosphere to sink into the asthenosphere at the subduction zones. On the other hand, since continents are obviously not sunk by this process, a mixed system exists at the present time in which parts of the lithosphere separate from the main convective flow and remain permanently at the surface. Allowing that the lithosphere of the early Earth might have had a different thickness and the crustal structure may have been quite different, it is certainly possible that a less viscous mantle convected underneath a more or less complete lithosphere at that time. Such a lithosphere would have been subject to considerable horizontal stress by the convective shear flows along its base.

Radiogenic Heat. A considerable part of the surface heat flow from continents is generated by the radioactive decay of K, U, and Th in the crust. These elements have moved during geologic time, but wherever they were, in the crust or in the mantle, they generated much more heat in Precambrian time. A rough rule of thumb for a typical rock composition is a factor of two times the present heat production at -2.5 Ga and a factor of three at -3.5 Ga. We can assume that rock chemistry has not altered much due to radioelement abundance changes, and therefore geologically similar granites formed in the Precambrian and recently should have similar heat productivities at the present time.

Studies of surface heat flow in many locations have shown directly that in crystalline provinces (i.e., areas that have undergone orogenesis) the heat-producing radiogenic elements are concentrated in the top of the crust (Lachenbruch, 1970). This is explained by the fact that K, U, and Th tend to travel together in crustal rocks in the granitic (sensu stricto) and pegmatitic phases, which are most abundant in the epizone and rare in the catazone. The heat flow data suggest that a good approximation for the average vertical variation of heat production is an exponential function with a decay constant of 5 to 10 km.

Gravity Tectonics. This paper is concerned with models of orogenesis, i.e., geologic processes where large deformations have occurred. Such deformations are clearly the result of long-sustained stresses that are not quickly relieved by small deformations or motions. The only source of such stresses is the force of gravity acting on density differences. What stress pattern this produces and what displacements and deformations result depends on the rheology of the materials, and whether any change in state of the materials changes the density pattern. In large-scale geology, long time intervals, and deformation at moderate to great depths, we can assume that plastic creep is the main rheological response to stress. Thus, we follow the guidelines of Ramberg (1967) in relating geological observations to possible mechanisms, and pay careful attention to what produces density changes and what affects the rheology of rocks.

From the physical point of view, large-scale orogenic deformation is a manifestation of *gravity tectonics,* a term geologists often use to mean merely gravity-driven allochthonous sliding or, in some cases, the rising and/or sinking of diapirs (usually called *vertical tectonics*). When some parts of a gravitationally stressed system are much more rigid than others, stresses may easily be transmitted laterally without noticeable deformation of the rigid member. Adjacent softer members may then deform by lateral compression or tension, transcurrent shear, etc., and yet there may be no local density differences causing the stress field. Nevertheless, such a situation is a manifestation of gravity tectonics in the broad sense.

An important corollary of gravity tectonics (sensu lato) is that virtually all geologic systems are in a state of non-hydrostatic stress at all times. The only density distributions that give rise to pure hydrostatic stress are those with a perfectly spheroidal density layering. Real geologic structures are only crudely and locally stratified, as they vary from place to place. Isostatic compensation between adjacent structures serves only to reduce, not eliminate, the non-hydrostatic stresses. As a consequence, orogenic deformation takes place whenever the viscosity of a part of the lithosphere drops enough for the stress to cause an important deformation in a reasonable period of time. It seems unlikely that regional non-hydrostatic stress levels vary by more than one or two orders of magnitude, while the effective viscosity of geologic materials can easily differ and change by a factor of 10^5 or even 10^{10}.

Rheology. At the present state of our knowledge, it is difficult to make many quantitative statements about rock rheology. The field is fraught with experimental difficulties. (See reviews by Weertman and Weertman, 1975; Carter, 1976; Holland and Lambert, 1969; Kerrich and Allison, 1978.)

A short summary of the field suffices here: Rocks in shear should be regarded roughly as Maxwell solids (spring in series with a viscous dashpot); most crystalline rocks will have (very roughly) similar elastic moduli that are not strongly dependent on pressure, temperature, and the environment; viscosities, however, vary enormously depending on lithology, temperature, chemical state, and stress level.

The steady-state creep viscosity of mono-mineralic rocks is reasonably well understood for sub-solidus conditions. It is stress dependent with an exponential variation in the range of 2 to 5, and has a rapid temperature dependence fitted by a simple thermal activation law (exp Q/kT). Table I gives a very rough guide to what values of viscosity are significant in various geologic circumstances. Also shown is the predicted viscosity variation of Westerly Granite from Goetze, 1971.

Volcanism. A fundamental feature of the Earth's mantle is that the melting point rises more steeply with increasing depth than does the temperature, the temperature being constrained by convective processes to lie close to an adiabat, except in the thermal boundary layers. Partial melting therefore occurs near the top. At the present time, partial melting beneath oceanic ridges (presumably located over mantle upwellings) liberates about 10 km^3 per year of basaltic partial melt. Whether all the partial melt formed actually gets to the surface, however, is unclear. Hot, partially melted upper mantle might simply re-freeze during lateral convection. Whatever the present situation, it is unlikely that less partial melt was created in the past. More likely, a hotter and more vigorously convecting mantle created more.

TABLE I
VISCOSITIES IN GEOLOGICAL SITUATIONS

Situation	Viscosity (poise)	Dimension (m)	Velocity (ms^{-1})
Permanent gravity features in shield	$>10^{28}$	10^4	$<10^{-15}$
Mantle, post-glacial rebound	$\approx 10^{22}$	10^4-10^6	$\approx 10^{-9}$
Mantle, convection	$\approx 10^{22}$	10^6	10^{-9}
Crust, gneiss dome	$\approx 10^{22}$	10^4	$10^{-9}-10^{-10}$
Sedimentary basin, salt dome	$\approx 10^{19}$	10^3	$\approx 10^{-9}$
Magma chamber, crystal settling	$<10^8$	10^{-2}	$>10^{-6}$
Surface, lava flow	10^1-10^3	$10-10^3$	>1

TEMPERATURE DEPENDENCE OF THE EFFECTIVE VISCOSITY
OF GRANITE

Temperature (°C)	300	400	500	600	700
Rel. viscosity (η_T/η_{700})	10^9	5×10^5	3×10^3	30	1

Estimated from measurements of transient creep of Westerly Granite by Goetze (1971). The value of η_{700} is estimated at 10^{22} poise with a tolerance of ± 2 in the exponent.

THE MODEL

In this research, my objective has been to find a model for the tectonic processes of the early Earth which is consistent both with the geophysical conclusions described above, and with the main features of Archean geology. The procedure has been to examine a great variety of possible tectonic processes and select those elements which could explain the geology and also could be linked together into a coherent history by means of a plausible physical argument. The model is first outlined without justification, and in the next section its various elements are explained.

Although there is no consensus on a tectonic model for the Archean, there is considerable agreement on the general nature of Archean geology. Many readers will be familiar with this subject. Others may find a good introduction in the compilation (with commentaries) of classic papers by McCall (1977) and in symposia volumes edited by Sutton and Windley (1973) and Windley (1976b). Important summaries of the Canadian Archean are given by Goodwin *et al.* for the Superior province, McGlynn and Henderson for the Slave province, and Davidson for the Churchill province in the review volume *Variations in Tectonic Styles in Canada* (Price and Douglas, eds., 1972). Young (1976) provides an interesting short review that compares the Canadian Archean with African and Australian cratons, and gives a good bibliography.

Outline of the Model

It is envisaged that by about -4.3 Ga the mantle had solidified and a primitive crust and lithosphere had formed on the Earth. This crust was heavily bombarded by meteorites and evolved volcanically under this influence. A primitive ocean and atmosphere existed from very early time.

By about -4.0 Ga, the meteoritic bombardment had abated sufficiently that the surface pattern of volcanism was controlled by mantle convection. The pattern changed frequently, since the hotter Archean mantle was even more unstable than the present one. The rate of volcanic effusion onto the early crust was higher than at present, the global accumulation averaging perhaps 50 to 100 km from -4.0 to -3.0 Ga. However, the volcanism was far from uniform; some areas of the Earth received vast quantities, while others were left in a relatively primitive state.

The Archean crust (archeocrust) did not simply accumulate the volcanics. The thermal gradient was steep, and as early volcanics and associated supracrustal rocks were buried, they were metamorphosed and partially melted. Granitic-pegmatitic partial melt was released at a relatively shallow level (about 20 km), leaving a residuum of mafic granulite. In turn, this residuum partially melted at greater depth (about 50 km) to produce tonalitic magma and an ultramafic residuum. At even greater depth (the bottom of the lithosphere), the ultramafic residuum softened and returned to the mantle convective flow. Due to their low density, the granitoid partial melts rose through the crust and accumulated at mid-crustal depth (5 to 15 km).

Wherever a granitoid middle crust was formed, any new thick volcanic pile acted as a thermal blanket, heating and softening the granitoid rocks and the basal parts of the volcanic pile. Due to the density inversion, a rapid diapiric overturn took place, with metavolcanic rocks sinking between rising domes of granite gneiss. This process, which occurred at many different locations and times, was responsible for much of the surface geology now observable on Archean shields.

As time progressed and heat flow from the mantle subsided somewhat, the lithosphere began to thicken from below, by retaining more cool, rigid, ultramafic material. The thickness was greatest where the lithosphere was coolest, i.e., where the upper lithosphere had not accumulated great quantities of volcanics and therefore had not built up a thick granitoid middle crust with its large radiogenic heating capacity.

As differences in lithospheric thickness became greater, they influenced the upper mantle's pattern of convection. Gradually, mantle volcanism and the consequent accumulation of granitoids in the crust became tied to specific parts of the surface. Paradoxically, the areas that were receiving most of the volcanics had a thinner crust and lithosphere than the more primitive parts, because the much higher thermal gradient produced by high mantle heat flow and high crustal radioactivity caused the chemical and physical processes that were controlling the layering to occur at shallower depths. Such areas of the lithosphere would, however, have accumulated a thicker granitoid middle crust than the more primitive areas. Thus, their average density became lower.

By about -3 Ga, the lithosphere in the most primitive areas had thickened sufficiently that the buoyancy of the crustal component was overcome by the excess density of the cold, lithified upper mantle. However, since the primitive regions were the strongest and most rigid parts of the lithosphere, it was difficulty for them to sink. Nevertheless, by -2.6 Ga, the situation had become so unstable that subduction of thick lithosphere took place locally, probably triggered by some strong fluctuation in the unstable mantle convection.

The effect of even a small amount of subduction was profound. Parts of the lithosphere were able to move laterally, and some new oceanic crust was created. Changes in the mantle convection pattern caused subduction of primitive parts of the lithosphere to become more widespread, and many of the soft, hot parts of the lithosphere that had recently been active were compressed laterally by the new convective system. These compressed areas became continental cratons.

As envisaged, the Kenoran cratonization event at -2.6 Ga created a large quantity of new oceanic crust and some large continental cratons, thereby disposing of one-half to two-thirds of the archeocrust. However, the geological evidence of the Proterozoic suggests that continuous subduction of new oceanic crust did not begin immediately. The Earth appears to have returned to a relatively fixed lithosphere type of convection pattern until the next episode of cratonization at -1.8 Ga. This event converted most of the remaining archeocrust into continental craton and began the continuous subduction of thick, dense (i.e., old oceanic) lithosphere observed in present-day plate tectonics.

Explanations and Details

Formation of Greenstone Belts. A remarkable feature of Archean cratons is that areas of similar regional metamorphic level have very similar geology. Metamorphic levels below the upper amphibolite facies are the most common, and such areas display the greenstone-granite (metavolcanic-plutonic) terrane for which the Archean is famous. Higher grade regions generally contain fewer identifiable volcanic rocks (mostly amphibolitic schlieren and tenuous greenstone belts), but a good case can be made that many of these areas are essentially similar to lower grade regions except that the level of exposure is deeper (e.g., Windley and Bridgwater, 1971; Goodwin, 1977).

Figure 1 illustrates the general, universal characteristics of the greenstone-granite terrane. Hypothetical maps and sections by Anhaeusser *et al.* (1969) and by Glikson (1978), based mainly on the Archean geology of Australia and Africa, are compared with the Sturgeon Lake-Savant Lake area of the Superior province, with which the author has had direct experience.

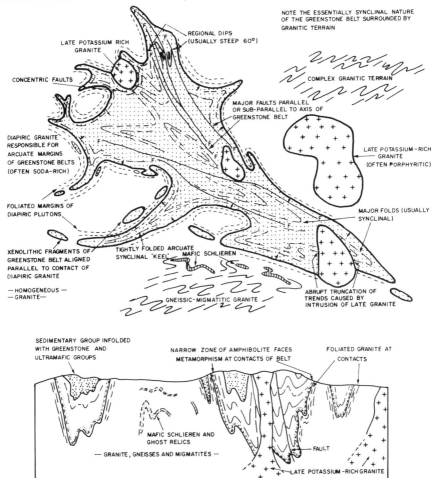

Figure 1a. A schematic map and section of typical Archean greenstone-granite terrane, based on the Barberton Mountain Land in Rhodesia (Anhaeusser *et al.*, 1969).

The typical metavolcanic-plutonic terrane consists mainly of steeply dipping, deformed metavolcanics and closely associated sediments which occur as large or small remnants between complex domes of granitoid gneiss. The metavolcanic assemblages (greenstone belts) are often broken and incomplete, but where better preserved they usually have a synformal aspect, with the inner parts less metamorphosed than the outer parts. Dips of bedding and foliation are almost always very steep. Strikes within a local area may be highly aligned, but over a large area their directions vary according to the pattern of granitoid domes. However, there is often some overall mean alignment of strikes throughout one craton or subprovince. Strains in the greenstone belts vary from modest to extremely large.

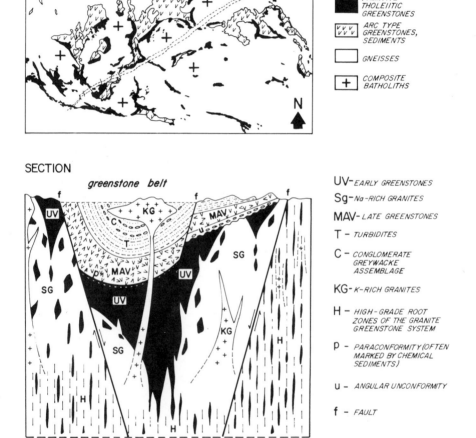

Figure 1b. A map of greenstone belts in the Rhodesian craton showing the primitive tholeiitic and arc-like greenstones surrounding composite batholiths, and a schematic cross-section showing the generalized stratigraphy (Glikson, 1978).

The emplacement of large quantities of mafic volcanics at the Earth's surface takes place in many different environments: on continents and in ocean basins, at ridges and trenches and on the plates. What is important in our context is determining the environment in which the volcanism occurred and the successor events that formed the volcanics into typical Archean greenstone belts squeezed between granitoid domes.

One of the oldest explanations for the greenstone-granite terrane is that the mafic volcanics were laid down on a pre-existing sialic granitoid crust, and being more dense than the granitoid substratum, they eventually sank into it while the granitoid rocks domed up (e.g., MacGregor, 1951; Anhaeusser et al., 1969). However, when Rb-Sr whole-rock age dating showed that most of the granitoid rocks exhibit low initial ratios and thus could not have had a long-pre-history in their

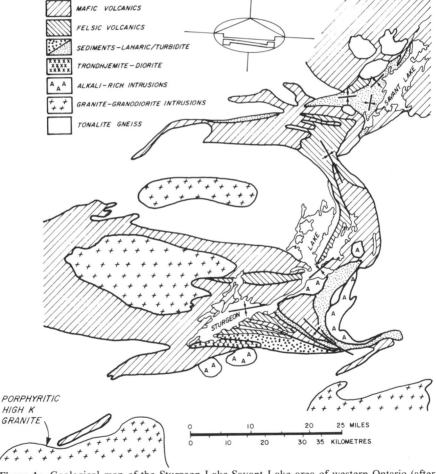

MAFIC VOLCANICS

FELSIC VOLCANICS

SEDIMENTS–LAHARIC/TURBIDITE

TRONDHJEMITE–DIORITE

ALKALI–RICH INTRUSIONS

GRANITE–GRANODIORITE INTRUSIONS

TONALITE GNEISS

PORPHYRITIC HIGH K GRANITE

0   10   20   25 MILES
0   10   20   30 35 KILOMETRES

Figure 1c. Geological map of the Sturgeon Lake-Savant Lake area of western Ontario (after Franklin, 1978, with additions from Goodwin, 1978, and Shegelski, 1976).

1. REST STATE

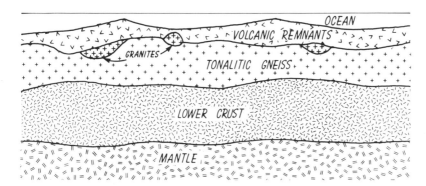

2. VOLCANISM-BASALT FROM MANTLE

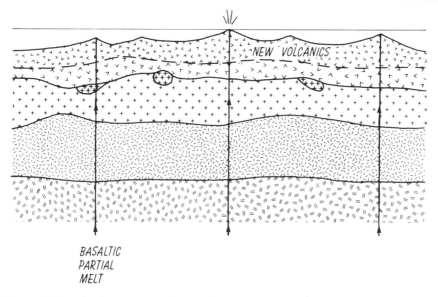

Figure 2. The volcanic-plutonic cycle, by which a newly emplaced volcanic pile is ingested by the crust: 1) The granitoid crust has a high radiogenic heat productivity contributing to a high temperature gradient, but it remains relatively cool and rigid because it is near the surface, 2) A mafic volcanic blanket 5 to 15 km thick is emplaced from a mantle source, 3) The temperature in the granitoid crust rises, metamorphism begins and viscosity falls drastically, 4) Diapirism begins 10-50 Ma after volcanism. The lower parts of the volcanic pile are ingested and granitoid domes rise, 5) Wet partial melting in the subcrust transports radioelements back towards the surface in granitic magmas. The crust cools slowly and returns to its rigid state.

3. ACTIVATION-THERMAL WAVE SOFTENS UPPER CRUST

4. DIAPIRISM - VOLCANICS PARTIALLY INGESTED

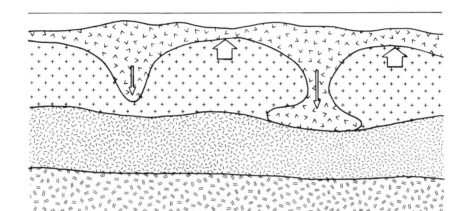

5. PLUTONISM - RADIOELEMENTS RETURN TO NEAR SURFACE

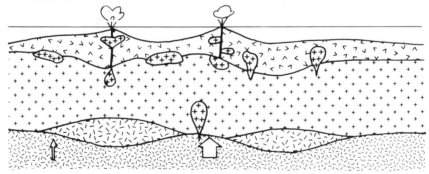

6. REST STATE AGAIN

present form, a variety of other theories were put forward. None, however have succeeded in being more convincing than the earlier model.

The sialic basement model has much to recommend it from a structural and tectonic point of view. It explains granitoid clasts in basal conglomerates beneath volcanic units (e.g., Baragar and McGlynn, 1976) and how the cratons ultimately stabilized by deep erosion. Furthermore, this model can be modified to overcome the Rb-Sr initial ratio problem.

A quantitative model study had been carried out by West and Mareschal (1979) and Mareschal and West (1980) of how a layer of volcanics may eventually deform and sink into a lighter substratum. The process is essentially that of vertical gravity tectonics as modelled by Ramberg (1967); the rate of diapirism is controlled by creep viscosity at depth, which itself is controlled by temperature. The modelling, which used analytical methods in one dimension and finite element methods in two dimensions, shows that a volcanic blanket 5 to 10 km thick on a crust of *high* surface heat flow can produce sufficient temperature rise and viscosity fall in the substratum to cause rapid diapirism 10 to 50 Ma after emplacement.

The Archean geologic record shows a mainly submerged crustal surface prior to cratonization, with frequent temporary emergence of volcanic islands and, much more rarely, of granitoid crust. Thus, granitoid crust of that era was not deeply eroded and should have contained rocks of the cata-, mezo-, and epizones, with granites having high concentrations of K, U, and Th in the epizone. Postulating a surface radio-element concentration similar to that in young, uneroded crust (i.e., $\approx 5\%$ K, 7 ppm U, 50 ppm Th), assuming an exponential depth dependence with decay constant of 10 km, and also extrapolating the radiogenic heat productivity back to -3.0 Ga provides a very potent heat source in the upper crust – capable of producing, on its own, a temperature gradient of about 35 K/km in rocks above it. Equilibrium temperatures in the lower crust will thus become excessively high whenever the radioactivity is buried significantly. Then when diapirism occurs, it will liberate pegmatitic partial melt from the overheated subcrust, transferring the radioelements back near the surface after their burial by the volcanics. Figure 2 qualitatively illustrates this thermally activated, diapiric process. Table II provides data from the modelling work.

Crustal Evolution. At the present time, the 10 km^3/year of mafic partial melt introduced into the crust at oceanic ridges appears to be approximately matched by a similar loss in the subduction process. It seems that only arc volcanism and continental plateau volcanism currently contribute much to the volumetric growth of continental sial. However, if the lithospheric subduction process did not operate in Archean time, the volume of sial could have grown very rapidly. A hotter-than-present mantle might well have liberated 30 km^3/year of basalt, which is enough to cover the present continental surface to a depth of 100 km in 150 Ma. Clearly, if volcanism occurred on such a scale, a mechanism for returning much of the volcanic material must also have existed.

Figure 3 shows a schematic view of what the crustal section might have looked like and what processes might have been taking place in it if the surface were subjected to regular increments of basalt at a rate of, say, 10 km per 100 Ma under the high heat productivity conditions of the Archean era. Supracrustal materials would

TABLE II
MODEL STUDIES OF THERMALLY ACTIVATED DIAPIRISM

Thermal Modelling

Model 1: Uneroded crust at -3.0 Ga, surface radioelement concentration 5%K, 50 ppm Th, 7 ppm U, exponential depth dependence with a 10 km natural logarithmic decrement.

Model 2: Eroded crust at -2.2 Ga, same as above but top 5 km removed.

Temperature Jump

Crust Model (no.)	Surface Heat Prod. ($\mu W\ m^{-3}$)	Mantle Heat Flow ($mW\ m^{-2}$)	Volc. Thick. (km)	Ultimate Temp. Jump (K)
1	11.8	40	5	320
2	6	30	5	190
1	11.8	40	10	650
2	6	30	10	385

Penetration of Transient (all models, approx.)

Depth below surface (km)	10	15	20	25	30
Time to $\Delta T/2$ (Ma)	3	7	13	22	30

Flow Modelling

Quantity	Model A	Model B
η_{700} (poise)	4×10^{21}	7.8×10^{22}
start at t=0 volc. thick. (km)	9.5-10.5	2.5-10.5
at t (Ma)	17.0	113
volc. thick. (km)	7.5-13.75	2 - 12.5
max. vert. vel. (cm/yr)	0.8	0.1
at t (Ma)	17.7	118.3
volc. thick. (km)	6.25-22	1.8 - 17
max. vert. vel. (cm/yr)	8	0.3

WEST

be increasingly metamorphosed and partially melted as they became deeply buried. Being light, the partial melts would rise and accumulate in the upper crust. The successive residues would become more and more mafic and refractory as they passed to greater and greater depths. The final residuum would be ultramafic, and at sufficient depth it would soften and could be swept into the convective flow of the mantle. According to this model, only a granitoid fraction of each basaltic increment

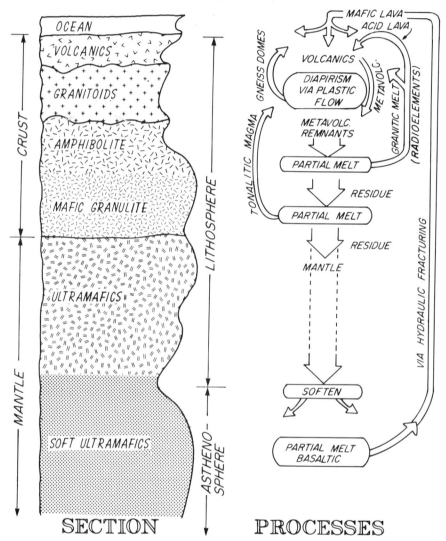

SECTION   PROCESSES

Figure 3. The evolutionary process by which part of the volcanic material emplaced on the surface eventually returns to the mantle and part accumulates in the upper crust as granitoid rocks.

accumulates in the crust. The dimensions of the other crustal layers reflect p, T conditions rather than the amount of volcanism.

The foregoing model can be fitted with the concept of volcanic-plutonic cycles discussed in the previous section and with several additional features of the geology. In Figure 3, I have suggested two main levels of partial melt separation: one where wet partial melting extracts the K, U, and Th in a pegmatitic granite phase, the other a drier partial melting where the tonalitic magma separates out. Such a model is capable of explaining the Rb-Sr initial ratio problem. The date obtained from a tonalite by the whole-rock isochron method would be the date at which it separated from its parent granulite.separation and aggregation of partial melt is an effective mixing process. Subsequent diapiric flows of the tonalite in the upper crust by solid creep would not disturb the whole-rock age much, although mineral ages might be reset. The initial Sr^{86}/Sr^{87} ratio of the tonalite's isochron would be only slightly different from the ratio in the parent basalt at the time of its release from the mantle, as most of the time between volcanism and decomposition to tonalite would be spent in the granulite layer where a Rb/Sr ratio of 0.03 is likely. Any high Rb material included with the supracrustals during their ingestion would rapidly have been regurgitated in the pegmatitic partial melts liberated during diapiric overturn rather than mixed with the solid residual phases in the lower crust.

To reach the evolved state shown in Figure 3, a total accumulation of many tens or even a hundred kilometres of volcanics is required. The whole lithospheric section must have been replaced at least once. I have not attempted to estimate the equilibrium thicknesses of the various layers under different thermal conditions, but it is clear that they would generally have increased with time. A crustal thickness of 40 km seems a reasonable estimate towards the end of the Archean.

Cratonization. The isostatic state of a crustal block can be altered in several ways to produce the required thickness for permanent surface uplift (Fig. 4). The simplest mechanism is lateral compression, where the whole lithosphere has become soft enough for deformation to proceed under the moderate regional stresses that might be available, for example opposing shear flows in the upper asthenosphere. A plate in such a soft state would essentially have delithified.

Other mechanisms involve changing the mean density of a plate by transport of material into or out of it. The crust might be underplated or intruded by a magma that will crystallize as a light rock. Besides basaltic partial melt from the mantle, the only obvious source for such material is a downgoing lithospheric slab, and it is unlikely that a wide zone could be underplated from such a source. Widespread underplating of, or intrusion into, the lower crust of large volumes of liquid magma is unsatisfactory from another viewpoint: there is little reason that the magma should stop its ascent at the base of the crust. However, a case could be made for the arrest by freezing of a high melting-point, light magma such as anorthosite. Alternatively, the lithosphere might drop a load from its base. If there is material in the lithosphere that is denser than the subjacent asthenosphere, e.g., the lower lithosphere has converted to a dense phase, its replacement by normal mantle would reduce the mean density.

None of the above mechanisms (nor other similar ones) can be ruled out *a priori*. However, we have been investigating a tectonic model where large-scale subduction

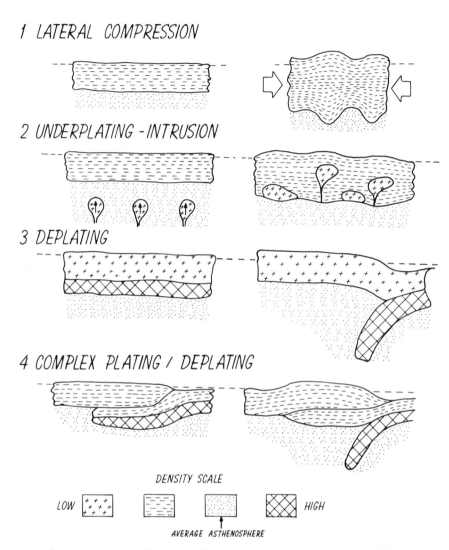

SCHEMATIC CRATONIZATION PROCESSES

1 LATERAL COMPRESSION

2 UNDERPLATING - INTRUSION

3 DEPLATING

4 COMPLEX PLATING / DEPLATING

DENSITY SCALE

LOW *HIGH*

AVERAGE ASTHENOSPHERE

Figure 4. Some methods by which the surface of the crust might be elevated so that a deep erosion will take place. 1) The lithosphere is laterally compressed, increasing the thickness of the crust. 2) Light material is introduced from below by magmatic process. 3) An excessively dense part of the lower lithosphere is subducted (deplating). 4) One plate overrides a second, barely buoyant plate. The change in pressure triggers deplating.

of lithospheric plates does not take place prior to -2.6 Ga because the hotter thermal regime does not permit the lithosphere to become cool and dense enough to sink the light crustal component attached to its top. Overheating and consequent delithification is a possibility in such circumstances. Thus, cratonization by lateral compression is a more attractive mechanism than the others.

Figure 5 shows how mantle convection would occur under a fixed lithosphere. Since the mantle temperature would be substantially higher in the upward flows than in the downward ones, the lithosphere would be thinner over an upwelling and would have a higher diffusive heat flow through it. Also, because basaltic partial melt is most easily produced in the mantle upwelling, we would expect the highest rate of volcanic effusion onto the crust to be located above it. If convection remains fixed relative to the lithosphere for a long period of time (about 300 Ma), the crust over the upwelling would be very actively evolving (Figs. 2 and 3), and there might be no lithified mantle remaining beneath the crust.

Although the lithospheric plate would be delithified over the upwelling, it would not immediately undergo regional deformation because it would be surrounded by rigid plate. Nevertheless, local vertical gravity tectonics would be very active in such

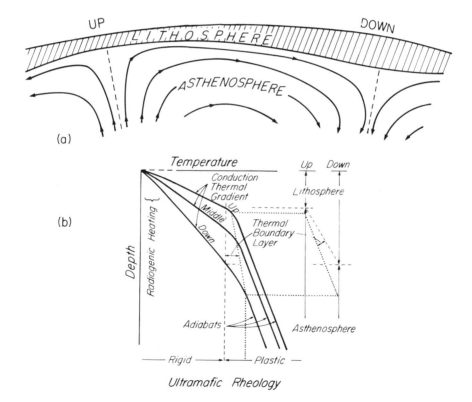

Figure 5. Convection of asthenosphere under an unsinkable lithosphere. The lithosphere adjusts its thickness to the convective system because the base of the lithosphere is approximately an isotherm. The "fluid" asthenosphere maintains a boundary layer beneath the lithosphere where thermal transport is largely conductive. Stress is applied to the lithosphere by the drag of the convection.

(a)

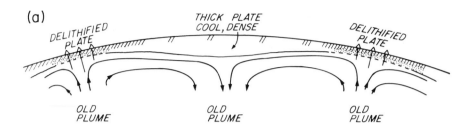

(b)

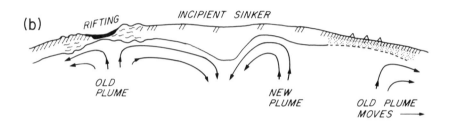

(c)

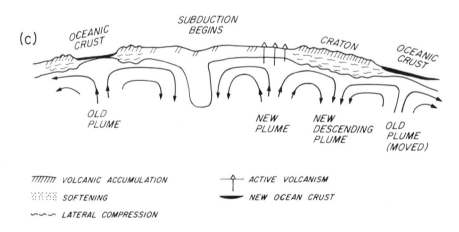

```
//////// VOLCANIC ACCUMULATION          ⟍⟋ ACTIVE VOLCANISM

SOFTENING                               ▬▬ NEW OCEAN CRUST

~~~~ LATERAL COMPRESSION
```

Figure 6. A scenario for the beginning of subduction and the formation of cratons. a) Large differences in the thickness of the lithosphere build up. b) Advent of a new plume produces a local sharp difference in thickness of the lithosphere. Deformation of the lithosphere begins. c) The onset of subduction triggers substantial alterations in convective pattern in the mantle. Soft areas of lithosphere are compressed and new oceanic crust is produced.

a softened region. If a large area delithified and the convective upflow remained steady, the central part might conceivably rift while the side zones thickened (cf. left hand side of Fig. 6). But the convective shear flow at the base would provide only modest stress because it would have little area to act on, so only minor cratonization could take place in this manner.

Local lithospheric thickness will be determined not only by the mantle convection pattern, but by the local crustal composition as well. Evolved crust will have a larger content of radioelements than less evolved crust, because it will contain a thick granitoid section, and for the same heat flow from the asthenosphere, it will have a thinner lithosphere. The most mafic crust would probably be located over a downflow in the asthenosphere, which would perpetuate and exaggerate the differences between more and less evolved crust. Thus an effective coupling would gradually arise between the pattern of mantle convection and lithospheric thickness.

As mantle convective activity declined and the lithosphere generally thickened, the more mafic parts of the lithosphere would become more dense than the asthenosphere, and therefore sinkable. Although these areas might have difficulty deforming enough to allow their surfaces to be 'flooded' and thus might remain metastably at the surface for a long time, the situation would be ripe for a catastrophic change. Figure 6 shows how sinking might begin, and how subsequent events would lead to a major cratonization period in which delithified areas would be thickened, much of the remaining thick, cold lithosphere would be subducted, and a quantity of oceanic crust would be rapidly produced to compensate for the loss of area. The key feature in this scenario is that it is triggered by a change in the pattern of mantle convection; once some subduction begins and some oceanic crust forms, what might have been a minor change in convective pattern becomes a global transition.

Summary. The graphs in Figure 7 show how the asthenosphere-lithosphere and mantle-crust systems are considered to have evolved through time.

The gradually cooling mantle would have had a falling temperature and convective rate. Partial melting at the top of convective upwellings should have been much more prevalent in earlier time than now (Fig. 7a) Basaltic magma generated in the mantle was transported to the surface via hydraulic fractures. Dyke injection through a crust that is not in tension requires a substantial over-pressure in the magma, which in turn requires that the magma reservoir be at a substantial depth. Thus it is envisaged that the magma delivery mechanism became more effective as the lithosphere thickened. Taken together with the falling rate of partial melt production, this suggests that the global rate of basaltic volcanism was possibly at a maximum at about -3.0 Ga. Other mechanisms besides dyke injection presumably operated during the formational and meteoritic impact phases of initial crustal development to bring fractionated mantle into the crust (Fig. 7b).

As time progressed, those areas which received substantial volcanics evolved according to the pattern shown in Figures 2 and 3, and the quantity of granitoids in the crust increased slowly. Because the mean crustal density of those areas gradually declined, they became more elevated, and islands emerged from the ocean cover. In the areas of most intense volcanism, the accumulated thickness of granitoids was probably limited by thermal considerations rather than by the supply of volcanics, as

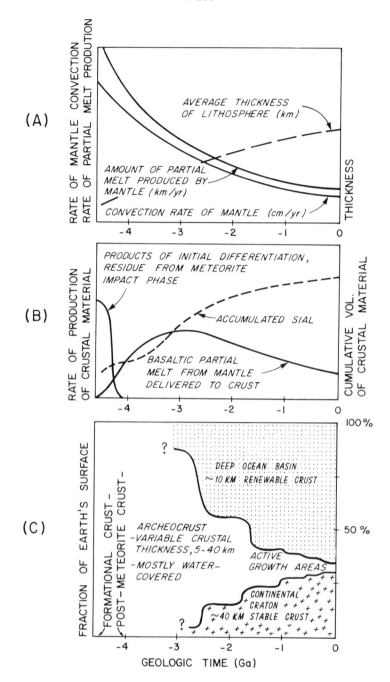

Figure 7. Crustal formation through geologic time.

a thick, radioactive, granitoid crustal layer could not have cooled enough to support itself and its surface loads.

As the contrast between less and more evolved parts of the lithosphere increased, the mantle convection pattern tended to couple to the differences in lithospheric thickness. In the evolving parts of the lithosphere, the temperature remained high, near the softening point of the various materials, while in the thicker, less evolved parts, it would have been lower, thereby keeping those areas more rigid. By -2.6 Ga, parts of the lithosphere were sufficiently thick, cold, and dense to sink, and eventually a major cycle of cratonization took place. Successive periods of cratonization completed the conversion of archeocrust into continent and ocean basin (Fig. 7c).

OBSERVABLE CONSEQUENCES

The objective of this section is to describe by example how various aspects of the model can be related to the observable geology of shields. The examples are mostly from the Canadian Archean, particularly the Superior Province.

Volcanic-Plutonic Terrane. The tectonic model requires that the exposed surface of an Archean shield be a somewhat laterally compressed and vertically dilated version of the structures shown in Figures 2 and 3. Most of the diapiric motion of the final volcanic-plutonic cycle should have been completed. Only relics of very slow motions would have been frozen into the crust during cooling following cratonization. Thus, we make the following correlations:

In areas of relatively low metamorphic grade which preserve large greenstone belts (such as the Abitibi region of the Superior Province), the volcanics are a surface layer (Fig. 2.5) which did not become quite hot and soft enough to participate fully in diapiric overturn. They have been intruded in the plutonic stage of the cycle, and distorted and compressed by cratonization and by the shear flows.

Areas of somewhat higher metamorphic grade such as the Savant-Sturgeon Lakes belt of Western Ontario shown in Fig. 1c exhibit a higher ratio of granitoid-greenstone exposure and represent a level of exposure just at the volcanic-granitoid interface.

Higher grade areas of the shield (such as the Berens River Sub-province) should be representative of the middle part of the granitoid zone. Any very large blocks of greenstones that reached such a depth must have moved vertically at considerable speed and should not be found there now. Only smaller, slowly sinking, thin bands of volcanics could have been trapped there by the erosion and cooling that followed the cratonization uplift.

The granulite terranes of the shields are considered to represent the lowermost part of the granitoid zone, not the mafic granulites of the lower crust. Continued cooking of the tonalitic gneiss and gradual removal of water with pegmatitic fractions could have raised the deeper parts to the granulite facies. However, the gneiss observed in the field is not usually sufficiently mafic and dense to prevent the sinking of large amphibolitic masses through it. Thus, it is not a representative of the lower crust.

Age of Greenstones. It is difficult to determine directly when the mafic volcanics of the greenstone belts were formed. Radiometric ages have largely been determined by Rb-Sr and Pb-Pb whole-rock isochron methods on the associated plutonic volcanics. Geological inference then places limits on the age of the greenstones.

Most Archean formational age determinations are in the range of 2.6 to 2.9 Ga, but there are many cases of well-established older ages. Particularly notable are the 3.8 Ga gneisses from the Minnesota River Valley (Goldrich and Hedge, 1974), which contain amphibolitic material, the 3.75 Ga Greenlandic Amitsoq gneisses which enclose Isua metavolcanics and iron formations (Moorbath *et al.*, 1977), the related rocks of similar age in Labrador, and the 3.6 Ga Rhodesian "basement granite gneiss" (Hickman, 1974). Ages between 3.3 and 3.0 Ga have now been found in most of the well-explored cratons.

The observed pattern of formational ages is consistent with the tectonic model. Since each volcanic-plutonic cycle tends to obliterate evidence of earlier cycles, most of the volcanics preserved in a craton will date from the last cycle. And since cratonization requires a hot crust, the last volcanic cycle will normally have begun just prior to cratonization.

Nevertheless, traces of earlier volcanic cycles should be found. Volcanic remnants of preceding cycles may lie uncomformably at the base of the stratigraphic section built by the latest cycle. Small remnants of any age might be found as xenoliths in and between the tonalitic gneiss domes.

In the granitoid rocks, one might expect newly formed molten tonalite from the lower crust to rise high into the granitoid crust above older deformed tonalite. Thus, the most obvious domes of relatively massive granitoids should exhibit the youngest whole-rock ages, while older ages might be found in deeper, relatively high-grade gneiss or in structurally remobilized gneiss brought up from deeper levels, as long as the metamorphism or remobilization was not so intense as to reset the isotopic system.

This picture fits quite well with observations in the Superior Province, where most volcanic belts appear to have been formed between -2.85 and -2.70 Ga, but volcanics as old as 2.96 Ga have been found (Nunes and Thurston, 1978). Also most of the Rb-Sr isochrons on granitoids yield Kenoran ages, but zircons older than 3.0 Ga have been found in some of the gneisses (Krogh *et al.*, 1976).

Two situations favour the preservation of very old rocks: early cratonization, or a long quiescent period between rock formation and cratonization. The extremely old rocks of West Greenland and Labrador seem more reasonably interpreted as the latter situation, while the part of the Southern African craton of pre-Kenoran argon age is likely an example of the former.

Composition of Greenstones. The composition of Archean volcanic rocks has attracted much interest and is central to several theories for the origin of greenstone belts. Mafic rocks predominate, with widespread evidence for submarine crystallization. Silicic volcanics occur mainly as fragmentals. Volcanics of both tholeiitic and calc-alkaline type occur and two different associations can be identified.

In one, the rocks are generally mafic tholeiites which occur as laterally extensive flow series and often include ultramafic members (komatiites). This association resembles modern ocean ridge volcanics and will be referred to as the "primitive tholeiitic suite."

The second association resembles the volcanics of modern island arcs and will be referred to as "arc-like volcanics." The rocks may be tholeiitic or calc-alkaline, and compositions run from basalt to dacite (and rare rhyolite). Although more common than the primitive theoleiites, the arc-like volcanics are often less extensive laterally and vary much more in composition. Series of flows are often found which exhibit a bimodal mafic-felsic alternation, varying from mainly mafic at the base of the stratigraphic sequence to mainly felsic near the top. Larger greenstone belts usually contain more than one flow series (mafic to felsic cycle).

Greenstone belts in southern Africa and Australia commonly contain volcanics of both primitive tholeiitic and arc-like suites, with the primitive suite being older. In Canada the arc-like suite predominates and, where the primitive tholeiites are found, they may not predate all the arc-like volcanics.

The tectonic model presented here does not provide two different settings for primitive and arc-like volcanics. However, it can still provide an explanation for the two suites. Primitive tholeiitic volcanism would occur when quiescent crust (i.e., in the rest state of Fig. 2) was newly exposed to a mantle upwelling. Once the partial melt in the mantle had become interconnected, it would begin to fracture its way to the surface via dykes. Hydraulic fracturing (Anderson, 1951) is the fracture mechanism. Because of the large excess pressure generated in the top of a light magma connected vertically for tens of kilometres through a denser crust, penetration would be very rapid through the upper crust where the magma temperature considerably exceeds the melting temperature of most of the wall rocks. Thus, any major contamination is avoided. Repeated fractures would feed repeated sets of flows.

In contrast, arc-like mixed volcanism would occur on crust that had recently experienced one or more volcanic-plutonic cycles, and whose internal temperature was at the softening point or possibly even above the solidus at some levels. Since hydraulic fracturing requires that the host medium react to stress as an elastic solid, not as a viscous quasi-fluid, dyke injection would have been inhibited and melt transport would have taken place by diapiric and stoping mechanisms. Because these are much slower, contamination of the magma en route through the crust is likely. Also, temporary magma chambers where further differentiation can take place are likely to form at crustal levels in any slow transport process. The magma arriving at the surface could thus be quite variable in composition and form rocks of the arc-like suite. According to this scenario, primitive tholeiites should usually be situated at the bottom of a stratigraphic succession, but their occasional presence at other levels is also explicable.

Ultramafic Volcanism. Ultramafic volcanic rocks, particularly those high in magnesium, appear to have been formed only in Precambrian time, mainly in the Archean. Since a substantially higher temperature is required to form ultramafic magma than the usual basaltic partial melt, the disappearance of ultramafics from the younger geologic record is usually ascribed to a steady decline in mantle temperature. This is certainly in accord with the model developed here, but an additional factor may have operated.

It is proposed that asthenospheric convection changed from a mode in which the lithosphere acted as a rigid upper boundary to one in which the lithosphere acts as the asthenosphere's own thermal boundary layer. In the former case, the asthenosphere must have had its own thermal boundary layer in the shear flow beneath the lithosphere, and there would have been a steep superadiabatic temperature gradient in it. Since the base of the lithosphere is the level where the viscosity of mantle material changes from a "solid" to a "fluid" value, it has a fixed temperature. Consequently, the bulk of the mantle must have been substantially hotter when a fluid boundary layer existed below the lithosphere than when the lithosphere began to convect with the rest of the flow and itself was the boundary layer. A sudden decline in ultramafic volcanism is therefore to be expected when the changeover occurred.

Crustal Structure. A geophysical puzzle is how to correlate the mapped geology of cratons, which can often be extrapolated to a depth of several kilometres, with the quasi-layered structure of the crust which is found by most geophysical investigations – particularly seismic crustal soundings. According to the theory presented here, layering is a natural consequence of metamorphism and gravitational segregation, and it should predominate from the amphibolitic layer of the crust downwards (Fig. 3). Such layering would be interrupted by scars from the vertical movement of magmas, but the metamorphism should have obliterated most of the supracrustal history of the rocks.

Many seismic surveys in shields indicate a relatively uniform crustal P-wave velocity of less than 6.5 km/s to a depth of about 20 km, below which a very noticeable increase occurs. Also, the velocity profile often shows a marked rising gradient as the M discontinuity is approached. To explain strong reflections from it, the M discontinuity is frequently interpreted as having a complex structure, with velocity interlayering. Traceable interfaces below it are also often found.

These seismic observations can be correlated with the tectonic model as follows: At higher crustal levels, the P-wave velocity in the amphibolite zone and lowermost granitoids is prevented from rising with depth by high pore water pressure, a relict of the metamorphic dewatering process. The increase in velocity at about 20 km indicates the top of the mafic granulites of the lower crust. Near the base of the crust, the velocity rises in the zone where tonalite-ultramafite segregation last took place. This partial melt segregation may have produced the crude interlayering seen at the M discontinuity. Stratification in the uppermost mantle may reflect slight compositional discontinuities where ultramafic material was successively added to the mantle from the lower crust or underplated onto the lithosphere as the Earth cooled and the lithosphere thickened.

Structural Imprint of Cratonization. The main cratonization events would have thickened several large areas, probably well separated on the globe. The dimensions of these regions would have been set by the amount of delithified plate existing at the time. Taking into account more recent fragmentation and suturing, the thickened areas must roughly correspond to the present Archean cratons. Within a craton, the strain accumulated during cratonization would have been mainly two-dimensional, with the principal shortening being horizontal, the principal elongation vertical. The horizontal shortening would have corresponded approximately to the alignment of the original volcanic piles, since both were controlled by the mantle convective pattern of the time.

In the mesozone, the level typically now exposed, lateral compression would have taken place by plastic flow. However, compression would certainly have produced buckling and thrust faulting in the more rigid, near-surface materials now mostly lost by erosion. Differential motion of large, near-surface rigid fault blocks would have dragged plastic material into heterogeneous deformation patterns, as would any large-scale irregularities in the rheology at any level. Thus, a structural pattern of lateral compression by ductile strain, possibly including transcurrent ductile tear faults, should be expected at the present surface, superimposed on the diapiric trough and dome pattern of greenstone belts and granitoid gneisses that was developed during the evolution of the archeocrust.

The foregoing is consistent with what is found in Archean greenstone-granite terranes. For instance, in the Ontario part of the Superior Province the direction of regional shortening is NNW-SSE to N-S, approximately normal to the regional alignment of greenstone belts (e.g., Schwerdtner et al., 1979). A complementary set of transcurrent fractures correlates with this shortening, with the NE-SW set being more obviously developed in western Ontario. East-West cataclastic fault zones also occur in association with regional east-west geologic boundaries (e.g., Quetico fault). Major transcurrent ductile tears are exemplified by the Kapuskasing feature, the Winisk fault, and the Miniss Lake fault. Halls (1978) has suggested that the Matachewan dykes emplaced just as cratonization ended also reflect a N-S compressional event. He suggests that a similar dyke pattern exists in several of the cratons.

Stabilization of Cratons. Seismic studies give crustal thicknesses of Archean cratons in the range of 30 to 50 km and gravity studies indicate good regional isostatic compensation. The cratonic surfaces of shields and platforms generally have only low or moderate topographic relief, and regional elevations vary from about +1 to −3 km asl.

The present level of exposure is generally mesozonal or catazonal. In some of the high-grade terrane, the metamorphic mineralogy demands that pressures of 6 to 7 km were reached. It is therefore reasonable to suppose that a low-grade area such as the Abitibi belt has lost about 5 km to erosion while high-grade terrane like that of central New Quebec may have lost 15 km or more. To be consistent with such figures, cratonization in the tectonic model is envisaged as beginning with a crust perhaps 40 km deep, thickening it to 45 to 70 km by a lateral contraction of 10 to 40%, then eroding the upper 5 to 20 km away to leave a crust 35 to 45 km deep.

Erosion following the thickening and uplift substantially cooled the crust, in two ways. Firstly, the temperature profile gradually shifted downwards on the depth axis by the amount of the erosion. Secondly, the most highly radioactive materials were removed, reducing the steep, near-surface temperature gradient. Thus, substantial cooling took place in the upper crust in about 10 Ma and reached the deeper levels of the lithosphere in about 300 Ma. Once cooled, the crust became rigid, and the rigidity could only be destroyed by a return to former temperature levels. This would require: 1) a new cover of sediment or volcanics, 2) replacing the lost radioactivity (or increasing the cover to make up for the loss), and 3) returning to the higher mantle heat flow conditions that likely existed prior to cratonization. The second and third requirements would have become harder to fulfil as geologic time progressed and radioactivity and mantle convective activity declined.

Thus, the stability of cratons is viewed as an immunity to softening by subsequent geologic events rather than as an absence of processes which might, in other environments, have caused orogeny. This seems consistent with the observations. For instance, the remnant Nipigon and Nippissing plates still lying on the Superior craton are evidence that substantial volcanic material was emplaced on the craton subsequent to cratonization, but no softening or ingestion occurred. Also, the numerous diabase dyke swarms that cross most cratons presumably fed flows or sills on the surface, yet the undistorted structure of the dykes indicates that no plastic deformation took place in the craton.

Metasedimentary-Migmatite Areas. In the Superior Province and the Slave Province, there are extensive areas of migmatized metasediments which are somewhat atypical of the usual Archean volcanic-plutonic terrane. In the Slave Province, these metasediments have a widespread occurrence but, except for their great areal extent, are similar to the usual basinal deposits found at the top of many greenstone sequences. In the western Superior Province, these metasediments occur in two approximately east-west trending belts (Quetico and English River) in which the fold axes are all subparallel to the strike of the belts. This is in marked contrast to the volcanic-plutonic terrane in between, where strikes are much more variable and tend to conform to the pattern of granitoid domes.

According to the tectonic model, the erosion of a craton would provide a large quantity of sediment, much of it containing high proportions of K, U, and Th, which would find its way into shallow oceans overlying remaining archeocrust and into new deep ocean basins. The thermal effect of a thick sedimentary cover on archeocrust would be similar to that of a volcanic pile, although since the cover would not be more dense than its basement, diapirism would not take place immediately. Only if such a thick sequence of sediment were accumulated that metamorphic reactions converted the basal sediment to a dense mafic gneiss, would much new diapiric activity arise. However, a sufficient quantity of sediment could certainly delithify highly evolved archeocrust.

The unusually extensive metasedimentary terrane of the Slave Province may be an example of delithification by sediment. It is possible that cratonization proceeded there as a kind of wave. Uplift and lateral compression may have begun in an initial soft zone. The erosion products of the uplifted part were then deposited on the adjacent archeocrust, facilitating its final softening, compression, and uplift. Such a process may also explain the English River and Quetico metasedimentary belts in the Superior Province. There, local rifting of the crust may have occurred prior to the cratonization event during a cycle of volcanic emplacement (Fig. 6, left-hand side). The rift would have filled rapidly with sediment once cratonization of the marginal area began. Within about 30 Ma, the crust under the rift would have softened and, presuming the main cratonization stresses were still active, a linear metasedimentary fold belt would have been created.

Proterozoic Cratons. In marked contrast to the Archean Superior and Slave Provinces, the Churchill Province, which was cratonized at −1.8 Ga, exhibits a great variety in its geology. It contains greenstone belts formed in Archean and Proterozoic time, exemplified respectively by the Kaminak and Flin Flon-Snow Lake Subprovinces (Wanless and Eade, 1975; Bell, *et al.*, 1975) and also geosynclinal belts such as the circum-Ungava Geosyncline and Wollaston Fold Belt along with many other structures whose origin is not yet clear (Davidson, 1972).

According to the tectonic model, much primitive archeocrust was subducted during the Kenoran cratonization. Presumably, a volcanic arc analogous to a modern island arc would have formed on the new ocean floor or the remaining archeocrust wherever the subduction occurred. Also, the Kenoran cratons would have provided a source of sediment to build continental shelves not only along the oceanic margins of the cratons but also along the oceanic margins of remaining archeocrust. Such features would have been incorporated into the continental crust when the remaining archeocrust was cratonized.

Oxygen in the Atmosphere. The amount of oxygen in the Archean atmosphere is a highly controversial subject. Several authors (e.g., Cloud, 1973; Schidlowski, 1976) point to the photosynthetic origin of oxygen and suggest that the early atmosphere may have been reducing. Others (e.g., Kimberley and Dimroth, 1976) point to geologic evidence in Archean rocks that indicates oxygen was present. The questions raised cannot be settled here. However, the Kenoran cratonization greatly increased the area of dry land and shallow water on the Earth. This might well have had an important influence on the global amount and type of photosynthetic activity and therefore on the rate or course of evolution.

CONCLUDING REMARKS

In this paper I have described a speculative theory for the origin of continental crust. Models of Archean geotectonics abound. What can be claimed for this one is that not only can it go some distance towards explaining the known observational data, but its various elements can be knit together by a plausible physical argument. Much attention has been paid to the roles of gravity, density, temperature, and rheology because to explain orogeny in continental rocks one must go beyond the paradigm of plate tectonics and look into how parts of the lithosphere can lose their strength, allowing the rocks at crustal level to behave in a plastic or quasi-fluid manner.

The more novel elements of the theory are the following: 1) The emphasis on cratonization as a distinct process; 2) the correlation of cratonization and the formation of ocean basins with a change in the nature of the asthenosphere's thermal boundary layer; 3) the connection between radioactivity in crustal granitoid rocks and the orogenic activity which produced greenstone belts; 4) the suggestion that both the lower crust and the upper mantle under continental cratons are the residues of long-metamorphosed supracrustal rocks.

ACKNOWLEDGEMENTS

The research on which this speculation is based was supported by the National Research Council of Canada and by the Department of Energy, Mines and Resources (Geotraverse Project). Several months spent with Cominco Ltd. in 1973 provided the initial opportunity to consider such matters. H. C Halls ably criticized a draft manuscript and discussions with W. Schwerdtner were very helpful. Drafting was done ably by K. Khan. A great deal of editorial assistance was received from P. Ohlendorf. All this help is gratefully acknowledged.

REFERENCES

Anderson, E.M., 1951, The Dynamics of Faulting and Dyke Formation with applications to Britain (Second Edit.): Oliver and Boyd.

Anhaeusser, C.R., Mason, R., Viljoen, M.J. and Viljoen, R.P., 1969, A reappraisal of some aspects of Precambrian Shield geology: Geol. Soc. America Bull., v. 80, p. 2175-2200.

Anhaeusser, C.R., 1973, The evolution of the early Precambrain crust of southern Africa: Phil. Trans. Royal Soc. London, v. 273A, p. 359-388.

Baragar, W.R.A. and McGlynn, J.C., 1976, Early Archean basement in the Canadian Shield, a review of the evidence: Geol. Survey Canada Paper 76-14, 20 p.

Baragar, W.R.A. and McGlynn, J.C., 1978, On the basement of Canadian greenstone belts: Discussion: Geosci. Canada, v. 5, n. 1, p. 13-15.

Baer, A.J., 1977, Speculations on the origin of the lithosphere, Precambrian Research, v. 5, p. 249-260.

Bell, K., Blenkinsop, J., and Moore, J.M., 1975, Evidence for a Proterozoic greenstone belt from Snow Lake, Manitoba: Nature, v. 258, no. 5537, p. 698-701.

Burke, K., Dewey, J.F., and Kidd, W.S.F., 1976, Dominance of horizontal movements, arc and microcontinental collisions during the later perimobile regime: in Windley, B.F., ed., The Early History of the Earth: Wiley-Interscience, p. 113-130.

Carter, N.L., 1976, Steady state flow of rocks: Reviews of Geophysics and Space Physics, v. 19, p. 301-360.

Cloud, P.E., 1973, Paleoecological significance of the banded iron-formation: Econ. Geol., v. 68, p. 1135-1143.

Condie, K.C., 1978, Origin and early development of the earth's crust (Abstr.): in Smith, I.E.M. and Williams, J.G., eds., Proceedings of the 1978 Archean Geochemistry Conference: Dept. Geology, Univ. Toronto, p. 337-8.

Davidson, A., 1972, The Churchill Province: in Price, R.A. and Douglas, R.J.W., eds., Variations in Tectonic Styles in Canada: Geol. Assoc. Canada Special Paper 11, p. 381-434.

Franklin, J.M., 1978, Petrochemistry of the south Sturgeon Lake volcanic belt: in Smith, I.E.M. and Williams, J.G., eds., Proceedings of the 1978 Archean Geochemistry Conference: Dept. Geology, Univ. Toronto, p. 161-180.

Glikson, A.Y., 1976, Stratigraphy and evolution of primary and secondary greenstones: significance of data from shields of the southern hemisphere: in Windley, B.F., ed., The Early History of the Earth: Wiley-Interscience, p. 257-278.

Glikson, A.Y., 1978, On the basement of Canadian greenstone belts: Geosci. Canada, v. 5, n. 1, p. 3-12.

Goetze, C., 1971, High temperature rheology of Westerly granites: Jour. Geophys. Research, v. 76, p. 1223-1230.

Goldich, S.S. and Hedge, C.E., 1974, 3800 Myr. granitic gneiss in southwestern Minnesota: Nature, v. 252, p. 467-468.

Goodwin, A.M., 1977, Archean basin-craton complexes and the growth of Precambrian shields: Canadian Jour. Earth Sci., v. 12, p. 2737-2759.

Goodwin, A.M., 1978, Archean crust in the Superior Geotraverse Area: Geologic overview: in Smith, I.E.M. and Williams, J.G., eds., Proceedings of the 1978 Archean Geochemistry Conference: Dept. Geol., Univ. Toronto, p. 73-106.

Halls, H.C., 1978, The structural relationship between Archean granite-greenstone terrains and late Archean mafic dikes: Canadian Jour. Earth Sci., v. 15, n. 10, 1978, p. 1665-1668.

Hargraves, R.B., 1976, Precambrain geologic history: Science, v. 193, p. 363-371.

Hickman, M.H., 1974, 3500 Myr-old granite in southern Africa: Nature, v. 251, p. 295-296.

Holland, J.G. and Lambert, R. St. J., 1969, Structural regimes and metamorphic facies: Tectonophysics, v. 7, n. 3, p. 197-217.

Kaula, W.M., 1979, Thermal evolution of Earth and Moon growing by planetesimal impacts: Jour. Geophys. Research, v. 84, n. B3, p. 999-1008.

Kerrich, R. and I. Allison, 1978, Flow mechanisms in rocks: Geosci. Canada, v. 5, n. 3, p. 109-118.

Kimberley, M.M. and Dimroth, E., 1976, Basic similarity of Archean to subsequent atmospheric and hydrospheric compositions as evidence in the distributions of sedimentary carbon, sulphur, uranium and iron: in Windley, B.F., ed., The Early History of the Earth: Wiley-Interscinece, p. 579-585.

Krogh, T.E., Harris, N.B.W., and Davis, J.L., 1976, Archean rocks from the eastern Lac Seul region of the English River Gneiss Belt, northwestern Ontario, Part 2 – Geochronology: Canadian Jour. Earth Sci., v. 13, no. 9, p. 1212-1215.

Lachenbruch, A.H., 1970, Crustal temperature and heat production: Implications of the linear heat flow relation: Jour. Geophys. Research, v. 75, p. 3291-3300.

MacGregor, A.M., 1951, Some milestone in the Precambrian of southern Africa: Proc. Geol. Soc. South Africa, v. 54, p. 27-71.

Mareschal, J-C. and West, G.F., 1980, A model for Archean tectonism, Part 2: Numerical models of vertical tectonism in greenstone belts: Canadian Jour. Earth Sci., v. 17, n. 1, p. 60-71.

McCall, G.J.H., ed., 1977, The Archean: Search for the Beginning: Dowden, Hutchinson and Ross.

McKenzie, D. and Weiss, N., 1975, Speculations on the thermal and tectonic history of the Earth, Geophys. Jour., v. 42, no. 2, p. 131-174.

Moorbath, S., Allaart, J.H., Bridgwater, D., and McGregor, V.R., 1977, Rb-Sr ages of early Archean supracrustal rocks and Amîtsoq gneisses at Isua: Nature, v. 270, p. 43-45.

Nunes, P.D. and Thurston, P.C., 1978, Evolution of a single greenstone belt over 220 million years – a zircon study of the Uchi Lake area, Northwestern Ontario: in Short Papers of the Fourth International Conference, Geochronology Cosmochronology, Isotope Geology, 1978; United States Geol. Survey Open-file Report 78-701, p. 313-315.

Peltier, W.R., 1980, Mantle convection and viscosity: in Dziewonski, A., and Bosci, E., ed., Proceedings of the Enrico Fermi International School of Physics (Course LXXVII) Societa Italiana di Fisica: New York-London, Academic Press.

Price, R.A. and Douglas, R.J.W., eds., 1972, Variation in Tectonic Styles in Canada: Geol. Assoc. Canada Spec. Paper 11.

Ramberg, H., 1967, Gravity, Deformation and the Earth's Crust: Academic Press, New York.

Schidlowski, M., 1976, Archean atmosphere and evolution of the terrestrial oxygen budget: in Windley, B.F., ed., The Early History of the Earth: Wiley-Interscience, p. 525-535.

Schwerdtner, W.M., Stone, D., Osadetz, K., Morgan, J., and Stott, G.M., 1979, Granitoid complexes and the Archean tectonic record in the southern part of north-western Ontario, Canadian Jour. Earth Sci., v. 16, no. 10, p. 1965-1977.

Sharpe, H.N. and Peltier, W.R., 1978, Parameterized mantle convection: Geophys. Research Letters, v. 5, no. 9, p. 737-740.

Shegelski, R.J., 1976, Coarse clastic facies of the Savant Lake and Sturgeon Lake greenstone belts: in Geotraverse Conference, 1976: Dept. Geology, Univ. Toronto, p. 28.1-28.26.

Sutton, J. and Windley, B.F., eds., 1973, A Discussion of the evolution of the Precambrian Crust: Phil. Trans. Royal Soc. London, Series A, v. 273, no. 1235, p. 315-581.

Wanless, R.K. and Eade, K.E., 1975, Geochronology of Archean and Proterozoic rocks in the southern District of Keewatin: Canadian Jour Earth Sci., v. 12, p. 95-114.

West, G.F. and Mareschal, J-C., 1979, A model for Archean tectonism, Part I: the thermal conditions: Canadian Jour. Earth Sci., v. 16, p. 1942-1950.

Weertman, J. and Weertman, J.R., 1975, High temperature creep of rock and mantle viscosity: Ann. Review Earth and Planetary Sci., v. 3, p. 293-315.

Windley, B.F., 1976a, New tectonic models for the evolution of Archean continents: in Windley, B.F., ed., The Early History of the Earth, Wiley Interscience, p. 105-112.

Windley, B.F., ed., 1976b, The Early History of the Earth: New York, Wiley-Interscience.

Windley, B.F. and Bridgwater, D., 1971, The evolution of Archean low and high grade terrains: Geol. Soc. Australia, Spec. Publ. 3, p. 33-46.

Wilson, J.T., 1949a, The origin of continents and Precambrian history: Trans. Royal Soc. Canada, Ser. 3, v. 43, p. 157-184.

Wilson, J.T., 1949b, Some major structures of the Canadian Shield: Canadian Mining Metal. Bull., v. 42, no. 451, p. 547-554.

Young, G.M., 1976, Some aspects of the evolution of the Archean crust: Geosci. Canada, v. 5, p. 140-149.

The Continental Crust and Its Mineral Deposits, edited by D. W. Strangway, Geological Association of Canada Special Paper 20

MAJOR DIAPIRIC STRUCTURES IN THE SUPERIOR AND GRENVILLE PROVINCES OF THE CANADIAN SHIELD

W. M. Schwerdtner
Department of Geology,
University of Toronto,
Toronto, Ontario M5S 1A1
S.B. Lumbers
Department of Mineralogy and Geology,
Royal Ontario Museum,
Toronto, Ontario M5S 2C6

ABSTRACT

Massive granitoid batholiths are more akin to horizontal sheets than deep-rooted igneous stocks. Because such batholiths are generally lighter than their overburden, they will rise diapirically if the supracrustal and granitoid materials become ductile and if the bulk ductility contrast between supracrustal rocks and granitoid material is relatively low. Depending on the time interval between plutonism and diapirism, this low ductility contrast can be achieved by: 1) cooling of granitoid magma and heating of the host rocks or 2) joint heating of granitoid and supracrustal rocks during regional metamorphism. The first process, which seems to have been active in the Superior Province of Ontario, requires that the large mass of a batholith is accumulated piecemeal by separate intrusion of small volumes of high-ductility magma. Overall cooling of the magma and associated lowering of ductility will occur during this accumulation until large-scale diapirism can commence. At the same time, the adjacent supracrustal rocks will be heated and become more ductile. The second process seems to be responsible for the tectonic deformation of many granitoid plutons in the Grenville Province of Ontario as well as the tight and recumbent folding of the enveloping supracrustal rocks. Unlike the hinge zones of typical buckle folds and related structures, the crestal regions of gneiss diapirs are characterized by subhorizontal flattening foliations and stretching lineations subparallel to dip. Bulk densities calculated from 678 handspecimens of plutonic and supracrustal rocks of the Grenville Province suggest that during diapirism, high strain levels caused major changes in the densities of the plutonic rocks. The density changes suggest important volume changes with concomitant mobility of material during diapirism, which could prove important in metamorphic studies and in concentration of mineral deposits.

RÉSUMÉ

Les batholites granitoïdes massifs sont plus susceptibles de former des couches horizontales que les stocks ignés aux racines profondes. Parce que de tels batholites sont généralement plus légers que les roches sus-jacentes, ils tendent à s'élever en diapirs si les roches superficielles et les matériaux granitoïdes deviennent ductiles et si le contraste global de ductilité entre les roches superficielles et les matériaux granitoïdes est relativement faible. Selon l'intervalle de temps écoulé entre le plutonisme et le diapirisme, ce faible contraste de ductilité peut se produire (1) par refroidissement du magma granitoïde et le réchauffement des roches encaissantes ou (2) par le réchauffement simultané des roches granitoïdes et superficielles durant le métamorphisme régional. Le premier processus, qui semble avoir été actif dans la province du Lac Supérieur en Ontario, requiert que la grande masse d'un batholite s'accumule morceau par morceau par l'intrusion séparée de petits volumes de magma très ductile. Le refroidissement global du magma et l'abaissement de la ductilité se produira durant cette accumulation jusqu'à ce que le diapirisme à grande échelle puisse débuter. Au même moment, les roches superficielles adjacentes se réchaufferont et deviendront plus ductiles. Le second processus semble responsable de la déformation tectonique de plusieurs plutons granitoïdes de la province de Grenville en Ontario aussi bien que des plis serrés et couchés des roches superficielles enveloppantes. A l'inverse des zones de charnière des plis bouclés typiques et des structures qui leur sont associées, les régions de crête des diapirs de gneiss se caractérisent par des foliations subhorizontales d'aplatissement et des linéations d'étirement subparallèles au pendage. Les densités globales calculées sur 678 échantillons de roches plutoniques et superficielles de la province de Grenville suggèrent que durant le diapirisme, les niveaux élevés de déformation ont causé des changements majeurs dans les densités des roches plutoniques; ce changement de densité des roches implique des changements de volume importants avec accroissement de la mobilité du matériel durant le diapirisme ce qui pourrait s'avérer important dans les études de métamorphisme et dans la concentration de dépôts de minéraux.

INTRODUCTION

Diapirism was originally defined as a process in which geological materials rise vertically from deeper levels of the Earth's crust, thereby piercing the overlying rocks and/or unconsolidated sediments (O'Brien, 1968, p. 2). Taken literally, *diapir* is synonymous with piercement structure.

This original definition is very broad and includes all discordant igneous intrusions (O'Brien, 1968). Two important elements are absent from the definition: 1) a specification of the original geometry of the diapiric medium, and 2) a provision for the vertical rise of buoyant material without piercement of the overburden. If the overburden is highly ductile, it will be stretched rather than pierced (Ramberg, 1967).

In this paper, the term diapir will be restricted to structures which, like evaporite domes, develop from subhorizontal tabular or sheet-like masses. Diapirs which do not pierce the overburden will be called concordant, whereas those which do, will be termed discordant. Both types can be either mature, i.e., mushroom-shaped, or immature, i.e., ridge-shaped or domical. Buoyant masses whose geometry at the onset of ascent does not correspond to subhorizontal layers or tabulae will be called intrusions. There are intrusive sheets, stocks, dykes, etc., all of which can but need not, be igneous. Good examples are the crescentic granitoid sheets (Schwerdtner, 1976a, 1978; Sutcliffe, 1977, 1978) which occur in greenstone terranes of the Superior Province of Ontario.

Within the Canadian Shield, particularly the Superior and Grenville Provinces, large areas of gneissic, predominantly granitoid, plutonic rocks abound and have been variously interpreted as basement complexes to intermingled supracrustal sequences, or primordial crust, mainly because of the highly metamorphosed state of the plutonic rocks. Our investigation of these gneissic granitoid terranes suggests that the granitoid material was metamorphosed while rising diapirically toward the supracrustal cover because the plutonic rocks are lighter than the supracrustal rocks and both rock groups became ductile with a low bulk ductility contrast. High ductility and low bulk ductility contrast can be achieved either during cooling of the granitoid magma and concomitant heating of the host supracrustal rocks, or by joint heating of granitoid and supracrustal rocks during regional metamorphism. Thus, none of the gneissic granitoid terranes represent primordial crust, but some may represent older basement to supracrustal cover rocks if regional metamorphism was the contributing factor in achieving low ductility contrast.

Because diapirism appears to be an essential factor in the tectonic and metamorphic history of the Canadian Shield, we will first briefly review the main concepts and principles involved in describing and recognizing diapirs, and then illustrate these by two examples, one from the western Superior Province where diapirism was apparently achieved by cooling of granitoid magma, and the other from the Grenville Province where diapirism occurred during regional metamorphism and involved the deformation of a regional unconformity.

RECOGNITION OF DIAPIRS COMPOSED OF METAPLUTONIC GRANITOID ROCKS

General Principles

The following criteria have been used to identify large diapiric structures: 1) An appropriate density contrast between the apparently diapiric material and the mantling rocks; 2) circular to oval outlines of apparent diapirs in map view; and 3) subhorizontal stretching fabrics in the crestal region of domes and antiforms (Schwerdtner et al., 1978a and 1979).

With some important modifications to be explained in the subsequent section, these principal criteria will be used throughout the present paper.

Given an original state of inverted density stratification with appropriate thickness ratio (Ramberg, 1967), a diapiric rise of light granitoid material cannot be avoided if its ductility is high and the overlying supracrustal rocks are not too strong. Therefore, criterion 1 should be modified to take account of the requirement that, besides an appropriate density contrast, there must be no significant difference in ductility between the diapiric material and the overlying rocks (Ramberg, 1970b). Such low ductility contrasts lead to similar magnitudes of coeval finite strain in granitoid diapirs and mantling rocks.

Unfortunately, the original density contrast can no longer be measured but must be deduced from petrologic evidence. In addition, the contemporaneity of finite strains in juxtaposed rock assemblages cannot always be demonstrated. These difficulties are somewhat lessened if the apparent diapirs are composed of metaplutonic granitoid gneisses whose parents were originally emplaced into a metavolcanic or metasedimentary overburden which in turn became the supracrustal mantle of the present granitoid structures. Given these conditions, the present density of the granitoid rocks represents a maximum value for the original density of the diapiric material. Moreover, both the granitoid and supracrustal rocks must have undergone at least one coeval ductile deformation.

Criterion 2 does not apply where the original supracrustal overburden had lateral variations in thickness, density and/or ductility (Ramberg, 1963a, 1967; Stephansson, 1972; Schwerdtner et al., 1978a; Schwerdtner and Troëng, 1978). In such cases, the specific non-circular outlines of first-order diapirs can be predicted from the original structure and lithology of the overburden. As will be discussed below, the horizontal outline of "parasitic" diapirs within multi-order structures becomes more nearly circular with increasing order. This geometric effect is highly valuable in identifying, major diapiric structures on geological maps.

The general reasoning behind criterion 3 has been presented by Schwerdtner et al. (1978a). Recent model studies have shown that both mature and immature concordant diapirs have quasi-concordant neutral surfaces above which the total horizontal strain is tensile (Fig. 1b). The magnitudes of total horizontal extension are relatively small in immature single-order diapirs (Dixon, 1974; Schwerdtner et al., 1978a), but become progressively larger as the diapirs mature and assume the shape of broad mushrooms (Dixon, 1975). These generalizations do not strictly apply to multi-order diapirs, in which the crestal region of immature higher-order domes are already characterized by relatively high magnitudes of total strain.

In contrast to realistic buckle folds (Fig. 1a), diapiric structures have total-strain patterns which are markedly different from those of interdiapiric depressions (Fig. 1b). Accordingly, the flattening foliation is sub-horizontal in the crestal region of domes and subvertical in the interdiapiric depressions. This difference in foliation attitude can also be accounted for by non-diapiric models, all of which are complex and involve repeated tectonism with passive superposition of finite strains and associated mineral fabrics (Figs. 1c and 1d).

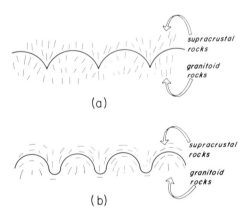

Figure 1. Schematic model of producing a contrast in foliation attitude between domical antiforms and synforms; (a) buckling of a batholithic roof at low ductility contrast, (b) diapirism, (c

To justify such complex models for granitoid domes of a given region, it must be demonstrated that both the granitoid rocks and supracrustal units possessed a flattening foliation prior to doming. Such foliation appears to have been absent in the Lake of the Woods region of the western Superior Province in northwestern Ontario (Schwerdtner *et al.*, 1979).

A fruitful investigation of dome structures in metamorphic terranes is impossible unless two important aspects of diapirism are considered: 1) the development of multi-order structures and 2) the growth of inclined domes and diapiric nappes. A general discussion of both phenomena follows.

Diapirs of Different Orders

By analogy with conventional folds (Ramsay, 1967, p. 354), different orders of diapirs are apt to form in a given low-density member. According to Ramsay's (1967) definition, structures of different orders need not evolve simultaneously. For example, during the buckling of multilayered sequences, higher-order folds develop prior to lower-order folds (Ramberg, 1963b, 1970a).

Using a theoretical model, Ramberg (1968) predicted that diapirs of different orders will grow simultaneously in Newtonian media. This prediction has not been borne out in model experiments with soft materials, where second-order diapirs have dome shapes and form invariably after the first-order diapirs. Moreover, the development of second-order diapirs requires that the first-order structures be ridge-

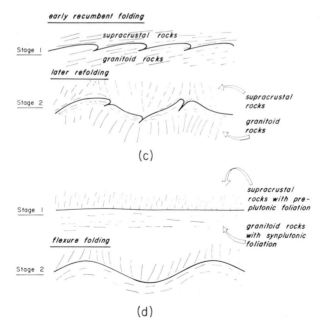

(c)

(d)

and d) repeated deformation. The refolding (c) is by buckling at low ductility contrast. Foliation is normal to the local direction of maximum total shortening.

shaped (Ramberg, 1967, Figs. 11, 13 and 17; Stephansson, 1972, Schwerdtner and Troëng, 1978).

Geological examples of second-order diapirs that emerge from diapiric ridges are found in many evaporite structures (Ahlborn and Richter-Bernburg, 1955; von Trusheim, 1960; Gould and DeMille, 1964; Schwerdtner and Morrison, 1974). But second-order diapirs appear to be equally common in oval salt domes (Kupfer, 1963; Schwerdtner and Clark, 1967), a phenomenon which cannot be explained on the basis of Ramberg's (1967) model experiments.

In gneissic granitoid terrane, more than two orders of diapirs are generally encountered. Such large structures will be called multi-order diapirs. Because first-order structures are usually difficult to recognize, smaller scale structures will be termed higher-order diapirs.

Nappe Structures and Inclined Domes of Metaplutonic Granitoid Rock

Stephansson (1972) and Talbot (1974) showed experimentally that incipient diapirs grow approximately normal to the original boundaries of the low-density member which contains the diapiric source material. Accordingly, the axis of diapiric domes can be markedly inclined, especially if higher-order diapirs emerge from sloping roofs of gentle lower-order structures.

The same experimenters showed that diapiric nappes can develop upon maturation of inclined immature diapirs (Stephansson, 1972, Fig. 23; Talbot, 1974, Figs.

Figure 2. Layered tonalite-granodiorite, eastern Irene-Eltrut Lakes granitoid complex (Fig. 9). Note shallow dip of foliation.

1-3). A decade earlier, Ramberg (1963a, 1967) had produced bilateral diapiric nappes of great complexity in centrifuged models and interpreted orogenic nappes in the Scandinavian Caledonides and elsewhere as diapiric structures.

As illustrated by Schwerdtner et al. (1978a), differentiation between inclined immature diapirs and inclined nappe structures can be difficult in regions of low topographic relief. In such regions, the inclination of diapirs is generally revealed by the dips of planar structures in the supracrustal envelope. If the diapirs are immature, then it is possible to apply the concepts developed in the previous section. As shown in the example of the Archean Ash Bay dome (Schwerdtner et al., 1978a), the centre of the exposed horizontal section through the crestal region of a moderately inclined diapir is defined by horizontal foliations. By contrast, the foliations in the supra-crustal envelope have a consistent dip in spite of internal folding (Blackburn, 1973 and 1976; Edwards and Lorsong, 1976; Edwards and Sutcliffe, 1978).

As illustrated by cumulative-strain models of upright mature diapirs (Dixon, 1975), an oblique section through the nappe-like crestal region will encounter folia-tion patterns which are similar to those of inclined immature diapirs. Because of a tendency to develop shallow attitudes of flattening foliations in the supracrustal rocks near large diapiric nappes, it can be difficult to distinguish mature diapirs from non-diapiric structures (cf. Fig. 1). If the rocks within and between high-order domes belong to the same magmatic suite and are part of the same batholith, then none of the flattening foliations can be preplutonic (Fig. 1d) and the diapiric origin of the domes may be established more readily.

Figure 3. Layered tonalite-granodiorite cut by various dykes, same general area in Figure 2.

CONCORDANT GNEISS DIAPIRS IN THE SOUTHERN SUPERIOR PROVINCE, NORTHWESTERN ONTARIO

The west-central sheet of the Ontario Geological Map (Ayres *et al.*, 1972) is dominated by the colour contrast between a network of green units (metavolcanics) and a pink sea of granitoid rocks. Grey metasedimentary units abound in two major zones, the Quetico Belt and the northern half of the English River Belt (Ayres *et al.*, 1972; Ayres, 1978). Between the metasedimentary zones lies a complexly structured domain of metavolcanic and granitoid rocks. This large domain, composed of the Wabigoon Belt and the southern half of the English River Belt (Ayres *et al.*, 1972; Ayres, 1978; Goodwin *et al.*, 1972) contains most of the concordant gneiss diapirs identified to date in the Archean terrane of northwestern Ontario (Schwerdtner *et al.*, 1978b). These major structures comprise various types of gneissic tonalite-granodiorite of the early plutonic suite (Schwerdtner *et al.*, 1979).

Figure 4. Dyke of gneissic tonalite-granodiorite with augen texture within a large greenstone enclave. Highway 17, 30 miles southeast of Ignace.

Because of their composite nature and the magmatic origin of the gneissic tonalite-granodiorite, the major diapirs have been called "granitoid complexes" (Sage *et al.*, 1975; Schwerdtner *et al.*, 1978a, 1978b, 1979). Some granitoid complexes are largely composed of a late plutonic suite of massive to foliated postdiapiric rocks ranging in composition from diorite to quartz monzonite, and have been described as batholiths (Ziehlke, 1973, 1974; Ayres, 1974).

Structural Features of the Gneissic Tonalite-Granodiorite

The evidence for a plutonic origin of the gneissic tonalite-granodiorite has been summarized by Ayres (1974, 1978) and Schwerdtner *et al.* (1978b, 1979). This evidence consists of: 1) igneous composition, 2) uniformity of composition of relatively large areas, 3) pretectonic fragmentation of mafic tonalitic phases by granodiorite or trondhjemite, 4) medium-grained dykes of gneissic tonalite-granodiorite within

Figure 5. Laminated tonalite-granodiorite, same location as Figure 2.

coarser grained gneissic tonalite-granodiorite of identical composition and texture, 5) relict igneous textures in the least deformed rocks and 6) the presence of euhedral to subhedral zircon characteristic of magmatic rocks.

On the basis of planar mineral fabrics and minor structures, three principal varieties of gneissic tonalite-granodiorite can be distinguished in the field: 1) the layered variety, 2) the laminated variety, and 3) the augen variety or clotty to polygonal wire-mesh variety. Transitions between these varieties are common in most areas. The rocks of all varieties can exhibit discordant veins of quartz-feldspar and postdiapiric dykes of massive diorite-quartzmonzonite (Figs. 3 and 8).

The layers of variety 1 are concordant veins (Fig. 2) or narrow supracrustal relicts (Fig. 3). Depending upon the grain size of feldspars and the total amount of mafic minerals, the foliated rocks of variety 3 exhibit feldspar augen (Fig. 4) or strained, clotty to polygonal wire-mesh textures of the mafic minerals. This variety

Figure 6. Massive quartz monzonite with a distorted xenolith of gneissic tonalite-granodiorite containing a melanocratic dyke. Eastern boundary of White Otter Batholith (Fig. 9).

grades into variety 2 with increasing magnitude of deformation (Schwerdtner and Mason, 1978). Variety 2 grades into variety 1 with increasing width and number of quartzo-feldspathic veinlets (Figs. 5 and 8).

The first author has found it difficult to map the three varieties on a scale of 1:15,000. Changes between different varieties occur on the scale of large outcrops, and the gradual transition between varieties hampers the mapping. In the east half of the Irene-Eltrut Lakes Complex (Schwerdtner *et al.*, 1979), there are many small bodies of early granodiorite or trondhjemite whose border zones involve crudely layered rocks that are highly strained. This type of layering may be partly due to plutonism, but cannot be easily separated from that of variety 1.

Minor, late-tectonic shear zones are common in the gneissic tonalite-granodiorite (Schwerdtner *et al.*, 1979). Together with minor folds, these shear zones contribute substantially to the complex structure of the early granitoid rocks. The intrusion of postdiapiric granitoid plutons has led to further distortion of the gneissosity (Fig. 6).

Contacts between Gneissic Tonalite-Granodiorite and Supracrustal Rocks

The large-scale interface between gneissic tonalite-granodiorite and adjacent supracrustal rocks is a preferred site for later granitoid plutons (Ayres, 1974; Schwerdtner *et al.*, 1978a, 1978b). Even where such postdiapiric plutons are absent, the original contact relationships have generally been modified by the tectonic deformation incurred during regional diapirism and metamorphism of the original tonalite-granodiorite (Schwerdtner *et al.*, 1978a, 1978b).

The syndiapiric ductility of the gneissic tonalite-granodiorite was similar to that of the supracrustal rocks. When deformation magnitudes are not extreme, unequivocal evidence of large synplutonic ductility contrasts is commonly preserved. This evidence consists of 1) quasi-concordant dykes and sheets of gneissic tonalite within greenstone and 2) angular xenoliths of greenstone within gneissic tonalite-granodiorite.

Because the deformation magnitudes are generally high near the boundary of greenstone belts, a distinction between concordant sheets and originally discordant dykes is often difficult. Flattened sheets of gneissic tonalite-granodiorite range in thickness from a few centimetres to several hundred metres. Good examples of thick sheets can be found along the southern boundary of the Phyllis Lake greenstone belt in the central Wabigoon Subprovince (Schwerdtner, 1976b). Here the total compressive strain due to diapiric deformation is over 50 per cent.

Deformed narrow sheets of veined tonalite-granodiorite within greenstone can be observed in the roadcuts of Highway 17, southeast of Ignace (Schwerdtner and Mason, 1978) and on Gulliver Lake (Schwerdtner *et al.*, 1978, Fig. 7). The sheets have sharp contacts, contain small xenoliths and display strained igneous textures (Fig. 4).

Near the boundaries of some greenstone belts, the gneissic tonalite-granodiorite contains numerous layers of amphibolite and hornblende-biotite gneiss, together with hornblende-rich mafic laminae (Fig. 3). These planar mafic elements appear to be remnants of supracrustal rocks, and the intervening tonalitic-granodioritic material has been injected or produced in situ. In either case, the gneissic tonalite-

Figure 7. Angular xenolith of mafic metavolcanics within laminated tonalite-granodiorite, same location as Figure 4.

Figure 8. Swarm of small mafic xenoliths in laminated tonalite-granodiorite, same location as Figure 4.

granodiorite would be younger than the adjacent supracrustal rocks of the greenstone belt.

Xenoliths of supracrustal rocks are common near greenstone belts or large greenstone enclaves within gneissic tonalite-granodiorite. In spite of large tectonic strain, the original angularity of greenstone xenoliths is locally preserved (Fig. 7). The xenoliths have metamorphic textures and structures which differ from those of metamorphic dykes.

Swarms of elongate xenoliths (Fig. 8) are typical of many areas in the central Wabigoon Belt that are rich in large greenstone enclaves (Schwerdtner, 1976b). These xenolith swarms have little affinity to boudinage structures, but are akin to tectonic breccias formed under conditions of high ductility contrast and chemical mobility in the evaporite domes of the Canadian Arctic Islands (Schwerdtner and Byers, 1965).

Gneiss Diapirs of Different Order

Each major gneiss diapir is composed of several higher-order diapirs which complicate the structural picture and hamper the recognition of major domes (Fig. 9). The higher-order dome structure is, in turn, obscured by numerous post-diapiric plutons of diorite-quartz monzonite, which cut the elliptical pattern of gneissosity or rim individual domes (Schwerdtner *et al.*, 1978a, 1978b, 1979).

The oval structure of some higher-order diapirs is discernible on topographic maps or aerial photographs. On this scale, compelling evidence of diapirism is furnished by the presence of low-density rocks (trondhjemite and leuco-granodiorite) with subhorizontal flattening foliations at or near the centre of lithologically-zoned, elliptical structures. Mineral lineations are also indicative of diapirism but so far have only been worked out for one higher-order diapir, the Ash Bay dome (Schwerdtner *et al.*, 1978a).

The higher-order diapirs are partly surrounded by arcuate greenstone arms or narrow amphibolite-rich septa which are connected with the main envelope of the major diapirs (Fig. 9, and Schwerdtner *et al.*, 1979). The inner gaps (Fig. 9) of the greenstone septa attest to the domical character of the major structure and the antiformal curvature of the enveloping surface to the higher-order diapirs. As revealed by the attitudes of foliation at the gneiss-greenstone interface, some higher-order diapirs are inclined or have nappe-like asymmetric shapes (Fig. 9).

Owing to the large diameter of individual major domes, which is commonly two or three times as great as the thickness of the present granitoid crust, the amplitude of the antiformal enveloping surface must be relatively small. The diapiric origin of these crudely elliptical domes is supported by several lines of indirect evidence. Firstly, large magnitudes of horizontal extension occur in the crestal region of the higher-order diapirs. These magnitudes are far too high for the crestal region of single-order immature diapirs (Schwerdtner *et al.*, 1978a), but seem appropriate for multi-order diapirism involving a coaxial superposition of finite strains. Secondly, the major low-amplitude domes are more elongate in an east-west direction than most of the oval higher-order diapirs. In addition, the major domes are commonly connected by narrow zones of gneissic tonalite-granodiorite, and resemble elliptical

Figure 9. Structure and lithology of the Irene-Eltrut Lakes granitoid complex and its supra-crustal envelope. Crosses indicate horizontal foliation or gneissosity (modified from Schwerdtner *et al*. 1979, Fig. 2).

second-order diapirs that coalesce to form first-order diapiric ridges (von Trusheim, 1960). The proportionality between size of structure and degree of relative elongation is typical for multi-order diapirs (cf. von Trusheim, 1960 and Ramberg, 1967, p. 55).

Owing to the low amplitude of the major diapirs, the minimal width of the original sheet-like batholiths which spawned the gneiss diapirs is only slightly greater than that of the diapirs themselves. Apparently, there were two major elongate batholiths located north-northwest and southeast of the Kenora-Savant Lake greenstone mass, Wabigoon Belt (Ayres et al., 1972).

To date, only two major gneiss diapirs have been mapped in detailed reconnaissance fashion: the Rainy Lake diapir (Schwerdtner et al., 1978a, 1978b) and the Irene-Eltrut Lakes diapir (Fig. 9). Parts of these and other diapirs have been studied in sufficient detail, but no first-order diapir has been analyzed entirely.

Some postdiapiric plutons associated with the gneiss diapirs were emplaced in a brittle fashion, and failed to deform the structural patterns in the gneissic tonalite-granodiorite (Schwerdtner et al., 1978a, 1979). Other plutons appear to have modified the original shape of second-order diapirs, but this general problem is still under investigation.

Evolution of the Gneiss Diapirs

Whereas the diapiric nature of the gneiss domes can be established by means of deformed mineral fabrics, their tectonic environment must be reconstructed on the basis of general petrologic principles. As pointed out by Ramberg (1967, 1970b), large concordant diapirs will not develop unless the ductility contrast between the original diapiric medium and its overburden is relatively low. Such low contrasts could be obtained during high-rank regional metamorphism. But as recently emphasized by Ayres (1978), the metamorphic zoning of typical greenstone belts in northwestern Ontario is not compatible with regional metamorphism of higher rank.

It appears that the low ductility contrast required for the diapirism was obtained by heating of the wall rocks and concomitant cooling of granitoid magma while large batholiths were being accumulated by piecemeal intrusion. The individual intrusive phases (magma) had high ductility, but their mass was too small to pierce the cold roof rocks of the incipient batholiths. These roof rocks seem to have acted as a large-scale barrier for the ascending granitoid magma until the mass of granitoid material was sufficiently large, and the ductility contrast between overburden and batholith sufficiently small. While the temperature and ductility of the overburden were increasing, the temperature and ductility of the granitoid material were decreasing.

Although this hypothesis of late-magmatic diapirism needs to be tested by quantitative geophysical modelling, it is strongly supported by negative geological evidence. Unless it can be shown that the common melanocratic dykes in the gneissic tonalite-granodiorite (Schwerdtner et al., 1979) are actually metagabbroic feeders rather than plutons of contaminated tonalite-granodiorite, there is no sound basis for invoking an early metavolcanic assemblage which was regionally metamorphosed together with the original tonalite-granodiorite before most of the present metavolcanics were laid down. Such an earlier generation of metavolcanics, unconformably overlain by a thick supracrustal assemblage, would now occur at the margin of large greenstone belts adjacent to the gneiss diapirs – a structural site at which no rocks have been dated so far.

THE ALGONQUIN BATHOLITH: A MAJOR GNEISS DOME FORMING THE CENTRAL PART OF THE GRENVILLE PROVINCE OF ONTARIO

Detailed and reconnaissance mapping of the Grenville Province of Ontario (Lumbers, 1971, 1976, in press) has revealed a major gneiss dome, the Algonquin batholith, which dominates the central part of the Province (Fig. 10). The batholith is incompletely mapped, and much more petrological and structural data are required to fully understand its history. Nevertheless, sufficient structural and density data have been obtained to show that the batholith became diapiric during late high-rank regional metamorphism. Moreover, a coarse clastic sequence, marking the base of a Late Precambrian supracrustal accumulation in the Grenville Province, was deformed by higher-order diapirism and was infolded into the batholith. This example is therefore more complex than that given above for the Superior Province, but it does provide an excellent illustration of the second major mechanism leading to diapiric activity, regional metamorphism. The batholith will be considered first in the context of its regional setting, and then from the viewpoint of its internal structures, lithology, and density relative to the enclosing supracrustal rocks. Because the diapirism was triggered by regional metamorphism, knowledge of the regional setting of the batholith is essential and provides data of a general nature concerning ductility levels required for the diapiric rise of the batholith.

The geologic setting of the Algonquin batholith is shown in Figure 10 and briefly described below; additional details are given elsewhere (Lumbers, 1978, in press). The batholith separates two major supracrustal accumulations of contrasting age and lithology. The older of the two accumulations, deposited between 2.5 and 1.8 Ga ago during Middle Precambrian time, is confined to the northern two-thirds of the Grenville Province. The base of the accumulation, marked by a coarse clastic sequence derived from submarine fan deposits (Lumbers, 1978), rests unconformably upon Archean rocks that extend into the Grenville Province from the Superior Province across the Grenville Front Boundary Fault, which is the northern boundary of the Grenville Province (Lumbers, 1978). The remainder of the accumulation consists of greywacke-argillite deposited as turbidites in deep water below wave-base (Lumbers, 1978) and shallow- to deep-water deposits derived from impure sandstone, shale, arkose, subarkose, orthoquartzite, and rare iron formation and marble (Lumbers, 1978, in press). The shallow- to deep-water deposits may represent the youngest part of the accumulation (Lumbers, in press), and were intruded by the Algonquin batholith.

The youngest accumulation, deposited between 1.5 and 1.25 Ga ago during Late Precambrian time (Lumbers, in press), includes rocks commonly referred to as the "Grenville Supergroup". The accumulation dominates the southern third of the Grenville Province (Fig. 10), and rests unconformably upon rocks of the Algonquin batholith (Lumbers, in press) A coarse clastic sequence marks the base of the accumulation and was deposited upon the southern and at least part of the western

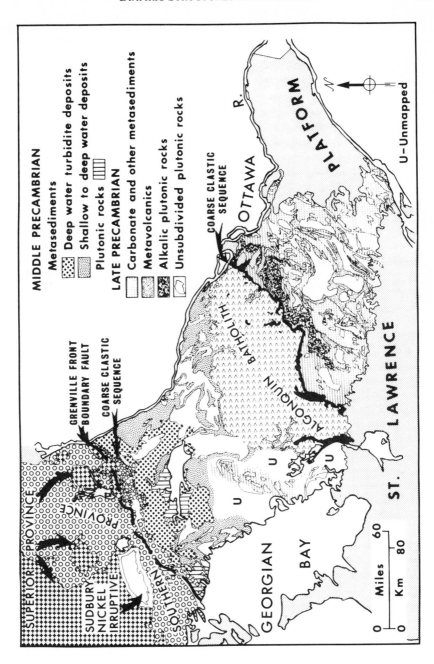

Figure 10. Sketch map showing some major geological features of the Grenville Province of Ontario. "U" denotes portions that are unmapped.

flanks of the batholith; small outliers of the sequence are also infolded within the batholith. Rocks of the sequence are derived largely from basal arkose, micaceous arkose, and arkosic conglomerate; local intercalations of siliceous calcitic marble, calcareous and siliceous shaley metasediments, and quartzose metasandstone are common (Lumbers, in press). The sequence undergoes a facies change southeastward across strike into carbonate rocks that dominate the youngest accumulation and reflect a major carbonate marine basin that can be traced southeastward across strike for over 240 km into the Adirondack region of New York State. Subordinate clastic siliceous metasediments and calcareous shaley metasediments are commonly intercalated with the carbonate rocks, and major volcanism occurred within the northwestern part of the basin (northwest of the Frontenac axis in Ontario) about 1.3 Ga ago (Lumbers, 1967, in press). Because no basement rocks have been unequivocally identified within the carbonate basin, the Algonquin batholith was apparently emplaced near the margin of a Middle Precambrian continental mass. Geochronological data indicate that the batholith was emplaced between 1.5 and 1.4 Ga ago (Lumbers, in press); it may have been part of a major magmatic event (Silver et al., 1977), extending from Labrador to southern California, which occurred during the same time. The oldest plutonic rocks identified within the carbonate basin are about 1250 Ma old; this age provides an upper time limit for the deposition of the accumulation. A second basin of supracrustal rocks, possibly contemporaneous with the rocks of the carbonate basin, extends northward from southeastern Georgian Bay to just south of Lake Nipissing (Lumbers, in press).

Most of the plutonic rocks that cut the older accumulation belong to the anorthosite suite and range in age from about 1.5 to 1.1 Ga (Lumbers, 1975; Lumbers and Krogh, 1977). The only older plutonic bodies consist of quartz monzonite and minor granodiorite about 1.7 Ga old, which were emplaced within the Grenville Front Tectonic Zone and the Middle Precambrian metasediments adjacent to the Tectonic Zone (Fig. 10; Lumbers, 1978). Between 1.3 and 1.0 Ga ago, a large variety of plutonic rocks were emplaced within the carbonate basin southeast of the Algonquin batholith (Lumbers, in press).

Nearly all the supracrustal and plutonic rocks were subjected to a high-rank regional metamorphism which culminated between about 1.3 and 1.0 Ga in the northern part of the Grenville Province and about 1.1 Ga in the carbonate basin (Lumbers, 1967, 1978, in press). This metamorphism converted most of the rocks into coarsely recrystallized and deformed gneisses under temperature and pressure conditions of the middle to upper almandine amphibolite facies. Commonly referred to as the "Grenvillian Orogeny", the metamorphism affected the entire Grenville Province of the Canadian Shield, allowing this province to be recognized as a unique, primary subdivision of the Canadian Shield. Within the Late Precambrian carbonate basin, only that region in which the exposed volcanic rocks are thickest has escaped the high-rank metamorphism (Lumbers, 1967). Mere vestiges of regional metamorphic events older than the high-rank event are present in both the older and younger supracrustal accumulations, but the older events were less intense than the high-rank metamorphism (Lumbers, 1978, in press). A few mafic and felsic igneous stocks scattered throughout the Grenville Province, and a large variety of dykes and alkalic rock-carbonatite complexes associated with the Ottawa-Bonnechere Graben postdate the high-rank regional metamorphism (Lumbers, in press).

The Algonquin batholith intruded supracrustal rocks of the older accumulation after these rocks were regionally metamorphosed, probably during the Penokean event that regionally metamorphosed rocks of the Southern Province near Sudbury and to the west between 2.16 and 1.8 Ga ago (Lumbers, 1978). Details of the mechanisms whereby the batholith was emplaced are poorly known, and primary intrusive relationships were somewhat obscured by tectonic deformation during the late high-rank regional metamorphism. Conceivably, the batholith underwent late-magmatic diapirism similar to that postulated above for gneiss diapirs in the Superior Province, but data are insufficient to prove this. Most of the tectonic and metamorphic features now preserved in the batholith and the surrounding rocks were formed during the late high-rank regional metamorphism when the batholith was reactivated and became diapiric toward the overlying supracrustal rocks. Most of the alkalic complex shown in Figure 10 immediately south of the batholith was emplaced during the latter stages of plutonic activity within the carbonate basin and is in part syn-metamorphic relative to the late high-rank metamorphism (Lumbers, in press). Gneissic foliation and stretching lineations in rocks of the complex, in the supracrustal host rocks of the carbonate basin, and in plutonic rocks of the batholith, have similar attitudes and high levels of strain (Lumbers, in press). Moreover, supracrustal rocks of the carbonate basin were at least partly deformed and metamorphosed prior to the diapirism because they underwent moderate regional metamorphism before the onset of the major plutonic activity within the carbonate basin (Lumbers, 1967, in press). Thus, as metamorphism increased within the basin, the supracrustal rocks decreased in porosity and increased in density to the levels now observed (Table II). These relationships also show that during diapirism the ductility contrast between the batholith and the enveloping rocks was low and the overall ductility was high.

Major Lithological and Structural Features

Only the eastern third and portions of the northern margin of the Algonquin batholith have been mapped in sufficient detail to determine the major lithological and structural features (Lumbers, 1976, in press); the western contact and portions of the southern contact of the batholith are incompletely mapped (Fig. 11). General reconnaissance work in the remainder of the batholith can be used only to outline the batholith and give crude estimates of the distribution of lithologies.

The batholith consists of anorthosite suite rocks (Lumbers, 1975), mainly gneissic quartz monzonite, pink to green monzonitic rocks, and quartz syenite; several units of tonalite and scattered small units of anorthosite and related mafic rocks are also present, especially in association with the monzonitic and syenitic rocks. Table 1 gives estimates of the percentages of the major rock types in the batholith, based upon work to date.

More detailed information on the petrography, distribution, and structure of the various rock types in the eastern third and northern parts of the batholith is given in Lumbers (1976, in press). The tonalite and the monzonitic and syenitic rocks are petrographically similar to the same rock types which have been mapped in plutons of anorthosite suite rocks to the north and northwest of the batholith (Lumbers, 1971, 1975). Quartz monzonite in the vicinity of tonalitic, anorthositic, and dioritic units is rich in xenoliths of amphibolite and tonalitic rocks, and contains up to 15 per cent amphibole as the chief ferromagnesian mineral. Such xenolith-rich phases also occur locally where large units of the mafic phases are absent. Cross-cutting relationships of the various phases suggest that the anorthositic and mafic rocks are the oldest, followed in order by the monzonitic and syenitic rocks and quartz monzonite.

Although all intrusive phases of the batholith are mainly gneissic and recrystallized, some partly metamorphosed phases with relict primary features indicative of a magmatic origin occur locally.

These features are: 1) relict igneous feldspars and, rarely, pyroxenes; 2) relict igneous textures; 3) igneous composition; 4) uniformity of composition of individual phases; 5) the preservation of xenoliths and cross-cutting dykes; and 6) the presence of euhedral to subhedral igneous zircons in tonalite and the felsic phases.

For all the rock types, in most large outcrops free of vegetal cover, small, irregularly developed areas of massive to only slightly gneissic phases are commonly preserved (Fig. 12), indicating that the batholith was once composed of predominantly massive rocks. Deformation and recrystallization produced three main varieties of gneiss in all the rock types: 1) augen gneiss (Fig. 13), in which a few igneous features are preserved; 2) laminated gneiss (Fig. 14), lacking relict primary igneous features and marked by thin, relatively continuous, sub-parallel layers alternating with thicker layers of contrasting mineralogy; 3) veined gneiss (Fig. 15), charac-

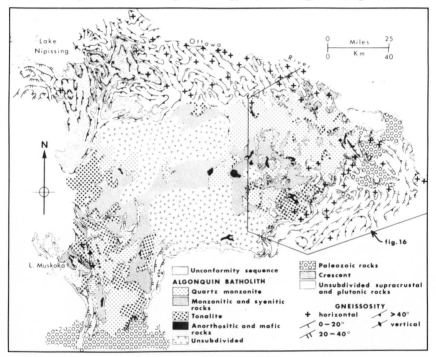

Figure 11. Sketch map showing the known distribution of lithologies within the Algonquin batholith and major structural trends in the surrounding rocks.

terized by numerous, coarse-grained, discontinuous, moderately thick felsic layers which generally are folded. All transitions between the three varieties of gneiss can be found in individual outcrops, but the laminated and veined varieties predominate in zones where the attitude of gneissic foliation is subhorizontal or shallowly dipping (<40°). Although the three varieties of gneiss occur in all the rock types, veined gneiss is rare in greenish monzonitic rocks. Strain measurements indicate that the laminated and veined varieties suffered greater deformation than the augen variety (Schwerdtner *et al.*, 1977); late tectonic shear zones are also common in them (Lumbers, in press).

In general, internal structures and structures of the enveloping rocks indicate that the Algonquin batholith is a large ridge-shaped body with numerous higher-order diapirs and that it narrows greatly to the northeast near Ottawa River (Fig. 11). Gneissosity trends within the batholith are subparallel to lithologic boundaries on a large scale; in the eastern third of the batholith where most of the mapping has been done, several domes are outlined by both gneissosity and lithology (Fig. 16). Air photographs and limited reconnaissance work also indicate that such domes are common elsewhere within the batholith. Maps of the eastern third of the batholith show that across the interface between the batholith and the supracrustal rocks of both the older and younger accumulations, the attitude of the gneissosity is similar, indicating a gross ridge-shaped pattern with the gneissosity dipping shallowly outward from the batholith. A prominent crescent-shaped mass of anorthosite suite rocks composed mainly of quartz monzonite and monzonitic rocks (Lumbers, 1971, 1976) lies near the northwestern margin of the batholith (Fig. 11). Gneissosity steepens markedly near the crescent, which is slightly younger and less deformed

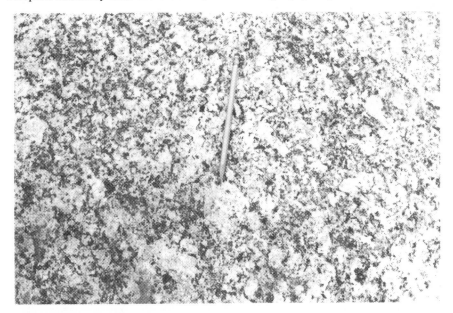

Figure 12. Massive monzonitic rock containing relict mesoperthite phenocrysts.

than rocks of the batholith (Lumbers, 1971). The crescent has a central septum of supracrustal rocks dividing it into two parts: the Powassan and Bonfield batholiths (Lumbers, 1971). (A similar crescent-shaped body, the Dashwa Lake pluton, has been mapped by Schwerdtner *et al.* [1979, Fig. 2] within a large gneiss dome in northwestern Ontario.) Other bodies of anorthosite suite rocks to the north of the main Algonquin batholith and east of the crescent (Fig. 10; Lumbers, 1976) show deformation similar to the batholith, and are separated from it by narrow zones of supracrustal rocks; these anorthosite suite bodies may be related to the batholith.

The higher-order diapirs mapped in the eastern third of the batholith (Fig. 16) are cored by quartz monzonite and are surrounded, at least in part, by steeper-dipping zones of either monzonitic and syenitic rocks, or mafic rocks which mark the termination of the higher-order diapirs. Following the culmination of the high-rank regional metamorphism, the subhorizontal foliation of the domes was gently folded by late warping during uplift. Such warping is common throughout the batholith (Fig. 17) and complicates structural interpretations. Laminated and veined gneiss dominate within the domes, indicating high finite strain. Other higher-order diapirs are evident near the northern margin of the batholith, where arcuate arms of supracrustal rocks extend into the batholith from the main contact (Fig. 11).

The long diameter of the batholith parallels the overall trend of the folded unconformity that marks the base of the younger supracrustal accumulation; this unconformity has also been warped by higher-order diapirism (Fig. 16). Rocks of the unconformity sequence are infolded into the batholith up to 48 km northwest of its exposed margin. This suggests that the batholith was once covered by at least part of the younger supracrustal accumulation.

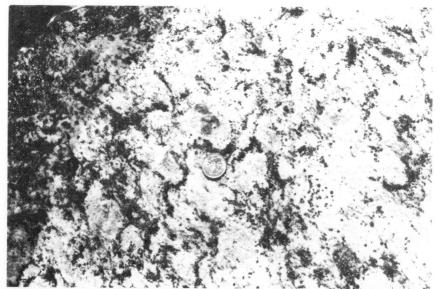

Figure 13. Typical augen gneiss developed in anorthosite of the Algonquin batholith. Some of the light grey plagioclase aggregates contain cores of primary plagioclase (medium-grey patch above one cent scale); dark grey matrix consists of garnet, biotite, and amphibole.

Figure 14. Typical laminated gneiss developed in anorthosite of the Algonquin batholith; thin, discontinuous, subparallel, mafic layers (dark grey) alternate with thicker plagioclase-rich layers.

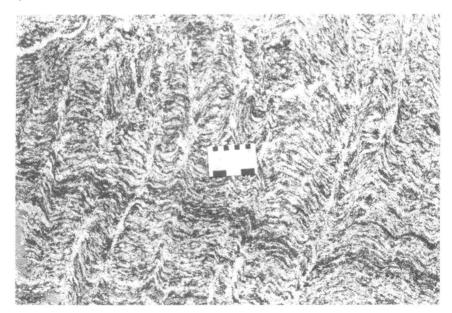

Figure 15. Typical veined gneiss developed in tonalite of the Algonquin batholith.

Bulk Densities of the Batholith and Supracrustal Rocks

If the Algonquin batholith underwent diapirism during the late high-rank regional metamorphism, its bulk density must have been less than the bulk densities of the two supracrustal rock accumulations prior to diapirism. In order to estimate bulk densities of these rock bodies, densities of 678 hand specimens of the batholith and surrounding rocks were measured. Results for rocks of the batholith are shown in Table 1. Average densities of the seven major rock types mapped within the batholith are tabulated together with the number of specimens measured for each type, the standard deviation, and the observed range in densities. For the felsic rocks and tonalite, average densities are also shown for the three main structural varieties of

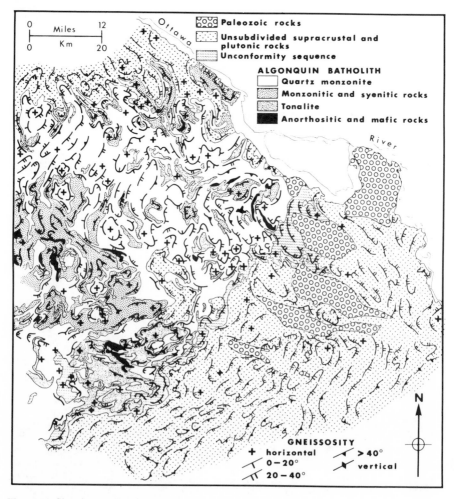

Figure 16. Sketch map showing major structural and lithological features of the eastern third of the Algonquin batholith (See Figure 11 for location).

these rocks: augen gneiss, and laminated and veined gneiss. Too few specimens of anorthositic, gabbroic and dioritic rocks were available to subdivide their densities in this manner. Average densities of quartz monzonite, quartz syenite, and the monzonitic rocks increase with increasing strain. The least strained variety of these rocks, augen gneiss, invariably gives the lowest density. For tonalite, the average density decreases with increasing strain. Petrographic studies in progress suggest that the observed changes in densities correlate with changes in mineralogy and the modal abundance of the minerals present. For example, in quartz monzonite, the modal content of biotite increases with increasing strain in the rock.

Abundances of each rock type within the batholith were estimated from lithologic maps used to compile Figure II; it was assumed that the relative proportions of rock types do not change in the central, unsubdivided portion of the batholith. Two averages were calculated, one for the densities in augen gneiss structural types, and the other for the densities of all the structural types. Because the augen gneiss structural type is the least strained and metamorphosed, its calculated density, 2.73, is the best approximation of the bulk density of the batholith prior to metamorphism and the onset of diapirism. The other density, 2.75, approximates the bulk density of the batholith following diapirism.

Average densities of the major rock types comprising the two supracrustal accumulations in the vicinity of the batholith are given in Table II. Abundances of the various rock types were estimated from lithologic maps compiled at a scale of 1:63,360 (Lumbers, 1971, 1976, in press). Again it was assumed that the relative

Figure 17. Cliff face about 9 m high showing late warping of subhorizontal gneissic layering in laminated to veined gneissic quartz monzonite of the Algonquin batholith; the late warping is accentuated by thin biotite-rich layers (dark grey).

proportions of the various rock types do not change near the western and southwestern flanks of the batholith where the supracrustal rocks are largely unmapped. The estimated bulk densities are 2.76 for the Middle Precambrian accumulation, and 2.82 for the Late Precambrian accumulation.

Although all these figures are only approximations, it is evident that the bulk density of the batholith was lower than the bulk densities of the supracrustal accumulations prior to diapirism. It increased during diapirism and approached the bulk density of the Middle Precambrian supracrustal accumulation.

Summary and Discussion

The main features indicating that the Algonquin batholith became diapiric during the late high-rank regional metamorphism are:

1) The bulk density of the batholith before the high-rank regional metamorphism was less than the bulk density of both the Middle and Late Precambrian supracrustal rocks.

2) Rocks of the batholith were mainly massive prior to diapirism, and were deformed during the late high-rank regional metamorphism to produce numerous domes that are characterized by subhorizontal gneissic foliation and terminate in zones of relatively steeply dipping gneissic foliation. The domes are outlined at least in part of contrasting lithology, with the least dense phase invariably found in the core of the domes.

3) The attitudes of lineations and gneissic foliation across the interface between the batholith and the enclosing rocks are similar, as is the level of strain in both rock types. Thus, during the high-rank regional metamorphism, there was little ductility contrast between rocks of the batholith and the enclosing rocks, and the overall ductility was high.

4) The coarse clastic sequence marking the unconformity at the base of the Late Precambrian accumulation was deformed by higher-order diapirism, and rocks of the sequence were infolded into the batholith.

Density determinations on rocks of the batholith suggest that the bulk density of the batholith increased as diapirism progressed and approached the bulk density of the Middle Precambrian metasediments. The increase in density, together with a concomitant tectonic thinning of the overlying supracrustal rocks were most likely the major factors in terminating the diapirism of the batholith.

Further work may show that the batholith is linked with similar batholiths to the south in Ontario and to the northeast in Quebec. If so, such a chain of major batholiths reactivated during the late high-rank regional metamorphism to form diapirs could explain the reclined to recumbent folding that characterizes supracrustal gneisses of much of the Grenville Province. Reclined to recumbent folding extends for tens of kilometres outward from the batholith. Other large zones of reclined to recumbent folding, such as in the vicinity of Lake Nipissing (Lumbers, 1971, 1975) also appear to be related to batholithic plutons that underwent diapirism during the late high-rank regional metamorphism. Thus, diapirism was perhaps a major controlling factor in the Late Precambrian structural and metamorphic history of the Grenville Province.

TABLE I
DENSITY OF ROCKS OF THE ALGONQUIN BATHOLITH

Rock Type	Density Average	No. of Specimens	Standard Deviation σn-1	σn	Range in Density	Estimated percentage of Batholith
Quartz Monzonite						35
Augen Gneiss	2.64	74	0.022	0.022	2.59-2.69	
Laminated Gneiss	2.68	13	0.031	0.030	2.62-2.73	
Veined Gneiss	2.70	16	0.028	0.027	2.65-2.76	
All types	2.66	103	0.033	0.033	2.59-2.76	
Quartz Syenite						8
Augen Gneiss	2.68	9	0.045	0.042	2.61-2.76	
Laminated Gneiss	2.72	9	0.037	0.035	2.65-2.78	
Veined Gneiss	2.71	4			2.63-2.74	
All types	2.70	22	0.045	0.044	2.61-2.78	
Pink and Grey Monzonite Rocks						20
Augen Gneiss	2.71	32	0.043	0.042	2.64-2.80	
Laminated Gneiss	2.75	26	0.044	0.043	2.68-2.85	
Veined Gneiss	2.81	12	0.099	0.095	2.72-3.05	
All types	2.74	70	0.070	0.069	2.64-3.05	
Green Monzonite Rocks						15
Augen Gneiss	2.83	24	0.152	0.149	2.68-3.37	
Laminated Gneiss	2.89	6	0.111	0.101	2.82-3.11	
All types	2.84	30	0.145	0.142	2.68-3.37	
Tonalite						15
Augen Gneiss	2.84	36	0.116	0.114	2.71-3.20	
Laminated Gneiss	2.81	22	0.062	0.060	2.72-2.96	
Veined Gneiss	2.78	20	0.065	0.063	2.71-2.98	
All types	2.81	78	0.094	0.093	2.71-3.20	
Anorthositic Rocks	2.87	10	0.065	0.061	2.78-3.00	5
Gabbro and Diorite	3.00	8	0.071	0.066	2.91-3.09	2

Estimated average density of batholith using augen gneiss structural types − 2.73
Estimated average density of batholith using all structural types − 2.75

TABLE II
DENSITIES OF SUPRACRUSTAL ROCKS

Middle Precambrian Accumulation

Rock Type	Density	No. of Specimens	Standard Deviation σn-1	σn-1	Range in Density	Estimated Percentage of accumulation
Calcareous shaley metasediments	2.98	17	0.079	0.077	2.83-3.16	10
Micaceous metasandstones	2.67	62	0.025	0.025	2.62-2.74	45
Siliceous shaley metasediments	2.78	45	0.070	0.069	2.66-2.97	35
Feldspathic and Quartzose metasandstones	2.64	20	0.018	0.018	2.61-2.67	10

Estimated Average Density of Accumulation − 2.76

Late Precambrian Accumulation

Rock Type	Density	No. of Specimens	Standard Deviation σn-1	σn-1	Range in Density	Estimated Percentage of accumulation
Siliceous calcitic marble	2.81	24	0.055	0.054	2.69-2.89	45
Calcareous shaley metasediments	2.97	34	0.084	0.083	2.73-3.16	15
Siliceous shaley metasediments	2.84	35	0.153	0.150	2.71-3.60	11
Calcitic marble	2.73	20	0.022	0.021	2.71-2.80	10
Micaceous metasandstone	2.73	23	0.053	0.052	2.64-2.82	7
Unconformity sequence	2.67	26	0.050	0.049	2.59-2.79	5
Feldspathic and Quartzose meta-sandstones	2.68	16	0.034	0.033	2.63-2.75	3
Dolomitic marble	2.84	10	0.022	0.021	2.79-2.86	2
Skarn	3.17	19	0.133	0.129	2.88-3.35	1.5
Siliceous Dolomitic marble	2.96	6	0.109	0.099	2.86-3.16	0.5

Estimated Average Density of Accumulation − 2.82

CONCLUSIONS

Our analysis of structures developed in gneissic batholiths in both the Superior and Grenville Provinces suggests that such bodies do not reflect early crustal remnants, but rather, diapirism of major proportions without involvement of a previously existing sialic basement. In the Superior Province, diapirism apparently occurred at a late plutonic stage, whereas in the Grenville Province, diapirism occurred possibly at a late plutonic stage, but unequivocally during high-rank regional metamorphism.

Diapirism cannot occur without appreciable folding of the supracrustal cover rocks. If diapirs were largely immature, then folding is upright to reclined as in the Superior Province. If diapirism was mature (involving mushrooming effects), then folding of the supracrustal rocks is reclined to recumbent as in the Grenville Province.

During diapirism, high strain levels caused major density changes in the various lithologies of the Algonquin batholith. If such density changes prove to be true for all diapiric batholiths, they will have to be considered in interpreting gravity data and in constructing crustal models. Moreover, the density changes imply important changes in volume with concomitant mobility of material during diapirism. This could prove important in metamorphic studies and in concentration of mineral deposits.

ACKNOWLEDGEMENTS

Work in the Superior Province portion was supported by the Department of Energy, Mines and Resources (Geological Survey of Canada) and the National Research Council of Canada. Thanks are due to Jack and Grace Switzer, Atikokan, and Dale and Jean Bryan, Ignace, for their generous assistance, helpful advice, and friendly hospitality. The first author benefitted greatly from numerous discussions with the staff of the Ontario Geological Survey, particularly C.E. Blackburn, G.R. Edwards, W.D. Bond, F.W. Breaks, W.O. Mackasey, R.P. Sage, P. Thurston and N.F. Trowell, as well as fellow members of the Superior Geotraverse Group.

Financial support in the field and in certain aspects of the laboratory and compilation work pertinent to the Grenville Province portion of this study was supplied to Lumbers by the Ontario Geological Survey from 1972 to 1978. V.M. Vertolli assisted in numerous ways in the Grenville Province portion of the study, including drafting of figures and compilation of data. R.S. Simon carried out many of the density determinations.

REFERENCES

Ahlborn, O. and Richter-Bernburg, G., 1955, Exkursion zum Salzstock von Benthe (Hannover), mit Befahrung der Kaliwerke Ronnenberg und Hansa: Zeitschrift der Deutschen Geologischen Gesellschaft, 105 (Jahrgang 1953), p. 855-865.

Ayres, L.D., 1974, Geology of the Trout Lakes Area, District of Kenora (Patricia Portion): Ontario Division of Mines, Geol. Rept 113.

——————, 1978, Metamorphism in the Superior Province of northwestern Ontario and its relationship to crustal development: in Metamorphism in the Canadian Shield: Geological Survey of Canada Paper 78-10, p. 25-36.

Ayres, L.D., Lumbers, S.B., Milne, V.G. and Robeson, D.W., 1972, Ontario Geological Map: Ontario Dept. Mines and Northern Affairs, Maps 2196 and 2199.

Blackburn, C.E., 1973, Geology of the Otukamanoam Lake area, districts of Rainy River and Kenora: Ontario Division of Mines, Geological Report 109, 42 p.

—————————, 1976, Geology of the Off Lake-Burditt Lake area, district of Rainy River: Ontario Division of Mines, Geol. Rept. 109, 62 p.

Dixon, J.M., 1974, A new method for the determination of finite strain in models of geological structures: Tectonophysics, v. 24, p. 99-114.

—————————, 1975, Finite strain and progressive deformation in models of diapiric structures: Tectonophysics, v. 28, p. 89-124.

Edwards, G.R. and Lorsong, L., 1976, Pipestone Lake area (south half), districts of Rainy River and Kenora: Ontario Division of Mines, Preliminary Map P1103, Geological Series (Scale 4″ = 1 mile).

Edwards, G.R. and Sutcliffe, R.H., 1978, Straw Lake area, districts of Kenora and Rainy River: Ontario Division of Mines, Preliminary Map, Geological Series (Scale 4 = 1 mile).

Gould, D.B. and Demille, G., 1964, Piercement structures in the Arctic islands: Canadian Soc. Petroleum Geol. Bull., v. 12, p. 719-753.

Goodwin, A.M., Ambrose, J.W., Ayres, L.D., Clifford, P.M., Currie, K.L., Ermanovics, I.M. I.M., Fahrig, W.F., Gibb, R.A., Hall, D.H., Innes, M.J.S., Irvine, T.N., McLaren, A.S., Norris, A.W., and Pettijohn, F.J., 1972, The Superior Province: in Variations in Tectonic Styles in Canada, Price, R.A. and Douglas, R.J.W., Geological Association of Canada Spec. Paper II, p. 575-585.

Kupfer, D.H., 1963, Structure of salt in Gulf Coast domes: in Bersticker, A.C., ed., First Symposium on Salt: Northern Ohio Geol. Soc. p. 104-123.

Lumbers, S.B., 1967, Geology and mineral deposits of the Bancroft-Madoc area: Guidebook, Geology of Parts of Eastern Ontario and Western Quebec, 1967 Ann. Meetings, Geol. Assoc. Canada-Mineral. Assoc. Canada, Kingston, Ontario, p. 13-29.

—————————, 1971, Geology of the North Bay Area, Districts of Nipissing and Parry Sound: Ontario Dept. Mines and Northern Affairs, Geol. Dept. No. 94, 104 p.

—————————, 1975, Geology of the Burwash Area, Districts of Nipissing, Parry Sound and Sudbury: Ontario Div. Mines, Geol. Rept. No. 116, 158 p.

—————————, 1976, Mattawa-Deep River Area, District of Nipissing and County of Renfrew: Ontario Div. Mines, Prelim. Map Nos. 1196 and 1197, Scale, 1 inch to 1 mile.

—————————, 1978, Geology of the Grenville Front Tectonic Zone between Sudbury and Timagami, Ontario: Guidebook, Ann. Meetings, Geol. Soc. America-Geol. Assoc. Canada-Mineral. Assoc. Canada, Toronto, 1978, p. 347-361.

—————————, in press, Summary report on the metallogeny of Renfrew County: Ontario Geol. Survey, Misc. Paper, in press.

Lumbers, S.B. and Krogh, T.E., 1977, Distribution and age of anorthosite suite intrusions in the Grenville Province of Ontario: Abstract, 1977 Ann. Meetings, Geol. Assoc. Canada-Mineral. Assoc. Canada, Vancouver.

O'Brien, G.D., 1968, Survey of diapirs and diapirism: in Braunstein, J. and O'Brien, G.D., eds., Diapirs and Diapirism: American Association of Petroleum Geologists, Memoir 8, p. 1-9.

Ramberg, H., 1963a, Experimental study of gravity tectonics by means of centrifuged models: Geol. Instit. of the University of Uppsala Bull., v. 42, p. 1-97.

—————————, 1963b, Fluid dynamics of viscous buckling applicable to folding of layered rocks: American Assoc. Petroleum Geol. Bull.

—————————, 1967, Gravity, Deformation and the Earth's Crust: London, Academic Press, 214 p.

—————————, 1968, Instability of layered systems in the field of gravity, I: Physics of the Earth and Planet: Interiors, v. 1, p. 427-447.

—————————, 1970a, Folding of laterally compressed multilayers in the field of gravity, II numerical examples: Physics of the Earth and Planet: Interiors, v. 4, p. 83-120.

_____, 1970b, Model studies in relation to intrusion of igneous bodies: in Mechanisms of Igneous Intrusion: Liverpool Geol. Soc. Spec. Publ. 2, p. 261-272.

Ramsay, J.G., 1967, Folding and fracturing of rocks: New York, McGraw-Hill, 568 p.

Sage, R.P. Blackburn, G.E., Breaks, F.W., McWilliams, G.M., Schwerdtner, W.M. and Stott, G.M., 1975, Internal structure and composition of two granite complexes in Wabigoon Subprovince: in the proceedings of the Geotraverse Workshop 1975, Dept. Geology, University of Toronto, p. 24-33.

Schwerdtner, W.M., 1976a, Crescent-shaped granitic bodies in the Scotch Lakes area, Wabigoon subprovince: Proceedings of the 1976 Geotraverse Conference, p. 34-41, Precambrian Research Group, University of Toronto.

_____, 1976b, Lithology and structure of the Irene-Eltrut Lakes granitic complex between Atikokan and Ignace, a progress report: Proceedings of the 1976 Superior Geotraverse Conference, Precambrian Research Group, University of Toronto, p. 22-27.

_____, 1978, Strain patterns within conformable granitoid bodies of the Superior and Grenville Provinces, northwestern and southeastern Ontario: in Abstracts with Programs, Geol. Assoc. Canada and Mineral. Assoc. Canada, v. 3, p. 488.

Schwerdtner, W.M. and Byers, P.N., 1965, Comparison of boudins with some similar tectonic inclusions: American Assoc. Petroleum Geol. Bull., v. 49, p. 613-616.

Schwerdtner, W.M. and Clark, A.R., 1967, Structural analysis of Mokka Fiord and South Fiord Domes, Axel Heiberg Island, Canadian Arctic: Canadian Jour. Earth Sci., v. 4, p. 1229-1245.

Schwerdtner, W.M. and Morrison, M.J., 1974, Internal flow mechanisms of salt and sylvinite in Anagance diapiric anticline near Sussex, New Brunswick: Coogan, A.H., ed., Fourth Symposium on Salt: Northern Ohio Geol. Soc. and Kent State University, p. 241-248.

Schwerdtner, W.M. and Mason, D.R., 1978, Wabigoon Subprovince – structure and field relationships within a metavolcanic-gneissic-plutonic terrain: in Smith, I.E.M. and Williams, J.G., eds., Proceedings of the 1978 Archean Geochemistry Conference: Toronto University Press, p. 18-22.

Schwerdtner, W.M. and Troëng, B., 1978, Strain distribution within arcuate diapiric ridges of silicone putty: Tectonophysics, v. 50, p. 13-28.

Schwerdtner, W.M., Bennett, P.J. and Janes, T.W., 1977, Application of L-S tectonite scheme to structural mapping and paleostrain analysis: Canadian Jour. Earth Sci., v. 14, p. 1021-1032.

Schwerdtner, W.M., Sutcliffe, R.H. and Troeng, B., 1978a, Patterns of total strain within the crestal region of immature diapirs: Canadian Jour. Earth Sci., v. 15, p. 1437-1447.

Schwerdtner, W.M., Morgan, J., Osadetz, K., Stone, D., Stott G.M. and Sutcliffe, R.H., 1978b, Structure of Archean rocks in western Ontario: in Smith, I.E.M. and Williams, J.G., eds., Proceedings of the 1978 Archean Geochemistry Conference: Toronto University Press, p. 107-125.

Schwerdtner, W.M., Stone, D., Osadetz, K., Morgan, J. and Stott, G.M., 1979, Granitoid complexes and the Archean tectonic record in the southern part of northwestern Ontario: Canadian Jour. Earth Sci., in press.

Silver, L.T., Bickford, M.E., Van Schmus, W.R., Anderson, J.L., Anderson, T.H. and Madaris, L.G., Jr., 1977, The 1.4-1.5 b.y. transcontinental anorogenic plutonic perforation of North America (Abst.): Geol. Soc. America, Abstracts with Programs, v. 9, no. 7, p. 1176-1177.

Stephansson, U., 1972, Theoretical and experimental studies of diapiric structures of Öland: Geol. Institut. of the University of Uppsala, Bull., New Series, v. 3, p. 163-200.

Sutcliffe, R.H., 1977, Geology and emplacement of the Jack Fish Lake pluton, a major intrusion in the Rainy Lake dome: in Proceedings of the 1977 Superior Geotraverse Conference, Precambrian Research Group, University of Toronto, p. 146-154.

—————————, 1978, Geology of the Rainy Lake granitoid complex, northwestern Ontario: in Smith, I.E.M. and Williams, J.G., eds., Proceedings of the 1978 Archean Geochemistry Conference: Toronto University Press, p. 235-244.

Talbot, C.J., 1974, Fold nappes as asymmetric mantled gneiss domes and ensialic orogeny: Tectonophysics, v. 24, p. 259-276.

von Trusheim, F., 1960, Mechanism of salt migration in northern Germany: American Assoc. Petroleum Geol. Bull., v. 44, p. 1579-1587.

Ziehlke, D.V., 1973, Aulneau batholith project, geological investigations – progress report: in Centre for Precambrian Studies, University of Manitoba, 1973 Annual Report, p. 60-73.

—————————, 1974, Aulneau batholith project, geology and petrology: in Centre for Precambrian Studies, University of Manitoba, 1974 Annual Report, Part 2, p. 150-184.

The Continental Crust and Its Mineral Deposits, edited by D.W. Strangway,
Geological Association of Canada Special Paper 20

GEOCHEMICAL VARIETY AMONG ARCHEAN GRANITOIDS IN NORTHWESTERN ONTARIO

I.E.M. Smith[1] J.G. Williams[2]
Department of Geology, University of Toronto,
Toronto, Ontario M5S 1A1
Present Addresses
[1]Department of Geology, University of Auckland, Private Bag, Auckland, New Zealand
[2]Department of Geology, University of Otago, Dunedin, New Zealand

ABSTRACT

Knowledge of the variety and distribution of granitoids in Archean terranes is critical to models of the evolution of the Earth's early crust. Representative sampling of granitoid bodies in four areas of the Superior Province in northwestern Ontario suggests that granitoids of the volcanic-plutonic belts (diorite-granodiorite) are compositionally distinct from those of the gneissic terranes (leucogranite, tonalite-trondhjemite). The variety among these Archean granitoids results from differences in the depth of magma generation, the availability of water, and the previous melting history of the source. Primary variation may also be imposed by the ultimate nature of the source, whether mantle-derived igneous material or reworked crustal metasedimentary material.

RÉSUMÉ

La connaissance de la variété et de la distribution des granitoïdes dans les régions archéennes est essentielle pour élaborer des modèles d'évolution de la croûte primitive de la Terre. Un échantillonnage représentatif des masses granitoïdes dans quatre régions de la province du Lac Supérieur dans le nord-ouest de l'Ontario suggère que les granitoïdes des zones volcanoplutoniques (diorite-granodiorite) ont des compositions distinctes de celles des régions surtout gneissiques (leucogranite-tonalite-trondhjémite). La variété observée dans ces granitoïdes archéens résulte des différences dans la profondeur de génération du magma, la disponibilité de l'eau et l'histoire antérieure de fusion de la source. On peut aussi imposer une variation primaire par la nature ultime de la source, que ce soit du matériel igné dérivé du manteau ou du matériel métasédimentaire remanié provenant de la croûte.

INTRODUCTION

The geochemical development of Archean crust can be modelled in terms of a primitive, largely mantle-derived mafic igneous component and an evolved siliceous component which at least in part represents recycled pre-existing crustal material. The granitoids of Archean terranes provide an example of the geochemically evolved crustal component, and their diversity indicates a variety of petrogenetic conditions and source compositions contributing to the crust-forming process. In order to understand this process it is clearly important to identify and quantify the different types of granitoid occurring in ancient crust. To this end we present field and compositional data on some granitoids from the Superior Province of the Canadian Shield.

On a broad scale the Superior Province is made up of two contrasting rock associations: 1) gneissic terranes characterized by medium-to high-grade metamorphic rocks together with granitoids, and 2) metavolcanic-plutonic terranes made up of predominantly low- to medium-grade metavolcanic supracrustal assemblages together with granitoid intrusions. The basis for this paper is a collection of representative rock samples from granitoid bodies within gneissic terranes in the Miniss Lake area of English River Subprovince (Williams, 1977) and the Crystal Lake area of Quetico Subprovince (Williams, 1978), and from discrete granitoid plutons intruding metavolcanic rocks in the Pickle Crow area of Uchi Subprovince and in the Shebandowan area of Wawa Subprovince (Fig. 1).

FIELD OCCURRENCE AND PETROGRAPHY

Granitoid bodies in each of the two main geologic settings of Superior Province appear to have distinctive composition and form.

Granitoids within the gneissic terranes occur as narrow bands interlayered with mafic gneiss, as tabular bodies up to hundreds of metres thick, and as discrete plutons. Contacts between larger granitoid bodies and surrounding country rock are broad (10^2 metres and upwards) and poorly defined.

In the Miniss Lake area Williams (1977) identified three distinct types of granitoid.

1. White massive leucogranite comprising the leucosome component of migmatized metasediments of the Miniss Series and also occurring as small bodies referred to as diatexite (Breaks and Bond, 1977). This rock type is made up mainly of plagioclase (albite to sodic oligoclase) and microcline (slightly perthitic in part); myremekitic quartz-plagioclase intergrowths are not uncommon. Minor almandine garnet is ubiquitous; biotite, muscovite, and cordierite occur sporadically; and in several outcrops in Miniss Bay on Lake St Joseph, abundant apatite is accompanied by minor beryl. The texture is characterized by uneven grain size and is commonly pegmatitic.

2. Pink massive leucogranite, commonly interbanded and mixed with orthogneiss, is the youngest known lithology in the Miniss Lake area. It is composed of microcline (perthitic in part) and plagioclase ($\sim An_{10}$) with subordinate quartz and accessory biotite and muscovite. Grain size is typically medium to coarse; pegmatitic patches are uncommon.

3. Trondhjemite, a white massive, or more commonly well-foliated rock type which intrudes the paragneiss around northern Churchill Lake and west of Miniss Lake. Intrusion of the trondhjemite predates at least some of the migmatization. The rock is made up of plagioclase (~An$_{30}$), quartz, and biotite with minor K-feldspar apatite, muscovite, opaque oxides, epidote, and zircon.

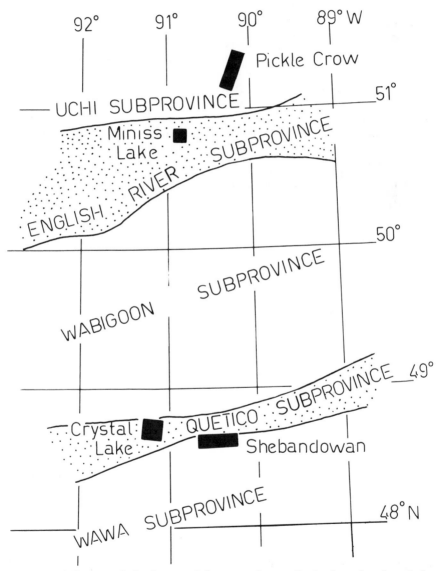

Figure 1. Locality map, indicating sampled areas and generalized subprovince boundaries. Gneissic terranes are stippled.

Metasedimentary rocks in the Crystal Lake area are intruded by two granitoid plutons. One of the plutons, the Eva Granite (Williams, 1978) is a white to locally pink, massive, coarse-grained (average grain size 2 to 5 mm) leucogranite. Typical specimens consist of microcline (perthitic in patches), plagioclase, and quartz with accessory mica, epidote, apatite, magnetite, and calcite. Two petrographically distinct phases are recognized within the Eva Granite: a biotite-rich phase in which the plagioclase is sodic oligioclase and a muscovite-rich phase containing calcic albite.

The second granitoid pluton, the Niobe Granite (Williams, 1978), is a compositionally and texturally heterogeneous leucogranite characterized by the presence of muscovite and garnet. It is always white and commonly well foliated; grain size is coarse to pegmatitic with K-feldspar megacrysts up to 30 cm across. The rock is composed of microcline (sparsely perthitic), plagioclase, and quartz, with accessory garnet, muscovite, biotite, and sphene; apatite and epidote occur sporadically. Garnet content is always low although variable on outcrop scale. Plagioclase composition is variable from albite (An 5) to albite-oligioclase so that in places the composition of the Niobe Granite is transitional toward that of the muscovite-bearing phase of the Eva Granite.

Monzodiorites occur as a 2-km wide belt of partially migmatized and assimilated inclusions within the Eva Granite. These consist of plagioclase and K-feldspar with subordinate hornblende and biotite, and accessory apatite, sphene, and zircon.

Tonalite rafts ranging up to outcrop size are scattered through the Eva and Niobe Granites. Although grain size and texture are variable, their composition is fairly constant. The most abundant mineral is plagioclase (oligoclase-andesine), accompanied by subordinate quartz, biotite and hornblende, and accessory K-feldspar,

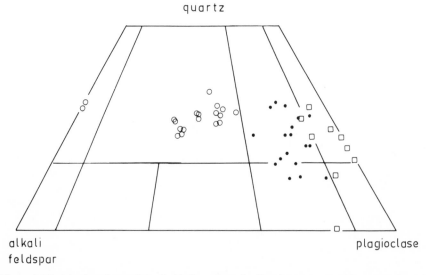

Figure 2. Modal mineralogy of granitoids from Superior Province expressed in terms of quartz, alkali feldspar (including albite <An₅), and plagioclase. Symbols as for Figure 3. Boundary lines of the IUGS granitoid classification scheme are given.

epidote, apatite, sphene, and opaque oxides with sparse muscovite, chlorite, and calcite.

Trondhjemite occurs as inclusions within both the Eva and Niobe Granites. These inclusions are typically less mafic and contain less K-feldspar than the tonalites but in other respects their mineralogy is similar.

Granitoids within the metavolcanic-plutonic belts of the Superior Province in northwestern Ontario occur as discrete rounded plutons partly or completely surrounded by metavolcanic rocks. In contrast to plutons in gneissic terranes, these bodies typically show well-defined contact zones (typically less than 100m wide) against the surrounding country rock. A sampling of plutons from the Pickle Crow and Shebandowan areas revealed compositions grouped within the diorite-monzodiorite-grandiorite fields (Fig. 2). These rocks are characteristically medium-grained and consist of plagioclase (typically oligoclase), microcline, or less commonly microcline microperthite and quartz together with biotite and muscovite or biotite and minor hornblende. Accessory mineral assemblages are generally complex; colourless epidote is ubiquitous, comprising up to six per cent of some specimens, sphene, apatite, opaque oxides, and zircon are common; allanite and fluorite are rare. Some of the plutons within metavolcanic-plutonic terranes (e.g., the Clearwater Lake intrusion in the Shebandowan area) show systematic compositional zoning toward more quartz-rich cores; others show little variation in composition throughout.

CHEMICAL COMPOSITIONS

Major and trace element analyses of the granitoid specimens from Superior Province are presented in Tables I to III. Table I contains analyses of leucogranite from the Miniss Lake and Crystal Lake areas; specimens were selected as representative of the granitoids exposed in these two areas, on the basis of detailed mapping (Williams, 1977, 1978). The analyses of granitoids from metavolcanic-plutonic belts presented in Table 2 are of specimens collected from a variety of small plutons using the 1 inch to 2 miles preliminary geologic maps of the Ontario Geological Survey as a base; in a number of cases poor outcrop limited the availability of material from these plutons. Analyses in Table III are of a gneissic trondhjemite from the Miniss Lake area and of diorite, tonalite, and trondhjemite inclusions and rafts from the Crystal Lake area.

Clearly our data base suffers from the limited availability of outcrop and also from the geographic restrictions of our mapping programme. However, we feel that the selection of representative samples with good field control from several areas of the western Superior Province has provided data which allow some general conclusions on the nature and origin of Archean granitoids.

In general the major element abundances of the Archean granitoids form a continuous trend. The granodioritic suite from the metavolcanic-plutonic belts have generally lower SiO_2 contents (65-74 wt %) than the leucogranites in the gneissic terranes (70-76 wt %), although there is some overlap. SiO_2 contents of the tonalitic suite (diorite-tonalite-trondhjemite) from the gneissic terranes mainly overlap with granodioritic samples and extend to lower values. Ferromagnesian and related major elements show a consistent negative correlation with SiO_2 throughout the sampled

TABLE I

LEUCOGRANITES FROM GNEISSIC TERRANES OF NORTHWESTERN ONTARIO

	MINISS LAKE AREA								CRYSTAL LAKE AREA						
	Biotite Bearing				Garnet Bearing				Biotite Bearing		Muscovite Bearing			Garnet Bearing	
No:	1	2	3	4	5	6	7	8	9	10	11	12	13	14	15
Wt. % SiO$_2$	70.55	71.07	74.71	73.34	74.58	75.07	76.17	71.82	74.07	74.26	74.21	74.94	74.96	74.73	75.09
TiO$_2$	0.06	0.15	0.14	0.05	0.02	0.05	0.03	0.23	0.06	0.06	0.06	0.07	0.05	0.02	0.03
Al$_2$O$_3$	14.23	14.34	13.61	14.39	14.02	14.55	13.97	14.50	14.32	13.58	14.14	14.08	13.78	14.74	14.63
Fe$_2$O$_3$	0.46	0.55	0.52	0.17	0.12	0.08	0.07	0.63	0.60	0.43	0.32	0.41	0.38	0.32	0.39
FeO	0.43	0.79	0.45	0.63	0.20	0.48	0.38	1.15	0.20	0.56	0.28	0.48	0.59	0.21	0.24
MnO	0.02	0.05	0.03	0.05	0.03	0.03	0.05	0.04	0.03	0.03	0.01	0.04	0.03	0.08	0.08
MgO	0.76	0.71	0.54	0.24	0.24	0.04	0.05	0.59	0.27	0.03	0.31	0.26	0.03	0.08	0.01
CaO	0.80	1.11	0.92	0.56	1.10	1.25	1.07	1.02	0.91	0.55	0.43	0.33	0.35	0.25	0.28
Na$_2$O	4.49	4.50	3.23	3.32	3.40	3.50	3.71	3.62	3.85	3.37	3.22	3.60	3.51	3.25	2.73
K$_2$O	6.69	5.44	5.45	5.59	4.15	4.73	4.73	5.24	4.79	5.35	6.33	4.63	5.06	5.41	5.81
P$_2$O^5	0.11	0.04	0.12	0.12	0.10	0.12	0.08	0.12	0.07	0.06	0.05	0.07	0.04	0.07	0.03
LOI	0.20	0.40	0.35	0.30	0.40	0.30	0.20	0.88	0.60	0.69	0.49	0.61	0.73	0.49	0.54
TOTAL:	98.80	99.15	99.95	98.57	99.80	99.62	100.59	99.84	99.77	99.81	99.85	99.81	99.81	99.64	99.86
ppm Ba	571	1170	493	193	710	307	520	653	555	608	254	297	163	<5	7
Rb	164	126	121	214	103	73	83	169	557	157	170	207	199	338	284
Sr	231	251	116	66	190	111	138	161	485	143	93	51	47	7	7
Pb	35	40	29	31	38	45	35	32	22	41	47	32	26	5	7
Zr	70	152	96	40	55	44	34	176	87	108	51	75	60	33	35
Nb	4.5	7.0	6.0	10.0	5.5	6.0	4.5	10.0	2.0	8.0	7.0	13.0	9.0	17.0	16.0
Y	9	7	9	5	6	13	10	14	7	9	10	13	18	5	5
V	5	12	<5	<5	<5	<5	<5	16	<5	<5	<5	<5	<5	<5	<5
Cr	9	7	5	7	7	12	9	9	5	6	<5	7	13	<5	<5
Ni	<5	<5	<5	<5	<5	<5	<5	<5	<5	<5	<5	<5	<5	<5	<5
Cu	9	5	5	7	8	7	9	10	14	8	10	7	7	<5	11
Zn	23	31	28	32	16	16	15	48	27	27	21	28	29	29	36
Ga	18	19	18	23	15	16	16	19	14	10	21	19	19	32	32
Modal Plagioclase	27.6	28.2	28.9	30.0	27.8	33.5	34.9	31.9	30.6	33.1	31.4	36.5	34.1	34.4	34.2
Quartz	26.4	28.0	28.7	29.7	40.4	37.2	29.4	36.2	29.9	30.6	25.1	32.7	32.7	46.0	46.8
K-feldspar	41.1	39.1	40.3	32.4	41.7	23.5	27.1	32.6	30.7	34.0	40.6	25.8	26.5	11.7	34.2
Biotite	4.5	3.6	1.8	3.3	tr	1.7	tr	4.5	tr	1.3	1.3	1.3	1.5	—	13.9
Muscovite	0.1	0.4	0.4	3.8	tr	tr	tr	tr	tr	tr	tr	8.3	5.1	7.3	1.0
Garnet	—	—	—	0.4	0.6	0.8	0.6	0.9	tr	tr	tr	tr	tr	0.5	0.8

tr = trace

Major elements by XRF (Norrish and Hutton, 1969) at University of Toronto, except for FeO (titration), Na$_2$O (flame photometry) and LOI (gravimetry).

Trace elements by XRF at Memorial University. Modes determined on 15 × 10 cm stained slabs.

TABLE II
GRANITOIDS FROM METAVOLCANIC-PLUTONIC BELTS OF NORTHWESTERN ONTARIO

		16	17	18	19	20	21	22	23	24	25	26	27	28
Wt. %	SiO_2	65.39	65.60	66.30	66.36	67.76	68.00	69.38	69.47	70.20	70.80	71.29	72.17	73.90
	TiO_2	0.57	0.56	0.51	0.54	0.37	0.55	0.29	0.31	0.27	0.27	0.24	0.27	0.10
	Al_2O_3	14.30	15.94	15.85	16.18	15.68	14.54	14.23	15.33	15.67	15.41	15.50	15.64	15.14
	Fe_2O_3	2.33	1.49	1.26	1.43	1.62	1.26	0.84	0.63	0.34	0.84	0.78	0.71	0.38
	FeO	3.31	1.88	1.81	1.93	1.09	2.54	1.37	1.13	1.63	1.04	0.55	0.81	0.27
	MnO	0.09	0.05	0.07	0.04	0.04	0.07	0.06	0.03	0.03	0.04	0.02	0.01	0.01
	MgO	2.22	1.34	1.37	1.41	1.06	1.73	1.43	0.66	0.82	0.69	0.41	0.30	0.04
	CaO	4.00	2.87	2.86	2.90	3.03	3.27	2.62	2.74	2.83	2.72	2.33	2.64	1.54
	Na_2O	4.42	5.48	2.94	5.00	4.40	3.91	4.34	4.70	5.00	5.04	5.11	5.16	5.23
	K_2O	1.97	2.70	2.94	2.86	2.10	2.55	3.59	2.35	1.87	1.93	2.43	1.97	3.18
	P_2O_5	0.13	0.22	0.18	0.20	0.17	0.19	0.11	0.11	0.10	0.10	0.06	0.08	0.01
	LOI	0.98	1.30	1.42	1.05	1.65	0.91	1.21	1.68	0.88	0.91	0.81	0.23	0.18
	TOTAL:	99.71	99.43	99.57	99.90	98.97	99.52	99.47	99.14	99.64	99.79	99.53	99.99	99.98
ppm	Ba	429	1285	1159	1198	589	654	616	879	1310	565	891	639	857
	Rb	55	66	86	84	54	110	94	63	85	61	65	51	99
	Sr	220	1128	1051	1058	502	250	664	744	871	480	576	546	590
	Pb	5	15	22	20	9	11	20	16	17	16	18	18	26
	Zr	214	204	190	201	128	200	109	153	178	117	121	139	90
	Nb	14	3	4	6	5	10	4	<2	6	4	2	3	3
	Y	19	15	12	12	10	24	9	9	13	8	8	9	5
	V	65	56	49	55	40	71	40	30	45	31	19	26	10
	Cr	61	12	16	20	15	36	46	16	8	6	10	9	10
	Ni	48	6	16	5	<5	24	24	<5	<5	<5	<5	<5	<5
	Ca	52	15	14	17	14	16	11	15	12	14	6	9	9
	Zn	69	65	67	65	54	66	43	67	60	51	51	55	43
	Ga	21	23	22	23	18	20	20	24	21	22	23	23	25
Modal	Plagioclase	48.7	56.7	57.4	57.7	45.8	41.3	51.1	49.8	53.9	46.9	50.8	53.4	47.6
	Quartz	18.7	13.3	16.8	13.1	24.9	28.2	16.6	25.2	19.0	26.4	26.7	27.5	26.6
	K-feldspar	8.2	17.4	14.8	15.1	3.8	8.1	18.2	11.9	17.3	7.0	14.0	10.8	22.5
	hornblende	16.7	2.3	3.6	3.9	—	—	6.2	—	—	—	—	—	—
	biotite	5.9	8.1	4.4	7.1	5.4	14.8	4.1	5.7	4.1	5.6	4.1	5.6	0.5
	muscovite	—	—	—	—	13.5	tr	tr	3.1	4.2	7.5	2.7	1.2	1.8
	epidote	1.5	1.6	1.8	2.0	4.4	6.5	2.5	2.1	0.8	6.0	1.1	0.9	0.7

Analytical methods as in Table I.

TABLE III
TONALITIC ROCKS FROM SUPERIOR PROVINCE

		29	30	31	32	33	34	35	36	37
Wt. %	SiO2	55.10	64.19	64.94	66.98	69.82	67.61	68.46	71.25	71.29
	TiO2	0.78	0.41	0.44	0.16	0.12	0.57	0.28	0.33	0.24
	Al2O3	16.60	16.28	15.97	18.39	17.40	15.77	16.30	14.94	15.62
	Fe2O3	2.22	1.53	1.15	0.58	0.17	0.88	1.12	0.77	0.85
	FeO	4.20	2.35	2.70	1.08	0.76	2.80	1.37	1.63	1.23
	MnO	0.09	0.05	0.07	0.04	0.02	0.04	0.06	0.02	0.02
	MgO	4.66	2.67	2.51	0.64	0.69	1.41	1.21	0.81	0.37
	CaO	5.85	4.05	4.10	3.13	3.30	3.28	4.15	3.37	2.67
	Na2O	5.27	4.54	4.32	6.79	5.94	4.50	4.56	4.33	4.55
	K2O	2.70	2.25	2.22	1.01	0.80	1.90	0.92	1.35	2.22
	P2O5	0.59	0.19	0.22	0.09	0.09	0.16	0.14	0.13	0.09
	LOI	1.40	1.38	1.03	1.02	0.73	0.28	0.13	0.37	0.52
	TOTAL:	99.46	99.89	99.67	99.91	99.84	99.20	98.70	99.30	99.67
ppm	Ba	1881	948	686	144	322	579	233	212	885
	Rb	89	65	112	54	32	63	22	58	53
	Sr	2502	629	702	493	411	465	443	224	308
	Pb	18	12	11	17	2	14	3	8	13
	Zr	272	140	139	55	58	163	124	176	110
	Nb	3	5	6	7	3	5	6	5	8
	Y	23	9	14	9	6	11	8	10	8
	V	114	66	67	18	8	78	24	23	20
	Cr	146	49	76	7	6	13	10	13	11
	Ni	103	31	31	4	9	6	3	11	—
	Ca	34	45	11	59	28	18	10	15	13
	Zn	103	62	68	45	28	73	42	54	45
	Ga	23	19	24	45	17	24	20	21	20
Modal	Plagioclase	56.2	59.2	53.6	74.1	72.8	60.1	67.5	54.3	56.1
	Quartz	—	12.2	22.5	19.6	23.2	25.9	25.7	32.8	30.3
	K-feldspar	10.0	5.9	6.7	0.3	0.3	1.8	tr	4.0	8.0
	hornblende	23.2	8.7	3.9	—	—	—	—	tr	—
	biotite	7.9	11.2	11.8	4.1	2.1	11.2	6.3	7.2	4.6
	muscovite	tr	—	0.2	1.4	1.4	—	—	0.9	0.9
	epidote	—	2.0	0.8	0.3	0.1	—	—	0.2	—

Analytical methods as in Table I

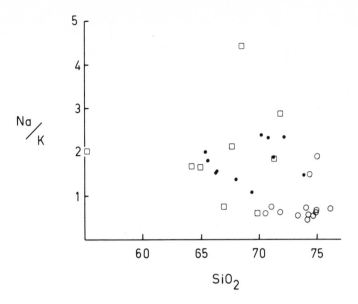

Figure 3. FeO (total) – SiO₂ plot for analyzed Superior Province granitoids. Open circles-leucogranites, open squares-tonalite suite, dots-granodiorites.

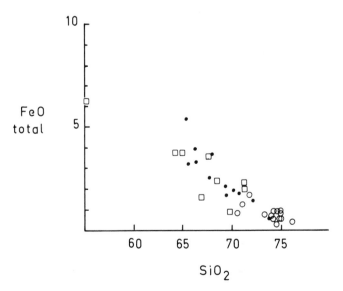

Figure 4. Na/K (atomic proportions) – SiO₂ plot for analyzed Superior Province granitoids. Symbols as for Figure 3.

granitoids (Fig. 3). However, marked differences in the relative abundances of Ca, Na, and K distinguish the granodioritic rocks (high Na, Ca) from the leucogranites (high K, low Ca); the tonalitic suite is high in Ca, with variable Na/K (Fig. 4).

Trace element abundances in the Superior Province granitoids vary both within and between compositional groups. Ferromagnesian trace elements (Ni, Cr, V) are consistently very low in the leucogranites. Biotite-bearing leucogranites are higher in Ba, but in other respects are not significantly different from most of the muscovite-garnet-bearing leucogranites. However, two garnet-bearing leucogranites from the Crystal Lake area (Table I, Analyses 14, 15) are distinctive because of very high Rb contents and low abundances of most other trace elements except for Ga, Zn, and Pb.

In contrast, the granodioritic granitoids are higher in ferromagnesian elements and Sr, and lower in Rb. There are no clear consistent trends within the suite and several elements, notably Ba, Sr, and Zr, show marked variability. In general, trace element characteristics in the tonalitic suite are similar to those in the grandioritic suite.

DISCUSSION

An important observation arising from this study is that the granitoids which intrude supracrustal rocks to form the metavolcanic-plutonic belts of the Superior Province are significantly different from the granitoids associated with high-grade metamorphic rocks in the gneissic terranes. Granitoids in the metavolcanic-plutonic belts form discrete well-defined plutons within the compositional range of diorite-monzodiorite-granodiorite. In contrast, granitoids within the gneissic terranes are either leucogranite ranging toward alkali-leucogranite, or tonalite-trondhjemite. Observations in other parts of Superior Province, indicating that tonalite is a major component of some Archean terranes, suggest a deficiency in our sampling programme.

Although differences in the form of contacts of the granitoid plutons in each geologic setting could be explained by the hypothesis that they represent sections at different levels through the crust, the compositional differences preclude such a simple explanation. For example, if the geochemically more evolved granitoids observed in the gneissic terranes are related to the granodioritic suite of the metavolcanic-plutonic belts by a single-stage process such as crystal fractionation, the more evolved granitoids would logically be found associated with, or at higher crustal levels than, the more primitive granitoids.

Field observations of widespread migmatites in Archean high-grade metamorphic terranes show clearly that metasedimentary rocks undergo partial melting to produce granitic melts. Such melts tend to be peraluminous (molecular $Al_2O_3/(CaO + Na_2O + K_2O) > 1.03$) and have been termed S-type granitoids (Chappell and White, 1974). Among the sampled granitoids from Superior Province most of the leuco-granites have S-type characteristics as defined by Chappell and White (1974) and Chappell (1978), whereas the granodioritic and tonalitic suites have igneous source characteristics (I-type of Chappell and White, 1974).

Our data indicate that there is a greater variety of granitoid compositions in Archean gneissic terranes than in the metavolcanic-plutonic belts; this is reflected in the trace element abundances and in the mineralogy. Such differences are emphasized by the extreme depletion in most elements of the two samples of Niobe Granite (Table I, Analyses 14, 15).

The purpose of this paper has been to point out what appear to be significant differences between granitoids in metavolcanic-plutonic belts and those in the gneissic terranes of Superior Province. Work in progress will, we hope, provide quantitative constraints on the processes responsibe for the observed variety of compositions. Our preliminary conclusions are: 1) that the igneous rocks of the metavolcanic-plutonic belts result from mantle differentiation followed by comparatively rapid reworking of the crustal material so produced, and 2) that the granitoids in the gneissic terranes result from a variety of source materials and melting conditions perhaps controlled by the availability of water. This model implies that the gneissic terranes represent an older crustal component which has undergone complex and prolonged reworking, whereas the metavolcanic-plutonic belts represent comparatively recent additions to the crust. In a general way the measured ages of rock types in the gneissic terranes ($\sim$3.0 Ga, Wooden, 1978) and the metavolcanic plutonic belts ($\sim$2.7 to 2.9 Ga, Krogh and Davies, 1971) bear this out.

ACKNOWLEDGEMENTS

This research was supported by grants from the Department of Energy, Mines and Resources, and the National Research Council to A.M. Goodwin and by a Negotiated Development Grant to the Department of Geology, University of Toronto.

REFERENCES

Breaks, F.W. and Bond, W.D., 1977, Manifestations of recent reconnaissance investigations in the English River subprovince, northern Ontario: Geotraverse Conference proceedings 1977, University of Toronto, p. 170-211.
Chappell, B.W., 1978, Granitoids from the Moonbi district, New England batholith, eastern Australia: Geol. Soc. Australia Jour., v. 25, p. 267-283.
Chappell, B.W. and White, A.J.R., 1974, Two contrasting granite types: Pacific Geol., v. 8, p. 173-174.
Krogh, T.E. and Davies, G.L., 1971, Zircon U-Pb ages of Archean metavolcanic rocks in the Canadian shield: Carnegie Institution of Washington Year Book, v. 70, p. 241-242.
Norrish, K. and Hutton, J.T., 1969, An accurate X-ray spectrographic method for the analysis of a wide range of geological samples: Geochim. Cosmochim. Acta, v. 33, p. 431-453.
Williams, J.G., 1977, Geology of Miniss Lake area, northern English River Gneiss Belt: Geotraverse Conference Proceedings 1977, University of Toronto, p. 109-116.
Williams, J.G., 1978, Geology of the Crystal Lake area, Quetico subprovince: in Smith, I.E.M. and Williams, J.G., eds., Proceedings of the 1978 Archean Geochemistry Conference, University of Toronto.
Wooden, J.L., 1978, Rb-Sr isotopic studies of the Archean rocks of the eastern Lake Seul and Kenora areas, English River Subprovince, Ontario: in Smith, I.E.M. and Williams, J.G., eds., Proceedings of the 1978 Archean Geochemistry Conference, Universtiy of Toronto.

APPENDIX 1: ANALYZED SPECIMENS

No.	U of T Catalogue	
1	585	Pink biotite-bearing leucogranite, Miniss Lake area
2	586	Pink biotite-bearing leucogranite, Miniss Lake area
3	587	Pink biotite-bearing leucogranite, Miniss Lake area
4	582	White garnet-bearing leucogranite, Miniss Lake area
5	584	White garnet-bearing leucogranite, Miniss Lake area
6	583	White garnet-bearing leucogranite, Miniss Lake area
7	581	White garnet-bearing leucogranite, Miniss Lake area
8	564	Biotite-bearing leucogranite, Eva Granite, Crystal Lake area
9	563	Biotite-bearing leucogranite, Eva Granite, Crystal Lake area
10	565	Biotite-bearing leucogranite, Eva Granite, Crystal Lake area
11	568	Muscovite-bearing leucogranite, Eva Granite, Crystal Lake area
12	567	Muscovite-bearing leucogranite, Eva Granite, Crystal Lake area
13	566	Muscovite-bearing leucogranite, Eva Granite, Crystal Lake area
14	570	Garnet-bearing albite granite, Niobe Granite, Crystal Lake area.
15	571	Garnet-bearing albite granite, Niobe Granite, Crystal Lake area.
16	653	Granodiorite, Badesdawa Lake Pluton, Pickle Crow area.
17	660	Monzodiorite, Burchell Lake Pluton, Shebandowan area.
18	661	Monzodiorite, Burchell Lake Pluton, Shebandowan area.
19	658	Monzodiorite, Burchell Lake Pluton, Shebandowan area.
20	651	Granodiorite, Orchig Lake Pluton, Pickle Crow area.
21	652	Granodiorite, Hooker Burkoski Stock, Pickle Crow area.
22	662	Monzodiorite, Greenwater Lake Pluton, Shebandowan area.
23	654	Granodiorite, Osnaburgh Lake Pluton, Pickle Crow area
24	659	Granodiorite, Burchell Lake Pluton, Shebandowan area.
25	650	Granodiorite, Orchig Lake Pluton, Pickle Crow area.
26	655	Granodiorite, Ace Lake Pluton, Pickle Crow area.
27	657	Granodiorite, Riach Lake Pluton, Pickle Crow area.
28	656	Granodiorite, Riach Lake Pluton, Pickle Crow area.
29	576	Monzodiorite inclusion, Crystal Lake area.
30	574	Tonalite inclusion, Crystal Lake area.
31	575	Tonalite inclusion, Crystal Lake area.
32	572	Trondhjemite inclusion, Crystal Lake area.
33	573	Trondhjemite inclusion, Crystal Lake area.
34	589	Gneissic trondhjemite, Miniss Lake area.

VERTICAL GEOMETRY
OF THE CRUST

The Continental Crust and Its Mineral Deposits, edited by D.W. Strangway,
Geological Association of Canada Special Paper 20

STRUCTURE OF THE CONTINENTAL CRUST: A RECONCILIATION OF SEISMIC REFLECTION AND REFRACTION STUDIES

M.J. Berry and J.A. Mair
Earth Physics Branch,
Department of Energy, Mines and Resources,
1 Observatory Crescent, Ottawa, Ontario K1A 0E4

ABSTRACT

The interpretation of seismic refraction surveys by modern techniques, including extensive use of synthetic seismograms, reveals structures having lateral variations, transition zones rather than sharp boundaries and low velocity zones often in the upper crust and sometimes in the lower crust. In contrast, recent very detailed deep crustal reflection surveys reveal highly complex structures with significant short wavelength variations and boundaries that are most successfully modelled as first order discontinuities or as complex laminae of alternating high and low velocity material. The reflection and refraction techniques are thus providing different views of reality. We suggest that the reflection technique reveals the structural mix of Earth materials that is the cumulative result of past orogenic cycles, whereas the refraction method reveals transitional boundaries that have been largely formed in response to the temperature and stress regime acting on the material of the crust. The reflection and refraction boundaries may be related or they may have cross-cutting relationships. We conclude that both reflection and refraction data are necessary for a fuller understanding of the nature and composition of the Earth's crust.

RÉSUMÉ

L'interprétation des levés de sismique-réfraction par les techniques modernes, y compris l'utilisation intensive de sismogrammes synthétiques, révèle des structures qui ont des variations latérales, des zones de transition plutôt que des limites nettes et des zones à faible vitesse, souvent dans la croûte supérieure et quelquefois dans la croûte inférieure. Par contraste, des études récentes très détaillées de réflexion à grande profondeur dans la croûte révèlent des structures très complexes avec des variations et des limites significatives de courtes longueurs d'ondes qu'on peut modeler avec beaucoup de succès comme des discontinuités de premier ordre ou comme des laminae complexes de matériel à vitesse alternativement élevée et faible. Les techniques de réflexion et de réfraction présentent donc des vues différentes de la réalité. Nous suggérons que la technique de réflexion met en relief le mélange structural des matériaux terrestres qui est le résultat de l'accumulation de cycles orogéniques passés, alors que les méthodes par réfraction révèlent les zones limites de transition qui se sont surtout formées en réponse au régime de températures et de contraintes qui ont agi sur le matériel de la croûte. Les limites déterminées par réfraction et par réflexion peuvent être apparentées ou présenter des relations discordantes. Nous concluons que les données de réflexion et de réfraction sont toutes deux nécessaires pour mieux comprendre la nature et la composition de la croût terrestre.

INTRODUCTION

Refraction studies of the crust date to the last century and the work of Mohorovicić, Conrad and others. The method has continued to be the principal probe for mapping the structure of the crust until very recently and, up to the late 1960s,

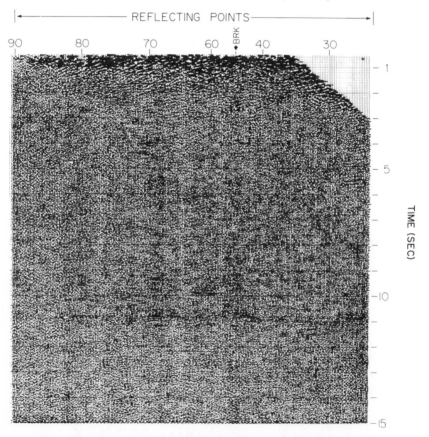

Figure 1. Vibroseis reflection section (1200% stack) from the Canadian Cordillera. Total length of section is approximately 20 km of surface coverage (after Mair and Lyons, 1976, with permission).

interpretations produced relatively simple layered models. Reflection techniques, developed to a high degree of sophistication by the oil industry, revealed at the same time that the upper crust was often highly complex, but these surveys were generally limited to sedimentary sequences. The application of theoretical seismograms in the late 60s and early 70s showed that the boundaries of the refraction models were in fact transition zones often of a complex layered nature. Recent very detailed deep crustal reflection surveys, using either exposive sources or the Vibroseis (registered trademark of the Continental Oil Company) technique, have revealed that complex structure exists throughout the crust at least to depths of the Morhorovicic boundary and probably deeper.

In this paper we will compare and seek to reconcile the models derived from reflection and refraction surveys of the crust and will suggest tectonic processes that may provide an explanation for the seemingly disparate results.

THE NATURE OF THE DATA

Modern seismic surveys of the crust range from extremely detailed near-vertical reflection Vibroseis profiles to very long-range refraction experiments, with detector spacing varying from 10 m to 5 km respectively. Figures 1 to 5 show representative

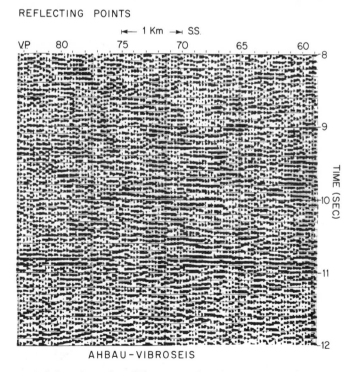

Figure 2. Expanded view of a section of Fig. 1. Note the coherent phase at about 11 seconds. The subsurface (ss) coverage scale is indicated (from Mair and Lyons, 1976, with permission).

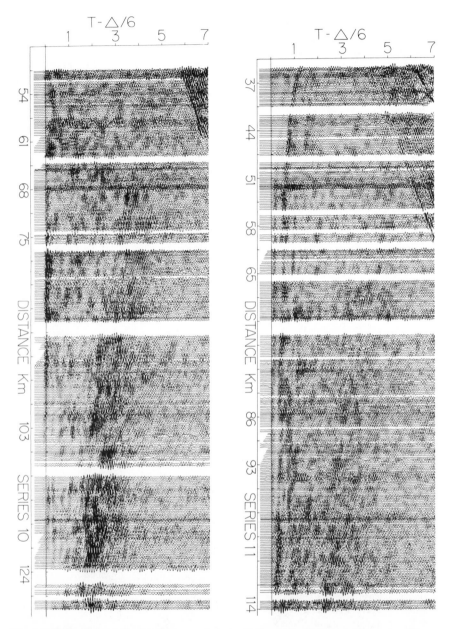

Figure 3. Reflection – refraction data recorded near Yellowknife, N.W.T., 1969, at two different offset distances by the same array (after Clee *et al.*, 1974, with permission). Note the coherent shallow reflection beginning at about 1.3 seconds at 35 km distance on the upper section and the reflected energy from the base of the crust at times of ~ 3.3 seconds (68 km) to ~ 1.5 seconds (124 km) on the lower section.

samples of seismic data from a number of experiments, and illustrate the very different nature of the data now available to studies of the Earth's crust.

Figure 1 shows a Vibroseis section that was recorded in 1973 in the Omineca crystalline belt of the Canadian Cordillera. Each trace represents a 1200% common reflection point stack and has had a variable gain function applied so that the event noticeable at 11 seconds is correlatable mainly through its phase coherence rather than its amplitude coherence. Figure 2 shows an expanded part of this section in the range of travel time from 8 to 12 seconds. An examination of the two figures shows that the seismic energy is more or less incoherent from the surface to two-way reflection times of 10 to 11 seconds where a laterally discontinuous horizon becomes evident.

In 1969 the Earth Physics Branch recorded a wide-angle reflection profile in the Slave Province in an attempt to study the detailed structure of a representative part of the Precambrian Shield. The upper section of Figure 3, recorded near Yellowknife, N.W.T. shows a highly correlatable shallow reflection which begins at about 1.3 seconds at 35 km and is apparently continuous over its 70-km length. The first-event refraction has a nearly constant velocity of 6 km/second. The lowest section of Figure 3 was recorded with the same recording locations but with the shot point shifted to give offset distances 15 km greater. On this section, there is little evidence of the shallow reflection, but excellent return of reflection energy from the base of the crust. Figure 4 is an expanded view of the lower section of Figure 3, showing the highly correlatable energy of this reflected phase over short segments of a few kilometres. The abrupt lateral changes in the character of the energy argue strongly for abrupt changes in the stratigraphy of the reflecting zone, and the duration of high amplitude energy return argues for a complex layered zone several kilometres in thickness.

Many long-range refraction profiles have been recorded with varying degrees of detail and uniformity of instrumentation. A recent high-quality profile recorded in Northern Britain (LISB-IV) is shown in Figure 5. In the upper intermediate range section the clear events (marked "a" and "d") are interpreted as refractions, while events "c" and "e" are identified as complex reflections from near-Moho depths (Bramford et al., 1978). Note that event "e" is correlatable over a distance greater than 100 km.

The lower seismic section of Figure 5 shows the full LISB-IV profile recorded along the length of the British Isles, to 700 km in offset distance. The long-range continuity of the refraction and wide-angle reflection events of this figure are in contrast to the discontinuous nature noted for the short-range, near-vertical reflected events of Figures 1 to 4.

Hirn et al. (1973) point out that the weak first arrival in the range from about 100 to 300 km, seen on Figure 5, is now commonly taken as the signature of the Mohorovicic Discontinuity. However this event is overtaken at greater distances by a much higher amplitude first arrival with approximately the same velocity and is frequently recorded out to 1000 km. It is this phase that was noted by Mohorovicic and taken by him to signify the crust – mantle transition.

It should be noted that there are no good examples in the literature where high-quality refraction data have been recorded along deep crustal near-vertical reflection profiles.

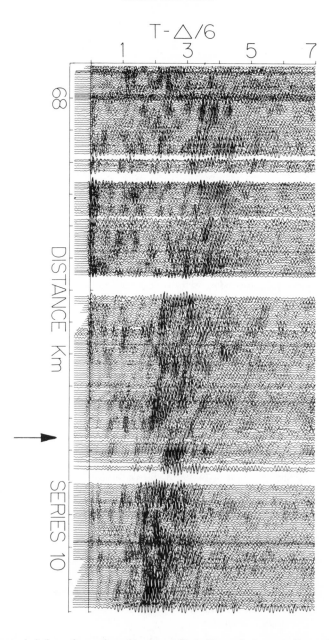

Figure 4. Expanded view of a section of lower profile in Figure 3. Arrow indicates location of a probable major fault in the crust-mantle transition zone (after Clee *et al.*, 1974, with permission).

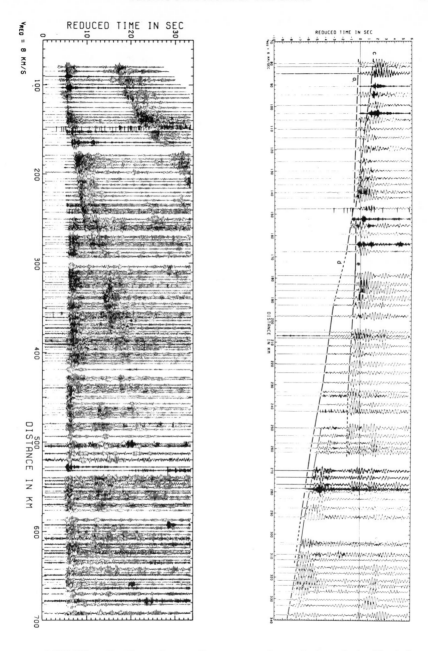

Figure 5. LISB-IV profile. Upper: intermediate range section, reducing velocity 6 km/s; lower: full profile, reducing velocity 8 km/s (from Bamford *et al.*, 1978, with permission).

SEISMIC MODELS OF THE CRUST

A quick glance at any geological map of Precambrian Shield regions shows considerable lateral heterogeneity in the rocks at the ground surface. This mapped structural complexity is generally thought to be typical of the crust at greater depth, and appears to be inconsistent with the results of long-range refraction experiments that are most simply interpreted in terms of a few subhorizontal layers.

Examples of long-range refraction results are shown in Figure 6. Three reversed refraction profiles, each about 400 km long, were recorded by the Earth Physics Branch in 1968 in the northeastern Canadian Shield, parallel to the Grenville Front, the boundary between the Grenville and Superior Precambrian Provinces (Berry and Fuchs, 1973). The six velocity-depth models represent the structure of the crust along a cone, starting at the shot point and extending down and along the profile, because the rays that bottom at greater depths are generally observed at greater distances. This feature of the seismic refraction method produces models that do not represent the velocity-depth function for any particular vertical section in the survey area. However, because the models have many features in common, we are confident that they are relevant to the real Earth.

These velocity-depth models from the Canadian Shield (Fig. 6) bear a striking resemblance to Mueller's (1977) generalized seismic model of the continental crust in Europe (Fig. 7, upper). His model is based mainly on seismic refraction work, but is supplemented and reconciled with a small amount of deep reflection data. The horizontal continuity implied by this model is partially qualified by Mueller (1977, p. 289):

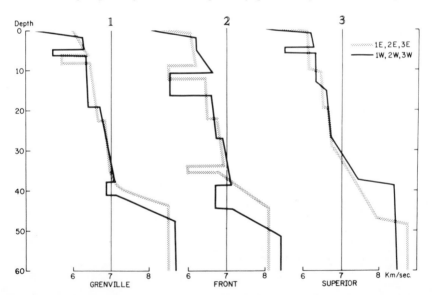

Figure 6. The six P-wave velocity models developed for the three reversed profiles of the Grenville-Superior refraction experiment. The solid lines represent profiles recorded by using eastern shot points, while the stippled lines represent profiles recorded by using western shot points (from Berry and Fuchs, 1973 with permission).

"Although not all block structures within each of the continents of the world show the same details in their velocity-depth function, the lateral consistency in structure is quite astonishing".

In contrast, Smithson's (1978) crustal model (Fig. 7, lower), based mainly on Vibroseis near-vertical reflection data from the United States, shows strong lateral and vertical changes. Three major zones are identified (Smithson, 1978, p. 750): "an upper zone of supracrustal rocks and granitic intrusions, a middle migmatitic zone, and a lower more mafic (andesitic) zone". The lateral heterogeneity revealed by this technique has led Smithson and his co-workers to discount any large-scale lateral continuity within the crust.

We suggest that these apparently contrary viewpoints are both realistic and reconcilable; both contribute to our understanding of the nature and evolution of the Earth's crust. The key to the reconciliation is the recognition that near-vertical reflected waves and long-range refracted waves are sensitive to different aspects of a three-dimensional structure. The acoustic impedance of a boundary or set of boundaries is a complex function involving, among other parameters, the angle of incidence and the wavelength of the impinging energy. Boundaries virtually opaque to one range of wavelengths or incident angles may be virtually transparent to others. The differences in field geometry between near-vertical seismic reflection and long-range refraction surveys will therefore, almost invariably, produce different information relevant to the nature of the Earth section traversed.

In the following section we will describe some of the more common features of the crust as revealed by refracted waves and will comment on the similarities and differences between these and the features detected by reflection surveys.

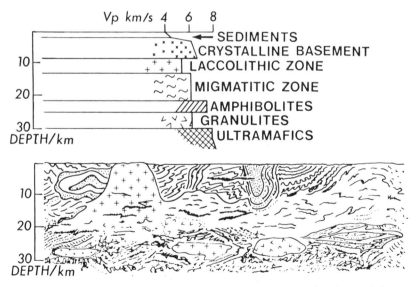

Figure 7. Upper: Crustal model of velocity vs. depth from seismic refraction work in Europe (after Mueller, 1977). Lower: Geological crustal section interpreted from Vibroseis reflection data in the United States (after Smithson, 1978).

Nature of the Crust

As we have written earlier (Berry and Mair, 1977), the seismic velocity structure of the crust is only of limited interest in itself, however detailed it may be. Its real value lies in what it can tell us about the tectonic history of a region, and about rock properties and other parameters such as temperature and stress in the crustal section.

The common features of continental crustal refraction models are: 1) the near-universal observation of a rapid velocity increase in the top 1 or 2 km, followed by a minor gradient to depths typically of 10 km; 2) the common observation of a low-velocity layer in the mid-upper crust, followed by a return to the velocities typical of the minor gradient; 3) the less common observation of a minor increase in velocity in the middle crust, over a depth interval of variable thickness, followed in some cases by a return to lower velocities; 4) the near-universal observation of a rapid velocity increase with depth to values in excess of 7.8 km/s at 30 to 60-km depth; this boundary, named after Mohorovicic, defines the crust-mantle interface; 5) the observation, where sufficient data have been gathered, that horizontal velocity anisotropy exists in the upper mantle.

The significance of some of these features is controversial, and we shall discuss them in turn.

1) the rapid velocity increase in the top 1 to 2 km of the crystalline crust corresponds very well with the velocity vs. pressure curves determined in the laboratory for granitic rocks, (e.g. Birch, 1961). The velocity increase is almost certainly caused by the closing of micro-fractures. We would expect that a reflection survey in, say, the Grenville-Superior region would not detect this significant feature, but would map instead a complex structure related to the geology seen at the surface. A refraction survey, on the other hand, finds the velocity gradient to be expressive and laterally coherent, though unable to resolve the more complex structure.

2) The low-velocity zone seen in the 5 to 10 km depth range in many parts of the Canadian Shield (Fig. 6) is commonly reported in such widely separated places as northern Germany, western Transvaal, Pennsylvania, southeastern Australia, central Japan, eastern Montana, the Baltic Shield and the Black Forest of West Germany (Berry and Mair, 1977). These interpretations are based partly upon reflections from the top and bottom of the layer. The Yellowknife section (Fig. 3, upper) provides a good example of two shallow reflections that may result from such a zone. (The strong phase beginning at 1.35 seconds at 35 km is preceded by a weaker phase about 0.5 seconds earlier.) Such zones cannot, in general, be unambiguously interpreted from refraction sections; however, their presence is occasionally revealed by their effects on secondary event amplitudes and by time delays apparent on first arrival events.

Mueller (1977) and Landisman et al. (1971) have argued that such low-velocity zones are most simply explained as widespread laccolithic granitic intrusions in the upper crust. Such intrusions undoubtedly exist, but it seems to us unlikely that they can provide a general explanation, given that the zone has been detected in such widely different tectonic settings. Berry (1972) has suggested that the combined effects of pore fluid and varying permeability with depth could provide an alternative explanation.

Figure 8 (from Todd and Simmons, 1972) shows that when the pore pressure within a rock sample equals the confining pressure, the P-wave velocity is reduced by as much as 15%. This leads naturally to the suggestion that the low-velocity zone is a region where the fluid pore pressure is equal to the confining pressure. This could be caused by the reduction of the effective permeability to zero by the closing of the microfractures in response to the lithostatic pressure. Short-range refraction profiles (e.g., Fig. 3) suggest that the upper boundary of the zone is well defined, which is reasonable as any degree of permeability will, over time, allow the pore pressure to assume hydrostatic levels. Only within a region of zero permeability would the pore pressure generally be equal to the confining pressure. In special cases one might hypothesize the existence of a cap-rock (as is found in many hydrocarbon reservoirs) to explain the trapped pore fluid, but this is unlikely to be a general explanation.

Short-range refraction data strongly suggest that the lower boundary of the low-velocity zone is also well defined and only several hundred metres thick. The experimental results of Nur and Simmons (1969) suggest that this may represent the transition from wet to dry rock, as depth increases in the upper crust. If such is the case the lower boundary represents the transition from rocks having free fluids in the pore spaces to those with effectively no pore spaces, and with the water chemically combined. In this case the depth of the transition is probably controlled by the temperature and stress fields as well as by the mineralogy.

Beneath a low-velocity zone the P-wave velocity may be slightly higher than that

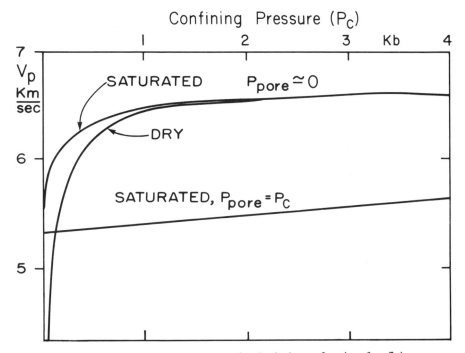

Figure 8. The effect of pore pressure on compressional velocity as a function of confining pressure in crystalline rocks. The upper curves are for zero pore pressure, while the lower line is for the case in which the pore pressure equals the confining pressure (after Todd and Simmons, 1972).

immediately above the layer. However, whether a velocity increase is present or not, the velocity gradient for the next several kilometres appears to be very similar to that detected in the upper crust. This seems to argue for a relatively uniform material at these depths, at least in a gross sense, the velocity being controlled largely by the pressure and temperature gradients.

3) Using the Grenville and Superior provinces (Fig. 6) as an example, we find a relatively sharp jump in velocity of about 0.3 km/s at depths of about 20 km. A similar feature has been detected by Kanasewich *et al.* (1969) beneath the Alberta Plains and by Hall and Brisbin (1965) beneath the Superior Province west of Lake Superior. Evidence of this feature within the Cordillera has been reported by Hales and Nation (1973), Berry and Forsyth (1975) and Cumming *et al.* (1979) from refraction data, but we can see no evidence of it in the reflection data of Figure 1. We have not detected evidence of this feature beneath the Slave Province from refraction data (Barr, 1971), nor from the refraction/reflection data shown in Figure 3.

The midcrustal refractor and sometime reflector may be related to the Conrad discontinuity in parts of Europe and is evident on Mueller's model (Fig. 7, upper) as the high velocity tooth at about 24 km-depth. In Canada it has been named the Riel discontinuity by some authors, partly to suggest that it may not represent the same feature in Canada as in Europe. Certainly, the refraction characteristics of the discontinuity appear to change significantly from region to region, with interpretations ranging from a major first-order discontinuity to a minor second-order transition in velocity. When it is present, it is usually interpreted as representing a change of rock type or bulk chemical composition. Mueller (1977) suggests it to be amphibolites overlying dry granulite facies rocks.

In modelling the Q-structure of the continental crust beneath Alberta, Clowes and Kanasewich (1970) conclude that the reflection properties of the Conrad (Riel) can be modelled by a complex sandwich of thin (about 200m) sills of alternating high and low-velocity material extending over a depth interval of less than 2 km. However, Smithson (1978) suggests that the seismic discontinuity, when reported, is simply due to the occurrence of a large body of intermediate to mafic rock that happens to be favourably located with respect to a seismic refraction profile. Certainly the "Conrad" appears to be both elusive and variable in nature when detected.

The proposal that the Conrad discontinuity in Europe represents amphibolites overlying granulites implies that it represents a metamorphic front rather than a simple change in bulk chemistry. In Alberta, the evidence of Clowes and Kanasewich (1970) that it may be a sill sandwich suggests that it was formed in a varying anisotropic stress field with minimum principal stress being vertical during sill emplacement.

There is occasional evidence that in the middle of the lower crust another low-velocity layer exists. Landisman *et al.* (1971) review the evidence for this and suggest that the zone can be explained in terms of certain dehydration reactions that release free water into the surrounding pore spaces. Such an explanation is similar to that favoured earlier for the low-velocity zone in the upper crust.

On the basis of the above we suggest that the lateral continuity of deep crustal horizons, as interpreted from seismic refraction surveys, is caused in large part by metamorphic processes involving water. Such horizons, being gradational, will be transparent to near-vertical reflection profiling. Lamellar sill sandwich boundaries will, on the other hand, appear highly reflective to reflection techniques while virtually transparent to refraction techniques. The sills may be a relatively localized phenomenon, indicating a history of the varying stresses and temperatures at these depths. Both features are clearly relicts of the tectonic history of the crustal section.

4-5) The significant velocity increase at depths of 30 to 60 km from velocities of about 7 km/s to values greater than 7.8 km/s is a nearly universal phenomenon which has been adopted, since Mohorovicic's discovery, as the operational definition of the base of the crust.

The model of the crust and the upper mantle shown in Figure 9 is a composite of Mueller's crustal model in the upper 30 km and the upper mantle model of Hirn *et al.* (1973) and D.P. Hill (pers. commun.) deduced from long-range European profiles such as those of Figure 5. We will restrict our discussion to the region depicted here by a sharp velocity increase at 30 km. We note, however, the interpretation of yet another low-velocity zone beneath this transition, i.e., that the Moho we usually detect with refraction profiles of modest length may be a fairly thin, high-velocity lid. This zone is usually a good refractor and sometimes an excellent reflector, although there is some evidence (Meissner, 1973) that the reflection may arise from somewhat shallower depths.

Petrologically there have been three principal hypotheses offered to explain the nature of the crust-mantle transition: 1) an isochemical phase change, namely basalt to eclogite (Lovering, 1958; Ito and Kennedy, 1971); 2) a chemical discontinuity, namely from intermediate granulite facies rocks to peridotite (Ringwood & Green, 1966; Green & Ringwood, 1967, 1972), or 3) a hydration reaction, namely serpentine to peridotite (Hess, 1962) especially under the oceans (Claque and Straley, 1977). The chemical

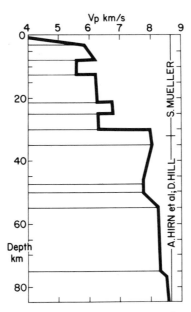

Figure 9. A composite velocity-depth model based on European data (after Mueller, 1977; Hirn *et al.*, 1973; D.P. Hill, pers. commun.).

discontinuity is the most commonly accepted concept at this time. However, to explain the seismic data we suggest a further hypothesis. 4) a seismic boundary resulting from crystallization or recrystallization in a present, or paleo, anisotropic stress field (Babuska, 1970). This does not necessarily exclude any of the first three suggestions.

Babuska's hypothesis has not received much attention or support except by Berry and Mair (1977) and more recently, we believe, by Phinney (1978). It certainly explains the observations of velocity anisotropy in the upper mantle and gives the rocks certain mechanical properties that we find reasonable in explaining much of the available data.

It has been proposed by many authors, including Fuchs (1969), Clowes and Kanasewich (1970), Meissner (1973), Clee et al. (1974), Dohr and Meissner (1975), and Cumming and Chandra (1975), that the crust-mantle transition zone is a many-layered complex of alternating high and low-velocity lamina having a total thickness of a few kilometres. Individual layers are postulated as thinner than the seismic wavelengths which are used to probe the Earth. The model is thought by many authors to explain best the observations that 1) near-vertical reflection energy returning from deep within the crust is of much higher amplitude than can be explained by a reasonable first-order discontinuity or by ramp functions of increasing velocity with depth, and 2) the reflected energy is limited in band-width, little of the detectable energy having frequencies below 10 Hz. It appears that the model applies to both Precambrian and younger regions.

We add to the above that the reflected energy returned from the zone should persist at relatively high amplitudes for up to 2-s duration in some areas. The Yellow-knife section (Fig. 3, upper), for example, shows a short-duration shallow reflection, indicating a short down-travelling pulse, but a long-duration signal is recorded from the crust-mantle transition (Fig. 4). In addition, the Vibroseis section from the Cordillera (Fig. 2) shows that for a second or so above and below the main arrival at 11 seconds there is a complex series of sub-parallel, reflected events, the assemblage being laterally discontinuous over distances greater than about 1 km.

Figure 10 models a laminated transition zone with a more or less random sequence of high and low-velocity beds of variable thickness, but all considerably less than typical wavelengths of recorded seismic energy. The down-going pulse (indicated at the top left), when reflected from the model at various angles of incidence, produces the changing pulse shapes shown. These signals contain contributions from all the multiples and converted waves generated by the interfaces. The calculations assume an incident plane wave-front for simplicity, but calculations using spherical waves would not change our conclusions. Numerical experiments indicate that the only effective way to increase the apparent duration of the reflected energy is to increase the total thickness of the transition layer or the duration of the incident pulse. In this example it is interesting to note that the highest-amplitude wavelet, apparent at $0°$ incidence, does not correlate with any specific interface, but is a result of constructive interference. Under typically noisy field recording conditions, which might obscure all events other than this wavelet, we might be inclined to interpret a sharp first-order transition, whereas we have contributions from a zone of over 3 km in thickness.

Berry and Mair (1977) have described some processes that might plausibly generate such theoretically appealing models. Some of the main points for the continental environment can be summarized as follows:

1) It appears that anisotropic stress fields and temperatures suitable for recrystallization with a preferred orientation of the crystal axes are common during the formation of the Earth's crust. In general, crustal material, particularly in the lower crust and upper mantle, will be anisotropic; isotropic behaviour is simply a convenient assumption when we lack the data to postulate otherwise.

2) Berry and Fuchs (1973) have described a weak dependence on P_n velocity on direction for the eastern Canadian Shield, and Bamford (1973) has found this same dependence in parts of Germany. The effect has been well documented in the uppermost-mantle under parts of the Pacific (Hess, 1964; Raitt *et al.*, 1969; Morris *et al.*, 1969; Keen and Barrett, 1971). Laboratory measurements support this interpretation. Birch (1960, 1961) has described the high elastic anisotropy found in some rocks, especially in dunites, and Babuska (1970) lists the velocity anisotropy of samples of ultra-mafic rocks with values of up to 20%.

3) Certainly velocity anisotropy should be a serious consideration in seismic models of crustal structure, as most of the observational data reveal only the horizontal component of velocity. Relating horizontal velocity to rock density without know-

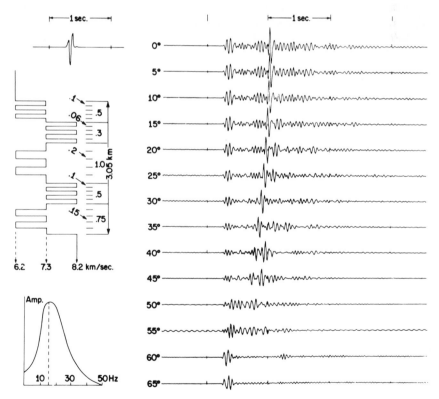

Figure 10. Theoretical reflection response of the indicated layered sequence to a plane wave pulse (upper left) at various reflecting angles. Spectrum of the pulse is shown at lower left (from Berry and Mair, 1977, with permission).

ledge of the vertical velocity could cause serious errors in gravity modelling. There is much laboratory and experimental data to suggest that the velocity anisotropy measured in the oceanic upper mantle is due to (re)crystallization of minerals in the anisotropic stress field that, predictably, occurs as the material cools and is transported away from oceanic ridges.

In and beneath the continental crust other mechanisms may be acting. Haxby and Turcotte (1976) have shown that erosion of the continental crust generates horizontal compressional deviatoric stress, whereas uplift and associated thermal stresses produce tension. On balance the combination leads to a predominantly horizontal deviatoric tension. During submergence and deposition the converse is true, and the crust is subjected to horizontal deviatoric compression. It seems plausible to us that a varying stress regime of this nature in the lower crust will, under favourable temperature conditions, lead to the development of velocity anisotropy, possibly at different levels as the process is repeated through time. The Cambrian sedimentary record of the Rocky Mountains provides an excellent example of a crustal section that has been subjected to such repeated cycles of emergence and submergence.

While direct evidence of stress is difficult to find, there is ample evidence of the resulting strains in the geological record. Figure 11 shows the scale of these strains in a cross-section of the Cordillera from the Omineca Crystalline Belt to the Plains of Alberta. There is reasonable geologic evidence, based on the metamorphic grade of exposed aluminium silicates, to indicate that the Cambrian rocks now exposed at elevations of 4 km above sea level (upper figure) were once buried beneath 10 km of Triassic-Jurassic sediments, as depicted in the lower part of the Figure (Campbell, 1973). Uplift of some 14 km, accompanied by a total crustal shortening of about 60 km, has occurred in the section over the last 160 Ma. In the eastern part of this section the style of faulting, well studied through exploration for hydrocarbons, is near-horizontal

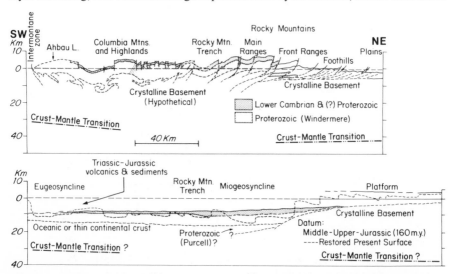

Figure 11. Structural cross-section, present (upper) and restored to a mid-upper-Jurassic datum (lower) of a section through the eastern Canadian Cordillera (after Campbell, 1973).

at depth, changing as the vertical load decreases, to near-vertical at the surface. In this environment old rocks may overlie younger ones, and the same formation may overlie itself several times. It seems reasonable, therefore, to believe that deep structures, including the Moho, have been involved, and that their mode of faulting has been near-horizontal under this stress. "Laminar plastic flow" may be a better term for the compensation for this stress at depths of 25 to 50 km and temperatures above 300°C (Turcotte, 1974).

We can thus reasonably explain the high amplitudes, the abrupt changes in reflection character and the duration of reflection energy returned from the deep crust in some areas. The lateral continuity over distances in excess of 100 km and the increase in apparent velocity of refracted energy, however, argue for a first-order transition in acoustic impedance or a transition layer several kilometres thick. A density and velocity change from granulites to peridotite seems a reasonable hypothesis to explain these observations. The reflection lamellae could lie above or be superimposed on this major transition. In some areas the lamellae may be absent (erased) or poorly structured, depending upon the most recent major tectonic events.

CONCLUSIONS

1) In general, four factors determine the velocity structure of the crust: a) the tectonic style of a region, including the geochemistry of the rocks and their orogenic history, b) the distribution of water in the crust, either chemically combined to yield hydrated minerals or free as pore fluid, c) the past and present states of stress, and d) the past and present temperature regimes.

2) Cross-cutting relationships between petrology and isograds of stress and temperature are evident at and near the surface and must exist at greater depths. Reflection and refraction techniques provide coherent data for first-order velocity transitions between thick layers, but otherwise diverge: reflection techniques tend to map variations of velocity related to fine-scale petrology, while refraction techniques tend to map the longer wavelength variations caused by metamorphism.

3) We find no reason to assume that deep seismic discontinuities, or even some shallow ones for that matter, are necessarily related in age to the exposed surface rocks. We consider them instead to be features whose seismic expression may be dependent to some extent upon petrology, but also upon the effects of past or present temperature and stress. We suggest that varying stress and temperature fields can erase and reform seismic boundaries, making them complex or simple, all within a generally gradational rock chemistry.

4) Water plays an important role in metamorphic processes. In addition we suggest it plays a key role in explaining low-velocity layers and zones of high conductivity in crustal sections of greatly varying age and tectonic style.

5) Reflection and refraction studies provide different data relevant to the Earth's structure. These are reconcilable, and both methods are necessary to improve our understanding of the nature and evolution of the crust.

Contribution from the Earth Physics Branch No. 812.

REFERENCES

Babuska, V., 1970, Elastic anisotropy of the upper mantle and the Mohorovicić Discontinuity: Studia Geoph. et Geod., v. 14, p. 296-309.

Bamford, S.A.D., 1973, Refraction data in western Germany – A time-term interpretation: Z. Geophys. v. 39, p. 907-927.

Bamford, S.A.D., Nunn, K., Prodehl, K. and Jacob, B., 1978, LISB-IV. Crustal Structure of Northern Britain: Geophys. Jour Royal Astron. Soc., v. 54, p. 43-60.

Barr, K.G., 1971, Crustal refraction experiment: Yellowknife 1966: Jour. Geophys. Res., v. 76, p. 1929-1947.

Berry, M.J., 1972, Low velocity channels in the earth's crust?: Comments Earth Phys. Geophys., v. 3, p. 59-68.

Berry, M.J. and Forsyth, D.A., 1975, Structure of the Canadian Cordillera from seismic refraction and other data: Canadian Jour. Earth Sci., v. 12, p. 182-208.

Berry, M.J., and Fuchs, K., 1973, Crustal structure of the Superior and Grenville provinces of the northeastern Canadian Shield: Seismol. Soc. Amer. Bull., v. 63, p. 1393-1432.

Berry, M.J. and Mair, J.A., 1977, The nature of the earth's crust in Canada: Heacock, J.G., ed., The Earth's Crust: Geophysical Monograph 20, Amer. Geophys. Union, Washington, D.C., p. 319-348.

Birch, F., 1960, Velocity of compressional waves in rocks to 10 kbar, I: Jour. Geophys. Res., v. 65, p. 1083.

_____, 1961, Velocity of compressional waves in rocks to 10 kbar, I: Jour. Geophys. Res., v. 66, p. 2199.

Campbell, R.B., 1973, Structural cross-section and tectonic model of the S.E. Canadian Cordillera: Amer. Jour. Earth Sci., v. 10, p. 1607-1620.

Claque, D.A. and Straley, P.F., 1977, Petrologic nature of the oceanic Moho: Geology, v. 5, p. 133-136.

Clee, T.E., Barr, K.G. and Berry, M.J., 1974, The fine structure of the crust near Yellowknife: Canadian Jour. Earth Sci., v. 11, p. 1534-1549.

Clowes, R.M. and Kanasewich, E.R., 1970, Seismic attenuation and the nature of reflecting horizons within the crust: Jour. Geophys. Res., v. 75, p. 66-6705.

Cumming, G.L. and Chandra, N.N., 1975, Further studies of reflections from the deep crust in southern Alberta: Canadian Jour. Earth Sci., v. 12, p. 539-557.

Cumming, W.B., Clowes, R.M., and Ellis, R.M., 1979, Crustal structure from a seismic refraction profile across southern British Columbia: Canadian Jour. Earth Sci., v. 16, p. 1024-1040.

Dohr, G.P., and Meissner, R., 1975, Deep crustal reflections in Europe: Geophysics, v. 40, p. 25-39.

Fuchs, K. 1969, On the properties of deep crustal reflectors: Z. Geophys., v. 35, p. 133-149.

Green, D.H. and Ringwood, A.E., 1967, An experimental investigation of the gabbro to eclogite transformation and its petrographic applications: Geochim. Cosmochim. Acta., v. 31, p. 767-833.

_____, 1972, A comparison of recent experimental data on the gabbro-garnet granulite – eclogite transformation: Jour. Geol. v. 80, p. 277-288.

Hales, G.L. and Nation, J.B., 1973, A Seismic refraction survey in the northern Rocky Mountains: more evidence for an intermediate crustal layer: Geophys. Jour. Royal Astron. Soc., v. 35, p. 381-399.

Hall, D.H. and Brisbin, W.C., 1965, Crustal structure from converted head waves in central western Manitoba: Geophysics, v. 30, p. 1053-1067.

Haxby, W.F. and Turcotte, D.L., 1976, Stresses induced by the addition or removal of overburden and associated thermal effects: Geology, v. 4, p. 181-184.

Hess, H.H., 1962, History of ocean basins: in Engel, A.E.J., James, H.L., and Leonard, B.F., eds., Petrologic Studies (Buddington volume); New York, Geol. Soc. America, p. 599-620.

_____, 1964, Seismic anisotropy of the uppermost mantle under oceans: Nature, v. 203, p. 629-631.

Hirn, A., Steinmetz, L., Kind, R. and Fuchs, K. 1973, Long range profiles in western Europe: II Fine structure of the lower Lithosphere in France (Southern Bretagne): Zeit. fur Geophys., v. 39, p. 363-384.

Ito, K. and Kennedy, G.C., 1971, An experimental study of the Basalt-Garnet granulite – eclogite transition, in the Structure and Physical Properties of the earth's crust: Heacock, J.G., ed., The Earth's Crust: Geophysical Monograph 20, Amer. Geophys. Union, Washington D.C., p. 303-314.

Kanasewich, E.R., Clowes, R.M. and McCloughan, C.H., 1969, A buried Precambrian rift in western Canada: Tectonophysics, v. 8, p. 513-527.

Keen, C.E. and Barrett, D.L., 1971, A measurement of seismic anisotropy in the northeast Pacific: Candian Jour. Earth Sc., v. 8, p. 1056-1064.

Landisman, M., Mueller, S. and Mitchell, B.J., 1971, Review of evidence for velocity inversions in the continental crust: The Structure and Physical Properties of the Earth's Crust: Heacock, J.G., ed., The Earth's Crust: Geophysical Monograph 20, Amer. Geophys. Union, Washington, D.C., p. 11-34.

Lovering, J.F., 1958, The nature of the Mohorovicic Discontinuity: Trans. Amer. Geophys. Union, v. 39, p. 947-955.

Mair, J.A. and Lyons, J.A., 1976, Seismic reflection techniques for crustal structure studies: Geophysics, v. 41, p. 1272-1290.

Meissner, R., 1973, The 'Moho' as a transition zone: in Geophysical Surveys, v. 1, p. 195-216, D. Reidel, Hingham, Mass.

Morris, G.B., Raitt, R.W. and Shor, G.G. Jr., 1969, Velocity anisotropy and delay-time maps of the mantle near Hawaii: Jour. Geophys., Res., v. 74, p. 300-4315.

Mueller, S., 1977, A new model of continental crust: Heacock, J.G., ed., The Earth's Crust: Geophysical Monograph 20, Amer. Geophys. Union, Washington, D.C., p. 289-317.

Nur, A. and Simmons, G., 1969, The effect of saturation on velocity in low porosity rock: Earth Planet. Sci. Letters, v. 7, p. 183-193,

Phinney, R.A., 1978, Interpretation of reflection seismic images of the lower continental crust (abs.): EOS, Trans. Amer. Geophys. Union, v. 59, p. 389.

Raitt, R.W., Shor, G.G., Jr., Francis, T.J.G. and Morris, G.B., 1969, Anisotropy of the Pacific upper mantle: Jour. Geophys. Res., v. 74, p. 3095-3109.

Ringwood, A.E. and Green, D.H., 1966, Petrologic nature of the stable continental crust, The Earth Beneath the Continents: Steinhart, J.S. and Smith, T.J., eds., Geophysical Monograph 10, Amer. Geophys. Union, Washington, D.C., p. 611-619.

Smithson, S.B., 1978, Modelling continental crust: structural and chemical constraints: Geophys. Res. Letters, v. 5, p. 749-752.

Todd, T. and Simmons, G., 1972, Effect of pore pressure on the velocity of compressional waves in low-porosity rocks: Jour. Geophys. Res., v. 77, p. 3731-3743.

Turcotte, D.L., 1974, Are transform faults thermal contraction cracks?: Jour. Geophys. Res., v. 79, p. 2573-2577.

The Continental Crust and Its Mineral Deposits, edited by D.W. Strangway,
Geological Association of Canada Special Paper 20

EARTHQUAKES AND OTHER PROCESSES WITHIN LITHOSPHERIC PLATES AND THE REACTIVATION OF PRE-EXISTING ZONES OF WEAKNESS

Lynn R. Sykes

Lamont-Doherty Geological Observatory and Department of Geological Sciences, Columbia University,
Palisades, New York 10964

ABSTRACT

The distribution of earthquakes and of igneous rocks postdating continental rifting is summarized and placed into a plate tectonic framework for about 15 continental areas within
lithospheric plates. In continents, intraplate earthquakes tend to concentrate along pre-existing
zones of weakness associated with the last major orogenesis that preceded the opening of the
present oceans. Many pre-existing zones of weakness (including fault zones, suture zones, failed
rifts and other tectonic boundaries), particularly those near continental margins, were reactivated
during the early stages of continental fragmentation. In contrast, intraplate shocks rarely occur
within the older oceanic lithosphere or within the interiors of ancient cratonic blocks of the
continents. In some instances, alkaline magmatism persisted for as long as 100 Ma along faults
that were reactivated during continental separation. The type of intraplate magmatism appears to
be related to the thickness of the lithosphere. A number of old zones of weakness have moved
repeatedly throughout geologic history in response to various plate tectonic events. The state of
stress appears to be different along some of these features that are now located within plates than
it was when the features were reactivated during the early opening of an ocean. Horizontal
compressive stresses appear to be present today in many of the pre-Mesozoic orogenic belts that
were reactivated by continental rifting. Large compressive stresses within plates appear to be
transmitted long distances through the lithosphere. During the fragmentation of a super-
continent, multi-branched rifting usually follows the youngest zone of previous orogenesis rather
than passing through old cratonic areas where the lithosphere is thick, cold and strong. Rift
junctions, which have been interpreted by some as hot spots related to mantle plumes, seem
instead to be strongly related to the pre-existing mosaic of cratons and younger belts of deformation. Thus, a number of hot spots in continents appear to be passive features rather than the
surficial expression of mantle plumes. During the early development of an ocean, the pre-existing
mosaic of structural elements within the thick continental lithosphere may result in large normal

forces across some plate margins, leaky transform faulting and localized stress concentrations. The early directions of sea-floor spreading and of transform faulting may be altered by these boundary forces and by the geometrical constraints imposed in separating old cratonic blocks.

RÉSUMÉ

On passe en revue la distribution des séismes et des roches ignées post-datant l'effondrement continental et on les replace dans un cadre de tectonique des plaques pour environ 15 régions continentales à l'intérieur de plaques lithosphériques. Dans les continents, les séismes entre les plaques tendent à se concentrer le long de zones de faiblesse pré-existantes associées à la dernière orogénèse majeure qui a précédé l'ouverture des océans actuels. Plusieurs zones de faiblesse pré-existantes (comprenant des zones de failles et d'autres limites tectoniques, des zones de suture, des fossés d'effondrement faillés), particulièrement celles qu'on retrouve près des bordures continentales, ont été réactivées durant les premiers stades de fragmentation continentale. Par contraste, les séismes dans les plaques se produisent rarement à l'intérieur de l'ancienne lithosphère océanique ou à l'intérieur d'anciens blocs cratoniques continentaux. Dans certains cas, le magmatisme alcalin a persisté pour une période allant jusqu'à 100 Ma le long de failles qui ont été réactivées durant la séparation continentale. Le type de magmatisme dans les plaques semble être relié à l'épaisseur de la lithosphère. Un certain nombre d'anciennes zones de faiblesse ont bougé de façon répétée au cours de l'histoire géologique en réponse à différents épisodes de tectonique des plaques. Le long de certaines de ces structures qui sont maintenant localisées à l'intérieur des plaques, l'état des contraintes semble avoir été différent de ce qu'il était quand ces structures ont été réactivées durant la phase initiale d'ouverture d'un océan. Des contraintes horizontales de compression semblent être présentes actuellement dans plusieurs des zones orogéniques d'avant le Mésozoïque qui ont été réactivées par l'effondrement continental. De grandes contraintes de compression dans les plaques semblent se transmettre sur de longues distances à travers la lithosphère. Durant la fragmentation d'un super-continent, l'effondrement se fait selon plusieurs embranchements et suit habituellement la zone la plus récente d'orogénèse antérieure plutôt que de passer par des régions cratoniques anciennes où la lithosphère est épaisse, froide et résistante. Les jonctions de fossés d'effondrement que certains ont interprétées comme des points chauds associés à des panaches du manteau semblent plutôt être fortement reliés à la mosaïque pré-existante de cratons et des zones de déformation plus récentes. Ainsi, un certain nombre de points chauds sur les continents semblent être des caractéristiques passives plutôt que l'expression super-ficielle des panaches du manteau. Au cours des premières phases de développement d'un océan, la mosaïque pré-existante d'éléments structuraux à l'intérieur de l'épaisse lithosphère continentale peut produire de grandes forces perpendiculaires à certaines bordures de plaques, des failles transformantes avec fuites de matériel, et des concentrations locales de contraintes. Les directions initiales de l'expansion des fonds océaniques et de la formation de failles trans-formantes peuvent avoir été altérées par ces forces aux limites et par les contraintes géométriques imposées par la séparation de vieux blocs cratoniques.

INTRODUCTION

This paper is an abbreviated and updated version of a longer paper (Sykes, 1978) which discussed in detail the distribution of earthquakes and alkaline magmatism for about 15 areas situated within large lithospheric plates. I undertook a review and synthesis of data on intraplate areas since so little was known about tectonic processes operating today in those regions; we now know a great deal about processes at most plate boundaries.

One of the central points of the 1978 paper was that many intraplate phenomena can be placed into a plate tectonic framework. Many of the present zones of intraplate deformation occur along old tectonic features that were reactivated during continental fragmentation and the development of new oceans. This paper will describe a model for the fragmentation of a super-continent, such as the separation of South America and Africa in the Mesozoic. The fragmentation of a super-continent does not usually occur along a single narrow zone; rather, it is a process of multi-branched deformation that generally follows zones of weakness inherited from previous orogenies. It appears that stresses can vary in a complicated manner both in space and time during continental fragmentation and during the early development of an ocean.

My ideas about intraplate tectonics and reactivation of old zones of weakness developed from two points made by Wilson in his papers on transform faulting and the history of ocean basins (Wilson, 1965, 1966). In his paper on transform faulting, Wilson pointed out that some of the major offsets of the Mid-Atlantic Ridge in the equatorial Atlantic have the same shape as the continental margins of Africa and Brazil, and he proposed that these major offsets may have developed along old lines of weakness in the two continents. In his 1966 paper, Wilson hypothesized that the present Atlantic Ocean developed in the Mesozoic along nearly the same zone in which a Proto-Atlantic Ocean closed in the Paleozoic. He depicted that zone of closure – either in the Appalachian fold belt or in its continuations into Greenland, Europe and West Africa – as a broad zone of weakness compared with the older orogenic belts on its flanks. In my work I emphasize the pronounced effect of the pre-existing tectonic fabric on the reactivation of old zones of weakness over a broad area during continental fragmentation and on the development of new ocean basins and major transform faults. Some of the multi-branched zones of deformation that were active during the early stages of continental fragmentation eventually become stranded within a plate when a single, throughgoing plate boundary develops.

EASTERN AND CENTRAL NORTH AMERICA

Although most of the world's larger earthquakes occur along plate boundaries, a small percentage of the total global seismic activity occurs within lithospheric plates. This, of course, demonstrates that the interiors of lithospheric plates are not inert and undeformed. Historically, several shocks of magnitude larger than 6 have occurred in a number of intraplate areas. Several of these have been quite damaging, including the large shocks in southeastern Missouri in 1811-1812, the Charleston earthquake in South Carolina in 1886, earthquakes off Cape Ann, Massachusetts in the 18th century, several large shocks near Montreal and Quebec during the last 400 years and the Grand Banks earthquake of 1929 (Fig. 1). In eastern North America an understanding of intraplate seismicity is crucial, since several historic shocks have occurred near what are now centers of population, and since most of the nuclear power reactors in the United States and Canada are siutated in an intraplate environment.

Intraplate earthquakes are usually characterized by extremely large felt areas for a given release of energy or for a given length of rupture. For example, the 1971 San Fernando earthquake in California and the Charleston shock of 1886 appear to have had nearly the same magnitudes and energy outputs. Nevertheless, the felt area of the Charleston earthquake was about 20 times that of the San Fernando shock (Fig. 2).

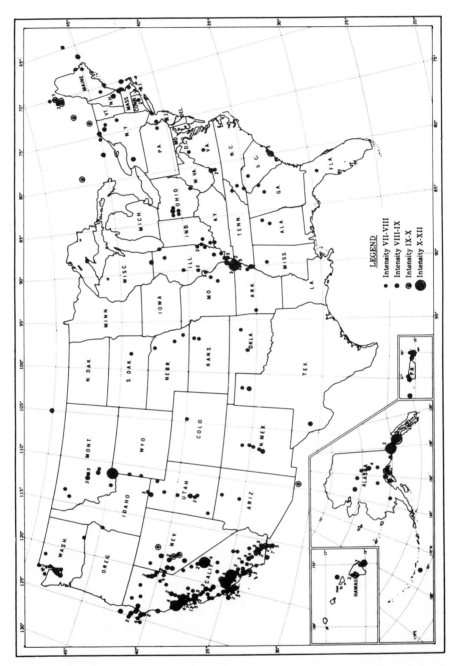

Figure 1. Damaging earthquakes of the United States through 1969 (after von Hake and Cloud, 1969).

This great difference is probably due to the much different rates of attenuation for waves of periods near one second in the two areas. Although the area affected by the New Madrid sequence of 1811 and 1812 was very great, the rupture zones and energy release in those shocks do not appear to have been as large as in great earthquakes situated along plate boundaries, such as the San Francisco earthquake of 1906. It is clear that the felt areas and regions of significant damage can be very appreciable for earthquakes of magnitude 5 to 7 within lithospheric plates.

Although a number of seismic zones or trends have been postulated within eastern North America (see Sykes, 1978), our understanding of intraplate earthquakes is still rudimentary. It is, of course, crucial to know whether large shocks like the Charleston earthquake of 1886 can be expected to occur at random anywhere in eastern North America, or whether shocks of that type are related to particular tectonic structures for which the long-term hazard is greater.

Figure 3 is a compilation after York and Oliver (1976) of reported earthquakes for eastern and central North America. A belt of earthquakes generally follows the Appalachian orogenic belt. With the exception of earthquakes in South Carolina, very few shocks can be seen in Figure 3 along the coastal plain south of Delaware. If we examine a compilation of moderate and large shocks that have been located instrumentally, the patterns are very similar to those in Figure 3. One of the surprising things to emerge from such a compilation is that most of the seismic activity is concentrated within the continent; little is found either within the older oceanic lithosphere or near

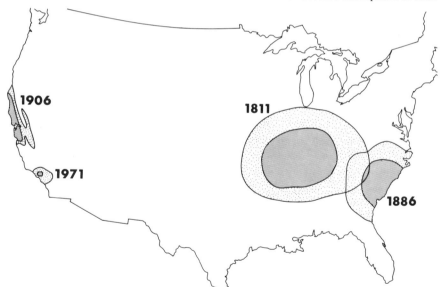

Figure 2. Areas sustaining a given level of intensity of shaking for earthquakes near New Madrid, Missouri, 1811; Charleston, South Carolina, 1886; San Francisco, California, 1906; San Fernando, California, 1971. Hatching denotes areas of intensity VII on Modified Mercalli (MM) scale; stipled pattern denotes areas affected by intensity VI. Note the much greater area affected by earthquake of 1886 than for shock of 1971 even though the two events were of approximately the same magnitude. From *Mosaic* (U.S. National Science Foundation), v. 7, p. 2-3, 1976.

the transition from oceanic to continental crust. The few focal mechanisms that are available for the western Atlantic indicate a predominance of thrust faulting (Sykes and Sbar, 1974). Likewise, about a dozen focal mechanism solutions for eastern North America also indicate a predominance of high-angle reverse faulting or thrust faulting such that the maximum compressive stress is nearly horizontal (Sbar and Sykes, 1973; Aggarwal and Sykes, 1978; Yang and Aggarwal, 1977). The similarity in the focal mechanisms suggests that horizontal compressive stresses may be transmitted long distances in the lithosphere. The present state of stress in much of eastern North America differs from that which existed near the Atlantic continental margin during the early separation of North America from Africa in the Triassic-Jurassic. One possible explanation of the state of stress and the occurrence of earthquakes in the Appalachian orogenic belt is that this area is weaker than the older oceanic lithosphere, because it contains a number of old fault zones and other major zones of weakness that have been inherited from Paleozoic and older orogenies.

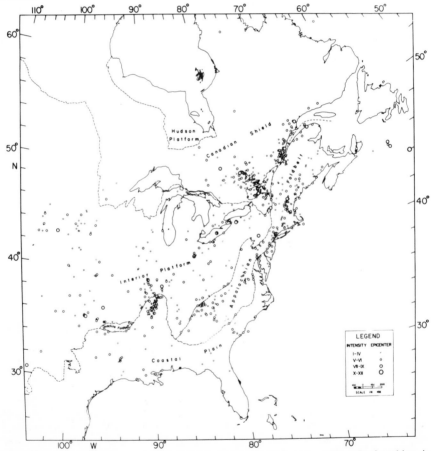

Figure 3. Distribution of reported earthquakes in eastern North America, 1928-1971, from historic and instrumental data (after York and Oliver, 1976).

Figure 4 shows the distribution of historic earthquakes in the southeastern United States, after Bollinger (1973). He concludes that seismic activity is present along two parts of the Appalachian orogenic belt, and delineates two other seismic trends that are nearly perpendicular to the Appalachians, in central Virginia and in South Carolina and Georgia. Activity in the southern Appalachian zone generally coincides with the faulted and folded portion of that belt (Fox, 1970).

Many of the major faults in South Carolina trend northeasterly. Seismic activity in central and western South Carolina is located in a zone that may not be continuous with activity farther to the southeast, near Charleston (Nishenko and Sykes, 1979). The epicentre of the Charleston earthquake of 1886 is located landward of a major fracture zone in the western Atlantic, the Blake Spur fracture zone. Fletcher *et al.* (1978) speculate that the South Carolina-Georgia seismic zone shown in Figure 4 may be a landward extension of the Blake Spur fracture zone. There is no indication on land, however, of major faults with a northwesterly orientation (Nishenko and Sykes, 1979). Thus, the tectonic setting of the Charleston earthquake is still uncertain.

Several seismograph networks have been installed during the last eight years in the northeastern United States and in the area of the New Madrid earthquakes in southeastern Missouri. Figure 5 shows a compilation of instrumentally located earthquakes for a recent seven-year period in the northeastern United States and in adjacent parts of Canada. The greatest concentration of activity is in a northwesterly trending zone that extends from the Adirondack region of northern New York to Kirkland Lake, Ontario. Yang and Aggarwal (1977) obtained about ten focal mechanism solutions for earthquakes in that zone, which are all very similar and indicate thrust faulting

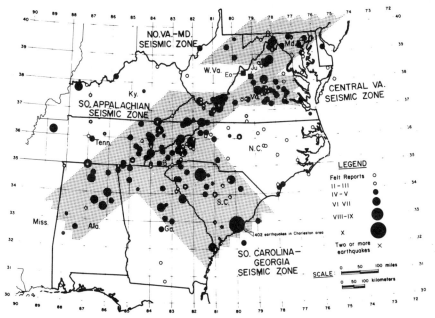

Figure 4. Historic seismicity for southeastern United States, 1754-1970 (after Bollinger, 1973, with additions). Eo and Ju are localities of alkaline rocks of Eocene and Jurassic age in Virginia.

along planes striking north-northwest, i.e., parallel to the seismic zone. Other concentrations of activity can be seen in Figure 5 in western New York and in the greater New York City area. A dense network of stations has been operating longer in New York and New Jersey than in New England. Hence, the actual rate of activity may be higher in eastern Massachusetts and in Maine than the figure suggests.

Some faults with a north-northwest strike can be seen in the Adirondack region of northern New York on the geologic map of the State or on photographs taken from space. A prominent set of older faults striking northeasterly does not appear to be seismically active. In western New York three focal mechanisms indicate motion along the Clarendon-Linden fault, a major fault that extends in a northerly direction from the Pennsylvania-New York border to Lake Ontario. The Clarendon-Linden fault appears to be a major old zone of weakness that was active during the Devonian as well as earlier during the Paleozoic.

One of the clearest examples of seismic activity being related to old fault systems can be seen in northern New Jersey and southern New York, within the greater New York City area (Figs. 5 and 6). Many of the earthquakes in this area, which have been recorded since a network began operating in 1974, are situated within the Precambrian terrane of the Hudson Highlands and Reading Prong. Very little activity has been observed within the Triassic Newark graben. Likewise, activity drops off abruptly to

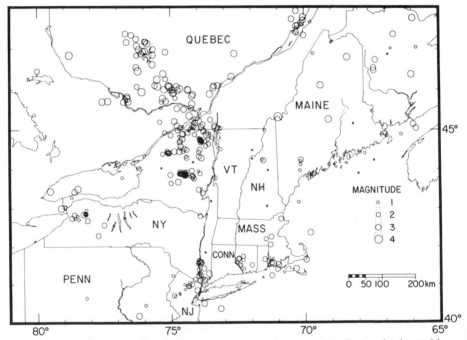

Figure 5. Epicenters of earthquakes from 1970 to 1977 in eastern North America located by various networks in the area (after Aggarwal and Sykes, 1978). For New York State and northern New Jersey the coverage is probably complete for magnitudes greater than 2; for New England the coverage is poorer until about 1975. Note the northeasterly alignment of earthquakes in northern New Jersey and southern New York. Asterisks denote unidentifiable events.

the northwest of the Hudson highlands (approximately a line connecting events of 7 and 25 in Fig. 6). Seismic activity is situated along the Ramapo and other faults within the Hudson highlands that strike northeasterly. Ratcliffe (1971) showed that the Ramapo fault was active in the late Precambrian, twice during the Paleozoic and in the late Triassic-early Jurassic during the fragmentation of North America and Africa. Some of the early deformation was in response to plate tectonic events when the region was near a plate boundary. The presence of mafic and ultramafic bodies, such as the Cortland complex, which must have had their source in the upper mantle during the Ordovician, indicates that the Ramapo fault zone is a major deep fault, much like those described by workers in Africa and the Soviet Union.

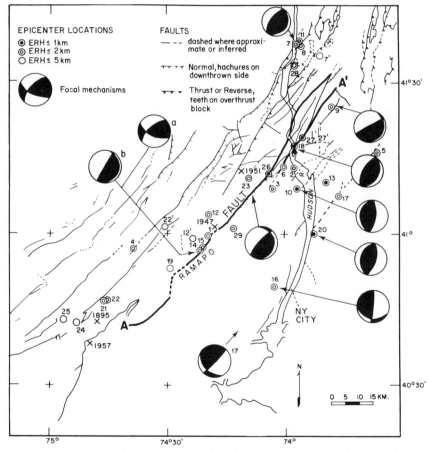

Figure 6. Fault map of southeastern New York and northern New Jersey showing epicenters (circles) of instrumentally located earthquakes from 1962-1977 (after Aggarwal and Sykes, 1978). Indicated uncertainties (ERH) in epicentral locations represent approximately two standard deviations. Focal mechanism solutions (FMS) are upper hemispheric plots; dark area represents the compressional quadrant. The Ramapo fault and two of its major branches (A-A') are shown by heavy lines; crosses denote locations for older events near the Ramapo fault. Triangle shows location of nuclear power reactors at Indian Point, New York.

The Ramapo fault was reactivated during the Triassic-Jurassic stage of continental fragmentation that led to the development of the present Atlantic Ocean. At that time the predominant motion on the Ramapo fault was one of normal faulting. Geological work done in conjunction with the Indian Point nuclear power plants (Dames and Moore, 1977; Ratcliffe, 1976) indicates that considerable strike-slip motion was present along the Ramapo fault some time later during the Jurassic. Several focal mechanism solutions for earthquakes in the region (Fig. 6) indicate that the present motion, however, is high-angle reverse faulting. Thus, the state of stress along the Ramapo fault has changed with time. Figure 7, a cross section showing seismic activity and focal mechanisms in the vicinity of the Ramapo fault, shows that the seismically active zone extends from the surface trace of the fault to a depth of about 10 km. The activity dips about 60° to the southeast, in agreement with the surface measurements of the dip. Other seismic activity in the Hudson highlands, as well as in the Manhattan Prong of Westchester County, New York, appears to be located along other northeast-trending faults that are also quite old.

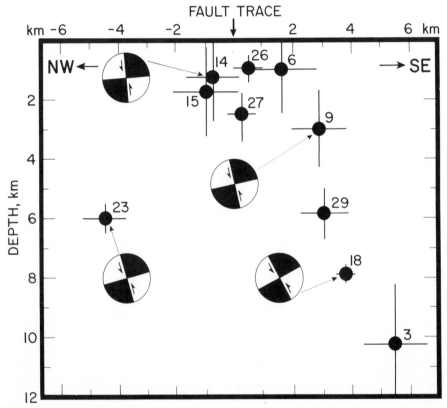

Figure 7. Composite (stacked) vertical cross section showing focal depths and focal mechanism solutions (FMS) for earthquakes within 10 km of trace of Ramapo fault (after Aggarwal and Sykes, 1978). Dark area is compressional quadrant. Only those events are plotted for which reliable focal depths could be determined. Bars represent one standard deviation.

The most active seismic region in eastern and central North America is situated near the northern end of the Mississippi Embayment near New Madrid, Missouri (Figs. 1 and 3). The activity in that region shows one of the clearest relationships with faults of Cretaceous or Cenozoic age. The Mississippi Embayment has been subsiding since the Cretaceous; the present embayment appears to be situated along an old tectonic feature of late Precambrian age, possibly an aulacogen (Ervin and McGinnis, 1975).

Alkaline rocks and diatremes occur at a number of localities in the midcontinental region of the United States, particularly in the Mississippi Embayment (Fig. 8). Many of these rocks are of middle Cretaceous age, coinciding with the beginning of subsidence in the Mississippi Embayment.

Some workers have proposed that since earthquakes seem to be concentrated near mafic and alkaline rock complexes such as those in Figure 8, a difference in the elastic moduli might be creating a stress concentration and hence the earthquakes. I believe, however, that the reason alkaline rocks and earthquakes are often found in the same regions is that they are both related to major zones of weakness that have been reactivated by continental fragmentation or continental collision. This will be discussed at greater length in the sections on Africa and South America

Zartman *et al.* (1967) and Zartman (1977) have dated a number of occurrences of alkaline rocks in the eastern and central United States, and conclude that these rocks

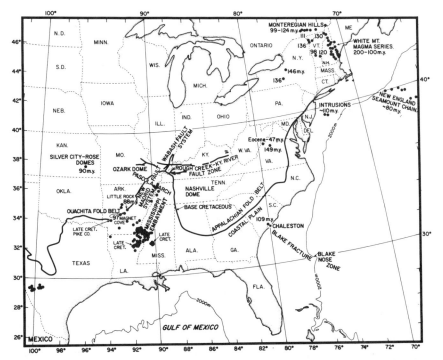

Figure 8. Alkaline igneous rocks of Jurassic and younger age in central and eastern United States and in southeastern Canada (after Sykes, 1978).

were probably emplaced along zones of crustal weakness. Several of these occurrences are younger than the late Triassic deformation associated with the separation of North America from Africa. In addition to the occurrences shown in Figure 8,
Zartman (1977) has recently obtained similar dates for alkaline rocks in southwestern
Pennsylvania and in northeastern Kansas. The occurrences in northeastern Kansas
are situated astride the Nemaha fault. Steeples *et al.* (1979) have located several small
earthquakes along this fault and believe that several of the historical earthquakes in
Kansas are associated with it. The Nemaha fault is another major structural feature of
great age that appears to have been reactivated since the Jurassic. It has been the site of
earthquakes and of alkaline magmatism.

One of the longest belts of alkaline magmatism that postdates the separation of
Africa from North America extends from the Monteregian Hills of southern Quebec
through the White Mountain Magma Series of New Hampshire and along the New
England Seamount Chain in the western Atlantic (Fig. 8). Radiometric ages indicate
that magmatism began in New Hampshire about 200 Ma ago, coinciding with initial
rifting between North America and Africa, and continued for 100 Ma. The western end
of the Monteregian province was also the site of carbonatitic intrusions during the
Cambrian (Rankin, 1976). A Paleozoic episode of alkaline magmatism has been identified by Zartman (1977) and others in eastern Massachusetts near the southeastern
end of the White Mountain Magma Series.

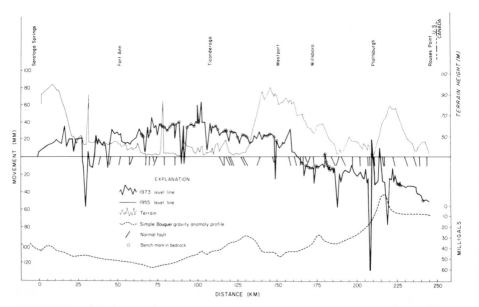

Figure 9. Geodetic leveling profile along the eastern flank of the Adirondack dome for the period
1955-1973 (after Isachsen, 1975). Note that the region of maximum uplift is rising at a rate of about
2 mm/yr relative to Saratoga Springs (and to sea level in New York City) and 4 mm/yr relative to
the United States – Canadian border. Another profile along the center of the Adirondacks
indicates uplift of about 3.7 mm/yr relative to Utica, New York, in the Mohawk Valley (Isachsen,
1976).

Alkaline magmatism of Paleozoic age has also been identified along the Rough Creek fault zone and the 38° lineament in West Virginia, Kentucky, southern Illinois, Missouri and eastern Kansas (Zartman, 1977). Similar occurrences of Paleozoic age are found near the head of the Mississippi Embayment in Missouri and southern Illinois. Thus, several of the regions that have been affected by Jurassic or more recent alkaline magmatism have a history of similar magmatism of Paleozoic or Precambrian age. This suggests that these rocks have been injected along major zones of weakness that have been periodically reactivated throughout geologic history.

A number of regions within lithospheric plates have been uplifted a considerable amount during the past few million years: large parts of the Appalachian Mountains, the Adirondacks, the White Mountains in New Hampshire, Caledonian zones of deformation in Norway and the Urals in the U.S.S.R. Very little is known about this process or processes. Levelling lines indicate that the Adirondack region of northern New York is being uplifted at a rate of about 2 to 3 mm per year (Fig. 9), which is sufficient to have produced the topography of the Adirondack region in about 1 million years. It seems unlikely that this uplift is related to glacial loading or unloading during the Quaternary, since comparable uplift is found in eastern Australia, where an extensive ice sheet did not form. The uplift is puzzling, since focal mechanism solutions for earthquakes in the Adirondack region indicate thrust faulting and horizontal compression. Although the region is moderately active seismically, the seismic belt shown in Figure 5 extends from the Adirondacks a considerable distance to the northwest beyond the area of uplift.

AFRICA

Seismicity and Reaction of Old Faults

Figure 10 shows the distribution of earthquakes in Africa since 1900. Two of the most active areas – the East African Rift system and the northernmost part of Africa (which is part of a broad plate boundary between Europe and Africa) – are not shown in the figure, since I was mainly interested in intraplate earthquakes. Seismic activity in southern Africa and along the East African Rift System is spread over a very large region. This may be contrasted with active plate margins in the oceans, which tend to be very narrow and single branched. The multi-branched character of the seismicity and tectonic structures in East Africa may be typical of the early fragmentation of a super-continent, such as the early fragmentation of Africa and South America. Presumably only one of the zones of deformation will eventually become a throughgoing plate boundary along which an ocean may develop.

Of greater relevance to this paper are two examples – from Ghana and Morocco – of old faults that have been reactivated and are now the sites of seismic activity. The region near Accra, Ghana, has experienced three earthquakes of magnitude 6 or larger within the last 120 years. Two of these shocks, in 1939 and 1906, are designated in Figure 10 by the numerals 39 and 06. They appear to have occurred not within oceanic crust, but within the continent, just landward of the end of the great Romanche fracture zone. Although the Accra area is located near the intersection of the Romanche fracture zone and a major tectonic boundary along the east side of the West African craton (Fig. 10), seismic activity in that area does not appear to join up with that along

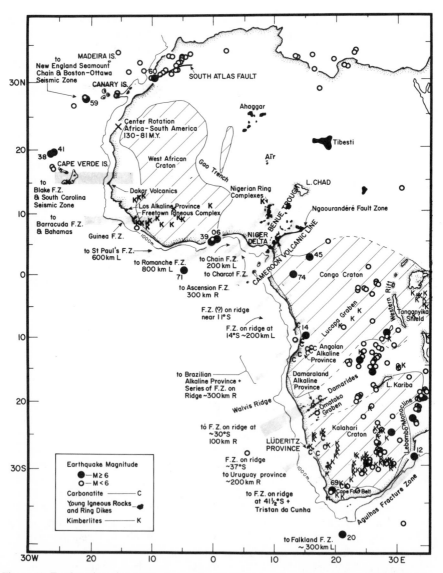

Figure 10. Earthquakes along Atlantic margin of northern and southern Africa, 1900-1975 (after Sykes, 1978). Carbonatites, Kimberlites and young igneous rocks shown are of Jurassic or younger age. Small circles (transform directions) are drawn about the center of rotation for either the movement of Africa with respect to South America from 125 to 81 m.y. (south of 5°N) or the movement of Africa with respect to North America from 81 to 180 Ma (north of 7°N). Small circles denoted by continuous lines are drawn along either mapped fracture zones, prominent offsets in the pattern of magnetic anomalies of Mesozoic age near the continental margin, offsets of continental margin or zones of recent volcanism in Canary and Cape Verde Islands. These transform directions have been extended a few hundred kilometers into continent. Small circles

plate boundaries, but instead is an isolated hot spot of seismic activity. The two tectonic features have quite different strikes and different ages.

Fletcher *et al*. (1978) argue that seismic activity near the ends of major oceanic fracture zones may be related to faults propagated inland from those oceanic fracture zones. I believe the following is a more reasonable explanation: Major oceanic fracture zones develop where the continental margin of the newly developing ocean changes strike, which most likely occurs near major zones of weakness inherited from previous orogenies within the continents. In this case the strike of the oceanic fracture zone would probably not be the same as the old zone of weakness. The strike of the fracture zone represents the direction of early relative movement between the two separating continents. During continental fragmentation a number of older faults are reactivated, which may be the sites of alkaline magmatism and of earthquakes for tens of millions of years after the region has ceased to be a plate boundary of either the transform or spreading type. Thus, the seismic activity near Accra appears to be occurring along an old major zone of weakness that was reactivated during the early separation of Africa and South America. Whether such a zone is seismically active or continues to be a locus of alkaline magmatism would depend upon the orientation of the fault relative to the stress tensor that is acting today.

Sykes (1978) has summarized evidence for Cenozoic deformation near Accra, Ghana and for other parts of the Gulf of Guinea. Wright (1976), Sykes (1978) and others discuss a number of older faults in that region that were reactivated in the Cretaceous and Cenozoic.

The South Atlas fault of Morocco (Fig. 10) has also undergone repeated deformation during the Earth's history. The westernmost part of the fault zone is seismically active today; a damaging earthquake occurred near Agadir in 1960. The South Atlas fault appears to be one of the major fault zones bounding the north side of the West African craton. Again, the strike of the South Atlas fault differs from that of the nearly east-west direction of transform faulting in the nascent Atlantic Ocean. A major fault in the Atlantic to the north of the Canary Islands and the magmatism of the Canary Islands themselves appear to have developed seaward of the South Atlas fault during the Mesozoic.

Intraplate Magmatism

The African continent contains a number of igneous bodies that were emplaced during or after the early separation of Africa from either North America or South America. Kimberlites, carbonatites, ring dikes, and various alkaline rocks of Jurassic or younger age are shown in Figure 10. Nearly all of the occurrences of Kimberlite are confined to the older cratons. In South Africa a great many kimberlites were injected

denoted by dashed lines are more questionable. Last two digits of date in years are shown beside larger earthquakes. The Canary Islands and Cape Verde Islands have conjugate points in North America near New England seamount chain and Charleston, South Carolina, in a reconstruction of Africa and North America that closes the Atlantic. The amount of apparent offset at the present ridge axis is indicated in kilometers as being either left (L) or right (R) lateral. Other features at conjugate points in North America and South America are indicated. Note that a number of Kimberlites, Carbonatites, other alkalic rocks and earthquakes are located near the ends of major oceanic transform faults. Cratonic areas older than 800 Ma are denoted by hatching.

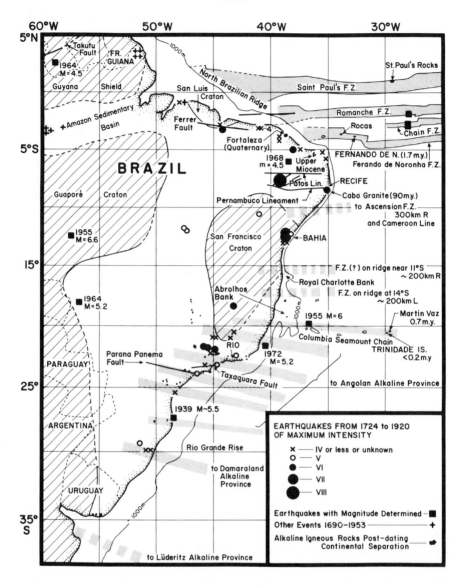

Figure 11. Seismicity, igneous rocks post-dating continental separation of South America from Africa and fracture zones near continental margin of Brazil (after Sykes, 1978). Intensities are for Rossi-Forel scale. Fracture zones near continental margin are denoted by continuous shaded lines and have been extended a few hundred kilometers into continent. Fracture zones inferred from offsets of Mid Atlantic ridge by appropriate finite rotations are indicated by dashed shaded lines. Hatching denotes cratonic areas older than 800 Ma.

into the Kalahari craton starting about 140 Ma ago and lasting until about 80 Ma ago (Sykes, 1978). The earlier age corresponds to the time of initial rifting between Africa and South America. The younger age corresponds to the time when the Falkland Plateau finally cleared South Africa along the Agulhas fracture zone. Hence, the timing of Kimberlites in South Africa appears to be intimately related to the process of continental fragmentation and early opening of the South Atlantic. One apparent enigma has been the restriction of kimberlites to the older cratonic areas. Assuming old zones of weakness were reactivated during continental fragmentation and early development of the Atlantic Ocean, I propose that kimberlites are emplaced where the lithosphere would be thickest, i.e., beneath cratonic areas that have not been disturbed by widespread deformation for hundreds of millions of years.

It is likely that the lithosphere is thinner beneath continental areas that have been affected by more recent orogenesis in either the Paleozoic of late Precambrian time. Prior to the development of the South Atlantic in the Mesozoic, the last major period of orogenic activity in Africa, the so-called Pan African event, occurred during the late Precambrian and early Paleozoic. Most of the younger magmatism that does not appear to have originated at such great depths as Kimberlites or carbonatites occurs within the areas affected by Pan African deformation. This includes the Nigerian ring complexes, the Cameroon Volcanic Line, and volcanic rocks near Dakar and in the Canary Islands. Thus, the type of intraplate magmatism appears to be governed by the thickness of the lithosphere. The emplacement of alkaline rocks seems to be controlled by old zones of weakness that were reactivated during the early separation of Africa from South America or from North America.

Since the distribution of intraplate magmatism is so strongly influenced by the presence of old zones of weakness and the pre-existing tectonic fabric, I do not favour the explanation that these igneous bodies indicate the presence of a mantle plume. Rather, I regard them more as passive features emplaced along old zones of weakness when the state of stress was favourable for these bodies to be injected into the lithosphere.

BRAZIL

A number of occurrences of alkaline rocks that are younger than the age of continental separation (about 125 Ma) are shown in Figure 11. As in Africa, most of these are found in the non-cratonic areas. Several of these bodies, e.g., the Cabo granites of 90 Ma, occur along major Precambrian faults that were reactivated during the separation of Africa and South America. Some have been emplaced as recently as the Quaternary (e.g., the Fortaleza body). The Pernambuco lineament, along which the Cabo granite was emplaced, appears to be continuous with the Ngaourandéré fault zone in equatorial Africa (Fig. 10). On the African side, the Cameroon Volcanic Line developed along this old zone of weakness.

AUSTRALIA

Figure 12 shows the distribution of earthquakes in Australia and surrounding areas, the distribution of fracture zones in the oceans and the configuration of the older cratonic blocks which make up much of the western two-thirds of Australia. Australia is one of the most active areas in the world for intraplate earthquakes.

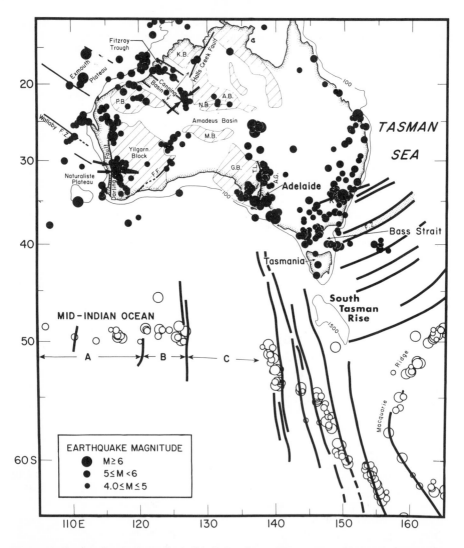

Figure 12. Earthquakes in Australia (solid circles) from 1897 to 1972 and in the eastern Indian Ocean from 1950 to 1966 (open circles) after Sykes (1978). Solid lines in oceans denote fracture zones. Horizontal arrows denote direction of maximum compressive stress inferred from well-determined focal mechanism solutions of earthquakes. Kimberlite occurrences of possible Quaternary age indicated by K. Hatching denotes cratonic areas experiencing their last major period of deformation more than 1 Ga ago. KB is Kimberley block; AB, Arunta block; MB, Musgrave block; PB, Pilbara block; GB, Gawler block; AG, Adelaide geosyncline; FF, Frasier fault; TL, Torrens lineament, and NB, Ngalia basin.

One of its most seismically active areas extends northward from Adelaide along 138°E: along the eastern side of the Gawler cratonic block and along the Adelaide geosyncline. Some of the major fracture zones in the Indian Ocean appear to have developed seaward of this zone (Fig. 12). It is quite clear that the seismic activity is largely confined to the continent, however, and does not extend along the older parts of the fracture zones in the oceans. The seismic activity in the oceans is largely confined to those parts of fracture zones that are located between spreading ridges.

In northwestern Australia a zone of seismic activity is situated between the Kimberley and Pilbara cratonic blocks. A number of occurrences of mafic, alkaline rocks of Mesozoic or Cenozoic age are found in that area. One focal mechanism solution indicates compression in a northeasterly direction, nearly perpendicular to the strike of the Canning Basin and Fitzroy Trough. This suggests that the two basins are being compressed between the two cratonic blocks. As in eastern North America, the focal mechanism solutions in Australia indicate a predominance of horizontal compression and thrust faulting.

THE EFFECT OF CRATONS ON THE OPENING OF THE SOUTH ATLANTIC

Figure 13 is a schematic illustration of the development of continental rifting and sea-floor spreading in the North and South Atlantic between the various cratonic blocks. I am indebted to Dr. C. Scholz for suggesting some of these ideas to me during an early reading of my 1978 paper. Several principles can be stated concerning the mode of continental fragmentation, the directions of transform faulting, the state of stress and the loci of new oceans:

1) Cratons resist being deformed. Hence, new oceans and other zones of deformation during the early separation of a super-continent are concentrated around the margins of cratons along younger orogenic belts such as the Appalachians or one of the belts of Pan African deformation. This is reasonable, since the lithosphere is likely to be thicker, older and stronger in the cratons than in the surrounding orogenic belts.

2) Continental fragmentation tends to be diffused in space and multi-branched. A present-day example of this might be the widespread distribution of deformation and seismic activity in East Africa. A number of workers have pointed out that the various branches of the East African Rift System tend to follow the youngest mobile belts, rather than passing through the older cratonic areas.

3) The configuration of the cratons will affect the ease with which a super-continent can be fragmented, the directions of spreading and the state of stress. It is unlikely, for example, that the strike of mobile belts between cratons is exactly the same as the directions the cratons would move apart in response to only the driving forces of plate tectonics. Large normal stresses could be transmitted across certain long transform faults during the early spreading history of an ocean, and then relax once the cratons have cleared one another. This clearing could lead to a major reorganization of the directions of transform faulting, and hence to changes in intra-plate magmatism and the state of stress.

Figure 13A shows multi-branched deformation developing on several sides of the West African craton from about 200 to 160 Ma ago. From 160 to 125 Ma ago the North Atlantic opens by sea-floor spreading (Fig. 13B). The Canary Islands lie near the

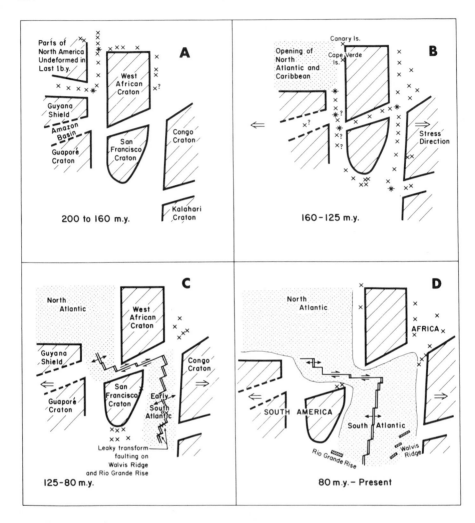

Figure 13. Schematic representation of continental rifting between cratons, of magmatism (crosses), and of movements between old cratonic nuclei (hatched areas) of Africa and North and South America as a function of time. Note that continental rifting, new oceans and magmatism develop along the youngest fold belts that predate rifting (blank areas) and avoid as much as possible passing through the older and thicker lithosphere of cratons. Large asterisks denote junctions of old fold belts that may become centres of unusually high rates of magmatism, i.e., hot spots. Magmatism during the early stages of continental fragmentation may be even more widespread along other mobile belts than is indicated here. A small piece of West African craton appears to have been broken off during the rifting and is now present in South America (San Luis craton, Fig. 11). After Sykes (1978) and including suggestions made to the author by C. Scholz (personal communication).

juncture of three older zones of deformation, and magmatism continues there long after the islands were located near a plate boundary.

Figure 13B also shows multi-branched deformation and magmatism occurring along the edges of several cratons in Africa and South America. About 125 Ma ago a throughgoing plate boundary develops into the nascent South Atlantic Ocean (Fig. 13C). As judged by the strike of the older transform faults, from about 125 to 80 Ma ago the directions of sea-floor spreading are northeasterly with respect to Africa. That direction may be dictated not only by the driving forces of plate tectonics, but also by the configurations of the cratons. A series of great fracture zones develops in the equatorial Atlantic to connect up deformation and sea-floor spreading in the North Atlantic with the development of the early South Atlantic between the Congo and San Francisco cratons.

About 80 Ma ago the directions of transform faulting change in the South Atlantic (Fig. 13D). This is about the same time that thick lithosphere is finally cleared on either side on some of the long transform faults in the equatorial Atlantic. Perhaps the directions of transform faulting during the last 80 Ma reflect the driving forces of plate tectonics rather than boundary forces across some of the large transform faults. About 80 Ma ago deformation in the Benue trough (Fig. 10), between the West African and Congo cratons, changes from minimal extension to horizontal compression. That change may be related to the clearing of thick lithosphere along the long transform faults in the equatorial Atlantic and to changes in the directions of transform faulting.

The Walvis Ridge and the Rio Grande Rise in the South Atlantic are two aseismic ridges that are still not well understood tectonically. Wilson and others propose that they represent the track of a hot spot or mantle plume that is presently located near the Mid-Atlantic Ridge at Tristan da Cunha. Most, if not all of these features, however, are located on crust that is older than 80 Ma. An exception is the western part of the Walvis ridge, which is located on crust younger than 80 Ma and is decidedly different from the central and eastern parts; it is composed mainly of seamounts and guyots (Dingle and Simpson, 1976). Portions of the Walvis Ridge and Rio Grande Rise are oriented along directions of transform faulting, but are not as long as some of the great transform faults in the equatorial Atlantic. They appear to have developed seaward of major orogenic belts on the south side of the Congo and San Francisco cratons, and may have acted as leaking transform faults from about 125 to 80 Ma ago, when the West African and San Francisco cratons were in contact with one another along the great equatorial fracture zones. Leaky transform faulting seems to have stopped or decreased considerably about 80 Ma ago along the Walvis Ridge and the Rio Grande Rise, when the direction of transform faulting changed in the South Atlantic.

Thus, we can imagine a complicated variation of stresses in both space and time during the early development of an ocean, when large normal stresses may be transmitted across some of the long transform faults. At that time there may still be major interaction between some of the cratons. When the stresses between cratons suddenly diminish, changes in the style of deformation, as along the Benue trough, and changes in intraplate magmatism may result.

ACKNOWLEDGEMENTS

I thank Dr. Christopher Scholz for making a number of suggestions about the effect of cratons on the development of new ocean basins and on the state of stress and directions of transform faulting. He and Dr. Klaus Jacob reviewed the manuscript. This study was partially supported by the Earth Sciences Section, National Science Foundation under NSF grant EAR 75-03640. Lamont-Doherty Geological Observatory Contribution 2885.

REFERENCES

Aggarwal, Y.P., and L.R. Sykes, 1978, Earthquakes, faults and nuclear power plants in southern New York-northern New Jersey: Science, v. 200, p. 425-429.

Bollinger, G.A., 1973, Seismicity and crustal uplift in the southeastern United States: American Jour. Science, v. 273-A, p. 396-408.

Dames and Moore, 1977, (Consultants, Cranford, New Jersey) Geotechnical investigation of the Ramapo fault system in the region of the Indian Point generating station: for Consolidated Edison of New York, Inc., v. 1 and 2, 28 March 1977.

Dingle, R.V., and E.S.W. Simpson, 1976, The Walvis ridge: A review: in Drake, C., ed., Geodynamics Progress and Prospects, American Geol. Instit., Washington, D.C., p. 160-176.

Ervin, C.P., and L.D. McGinnis, 1975, Reelfoot rift: Reactivated precursor to the Mississippi embayment: Geol. Soc. America Bull. v. 86, p. 1287-1295.

Fletcher, J.B., M.L. Sbar, and L.R. Sykes, 1978, Seismic trends and travel-time residuals in eastern North America and their tectonic implications: Geol. Soc. America Bull. v. 89, p. 1656-1676.

Fox, F.L., 1970, Seismic geology of the eastern United States: Bull. Assoc. Eng. Geol., v. 17, p. 21-43.

Isachsen, Y.W., 1975, Possible evidence for contemporary doming of the Adirondack mountains, New York, and suggested implications for regional tectonics and seismicity: Tectonophysics, v. 29, p. 169-181.

──────────────, 1976, Contemporary doming of the Adirondack Mountains, New York (abstract): EOS Trans. Amer. Geophys. Union, v. 57, p. 325.

Nishenko, S.P., and Sykes, L.R., 1979, Fracture zones, Mesozoic rifts and the tectonic setting of the Charleston, South Carolina earthquake of 1886 (abstract): EOS Trans. Amer. Geophys. Union, v. 18, p. 310.

Rankin, D.W., 1976, Appalachian salients and recesses: Late Precambrian continental breakup and the opening of the Iapetus Ocean: Jour. Geophys. Res., v. 81, p. 5605-5619.

Ratcliffe, N.M., 1971, The Ramapo fault system in New York and adjacent northern New Jersey: A case of tectonic heredity: Geol. Soc. America Bull., v. 82, p. 125-142.

──────────────, 1976, Final report on major fault systems in the vicinity of Tompkins Cove-Buchanan, New York: for Consolidated Edison of New York, Inc., 24 June 1976.

Sbar, M.L., and Sykes, L.R., 1973, Contemporary compressive stress and seismicity in eastern North America: An example of intra-plate tectonics: Geol. Soc. America Bull., v. 84, p. 1861-1882.

Steeples, D.W., DuBois, S.M., and Wilson, F.W., 1979, Seismicity, faulting, and geophysical anomalies in Nemaha County, Kansas: Relationship to regional structures: Geology, v. 7, p. 134-138.

Sykes, L.R., 1978, Intraplate seismicity, reactivation of preexisting zones of weakness, alkaline magmatism and other tectonism postdating continental fragmentation: Reviews of Geophysics and Space Physics, v. 16, no. 4, p. 621-688.

Sykes, L.R., and Sbar, M.L., 1974, Focal mechanism solution of intraplate earthquakes and stresses in the lithosphere: in Kristjansson, L., ed., Geodynamics of Iceland and the North Atlantic Area, D. Reidel, Hingham, Mass., p. 207-224.

von Hake, C.A., and Cloud, W.K., 1969, United States Earthquakes 1969, 80 pp., National Oceanic and Atmospheric Administration, Rockville, Maryland.

Wilson, J.T., 1965, A new class of faults and their bearing on continental drift: Nature, v. 207, p. 343-347.

——————————, 1966, Did the Atlantic close and then re-open?: Nature, v. 211, p. 676-681.

Wright, J.B., 1976, Fracture systems in Nigeria and initiation of fracture zones in the South Atlantic: Tectonophysics, v. 34, p. T43-T47.

Yang, J.P., and Aggarwal, Y., 1977, Seismotectonics of eastern North America, 2, northern New York and southern Quebec region (abstract): EOS Trans. Amer. Geophys. Union, v. 58, p. 432.

York, J.E., and Oliver, J.E., 1976, Cretaceous and Cenozoic faulting in eastern North America: Geol. Soc. America Bull., v. 87, p. 1105-1114.

Zartman, R.E., 1977, Geochronology of some alkalic rock provinces in eastern and central United States: in Annual Reviews of Earth Planet. Sci., v. 5, p. 257-286.

Zartman, R.E., Brock, M.R., Heyl, A.V., and Thomas, H.H., 1967, K-Ar and Rb-Sr ages of some alkalic intrusive rocks from central and eastern United States: Amer. Jour. Science, v. 265, p. 848-870.

The Continental Crust and Its Mineral Deposits, edited by D.W. Strangway,
Geological Association of Canada Special Paper 20

MAGNETOMETER ARRAYS AND LARGE TECTONIC STRUCTURES

D. Jan Gough
Institute of Earth and Planetary Physics,
University of Alberta,
Edmonton, Alberta T6G 2J1

ABSTRACT

Observations of time-varying magnetic fields in the period range 15 to 150 minutes, by means of large two-dimensional arrays of recording magnetometers, have led to the discovery and delineation of several large, highly conductive structures in the crust and upper mantle. In the western United States the observations are well fitted by induced currents flowing in the upper mantle, and the regions of high conductivity correlate well with high heat flow and with seismological evidence of partial melting. In contrast to these upper mantle structures, related to partial melt, a narrow conductor at crustal depths joins the northern end of the Southern Rockies in Wyoming to the Black Hills of South Dakota and to the Wollaston Domain of north central Saskatchewan. This conductor may be associated with an interplate boundary of Proterozoic age, and the high conductivity with graphite and possibly other metamorphic minerals. In southern Africa, two major crustal conductors have been well mapped by four array studies. One conductor revealed a hitherto unrecognized southwestward extension of the Luangwa-Zambezi Rift and saline water in fractures may provide the high conductivity. The other conductor underlies the tip of the continent south of 30°S. This body correlates with a major magnetic anomaly, with an isostatic gravity anomaly and with geochronologic and geologic information, in a manner consistent with a hypothetical underthrust of oceanic crust in Proterozoic time. Here, on this interpretation, hydrated ophiolitic rocks would constitute the good conductor.

These examples illustrate both the exploratory potential of magnetometer arrays and the various possible causes of high conductivity. Correlations with other geophysical and geologic data sometimes allow discrimination between partial melting in the mantle, conductive minerals in the basement, and saline water in the upper crust.

RÉSUMÉ

Les observations sur les champs magnétiques variables dans le temps pour une gamme de périodes de 15 à 150 minutes au moyen de grands réseaux bi-dimensionnels de magnétomètres enregistreurs ont mené à la découverte et à la délimitation de plusieurs grandes structures très conductrices dans la croûte et dans le manteau supérieur. Dans l'ouest des Etats-Unis, on explique les observations par des courants induits circulant dans le manteau supérieur, et les régions à conductivité élevée correspondent bien aux zones de flux thermique élevé et avec l'évidence sismologique de fusion partielle. En contraste avec ces structures du manteau supérieur qui se rattachent à la fusion partielle, un conducteur étroit dans la croûte joint la partie nord du sud des Rocheuses dans le Wyoming aux Black Hills du Dakota Sud et au domaine de Wollaston du centre nord de la Saskatchewan. Ce conducteur pourrait correspondre à une limite entre deux plaques d'âge Protérozoïque, et la forte conductivité serait causée par le graphite et peut-être d'autres minéraux métamorphiques. En Afrique du Sud, on a cartographié deux conducteurs majeurs dans la croûte à l'aide de quatre études sur des réseaux. Un conducteur a révélé une extension jusque là inconnue du rift de Luanga-Zambezi vers le sud-ouest; l'eau salée dans les fractures pourrait expliquer la conductivité élevée. L'autre conducteur se situe en dessous de la pointe du continent au sud de 30° S. A ce conducteur correspondent une anomalie magnétique importante, une anomalie de gravité isostatique et d'autres données géochronologiques et géologiques d'une manière compatible avec l'hypothèse d'un sous-chevauchement de la croûte océanique au Protérozoïque. Ici, dans cette interprétation, les roches ophiolitiques hydratées constitueraient le bon conducteur.

Ces exemples illustrent le potentiel pour l'exploration des réseaux de magnétomètres et les différentes causes possibles de conductivité élevée. Les corrélations avec d'autres données géophysiques et géologiques permettent quelquefois une discrimination entre la fusion partielle dans le manteau, les matériaux conducteurs dans le socle et les eaux salées dans la croûte supérieure.

INTRODUCTION

Systems of electric current in the magnetosphere and ionosphere are frequently disturbed as a result of events on the Sun. Such disturbances generate time-varying magnetic fields at the Earth's surface, with amplitudes of tens, hundreds or thousands of nanoteslas in the period range 10 to 1440 minutes. Magnetic fields in this period range penetrate the Earth's crust and upper mantle, and there induce secondary currents which flow preferentially in the more conductive rocks. Although magnetometers at the Earth's surface observe the sum of the fields of external and internal currents, it is usually possible to ascribe spatial anomalies in these fields to either external (primary) currents or internal (induced) currents. If data are available from a two-dimensional array of magnetometers, each recording three components of time-varying fields, one may be able to map concentrations of electric current and so, by inference, conductive structures in the Earth. The resolution of such maps is determined by the magnetometer spacing in the array: the arrays used in work considered in this paper involved 24 to 56 magnetometers at spacings of 50 to 150 km, so that large conductors in the crust or upper mantle are delineated with a precision of order 50 km on the map. Depth resolution is less satisfactory, and long-line four-electrode resistivity sounding or magnetotelluric sounding are superior in this regard. In a few cases two-dimensional forward model calculations can be used in interpretation, but the crust and upper mantle are electrically extremely heterogeneous and an array may observe concentrations, by conductive structures under the array, of currents induced elsewhere by unknown fields in conductors of unknown shapes. In such cases quantitative modelling is impossible, but the current concentrations can still be mapped.

Magnetometer array studies are subject to a depth limit of about 400 km, because there is, in the depth range 400 to 600 km, a general, planetary scale rise in conductivity

to values in excess of 1 S/m (Banks, 1972) which effectively limits further penetration
of magnetic fields with periods of one day or less. High electrical conductivities occur
in this near-surface shell in several ways. In the top few kilometres, where rock pores
or cracks are connected, saline water may produce high conductivities. At all crustal
depths, conductive minerals such as graphite or sulphides may occur along metamorphic belts. Hydrated minerals in ophiolitic rocks may similarly provide compositional
conductors. In the upper mantle, partial melting of one to ten percent may raise the
conductivity two orders of magnitude, if the melt is in interconnected spaces. This list
of possibilities is not complete, but shows that a high-conductivity anomaly is an
ambiguous object. Usually other geophysical or geological evidence is invoked to
reduce the possibilities, sometimes to one. By contrast, a region of conductivity of
order 10^{-4} S/m can only consist of dry crystalline rock at temperatures well below its
solidus Temperature.

   To conclude this brief description of geophysical exploration with magnetometer

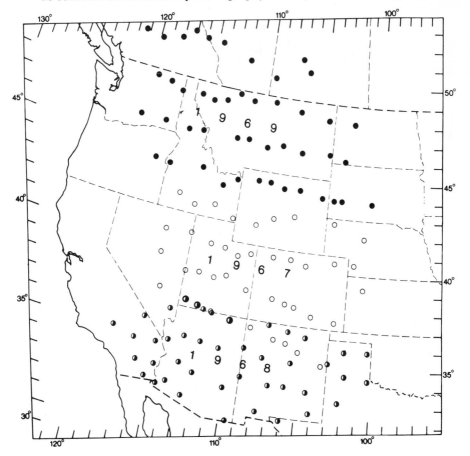

Figure 1. Magnetometer locations in the array studies of 1967, 1968 and 1969.

arrays, one may note that the range of conductivities found in the outermost 400 km of this planet is from 10^{-4} S/m for dry, crystalline crustal rocks to 4 S/m for seawater.

Reviews of magnetometer array methods, and results up to 1973, are given by Gough (1973) and Frazer (1974).

WESTERN NORTH AMERICA

Three large array studies, each with about 50 magnetometers deployed in four east-west lines, were made by the University of Texas at Dallas and the University of Alberta in the years 1967, 1968 and 1969 (Fig. 1). These arrays extended from the Great Plains, which lie on the craton, across much of the western superprovince with its Tertiary and later volcanics and diverse evidence of recent tectonic activity. Results of

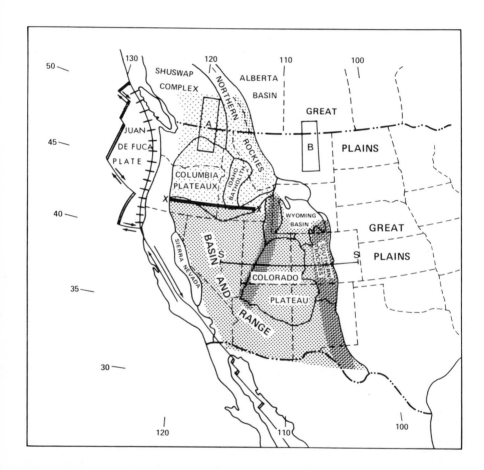

Figure 2. Distribution of electrical conductivity in the upper mantle and tectonic provinces of western North America. The density of stippling gives a qualitative indication of variations of conductivity.

these arrays revealed much structure in the electrical conductivity west of the Great Plains, as represented schematically in Figure 2. Highly conductive upper mantle underlies the Basin and Range tectonic province, with still greater thickness (or higher conductivity) of the conductive rock under the Wasatch Front and Southern Rocky Mountains, both of which have been uplifted in post-Cretaceous time. The anomalies across these tectonic provinces were the only ones the writer has encountered, in some ten array studies, in which the currents observed by the array were, in the main, induced directly in the underlying structures, so that two-dimensional modelling could be used to fit the observed magnetovariational anomalies. (These results come from the first large two-dimensional array study: had we stopped then, we might have preserved hopes of fitting two-dimensional anomalies in general.) One conductivity model which fits the array data along latitude 38.5°N (SS' in Fig. 2) is shown in figure 3 (Porath, 1971). In this model the depth of the conductor, which is indeterminate from the magnetovariational data, has been fitted to the low-velocity layer of seismology. A point made by this model is the thickness of the conductive layer, which is 100 km or

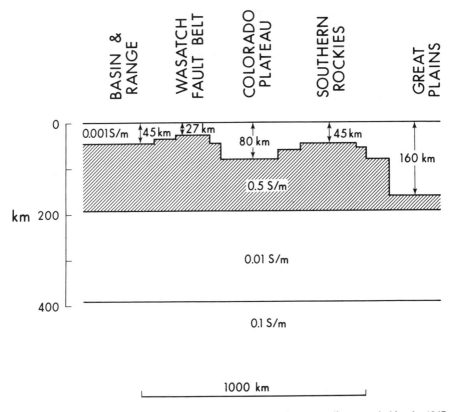

Figure 3. A model of conductivity which fits magnetic variation anomalies recorded by the 1967 array. The section is at latitude 38.5°N., along SS in Fig. 2. Conductivities are in Siemens/metre. After Porath, 1971.

more everywhere under the western region. This thickness assumes a high conductivity of 0.5 S/m: with a lower conductivity the thickness becomes still greater. The structure has to be in the mantle because the crust cannot contain it.

Since the conductive structures revealed by the magnetometer arrays correlate well with high heat flow and with the areas where there is seismological evidence of partial melting (Gough, 1974), the stippling in Figure 2 can be regarded as a first-order map of partial melting. It is worth noting that the Yellowstone plume is at the northern end of our Wasatch Front conductive ridge in the uppermost mantle. Geothermal exploration along the whole Wasatch fault belt might be productive, for instance by means of closely-spaced magnetometer arrays of higher resolution.

An important feature of the conductive structure is the abrupt reduction of the thickness-conductivity product at the northern boundary of the Basin and Range Province, near 42°N (*XX'* in Fig. 2). North of this boundary of the conductive layer thickness falls to about 10 km (at constant conductivity). Presumably the partial-melt structure under western North America is related to the complex pattern of subduction and associated ascent of melt in Tertiary and Quaternary times, in ways difficult to unravel. It is a little unexpected to find an order of magnitude less partial melt above the still-active subduction of the northwestern United States and southwestern Canada, than under the western U.S.A., where underthrusting ceased 10 to 20 Ma ago. One could speculate about combination of two or more subduction systems with suitable relaxation times for ascent of melt, or one might consider a possible upward flow in mantle convection, under the Basin and Range, but the total geophysical-geological information leaves the origin of the partial-melt structures under western North America as a fascinating enigma.

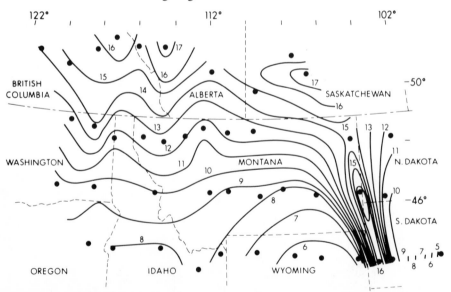

Figure 4. Anomaly in the amplitude (nT) of the east-west horizontal field *Y* of a polar magnetic substorm, at period 48 min., recorded by the 1969 array, above the North American Central Plains crustal conductive structure. After Gough and Camfield, 1972.

THE NORTH AMERICAN CENTRAL PLAINS CONDUCTIVE ANOMALY

In the 1967 array study we encountered a very large anomaly in the variation fields in the north-east corner of the array. It was clear that there was a concentration of

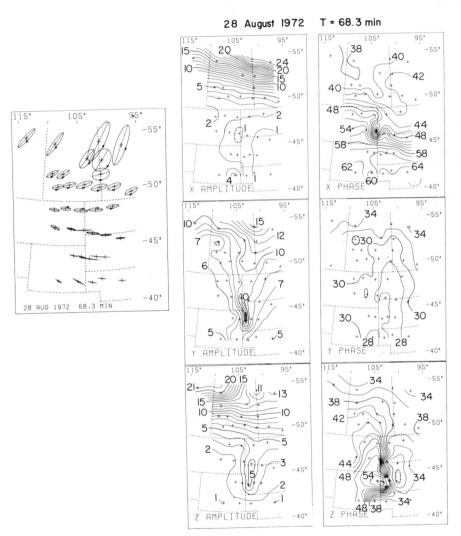

Figure 5. Fourier transform amplitudes (nT) and phases (min) at period 68 min from a substorm on 1972 August 28, near the North American Central Plains conductive structure. Polarizations of the horizontal components are shown at left. After Alabi, Camfield and Gough, 1975.

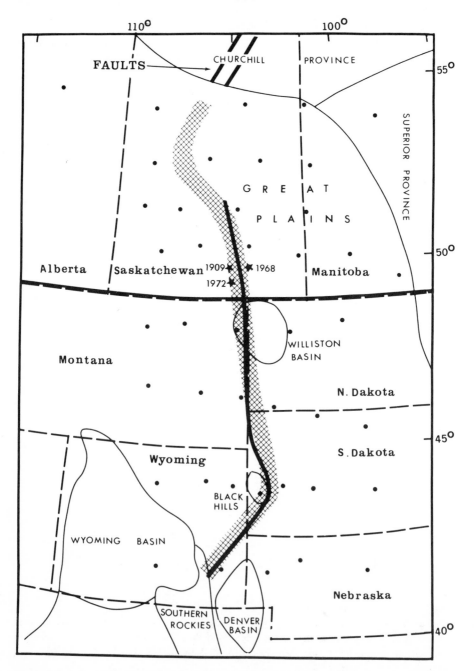

Figure 6. Location of the North American Central Plains conductive structure, from the results of the 1972 array. Stars locate earthquake epicentres. After Alabi *et al.*, 1975.

current in the general area of the Black Hills of South Dakota. The 1969 array, which was intended mainly to investigate the Cordillera of northwestern U.S.A. and southwestern Canada, was stretched eastward to cover the Black Hills. As a result, one of the largest known anomalies in magnetic variation fields was discovered (Fig. 4). In South Dakota the currents and the conductive structure must be less than 38 km deep, because the width at half-amplitude of the anomaly gives this depth for a line current. The conductive body is therefore crustal, and may well be near the top of the crystalline basement. Some induced potential anomalies near the Black Hills are believed to be associated with graphite in the basement, and a basement map of South Dakota compiled by Lidiak from other geophysical and borehole data shows a metamorphic belt which coincides with the conductive structure (Gough and Camfield, 1972). An array study was made in 1972 by Earth Physics Branch and the University of Alberta, with 40 magnetometers deployed to map the conductor northward and southward (Alabi *et al.*, 1975). Figure 5 shows a set of maps of amplitudes and phases of Fourier transforms, at a period of 68 minutes, of the magnetic field of a polar magnetic substorm. The external current system in this type of event includes an electrojet along the auroral zone, which is just north of the array, and the maps of

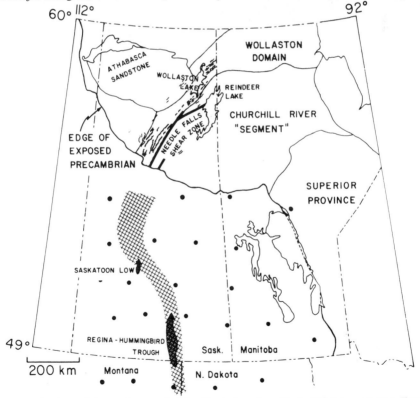

Figure 7. Geology of the northern end of the North American Central Plains conductor. Trend lines in the Wollaston Domain join conductors mapped in airborne electromagnetic surveys. After Camfield and Gough, 1977.

amplitudes of X (northward horizontal) and Z (vertical) components, north of the Canadian-U.S. border, show mainly this auroral electrojet. The Y (eastward) amplitude and Z phase maps however show the currents channelled in the North American Central Plains (N.A.C.P.) conductor. From several such sets of maps, from other maps of induction arrows representing transfer functions, and from inspection of the original magnetograms, the course of the conductive zone was traced to the northern limit of the array, 50 km south of the exposed shield of northern Saskatchewan, and southwestward from the Black Hills to the northern end of the Southern Rockies in Wyoming (Figs, 6,7,8).

Camfield and Gough (1977) have examined geological and geophysical information of all available kinds to attempt an interpretation of the conductive structure. In the exposed shield (Fig. 7) the Wollaston Domain, within the Churchill Province of the Shield, lies on strike and in line with the northern end of the mapped part of the N.A.C.P. conductor. The intensely, often isoclinally, folded schists and gneisses of the Wollaston Domain contain many conductivity anomalies found in airborne electro-magnetic surveys. These conductivity anomalies strike northeast-southwest, along the fold belt and parallel to the N.A.C.P. conductor; and many have been identified with graphitic garnet-biotite gneiss units (Camfield and Gough, 1977). Well-

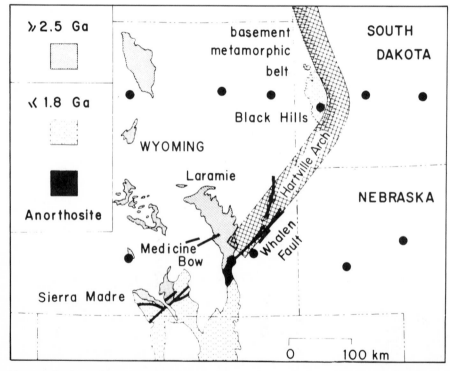

Figure 8. Geology of the N.A.C.P. conductive body south of the Black Hills. A geochronologic boundary is marked by the fault zone shown. There are faults, not shown, with other strikes. After Camfield and Gough, 1977.

developed sheets of graphite in nearly vertical limbs of folds are seen in outcrop in the Wollaston Domain. Graphite had already been suspected as a conductive mineral in the Black Hills section of the conductor, and may be involved along much of the 1800 km length of the N.A.C.P. conductor.

At the southern end the conductor skirts the east side of the Black Hills and turns southwest to run along the Hartville Arch and the Whalen Fault to the southern limit of the array in the Laramie Mountains at the northern end of the Southern Rockies (Fig. 8). Exposures of Precambrian basement occur along the crest of the Hartville Arch, some showing graphite. Geochronologic work by Houston and others, referenced by Camfield and Gough (1977), together with geological data, indicate an age boundary along the line of the Whalen Fault (and the N.A.C.P. conductor). Continental granitic rocks of ages 2500 Ma or more are found in central Wyoming to the northwest of the boundary, and intrusives of ages 1800 Ma or less are the oldest rocks known southeast of it. The boundary is marked by a major fracture zone across the Medicine Bow and Sierra Madre fingers of the Southern Rockies province. Hills *et al.*(1975) have proposed a Proterozoic plate boundary along the fracture zone and geochronologic boundary, with subduction under Nebraska and Colorado along the Whalen Fault stopping with an intercontinental collision about 1700 My ago. Camfield and Gough (1977) extended this hypothesis with the suggestion that the whole N.A.C.P. conductor marks a large fracture zone which may be a proterozoic plate boundary, though without implying that subduction was involved along its whole length. Graphite is common in major fracture zones, and has been observed at the northern end in the Wollaston Domain and at places already noted in South Dakota and Wyoming.

It is quite possible, though unproven, that the N.A.C.P. conductive structure continues along the Wollaston Domain to Hudson Bay, and the southern end down the subducted plate of Hills *et al.*(1975) to the conductive ridge due to partial melt in the upper mantle beneath the Southern Rockies. Currents concentrated in it may conceivably be induced partly by auroral-zone and polar-cap external fields in Hudson Bay, and partly by mid-latitude fields in the upper mantle under Colorado. The problems of modelling induction processes of such baroque complexity are far beyond present computers, even if the input parameters could be specified. It is no surprise that Porath *et al.*(1971) were unsuccessful in modelling the observed anomalous fields above the N.A.C.P. conductor, in a two-dimensional calculation for induction in the crustal conductor itself.

A RIFT ANOMALY IN SOUTHERN AFRICA

Well-known magnetovariational and magnetotelluric work had been done on the Gregory Rift of Kenya, by Banks and his associates and by Hutton (1976) and hers. The present discussion considers results of two investigations with two-dimensional magnetometer arrays in southern Africa which have discovered an extension of the East African rift system across the continent to the Atlantic coast. The first array study, in 1972 (de Beer *et al.*, 1976), employed 25 magnetometers in South-West Africa, Botswana and Rhodesia to look for possible conductivity anomalies associated with the Damara Geosyncline in South-West Africa, and with the seismically active belt running southwestward from the Middle Zambezi rift (in which Lake Kariba lies) to the Okavango Delta of Northern Botswana. Figure 9 from de Beer *et al,* (1975), shows the position of a *single* conductor linking the Damara Geosyncline to the Okavango Delta and Middle Zambezi. This unforseen result of the 1972 array was combined by de Beer *et al.* (1975) with mapped fractures and lineations, and with the seismicity of the region, in the hypothesis that the Luangwa/Middle Zambezi Rift continues south-westward to the Okavango Delta and thence westward across South-West Africa to the Damara. Earthquakes of magnitudes up to 6.7 have been recorded from the Okavango Delta since the inception of the South African seismological observatory system in 1949. In the Middle Zambezi, Lake Kariba triggered a large swarm of earthquakes (Gough and Gough, 1970) , the main shock having a normal-faulting mechanism (Sykes, 1967). Scholz *et al.* (1976) have used micro-earthquake locations and mechanisms to show that rifting is in progress, with normal-faulting seismicity, along the Tanganyika-Luangwa-Middle Zambezi-Okavango Rift.

The high conductivity observed along the Zambezi-Okavango Rift is probably at crustal depth (de Beer *et al.*, 1976) and may well be associated with saline water in rift fractures. De Beer *et al.* (1975) suggested that its westward extension might mark an extension of the rift structure along old weak zones in the lithosphere.

A second array was deployed in 1977 over the northern half of South-West Africa to investigate the western end of the anomaly at closer spacing and to the coast. A preliminary report of this investigation has been given by de Beer (1979). This work

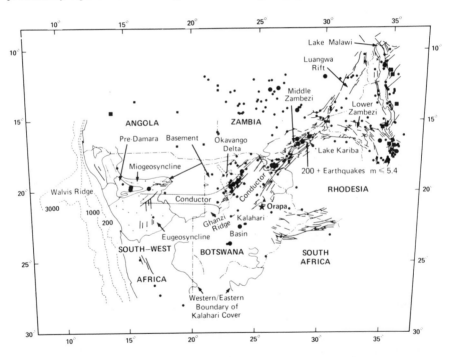

Figure 9. The conductivity anomaly along the Zambezi-Okavango-Damara rift, in relation to faults, fold axes and earthquake epicenters. Small dots, $2 \leq m \leq 4$; large dots, $m > 4$. After de Beer, Gough and van Zijl, 1975.

confirmed the results of the 1972 array east of 18°E., the western limit of the earlier array, but showed that the conductor bends to the southwest between 17° and 18°E. and then runs parallel to the fractures in the Damara Geosyncline. Figure 10 shows the position of the conductive anomaly, as revealed by the two arrays, in relation to the African rifts and shields.

Schlumberger soundings over the conductive zone, made by the Geophysics Division of the South African National Physical Research Laboratory, show a drop in resistivity from 700 ohm.m to 20 ohm.m at a depth of only 3 km. The high conductivity is therefore unlikely to be related to high temperatures. It could be associated with saline water in fractures or with conductive minerals. De Beer (1979) points out that the conductive belt joining the Damara and Zambezi belts could be a result of a suture containing oceanic lithosphere wedged between continental lithosphere slabs, on the model proposed by Burke *et al.* (1977), but at a different site from that proposed by those authors. It would be reasonable to suppose that under a different stress field, the continent would later fracture along the old suture. Where the rifting is more active,

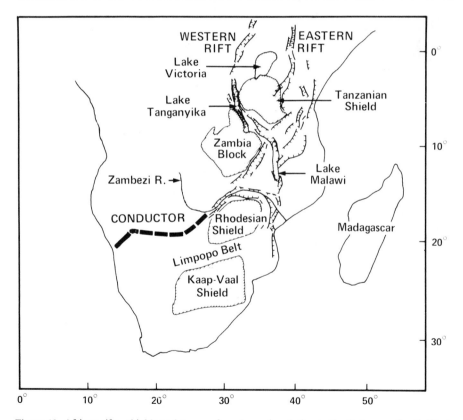

Figure 10. African rifts, shields and zones of weakness in relation to the Botswana-South-West Africa conductive structure. From de Beer, Gough and van Zijl, 1975 modified at the west end from de Beer, 1979.

part of the induced current could flow in the oceanic rocks and part in groundwater in the new fractures.

No less than 15 of 16 known calc-alkaline igneous complexes lie within the conductivity anomaly, 14 in the Damara Geosyncline and one in northwestern Rhodesia. This supports the argument for a zone of weakness, following the discussion by Sykes (1978). Sykes also suggests a relation between this zone of weakness and transform faults in the Atlantic floor.

THE SOUTHERN CAPE ANOMALY

An array study in South Africa in 1971, intended to look for a possible change in crustal or upper mantle conductivity at the boundary between the 2500 Ma-old Kaap-vaal craton and the 1000 Ma-old Namaqua-Natal mobile belt, discovered instead a very large conductivity anomaly elongated east-west and underlying the southern limit of the Karoo Basin and the Cape Fold Belt (Gough *et al.*, 1973). Only one station, at the southern corner of the triangular array of 1971, lay above the conductor. In 1977 an array of 53 magnetometers was placed over the southern end of the continent, south of 30°S., in order to map and study this Southern Cape anomaly. Analysis of the results of this investigation is in progress. The conductive structure has been delineated in some detail from the west coast to the south-east coast. It correlates closely with the well-known Beattie anomaly in the static magnetic field, and is related to the southern boundary of the Namaqua-Natal mobile belt in a way that suggests that the conductor may consist of ophiolitic rocks incorporated in a marginal geosyncline along the Proterozoic continental edge of Gondwanaland. Other geologic and geochronologic evidence supports this hypothesis.

CONCLUSIONS

These examples show both the capabilities and the limitations of magnetometer arrays. As a mapping technique the method is excellent and has revealed many structures of both present and past tectonic significance. Depth estimation and quan-titative modelling are difficult and uniqueness far off, but an array study, involving two or three scientists for a year or so, has usually proved to reveal interesting structure in the top 400 km of the Earth, even though the structure discovered may be unexpected and unrelated to that the array was intended to investigate. Other methods can then be used to estimate depths. The ambiguity in the cause of high electrical conductivity in rocks is great, but can usually be resolved in favour of some one of the alternatives, as these examples may have shown, by recourse to other geophysical, geochemical, geochronologic and geological information.

ACKNOWLEDGEMENTS

In array studies over the last fourteen years I have benefitted from working with my colleagues J.S. Reitzel, H. Porath, C.W. Anderson III, D.W. Oldenburg, P.A. Camfield, A.O. Alabi, F.E.M. Lilley, M.W. McElhinny, J.H. de Beer, J.S.V. van Zijl, V.R.S. Hutton, J.M. Sik, G. Rostoker and J.R. Bannister, some of whom have been my students and all of whom have contributed to my understanding of the phenomena. J.H. de Beer has kindly read this paper and suggested improvements.

REFERENCES

Alabi, A.O., Camfield, P.A. and Gough, D.I., 1975, The North American Central Plains conductivity anomaly: Geophys. Jour. Royal Astron. Soc., v. 43, p. 815-833.

Banks, R.J., 1972, The overall conductivity distribution of the Earth: Jour. Geomag. Geoelectr., v. 24, p. 337-351.

Burke, K., Dewey, J.F. and Kidd, W.S.F., 1977, World distribution of sutures- the sites of former oceans: Tectonophysics, v. 40, p. 69-99.

Camfield, P.A. and Gough, D.I., 1977, A possible Proterozoic plate boundary in North America: Canadian Jour. Earth Sci., v. 14, p. 1229-1238.

de Beer, J.H., 1979, The tectonic significance of geomagnetic induction anomalies in Botswana and South-West Africa: Proceedings of a Conference on "The Role of Geophysics in the Exploration of the Kalahari", Lobatse, Botswana, Feb. 1979.

de Beer, J.H., Gough, D.I., and van Zijl, J.S.V., 1975, An electrical conductivity anomaly and rifting in southern Africa: Nature, v. 225, p. 678-680.

de Beer, J.H., van Zijl, J.S.V., Huyssen, R.M.J., Hugo, P.L.V., Joubert, S.J. and Meyer, R., 1976, A magnetometer array study in South-West Africa, Botswana and Rhodesia: Geophys, Jour. Royal Astron. Soc., v. 45, p. 1-17.

Frazer, M.C. 1974, Geomagnetic deep sounding with arrays of magnetometers: Rev. Geophys. and Space Phys., v. 12, p. 401-420.

Gough, D.I., 1973, The interpretation of magnetometer array studies, Geophys. Jour. Royal Astron. Soc., v. 35, p. 83-98.

Gough, D.I., 1974, Electrical conductivity under western North America in relation to heat flow, seismology and structure: Jour. Geomag. Geoelectr., v. 26, p. 105-123.

Gough, D.I. and Camfield, P.A., 1972, Convergent geophysical evidence of a metamorphic belt through the Black Hills of South Dakota: Jour. Geophys. Res., v. 77, p. 3168-3170.

Gough, D.I. and Gough, W.I., 1970, Load-induced earthquakes at Lake Kariba-II: Geophys. Jour. Royal Astron. Soc., v. 21, p. 79-101.

Gough, D.I., de Beer, J.H. and van Zijl, J.S.V. 1973, A magnetometer array study in southern Africa: Geophys. Jour. Royal Astron. Soc., v. 34, p. 421-433.

Hills, F.A., Houston, R.S. and Subbarayuda, G.V., 1975, Possible Proterozoic plate boundary in southern Wyoming: Geol. Soc. America, Abst. with Programs, v. 7, p. 614.

Hutton, R., 1976, Induction studies in rifts and other active regions, Acta Geodaet, Geophys. Montanist., v. 11, p. 347-376.

Porath, H., 1971, Magnetic variation anomalies and seismic low-velocity zone in the western United States: Jour. Geophys. Res., v. 76, p. 2643-2648.

Porath, H., Gough, D.I. and Camfield, P.A. 1971, Conductive structures in the northwestern United States and southwestern Canada: Geophys. Jour. Royal Astron. Soc., v. 23, p. 387-398.

Scholz, C.H., Koczynski, T.A. and Hutchins, D.G., 1976, Evidence for incipient rifting in southern Africa: Geophys. Jour. Royal Astron. Soc., v. 44, p. 135-144.

Sykes, L. R., 1967, Mechanism of earthquakes and nature of faulting on the mid-oceanic ridges; Jour. Geophys. Res., v. 72, p. 2131-2153.

_____, 1978, Intraplate seismicity, reactivation of zones of weakness, alkaline magnetism and other tectonisms post-dating continental fragmentation: Rev. Geophys. Space Phys., v. 16, p. 621-688. (See also contribution in D. W. Strangway, ed., The Continental Crust and Its Mineral Deposits, Geol. Assoc. Canada Spec. Paper 20, 1980.)

The Continental Crust and Its Mineral Deposits, edited by D.W. Strangway, Geological Association of Canada Special Paper 20

CRUSTAL AND UPPER MANTLE ELECTRICAL CONDUCTIVITY STUDIES WITH NATURAL AND ARTIFICIAL SOURCES

R.N. Edwards, R.C. Bailey and G.D. Garland
Department of Physics,
University of Toronto,
Toronto, Ontario M5S 1A7

ABSTRACT

This paper attacks the problem of the location and significance of a zone of enhanced electrical conductivity which has been inferred at lower crustal depths below many regions of the Earth with different geological environments. Lower crustal conductivities have been indicated which are much above those predicted by the extrapolation of laboratory measurements made on probable rock types. Initially suggested by magnetotelluric measurements, the zone has also been indicated by controlled-source experiments and geomagnetic depth sounding. Our own observations include measurements with the former technique in the Canadian Shield and with the latter in the Appalachians.

A unified statistical approach employing the eigenvector decomposition of generalized linear inverse theory is applied to the results obtained by various methods in different places, in an attempt to improve the bounds on conductivity and depth.

RÉSUMÉ

Cet article s'attaque au problème de la localisation et de la signification d'une zone de conductivité électrique plus élevée qu'on semble déceler dans la partie inférieure de la croûte en dessous de plusieurs régions de la Terre dans les milieux géologiques différents. Les conductivités de la croûte inférieure semblent excéder de beaucoup celles qu'on peut prédire par extrapolation de mesures de laboratoire faites sur des types probables de roches. Comme le suggéraient initialement les mesures magnétotelluriques, cette zone a été mise en évidence par des expériences à source contrôlée et par des sondages géomagnétiques profonds. Nos propres observations sont faites à partir de mesures avec la première technique dans le bouclier canadien et avec la seconde dans les Appalaches.

On a appliqué une approche statistique unifiée utilisant la décomposition des eigenvecteurs de la théorie linéaire inverse généralisée aux résultats obtenus par différentes méthodes en différents endroits, dans le but de préciser les limites de conductivité et de profondeur.

INTRODUCTION

This paper is divided into two main sections. In the first, we review our results obtained over several years from studies on the electrical structure of the crust in eastern North America. The details of experimental and interpretive techniques are not given, because they have been discussed elsewhere. In the second part of the paper the chief features of the models are discussed, and the certainty with which depth and conductivity may be determined by magnetotelluric, long-wire controlled-source and Schlumberger resistivity sounding is investigated.

ELECTRICAL CRUSTAL STUDIES IN EASTERN NORTH AMERICA

Our investigations of the electrical structure of the crust in eastern North America began with measurements at single stations distributed through southern Quebec, the Maritimes, and Newfoundland (Fig. 1). Measurements of the time variations of three components of the magnetic field, and two of the electric field, were made at most of these stations. The magnetic fields were analyzed in terms of transfer functions or induction arrows by Bailey *et al.* (1974), and the magnetotelluric relationships by Kurtz and Garland (1976).

Induction arrows for a given period of magnetic disturbance are obtained from the statistical study of three magnetic components. The arrows are drawn to show the azimuth of horizontal magnetic field in phase with the vertical field (or, for the "imaginary" arrows, in quadrature with the vertical field). The length of an arrow is proportional to the ratio of the correlated parts of the vertical to horizontal amplitudes. The analysis of induction arrows does not by itself yield a model of the variation of conductivity with depth. Rather, the arrows can delineate lines of electric current concentration within the Earth by virtue of the fact that they tend to point towards (and perpendicular to) adjacent current concentrations. In eastern Canada, the principal result was that at periods greater than 1220 seconds, the in-phase arrows pointed predominantly south, suggesting a source to the south of the study area (Fig. 1). It is true that a bias in direction can be produced by a source effect, for a southward-pointing arrow is produced by an external source current, that is, one above the earth, to the north of a station. The possibility existed, therefore, that this overall southerly direction of arrows at the longer periods was simply a reflection of the fact that the sources of most of the events analyzed lay in the region of the auroral electrojet to the north. We shall return to this point shortly. At shorter periods (down to 154 seconds), the arrows at most stations rotated, showing the influence of more local current concentrations. In particular, Logan's Line, marking the northern limit of the Appalachians, appeared to be characterized in this way. The seismically-active region near the intersection of this line with the Saguenay River was marked by rapid changes in direction, particularly of the imaginary arrows, which could be produced by leakage of near-surface currents to a conductor at depth.

The magnetotelluric method involves the calculation of a frequency-dependent

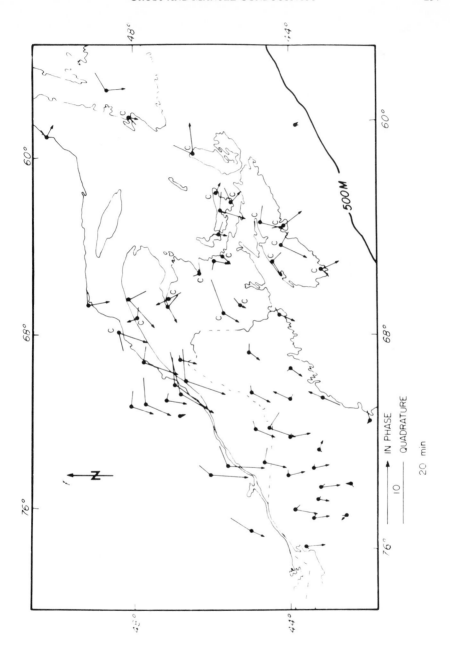

Figure 1. Transfer functions of eastern Canadian and northeastern United States stations for a period of 20 minutes. Scale at bottom indicates the vertical to horizontal ratio; stations marked "C" are from Cochrane and Hyndman (1974).

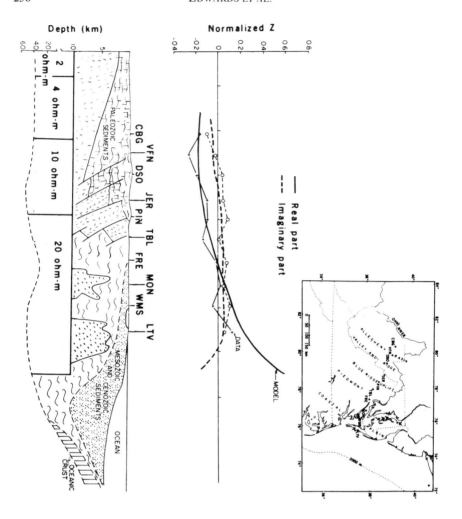

Figure 2. Vertical field variations at a period of 1 hour, expressed as fractions of the amplitude near a ''normal'' coastline, as obtained on a traverse (inset) across the southern Appalachians by Edwards and Greenhouse (1975). At bottom is shown the resistivity structure required to give the model curves.

impedance computed from the ratio of the time-varying electric field, as given by the recorded voltage drop between two electrodes, to the time-varying magnetic field. In principle, when conditions are ideal, an interpretation of conductivity as a function of depth may be made. In our measurements, the interpretation was greatly complicated by the local distortion of the electric field at nearly all stations. This was evidenced by a high degree of polarization in the field, often relatable to local contacts or fracture zones in the exposed rocks, and by statistical tests on the field relationships, which showed that the ideal, uniform, magnetotelluric conditions did not exist. In spite of these difficulties, reasonable two-dimensional models of crustal conductivity were obtained to fit most stations. In general, two different models of conductivity were found to characterize the region. For stations on the Shield, a relatively resistive (600-100 Ω-m) upper crust was found to be underlain by a more conductive (200 Ω-m) lower crust, itself underlain by a resistive (5000 Ωm) upper mantle. By contrast, stations in the northern Appalachians appeared to be characterized by a uniformly more resistive (1000 Ωm) crystalline crust and a more conductive (100 Ωm) upper mantle. It is interesting that two models similar to these have recently been proposed by Jones and Hutton (1979) for stations in Britain, one north and one south of the Southern Uplands Fault.

To return to the analysis of induction arrows, the indication of a source to the south of eastern Canada showed the importance of making observations over the more

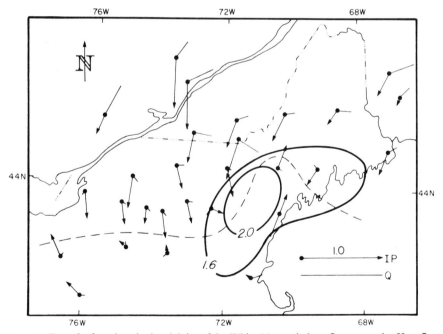

Figure 3. Transfer functions in the vicinity of the White Mountain heat flow anomaly. Heat flow contours are in heat flow units (1 hfu = 41.9 S.I. units); the broken line indicates the axis along which the direction of the transfer function arrows (real part) traverses.

southerly Appalachians. Edwards and Greenhouse (1975) established a traverse across the Appalachians, reaching the coast near the latitude of Washington, D.C. All of the vertical fields measured by them were small, suggesting the presence of a conductor beneath the region. In particular, the expected enhancement of the vertical field near the coast, due to electric currents induced in the ocean, was absent. The explanation (Fig. 2) was that the oceanic effect was balanced by extremely conductive material within the lower crust beneath the Appalachians. Edwards and Greenhouse proposed that the conductivity was provided by a very high degree of hydration; recent

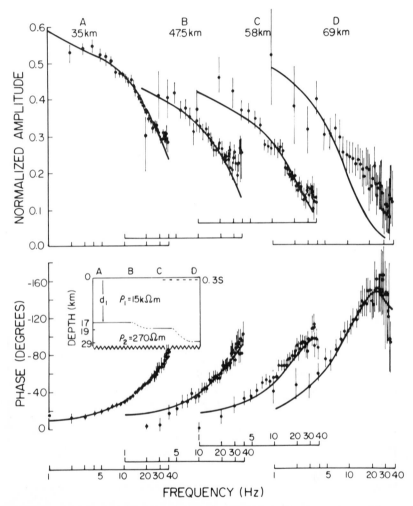

Figure 4. Normalized vertical field amplitude, as a function of frequency, for 4 stations, at the distances shown, west of the Timmins controlled-source transmitter. Solid curves are computed for the crustal models indicated.

seismological evidence (Oliver, 1980) suggests that this may lie in a mass of concealed, underthrust sedimentary rocks.

The region between eastern Canada and the profile of Edwards and Greenhouse has been studied by Bailey et al. (1978). In this area (Fig. 3) the induction arrows show a definite reversal, from southward-pointing to northward-pointing at the most southerly stations. This sudden change in direction cannot be a source effect; a concentrated zone of current flow in the lower crust or upper mantle is definitely indicated. This zone, which presumably marks the northern limit of the unusually conductive crust of the southern Appalachians, trends across the strike of the mountains, approximately along the Mohawk Valley, but is warped to the north in the vicinity of the White Mountains. Bailey et al. proposed that higher-than-normal temperatures in the lower crust or upper mantle, which could be produced by the upper crustal concentrations of radioactive minerals responsible for the White Mountain heat flow anomaly (Fig. 3), could enhance the conductivity sufficiently to produce this warping.

This review of investigations with natural field variations in eastern North America has indicated some of the limitations inherent in the available methods. Geomagnetic depth sounding through the use of induction arrows provides a sensitive indication of the location of anomalous, induced currents, for which models such as that of Figure 2 can be constructed, but the method does not yield a conductivity-depth model in a straightforward way. Furthermore, in those cases where the arrows over an entire region show a systematic trend in direction, source effects may be suspected. By contrast, the magnetelluric approach gives a first estimate of conductivity as a function of depth, but it suffers from the local distortion of the electric field in regions where the near-surface rocks are laterally heterogeneous in electrical properties. For these reasons, a controlled-source experiment has been carried out (Duncan et al., 1979) near Timmins, Ontario, specifically to sound to lower crustal depths. The source was an abandoned power line, 21 km long, grounded at both ends and energized with a commutated D.C. signal of variable pulse width. Magnetic field components were measured at distances up to 69 km from the wire, and the signal was extracted by means of a cross-correlation technique. In general this signal includes the magnetic fields of the current in the wire, of the normal return current in the earth and of any currents induced in conductors at depth. The measured fields at known distances, as a function of frequency, are inverted to give a conductivity structure. At Timmins, the most useful results were obtained to the west of the wire, in an area of essentially granitic rock. To the east, an extensive greenstone terrain provided local near-surface conductors which limited the possibility of sounding to depth. The results for four western stations are shown in Figure 4. Stations 19 and 58 give clear indication of a sharp reduction in crustal resistivity, from 15,000 Ωm to 270 Ωm, at a depth of 17 km. The more distant stations 57 and 55 require the conductor, if present, to be considerably deeper. Duncan et al. concluded that there was evidence for a conductive lower crust in the Timmins area, but that the conductive region was variable in depth, and possibly discontinuous laterally.

LIMITS ON MODEL PARAMETERS

All of the investigations mentioned above include indications of relatively high conductivity at some depth within the continental crust. Similar indications have been found in other parts of the world by the same methods (Nekut *et al.*, 1977; Jones and Hutton, 1979), and also by large-scale four-electrode Schlumberger resistivity sounding (Van Zijl and Joubert, 1975; Blohm *et al.*, 1977). Because the existence of these high conductivities must imply restraints on the composition or hydration state of the region concerned, it is important to determine the limits within which the depth and conductivity are established by each method. In particular, it is desirable to establish whether the indicated conductive region lies within the lower crust or uppermost mantle. The approach has been to adopt a model for the conductivity structure of the upper 200 km of the earth, and, assigning typical errors for field observations by the magnetotelluric, controlled-source and Schlumberger methods, investigate the degree of certainty with which the model parameters could have been obtained in each case. The model adopted was in fact that of Blohm *et al.* (1977) for Southern Africa, shown in Figure 5. It is not dissimilar to those indicated for eastern North America, consisting of a resistive upper crust, a lower crust in which the resistivity drops to 50 Ωm, a resistive upper mantle, and a decrease in resistivity at depth within the mantle.

    Standard methods exist for fitting parameterized models to measured data (e.g., Lawson and Hanson, 1974) by least-squares techniques. These compute the values of the parameters of the best fitting model and their errors. Having obtained a final model, a useful statistical procedure is to perform an eigensolution analysis based on the

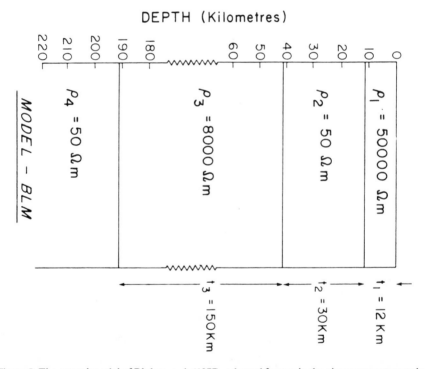

Figure 5. The crustal model of Blohm *et al.* (1977), adopted for use in the eigenparameter study.

inverse methods of Wiggins (1972) and Jackson (1972). This approach allows the inherent ambiguities of the model-fitting procedure to be described in a more physical way than would a simple statement of the errors of the model parameters. For example, it is often the case that a conductive layer significantly influences the measurements only through its conductivity-thickness product; a wide range of actual resistivity and thickness values for the layer is acceptable as long as the conductivity-thickness product has an appropriate value. As a result, the computed errors for these parameters are large even when a certain combination of the parameters is well determined. Although inspection of the parameter correlation matrix (Inman, 1975) can reveal that a parameter combination is well determined even when the individual parameters are not, this is often very difficult if correlations exist between many parameters. It is also difficult to describe succinctly which parameter combinations can be varied without affecting the fit of the model. In the eigensolution analysis, the model is allowed to determine for itself which independent parameter combinations are constrained by the data and to what degree.

Let the best fitting model have parameters p_j, $j=1,N$. The p_j will be the layer thicknesses and resistivities. For small variations, dp_j in these parameters, the expected changes in the measurements Y_i ($i=i,M$) will be given by

$$dY_i = \text{sigma } (j=1,N) \text{ of } A_{ij} * dp_j$$

where each coefficient A_{ij} is simply the sensitivity of datum Y_i to a change in parameter p_j; i.e., A_{ij} is the partial derivative dY_i/dp_j. These can be calculated by solving the forward problem with slightly changed model parameters. It is possible to choose new linear combinations of both the old data and the old parameters so that the complicated relationships described by equations (1) become very much simpler. Such combinations are found by applying what is called singular-value decomposition to the matrix A_{ij} of partial derivatives, as described, for example, by Wiggins (1972). Standard computer programmes exist for this purpose. Now, if we describe these by

$$Y'_i = \text{sigma } (j=1,M) \text{ of } V_{ij} * Y_j \qquad i = 1,M$$

and

$$p'_k = \text{sigma } (l=1,N) \text{ of } U_{kl} * p_l \qquad l = 1,N$$

where the coefficients (weights) V_{ij} and U_{kl} are those found by the singular-value decomposition, and if we call the quantities Y'_i and p'_k eigendata and eigenparameters respectively, equations (1) above simplify to the set of simple relationships given by

$$Y'_i = L_i * p'_i \qquad (i=1,N).$$

Only one set of weights V_{ij} and U_{kl} will permit this simplification. The terms L_i are the eigenvalues of A obtained by the singular-value decomposition. We now have parameter combinations p'_i, each of which is determined independently by the data through the Y'_i. There is no longer any need to worry about parameter correlations if we describe the model in terms of these eigenparameters and their errors.

This procedure clearly has limited appeal if we cannot identify the eigenparameters physically in our model. Fortunately, this physical identification is often quite straightforward, particularly if models are described from the outset in terms of the logarithms of the resistivities and thicknesses rather than the resistivities and thicknesses themselves. As an example, consider an eigenparameter which is a combination of two original parameters p_1 and p_2 with equal but opposite unit weights, where p_1 and p_2 are the logarithms of the thickness and the resistivity of a particular layer. Then

$$p' = p_1 - p_2 = \log(\text{thickness}) - \log(\text{resistivity})$$

so that $\quad p' = \log(\text{thickness*conductivity})$

Thus we can identify p' as effectively describing the logarithm of the conductivity-thickness product of the layer.

The error in each eigenparameter can be calculated simply (Wiggins, 1972; Jackson, 1972). If the data have been scaled initially so as to have independent errors of unit standard deviation, then the standard deviation of each eigenparameter is simply the reciprocal of L_j, the corresponding eigenvalue. In the example given above, the computed standard deviation of p' is just the fractional error in the conductivity-thickness product. The errors of the eigenparameters are uncorrelated with each other, so that description of model ambiguities in terms of the eigenparameters and their errors is straightforward.

This procedure is used to produce the results shown on Tables I, II and III. The eigenparameters are ranked from top to bottom in order of decreasing eigenvalue (increasing standard deviation). The row of coefficients gives the weights with which each original parameter is included in the combination of parameters for that eigenparameter. For example, for weights $U_1, U_2 . . .$, changes in the eigenparameter p' are related to changes in the original parameters p by

$$\delta p'/p' = U_1 \, \delta p_1/p_1 + U_2 \, \delta p_2/p_2 + ...$$

TABLE I

Statistics of the eigenparameter analysis for the Schlumberger sounding case. Symbols t and ρ refer to layer thickness and resistivity of the adopted model; figures in bold type indicate the weights of the most significant parameters contributing to each eigenparameter. The fractional standard deviation of eigenparameters is shown in the right-hand column. ·

STATISTICS - SCHLUMBERGER

EIGENPARM	$\delta t_1/t_1$	$\delta t_2/t_2$	$\delta t_3/t_3$	$\delta \rho_1/\rho_1$	$\delta \rho_2/\rho_2$	$\delta \rho_3/\rho_3$	$\delta \rho_4/\rho_4$	PHYSICS	$\delta p'/p'$
1	**-.96**	.08	-.00	-.26	-.09	-.03	-.00	Thickness Layer 1	0.01
2	0.11	**.70**	-0.04	.05	**-.70**	-.05	-.00	Conductivity - Thickness Layer 2	0.05
3	-.26	-0.02	.04	**.96**	.01	.04	.00	Resistivity Layer 1	0.19
4	.02	.05	**.70**	-.05	-.04	**.71**	.02	Resistivity - Thickness Layer 3	0.52
5	-.00	**.70**	-.03	.00	**.71**	.02	-.05	Resistivity - Thickness Layer 2	Greater
6	.00	-.04	**-.61**	-.00	-.05	**.62**	**-.49**	Conductivity - Thickness Layer 3	Than
7	-.00	.02	**-.36**	-.00	.01	**.33**	**.87**	Resistivity Layer 4	1

Each row of weights is normalized so that the sum of the squares of the weights is one (which is to be expected, as each row is obtained as an eigenvector of the matrix A_{ij} by the singular-value decomposition). Finally, the approximate physical interpretation of each eigenparameter is given, based on the weights in the row.

We have applied the eigenparameter analysis to the hypothetical sounding of the section shown in Figure 5 by three methods: Schlumberger, magnetotelluric and controlled bipole source. It is apparent from the above description of the theory that it is necessary to perform the forward calculations on the model to yield the quantities which would actually be observed in each case (i.e., apparent resistivity for the first

TABLE II

As for Table I, for magnetotelluric sounding

$$STATISTICS - MAGNETOTELLURICS \ (\nu = 3000 \ km^{-1})$$

EIGENPARM	$\delta t_1/t_1$	$\delta t_2/t_2$	$\delta t_3/t_3$	$\delta\rho_1/\rho_1$	$\delta\rho_2/\rho_2$	$\delta\rho_3/\rho_3$	$\delta\rho_4/\rho_4$	$\delta(\nu^2)/\nu^2$	PHYSICS	$\delta\rho'/\rho'$
1	**-.99**	.03	-.00	.04	-.15	-.00	.00	-.00	Thickness Layer I	0.03
2	.14	**.64**	-.12	-.00	**-.75**	-.00	.01	.00	Conductivity - Thickness Layer 2	0.03
3	-.03	-.03	.03	**-1.0**	-.03	.00	.00	-.00	Resistivity Layer I	0.12
4	.03	-.05	**.96**	.03	-.19	.01	.17	-.00	Thickness Layer 3	0.15
5	-.07	**.77**	.17	-.03	**.61**	-.01	-.06	-.00	Resistivity-Thickness Layer 2	0.21
6	.00	-.04	-.18	.00	.01	-.01	**-.98**	-.01	Resistivity Layer 4	Greater
7	-.00	-.01	.01	-.00	-.00	**-1.0**	-.01	.08	Resistivity Layer 3	Than
8	.00	.00	-.00	-.00	.00	.08	.01	**1.0**	(Wavenumber)^2	1

TABLE III

As for Table I, for controlled bipole source sounding

$$STATISTICS - CONTROLLED \ SOURCE$$

EIGENPARM	$\delta t_1/t_1$	$\delta t_2/t_2$	$\delta t_3/t_3$	$\delta\rho_1/\rho_1$	$\delta\rho_2/\rho_2$	$\delta\rho_3/\rho_3$	$\delta\rho_4/\rho_4$	PHYSICS	$\delta\rho'/\rho'$
1	**.99**	.05	.01	-.00	.13	-.00	-.00	Thickness Layer I	0.02
2	.11	.31	-.02	.02	**-.94**	-.00	.00	Resistivity Layer 2	0.04
3	-.09	**.95**	.02	-.04	.30	-.00	-.00	Thickness Layer 2	0.12
4	-.00	.03	.03	**1.00**	-.03	-.00	-.00	Resistivity Layer I	0.28
5	-.00	-.02	**1.00**	-.03	-.03	-.00	-.01	⎤	Greater
6	-.00	.00	.01	.00	.00	.05	**1.00**	⎬ NOISE	Than
7	.00	-.00	-.00	.00	.00	**-1.00**	.05	⎦	1

two, amplitude and phase of a magnetic field component for the third), and to assign
the errors with which these could reasonably be obtained. Figure 6, from Blohm *et al.*
(1977), shows the actual Schlumberger resistivity obtained in southern Africa, with
uncertainties of 25% in the apparent resistivities, as given by the authors. We have
adopted this error for the Schlumberger case, and also for the apparent resistivity as
determined in a hypothetical magnetotelluric sounding over the same section (Fig. 7).
Magnetotelluric observations also yield a phase difference between magnetic and
electric fields; as shown in the lower part of Figure 7, a constant realistic standard error
of 3° (Chaipayunzpun and Landisman, 1977) has been adopted for phase. The range of
periods chosen for the magnetotelluric calculations is such that the apparent resistivity
curve (upper part of Fig. 7) contains essentially the same information as does the
Schlumberger resistivity in Figure 6. For the controlled-source calculations, a method
based on that of Anderson (1974) has been used. Calculations have been carried out for
the field at a single station, (Fig. 8) 70 km from a bipole source (assumed to be short
compared to this distance). A frequency range of 0.01 to 100 Hz has been chosen, to

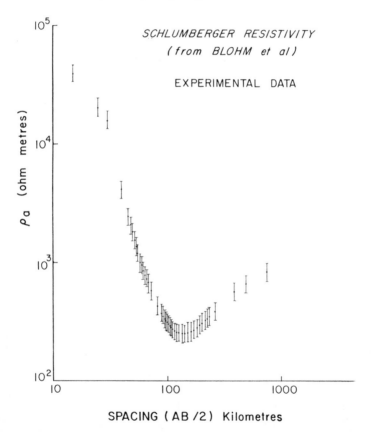

Figure 6. Schlumberger resistivity as a funciton of electrode spacing, obtained by Blohm *et al.*
(1977) in southern Africa.

correspond to that over which our own system can measure. The error in measured vertical magnetic field has been taken as 2% (and in phase 2°) on the basis of repeated measurements made during the Timmins experiment described above. This error is smaller than that assigned to the apparent resistivity for either the Schlumberger or magnetotelluric methods, because much of the uncertainty in the latter cases arises from the measurement of the electric field.

The statistical results for the eigenparameter analysis with these assumptions are shown on Tables I, II and III. Fractional errors in the eigenparameters are given in the right-hand column. All methods give the thickness of layer 1, that is, the depth to the first conductor, with remarkably small fractional error (0.01 to 0.03). Thereafter, the order of certainty with which different parameters can be determined varies between methods.

Both magnetotelluric and Schlumberger analyses give the conductivity-thickness product of the conductor as the second-best determined parameter. Thus, in the magnetotelluric case, the particular formula of the general equation for the eigenparameter is:

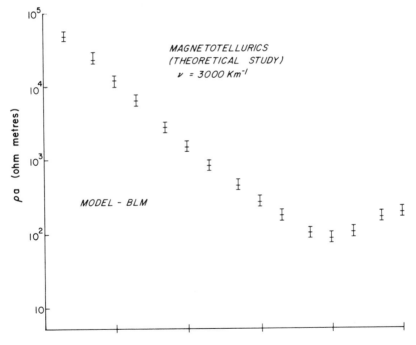

Figure 7. Computed magnetotelluric curves, for a source wavelength of 3,000 km, over the adopted model.

$$0.14 \ \frac{\delta t_1}{t_1} \ + 0.64 \ \frac{\delta t_2}{t_2} \ - 0.12 \ \frac{\delta t_3}{t_3} \ - 0.00 \ \frac{\delta \rho_3}{\rho_1}$$

$$- \ 0.75 \ \frac{\delta \rho_2}{\rho_2} \ - 0.00 \ \frac{\delta \rho_3}{\rho_3} \ + 0.01 \ \frac{\delta \rho_4}{\rho_4} \ + 0.00 \ \frac{\delta (v^2)}{v^2} = 0.03$$

The eigenparameter is dominated by the weights for $\frac{\delta t_2}{t_2}$ and $\frac{\delta \rho_2}{\rho_2}$; as explained above, the fact that the second of these is negative implies that it is the conductivity-thickness product which is determined. When two weights almost equally predominate, they should each be equal to $^1/\sqrt 2$ or 0.71. This condition is approximated for

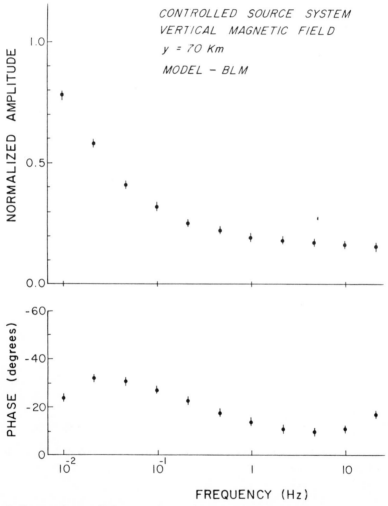

Figure 8. Computed controlled source observations over the adopted model.

eigenparameters 2 and 5 of Table II, and very closely achieved for eigenparameter 4 of Table I. Continuing with the comparison of the magnetotelluric and Schlumberger methods, both give the resistivity of the surface layer as eigenparameter 3, but thereafter they differ. Schlumberger sounding yields the product $\rho_3 t_3$ (equal weights for $\delta t_3/t_3$ and $\delta \rho_3/\rho_3$, both positive), but magnetotelluric sounding give t_3 alone. The difference arises from the fact that the former case current is forced through the relatively resistive third layer, whereas in the magnetotelluric situation virtually no current is induced in resistive layers. Magnetotellurics give the product $\rho_2 t_2$, with fractional error 0.21, as the fifth signficant eigenparameter, so that in principle the thickness and conductivity of layer 2 can be separated. The inverse of source scale length, v, has a vanishingly small weight in all magnetotelluric eigenparameters.

Turning to controlled-source sounding, we find that the eigenparameters are very different. All of the four significant ones are determined by single parameters rather than products. It would obviously be advantageous to sound at the same location with both controlled-source and either Schlumberger or magnetotelluric methods, for ρ_2 and t_2 could then be separated with a high degree of certainty. This is an extension of the result of Vozoff and Jupp (1975) who showed, for a different model, the advantage of jointly inverting magnetotelluric and Schlumberger measurements. When the fractional error exceeds unity, the quantities are assumed to be undetermined; in the case of the controlled-source calculations, parameters relating to layers 3 and 4 are labelled "noise", because it can be shown that, for the station distance chosen, these could never be determined.

The important conclusions of this analysis are 1) that the presence and depth of a relatively highly conducting layer, indicated by any of the three methods studied, may be accepted with considerable certainty and 2) where a conductor can be shown to lie at lower crustal depths, it is almost certainly in the crust, not in the upper mantle. Most regions have been investigated by only a single method, and in these cases it is the conductivity-thickness product of the conductor which is established, but there is certainly a good possibility for the future of using joint inversions to estimate conductivity and thickness separately.

SIGNIFICANCE OF THE RESULTS OF THE STUDY OF THE CRUST

The indication of a highly conducting layer or zone (resistivity 200 ohm m or less) in the lower crust below many regions of different tectonic character has provoked great interest in the laboratory study of the resistivity of basalt under lower crustal conditions. Brace (1971) carried out detailed measurements on samples of typical basalt, studying independently the effects of varying temperature on mineral conduction, and of varying pressure on conduction through pores. Increase of temperature increases the conductivity of minerals which behave as semiconductors, while increase in pressure decreases pore volume and inhibits conduction through the pores. Applying his results to the estimated temperature gradient in the eastern United States, Brace estimated that in the depth range of 15 to 35 km, both mechanisms could contribute significantly to the conduction, but the resistivity could hardly be less than 10^4 Ω-m.

More recent work has suggested two factors which could reduce this estimate. First, Stesky and Brace (1973) and Drury (1979) have reported lower resistivities in serpentinized rocks from the ocean floor, and it is possible that hydrated minerals, whose effect on the enhancement of conductivity in young crust was emphasized by Hyndman and Hyndman (1968), are more abundant in the lower continental crust than was formerly assumed. Secondly, it has been pointed out by Nekut *et al.* (1977) and F.J. Vine (pers. commun., 1979) that Brace's method of investigation underestimated the role of pore conduction, by not providing for the increase in salinity of pore fluids with increasing temperature. It is possible that with provision for this effect, the laboratory measurements on normal basalt would indicate resistivities of 10^3 Ωm or less under lower crustal conditions. To reduce the resistivity significantly below 10^3 Ωm, as required by the results of many soundings, including our own for the northern Appalachians and the Canadian Shield at Timmins, it is probably necessary to invoke the presence of hydrated minerals. The extremely high conductivities discovered beneath the southern Appalachians might be explained by concealed wedges of sedimentary rock, as mentioned above.

ACKNOWLEDGEMENTS

The Negotiated Development Grant of the National Research Council of Canada provided funds for the data acquisition systems and the controlled-source. Field work was supported through the Negotiated Development Grant and operating grants of the National Research Council and Research Agreements of the Department of Energy, Mines and Resources.

REFERENCES

Anderson, W.L., 1974, Electromagnetic fields about a finite electric wire source: United States Geol. Survey Report GD-74-04, Springfield, Nat. Tech. Inf. Service.

Bailey, R.C., Edwards, R.N., Garland, G.D., Kurtz, R. and Pitcher, D., 1974, Electrical conductivity studies over a tectonically active area in eastern Canada: Jour. Geomag. Geoelect., v. 26, p. 125-146.

Bailey, R.C., Edwards, R.N., Garland, G.D. and Greenhouse, J.P., 1978, Geomagnetic sounding of eastern North America and the White Mountain heat flow anomaly: Geophys. Jour. Royal Astron. Soc., v. 55, p. 499-502.

Blohm, E.K., Worzyk, P. and Scriba, H., 1977, Geoelectrical deep soundings in Southern Africa using the Cabora Bassa power line: Jour. Geophys., v. 43, p. 665-679.

Brace, W.F., 1971. Resistivity of saturated crustal rocks to 40 km. based on laboratory measurements: in Heacock, J.G., ed., The Structure and Physical Properties of the Earth's Crust: Amer. Geophys. Union, Monograph 14.

Chaipayungpun, W. and Landisman, M., 1977, Crust and upper mantle near the western edge of the Great Plains: in Heacock, J.G., ed., The Earth's Crust: Amer. Geophys. Union, Monograph 20.

Cochrane, N.A. and Hyndman, R.D., 1974, Magnetotelluric and magnetovariational studies in Atlantic Canada: Geophys. Jour. Royal Astron. Soc., v. 39, p. 385-406.

Drury, M.J., 1979, Electrical resistivity models of the oceanic crust based on laboratory measurements on basalts and gabbros: Geophys. Jour. Royal Astron. Soc., v. 56, p. 241-253.

Duncan, P.M., Hwang, A. Edwards, R.N., Garland, G.D. and Bailey, R.C., 1979, The development of a wide-band E.M. prospecting system: Geophysics, in press.

Edwards, R.N. and Greenhouse, J.P., 1975, Geomagnetic variations in the eastern United States: evidence for a highly conducting lower crust?: Science, v. 188, p. 726-728.

Hyndman, R.O. and Hyndman, D.W., 1968, Water saturation and high electrical conductivity in the lower continental crust: Earth Planet Sci. Letters, v. 4, p. 427-432.

Inman, J.R., 1975, Resistivity inversion with ridge regression: Geophysics, v. 40, p. 798-817.

Jackson, D. D., 1972, Interpretation of inaccurate, insufficient and inconsistent data: Geophys. Jour. Royal Astron. Soc., v. 28, p. 97-109.

Jones, Alan G. and Hutton, Rosemary, 1979, A multi-station magnetotelluric study in southern Scotland II. Monte-Carlo inversion of the data and its geophysical and tectonic implications: Geophys. Jour. Royal Astron. Soc., v. 56, p. 351-368.

Kurtz, R.D. and Garland, G.D., 1976, Magnetotelluric measurements in eastern Canada: Geophys. Journ. Royal Astron. Soc., v. 45, p. 321-347.

Lawson, C.L. and Hanson, R.J., 1974: Solving Least-Square Problems: Prentice-Hall.

Nekut, A., Connerney, J.E.P. and Kuckes, A.F., 1977, Deep crustal electrical conductivity; evidence for water in the lower crust: Geophy. Res. Letters, v. 4, p. 239-242.

Oliver, J., 1980, Seismic reflection profiling and deep crustal structures. This volume.

Stesky, R.M. and Brace, W.F., 1973, Electrical conductivity of serpentinized rocks to 6 kilobars: Jour. Geophys. Res., v. 78, p. 7614-7621.

Van Zijl, J.S.V. and Joubert, S.J., 1975, A crustal geoelectrical model for South African Precambrian granitic terrains based on deep Schlumberger soundings: Geophysics, v. 40, p. 657-663.

Vozoff, K. and Jupp, D.L.B., 1975, Joint inversion of geophysical data: Geophys. Jour. Royal Astron. Soc., v. 42, p. 977-991.

Wiggins, R.A., 1972, The general linear inverse problem: implication of surface waves and free oscillations for earth structure: Rev. Geophys. and Space Phys., v. 10, p. 251-285.

The Continental Crust and Its Mineral Deposits, edited by D.W. Strangway,
Geological Association of Canada Special Paper 20, 1980

SHALLOW ELECTRICAL SOUNDING
IN THE PRECAMBRIAN CRUST
OF CANADA AND THE UNITED STATES

D.W. Strangway and J.D. Redman
Department of Geology and Department of Physics,
University of Toronto, Toronto, Ontario M5S 1A1
D. Macklin
G.T.E. Sylvania, 189B St. Needham, Mass.

ABSTRACT

Audio frequency magnetotelluric (AMT) sounding observations have been conducted in Wisconsin, Michigan (upper peninsula), northwestern Ontario, eastern Manitoba, and north central Ontario. These observations show that in regions where there are only moderate lateral variations in resistivity the Shield has one or two thin electrical surface layers. These layers are at most a few hundred metres thick, and contain a low resistivity zone presumably associated with overburden and fracturing in the bedrock. Beneath this is a very thick, very resistive zone at least 5 kilometres and probably more than 10 kilometres thick, with a resistivity often 10,000 ohm-m to 50,000 ohm-m. Our data support other studies, which indicate a very conductive zone 15 to 25 kilometers beneath this.

In some portions of the shield it is not possible to conduct deep penetration experiments using electrical methods. There may be extensive near-surface faulting or highly conductive glacial clays near the surface. The presence of conductive graphites and/or serpentinites also makes crustal studies difficult using electrical methods. It is thus desirable to map out the near surface electrical structure using a method such as AMT before attempting to conduct deep crustal soundings by large electrode array systems or by using low frequency magnetotellurics.

RÉSUMÉ

On a fait des observations en utilisant des sondes magnétotelluriques avec des fréquences audio dans le Wisconsin, le Michigan (partie supérieure de la péninsule), le nord-ouest de l'Ontario, l'est du Manitoba et le centre nord de l'Ontario.Ces observations montrent que dans une région où il y a seulement des variations latérales modérées dans la résistivité, le bouclier possède une ou deux minces couches électriques en surface. Ces couches ont une épaisseur qui ne dépasse pas quelques centaines de mètres et elles contiennent une zone de faible résistivité qui correspond probablement au mort-terrain et à la fracturation du roc. En dessous, on observe une zone très épaisse, de résistivité très élevée d'au moins 5 km et probablement plus de 10 km d'épaisseur, avec une résistivité souvent de 10 000 ohm-m à 50 000 ohm-m. Nos données supportent d'autres études qui indiquent la présence d'une zone très conductrice de 15 à 25 km plus profonde.

Dans certaines parties du bouclier, il n'est pas possible d'effectuer des essais à grande profondeur en utilisant les méthodes électriques. C'est souvent causé par la présence de nombreuses failles près de la surface ou d'argiles glaciaires superficielles très conductrices. La présence de graphite, de serpentinites ou de ces deux conducteurs rend aussi les études dans la croûte difficiles avec les méthodes électriques. Il est donc avantageux de cartographier les structures électriques près de la surface en utilisant une méthode comme les sondages magnétotelluriques de fréquence audio avant de tenter des sondages électriques profonds en utilisant de grands réseaux d'électrodes ou en utilisant les méthodes magnétotelluriques à basse fréquence.

INTRODUCTION

This paper summarizes the results of a number of surveys conducted in the Precambrian Shield of Canada and the United States using the audio frequency magnetotelluric method (AMT). Magnetotelluric methods were first described by Cagniard (1953), who showed that natural magnetic fluctuations which were uniform over a sufficiently large region to behave approximately as plane waves could be used as sources for probing the electrical structure of the Earth. Using this approach, it was possible to measure the ratio of the electric field (E in volts/m) to the magnetic field (H in ampturns/m) and thus to determine the apparent resistivity. The apparent resistivity (ρ_a) is defined using the relationship $\rho_a = \frac{1}{\mu\omega}$ (E/H)^2. Since in a typical conductive Earth case, electromagnetic energy diffuses into the ground, the depth of penetration or skin depth is an inverse function of frequency. The skin depth is given as $\delta = \sqrt{2\rho/\mu\omega}$ where ρ is the resistivity in ohm-m, μ is the magnetic permeability (usually 4 π x 10^{-7} in henries/m), and ω is the rotational frequency (2 π f). Instrumentation to operate in the audio frequency range, i.e., from 10 hz to 20 khz, whose development was based on earlier research sponsored by Kennecott Copper Corp, was first described by Strangway and Vozoff (1969) and by Strangway et al. (1973). This instrumentation has since been improved, and in the past few years we have conducted several surveys in various parts of North America. Some of these have been documented in a series of papers (permafrost – Koziar and Strangway, 1975, 1978a; greenstone and gneissic belt in northwestern Ontario – Koziar and Strangway, 1978b; massive sulphides – Strangway and Koziar, 1979). In the present paper, we report on AMT results collected in Wisconsin and in Michigan in the Southern province, in southern Manitoba and in northwestern Ontario in the Superior province, and across the Grenville/Superior boundary near Sudbury, Ontario and at Chalk River, Ontario in the Grenville province (Fig. 1). In these surveys we have been able to find portions of the shield where sounding of the crust to depths of several kilometres is possible.

Our apparent resistivity measurements were made at the discrete frequencies of 12, 22, 35, 80, 210, 470, 860, 2100, 4300, 8700 and 10,000 hertz. At the present time, we measure only the magnitude of the apparent resistivity with observations usually taken in orthogonal directions. The electric field sensor is a cable, usually 50 metres long, and the magnetic field sensor is an induction coil. There have been a series of recent papers describing AMT work done by other groups. Hoover *et al.* (1976) describe a system developed at the United States Geological Survey. Hoover *et al.* (1978) and Mabey *et al.* (1978) summarize the results of extensive surveys in the Basin and Range province and in the Columbia Plateau basalts. Ngoc *et al.* (1978) report on a system they developed which operates at 1, 8, 145 and 3,000 hertz, and Benderitter *et al.* (1978) report on a system operating at 8, 17, 37, 80, 170, 370, 800 and 1700 hertz as used on a survey in Finland. This same equipment was used by Dupis and Iliceta (1974) and Dupis *et al.* (1974) to make measurements at the Lardarello geothermal site in Italy. We report now on the results of our individual surveys.

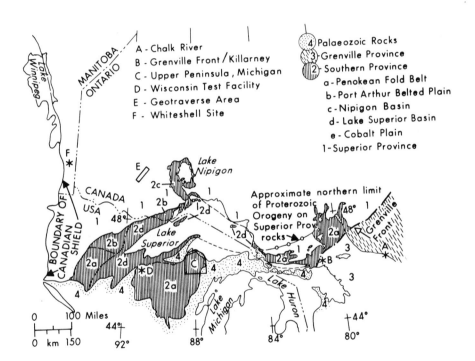

Figure 1. Location map showing the southern Superior province and the Grenville province of Ontario and the north central United States and locations of surveys.

WISCONSIN TEST FACILITY

The results described here are based on a set of measurements at the Wisconsin Test Facility (WTF) near Clam Lake, Wisconsin. This is the site of a powered antenna 22.5 km long in two orthogonal directions (Fig. 2) which is designed to operate at 45 and 75 hz. As this facility was set up as a preliminary test area for designing a submarine communication system, many previous electrical measurements have been made as an aid to characterizing the transmitter. Some of these reports include d.c. resistivity

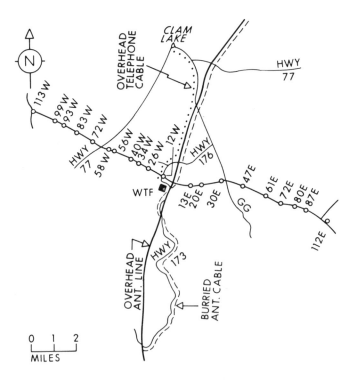

Figure 2. Sketch map showing the N-S and E-W arms of the Wisconsin Test Facility antenna near Clam Lake, Wisconsin. AMT soundings were conducted on the E-W arm at the locations shown.

measurements using a Schlumberger array (Davidson *et al.*, 1974). This sounding is shown in Figure 3 and has been interpreted to give the following layers: a thin layer 40 m thick with a resistivity of 450 ohm-m, followed by a layer 200 m thick of 13,700 ohm-m, then a layer 100 m thick of 300 ohm-m overlying a resistive half-space of 6350 ohm-m. An anisotropy of 1.33 was found in the basement layer, the larger value being in the N-S direction. The maximum electrode spacing was 16,000 metres.

There have been a number of other studies at the site using the antenna itself. Bannister and Williams (1974), for example, summarize measurements of the magnetic field (H) pattern of the antenna with a current (I) at ranges up to 75 kilometres from the antenna using the antenna frequencies of 45 and 75 hz. This approach, referred to as the H/I method, gives a regional resistivity value at each of these frequencies. These values show that the antenna pattern is skewed several degrees (clockwise from the E-W antenna and counterclockwise from the N-S antenna), indicating local anisotropy in the electrical structure. The apparent resistivities determined in this way are as follows:

45 hz − E.W. 3400 ohm-m NS 5800 ohm-m
75 hz − E.W. 3100 ohm-m NS 4500 ohm-m

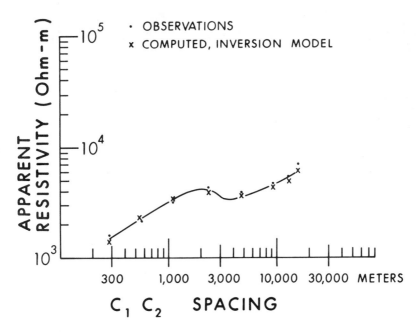

Figure 3. Schlumberger expanding electrode data roughly along the E-W, WTF antenna line (after Davidson *et al.*, 1974). The observations are shown with dots. The stars are a four layer model with the following parameters:

1. 40m   450 ohm-m
2. 200m   13,700 ohm-m
3. 100m   300 ohm-m
4. half-space   6,250 ohm-m

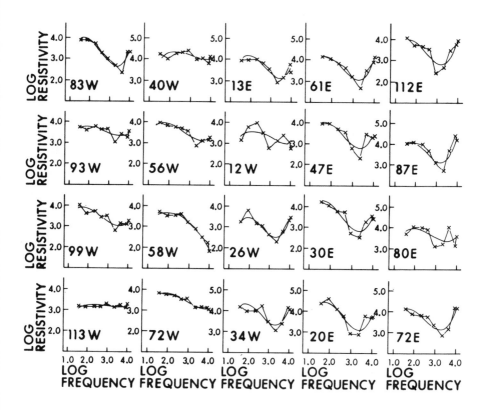

Figure 4. Plots of log resistivity versus log frequency for the 20 sounding stations shown in Figure 2. Frequency range was from 35 hertz to 20 khz. The smooth curves are a least squares fit to the data using a third order polynomial. These observations were taken with the electric field sensors oriented north-south. Similar data were also taken with the electric field sensor oriented east-west.

It should be emphasized that these are regional averages based on long traverses away from the antenna.

Figures 4 and 5 show the results of 20 AMT soundings conducted along the east-west antenna leg at roughly equal intervals. Figure 4 contains a set of representative plots of log frequency versus log resistivity. These data represent soundings at the stations shown in Figure 2 with the electric field sensor oriented N-S. The set of smooth curves in Figure 4, representing a least-squares, third-order polynomial fit to the observations, have been used to derive the pseudosections shown in Figure 5. The pseudosection is a plot of the contoured resistivity value (in logarithmic intervals) in a plane of horizontal position versus frequency. Interpolation between non-uniformly spaced stations was done on the basis of the square root of the resistivity (or the electric field), roughly simulating the effect of a long electric field measurement. This display emphasizes the lateral variations along the line, and in general lower frequencies correspond to greater sounding depths.

We have obtained mean values for the complete line at each frequency in an attempt to estimate an average electrical resistivity for it. These mean values are plotted in Figure 6, which shows the overall sounding curve for the line. Table I shows the same results, as well as the comparison with the H/I work at 45 and 75 hertz. This

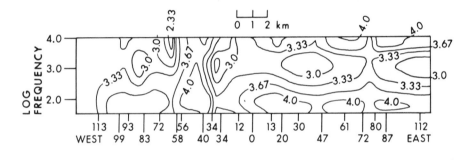

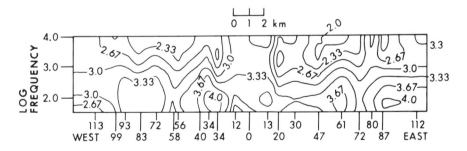

Figure 5a, b. Pseudosections showing the station locations along the horizontal axis and the frequency on the vertical axis. Data are log resistivity values contoured in units of 0.33 log ρ_a. The values are based on the fitted polynomial curve and the results are machine contoured by interpolating between stations using the square root of the resistivity. (a shows the data using E-W electric field sensors, and b shows the data using N-S electric field sensors).

TABLE I

		Papp ohm-m			
	N-S	Average		E-W	
Frequency	AMT (20 station interpolation)	AMT (20 station interpolation)	AMT (20 station interpolation)	AMT using full 20 km antenna for E-field sensor	H/I
10,200	4,790	2,090	467	809	—
8,680	3,470	1,660	437	1,932	—
4,290	1,510	977	447	186	—
2,140	1,320	977	617	72	—
858	1,952	1,510	1,200	250	—
473	3,241	2,510	2,000	501	—
213	5,501	4,170	3,090	3,204	—
83	8,910	6,610	4,170	5,521	3,100
35	7,760	5,370	3,090	2,120	3,500

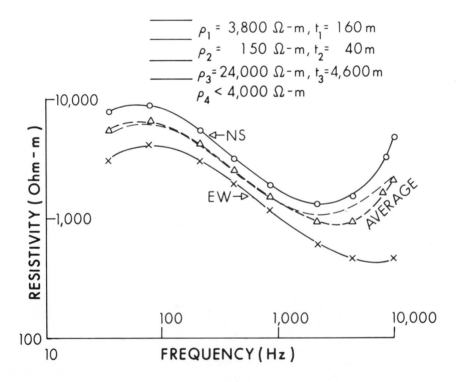

$$\rho_1 = 3,800 \ \Omega\text{-m}, \ t_1 = 160 \ \text{m}$$
$$\rho_2 = 150 \ \Omega\text{-m}, \ t_2 = 40 \ \text{m}$$
$$\rho_3 = 24,000 \ \Omega\text{-m}, \ t_3 = 4,600 \ \text{m}$$
$$\rho_4 < 4,000 \ \Omega\text{-m}$$

Figure 6. Averaged AMT results over the east-west line at the Wisconsin test facility (average based on interpolating square root of resistivity). Long dashes correspond to resistivity model shown.

comparison is reasonably good, considering the difference in the nature of the two methods, and is generally consistent with the Schlumberger sounding.

It should be noted that there is also strong anisotropy in the AMT data. When the current flow is in the N-S direction, the resistivities are significantly larger than when it is in the E-W direction. This is consistent with the anisotropy inferred from the antenna studies; it may reflect a true anisotropy in the resistivity of the rocks themselves, or perhaps a geometric factor caused by a more conductive zone north of and roughly parallel to the east-west antenna line. Since the effect seems larger at high frequencies, it may be related to shallow bedrock-surface topography controlled by such features as fractures.

It is possible to interpret these observations by means of a four-layer model with the following characteristics: 1) Top layer – highly resistive glacial sands and gravels a hundred metres or so thick, with anisotropy, 2) Second layer – about 40 metres thick with a resistivity of 150 ohm-m, 3) Third layer – resistivity of about 24,000 ohm-m and a thickness of about 5 kilometres. 4) Fourth layer – If the resistivity decreases at depth, this highly conductive zone occurs below about 7 kilometres. It must be at least this deep, or it would have affected the large array Schlumberger soundings and strongly affected the apparent resistivity values at frequencies above 35 hz.

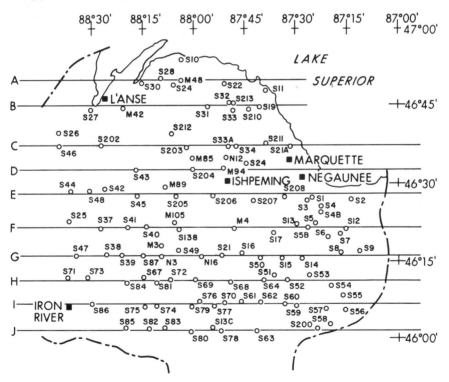

Figure 7. Location map of the upper peninsula of Michigan showing position of AMT sounding stations and profiles.

It appears that the upper crust near the Wisconsin Test Facility is highly resistive to depths of at least 7 kilometres. This is consistent with interpretations based on other regional studies of the area (See review by Sternberg and Clay, 1977). Dowling (1970) reported on magnetotelluric measurements on a regional scale over the Wisconsin arch at long periods of 1000 seconds to frequencies of 20 hz, almost overlapping into our present frequency range. His highest frequency resistivity values range from a few 100 ohm-m to nearly 10,000 ohm-m and then roll over to lower values between 1 and 5 hertz, suggesting a conductor at a depth of 10 kilometres or more.

Thus, combining the lower frequency range used by Dowling (1970) and by Sternberg and Clay (1977), the existence of a uniform, thick, high resistivity layer is confirmed; furthermore, at frequencies lower than our range there is clear evidence of a conductive layer at middle to lower crustal depths. In the study by Sternberg and Clay (1977) and elaborated on by Sternberg (1979), the "Flambeau anomaly" is also reported. This is a zone of conductive graphites and schists such as those frequently encountered in shield terranes.

MICHIGAN-UPPER PENINSULA

We ran a similar survey on the upper peninsula of Michigan, primarily to see if it would be an appropriate place to install a large antenna array for submarine communication. The same procedures and instrumentation were used for 102 stations spread

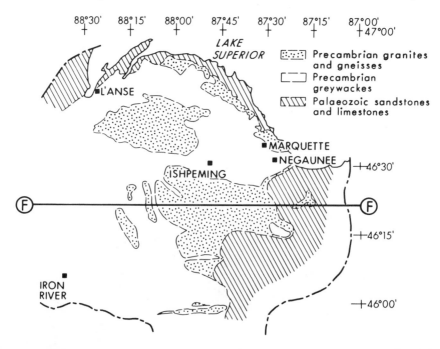

Figure 8. Generalized geological sketch map of the upper peninsula of Michigan showing Precambrian gneisses and granites, Precambrian greywackes, and Paleozoic limestones and sandstones.

over an area of 3400 square miles (Fig. 7). The overall observations are very similar to those in the Wisconsin area, although some regional trends are now apparent. Figure 8 is a geologic sketch map in which the following major units can be identified: 1) Ordovician and Cambrian sedimentary cover in the eastern part of the area, 2)

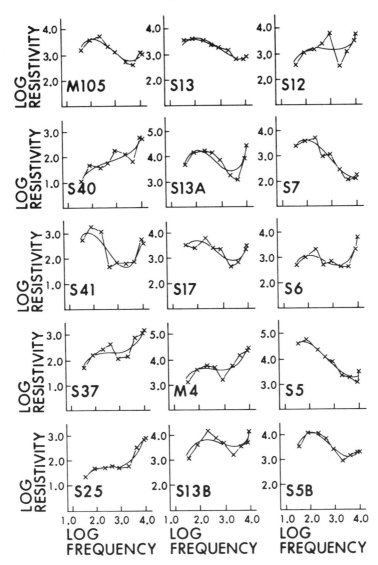

Figure 9. Log resistivity versus log frequency plots for Line F in the upper peninsula of Michigan (see Fig. 7). These plots are based on the average value of the two components measured N-S and E-W (averaged using the square root of the resistivity).

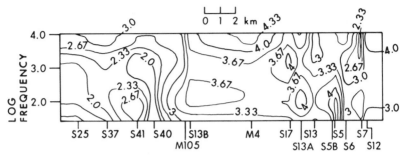

Figure 10. Pseudosection corresponding to line F. Vertical axis is log frequency and horizontal axis is distance along the profile. Contour interval is 0.33 log ρa.

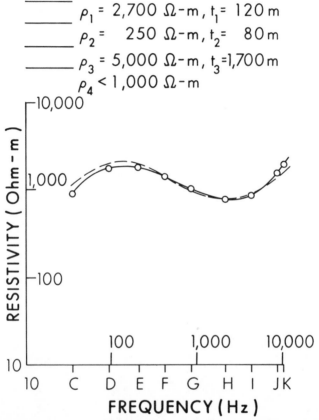

_____ $\rho_1 = 2,700 \ \Omega\text{-m}, t_1 = 120 \ m$

_____ $\rho_2 = \ \ 250 \ \Omega\text{-m}, t_2 = \ \ 80 \ m$

_____ $\rho_3 = 5,000 \ \Omega\text{-m}, t_3 = 1,700 \ m$

$\rho_4 < 1,000 \ \Omega\text{-m}$

Figure 11. Average log resistivity versus log frequency plot for the complete area surveyed. Mean values determined by averaging the square roots of the resistivity and interpolating these to equal intervals along lines A to J. Each line value determined in this way (see Table II) was then weighed according to the length of the line and then reaveraged. Dashed line corresponds to theoretical model shown in Figure.

TABLE II

AVERAGED APPARENT RESISTIVITIES FOR EACH TRAVERSE LINE UPPER PENINSULA, MICHIGAN

Frequencies = Hz

Line	Length of Interpolation (Miles)	35	95	213	430	864	2160	4310	8650	10,000
A	44	25.7	30.9	32.4	33.1	35.5	46.8	74.1	154.9	190.5
B	56	316.2	354.8	199.5	117.5	75.9	64.6	97.7	316.2	467.7
C	68	1548.8	2884.0	2570.4	1778.3	1230.3	955.0	1047.1	1862.1	2344.2
D	46	1349.0	2691.5	2511.9	1737.8	1288.2	933.3	1000.0	2089.3	2818.4
E	95	1659.6	3801.9	4786.3	4073.8	3235.9	2454.7	2344.2	3311.3	3801.9
F	94	1548.8	2884.0	2691.5	1949.8	1380.4	1148.2	1548.8	3630.8	4897.8
G	98	1412.5	2630.3	2511.9	1819.7	1202.3	812.8	851.1	1621.8	2089.3
H	94	831.8	1659.6	1862.1	1479.1	1023.3	660.7	602.6	851.1	1000.0
I	86	489.8	955.0	1071.5	891.3	691.8	549.5	588.8	955.0	1148.2
J	71	295.1	645.7	812.8	741.3	631.0	575.4	676.1	1258.9	1548.8
OVERALL WEIGHTED AVERAGE		909.0	1761.8	1817.3	443.4	1030.5	786.3	855.3	1547.7	1951.2

Cambrian, Jacobsville sandstone in the western and northern part of the area, 3) Precambrian slates, greywackes, iron formations and other sediments, 4) Precambrian gneisses and granites.

Representative AMT results are shown in Figure 9. These are a series of log ρ – log f plots similar to those from the Wisconsin region to the east along line F. Again we have drawn a number of pseudosections, one of which is illustrated in Figure 10. This represents a cross section through the region and shows quite clearly that the resistivity values are higher in the centre where the Precambrian granites and geisses are dominant, than at the edges where these rocks give way to greywackes and other metasediments. In the centre of the region, corresponding to the granites and gneisses, the resistivity values are at least 10,000 ohm-m. The values drop both over the Precambrian sediments to the west and over the Paleozoic cover to the east. This Paleozoic cover is thin, and it is probable that the meaaured resistivity is only slightly influenced in this area by the sediments.

With such sharp lateral variations, there is a question as to what "averaged values" mean. Nevertheless, many low frequencies use long electrode arrays which will average these variations, and it is interesting to consider the implications of regional patterns. The averaged values along lines A through J are tabulated in Table II, which shows both the line averages and a regional weighted average. This overall, averaged profile is shown in Figure 11. For what it is worth, this result implies that there is a near-surface layer with a resistivity of several thousand ohm-m, followed by

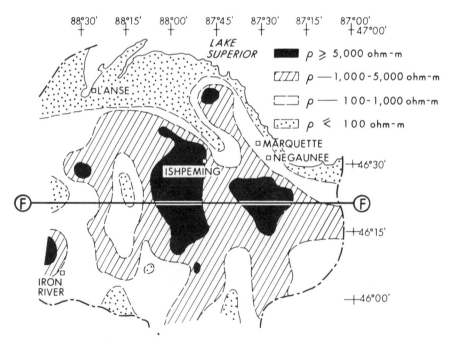

Figure 12. Contour of the apparent resistivity values as measured at 95 hertz over the upper peninsula of Michigan. Based on averaged N-S and E-W components.

a layer with a lower resistivity and a thickness of roughly 80 metres, and beneath this, a very resistive layer about 2 kilometres thick with a resistivity of more than 2000 ohm-m.

In Figure 12 we have shown the regional variations in a contoured plot of the apparent resistivity value determined at 95 hertz. This Figure shows a remarkable correlation with the mapped geology. The highest resistivities correspond reasonably well with the presence of granites and gneisses. These high values are quite similar to those reported earlier from the Geotraverse area in northern Ontario (Koziar and Strangway, 1978b), and to those seen in the Wisconsin area just discussed in this paper.

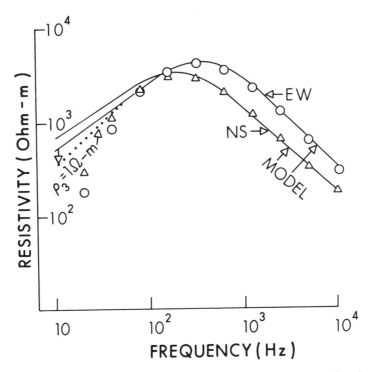

Figure 13. Average resistivity plot along a 15 mile profile in the Geotraverse area of northwestern Ontario. The points represent the observations and the smooth lines are three layer model fits as follows:

E-W component   1) $t/\rho = 0.19$ mho
                         2) $t = 1.7$ km; $\rho = 50{,}000$ ohm-m
                         3) half-space $\rho = 100$ ohm-m.
N-S component   1) $t/p = 0.25$ mho
                         2) $t = 2.0$ km; $\rho = 15{,}000$ ohm-m
                         3) half-space $\rho = 100$ ohm-m.
                         (from Koziar and Strangway, 1978b)

GEOTRAVERSE

Koziar and Strangway (1978b) reported on a detailed study of a portion of the Superior province referred to as the Geotraverse region. The overburden in this region generally consisted of sandy, glacial outwash and no clays. The resistivity values were in general very high, which suggests the absence of significant faulting. The electrode arrays were 275 m, considerably longer than in the previously discussed studies, and the stations were spaced 450 to 600 m apart. A total of 46 stations were occupied, mostly over granite and gneissic terrane. A local conductor representing a fault was detected at one station, but apart from this, the pattern of resistivity structure was remarkably uniform from station to station. A line average over a section 15 miles in length produced the pattern shown in Figure 13.

This can be interpreted quite simply as representing a three-layer Earth. The top layer is at most a few tens of metres thick and has a ratio of thickness/resistivity (t/ρ) of about 0.20 mho. Fluid in the thin, sandy, soil cover and cracks and fractures in the uppermost bedrock are probably responsible for the conductivity of this layer; extensive glaciation, which has removed most of the weathered material normally found in unglaciated regions, probably explains its thinness. The bedrock fractures close quickly with depth, and below this is a thick, highly resistive layer at least two kilometres thick. We can only determine a minimum value for its resistivity (which could easily be 100,000 ohm-m), suggesting a massive unit with very few cracks or fractures and essentially no fluids in pore spaces. At a depth of several kilometres the resistivity drops off quite sharply, suggesting the presence of a conductor at fairly shallow crustal depths. We will review possible explanations for such a zone in the discussion section of this paper.

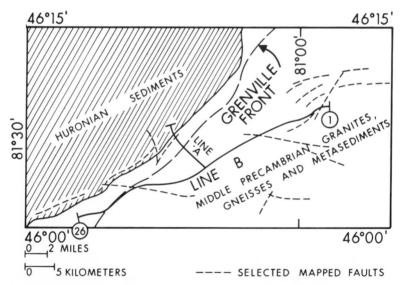

Figure 14. Sketch map of the Grenville Front region showing profiles A and B and locating the Grenville Front, the Huronian sediment contact and some of the mapped faults.

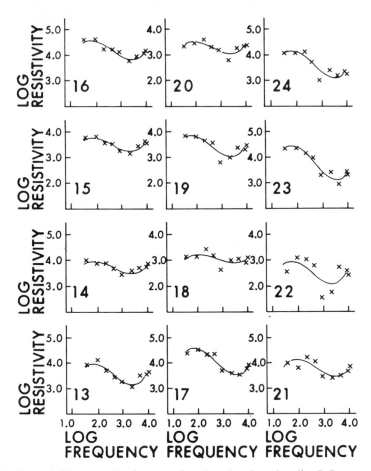

Figure 15. Log resistivity versus log frequency for selected stations along line B. Data used are the averages of the N-S and E-W measurements.

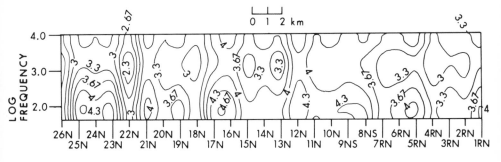

Figure 16. Pseudosection along line B showing resistivity contours in units of 0.33 log ρa. Horizontal scale is one mile between stations.

GRENVILLE FRONT

We conducted another survey in the vicinity of the Grenville Front near Sudbury, Ontario (Fig. 14). This consisted of two profiles, one (line B) along a road that crosses the Grenville Front, and a shorter one perpendicular to it (line A), also crossing the Grenville Front but ending up on Huronian sediments. The last station of line A was on

TABLE III

SUDBURY AVERAGE ρ OHM-M

Frequency	Line A	Line B
10,000	4,900	17,000
8,650	4,100	14,000
4,310	2,300	7,800
2,140	2,100	7,200
858	2,900	10,000
473	4,600	17,000
213	7,100	26,000
83	10,000	36,000
35	8,300	27,000

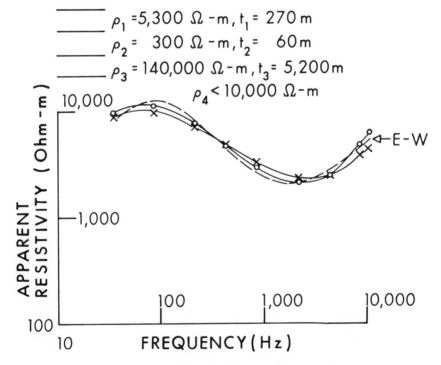

Figure 17. Mean apparent resistivity along line B showing the E-W value, the N-S value, and model fitted to these curves (dashed line).

the quartzite of the Cobalt group, which consists largely of quartz sandstone. The rest of the region is underlain mainly by felsic intrusions and gneissic rocks with small amounts of gneissic metasediments (Card, 1978). A representative set of soundings is shown in Figure 15 for stations 13 to 24 along line B. Readings were taken at one-mile intervals using a 50 m electrode array, and both N-S and E-W orientations were used. The pattern along both lines is remarkably uniform and quite similar to the patterns we have seen in the other Precambrian regions studied, although this is in the Grenville rather than in the Superior or the Southern Province.

Figure 16 shows a pseudosection along line B, again illustrating lower resistivities near the surface, with higher values at the lower frequencies frequently exceeding 20,000 ohm-m. There is no systematic variation observed across the Front. Local bands of lower resistivity are present at a few places along the line, at stations 11, 18, 22 and 26. In Figure 14 we have shown a few of the faults which have been mapped in this area. It is probable that these zones are associated with water-filled fractures. Because of the one mile sampling interval used, there are probably many more faults than we detected.

The overall resistivity corresponding to the line average of the electric fields is shown in Figure 17 and in Table III, and suggests a three-layer Earth. These data have been fitted to a model curve which has the following parameters: 1) ρ_1 = 5300 ohm-m; t_1 = 270 m; 2) ρ_2 = 300 ohm-m; t_2 = 60 m; 3) ρ_3 = 14,000 ohm-m; t_3 = 5200 m; ρ_4 < 10,000 ohm-m. Where line A intersects the quartzite there is a very high resistivity value (about 100,000 ohm-m), suggesting that pure, metamorphosed quartzite is highly resistive; it contains no water in pore spaces and is essentially fracture-free. This quartzite is a very massive unit that forms steep cliffs, which is consistent with our interpretation.

In general, the electrical resistivity in the typical metamorphosed Grenville terrane of granitic and gneissic rocks is like that in Wisconsin, the upper peninsula of Michigan, and northwestern Ontario. The representative pattern seems to be a near-surface low resistivity zone followed by a very high resistivity zone extending to depths of several kilometres, without any electrical signature associated with major faulting.

CHALK RIVER, ONTARIO

We have had an opportunity to carry out AMT sounding at two sites being studied by the Geological Survey of Canada and the Earth Physics Branch in connection with the nuclear waste disposal program. These sites illustrate two of the common problems that are encountered in mapping the Precambrian shield using electrical methods. The first of these, evident at Chalk River, Ontario, is the presence of extensive, water-filled fractures which introduce rapid local changes in the resistivity. This makes regional interpretations difficult and, as discussed by Keller and Furgerson (1977) makes depth interpretations ambiguous. Figure 18 shows the location of the survey and the grid that was occupied at 100 metre intervals (occasionally 50 m). A nearby power transmission line created enough interference to make it impossible to use the natural source fields. Accordingly, observations were made at 60 hertz and some of its harmonics (180 hz, 300 hz and 660 hz). The source field of the transmission line was sufficiently uniform

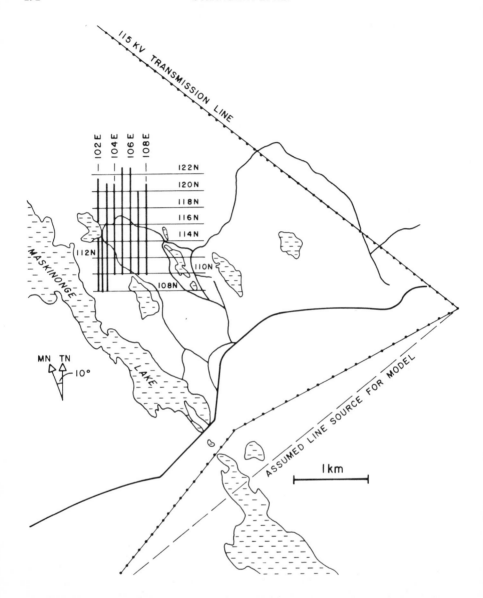

Figure 18. Location map showing the grid surveyed in the Chalk River, Ontario area. Power lines were the signal source used for the AMT soundings.

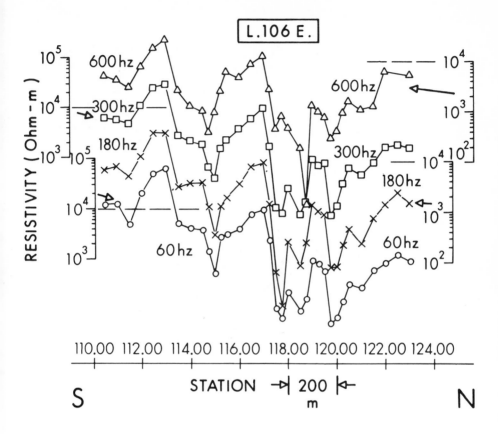

Figure 19. Apparent resistivity profiles along L. 106E at frequencies of 60, 180, 300 and 660 hz. Each curve is offset from the others so that the individual profiles can be followed. Line numbers are in units of 100's of metres.

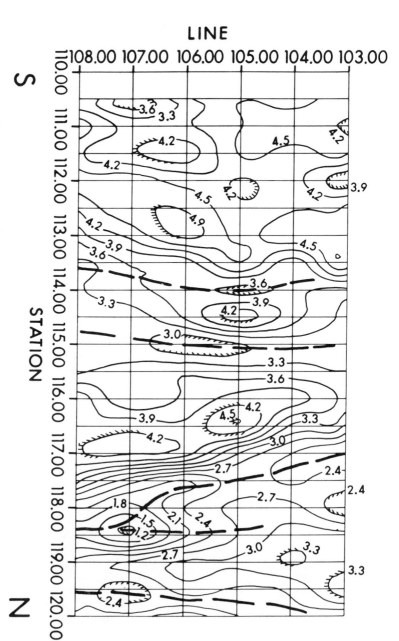

Figure 20. Contour map of the apparent resistivity over the grid area at 60 hertz. Contour interval is 0.33 log ρa. Dashed lines mark the location of possible conductive faults. Line numbers are in units of 100's of metres.

over the survey area to permit an electrical sounding to be made at the power line frequencies. A set of profiles along L. 106 illustrate the rapid variations over very short distances (Fig. 19). The typical apparent resistivity values are around 1000 ohm-m to 10,000 ohm-m, but there are local spots where the values drop as low as 100 or even 10 ohm-m. These low values are recognized from line to line, and no doubt represent the water-filled fracture zones which have been discovered by drilling in the area.

Figure 20 is a contour map of the apparent resistivity measured over this grid at 60 hertz. Although local power line sources do not permit sounding to depths as great as natural sources, the presence of the faults, clearly seen in Figure 20, implies that crustal sounding to depths of several kilometres would not be possible in such a region anyway. These conductors strike generally east-west across the region in accord with the regional geology, and the apparent resistivity values range from 10 ohm-m to 60,000 ohm-m over very short distances. Such rapid variations mean that the mag-

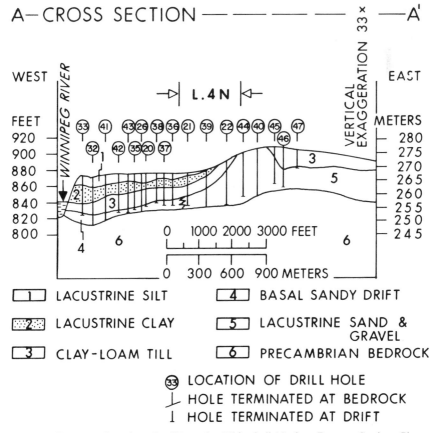

Figure 21. Cross section along L. 4N at the Whiteshell Nuclear Reactor Station, Pinawa, Manitoba. The overburden is based on extensive drilling as shown. This figure is from Grisak and Cherry, 1975.

netotelluric method is very sensitive to lateral changes and hence is an excellent tool for exploring for water-filled fractures. When these features are present, however, they make it almost impossible to conduct regional crustal studies by electrical methods, and incidentally mean that the region is not suitable for nuclear waste disposal. It must therefore be established whether or not such features exist in a region before conclusions about the crustal resistivity distribution can be drawn using low frequency magnetotellurics or large electrode array resistivity methods.

PINAWA, MANITOBA

A second problem often encountered in the Precambrian shield is the presence of highly conductive near-surface layers, which mask the subsurface at least in the frequency range studied here. If not recognized, they will affect depth interpretations by the lower frequency methods. These are the glacial clays left by the ancient lakes;

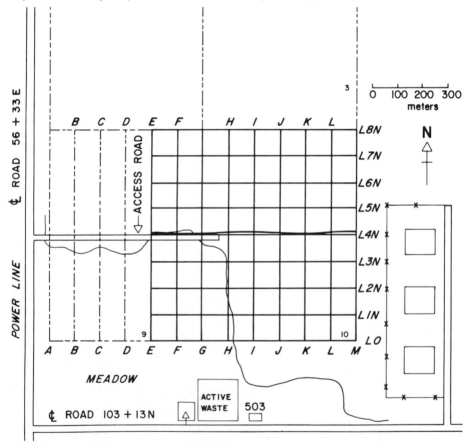

Figure 22. Grid surveyed at the Whiteshell Nuclear Reactor Station. 100-metre intervals were used with the electric field sensor oriented N-S.

since they often contain saline solutions, their resistivities are frequently as low as 1 ohm-m. This problem has plagued electromagnetic prospecting methods for many years, and many attempts have been made to devise techniques for mapping the electrical properties of the bedrock beneath. This has generally been frustrating, since it is difficult to separate the effect of a thickening or laterally varying clay from a

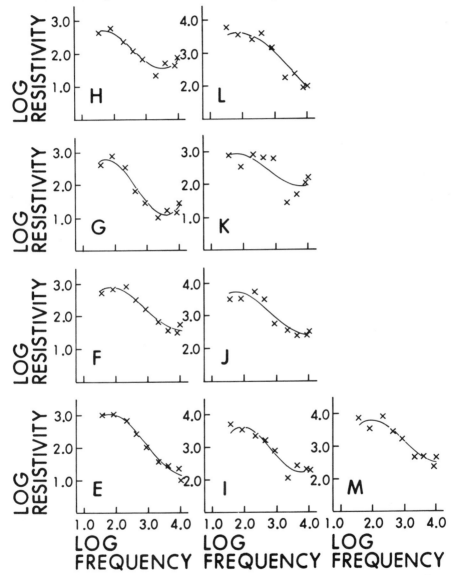

Figure 23. Log resistivity versus log frequency plot for the stations located along L. 4N from intersecting lines E to M.

variation in the bedrock. An example of such an attempt was discussed by Goldstein and Strangway (1975), where an artificial source AMT survey was done near Timmins, Ontario. The Pinawa region studied was on the Whiteshell Nuclear Reactor site in Manitoba. It is extensively covered by clay left by glacial Lake Agassiz. A cross section from the Whiteshell nuclear reactor station (Fig. 21) shows an extensive, lacustrine clay unit which thickens toward the west (the Winnipeg River). The overburden containing this clay is typically about 40 feet thick in the area.

Figure 22 shows the region that was surveyed using a grid of stations spaced at 100 m intervals. A typical set of log resistivity versus log frequency plots is shown in Figure 23. The near-surface apparent resistivity is typically as low as 10 ohm-m, reflecting the presence of conducting clay quite clearly. The resistivity then rises sharply to values of 1000 ohm-m to 10,000 ohm-m, reflecting the bedrock a few tens of metres down. Even this thin layer of clay affected our ability to penetrate to great depths, and restricted us to mapping the clay resistivity and thickness and the upper few kilometres of the bedrock. A pseudosection along L, 40 (Fig. 24) again illustrates the very low resistivity, near-surface conductor followed by the high resistivities of the bedrock at lower depths. Model fitting to these results suggests that bedrock is highly resistive until a depth of at least 1.5 to 4.5 kilometres is reached. The pseudosection also shows that the resistivity decreases to the west as the clay thickens. These results show that the AMT method can be used to see through conductive clays, although it is difficult to sort out the lateral variations in the bedrock from those within the conductive clay unit. Other electrical methods have more difficulty in seeing through.

There are many other causes of local resistivity variations which may occur near the surface and make deep penetration studies difficult to carry out. These have long been recognized by exploration geophysicists in the search for minerals, and include the effects of extensive conductive zones associated with graphite and serpentine. Sternberg and Clay (1977) have accidentally found such a region in Wisconsin, which they refer to as the Flambeau anomaly. Gough (1980) has reported on tracing such a zone into the subsurface beneath sedimentary cover in western Canada. Many exploration files document abundant zones of this type in the Canadian Shield.

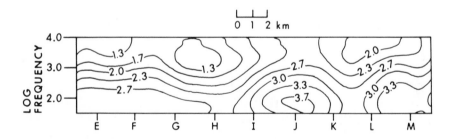

Figure 24. Pseudosection along line 4. The horizontal scale is 100 meters between stations. Contour interval is 0.33 log ρa.

DISCUSSION

Through the data presented in this paper we have attempted to show that in some portions of the glaciated Shield there are very high and relatively uniform resistivities beneath a thin, surface layer of overburden and fractured bedrock. This surface layer is frequently conductive relative to the bedrock, but varies considerably from place to place. Using a simple and rapid technique such as AMT, it is possible to map out such regions and locate portions of the Precambrian crust that are relatively uniform in their electrical resistivity. When this is done, it is possible to find regions which have large volumes of rock with high apparent resistivities and perhaps true resistivities of 50,000 ohm-m or more. These large, uniform regions provide the most useful sites for deep sounding techniques. Even by using audio frequencies in such areas, we have been able to map high resistivities to depths as great as 10 kilometres (Kryzan and Strangway, in prep).

We have also shown examples where deep sounding techniques have serious limitations. The extensive faulting in regions such as Chalk River poses an effective barrier to deep penetration studies. The highly conductive clays commonly occurring in the Precambrian Shield introduce masking conductors which may vary rapidly laterally, again making the deep sounding results ambiguous. Finally, graphitic shales, often extending over tens of miles of strike length, are frequently encountered in Precambrian areas.

If precautionary steps are taken to ensure that studies are done in regions of fairly uniform near surface resistivities, i.e., in the top 10 kilometres, there is little doubt that effective crustal sounding can be done with electrical methods. As shown in this paper, a rapid survey technique such as AMT is effective for this purpose. Most studies show that when depths of 15 to 20 kilometres are reached, the resistivity drops sharply from the high values reported here to much lower values of around 100 ohm-m or less. Edwards *et al.* (1980) reports on such low resistivities in the Appalachians and in the Precambrian shield near Timmins, Ontario. In the Precambrian near Timmins, these authors report that at depths from 19 to 29 kilometres the resistivity drops from 15,000 ohm-m to 270 ohm-m. The high values over a considerable thickness are consistent with our Precambrian measurements. Similar results have been reported by Nekut *et al.* (1977) in the Adirondacks and by Van Zijl (1977) in the South African Shield.

The cause of extremely high resistivities in the top 5 to 10 kilometres of gneissic and granitic rocks in the shield is the lack of significant fluids and fractures in this material. The drop in resistivity beneath is likely due either to the presence of pore space fluids in very small amounts or the presence of hydrated minerals as reviewed by Van Zijl (1977).

ACKNOWLEDGEMENTS

These observations have been compiled over a period of several years in connection with several different projects. Kennecott Copper graciously provided the basic interference analyzer, while the National Research Council of Canada has supported our continuing research. Energy, Mines and Resources, both directly and through contracts with Atomic Energy of Canada Ltd., provided the support for field studies in Ontario and Manitoba. Discussions with R. Niblett, W. Scott and L. Collett were very helpful. G.T.E.-Sylvania was responsible for a subcontract from the U.S. Navy for the Wisconsin and Michigan study.

Many individuals supported various aspects of this work. Dave Davidson of G.T.E.-Sylvania and E. Wolkoff of the U.S. Navy provided much support in interpreting the Wisconsin and Michigan results. A. Koziar did the Geotraverse work and A. Kryzan supported the field work and analysis of the Wisconsin and Michigan work. Scott Holladay and Chris Horne assisted on the Pinawa work, A. Hamud on the Chalk River work, and Susan Strangway on the Grenville Front work. Drafting was done by S. Shanbhag.

REFERENCES

Bannister, P.R. and Williams, F.J., 1974, Results of the August 1972 Wisconsin Test Facility effective earth conductivity measurements, Jour. Geophys. Res. V. 79, p. 725-732.

Benderitter, Y., Herisson, C., Korhonen, H. and Pernu, T., 1978, Magneto-telluric experiments in Northern Finland, Geophysical Prospecting, V. 26, p 565-571.

Cagniard, L., 1953, Basic theory of the magnetotelluric method of geophysical prospecting, Geophysics, V. 18, p. 605-635.

Card, K.D., 1978, Geology of the Sudbury-Manitoulin Area, Districts of Sudbury and Manitoulin, Ontario Geological Survey, Report 166, 238 p.

Davidson, D., Macklin, D.N. and Vozoff, K., 1974, Resistivity surveying as an aid in Sangiune site selection, IEEE Trans. on Comm., V. COM-22, p. 389-393.

Dowling, F.L., 1970, Magnetotelluric measurements across Wisconsin arch, Jour. Geophys. Res., V. 75, p 2683-2698.

Dupis, A. and Iliceto, V., 1974, An example of rapid magnetotelluric investigation of faulted structures: the Carboli area (Lardarello-Italy), Boll. di Geofis Teorica ed Applicata, V. 16, p. 125-136.

Dupis, A., Iliceto, V. and Norinelli, A., 1974, First magnetotelluric measurements on Lardarello site, Boll. di Geofis Teorica ed Applicata, V. 16, p. 137-152.

Edwards, R.N., Bailey, R.C. and Garland, G.D., 1980, Crustal and Upper Mantle electrical conductivity studies with natural and artificial sources: in Strangway, D.W., ed., The Continental Crust and Its Mineral Deposits: Geol. Assoc. Canada Spec. Paper 20, in press.

Goldstein, M.A. and Strangway, D.W., 1975, Audio frequency magnetotellurics with a grounded electric dipole source, Geophysics, V. 40, p. 669-683.

Gough, D.I., 1980, Magnetometer arrays and large tectonic structures: in Strangway, D.W., ed., The Continental Crust and Its Mineral Deposits: Geol. Assoc. Canada, Spec. Paper 20, in press.

Grisak, G.E. and Cherry, J.A., 1975, Hydrologic characteristics and response of a fractured till and clay confining a shallow aquifer, Can. Jour. Earth Sci, Vol. 12, p. 23-43.

Hoover, D.B., Frischknecht, F.C. and Tippins, C.L., 1976, Audio magnetotelluric sounding as a reconnaissance exploration technique in Long Valley, California, Jour. Geophys. Res., V. 81, p. 801-809.

Hoover, D.B., Long, C.L. and Senterfit, R.M., 1978, Some results from audiomagnetotelluric investigations in geothermal areas, Geophysics, V. 43, p. 1501-1514.

Keller, G.R. and Furgerson, R.B., 1977, Determining the resistivity of a resistant layer in the crust, AGU monograph 20, The earth's crust, p. 440-469.

Koziar, A. and Strangway, D.W. 1975, Magnetotelluric sounding of permafrost Science, V. 190, p. 566-568.

Koziar, A. and Strangway, D.W., 1978a, Permafrost mapping by audio frequency magnetotellurics, Can. Jour. Earth Sci, V. 15, p. 1539-1546.

Koziar, A., and Strangway, D.W., 1978b, Shallow crustal sounding in the Superior province by audio frequency magnetotellurics, Can. Jour. Earth Sci., V. 15, p. 1701-1711.

Kryzan, A. and Strangway, D.W., in prep. Audio frequency magnetotelluric sounding in the granites and greenstones of the Geotraverse area.

Mabey, D.R., Hoover, D.B., O'Donnell, J.E. and Wilson, C.W., 1978, Reconnaissance geophysical studies in the geothermal system in southern Raft River Valley, Idaho, Geophysics, V. 43, p. 1470-1484.

Nekut, A., Connerney, J.E.P. and Kuckes, A.F., 1977, Deep crustal electrical conductivity: evidence for water in the deep crust, Geophys. Res. Lett., V. 4, p. 239-242.

Ngoc, Pham Van, Boyer, D. and Choteau, M, 1978, Cartographic des "pseudo-resistivites apparentes" par profilage tellurique-tellurique associe a la magneto-tellurique, Geophysical Prospecting, V. 26, p. 218-246.

Sternberg, B.K. 1979, Electrical resistivity structure of the crust in the southern extension of the Canadian shield-layered earth models, Jour. Geophys. Res., V. 84, p. 212-228.

Sternberg, B.K. and Clay, C.S., 1977, Flambeau anomaly: a high-conductivity anomaly in the southern extension of the Canadian Shield, AGU monograph 20, The earth's crust, p. 501-530.

Strangway, D.W. and Koziar, A., 1979, Audio frequency magnetotelluric soundings — a case history at the Cavendish geophysical test range, Geophysics, V. 44, p. 1429-1446.

Strangway, D.W. and Vozoff, K., 1969, Mining exploration with natural electromagnetic fields, in Mining and Groundwater geophysics, GSC Econ. Geol. Rep., 26, p. 109-122.

Strangway, D.W., Swift, C.M. and Holmer, R.C., 1973, The application of audio frequency magnetotellurics (AMT) to mineral exploration, Geophysics, V. 38, p. 1159-1175.

Van Zijl, J.S.V., 1977, Electrical studies of the deep crust in various tectonic provinces of southern Africa, AGU monograph 20, The earth's crust, p. 470-500.

CRUSTAL MOTIONS

The Continental Crust and Its Mineral Deposits, edited by D.W. Strangway,
Geological Association of Canada Special Paper 20

ROTATION OF MICROPLATES
IN WESTERN NORTH AMERICA

Allan Cox
Department of Geophysics, Stanford University,
Stanford, California 94305

ABSTRACT

The widespread clockwise rotation of microplates and scholles along the western margin of North America has now been firmly established by paleomagnetic research. Among the models advanced to account for the observed rotation are the following: 1) Continuum tectonics in the form of penetrative shear, 2) Rotation of scholles loosened by extension in pull-apart basins, 3) Rotation of conjugate shear faults, 4) Distributed shear in the ductile subcrustal mantle, 5) Rotation during break-up and accretion of oceanic plates, 6) Differential back-arc spreading and basin and range extension, 7) Translation and rotation of continental microplates.

Although it is unlikely that only one mechanism produced all of the observed rotations, the fact that almost all of the rotations are clockwise suggests that they are linked in one way or another to dextral shear along the western boundary of North America. This could reflect continuing dextral motion along transforms or continuing dextral components of oblique subduction.

RÉSUMÉ

La rotation positive très fréquente des microplaques et des scholles le long de la bordure ouest de l'Amérique du Nord est maintenant solidement établie par les recherches en paléomagnétisme. Parmi les modèles proposés pour expliquer ce type de rotation, notons les suivants:

(1) La tectonique du continu sous forme de cisaillement pénétratif.
(2) La rotation de scholles dégagées par l'extension de bassins.
(3) La rotation de failles de cisaillement conjuguées.
(4) Le cisaillement réparti dans le manteau ductile sous la croûte.
(5) La rotation durant l'accrétion et la rupture de plaques océaniques.

(6) L'expansion différentielle derrière les arcs et l'extension des bassins et vallées de failles.

(7) La translation et la rotation de microplaques continentales.

Bien qu'il soit peu probable qu'un seul de ces mécanismes ait produit toutes les rotations observées, le fait que la plupart des rotations soient positives suggère qu'elles sont liées d'une façon ou d'une autre au cisaillement à droite le long de la bordure ouest de l'Amérique du Nord. Ceci pourrait refléter la continuation du mouvement vers la droite le long des failles transformantes ou l'extension vers la droite des composantes de subduction oblique.

INTRODUCTION

A key idea in plate tectonics is that most of the strain accompanying global tectonics occurs in narrow orogenic zones between major plates. Broad and complex orogenic zones (such as the Alpine-Himalayan system) that mark the boundary between converging continental plates have long been recognized as a major exception to this generalization. The accretionary terrane of western North America also appears to be a broad zone of orogenic activity related to an active plate boundary, but in this case the tectonic style is quite different from that produced by continental collision.

On a global scale the boundary between the North American and Pacific plates appears to be much simpler than the Alpine-Himalayan convergence. During the Cenozoic the boundary was characterized by the subduction of oceanic plates beneath the continent, generally with a component of dextral shear. Oblique subduction was followed in California by dextral shear along the San Andreas transform. On the basis of classical plate tectonics, little deformation would be expected along the edge of a continent facing a plate boundary of this type. However, in North America this is not the case. Atwater (1970) noted that "the idea that the San Andreas fault constitutes a simple boundary between two large, perfectly rigid plates is almost certainly too simplistic a view. Other active and inactive faults in California lie parallel to the San Andreas and probably have taken up some motion . . . We might take a still broader view and include the later Tertiary deformation of Oregon, Washington, Idaho, and the Basin and Range province, since this deformation has been described as a megashear in the San Andreas direction and sense." This interpretation is supported by recent paleomagnetic evidence for dextral shear, recorded as the widespread clockwise rotation of microplates or scholles.

The term "microplate" was introduced by paleomagnetists studying the tectonics of southern Europe to describe the rotation of regions the size of Italy and the Iberian Peninsula (Zijderveld and Van der Voo, 1973). Subsequently the term has been used rather loosely to describe displaced terranes of various dimensions. In the present study we will regard microplates as similar to megaplates in being segments of lithosphere that have been displaced relative to adjacent plates along a complete set of boundary faults that penetrate to the asthenosphere. It seems unlikely that the length or width of a microplate so defined would be much less than the thickness of the lithosphere.

The term "scholle" was introduced by Dewey and Sengör (1979) in their analysis of the complex tectonics of the Aegean region, where a key issue is whether the intracontinental convergence can be resolved into the relative motion of a number of small microplates, or whether continuum tectonics better describes the convergent zone (Tapponnier and Molnar, 1979). If the former is true, the large finite strain of convergence would produce an exceedingly heterogeneous strain distribution, with high strain zones bounding scholles within which strains are very small. The term "scholles" is used in this context to describe "the crustal fragments and splinters that form as a result of complex strain patterns during continental collision, and by virtue of their tectonic nature and behavior cannot be termed plates" (Dewey and Sengör, 1979). Scholles might be of any size, with an upper limit roughly equal to the thickness of the lithosphere, although some thrust sheets, which are scholles by the above definition, may have lengths exceeding this limit. We will see that the idea of scholles is also useful in considering the Pacific-North American plate boundary, where the dominant tectonic style is one of shear rather than convergence.

The first paleomagnetic evidence for a tectonic rotation (although I didn't recognize it as such at the time) was my discovery (Cox, 1957) that the early Eocene Siletz River Volcanics in the Coast Range of western Oregon record a paleomagnetic field direction 60° clockwise from north. As a minor historical note, this study was undertaken to obtain a pole from North America to compare with Irving's (1956) and

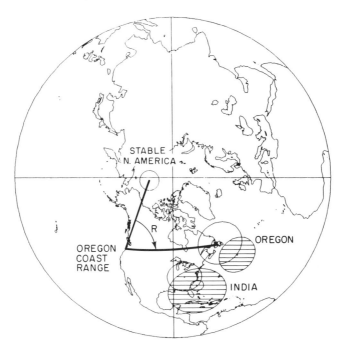

Figure 1. Early Tertiary paleomagnetic poles from the Oregon Coast Range, India, and the tectonically stable part of North America. The Oregon Coast Range has rotated by the angle R since the early Tertiary. Shaded Oregon pole, Cox (1957). Unshaded Oregon pole, Simpson and Cox (1977). Shaded Indian pole, Irving (1956). Unshaded Indian poles, Klootwijk (1976). Stable North American pole, Simpson and Cox (1977).

Deutch *et al.*'s (1958) early Tertiary pole from India located in the mid-Atlantic. The paleomagnetic pole for Oregon agreed quite well with the pole for India (Fig. 1). At the time it was unclear whether the principle of Occam's razor, properly applied, pointed to the conclusion that the pole had wandered rapidly during the Tertiary, or to the conclusion that India had moved northward and Oregon had rotated about a local axis in such a way as to bring the two poles into fortuitous agreement. Although Irving (1964) was quick to recognize that the latter explanation was the correct one, this was not widely accepted at the time and for more than a decade the Oregon result stood out as a rather embarrassing pole in the otherwise coherent body of data used to define the apparent polar wander path of North America.

Beck (1976) was the first to notice that as the amount of paleomagnetic data for North America had increased through the years, the Oregon result had been joined by a sizable number of anomalous poles. Moreover, most of these anomalous poles were from sampling sites in the western Cordillera. Beck further noted that most of these data were anomalous in the sense that the inclinations were shallower than expected or the declinations were clockwise from the expected direction or both. He pointed out that these data were consistent with a tectonic style in which mircroplates had been translated northward and rotated dextrally along the western plate boundary of North America. The most recent summary of paleomagnetic data from the Cordillera (Irving, 1979) lists 29 anomalous poles from western North America. Most are anomalous in the sense described by Beck (1976). The differences between the poles from stable North America and those from the Cordillera are too large and too systematic to be attributable to paleomagnetic noise. They clearly reflect regional tectonic processes.

DETERMINING ROTATIONS FROM PALEOMAGNETIC DATA

The results of a paleomagnetic study of a rock unit can be expressed by three numbers: I, the mean inclination below the horizontal; D, the mean declination measured clockwise from north; and a, the semi-radius of a cone containing the true paleofield vector with 95% probability. Alternatively, the I and D values may be transformed into paleomagnetic pole coordinates, three numbers again being required to describe the data: λ, the latitude of the pole; ϕ, the longitude of the pole; and A, the 95% confidence circle about the pole. All of the contemporaneous paleomagnetic data from one plate can be combined by taking an average of the paleomagnetic poles for a given interval of geologic time, avoiding data from tectonically disturbed regions like the Alps and the Cordillera of North America. A sequence of poles for successive time intervals is a so-called apparent polar wander path, which reflects the motion of the stable part of the plate relative to the Earth's rotational axis. A set of paleopoles for stable North America averaged over 30-million-year time intervals has been found by Irving (1979) for the past 300 Ma. Most of the poles from the Cordillera are discordant with this set of reference poles.

Paleopole plots provide a straightforward analytical method for determining whether the paleomagnetic data from a given formation in an orogenic zone is concordant with data from the stable part of the plate. If the standard error circle around the formation's paleomagnetic pole intersects the error circle around the expected pole of corresponding age as found from data from the stable part of the plate, then the data from the formation are consistent with those from the rest of the plate. If the circles do not intersect, the microplate or scholle carrying the sampling area has probably been displaced relative to the stable part of the plate. This motion can be resolved into two components: motions toward or away from the expected paleopole, recorded as inclination anomalies, and rotations of the microplate or scholle about a local Euler pole, recorded as declination anomalies. The former can best be assessed by converting the paleomagnetic inclination I of the formation to the paleo-colatitude p by the equation

$$p = \cot^{-1}(\tfrac{1}{2} \tan I), \ 0° \leqslant p \leqslant 180° \tag{1}$$

If p_0 is the present distance from the sampling site to the expected paleopole, then the amount of motion d toward the expected paleopole is given by

$$d = p - p_0 \tag{2}$$

where positive values of d indicate the microplate has moved toward the expected paleopole.

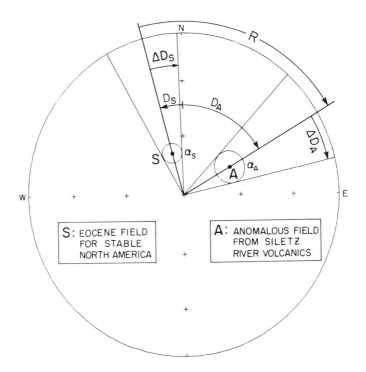

Figure 2. Paleofield directions corresponding to the paleopoles of Figure 1. The amount of rotation recorded by the Siletz River Volcanics is given by $R = D_A - D_S$. The error in R at 95% confidence is given by $\Delta R = (\Delta D_A^2 + \Delta D_S^2)^{1/2}$.

The uncertainty Δd in the displacement d can be estimated by combining the two confidence circles, A and A_0, for the formation and expected paleopoles,

$$\Delta d = (A^2 + A_0^2)^{1/2} \tag{3}$$

where Δd, the uncertainty in d at the 95 per cent confidence level, corresponds approximately to two standard errors of the mean. If $| d | > \Delta d$, the paleomagnetic data indicate a significant displacement toward or away from the expected paleopole. In the Cordillera there are several well-documented examples of microplates that have been displaced northward several thousand kilometres (for review see Irving, 1979).

The present study focuses on rotations about local Euler poles, which can be measured by comparing the observed paleomagnetic vector from a formation with the contemporaneous "expected" paleofield vector obtained by inverting the expected paleopole to give the paleofield direction that would have existed at the sampling site if the site had remained fixed relative to the stable part of the plate. To obtain a confidence interval for the expected paleofield direction, the confidence circle A_0 about the expected paleopole is mapped into a corresponding oval around the expected paleofield vector. The semi-radius of this oval in the horizontal direction corresponds to the following uncertainty in D,

$$\Delta D_0 = \sin^{-1} \frac{\sin A_0}{\sin p_0} \tag{4}$$

ΔD_0 depends both on the uncertainty A_0 in the expected paleopole and on the distance p_0 from the sampling site to the expected paleopole, and is much larger for sites near the paleopole. The amount of clockwise rotation R is found by subtracting the expected paleomagnetic declination D_0 from the mean paleomagnetic declination D for the formation,

$$R = D - D_0 \tag{5}$$

To estimate the uncertainty in R, it is first necessary to find the uncertainty ΔD in the measured paleomagnetic declination. If the paleopole for the formation and the confidence limit A are known, ΔD can be found from equation (4). Alternatively, if the mean paleomagnetic inclination I, declination D, and confidence circle a are known ΔD is given by

$$\Delta D = \sin^{-1} \frac{\sin a}{\cos I} \tag{6}$$

The uncertainty ΔR in the rotation angle at the 95% confidence level is then given by

$$\Delta R = (\Delta D^2 + \Delta D_0^2)^{1/2} \tag{7}$$

If $| R | > \Delta R$, the paleomagnetic data indicate a significant rotation of the sampling area relative to the stable part of the plate.

Figures 1, 2, and 3 summarize paleomagnetic data which indicate that a microplate or scholle has rotated. The paleopole for the lower Eocene Siletz River Volcanics shown in Figure 1 was based on a collection of more than 200 oriented samples collected over a broad sampling area in the Oregon Coast Range from 33 extrusive flows and dykes. The mean paleomagnetic direction for this formation is plotted in Figure 2, together with the 95% confidence interval. Also shown are the expected lower Eocene paleofield and the confidence circle as determined from the mean Eocene paleopole for the stable part of the North American plate. Because of the steep inclination, the uncertainties in the declinations are quite large. However, the rotation angle as shown in Figure 3 clearly indicates that the sampling area has undergone significant clockwise rotation. This is confirmed by similar paleomagnetic studies of nearby Eocene sediments and upper Eocene lava flows in the Oregon Coast Range which record clockwise rotations of the sampling area of from 68 to 51°. In this and in other studies to be discussed, the statistical signficance of the calculated rotations and the internal consistency of results from different formations establishes that the rotations are real; they are due to some tectonic process rather than to experimental noise.

MECHANISMS OF ROTATION

The rotations that occur during tectonic processes can be resolved into components about three orthogonal axes, two of them horizontal and one vertical. The dips and strikes of paleohorizontal surfaces, (ie., bedding planes) record the effects of rotations about the two horizontal axes. Techniques for analyzing this information comprise a major part of structural geology. On the other hand, information about rotations around vertical axes as recorded in paleonorth directions has only recently become available through paleomagnetic studies. So it is not surprising that our models for interpreting this information are still rather primitive and speculative. Still unanswered is the fundamental question of the style of deformation. Is it one of extremely heterogeneous strain with high strain zones bounding scholles within which strain is very small, as suggested by Dewey and Sengör (1979) for intracontinental convergence, or is the style one of continuum tectonics?

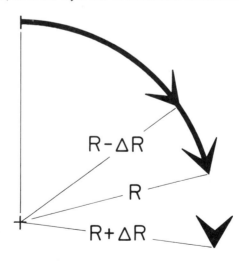

Figure 3. Symbol used on maps to describe a rotation of R ± ΔR, where the extreme arrows span the 95% confidence interval.

Penetrative Shear

The main difference between the tectonic styles of the Alpine-Himalayan and Cordilleran orogenic zones is that the former is characterized by compressional foreland shortening and crustal thickening due to the convergence of two continental plates, while the latter is characterized by dextral shear. The key question is how this regional shear produces the observed paleomagnetic rotations. At one end of the spectrum between continuum tectonics and microplate motions is the model of penetrative shear distributed uniformly and continuously throughout the orogenic zone. In this case the paleomagnetic vector would rotate with the magnetic grains as they themselves rotate as minute scholles in the penetratively shearing rock. Although this mechanism may have operated in a few highly deformed rocks, most of the formations in the Cordillera with large rotations are gently folded strata without

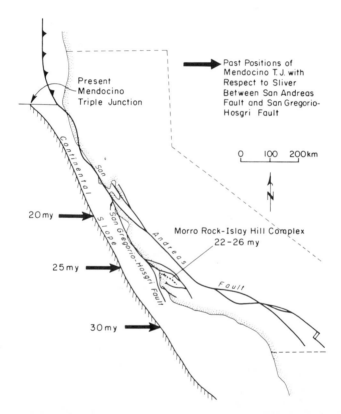

Figure 4. Morro Rocks – Islay Hill intrusive complex, located between the San Andreas and San Gregorio-Hosgri Faults, has rotated 49° since the time of intrusion 22 to 26 Ma ago. The Mendocino triple junction migrated northward past this site slightly before the time of intrusion. Since emplacement, these intrusives have been in a tectonic environment of dextral shear (Figure from Greenhaus and Cox, 1979).

extensive internal deformation. An enormous amount of penetrative shearing would be required to produce the rotations of more than 50° so commonly observed. Petrologic studies have not found evidence of this shearing.

Card Deck Models of Shear

The "mega-card deck" model of Livaccari (1979) for the Cordillera is similar to the earlier models of Beck (1976) and Wise (1963) in proposing that dextral shear in western North America is not penetrative in the sense described above, but rather is concentrated within northwest-trending strike-slip faults like the San Andreas that bound elongate microplates or scholles. Without question the dextral shear along the western margin of North America has been accommodated along a series of parallel or subparallel faults rather than along a single fault, so in a sense this model is simply a description of a known pattern of faulting. Moreover, it seems very likely that large translations of microplates along these faults are responsible for the anomalously shallow paleomagnetic inclinations that are common in the Cordillera (Beck, 1976; Irving, 1979). It is important to note, however, that this model would produce rotation only within the narrow fault zones themselves, while the paleomagnetically observed rotations are generally from rock units not in the fault zones but rather on blocks between the faults (Simpson and Cox, 1977; Kamerling and Luyendyk, 1979; Greenhaus and Cox, 1979). Thus, although the card deck model describes part of the tectonic style along the western continental margin, it does not account for the observed rotations.

Pull-Apart Basins

Many possible mechanisms for the rotation of scholles emerge if the bounding faults of the card deck model are not perfectly continuous and parallel. Perhaps the simplest is based on the pull-apart basins which develop between offset strands of

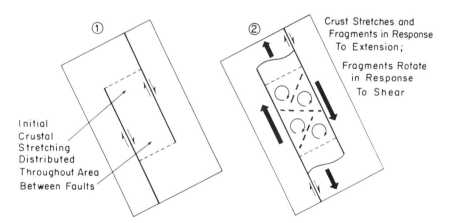

Figure 5. Pull-apart basin. Crustal blocks are formed by stretching of the brittle crust during initial extension, then are rotated clockwise in response to dextral shear in the basin.

strike-slip faults (Crowell, 1974). The following model was advanced to account for the paleomagnetically observed 49° clockwise rotation of Oligocene intrusives at Morro Rock in a pull-apart basin north of the Transverse Ranges in southern California (Fig. 4). In our model (Greenhaus and Cox, 1979) the pull-apart structure began with overlapping en echelon right-slip faults (Fig. 5). This fault pattern is observed quite commonly in the San Andreas transform zone. Initial extension in this basin during the Oligocene produced the intrusives that were sampled. After initial crustal fragmentation and volcanism in the basin, continued right-slip on the faults and accompanying extension loosened crustal fragments or scholles and allowed them to rotate in response to dextral shear exerted by the bounding faults and perhaps also by shear flow in ductile rocks between and beneath the intrusives. The scholles would have undergone different amounts of rotation if, as seems likely, the crust was originally heterogeneous and the initial pattern of crustal fragmentation irregular. This would account for the rather wide range of paleomagnetic declinations, all of which indicate substantial clockwise rotation but by varying amounts. Whether or not this model is correct in detail, the presence of a broad zone of subparallel en echelon faults in coastal California opens the possibility that rotations may occur quite commonly in extensionally loosened scholles.

Rotation of Conjugate Shear Faults

Under shear in the laboratory, materials like clay and plasticine commonly develop a conjugate set of shear faults termed Riedel faults. With continued shear deformation, this pattern of shear faults undergoes a distortion that rotates the faults and the blocks bounded by them. In reviewing these experiments, Freund (1974) pointed out that the broad network of faults spanning much of southern California appears to be such a system of conjugate shear faults that has undergone a large amount of rotation.

Paleomagnetic results from late mid-Miocene volcanic rocks from the western Transverse Ranges and continental borderland in southern California (Fig. 6) show clockwise rotations of more than 60° (Kamerling and Luyendyk, 1979). These well-documented rotations are consistent with Freund's (1974) model of rotated conjugate shears. However, Kamerling and Luyendyk (1979), pointing out that the paleomagnetic inclinations are anomalously shallow, have developed an alternative model in which the sampled areas are part of a microplate that originated about 10 degrees to the south. In this regard it is interesting to note that the results from Morro Rock (M in Fig. 6), which is not on the proposed microplate, also indicate clockwise rotation.

Distributed Shear in the Ductile Subcrustal Mantle

What is the mechanism by which shear stress propagates through a broad orogenic zone? Model experiments suggest two possibilities. In some experiments, a layer of clay or plasticine resting on a nearly frictionless base develops shear faults when strained from the sides, the main boundary condition being a constant rate of shear displacement on the two sides. In other experiments, shear faults develop in a layer of clay tightly coupled to a base that undergoes shear displacement produced by some mechanical device. If v_x is the velocity in the direction of shear at some point (x,y) at the base of the layer, the boundary conditions are that $v_y = 0$, $\partial v_x/\partial x = 0$, $\partial v_x/\partial y = $ constant. In the case of microplates being dispaced along faults cutting the lithosphere, the first model is probably appropriate since there is little viscous coupling between the lithosphere and the aesthenosphere (Forsythe and Uyeda, 1975). In regions like southern California, however, the close spacing of faults suggests that the blocks bounded by them are scholles, since it is unlikely that the boundary faults extend through to the aesthenosphere. Scholles of brittle upper crust appear to rest on more ductile lower crust and mantle, the boundary between the two probably being delineated by the drop-off in focal depths of earthquakes between 15 and 20 kilometres. If we consider the velocity fields in the ductile layer adjacent to the San Andreas transform, we can imagine two cases: constant velocities in both plates with a sharp jump across the transform, or a continuous change in velocity over a boundary zone (Fig. 7). The transition is probably sharp where the surface trace of the fault is a single straight fault, as in parts of northern California. However, where the trend of the main fault is curvilinear and the zone of surface faulting is broad, as in southern California, the velocities in the ductile zone may change over a correspondingly broad zone. It seems particularly unlikely that the velocities throughout the lithosphere follow the trace of the San Andreas fault at its major bend through the Transverse Ranges. On the basis of their seismic studies, Hadley and Kanamori (1977)

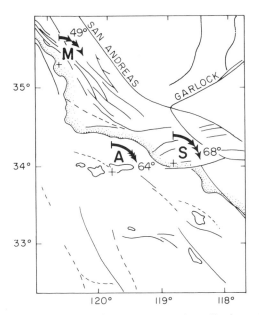

Figure 6. Clockwise rotations inferred from paleomagnetic studies in southern California. S and A: Conejo Volcanics with ages from 16 to 13 Ma exposed on Anacapa Island (A) and in Santa Monica Mountains (S) (Kamerling and Luyendyk, 1979). M: Morro Rock-Islay Hill intrusives (Greenhaus and Cox, 1979).

have argued that the trace of the fault through the ductile layer, may, in fact, be substantially different from the surface trace. In places where the difference in plate velocities is distributed through a broad transform zone in the ductile layer, the crustal scholles would experience torques proportional to the curl of the velocity field in the underlying ductile layer. If this model is correct, the dimensions of the scholles would be approximately equal to the thickness of the brittle layer, roughly 10 to 20 kilometres; this corresponds reasonably well to the spacing of the larger faults in southern California.

Rotation During Break-up and Accretion of Oceanic Plates

Menard (1978) has made a strong case for the break-up of the Farrallon plate as the East Pacific rise approached the trench along the coast of North America in the Tertiary. During the break-up, small oceanic microplates appear to have commonly rotated about local Euler poles (Menard, 1978). Additional rotation might be expected during accretion of the plates to the adjacent continent if, as was true during the Tertiary, the subduction was oblique.

This was the basis for one of the models advanced by Simpson and Cox (1977) to account for the observed post-Eocene paleomagnetic rotation of the Coast Range of Oregon (Fig. 8). In this model, the present Oregon Coast Range originated as ocean floor that rotated clockwise as it accreted to North America.

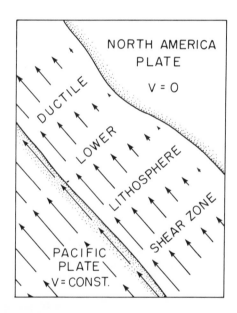

Figure 7. Velocity field in a hypothetical ductile shear zone between the North American and Pacific plates. Blocks of brittle crust above this zone would experience clockwise torques and rotations.

Differential Back-Arc Spreading and Basin and Range Extension

During the development of marginal seas behind the island arcs, a rifting arc may be rotated or bent if the rate of back-arc spreading varies along the length of the marginal sea. During the opening of the Japan Sea in the Tertiary, not only was the island arc rifted away from the continent, but, on the basis of paleomagnetic data (Kawai *et al.*, 1969), there appears to have been differential rotation between the northern and southern parts of the arc. Similar back-arc spreading may account for the observed rotation of the Blue Mountains in eastern Oregon (Wilson and Cox, 1980).

Basin and range extension may produce similar rotations if the amount of extension varies along strike. Hamilton (1969) proposed that the Klamath Mountains were originally aligned with the Sierra Nevada and moved westward to their present position during basin and range extension. As an alternative model to that shown in Figure 8, Simpson and Cox (1977) suggested that the Oregon Coast Range may have rotated clockwise about a Euler pole located north of the range (Fig. 9). The rotation

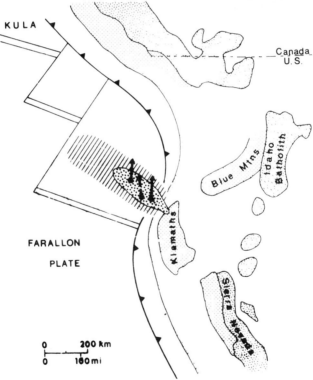

Figure 8. Model for rotation in which rocks that are to form the Oregon Coast Range (shaded area) originally are formed on the ocean floor, which rotates during accretion to the North American plate. From Simpson and Cox (1977).

of the Oregon Coast Range would, by this modle, reflect greater extension at the latitude of the Klamaths than at more northerly latitudes.

Although it is still too soon to know which, if either, of the two models of Simpson and Cox (1977) for the evolution of the Coast Range is correct, the growing body of paleomagnetic data from western Oregon (Fig. 10) strengthens the observational basis for the existence of large rotations. The discovery of rotations in subaerial lavas (sites Y, T, and G in Figure 10) suggests that the mechanism responsible for the rotation does not involve ocean floor only and that it may, at least in part, be due to differential basin and range extension.

Translation and Rotation of Microplates

Quite commonly where anomalously shallow inclinations have been observed in the Cordillera (indicating translation of a microplate from the south), the declinations

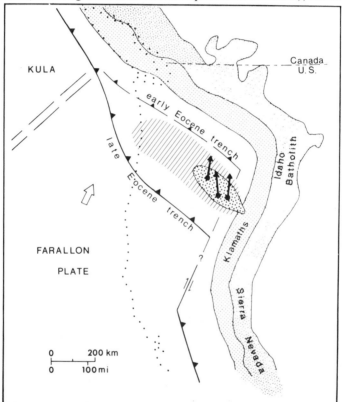

Figure 9. Model for rotation in which rocks that are to form the Oregon Coast Range (shaded area) originally form on ocean floor in eastern Oregon. The Coast Range then rotates to its present position along the Oregon coast, accompanied by basin and range extension that is greater at the latitude of the Klamath Mountains than it is farther to the north. From Simpson and Cox (1977).

have also been anomalous (Irving and Yole, 1972; Packer and Stone, 1974; Hillhouse, 1977; Kamerling and Luyendyk, 1979). There is no basis for determining whether the rotations occur during translation or during docking of the microplate and, in fact, both processes probably contribute.

CONCLUSIONS

The widespread occurrence of rotation within the Cordillera constitutes a new set of observations that have not yet been fully incorporated into either palinspastic reconstructions or tectonic theory. Since the rotations are found to occur at different times and in different tectonic settings, it is unlikely that all have been produced by any one mechanism. However, since almost all of the rotations have occurred in a clockwise sense, they probably are linked in some way to dextral shear along the western boundary of the Northern American plate, reflecting continuing dextral motion along transforms or oblique subduction zones.

Figure 10. Clockwise rotations inferred from paleomagnetic studies in western Oregon. S, TF, and Y: Eocene Siletz River Volcanics, Tyee-Fluornoy, and Yachats formations (Simpson and Cox, 1977). T: Eocene Tillamook Formation (Cox and Magill, 1977). G: Upper Eocene to Oligocene Gobel Volcanics (Beck and Burr, 1979). C: Upper Eocene Clarno Formation (Beck *et al.*, 1978).

320   COX

REFERENCES

Atwater, T., 1970, Implications of plate tectonics for the Cenozoic tectonic evolution of western North America: Geol. Soc. American Bull., v. 81, p. 3512-3536.

Beck, M.E., 1976, Discordant paleomagnetic pole positions as evidence of regional shear in the Western Cordillera of North America: American Jour. Sci., v. 276, p. 694-712.

Beck, M.E., Engebretson, D.C., Jr., Grommé, C.S., Taylor, E.M., Whitney, J.N., 1978, Paleomagnetism of the middle Tertiary Clarno Formation, north-central Oregon: Constraint on models for tectonic rotation: EOS, v. 59, p. 1058.

Beck, M.E. and Burr, C.D., 1979, Paleomagnetism and tectonic significance of the Goble volcanic series, Southwestern Washington: Geology, v. 7, p. 175-179.

Cox, A., 1957, Remanent magnetization of lower to middle Eocene basalt flows from Oregon: Nature, v. 179, p. 685-686.

Cox, A. and Magill, J. 1977, Tectonic rotation of the Oregon Coast Range (Abst.): EOS, v. 58, p. 1126.

Crowell, J.C., 1974, Origin of later Cenozoic basins in Southern California: Tectonics and Sedimentation: Society Economic Paleontology and Mineralogy, Spec. Paper 22, p. 190-204.

Deutsch, E.R., Radakrishnamurity, C., and Sahasrabudhe, P.W., 1958, The remanent magnetism of some lavas in the Deccan Traps: Philosophical Magazine, v. 3, no. 26, p. 170-184.

Dewey, J.F. and Sengör, A.M.C., 1979, Aegean and surrounding regions: Complex multiplate and continuum tectonics in a convergent zone: Geol. Soc. America Bull., v. 90, p. 84-92.

Forsyth, D. and Uyeda, S., 1975, On the relative importance of the driving forces of plate motion: Geophys. Jour. Royal Astron. Soc., v. 43, p. 163-200.

Freund, R., 1974, Kinematics of transform and transcurrent faults: Tectonophysics, v. 21, p. 93-134.

Greenhaus, M.R. and Cox, A., 1979, Paleomagnetism of the Morro Rock – Islay Hill complex as evidence for crustal block rotations in central coastal California: Jour. Geophys. Research, v. 84, p. 2393-2400.

Hadley, D. and Kanamori, H., 1977, Seismic structure in the Transverse Ranges, California: Geol. Soc. America Bull., v. 88, p. 1469-1478.

Hamilton, W., 1969, Mesozoic California and the underflow of Pacific mantle: Geol. Soc. America Bull., v. 80, p. 2409-2430.

Hillhouse, J.W., 1977, Paleomagnetism of the Triassic Nikolai greenstone, McCarthy quadrangle, Alaska: Canadian Jour. Earth Sci., v. 14, p. 2578-2592.

Irving, E., 1956, Paleomagnetic and paleoclimatological aspects of polar wandering: Geofisica Pura e Applicata, v. 33, p. 23-41.

_____, 1964, Paleomagnetism and Its Application to Geological and Geophysical Problems: New York, John Wiley and Sons, Inc., 399 p.

_____, 1979, Paleopoles and paleolatitudes of North America and speculations about displaced terrains: Canadian Jour. Earth Sci., v. 16, p. 669-694.

Irving, E. and Yole, R.W., 1972, Paleomagnetism and the kinematic history of mafic and ultramafic rocks in fold mountain belts: Earth Physics Branch, Dept. Energy Mines and Resources, Ottawa, v. 42, p. 87-95.

Kamerling, M.J. and Luynedyk, B.P., 1979, Tectonic rotation of the Santa Monica Mountains region, western Transverse ranges, California, suggested by paleomagnetic vectors: Geol. Soc. America Bull., Part I, v. 90, p. 331-337.

Kawai, N., Hirooka, K., and Nakajima, T., 1969, Paleomagnetic and potassium-argon age informations supporting Cretaceous-Tertiary hypothetic bend of the main island Japan: Paleogeog., Paleoclim., Paleoecol., v. 6, p. 277-282.

Klootwijk, C.T., 1976, The drift of the Indian subcontinent; an intrepretation of recent paleomagnetic data: Sonderdruck aus der Geologischen Rundschau, v. 65, p. 885-909.

Livaccari, R.F., 1979, Late Cenozoic tectonic evolution of the western United States: Geology, v. 7, p. 72-75.

Menard, H.W., 1978, Fragmentation of the Farallon plate by pivoting subduction: Jour. Geol., v. 86, p. 99-110.

Packer, D.R. and Stone, D.B., 1974, Paleomagnetism of Jurassic rocks from southern Alaska, and the tectonic implications: Canadian Jour. Earth Sci., v. 11, p. 976-997.

Simpson, R.W. and Cox, A., 1977, Paleomagnetic evidence for tectonic rotation of the Oregon Coast range: Geology, v. 5, p. 585-589.

Tapponnier, P. and Molnar, P., 1979, Active faulting and Cenozoic tectonics of the Tien Shan, Mongolia, and Baykal regions: Jour. Geophys. Research, v. 84, p. 3425-3459.

Wilson, D. and Cox, A., 1980, Paleomagnetic evidence for tectonic rotation of Jurassic plutons in Blue Mountains, eastern Oregon: Jour. Geophys. Research, in press.

Wise, D.U., 1963, An outrageous hypothesis for the tectonic pattern of the North American Cordillera: Geol. Soc. America Bull., v. 74, p. 357-362.

Zijderveld, J.D.A. and Van der Voo, R., 1973, Paleomagnetism in the Mediterranean area: in Implications of Continental Drift to the Earth Sciences, v. 1, NATO Advanced Study Inst., April 1972, New York, Academic Press, p. 133-161.

The Continental Crust and Its Mineral Deposits, edited by D.W. Strangway,
Geological Association of Canada Special Paper 20

STRESS FIELD, METALLOGENESIS AND MODE OF SUBDUCTION

Seiya Uyeda
Department of Earth and Planetary Sciences, Massachusetts Institute of Technology (on leave from
Earthquake Research Institute, University of Tokyo, No. 1-1, Yayo 1-chome,
Bunkyo-ku, Tokyo 113, Japan)

Chikao Nishiwaki
Institute for International Mineral Resources Development, Fujinomiya-shi, 418-02, Japan

ABSTRACT

The world's major endogenic copper ores are in the form of either porphyry copper or massive sulphide deposits. Most younger deposits of both types occur in the landward plates near present-day convergent plate boundaries, but their distributions are uneven: porphyry copper deposits are abundant in the Pacific margin of the North and South American continents, but not in the western Pacific island arcs. The typical copper deposits in Japan are massive sulphides (Kuroko). The proposed explanation for this uneven distribution is that porphyry copper mineralization and massive sulphide mineralization require compressional and extensional regional stress environments respectively, which are characteristic of the Chilean-type and the Mariana-type subduction zones. Porphyry copper deposits recently discovered in the southwestern Pacific (Philippines and Melanesia) may have been formed under the compressive stress regime resulting from collision tectonics.

RÉSUMÉ

Les minerais de cuivre endogéniques les plus importants au monde se présentent sous la forme de dépôts de cuivre porphyrique ou de sulfures massifs. La plupart des dépôts récents des deux types se trouvent du côté continental des plaques près des limites actuelles de plaques convergentes, mais leur distribution est inégale; les dépôts de cuivre porphyrique sont abondants dans la bordure pacifique des continents nord- et sud-américains mais non dans les arcs insulaires de l'Ouest du Pacifique. Au Japon, les dépôts de cuivre typiques sont formés de sulfures massifs (Kuroko). L'explication que nous proposons pour cette distribution inégale est que la minéralisation de cuivre porphyrique et de sulfures massifs requiert des conditions de contraintes régionales de compression et d'extension respectivement; les conditions sont caractéristiques des zones de subduction du type Chilien et du type des Marianes. Les dépôts de cuivre porphyrique découverts récemment dans le sud-ouest du Pacifique (Philippines et Mélanésie) ont pu se former sous un régime de contraintes de compression résultant d'une tectonique de collision.

INTRODUCTION

Since the establishment of the basic concepts of plate tectonics in the late 1960s (Wilson, 1965; McKenzie and Parker, 1967; Morgan, 1968; Le Pichon, 1968), the theory has been used to explain many geological phenomena (e.g., Dewey and Bird, 1970) including metallogenesis. Many of the important contributions to the plate tectonic models of metallogenesis up to the end of 1975 are included in Wright (1977). In these models, major sites for metallogenesis are assigned to the world's mid-oceanic ridge-rift systems (divergent plate boundaries) and to the continental and island arcs (convergent plate boundaries). Since metallogenesis is considered an integral part of magmatic activity, it is natural that the major sites of metallogenesis are found on the two types of the world's major interplate volcanic zones.

Known economically workable ore deposits generated at divergent plate boundaries are relatively few, Cyprus being an outstanding example (Sillitoe, 1972a; Spooner, 1979). However, the spectacular discoveries of hot brine on the bottom of the Red Sea (Degens and Ross, 1969), as well as hydrothermal activities on the Galopagos spreading centre (Williams et al., 1974; Corliss et al., 1979) and the East Pacific Rise (Geotimes, 1979; Francheteau et al., 1979), attest to the fact that active metallogenesis is indeed in progress at these diverging plate boundaries. On the other hand, most of the known young metallogenic provinces of endogenic type are located close to the present-day convergent plate boundaries (Guild, 1972). By integrating the available information within the plate tectonic framework, fairly detailed models have been proposed for the two major forms of sulphide copper ore deposits, namely the porphyry copper and the massive sulphide (Kuroko) type deposits (Sillitoe, 1972b, c, 1973; Mitchell and Garson, 1972; Sawkins, 1972; Hutchinson, 1973, 1979; Mitchell and Bell, 1973; Ishihara, 1974; Horikoshi, 1976).

As noted by these authors, porphyry copper deposits tend to occur in continental arcs and massive sulphides in island arcs. From these observations it has been suggested that porphyry copper mineralization is favoured by the existence of continental crust, while massive sulphide mineralization is not. However, as will be mentioned later, porphyry copper deposits have recently been discovered in island arc settings, especially in the southwestern Pacific, and Kesler (1973) pointed out that there is a significant difference in molybdenum and gold content in porphyry copper deposits in continental and island arc settings. Although we will not discuss such differences, we note that the nature of the crust may not be the only controlling factor for copper mineralization. It has been suggested, in fact, that porphyry copper mineralization takes place under compressive regional tectonic stress and massive sulphide mineralization under an extensional stress regime (Sillitoe, 1972a). In this paper we will re-examine this idea in terms of the two contrasting modes of subduction recently developed by Uyeda and Kanamori (1979).

DISTRIBUTIONS OF PORPHYRY COPPER AND MASSIVE SULPHIDE DEPOSITS

Figures 1 and 2 show the world distributions of post-Mesozoic porphyry copper and massive sulphide deposits respectively. Most of these deposits are located on the upper (landward) plates near the present convergent boundaries, which indicates that subduction of the oceanic plate has played an essential role in their formation. Note that the western North American coast was a subduction boundary until Miocene time (Atwater, 1970).

Figure 1 also shows an apparent east-west asymmetry in the occurrence of porphyry copper deposits in the circum-Pacific region. They are abundant in both the North and South American coastal regions, but are distinctly scarce in the western Pacific island arcs of the Aleutians, Kuriles, Japan, Ryukyu, Mariana and Tonga. In addition to the lack of porphyry copper deposits, there are a number of massive sulphides (mostly Kuroko-type) in Japan. The absence of porphyry copper deposits in the western Pacific island arcs, notably in the well-developed and extensively studied island arc of Japan, has long been an enigma to exploration geologists. In his pioneering paper on the plate tectonic model of prophyry copper deposits, Sillitoe (1972b) stated that one of the most obvious possible porphyry copper provinces is Japan, and that many other western Pacific arcs are also promising targets. In the following section, however, we propose that the difference in the form of sulphide ore deposits is largely, if not entirely, due to the difference in the stress regime at the time of mineralization and that the chances of discovering major porphyry copper ores in a region that was in an extensional regional stress regime at the time of mineralization are slim. Similarly, it is unlikely that major massive sulphide ores would be found in a region that was in a compressional regime at the time of mineralization.

A number of porphyry copper deposits have been discovered in the Philippines and in the Melanesian borderland region as a result of intensive exploration during recent years (Gustafson and Titley, 1978). We will show that the sites of these deposits are also considered to be under compressional stress.

STRESS REGIME AND THE MODE OF SUBDUCTION

Uyeda and Kanamori (1979) have recently discussed the stress regimes at subduction plate boundaries, with special reference to the mechanism of back-arc opening. Back-arc opening is a puzzling phenomenon because a compressional rather than extensional stress regime is more naturally expected to prevail in the back of a subduction plate boundary. Therefore, various ideas have been proposed to explain the possible origin of extensional strain in the back-arc regions (McKenzie, 1969; Hasebe et al., 1970; Karig, 1971; Packham and Falvey, 1971; Sleep and Toksöz, 1971; Moberly, 1972). Uyeda and Kanamori (1979) recognized the importance of the fact that not all back-arc basins are actively opening at present, and that some arcs have no back-arc basins at all. Apparently subduction is not a sufficient condition for the formation of a back-arc basin, although it may be a necessary one.

Uyeda and Kanamori (1979) classified arc-tranch systems according to the nature of their back-arc regions, and concluded that two basically different modes of subduction give rise to two different back-arc stress situations: ''continental arcs'' and

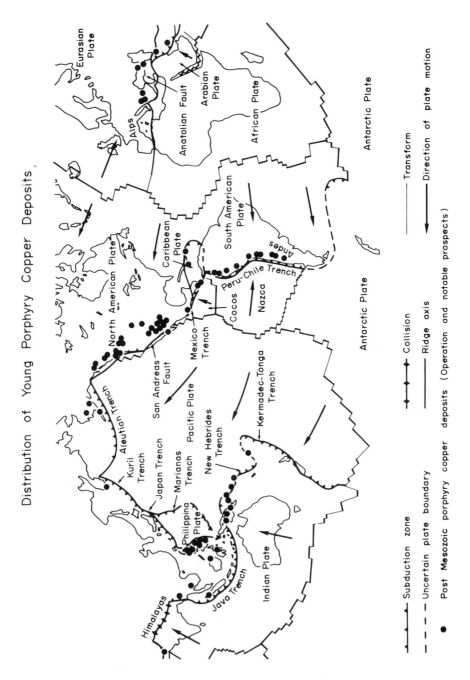

Figure 1. Distribution of young porphyry copper deposits. (Base map is after Dewey, 1972)

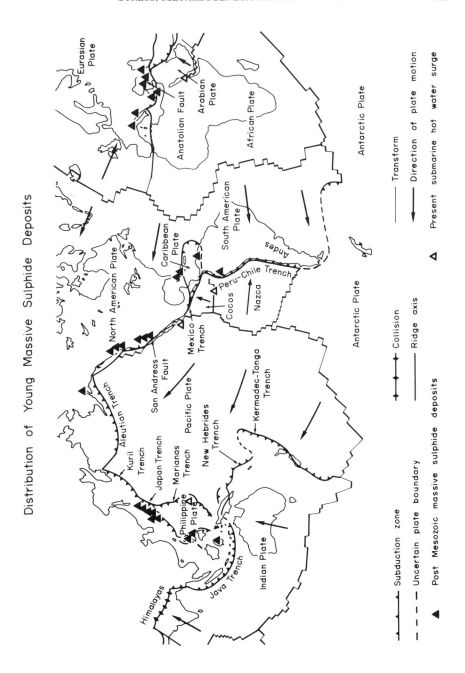

Figure 2. Distribution of young massive sulphide deposits. (Base map is after Dewey, 1972)

"island arcs." By definition, the former have no back-arc basins and the latter do. Island arcs are further sub-divided into back-arc inactive and back-arc active types. The inactive back-arc basins may have been formed by either an active opening process in the past or by a trapping of old ocean. Back-arc active arcs may also be further classified into two, depending on the process of the opening, either the typical back-arc spreading type, or what may be called the "leaky transform" type between mega-plates. This scheme of classification is illustrated in Table I (p. 337), with typical examples and possible tectonic regimes.

Of course, such a classification cannot be applied very strictly because there are arcs of intermediate or hybrid nature. For instance, western North America was probably a continental arc before the ridge collision, but then became a leaky transform type as evidenced by the extensional tectonics of the Basin and Range Province, Rio Grande Rift, and the Gulf of California. Arcs around the Caribbean plate are also complex. The Caribbean plate may have been a part of the Pacific Ocean bounded to the east by the Antilles arc, which produced the active Grenada back-arc basin. The Caribbean plate was trapped by the emergence of the Middle American arc in early Oligocene time, according to Malfait and Dinkelman (1972). Today, the back-arc of the Antilles appears to be inactive, whereas some parts of the back-arc of the Middle American trench show extensional tectonics (Plafker, 1976).

In the western Pacific, the Philippine Sea is believed to consist of a typical series of back-arc basins, with the Mariana Trough one of the most actively spreading back-arc basins and the Shikoku-Parece Vela basins inactive ones. The western Philippine Basin, however, is problematical. Uyeda and Ben-Avraham (1972) postulated it to be a part of the Pacific Ocean which was trapped by the Kyushu-Palau paleo-arc. The Okinawa Trough behind the Ryukyu arc may be an active back-arc basin, but a recent survey indicates that spreading might have become much subdued since the Miocene (Herman et al., 1979). The back-arc area of the Java-Sumatra arc consists of shallow seas; since it is assumed to be underlain by continental crust, the Indonesian arc may be classified as continental. However, to our knowledge, there have been no refraction seismic studies to confirm the continental crustal structure of the area. Since there are several thousand metres of sediment in these seas, shallowness of water in itself does not necessarily rule out the existence of oceanic crust.

Most enigmatic of all is the relationship between the New Herbrides arc and the Fiji Plateau. The small basin directly behind the New Hebrides arc does not appear to be spreading at present (Chung and Kanamori, 1978), whereas the Fiji Plateau may well be a basin generated in recent times (Chase, 1971).

Uyeda and Kanamori (1979) demonstrated that:

1) The present-day stress in the back-arc areas (deduced from the source mechanisms of intraplate earthquakes within the landward plates) is indeed compressional for most of the arcs, except the ones that have actively opening back-arc basins. The latter appear to show extensional source mechanisms. Stress regimes deduced from active faults and from the orientation of parasitic cones of active volcanoes (Nakamura, 1977) are also in agreement with these conclusions, indicating that the present-day stress state has persisted for a period exceeding the length of the time covered by seismic observations.

2) Although interplate thrust-type earthquakes occur at every subduction bound-

ary, truly great earthquakes of the new magnitude Mw (Kanamori, 1977) significantly greater than 8 occur almost exclusively in areas where there is no active back-arc spreading. Except for one shock in Assam, the 10 greatest earthquakes from 1904 to 1976, which accounted for more than 90 per cent of the total global seismic energy release, occurred at such subduction boundaries; none occurred at "back-arc active" boundaries (Fig. 3).

Based on these observations, Uyeda and Kanamori (1979) proposed that there are two fundamentally different modes of subduction. In one, two converging plates press hard against each other and in the other the subducting plate sinks without exerting strong compression on the upper plate. (There are, of course, modes of intermediate nature.) Using the names of the most typical end members of the two modes, the former are called the Chilean-type and the latter the Mariana-type modes of subduction.

These authors further showed that the difference in the mode of subduction is reflected in other important arc features: the Benioff-Wadati zone is generally steeper in the Mariana-type and shallower in the Chilean-type; the trenches themselves are deeper for the Mariana-type than for the Chilean-type. These and other observations indicate some far-reaching role of the difference in the subduction mode in the tectonics of subduction zones. The proposed two modes are illustrated in Figure 4.

Another feature of possible importance to the present problem of metallogenesis is the difference in the nature of arc volcanism. Nakamura (1977) noted that volcanism under extensional tectonics usually takes the form of fissure eruption or numerous small monogenetic cones, whereas under a compressional scheme volcanic eruption occurs repeatedly through fewer vents, to form the high edifices of polygenetic strato-volcanoes. Although this distinction in the styles of volcanism has been noted for the rift-type and arc-type environments, the same distinction, though still to be confirmed, seems applicable to the volcanism at the Mariana-type and the Chilean-type subduction zones. Thus, the chances for submarine volcanism are higher for the Mariana-type zones than for the Chilean-type.

It is well known that the abundance of andesites varies from one arc to another. Sugisaki (1972) attributed this variation to the difference in the convergence rate, while Miyashiro (1975) related it to the "maturity" of the arcs. Indeed, there is a clear tendency for andesitic rocks to be more abundant in arcs where there is a well-developed continental crust. Meanwhile, Lipman et al., (1972) showed that andesitic volcanism dominates in compression or subduction environments while bimodal (basalt-rhyolite) volcanism dominates in extensional or rift environments. As will be discussed more fully elsewhere, we propose that this distinction also holds true for the two types of subduction zones.

Three possibilities have been proposed to account for the difference in modes of subduction:

1) The two modes represent different stages in the evolution of subducting boundaries. Subduction starts with the Chilean-type and evolves into the Mariana-type (Kanamori, 1971).

2) To some extent, the age of the subducting oceanic plate may control the mode. A younger plate, being less dense, tends to subduct in the Chilean-type mode, the older and denser plate in the Mariana-type mode (Molnar and Atwater, 1978).

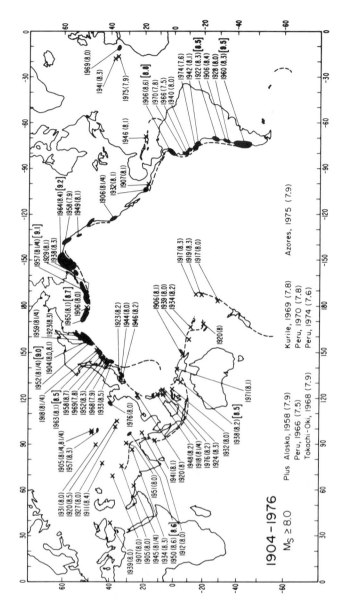

Figure 3. Great earthquakes during 1904-1976 (after Kanamori, 1978). Conventional magnitudes are in parentheses; new magnitudes are in brackets. Black areas are rupture zones.

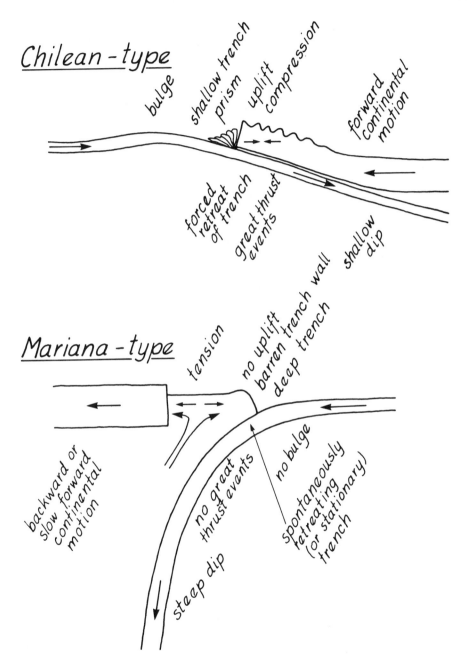

Figure 4. Two types of subduction (after Uyeda and Kanamori, 1979).

3) The motion of the upper plate also affects the mode of subduction. If the upper plate advances toward the trench, as the South America plate does, the Chilean-type mode occurs; if the upper plate retreats, as the Philippine plate does, the Mariana-type mode occurs. (Scholz et al., 1971; Wilson and Burke, 1972; Chase, 1978; Uyeda and Kanomori, 1979; Dewey, 1979).

Although this problem is very important, we will not discuss it further here. For the purpose of the present paper, recognition of these two modes of subduction and their attendant types of volcanism will suffice.

STRESS REGIME AND METALLOGENESIS

Porphyry copper deposits are stockwork and disseminated copper-iron sulphide accumulations associated with medium-felsic porphyries. In most cases, these porphyritic rocks are members of the calc-alkaline suite of arc-associated magmatism. Mainly because of their occurrence at subduction boundaries, it has long been considered that porphyry copper deposits were formed in an environment of compressive tectonic stress. In this paper, however, we have emphasized that not all subduction boundaries are under compression. The reason for suspecting a compressional setting for porphyry copper mineralization should now be re-examined. In our view, a compressional regime is needed for the following reasons:

1) The association of porphyry copper deposits with intrusive porphyritic rocks suggests that the mineralization took place in an environment where the magma was not allowed to extrude freely, but gradually cooled to develop the porphyritic textures until some critical point was reached for final rapid crystallization.

2) The depth of mineral emplacement was frequently very shallow, sometimes less than 1500 m (e.g., Fournier, 1968). Both the existence of very young (about 1 Ma) porphyry ore bodies and the rate of erosion testify that the depth of formation need not have been great.

In contrast to an extensional tectonic setting, horizontal compressional stress combined with the normal lithostatic pressure may inhibit the easy eruption of magma, sustaining it at a high level for a longer time. Such a compressional setting would satisfy the conditions proposed by Gustafson (1978): the separation of the volatiles is prevented until the magma rises to shallow reservoirs in the crust.

On the other hand, the Kuroko deposits in Japan are regarded as the "type" volcanogenic massive sulphides (Tatsumi, 1970; Ishihara, 1974; Horikoshi, 1976). It is believed that the Kuroko deposits were formed by hydrothermal activity associated with the Miocene acidic submarine volcanism, generally called the Green-tuff volcanism. This volcanism started about 30 Ma ago in the back-arc region of the Northeast Japan Arc and was accompanied by regional subsidence. The rate of eruption was very high in the beginning and has decreased gradually during the following 20 Ma. Since the Kuroko deposits and related volcanics are found in a belt consisting of numerous grabens or calderas bounded by normal faults, the shallow tectonic stress in the area at the time of mineralization was clearly extensional. This means that the Northeast Japan Arc was then the Mariana-type (now it is the Chilean-type). It is tempting to identify this period with the time of the formation of the Japan Sea, or at least a part of it.

According to Horikoshi (1960), the Kuroko mineralization was confined to the upper Nishikurozawa stage 13 Ma ago. Thus, the ore formation apparently took place at a later stage of the Green-tuff volcanism. The reason that the Kuroko ores did not form during the earlier stage is not clear in the literature. However, two observations have drawn our attention: 1) the volcanism during the Nishikurozawa stage (17 to 13 Ma ago) was the basalt-rhyolite bimodal type (Konda, 1974), which is regarded as characteristic of volcanism with extensional tectonics, and 2) the intensive subsidence of the grabens containing the Kuroko deposits started only at the Nishikurozawa stage (Sato, 1978). Both observations indicate that the stress state of the areas generating Kuroko deposits became strongly extensional only a few million years prior to Kuroko formation.

Horikoshi (1975), however, interprets the late occurrence of the Kuroko deposits in exactly the opposite way, stating that the stress changed from extensional to compressional at the Nishikurozawa stage, with the resultant squeezing out of previously stored metalliferous fluid giving rise to Kuroko ore formation. Although more studies are needed for a thorough understanding of the process of Kuroko formation, the evidence in favour of an extensional environment appears overwhelming.

As stated in the introduction, many spectacular discoveries are being made of hydrothermal activity at mid-oceanic spreading centres; these discoveries are valuable in elucidating the mineralization processes in the ocean floor. Recently, an almost equally spectacular hydrothermal activity has been discovered in the Mariana Trough, an actively opening back-arc basin (Hobart et al., 1979). An example of closely spaced heat flow measurements in the Mariana Trough is shown in Figure 5. Enormously variable heat flow within a few kilometres, as shown in the figure, can only be explained by active hydrothermal circulation in the crust. A thorough geochemical investigation of hydrothermal fluid in this area would be most valuable in understanding the metallogenesis in the back-arc region under a extensional regime, although the possible difference in the water-depth and the nature of the crust might preclude direct identification of the process in the Mariana Trough with the Kuroko generating process. From the above considerations, it may be postulated, at least as a working hypothesis, that the distributions of young porphyry copper deposits and massive sulphide deposits in the world are another manifestation of the two different modes of subduction, i.e., the Chilean-type and the Mariana-type subductions.

It is quite possible that some porphyry ore-genesis may be taking place beneath the subaereal volcanic zone in Japan today, because Japan has been a Chilean-type arc for the last several million years. For the same reason, massive sulphide ore-genesis may be in progress in the submarine volcanic areas, such as the Andaman Sea, the Mariana Trough, and the Gulf of California, although the last example is not in an arc situation at present.

PORPHYRY COPPER DEPOSITS IN THE SOUTHWESTERN PACIFIC OCEAN

As was noted earlier, there are a number of porphyry copper deposits in the southwestern Pacific region, notably in the Philippines, New Guinea, New Britain, and the Solomons. How does this fact fit into our general picture? These areas are known to be composed of extremely complicated sets of platelets, and the recent tectonics has involved subduction, sea-floor spreading, transform motion, collision, and probably arc polarity-reversal.

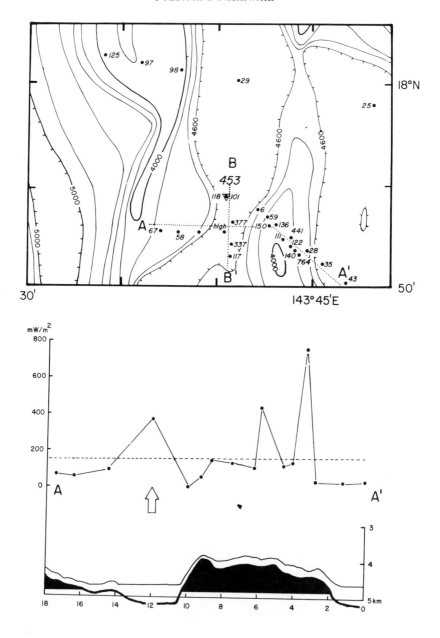

Figure 5. Examples of heat flow profiles in the Mariana Trough (after Hobart *et al.*, 1979). A-A', B-B' in the lower figures correspond to the survey lines shown in the top figure. Heat flow is in

In the Philippine region, porphyry copper deposits occur mainly in the so-called Philippine Mobile Belt bounded by the west-dipping Philippine and Mindanao Trenches and the east-dipping Manila, Visayan, and north Celebes Trenches. (Fig. 6) The west-dipping Philippine Tranch off northern Luzon is believed to have been generating the Sierra Madre del Norte by active volcanism until some disputed time in the Tertiary, ranging from Oligocene (Bowin *et al.*, 1978) to Pliocene (Murphy, 1973). However, no porphyry copper ore has been confirmed in this mountain range.

When the northern Philippine Trench became inactive, possibly due to the arrival of the buoyant Benham Rise, a new east-dipping subduction started at the Manila Trench. This subduction might have caused the magmatism by which the High Cordillera Central was formed. Porphyry copper deposits of Miocene age have been discovered in the Central Cordillera. Further south, the geology is even less well known, but the occurrences of porphyry copper, some of which are Oligocene age, are also found within the Mobile Belt bounded on both eastern and western sides by trenches.

In the New Guinea-Solomon region, almost equally complex tectonic processes appear to have taken place. Since the middle Miocene, the New Guinea Mobile Belt has been a collision zone between the northward-drifting Australian continent and the island arc that previously formed on a north-dipping subduction zone (Crook, 1978). The New Britain arc may be an eastern extension of that arc. In the Solomons, on the other hand, subduction zone was probably south-dipping until the buoyant Ontong

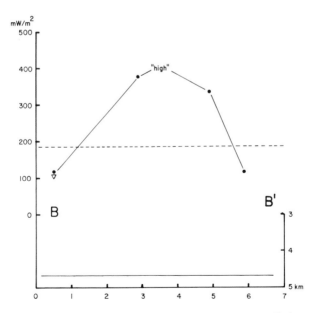

mw/m². 453 in the top figure is a DSDP site. The results are based on preliminary shipboard data reduction.

Java Plateau collided with it. Then the polarity of subduction was reversed. During this period, the Australian and Pacific plates were converging obliquely (Chase, 1971). Although the exact time relationships between these tectonic events and mineralization have to be investigated carefully, it appears that those southwestern Pacific areas with porphyry copper deposits are characterized by complex collision tectonics that would naturally call for high compressive stresses. In this regard, it is worth noting that the mode of subduction after a polarity reversal will generally be of the Chilean-type, because the former subducting (and now over-riding) plate would retain its trenchward motion even after the reversal.

In order to test the hypothesis that the southwestern Pacific porphyry copper ores were also formed under compressional regional stresses, we have compiled, in Figure

Distribution of Porphyry Copper Deposits in the Southwestern Pacific

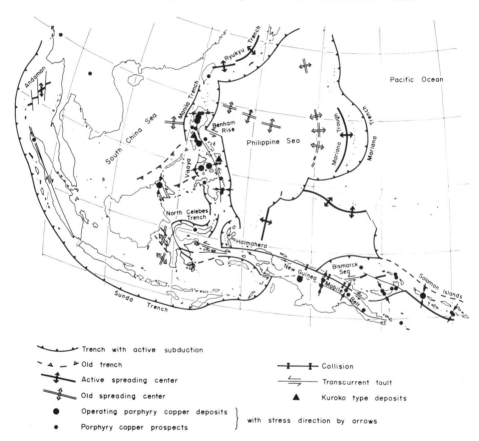

Figure 6. Distribution of porphyry copper deposits in the southwestern Pacific with estimated directions of maximum horizontal compressive axes at the time of mineralization (after Nishiwaki and Uyeda, in preparation).

6, the directions of the axes of maximum horizontal compression at several porphyry copper mines. These directions were deduced from the orientations of dykes and fracture patterns associated with the porphyry copper mineralization as documented by various investigators. A detailed account of the compilation will be published elsewhere (Nishiwaki and Uyeda, in prep.), but it may be noted here that the deduced directions of compression are in general agreement with the directions of convergence and collision expected from the regional plate tectonics. In fact, this agreement has already been noted as a characteristic feature by Titley and Heidrick (1978), from which much of the data in Figure 6 has been drawn.

TABLE I
CLASSIFICATION OF ARCS

CLASSIFICATION			TECTONIC REGIME	TYPICAL EXAMPLES
Continental Arc (without back-arc basins)			Compressive (or neutral)	Peru-Chile Alaska
Arc	Island Arc (with back-arc basin)		Back-arc inactivated	Compressive (or neutral)	Kuril, Japan Shikoku, Parece Vela Basins
		Back-arc inactive	Back-arc trapped	Compressive (or neutral)	Bering Sea
		Back-arc active	Back-arc spreading	Extensional	Mariana Trough Scotia Sea Lau Basin
			Leaky Transform	Extensional with Shear	Andaman Sea

CONCLUSION

The uneven distributions of young porphyry copper and massive sulphide deposits near convergent (subduction) plate boundaries can be explained if the mineralization of porphyry copper and massive sulphide requires compressional and extensional stress regimes respectively. The different distributions of these two types of ore deposits conform to the independent classification of the subduction boundaries based on the stress regime in the back-arc regions. Some porphyry copper deposits, notably in the southwestern Pacific region, are inferred to have been formed under a compressive stress generated or enhanced by the collision of various plates and platelets. The argument developed in this paper deals mainly with the young copper deposits. As for the older copper deposits, the time relationships of both porphyry and massive sulphide mineralizations with tectonic stress regimes are insufficiently known to test the present hypothesis. The alkaline porphyry copper deposits in British Columbia of Triassic to Jurassic age, for which an extensional regime has been suggested (Nielsen, 1976), would be one of the most interesting cases for future investigation.

ACKNOWLEDGEMENTS

The authors are grateful to Drs. S. Aramaki and K. Nakamura for discussions and Drs. P. Hurley, J. Edmond, and P. Molnar for valuable comments on the manuscript. Drs. S. Scott and R. L. Armstrong are also acknowledged for their critical reviews.

REFERENCES

Atwater, T.M., 1970, Implications of plate tectonics for the Cenozoic tectonic evolution of western North America, Geol. Soc. America Bull., v.81, p. 3513-3536.

Bowin, C., Lu, R.S., Lee, C.S., and Shouten, H., 1978, Plate convergence and accretion in Taiwan-Luzon region: Amer. Assoc. Petrol. Geol., v. 62, p. 1645-1672.

Chase, C.G., 1971, Tectonic history of the Fiji Plateau: Geol. Soc. Amer. Bull., v. 82, p. 3087-3110.

, 1978 Extension behind island arcs and motions relative to hot spots: Tour. Geophys. Res., v. 83, p. 5385-5387.

Chung, W.T. and Kanamori, H., 1978, A mechanical model for plate deformation associated with aseismic ridge subduction in the New Hebrides arc: Tectonophysics, v. 50, p. 29-40.

Corliss, J.B., Dymond, J., Gordon, L.I., Edmond, J.M., et al., 1979, Submarine thermal springs on the Galapogos Rift: Science, v. 203, p. 1073-1083.

Crook, K.A.W., 1978, Stage maps to illustrate the development of the southwest Pacific, 90 m.y. to present: a consequence of earth rotation? Australian Soc. Explor. Geophys. Bull., v. 9, p. 152-156.

Degens, E.T. and Ross, D.A., eds., 1969, Hot Brines and Recent Heavy Metal Deposits in the Red Sea: Springer Verlag, Berlin, 600 p.

Dewey, J.F., 1972, Plate tectonics: Sci. Amer., May, p. 56-58.

_____, 1980, Episodicity, sequence and style at convergent plate boundaries: in Strangway, D.W., ed., The Continental Crust and Its Mineral Deposits: Geol. Assoc. Canada Spec. Paper 20, in press.

Dewey, J.F. and Bird, J.M., 1970, Mountain belts and the new global tectonics: Jour. Geophys. Res., 75, p. 2615-2647.

Fournier, R.O., 1968, Depth of intrusion and conditions of hydrothermal alteration in porphyry copper deposits: Geol. Soc. Amer. Ann. Mtg. Abstracts, Mexico City, p. 101.

Francheteau, F., Needham, H., Coukroune, P., Juteau, T., et al., 1979, Massive deep-sea sulfide ore deposits discovered on the East Pacific Rise, Nature, 277, p. 523-528.

Geotimes, 1979, Vents yield surprises: v. 24, no. 7, p. 36.

Guild, P.W., 1972, Metallogeny and the new global tectonics: 24th Intern. Geol. Cong. Proc., v. 4, p. 17-24.

Gustafson, L.B., 1978, Some major factors of porphyry copper genesis, Econ. Geol., v. 73, p. 600-607.

Gustafson, L.B. and Titley, S.R., eds., 1978, Spec. Issue on Porphyry Copper Deposits of the Southwestern Pacific Islands and Australia: Econ. Geol., v. 73, no. 5.

Hasebe, K., Fuji N., and Uyeda, S., 1970, Thermal processes under island arcs: Tectonophysics, 10, p. 335-355.

Herman, B.M., Anderson, R.N., and Trachan, M., 1979, Extensional tectonics in the Okinawa Trough, Amer. Assoc. Petrol. Geol. Mem. 29, 199-208.

Hobart, M.A., Anderson, R.N., and Uyeda, S., 1979, Heat transfer in the Mariana Trough: EOS, 60, p. 383.

Horikoshi, E., 1960, The stratigraphic horizon of the "Kuroko" deposits in Hanaoka-Kosaka area, Green Tuff district of Japan: Min. Geol., (Soc. Min. Geol. Japan), v. 10, p. 300-310. (in Japanese).

, 1975, Genesis of Kuroko-stage deposits from the tectonic point of view: Jour. Volcan. Soc. Japan, v. 20, p. 341-353 (in Japanese with English Abstract).

, 1976, Development of Late Caenozoic petrogenic provinces and metallogeny in northeast Japan: in Strong, D.F., ed., Metallogeny and Plate Tectonics: Geol. Assoc. Canada Spec. Paper 14, p. 121-142.

Hutchinson, R.W., 1973, Volcanogenic sulfide deposits and their metallogenic significance: Econ. Geol., v. 68, p. 1223-1243.

_____, 1980, Massive base metal sulfide deposits as guides to tectonic evolution: in Strangway, D.W., ed., The Continental Crust and Its Mineral Deposits: Geol. Assoc. Canada, Spec. Paper 20, in press.

Ishihara, S., ed., 1974, Geology of Kuroko Deposits: Min. Geol., Spec. Issue, 6, 435 p.

Kanamori, H., 1971, Great earthquakes at island arcs and the lithosphere: Tectonophysics, v. 12, p. 187-198.

_____, 1977, The energy release in great earthquakes: Jour. Geophys. Res., v. 82, p. 2981-2987.

_____, 1978, Quantification of earthquakes: Nature, v. 271, p. 411-414.

Karig, D.E., 1971, Origin and development of the marginal basins in the western Pacific: Jour. Geophys. Res., v. 76, p. 2542-2561.

Kesler, S.E., 1973, Copper molybdenum and gold abundances in porphyry copper deposits: Econ. Geol., v. 68, p. 106-112.

Konda, T., Bimodal volcanism in the Northeast Japan Arc: Geol. Mag., Geol. Soc. Japan, v. 80, p. 81-89 (in Japanese).

LePichon, X., 1968, Sea-floor spreading and continental drift: Jour. Geophys. Res., v. 73, p. 3661-3697.

Lipman, P.W., Prostka, H.J., and Christiansen, R.L., 1972, Cenozoic volcanism and plate tectonic evolution of the western United States, Early and Middle Cenozoic: Phil. Trans. Royal. Soc. London, Ser. A, v. 271, p. 211-248.

McKenzie, D.P., 1969, Speculations on the consequences and causes of plate motions: Geophys. Jour, Royal Astron. Soc., v. 18, p. 1-32.

Nielsen, R.L., 1976, Recent developments in the study of porphyry copper geology - a review: Sato, T., 1978, The Kuruko Deposits - their origin and development: Kagaku, v. 48, p. 157-165 (in

Packham, G.H. and Falvey, D.A., 1971, An hypothesis for the formation of marginal seas in the western Pacific; Tectonophysics, v. 11, p. 79-110.

Plafker, G., 1976, Tectonic aspects of Guatemala earthquake of February 4, 1976: Science, v. 193, p. 1201-1208.

Sato, T., 1978, The Kuroko Deposits - their origin and development: Kagaku, v. 48, p. 157-165 (in Japanese).

Sawkins, F.J., 1972, Sulfide ore deposits in relation to plate tectonics: Jour Geol., v. 80, 4, 377-397.

Scholtz, C.H., Barazangi M., and Sbar, M., 1971, Late Cenozoic evolution of the Great Basin, western United States as an ensialic inter-arc basin: Geol. Soc. America Bull., v. 82, p. 2979-2990.

Sillitoe, R.H., 1972a, Formation of certain massive sulfide deposits at sites of sea-floor spreading: Trans. Instit. Min. Metall., v. 81, p. B141-B148.

_____, 1972b, A plate tectonic model for the origin of porphyry copper deposits: Econ. Geol., 67, 186-197.

_____, Relation of metal provinces in Western America to subduction of oceanic lithosphere: Geol. Soc. America Bull., v. 83, p. 813-817.

_____, 1973, Environments of formation of volcanogenic massive sulfide deposits: Econ. Geol., v. 68, p. 1321-1325.

The Continental Crust and Its Mineral Deposits, edited by D.W. Strangway,
Geological Association of Canada Special Paper 20

PLATE TECTONICS AND KEY PETROLOGIC ASSOCIATIONS

William R. Dickinson
Department of Geosciences,
University of Arizona,
Tucson, Arizona 85721

ABSTRACT

Key petrotectonic assemblages are rock associations that form during either persistent or transient plate interactions along plate boundaries. Divergent interactions include transient rifting and persistent sea-floor spreading. Convergent interactions include persistent plate consumption of oceanic lithosphere to produce arc-trench systems, and transient collision events that convert subduction zones to suture belts. Non-orogenic magmas erupted along spreading centres and rift belts are derived from partial melting of upwelling asthenosphere as pressure is released. Orogenic magmas erupted and injected along magmatic arcs are derived mainly from fluxing of asthenosphere where volatiles are introduced from descending slabs of subducted lithosphere. Irregular variations in isotopic ratios and potassium contents among arc igneous suites may partly reflect varying degrees of sediment subduction beneath the arcs. Subsidence of sedimentary basins and uplift of sediment sources are controlled by the plate interactions that govern changes in crustal profiles, thermotectonic effects, and flexures of plates under sedimentary or tectonic loads. Compositions of clastic sediments display systematic patterns that reflect plate tectonic settings of the depositional basins and their provenance terranes.

RÉSUMÉ

Les assemblages pétrotectoniques clés sont des associations de roches qui se forment durant les interactions persistantes ou transitoires des plaques le long de leurs bordures. Les interactions divergentes comprennent l'effondrement transitoire et l'extension persistante du fond de l'océan. Les interactions convergentes comprennent la consommation persistante par les plaques de la lithosphère océanique pour produire des systèmes arc-fossé, et des collisions transitoires épisodiques qui convertissent les zones de subduction en zones de suture. Les magmas non orogéniques ont fait éruption près des centres d'expansion et les zones d'effondrement proviennent de la fusion partielle dans les voûtes d'asthénosphère lorsque la pression se relâche. Les magmas orogéniques qui ont fait éruption et se sont injectés le long d'arcs magmatiques proviennent surtout de la fusion de l'asthénosphère lors de l'introduction de substances volatiles apportées par des lambeaux descendants de lithosphère subissant la subduction. Des variations irrégulières dans les rapports d'isotopes et les teneurs en potassium dans les suites ignées d'arc pourraient refléter partiellement les différents degrés de subduction de sédiments sous les arcs. La subsidence de bassins sédimentaires et le soulèvement des régions sources de ces sédiments sont contrôlés par les interactions de plaques qui gouvernent les changements dans le profil de la croûte, les effets thermotectoniques et la flexion de plaques sous les charges sédimentaires ou tectoniques. La composition des sédiments clastiques montre des patrons systématiques qui reflètent la mise en place par la tectonique des plaques des bassins de sédimentation et des régions de provenance des sédiments.

INTRODUCTION

Petrotectonic assemblages (Dickinson, 1971, 1972) are diagnostic rock associations that form characteristically under certain plate tectonic regimes. In general, plate settings include divergent, convergent, and transform boundaries or junctures between plates, and varied intraplate environments within plates. However, the most distinctive and voluminous petrotectonic assemblages form along the divergent and convergent plate boundaries, where lithosphere is created or consumed.

Both transient and persistent plate interactions are important for petrogenesis. Transient interactions are geologic processes whose time span is inherently limited, and whose behaviour is therefore discontinuous. By contrast, persistent interactions are long-term geologic processes which operate in either a steady-state or evolutionary mode.

This paper first reviews the major plate interactions that give rise to key petrotectonic assemblages. The origins of magmas related to plate boundaries are then discussed against that background. Finally, variations in the compositions of clastic sedimentary suites are related to differences in provenance, which are controlled by the plate settings of varied sedimentary basins.

KEY PLATE INTERACTIONS

Along divergent plate boundaries, sea-floor spreading to form an ophiolite sequence (Table I) atop new oceanic lithosphere is the persistent mode of plate interaction. Transient phases include rift separations of continental blocks or island arcs. Ophiolite sequences are most commonly interpreted as normal oceanic crust formed either at a mid-oceanic rise crest or within an inter-arc basin. However, the volcanic components of some ophiolites may well include island-arc volcanics (Miyashiro, 1973). The volcanogenic piles of intra-oceanic island arcs may rest upon either 1) pre-existing layered oceanic crust, perhaps modified by arc-related plutonism, or 2) some unique type of plutonic sub-stratum, also ophiolitic in a broad sense, that might be generated in the deep roots of an island arc massif during crustal extension associated with the opening of an inter-arc basin. In concept at least, the igneous rocks of a given ophiolite sequence might thus have been created by sea-floor magmatism or arc magmatism, or by both in sequence or combination.

Two main types of orogenic belts can develop along convergent plate bound-

TABLE I
KEY CRUSTAL ELEMENTS IN OPHIOLITE SEQUENCES OF OCEANIC CRUST

Oceanic Layers	Typical Thickness	Inferred Lithologic Types
1	0.5 km	Turbidite cover (variable thickness) over pelagic chert and/or micrite with intercalated hemipelagic argillite and basal horizons of metalliferous deposits
2A	~ 1 km	Pillow lavas, with associated pillow breccias and aquagene tuffs, cut locally by dikes and sills but porous overall
2B	~ 1 km	Massive altered lavas and sheeted dolerite dike complex
3A	~ 2.5 km	Metagabbro and microgabbro cut by altered dolerite dikes and local felsic intrusions
3B	~ 2.5 km	Massive and cumulate gabbro with uncertain intermixture of serpentinized peridotite
4	mantle	Olivinic peridotite of tectonitic fabric with uncertain intermixture of cumulate peridotite

after Peterson et al., 1974; Moores and Jackson, 1974; Christensen and Salisbury, 1975; Clague and Straley, 1977; Christensen, 1978

aries. Arc orogens occur where persistent consumption of oceanic lithosphere forms arc-trench systems. Collision orogens occur where attempted subduction of buoyant crustal blocks arrests plate consumption. Crustal collisions are thus transient events that convert subduction zones into deformed suture belts between blocks of continental crust. Continental collisions can also be viewed as the closing stage of the Wilson Cycle (Fig. 1).

Crustal collisions inhibit arc magmatism by interfering with further subduction of oceanic lithosphere, but a special type of anatectic magmatism may be fostered by collision events. Beneath suture belts, subterranean plate rupture may occur to allow subducted portions of the contiguous slab containing the buoyant crustal block to continue descent into the mante. Upwelling of asthenosphere in the vicinity of the rupture may expose crustal rocks of the collision orogen to abnormally high geotherms after the transient collision event (Toksoz and Bird, 1977; Bird, 1978).

The spectrum of arc-trench systems includes two end members (Fig. 2): 1) intra-oceanic island arcs where the back-arc area of marginal seas is expanded by sea-floor spreading; this occurs in the inter-arc basins that form sequentially as the arc massif is split by episodes of extensional rifting, and 2) continental-margin arcs where the back-arc area of terrestrial uplands is contracted by folding and thrusting; this takes place where the foreland fringe of a rigid continental block is underthrust beneath the rear flank of the arc orogen. In some arc-trench systems, both island arcs and terrestrial ones, neither spreading nor thrusting occurs in the back-arc region. The key factor that controls back-arc behaviour is probably the lateral motion (relative to back-arc lithosphere) of the trench hinge in the descending slab at the subduction zone.

In general, the steady retreat of the line of flexure at the trench hinge into the interior of a descending oceanic plate is due to the tendency of cold, dense lithosphere to sink into hot asthenosphere of lower density. Negative buoyancy is enhanced once subduction is well underway; phase transformations in the cold de-

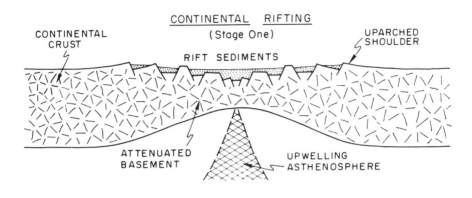

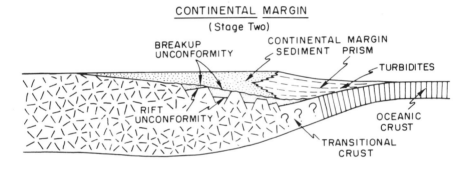

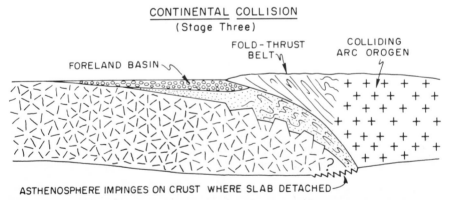

Figure 1. Sequential diagrams of key plate interactions in the Wilson Cycle, from transient continental rifting (top) to transient continental collision (bottom); intervening middle panel shows persistent intraplate sedimentation of rifted-margin prism along continent-ocean interface.

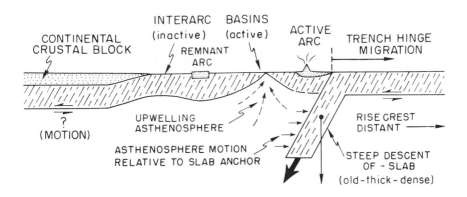

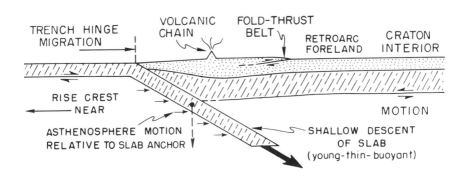

0   100   200   300
K M

Figure 2. Schematic geodynamics of island arcs with back-arc spreading (top) and terrestrial arcs with back-arc thrusting (bottom); arcs lacking back-arc deformation (see Fig. 3) are not shown. See text for discussion

scending slab are displaced upwards as much as 100 km relative to the surrounding asthenosphere (Ringwood, 1972). The resulting dense phases at anomalously shallow levels in the subducted slab act as sinkers. The angle at which the inclined slab descends into the mantle can be viewed as a resultant that balances the lateral motions of sea-floor spreading and plate convergence with the vertical motion of slab sinking (Luyendyk, 1970).

The direction of migration of the trench hinge with respect to back-arc lithosphere depends upon the net motion vector between the asthenosphere and the over-riding plate at the subduction zone (Wilson and Burke, 1972). Where back-arc lithosphere moves toward the plate flexure at the trench hinge, contractional tectonics may operate across the whole arc-trench system (Burchfiel and Davis, 1976). Retro-arc fold-thrust belts then develop behind the arc as a kind of secondary and antithetic subduction zone (Coney, 1973). Where oceanward motion of the overriding plate does not compensate for the retreat of the trench hinge, the crustal extension required within the arc-trench system concentrates at spreading centres within successive inter-arc basins. These form as the overriding plate splits repeatedly along the thermal axis of the magmatic arc.

The overall geodynamics of such arc-trench systems are complex (Uyeda and Kanamori, 1979). For example, the behaviour of the trench hinge depends partly upon the age of the oceanic lithosphere being subducted. Old oceanic lithosphere that is cold and dense will sink more readily than younger and hotter lithosphere (Molnar and Atwater, 1978). Retrograde migration of the trench hinge away from the overriding plate, which allows extensional inter-arc basins to develop behind the arc, is thus favoured where descending lithosphere is relatively old. On the other hand, where the descending slab is younger and more buoyant, subduction is more difficult and the trench hinge may press laterally against the flank of the arc-trench system, inducing widespread contractional deformation of the overriding plate.

Present-day arc-trench systems displaying back-arc spreading generally face east, while those showing back-arc thrusting generally face west (Fig. 3). Most of these systems lie around the circum-Pacific periphery, whose characteristics thus dominate the data. The marked asymmetry in age of the Pacific sea-floor means that east-facing systems with inter-arc basins mostly subduct the old sea-floor of the western Pacific, while the west-facing systems with retro-arc fold-thrust belts mostly subduct the young sea-floor of the eastern Pacific (see Fig. 2). No one knows yet whether the east-west asymmetry of Pacific sea-floor and circum-Pacific subduction systems is coincidental, or reflects a generic east-west bias in geotectonics (Dickinson, 1978).

MAGMATIC ORIGINS

The largest volume of igneous rocks is erupted along either convergent or divergent plate boundaries (Martin and Piwinskii, 1972). Processes associated with plate convergence generate a broad spectrum of orogenic magmas, which include both the arc volcanics and their co-genetic batholiths or stocks. Magmas erupted along divergent plate boundaries are contrasting non-orogenic types, of which the oceanic tholeiites of the sea-floor are most voluminous (Engel et al., 1965). Other non-orogenic suites occur locally at intraplate sites, especially among oceanic islands and on continental plateaus.

Nonorogenic Magmas

Parent non-orogenic magmas generated along spreading centres at divergent plate boundaries are derived from partial melting of asthenosphere. Upwelling of asthenosphere to fill the space created by receding edges of lithosphere triggers the requisite pressure release. Intraplate magmatism stems from analogous mantle motions, although the relative roles of two complementary phenomena are uncertain: 1) actively rising convective plumes in the asthenosphere (Briden and Gass, 1974), and 2) tensional fracturing of rigid plates of lithosphere (Turcotte and Oxburgh, 1978).

The evolution of magmas rising from the mantle probably involves progressive fractionation by partial crystallization at various pressure levels (O'Hara, 1965). Induced melting at crustal horizons affected by advective heating is also possible where crust is thick. The overall magmatic system is complex even under mid-

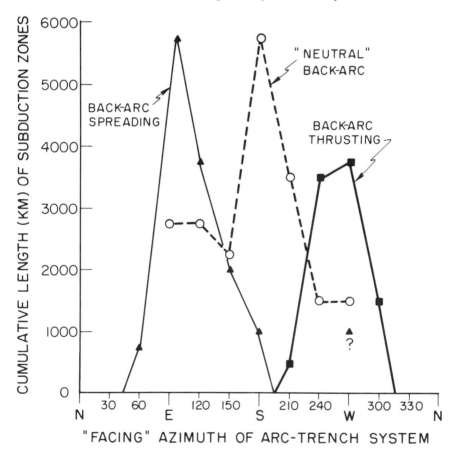

Figure 3. Plot showing systematic relationship between current back-arc tectonics and geographic orientation of modern arc-trench systems (after Dickinson, 1978). "Facing" azimuth is direction from arc toward trench normal to arc-trench trend.

oceanic rise crests where both crust and lithosphere have minimal thickness. Current models for the formation of oceanic crust at a spreading centre commonly assume the presence of only one magma chamber at a shallow level (Cann, 1974; Sleep, 1975; Kidd, 1977). However, this chamber is simply the one that congeals to form the gabbros and associated intrusives of the lower oceanic crust; sheeted dyke complexes of the mid-crust are emplaced into the roof of this expanding magma chamber and feed the sea-floor eruptions that produce the pillow lavas of the upper oceanic crust (Dewey and Kidd, 1977). The full history of the magmas that continuously or intermittently replenish the crustal magma chamber as it is evacuated by volcanism and crystallized by cooling remains speculative. However, the compositions of abyssal tholeiites generally cluster near low-pressure cotectics, and thus cannot be primary magmas produced by high-pressure melting at depth in the mantle (O'Hara, 1968). Repeated equilibration in multiple magma chambers lying at intermediate depths seems to be implied by this observation. The truly parental magmas are inferred to contain significantly more magnesium than the observed sea-floor basalts (Elthon, 1979).

The volcanic suites erupted along intracontinental rifts at incipient spreading centres are commonly much more alkalic than the abyssal tholeiites of the ocean floor. This difference may reflect the final equilibration of parent melts with a refractory mantle residuum at different pressure levels (Green, 1971). If the level of last equilibration is taken crudely as the base of the lithosphere, then the depth will clearly be greater under continental lithosphere that has cracked locally, but not yet separated, than under an oceanic spreading centre, where new lithosphere is just being formed by cooling of upwelling asthenosphere. McBirney and Gass (1967) suggested that the volcanic suites of oceanic islands may be systematically more alkalic in proportion to their distance from a mid-oceanic rise crest. Although Middlemost (1973) challenged the validity of their correlation, such a relation would be in harmony with the gradual thickening of lithosphere as a plate moves away from a spreading centre. Perhaps the correlation should be tested again using the square root of the age of the seafloor as the key value against which to plot suitable petrologic parameters.

Abundant isotopic data now indicate that the magmas of the ocean floor and oceanic islands are derived from quite heterogeneous mantle (O'Nions and Pankhurst, 1974). Mantle isochrons suggest that the heterogeneity stems from episodes of mantle differentiation and depletion reaching back into the Precambrian (Brooks et al., 1976). On balance, typical abyssal tholeiites of the seafloor reflect derivation from more depleted sources than the lavas of most oceanic islands. A favoured explanation for this fact has been the notion that depleted asthenosphere, which feeds rise-crest volcanism to build the oceanic crust, is underlain by less depleted mesosphere. At intervals along rise crests and elsewhere, hot plumes or blobs of this undepleted material are envisioned as rising through the depleted asthenosphere to construct the oceanic islands (Schilling, 1973). However, recent data obtained from drillholes penetrating deep into the seafloor indicate that the pattern of heterogeneity is both too complex petrochemically and too persistent through time to be explained well by the plume model (Wood et al., 1979). Moreover, the available data can be used to support a diametrically opposed model (Hedge, 1978), in which a relatively undepleted layer lying along the low-velocity zone at the top of the asthenosphere feeds island volcanism. Rise-crest volcanism to build normal seafloor is then viewed as a response to strong upwelling of more depleted asthenosphere beneath spreading centres. This alternative model has the twin virtues of allowing uniform upward concentration of incompatible elements within the asthenosphere, and of calling for the most forceful perturbation of the asthenosphere beneath mid-oceanic rise crests where divergent plate motions induce strong upwelling.

Orogenic Magmas

The vertical profile of crust and mantle beneath a magmatic arc includes the following petrologic elements from which magmas might be derived: 1) Granitoid and gneissoid rocks of the upper crust; these are the continental basement rocks or plutons and metamorphics of the arc massif exposed in uplifted terranes; 2) gabbroic and granulitic rocks of the lower continental crust whose exact nature is still uncertain (Smithson and Brown, 1977), although some exposures are available locally in mountain belts (Fountain, 1976); 3) ultramafic rocks of the subcontinental lithosphere; these are mainly refractory rocks unlikely to yield melts, but minor amounts of enriched components may be present (Brooks *et al.*, 1976); 4) asthenosphere that forms a wedge between the lithosphere of the over-riding plate and the lithosphere of the descending slab; and 5) the oceanic crust of the subducted lithosphere; this layer may be composed mainly of eclogites formed at elevated pressures from the igneous components of seafloor successions.

The processes that sustain arc magmatism are probably set in motion when the descending slab of lithosphere penetrates the asthenosphere below the arc massif. This inference is supported by relations in the modern Andes (Barazangi and Isacks, 1976). Active volcanism occurs only where plate descent is steep enough for the subducted slab of lithosphere to penetrate into the asthenosphere. Where the angle of plate descent is anomalously shallow, and the subducted plate slips subhorizontally beneath the over-riding plate, little or no volcanism occurs. This apparent relationship between arc magmatism and the upper part of the asthenosphere is confirmed by the fact that arc volcanic chains most typically stand about 100 to 150 km vertically above the inclined seismic zone beneath the arc. In effect, the arcuate volcanic chain is the surface projection of the curvilinear intersection between the inclined seismic zone marking the cold upper part of the descending slab and the low-velocity zone near the top of the asthenosphere (Dickinson and Hatherton, 1967).

Melting near the interface between the descending slab and the overlying mantle wedge might involve either the oceanic crust of the subducted slab or adjacent asthenosphere in the low-velocity zone. Magmas from either source would need to undergo extensive modification by fractionation and contamination during ascent to resemble closely the igneous suites observed at the surface (Ringwood, 1977). Water released from the thermal dehydration of hydrous minerals contained within the downgoing slab probably serves as an effective flux to trigger the melting process (Anderson *et al.*, 1976). For some arcs, details of elemental and isotopic geochemistry are compatible with derivation of the parent magmas principally from subducted oceanic crust (Meijer, 1976), but in other cases such a single source is unlikely (Gill, 1974). The most attractive general model proposes that partial melts are derived from

the underthrust oceanic crust, and move upward to mix with overlying asthenosphere, where they cause further diapirism and melting in the modified mantle materials (Kay, 1977). Alternatively, released water that carries contaminants derived
from the downgoing oceanic crust may move alone into the overlying mantle wedge,
there to flux asthenosphere directly (Anderson et al., 1978). Experimental work
indicates that hydrous mantle is a feasible source material for arc magmas (Mysen
and Boettcher, 1975), and petrologic relations in the central Andes suggest that the
mantle wedge above the descending slab is the major source for the partial melts that
feed arc volcanism there (Noble et al., 1975).

Further evidence suggests that contributions from the lower crust of the arc
massif are also present in some arc magmas. Melting in the roots of the arc could be
promoted by advective heat brought upward from the mantle by rising magmas.
Some intermediate andesitic and dacitic rocks are hybrids developed through the
mixing of mafic basaltic melts from the mantle and felsic rhyolitic magmas of probable crustal origin (Eichelberger, 1978). Moreover, granitic batholiths probably contain anatectic crustal components as well as magmatic contributions from the mantle
below (Wyllie et al., 1976). Low strontium isotope ratios in typical batholiths preclude significant contributions from the radiogenic upper crust, but allow the incor-
poration of materials from the lower crust, which may be composed largely of mantle
derivatives (Brown, 1977). Arc massifs probably evolve by means of the systematic
upward migration of incompatible elments as partial melting proceeds in the hot roots
of a gradually thickening crust (Jakes and White, 1971).

Sediment Subduction

Data on strontium isotopes (Table II) offer further insight into the petrogenesis
of orogenic magmas. Lavas from the seafloor and most oceanic islands display ratios
well below 0.7050. Volcanics from most intra-oceanic island arcs and many
continental-margin arcs display ratios nearly comparable to these values, but well
below those for selected oceanic islands with ratios above 0.7050. For these orogenic
magma suites, contributions from non-oceanic crustal sources seem to be minimal.
They include suites erupted through thick crustal profiles beneath the Cascades,
Mexico, Central America, Ecuador, Chile, and Japan.

However, other continental-margin suites from the central Andes in Peru and
Chile display ratios consistenty above 0.7050. Although some authors have attributed
these elevated ratios to assimilation or anatexis of continental crust in the arc roots, a
full evaluation of petrochemical trends within the region implies that contributions
from subjacent crust are minimal (Thorpe et al., 1976). Moreover, comparably high
ratios in the range of 0.7050 to 0.7075 are known in suites from the Banda island arc
where crust is thin. Indeed, the Banda values are consistently higher than values for
the nearby Sunda island arc, a submerged continental-margin arc with thicker crust
but low ratios that are normal for arc suites. The contrast between Banda and Sunda
leads to the idea that the higher strontium isotope ratios for Banda reflect incipient
crustal collision between the Banda arc and Australia in the Timor region (Whitford
et al., 1977). Sediments of continental derivation that spread northward from Australia may have been subducted deep beneath the Banda arc, but not beneath the
Sunda arc.

TABLE II
REPRESENTATIVE STRONTIUM ISOTOPE RATIOS IN VOLCANIC
SUITES FROM DIFFERENT TECTONIC SETTINGS

Tectonic Setting	Suites (N)	Total Range All Suites	Range for Suite Medians
Rise crests and seafloor	5	.7020-.7043	.7026-.7034
Oceanic islands (typical examples)	12	.7024-.7049	.7028-.7037
Oceanic islands (Samoa, Tristan)	2	.7050-.7066	.7051-.7057
Intra-oceanic island arcs	10	.7026-.7054	.7034-.7045
Continental-margin arcs (low set)	7	.7029-.7059	.7037-.7046
Continental-margin arcs (high set)	4	.7051-.7079	.7063-.7073
Banda arc (low)	1	.7054-.7059	.7057
Banda arc (high)	1	.7065-.7091	.7072

The subduction of continent-derived sediment to mantle depths is an attractive mechanism to explain variations in isotopic ratios for arc suites elsewhere (Armstrong, 1971). Although the low density of sediments would tend to resist subduction, significant quantities might be trapped within grabens or locked between lavas and sills in the oceanic crust undergoing subduction. If present along the inclined seismic zone, sediments would readily enter the melt fraction as magma generation began, or would enter an aqueous fluid if dehydration of the slab were to proceed without melting. The varying concentrations of radiogenic components in arc suites might then be controlled by varying degrees of sediment subduction. Recent interpretations of elemental and isotopic data for the central Andes (James, 1978) suggest that such bulk recycling of subducted sediments may be an important factor in the petrogenesis of many continental-margin arc suites. The consistently low strontium isotope ratios for most intra-oceanic island arcs presumably reflect relatively low sediment volumes in all those arc-trench systems. However, even minor sediment subduction may produce subtle but detectable isotopic effects (Church, 1973, 1976).

The concept of sediment subduction marks a return to ideas advanced two decades ago. Stille (1955) suggested that the faulting of crustal materials downward along the inclined seismic zone to levels near the low velocity zone would allow melted sediments to feed arc magmatism. He called this process 'palingenesis by underthrusting'' to distinguish it from ''palingenesis by subsidence'', which was then thought to occur along the keels of geosynclines. Later, Coats (1962) appealed to the dragging of sediments down along the inclined seismic zone to produce the arc magmas of the Aleutians. He proposed this process as means of contaminating mantle melts with crustal materials in an arc where no continental crust is present.

K-h Correlation

Despite the complexity of petrogenesis in magmatic arcs, an examination of the global data reveals a systematic relationship between the potassium content of arc lavas from active volcanic chains and the vertical depth of the inclined seismic zone in the mantle beneath (Fig. 4). This empirical correlation supports the inference that processes near the upper surface of the descending slab trigger arc magmatism. The considerable scatter of points on the plot of potassium content (K) at a given nominal silica percentage (57.5) versus depth (h) probably reflects real variations in petrogenesis from arc to arc, rather than random noise from imprecise data. Correlation coefficients for data from individual arcs are commonly higher than the inherent scatter in the global plot implies. In general, the increase in the potassium content of lavas across most broad arc terranes is striking and well known.

Recent detailed work along a transect of the Sunda arc (Whitford and Nicholls, 1976) indicates that the K-h correlation can be improved significantly by restricting attention to a subset of the local lavas. The preferred rocks are those whose strontium isotope ratios are low enough to reflect principle derivation from either the oceanic crust of the descending slab or the overlying wedge of asthenosphere, or both. The appropriate values preclude significant contributions to the volcanic magmas from continental crust, whether as subducted sediment at mantle depths or in the lower crust of the arc massif. When data from all twenty volcanic suites studied along the transect are plotted without selectivity, the K-h correlation coefficient is less than 0.85. By rejecting the suites with strontium isotope ratios of 0.7055 or more, and further restricting the plot to volcanoes where the basic K_2O/SiO_2 correlation is significant at the 99.9 per cent level, the correlation coefficient for the eight remaining suites improves to nearly 0.95. This exercise supports the fundamental mantle origin of arc magmas.

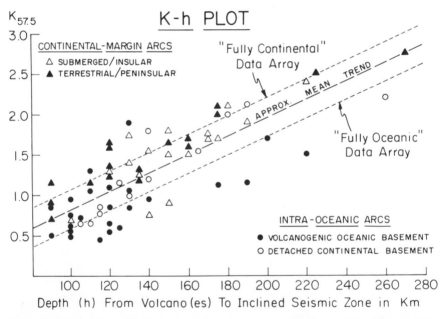

Figure 4. Mean global correlation (K-h plot) of level of potassium content in Quaternary volcanic suites from active magmatic arcs with depth to inclined seismic zone beneath the volcanic chains. Each of the 70 points represents a single volcano or volcano cluster. Ordinate ($K_{57.5}$) is per cent K_2O at 57.5 per cent SiO_2 on Harker variation diagram plotted volatile-free. Abscissa (h) is depth from surface to top of seismic envelope estimated from transverse hypocenter profiles published for selected arc transects. Modified after Dickinson (1975).

The main reason for the K-h correlation is unclear (Dickinson, 1975). Phase relations during partial melting at varying depths may modulate the potassium content of the parent magmas at the sites of initial magma generation. The buffering of potassium content in the melt phase by sanidine under different pressures is the most likely control (Marsh, 1976; Marsh and Carmichael, 1974). Both amphibole and phlogopite are also often mentioned as reservoirs for potassium at shallow mantle depths. Alternatively, the potassium contents of the magmas finally evolved at the surface may be governed by the progressive scavenging of potassium from mantle rock by hydrous fluids, in amounts proportional to the length of the travel path to the surface (Best, 1975). The scavenging agents might be both aqueous solutions derived from dehydration of the downgoing slab and hydrous magmas developed in the overlying wedge of asthenosphere. Regardless of its cause, the K-h correlation can be used successfully to reconstruct the approximate geometry of past arc-trench systems (Keith, 1978).

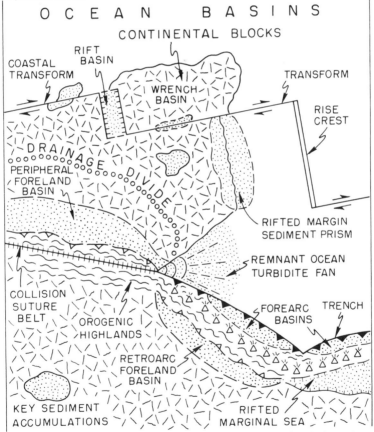

Figure 5. Sketch map showing main sites of major sedimentary basins in relation to key types of plate boundaries and associated sources for detrital sediment. Stipples indicate areas of sedimentation.

The mean K-h curve for continental-margin arcs shows distinctly higher potassium contents for a given depth than does the mean K-h curve for intra-oceanic arcs (see Fig. 4). Previously, this difference was explained by the participation of the lower crust of the arc massif in the petrogenesis of arc magmas (Dickinson, 1975). In terms of sediment subduction, the difference might be attributed to the tendency for greater volumes of sediment to be available for subduction along continental margins than in ocean settings.

SEDIMENTARY SUITES

Plate tectonics offers sound explanations for the subsidence of sedimentary basins (Dickinson, 1974), and provides a useful guide to the main categories of provenances from which detrital sandstone suites are derived (Dickinson and Suczek, 1979). To a considerable extent, earlier geosynclinal classifications of depositional sites have been supplanted by interpretations based on plate tectonic settings. The major motions of plates are horizontal. However, plate interactions produce uplift belts, which serve as major sources of sedimentary detritus, and belts of subsidence, where thick piles of sediment can accumulate.

Sediment Sources

Major uplifts occur in association with plate boundaries (Dewey and Bird, 1970). Along intracontinental rifts marking divergent plate boundaries, thermotectonic doming forms broad uplands. Steep relief then develops locally where crustal extension has produced block faulting. Wrench deformation along intracontinental transforms also creates local highlands when deforming plate edges warp and break (Wilcox *et al.*, 1973). However, the mountain belts of major orogenic systems occur in association with convergent plate boundaries (Fig. 5). Arc orogens may pass along strike into collision orogens as continental blocks are jammed against subduction systems consuming oceanic lithosphere.

There are four main kinds of orogenic highlands, based on the types of sedimentary detritus shed from each:

1) Uplifted subduction complexes, composed of ophiolitic rocks and oceanic sediments that were detached from the descending slab and gradually raised as an imbricate stack of tectonic slices beside the trench;

2) elevated magmatic arcs, composed of volcanogenic strata capping an arc massif whose major crustal components are plutons also emplaced by arc magmatism;

3) foreland fold-thrust belts, formed by structural telescoping of continental strata by back-arc contraction behind some continental-margin arcs, and;

4) complex collision sutures along which subduction complexes may be juxtaposed tectonically with foreland fold-thrust belts of partly subducted continental strata and uplifted basement rocks of the colliding continental block and eroded arc roots.

Sedimentary Basins

Major subsidence, which is necessary for thick sediment accumulation, occurs in three main ways: 1) by extensional thinning of crust during continental rifting along divergent plate boundaries, 2) by thermotectonic subsidence of cooling lithosphere within oceanic basins or along rifted continental margins and aborted rift troughs, and 3) by broad isostatic flexure of plate surfaces in response to sedimentary or tectonic loads nearby. Most basins can be grouped conveniently as either rifted basins or orogenic basins, according to the relative importance and the sequence of these three influences during their history. The evolution of rifted basins is controlled mainly by plate interactions along divergent plate boundaries, whereas orogenic basins evolve near convergent plate boundaries.

Rifted basins first undergo crustal thinning by extensional deformation of the substratum along a divergent plate boundary. This initial step creates the potential for eventual subsidence, which may be delayed by the effects of thermotectonic uplift during the progress of rifting. Later thermotectonic subsidence induces local sedimentation by attracting drainage from adjacent high-standing continental blocks. The growing sediment load may produce regional downflexure of lithosphere, which enlarges the surface area of the basin. Rifted basins include both 1) intracontinental troughs and downwarps flanked by undisturbed continental blocks, and 2) rifted continental margins with a continental block on one side but an ocean basin on the other. Rifted-margin sediment prisms (see Fig. 1) taper seaward as well as landward simply because no effective sediment feed comes from the ocean side.

Orogenic basins include several disparate types for which downflexure of lithosphere under the tectonic loads of stacked thrust sheets is a key factor in subsidence. Most characteristic are foreland basins, located between orogenic belts and stable cratons in adjacent continental blocks. Foreland basins are strongly asymmetric, with wedge edges on the cratonic side and deep keels adjacent to the orogenic belt. They develop as the surface of the continental block is warped downward beneath the thrust sheets of a fold-thrust belt along the flank of the adjacent orogen. Two distinct types of foreland basin are known: 1) retro-arc basins which occur behind active magmatic arcs undergoing contraction, and 2) peripheral basins, which form in front of suture belts in the pre-collision trench position (see Fig. 5).

Other orogenic basins include the trench, where little sediment can accumulate without immediate deformation, and fore-arc basins located between the magmatic arc and the trench (Dickinson and Seely, 1979). Basins associated with wrench deformation along transform plate boundaries can be either orogenic or rifted, depending upon whether compression or extension occurs locally along the transform trend.

Provenance Relations

Connections between sediment sources and detrital accumulations can be inferred from sandstone petrology. Characteristic species of mineral grains and varieties of rock fragments in sandstone frameworks are a reliable guide to the general nature of the provenance (Fig. 6). Three general classes and several subclasses of provenance can be defined.

1) Sources on continental blocks include the broad surfaces of stable cratons, from which mature quartzose sands are derived, and uplifted basement blocks, from which immature quartzo-feldspathic or arkosic sands are derived. The ultimate sources of both quartzose and arkosic sands are mainly granitoid and genissoid

basement rocks. Quartzose sands of cratonic derivation are deposited mainly in platform successions on the continental blocks, within mature intracontinental basins, and as parts of rifted-margin sediment prisms. Calcarenites are common associates. The basement uplifts from which the arkosic sands are derived include fault blocks of incipient rift belts, zones of wrench deformation along transforms, and some fractured foreland regions. The arkosic sands are deposited mainly in local rift

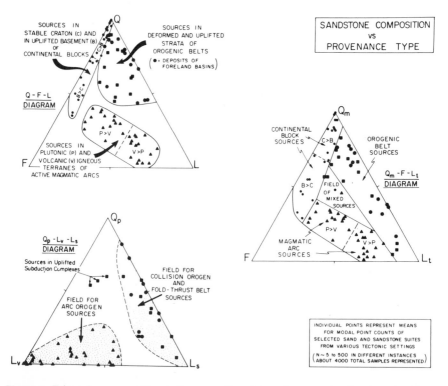

Figure 6. Triangular compositional diagrams of sandstone frameworks showing distinctions between detritus from different provenances as controlled by plate tectonic setting (after Dickinson and Suczek, 1979). Symbols:

Q   − Total Quartzose Grains
Qm − Monocrystalline Quartz Grains
Qp − Polycrystalline Quartz (~Chert)
F   − Total Feldspar Grains
L   − Unstable Lithic Fragments
Lt − Total Lithic Fragments
Lv − Volcanic-Metavolcanic Lithics
Ls − Sedimentary-Metasedimentary Lithics

Note that $Q = Qm + Qp$, $Lt = L + Qp$, and $L = Lv + Ls$, hence

$L = Qp + Lv + Ls$ and $Qp − Lv − Ls$ plot includes lithic fragments only.

troughs, including those later buried beneath rifted-margin sediment prisms and some along belts of wrench tectonism.

2) Sources in magmatic arcs include the surficial volcanic chains, from which lithic sands of volcaniclastic character are derived, and the subjacent plutonic terranes, from which more feldspathic sands are derived. Sands from these source rocks are commonly mingled in arc-derived detritus. Sands from deeply eroded arcs with exposed plutons have compositions gradational to the arkosic sands derived from basement uplifts of continental blocks. This relation is to be expected, for the deep roots of magmatic arcs represent continental crust in the making. Arc-derived sands are deposited mainly in trenches, where they are deformed to become parts of growing subduction complexes, and in either fore-arc or inter-arc basins. Some arc-derived detritus finds its way into foreland basins, particularly of the retro-arc type, but drainage divides along foreland fold-thrust belts commonly block the access of arc-derived sands to continental blocks. Conversely, sands from continental blocks seldom cross the drainage divides along the trends of magmatic arcs.

3) Sources in deformed and uplifted strata of orogenic belts include subduction complexes of oceanic strata, foreland fold-thrust belts of continental strata, and combinations of oceanic and continental strata uplifted along the suture belts of collision orogens. All these terranes are sources for recycled detritus. The sands are generally low in feldspar but otherwise range from quartzose to lithic varieties. The more quartzose examples are gradational to sands of cratonic derivation and largely represent recycled sand having that ultimate origin. Lithic fragments are mainly chert and argillite or slate from sedimentary and metasedimentary terranes. This population of lithic grains derived from collision orogens contrasts markedly with the volcanic and metavolcanic lithic fragments that are dominant in debris derived from arc orogens. The recycled orogenic sands are deposited mainly in foreland basins or in remnant ocean basins lying off the end of closing suture belts (see Fig. 5). Special variants also occur locally in trenches and fore-arc basins.

The characteristic flysch sequences incorporated into the heart of many mountain systems are probably largely remnants of oceanic turbidites shed longitudinally from collision orogens (Graham et al., 1975, 1976). The typical longitudinal paleocurrents observed in the beds are compatible with this setting, and deformation soon after deposition can be expected as the suture belt closes sequentially. The flysch in each segment of an orogenic belt thus represents the sedimentation on shrinking ocean floor or depressed continental edge just prior to final oceanic closure by collision. The overlying molasse is deposited during and immediately following the collision orogeny. Properly understood, the traditional tandem of flysch and molasse thus becomes a firm sedimentary record of the closing phases of the Wilson Cycle.

REFERENCES

Anderson, R.N., DeLong, S.E., and Schwarz, W.M., 1978, Thermal model for subduction with dehydration in the downgoing slab: Jour. Geol., v. 86, p. 731-739.

Anderson, R.N., Uyeda, Seiya, and Miyashiro, Akiho, 1976, Geophysical and geochemical constraints at converging plate boundaries – Part I: dehydration in the downgoing slab: Geophys. Jour. Royal Astron. Soc., v. 44, p. 333-357.

Armstrong, R.L., 1971, Isotopic and chemical constraints on models of magma genesis in volcanic arcs: Earth and Planet. Sci. Letters, v. 12, p. 137-142.

Barazangi, Muawia, and Isacks, B.L., 1976, Spatial distribution of earthquakes and subduction of the Nazca plate beneath South America: Geology, v. 4, p. 686-692.

Best, M.G., 1975, Migration of hydrous fluids in the upper mantle and potassium variation in calc-alkalic rocks: Geology, v. 3, p. 429-432.

Bird, Peter, 1978, Initiation of intracontinental subduction in the Himalaya: Jour. Geophys. Research, v. 83, p. 4975-4987.

Briden, J.C., and Gass, I.G., 1974, Plate movement and continental magmatism: Nature, v. 248, p. 650-653.

Brooks, Christopher, Hart, S.R., Hofmann, A., and James, D.E., 1976, Rb-Sr mantle isochrons from oceanic regions: Earth and Planet. Sci. Letters, v. 32, p. 51-61.

Brooks, Christopher, James, S.E., and Hart, S.R., 1976, Ancient lithosphere: its role in young continental volcanism: Science, v. 193, p. 1086-1094.

Brown, G.C., 1977, Mantle origin of Cordilleran granites: Nature, v. 265, p. 21-24.

Burchfiel, B.C., and Davis, G.A., 1976, Compression and crustal shortening in Andean-type orogenesis: Nature, v. 160, p. 693-694.

Cann, J.R., 1974, A model for oceanic crustal structure developed: Geophys. Jour. Royal Astron. Soc., v. 39, p. 169-187.

Christensen, N.I., 1978, Ophiolites, seismic velocities and oceanic crustal structure: Tectonophysics, v. 47, p. 131-157.

Christensen, N.I., and Salisbury, M.H., 1975, Structure and constitution of the lower oceanic crust: Rev. Geophys. Space Phys., v. 13, p. 57-86.

Church, S.E., 1973, Limits of sediment involvement in the genesis of orogenic volcanic rocks: Contr. Mineral. Petrol., v. 39, p. 17-32.

Church, S.E., 1976, The Cascade Mountains revisited: a re-evaluation in light of new lead isotopic data: Earth and Planet. Sci. Letters, v. 29, p. 175-188.

Clague, D.A., and Straley, P.F., 1977, Petrologic nature of the oceanic Moho: Geology, v. 5, p. 133-136.

Coats, R.R., 1962, Magma type and crustal structure in the Aleutian arc: in MacDonald, G.A., and Kuno, Hisashi, eds., The Crust of the Pacific Basin: Amer. Geophys. Union Mono. 6, p. 92-109.

Coney, P.J., 1973, Plate tectonics of marginal foreland thrust-fold belts: Geology, v. 1, p. 131-134.

Dewey, J.F., and Bird, J.M., 1970, Mountain belts and the new global tectonics: Jour. Geophys. Research, v. 75, p. 2625-2647.

Dewey, J.F., and Kidd, W.S.F., 1977, Geometry of plate accretion: Geol. Soc. America Bull., v. 88, p. 960-968.

Dickinson, W.R., 1971, Plate tectonics in geologic history: Science, v. 173, p. 107-113.

Dickinson, W.R., 1972, Evidence for plate-tectonic regimes in the rock record: Amer. Jour. Sci., v. 272, p. 551-576.

Dickinson, W.R., 1974, Plate tectonics and sedimentation: in Dickinson, W.R., ed., Tectonics and sedimentation: Soc. Econ. Paleontol. Mineral. Spec. Publ. 22, p. 1-27.

Dickinson, W.R., 1975, Potash-depth (K-h) relations in continental margin and intra-oceanic magmatic arcs: Geology, v. 3, p. 53-56.

Dickinson, W.R., 1978, Plate tectonic evolution of North Pacific rim: Jour. Phys. Earth, v. 26 (supplement), p. S1-S20.

Dickinson, W.R., and Hatherton, Trevor, 1967, Andesitic volcanism and seismicity around the Pacific: Science, v. 157, p. 801-803.

Dickinson, W.R., and Seely, D.R., 1979, Structure and stratigraphy of forearc regions: Amer. Assoc. Petroleum Geol. Bull., v. 63, p. 2-31.

Dickinson W.R., and Suczek, C.A., 1979, Plate tectonics and sandstone compositions: Amer. Assoc. Petroleum Geol. Bull., v. 63, p. 2164-2182.

Eichelberger, J.C., 1978, Andesitic volcanism and crustal evolution: Nature, v. 275, p. 21-27.

Elthon, Don, 1979, High magnesia liquids as the parental magma for ocean floor basalts: Nature, v. 278, p. 514-518.

Engel, C.G., Engel, A.E., and Havens, R.G., 1965, Chemical characterisitcs of oceanic basalts and the upper mantle: Geol. Soc. America Bull., v. 76, p. 719-725.

Fountain, D.M., 1976, The Ivrea-Verbano and Strona-Ceneri Zones, northern Italy: a cross-section of the continental crust – new evidence from seismic velocities of rock samples: Tectonophysics, v. 33, p. 145-165.

Gill, J.B., 1974, Role of underthrust oceanic crust in the genesis of a Fijian calc-alkaline suite: Contr. Mineral. Petrol, v. 43, p. 29-45.

Graham, S.A., Dickinson, W.R., and Ingersoll, R.V., 1975, Himalayan-Bengal model for flysch dispersal in the Appalachian-Ouachita system: Geol. Soc. America Bull., v. 86, p. 273-286. 273-286.

Graham, S.A., Ingersoll, R.V., and Dickinson, W.R., 1976, Common provenance for lithic grains in Carboniferous sandstones from Ouachita Mountains and Black Warrior Basin: Jour. Sed. Petrol., v. 46, p. 620-632.

Green, D.H., 1971, Composition of basaltic magmas as indicators of conditions of origin with application to oceanic volcanism: Royal Soc. London Philos. Trans., v. 268A, p. 707-725.

Hedge, C.E., 1978, Strontium isotopes in basalts from the Pacific ocean basin: Earth and Planet. Sci. Letters, v. 38, p. 88-94.

Jakes, Peter, and White, A.J.R., 1971, Composition of island arcs and continental growth: Earth and Planet. Sci. Letters, v. 12, p. 224-230.

James, D.E., 1978, On the origin of the calc-alkaline volcanics of the central Andes; a revised interpretation: Carnegie Instn. Washington Year Book 77, p. 562-789.

Kay, R.W., 1977, Geochemical constraints on the origin of Aleutian magmas: in Talwani, Manik, and Pitman, W.C., III eds., Island arcs, deep-sea trenches, and back-arc basins: Amer. Geophys. Union Maurice Ewing Ser. 1, p. 229-242.

Keith, S.B., 1978, Paleosubduction geometries inferred from Cretaceous and Tertiary magmatic patterns in southwestern North America: Geology, v. 6, p. 516-521.

Kidd, R.G.W., 1977, A model for the process of formation of the upper oceanic crust: Geophys. Jour. Royal Astron: Soc., v. 50, p. 149-183.

Luyendyk, B.P., 1970, Dips of downgoing lithospheric plates beneath island arcs: Geol. Soc. America Bull., v. 81, p. 3411-3416.

McBirney, A.R., and Gass, I.G., 1967, Relations of oceanic volcanic rocks to mid-oceanic rises and heat flow: Earth and Planet. Sci. Letters, v. 2, p. 265-276.

Marsh, B.D., 1976, Some Aleutian andesites: their nature and source: Jour. Geol., v. 84, p. 27-45.

Marsh, B.D., and Carmichael, I.S.E., 1974, Benioff zone magmatism: Jour. Geophys. Research, v. 79, p. 1196-1209.

Martin, R.F., and Piwinskii, A.J., 1972, Magmatism and tectonic settings: Jour. Geophys. Research, v. 77, p. 4966-4975.

Meijer, Arend, 1976, Pb and Sr isotopic data bearing on the origin of volcanic rocks from the Mariana island-arc system: Geol. Soc. America Bull., v. 87, p. 1358-1369.

Middlemost, E.A.K., 1973, Evolution of volcanic islands: Lithos, v. 6, p. 123-132.

Miyashiro, Akiho, 1973, The Troodos ophiolitic complex was probably formed in an island arc: Earth Planet. Sci. Letters, v. 19, p. 218-224.

Molnar, Peter, and Atwater, Tanya, 1978, Interarc spreading and Cordilleran tectonics as alternates related to the age of subducted oceanic lithosphere: Earth Planet. Sci. Letters, v. 41, p. 330-340.

Moores, E.M., and Jackson, E.D., 1974, Ophiolites and oceanic crust: Nature, v. 250, p. 136-139.

Mysen, B.O., and Boettcher, A.L., 1975, Melting of a hydrous mantle: Jour. Petrol., v. 16, p. 549-593.

Noble, D.C., Bowman, H.R., Hebert, A.J., Silberman, M.L., Heropoulous, C.E., Fabbi, B.P., and Hedge, C.E., 1975, Chemical and istopic constraints on the origin of low-silica latite and andesite from the Andes of central Peru: Gec'ogy, v. 3, p. 501-504.

O'Hara, M.J., 1965, Primary magmas and the origin of basalts: Scottish Jour. Geol., v. 1, p. 19-40.

O'Hara, M.J., 1968, Are ocean floor basalts primary magma: Nature, v. 220, p. 683-686.

O'Nions, R.K., and Pankhurst, R.J., 1974, Petrogenetic significance of isotope and trace element variations in volcanic rocks from the mid-Atlantic: Jour. Petrol., v. 15, p. 603-634.

Peterson, J.J., Fox, P.J., and Schreiber, E., 1974, Newfoundland ophiolites and the geology of the oceanic layer: Nature, v. 247, p. 194-196.

Ringwood, A.E., 1972, Phase transformations and mantle dynamics: Earth Planet. Sci. Letters, v. 14, p. 233-241.

Ringwood, A.E., 1977, Petrogenesis in island arc systems: In Talwani, Manik, and Pitman, W.C., III, eds., Island arcs, deep-sea trenches, and back-arc basins: Amer. Geophys. Union Maurice Ewing Ser. 1, p. 311-324.

Schilling, J.G., 1973, Iceland mantle plume: geochemical study of Reykjanes Ridge: Nature, v. 242, p. 565-575.

Sleep, N.H., 1975, Formation of oceanic crust: some thermal constraints: Jour. Geophys. Research, v. 80, p. 4037-4042.

Smithson, S.B., and Brown, S.K., 1977, A model for lower continental crust: Earth and Planet. Sci. Letters, v. 35, p. 134-144.

Stille, Hans, 1955, Recent deformations of the earth's crust in the light of those of earlier epochs: in Poldervaart, Arie, ed., Crust of the Earth: Geol. Soc. America Spec. Paper 62, p. 171-192.

Thorpe, R.S., Potts, P.J., and Francis, P.W., 1976, Rare earth data and petrogenesis of andesite from the North Chilean Andes: Contr. Mineral. Petrol., v. 54, p. 65-78.

Toksoz, M.N., and Bird, Peter, 1977, Modelling temperatures in continental convergence zones: Tectonophysics, v. 41, p. 181-193.

Turcotte, D.L., and Oxburgh, E.R., 1978, Intra-plate volcanism: Phil. Trans. Royal Soc. London, v. 288A, p. 561-579.

Uyeda, Seiya, and Kanamori, Hiroo, 1979, Back-arc opening and the mode of subduction: Jour. Geophys. Research, v. 84, p. 1049-1062.

Whitford, D.J., and Nicholls, I.A., 1976, Potassium variation in lavas across the Sunda arc in Java and Bali: in Johnson, R.W., ed., Volcanism in Australasia: Elsevier, New York, v. 63-76.

Whitford, D.J., Compton, W., Nicholls, I.A., and Abbott, M.J., 1977, Geochemistry of late Cenozoic lavas from eastern Indonesia: role of subducted sediments in petrogenesis: Geology, v. 5, p. 571-575.

Wilcox, R.E., Harding, T.P., and Seely, D.R., 1973, Basic wrench tectonics: Amer. Assoc. Petroleum Geol. Bull., v. 57, p. 74-96.

Wilson, J.T., and Burke, Kevin, 1972, Two types of mountain building: Nature, v. 239, p. 448-449.

Wood, D.A., Tarney, J., Varet, J., Saunders, A.D., Bougault, H., Joron, J.L., Treuil, M., and Cann, J.R., 1979, Geochemistry of basalts drilled in the North Atlantic by IPOD Leg 49: implications for mantle heterogeneity: Earth and Planet. Sci. Letters, v. 42, p. 7-97.

Wyllie, P.J., Huang, W-L., Stern, C.R., and Malloe, Sven, 1976, Granitic magmas: possible and impossible sources, water contents, and crystallization sequences: Canadian Jour. Earth Sci., v. 13, p. 1007-1019.

The Continental Crust and Its Mineral Deposits, edited by D.W. Strangway,
Geological Association of Canada Special Paper 20

LATITUDINAL VARIATIONS IN ENCYSTMENT MODES
AND SPECIES DIVERSITY IN JURASSIC
DINOFLAGELLATES

Edward H. Davies
Atlantic Geoscience Centre
Geological Survey of Canada
Bedford Institute of Oceanography
Dartmouth, Nova Scotia B2Y 4A2

Geoffrey Norris
Department of Geology,
University of Toronto,
Toronto, Ontario M5S 1A1

ABSTRACT

Lower, Middle, and Upper Jurassic dinoflagellate cyst assemblages from Arctic Canada are described in terms of diversity and encystment modes and compared with those from offshore eastern Canada and from southern and other northern hemisphere localities. Maximum diversities in the Sverdrup Basin occur in the Callovian-Oxfordian and correspond with a marine transgressive peak. Deltaic progradation in the late Jurassic in the Sverdrup Basin ultimately reduced dinoflagellate diversity compared with other regions. In general, boreal Jurassic dinoflagellate assemblages are of relatively low diversity at the species level and are dominated by proximate cysts.

Prior to the Middle Jurassic, however, maximum species diversity appears to occur in the high latitude assemblages, whereas in the Middle and Upper Jurassic assemblages maximum diversities occur in the mid-latitudes where chorate cysts also start to become a more diverse component of the flora. A simple gradational relationship between paleolatitude and dinoflagellate encystment modes and diversity is not substantiated for the Jurassic; facies relationships and other factors are also important controls on assemblage characteristics.

RÉSUMÉ

On décrit des assemblages de kystes de dinoflagellés datant du Jurassique inférieur, moyen et supérieur dans l'Arctique canadien en termes de la diversité et des modes d'enkystement en les comparant avec ceux des localités au large de l'est du Canada et des hémisphères nord et sud. Le maximum de diversité dans le bassin de Sverdrup se situe au Callorien-Oxfordien et correspond à un pic dans la transgression marine. La progradation deltaïque à la fin du Jurassique dans le bassin de Sverdrup y a finalement réduit la diversité des dinoglagellés par rapport à d'autres régions. En général, les assemblages de dinoflagellés boréaux du Jurassique sont de diversité relativement faible au niveau de l'espèce et sont dominés par les kystes proximaux.

Toutefois, avant le Jurassique moyen, le maximum de diversité des espèces semble se produire dans les assemblages de latitude élevée, alors qu'au Jurassique moyen et supérieur, le maximum de diversité dans les assemblages se retrouve à des latitudes moyennes où les kystes choroïdes semblent aussi devenir une composante plus diverse de la flore. Une relation simple de gradation entre la paléolatitude et l'enkystement et la diversité des dinoflagellés ne semble pas exister pour le Jurassique; les relations de faciès et d'autres facteurs sont aussi des moyens de contrôle importants pour les caractéristiques des assemblages.

INTRODUCTION

Dinoflagellates are common components of the marine plankton, generally occupying the upper part of the photic zone in marine waters, although other habitats are also occupied. They are minute single-celled algae (Division Pyrrhophyta) often having two stages in the life cycle. The motile flagellated stage is generally not preserved because the wall consists of cellulose which decays after death. Under certain conditions (sexual reproduction or seasonal climatic changes), however, many species of dinoflagellates form cysts with resistant organic walls which fossilize relatively easily. The cysts range in size from approximately 50 μm to 150 μm. Leading references to and discussions of the features of living and fossil dinoflagellates may be found in Sarjeant (1974).

Dinoflagellates appear to have had a long evolutionary history with putative origins in the Precambrian (see Norris, 1978, for discussion and references). Indisputable fossil dinoflagellates occur only in the Triassic and later. The earliest dinoflagellates are marine, as are the majority of Mesozoic and Cenozoic dinoflagellates. Migration into freshwater environments probably occured in the early Tertiary, although abundant fossil dinoflagellates are known from Mesozoic sediments deposited in both hypersaline and hyposaline environments (see Norris and Hedlund, 1972, and Dörhöfer and Norris, 1977, for further discussion).

Temporal changes in dinoflagellate cyst diversity on a global scale have been plotted by Bujak and Williams (1979) suggesting minimum species diversities in the Lower Jurassic and maxima in the middle and upper Cretaceous with a subsequent decline through the Tertiary. Present low diversities of extant cysts are not representative of all dinoflagellates because some do not now produce resistant cysts.

Dinoflagellate encystment modes show large variations. Proximate cysts have little if any ornament and mimic the parent thecal body. The inner wall layer (endophragm) may form a distinct body in cavate cysts, whereas it is closely adpressed to the outer wall layer (periphragm) in acavate cysts. Apteate cysts have a cover of relatively short processes. Chorate cysts bear long processes which suggest that the protoplasm in the theca underwent strong contraction prior to cyst formation. These and other encystment modes are illustrated in Artzner et al. (1979).

An empirical relationship between encystment modes and paleolatitude was discerned in the uppermost Jurassic and the Cretaceous by Dörhöfer (1979) and Norris and Dörhöfer (1977) and has been investigated in detail in the Albian by Dörhöfer (1980). In waters of normal salinity the proximate mode is most common in high latitudes, while the chorate mode is common to dominant in lower latitudes; higher diversities are also generally recorded in lower latitudes. Critical testing of these relationships has not been possible, however, because of the lack of comprehensive data on boreal dinoflagellates and the sporadic nature of available data. If proportions of encystment modes prove to bear a general relationship to latitude, this may be of use in continental reconstructions. Diversity gradients from equator to pole are well documented in some faunal groups and have been discussed extensively and used successfully. For example, Waterhouse and Bonham-Carter (1975) used brachiopod data to identify Permian rotational poles and paleolatitudes. Diversities of dinoflagellate floras have not been investigated systematically, partly due to the lack of comprehensive information.

The recent work by Bujak and Williams (1977) and Davies (1979) has provided comprehensive information on Lower, Middle, and Upper Jurassic dinoflagellates from offshore eastern Canada and the Sverdrup Basin, northern Canada respectively. Information from these papers is summarized below in terms of both diversity and encystment modes and evaluated for paleolatitudinal significance. Other studies from more limited stratigraphic intervals are also considered. Important contributions to Jurassic dinoflagellate stratigraphy of the Sverdrup Basin and summaries of previous work have been made by Tan and Hills (1978).

THE JURASSIC BOREAL DINOFLAGELLATE CYST FLORAS

General Character

The Jurassic boreal dinoflagellate floras from the Sverdrup Basin (Davies, 1979) are typically low in species diversity and consist of predominantly acavate proximate cyst growth modes (Fig. 1). A large number of barren samples is apparently due to an abundance of thermally degraded samples from Axel Heiberg Island. Excluding these thermally degraded barren samples, 90 % of the samples had 13 dinoflagellate cyst species or less per sample, 95% had 15 species or less, and 99% had 18 species or less per sample. The mean was six species per sample and mode was five species per sample.

The maximum number of species per sample was 24, occurring in the Upper Savik Formation on Ellef Ringnes Island, dated as Late Callovian to Early Oxfordian. With respect to the whole basin, a maximum of 45 species existed at an early Oxfordian chronohorizon (using Shaw's (1964) graphic correlation technique), corresponding to the same stratigraphic horizon containing the maximum number of species per sample. Twelve Jurassic Oppel-zones were recognized by Davies (1979); a maximum number of 49 species per Oppel-zone (Fig. 3) occurred in the Late Callovian to Early Oxfordian interval.

The encystment mode of these high latitude assemblages is predominantly acavate-proximate and represents 59% of all dinoflagellate cyst species from the Early Jurassic to Valanginian of the Sverdrup Basin (Fig. 2), although the actual proportion of encystment modes varies temporally and with facies as discussed later.

Thus, the overall character of the Jurassic dinoflagellate cyst assemblages of the Sverdrup Basin is one of low diversity (although preservation and recovery were generally excellent), with acavate proximate cysts being dominant.

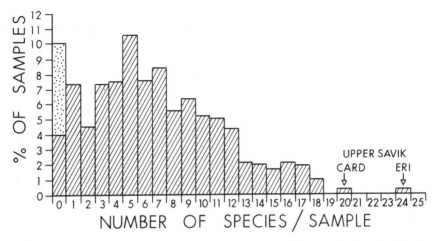

Figure 1. Species diversities within Lower Jurassic to Valanginian samples from the Sverdrup Basin (after Davies, 1979). The maximum number of species per sample (24) is found in the upper Savik Formation at Reindeer Peninsula (ERI). The second highest maximum number of species per sample (20) also occurs in the upper Savik Formation at Central Amund Ringnes Dome (CARD). The stippled pattern represents thermally degraded samples in which palynomorphs have apparently been totally destroyed. Frequencies of diversities are expressed as percentages of the 672 samples examined whose stratigraphic positions are detailed in Davies (1979).

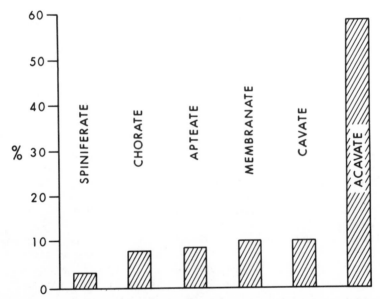

Figure 2. Relative percentages of species in encystment modes. Calculations are made on a total of 128 dinoflagellate cyst species found within the Jurassic to Valanginian of the Sverdrup Basin (after Davies, 1979).

Temporal Changes

The Early to Middle Jurassic is characterized by pulsating increases in species diversity across the basin (Fig. 3), originating with an impoverished flora of a minute *Lithodinia* sp. and *Nannoceratopsis* spp., expanding and diversifying into the phallocystacean flora of the Toarcian and Bathonian with typically *Susadinium scrofoides* Dörhöfer and Davies, *Phallocysta eumekes* Dörhöfer and Davies, *Comparodinium aquilonium* Dörhöfer and Davies, *Dapcodinium* spp., and abundant *Nannoceratopsis* spp. This succession is linked to the Savik shale transgression over the Sverdrup Basin. The phallocystacean flora has almost exclusively acavate proximate cysts (81% to 96% of the total species, or 10 to 21 species). The remainder is composed of membranate species of *Comparodinium* and *Chlamydophorella*. (4% to 19%, or 1 to 2 species)

The first pareodiniacean cysts appear as *Glomodinium* with the onset of the peak in diversity; thereafter, pareodiniacean cysts continue as an important and often dominant component throughout the Middle and Late Jurassic.

In the middle Middle Jurassic the phallocystacean flora declines and is replaced by lithodiniacean assemblages. This decline is paralleled by a decline in species diversity reaching a minimum of 12 species in the Bathonian and by a minor regression

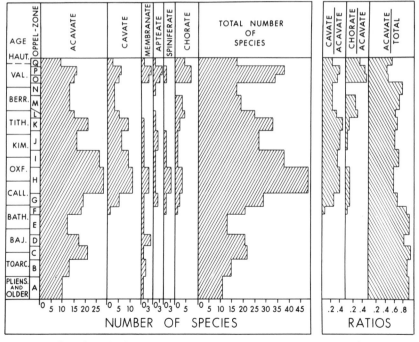

Figure 3. Number of species in various encystment modes are shown together with total species diversity in each Oppel Zone recognized in the Lower Jurassic-Valanginian interval, Sverdrup Basin. Ratios of selected encystment modes in each zone are also plotted. (Data derived from Davies, 1979.)

within the Sverdrup Basin. The lithodiniacean succession begins at the onset of another pulse of the Middle Jurassic transgression. As the transgression resumes in the Late Bathonian, species and growth mode diversity increase. Four new growth modes are established: cavate, apteate, spiniferate, and chorate (Fig. 3). Cavate species displace the membranate forms as secondary in importance.

The families Lithodiniaceae, Gonyaulacystaceae, Pareodiniaceae, Scriniocas-saceae, Cleistosphaeridiaceae, Spiniferitaceae, and Endoscriniaceae become well established by the time of the peak in transgression, and maxima in basinal and sample species diversity at the Middle/Late Jurassic boundary.

The extensive marine transgression was followed by a minor regression, a decline in cyst diversity (27 species), and an increase in the proportions of acavate species (Fig. 3). The major regression in the Kimmeridgian is marked by a subtle increase in the percentage of apteate cysts, reflecting an influx of *Batiacasphaera* species.

Extensive deltaic progradation is initiated in the Kimmeridgian across the Sver-drup Basin, and appears to have spurred species diversity perhaps through environ-mental stress and the expanded development of niches. As the coarser deltaic clastics became more widespread through the Tithonian to Berriasian, a flora develops that is characterized by a decrease in species and cyst growth mode diversity, and an increase in acavate and more markedly chorate cyst species such as *Prolixosphaeridium* sp. and *Tanyosphaeridium* sp. (see Fig. 3).

The earliest Cretaceous (Berriasian to Valanginian) is marked by a rapid, exten-sive transgression, and a decline in the proportions of acavate species (although they remain the dominant form). The continous increase in the proportions of chorate cyst species might reflect warming trends in the Lower Cretaceous. A rapid regression in the Late Valanginian is paralleled by a decrease in species diversity.

Latitudinal Comparisons

Bujak and Williams' (1977) data on the Jurassic dinoflagellate biostratigraphy of offshore eastern Canada is based on comprehensive work for a large stratigraphic interval that may be compared to the dinoflagellate succession in the Sverdrup Basin, although different depositional conditions at certain times must be taken into account. A summary of the encystment modes is presented in Figure 4.

In the Lower through Middle Jurassic there is a steady increase in the number of dinoflagellate cyst species. As in the Sverdrup Basin, the cyst flora is dominated by acavate cysts; however, the diversity is lower and the Toarcian-Bajocian diversity peak is not present. Wille and Gocht (1979) have recorded 18 species from the Pliensbachian and Toarcian of Germany. The seven species of *Comparodinium* described by them have growth modes ranging from acavate through apteate to chorate. The assemblages are dominated by acavate cyst species. Filatoff (1975) recorded a similar diversity peak (8 species) of acavate phallocystaceans in the lowest Middle Jurassic of Western Australia.

With the advent of the Bathonian, other cyst modes are introduced into the North proto-Atlantic. Acavate cysts continue to dominate throughout the Middle and Late Jurassic, but they steadily decline from a maximum in the Callovian.

The similarity in dinoflagellate cyst characters between the Sverdrup Basin (60° to 65°N paleolatitude) and offshore eastern Canada (25° to 30°N paleolatitude) begins to decrease in the late Middle Jurassic and continues through the Late Jurassic. In the high latitudes acavate cyst species continue to dominate) 57% to 68% of the total number of cysts are acavate in the Callovian to Tithonian), whereas in the lower latitudes they decline to less than 50% in the Late Jurassic (51% to 55% in the Callovian; 34% to 46% in the Late Jurassic).

The peak in cavate species diversity occurs in the Kimmeridgian in lower latitudes. The peak in the high latitudes occurs in the Late Callovian to Early Oxfordian.

Cavate species in the lower latitudes have a higher diversity than cavate species in high latitudes in the late Middle to Late Jurassic. It appears that the chorate diversity curve of the Sverdrup Basin is bimodal, flanking the single peak of diversity of chorate cysts in eastern Canada.

In offshore eastern Canada acavate diversity begins to decrease at the end of the Callovian (Fig. 4) but it increases until the Oxfordian in the Sverdrup Basin (Fig. 3).

Figure 5 illustrates the differences between the number of Jurassic species recorded by Bujak and Williams (1977) and Davies (1979) in low and high latitudes respectively. It is clear that diversity in dinoflagellate cysts is not inversely proportional to paleolatitude. Other ecological factors unrelated to latitude must come into play. Throughout the late Middle Triassic to Early Jurassic (Fig. 6) the Arctic maintained the highest dinoflagellate species diversities recorded in this interval (1 to 23 species of acavate chorate and membranate cysts). Most other records of Triassic dinoflagellate cysts in lower latitudes record only *Rhaetogonyaulax rhaetica* Sarjeant (Lund, 1977; Sarjeant, 1963; Harland *et al.*, 1975; Fisher, 1972). This may be a monographic artifact or it may be the result of true biogeographical differences. Although most Triassic and Early Jurassic studies deal with terrestrial palynomorphs,

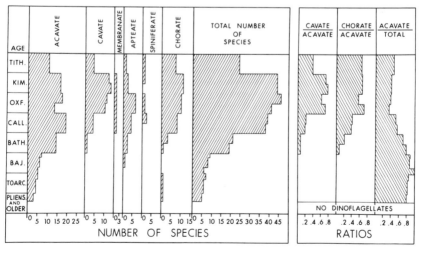

Figure 4. Number of species in various encystment modes are shown together with total species diversity in each stage from offshore eastern Canada (data derived from Bujak and Williams, 1977). Ratios of selected encystment modes in each zone are also plotted.

some have also recorded acanthomorphic acritarchs (Wall, 1965; Jansonius, 1962; Schurrman, 1977; Filatoff, 1975). These might be related to the earliest dinoflagellates such as *Suessia, Rhaetogonyaulax, Noricysta, Dapcodinium,* and *Sverdrupiella,* but this is speculative. Morbey (1975) recorded 6 dinoflagellate species (33% acavate and 67% chorate) from the Rhaetian of Austria. A similar chorate dinoflagellate assemblage has been recorded by Dörhöfer and Davies (1980) from the Early Jurassic of the Sverdrup Basin. Diversification of acavate and cavate cysts may well have started in the Arctic and spread southward in the Toarcian-Bathonian as the seaway through the North Atlantic developed. This diversification may have begun with the northward movement of North America through the Triassic, which culminated in the Middle Jurassic. The added stress of an increasingly continental climate placed on the forms within the Arctic Basin (Dörhöfer and Davies, 1980) may have stimulated the production of proximate cysts from acritarchous ancestors.

The Bathonian appears to have been a time of change from high boreal diversities

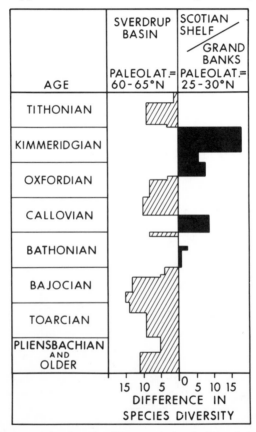

Figure 5. Changes in the species diversity difference during the Jurassic between the Sverdrup Basin (60° to 65°N paleolatitude) and offshore eastern Canada (25° to 30°N paleolatitude). The difference in total number of species per stage is plotted under the region with greatest diversity.

to high diversities in low latitudes. Diversity differences shift between the two regions throughout the Jurassic, with the maximum of the differences for the boreal becoming less and the maximum of the differences of the lower latitudes becoming greater (See Fig. 5).

Table I indicates the number of species in acavate, cavate, and chorate encystment modes, as well as the total number of cysts (including the less important apteate, spiniferate and membranate cysts) as a function of time and geographical extent.

TABLE I

Latitudinal variation in encystment mode diversities at the species level during the Callovian to Kimmeridgian transgression. Number of species from each region are entered for each stage as acavate, cavate, chorate, and total numbers. Discrepancies between numbers listed as "total" and the sum of acavate, cavate, and chorate cysts are due to the presence of other encystment modes (spiniferate, apteate, etc.).

REGION	PALEO-LATITUDE	EARLY CALLOVIAN	EARLY OXFORDIAN LATE CALLOVIAN	EARLY KIMMER-IDGIAN LATE OXFORDIAN	EARLY-MIDDLE KIMMER-IDGIAN	CYST GROWTH MODES
SVERDRUP BASIN CANADA (DAVIES, 1979)	65°± 60° N	19	28	26	16	ACAVATE
		5	11	9	6	CAVATE
		2	3	1	1	CHORATE
		26	42	36	23	TOTAL
SCOTLAND *(LAM & PORTER,1977) **(GITMEZ & SARJEANT, 1972)	40° N	14*	6*	19*	11**	ACAVATE
		2	2	10	14	CAVATE
		6	5	11	10	CHORATE
		22	13	40	35	TOTAL
ENGLAND *(SARJEANT,1962) **(IOANNIDES et al.,1977) ***(GITMEZ & SARJEANT,1972)	35° N	5*	6*	18**	44***	ACAVATE
		5	3	7	24	CAVATE
		6	1	11	18	CHORATE
		16	10	36	86	TOTAL
FRANCE *(SARJEANT,1962) **(GITMEZ & SARJEANT, 1972)	35° N		19		29**	ACAVATE
			7		23	CAVATE
			11		11	CHORATE
			37		63	TOTAL
OFFSHORE EASTERN CANADA (BUJAK & WILLIAMS, 1977)	30°± 25° N	21	14-20	17-18	17	ACAVATE
		4	4-10	11-13	12	CAVATE
		8	8-10	10-11	11	CHORATE
		33	39-41	44-46	60	TOTAL
ROMANIA (BEJU,1971)	20° N	14	14			ACAVATE
		7	8			CAVATE
		1	4			CHORATE
		22	26			TOTAL
AFGHANISTAN (ASHRAF,1979)	15°± 10° N	4	8	7	12	ACAVATE
		0	3	5	3	CAVATE
		2	5	4	6	CHORATE
		6	16	16	21	TOTAL
OFFSHORE SOUTHERN ARGENTINA (HARRIS,1978) (HEDLUND & BEJU, 1976)	60° S	2	7			ACAVATE
		0	3			CAVATE
		0	2			CHORATE
		2	12			TOTAL

Monographic differences are evident, for example, in the Kimmeridgian data from England (Ioannides et al., 1977; Gitmez and Sarjeant, 1972), which indicate large differences in the number of species found in adjacent zones. These differences are difficult to reconcile, and underline the problems of synthesizing data on this scale.

The lowest latitudes appear to have fewer species than the mid-latitudes through the Callovian to Kimmeridgian. In the early Callovian the mid-latitudes have more species than the high latitudes. This is reversed in the Late Callovian-Early Oxfordian and reversed back again in the Kimmeridgian. The greatest differences appear to be during the Kimmeridgian in Europe. This location would coincide with the up-welling of Tethyan waters on the continental shelf where mixing with the high nutrient, less saline, cooler polar water might have occurred.

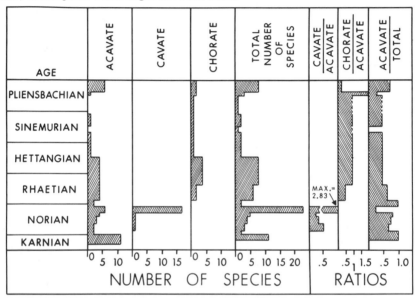

Figure 6. Triassic and Early Jurassic dinoflagellate cyst diversities by encystment modes in northern Canada and Alaska (after Fisher and Bujak, 1975; Bujak and Fisher, 1976; Wiggins, 1976). Ratios of selected encystment modes in each stage are also plotted.

CONCLUSIONS

1) The Jurassic boreal dinoflagellate cyst floras are characterized by low assemblages diversity with a predominantly acavate proximate character.

2) The pulsating diversity throughout the Jurassic parallels the pulsating transgression within the Sverdrup Basin. The Callovian-Oxfordian increase in dinoflagellate diversity appears to be correlated with the worldwide major Callovian-Oxfordian transgression.

3) Within the limitations of published data on Jurassic dinoflagellate cysts there appears to be:

a) a domination of proximate cysts species diversity over other encystment mode diversities throughout the Jurassic;

b) a pulsating shift in species diversity distribution from a maximum in the Boreal region from the Late Triassic to middle Middle Jurassic, to a mid-latitudinal maximum, preceding a world-wide diversification associated with a transgression in the Callovian to Kimmeridgian;

c) an increase in the chorate/acavate and the cavate/acavate cyst ratios towards the mid-latitudes during the Callovian to Kimmeridgian, as well as an overall increase of proximate species numbers through the Jurassic. Chorate species become common beginning in the Callovian, and attain higher diversities in the Late Jurassic in mid-latitudes than in high latitudes;

d) a probable radiation of dinoflagellate cysts from the proto-Arctic ocean during the Triassic and Early Jurassic, and an invasion of the lower latitudes in the Middle Jurassic.

4) It is clear, therefore, that a uniform gradational relationship between latitudinal position and the character of dinoflagellate assemblages cannot be assumed for purposes of Jurassic paleogeographic placement. For example, on the basis of available data, mid-latitudes, show the highest dinoflagellate diversities in the Middle and Late Jurassic, compared with those in sub-equatorial and polar regions. Comprehensive data on Jurassic dinoflagellates from critical areas are, however, still lacking. Chorate and spiniferate cysts become more important components of the dinoflagellate flora in the Cretaceous, and their relationship to latitude may be different compared with those in the Jurassic.

5) Facies relationships must be considered in assessing differences in diversities and encystment modes between regions.

ACKNOWLEDGEMENTS

This work was supported by a National Research Council of Canada operating grant (G. Norris, principal investigator), an NRC negotiated development grant to the Department of Geology, University of Toronto, and a research agreement with the Department of Energy, Mines and Resources (G. Norris, principal investigator). We are grateful to the Institute of Sedimentary and Petroleum Geology, Calgary for logistical support in the field and for providing some samples from northern Canada, and to Phillips Petroleum Company for the technical services in the production of the manuscript.

REFERENCES

Ashraf, A.R., 1979, Die räto-jurassischen Floren des Iran und Afghanistans. 6. Jurassische and unterkretazische Dinoflagellaten und Acritarchen aus Nord-Afghanistan: Palaeontographica, Abt. B, 169, p. 122-158.

Artzner, D., Davies, E.H., Dörhöfer, G., Fasola, A., Norris, G., and Poplawski, S., 1979, A systematic illustrated guide to fossil organic-walled dinoflagellate genera (Dinophyceae, Triassic-Quaternary): Royal Ontario Museum Life Sci. Ser.

Beju, D., 1971, Jurassic microplankton from the Carpathian Foreland of Romania: Annls. Instit. Geol. Publ. Hung. 54, 275-302, pl. 1-8.

Bujak, J.P., and Fisher, M.J., 1976, Dinoflagellate cysts from the Upper Triassic of Arctic Canada: Micropaleontology, v. 22, p. 44-70.

Bujak, J. P. and Williams, G.L., 1977, Jurassic palynostratigraphy of offshore eastern Canada: in Swain, F.M., ed., Stratigraphic Micropaleontology of Atlantic Basin and Borderlands: Elsevier Scientific Publ. Co., p. 321-339.

——————————, 1979, Dinoflagellate diversity through time: Marine Micropaleo., v. 4, p. 1-12.

Davies, E.H., 1979, Jurassic and Lower Cretaceous dinoflagellate cysts of the Sverdrup Basin, Arctic Canada: taxomony, biostratigraphy, chronostratigraphy: Ph.D Thesis, University of Toronto, Toronto, Ontario, Canada, 591 p.

Dörhöfer, G., 1976, Dinoflagellate cysts from the German Albian: 9th Ann. Meeting, Amer. Assoc. Strat. Palyn. Abstracts, p. 218.

——————————, 1980, Principles of Albian dinoflagellates cyst provincialism: Marine Micropaleo., in press.

Dörhöfer, G. and Davies, E., 1980, Evolution of archeopyle and tabulation in rhaetogonyaulacinean dinoflagellate cysts: Royal Ontario Museum Life Sci. Ser.

Dörhöfer, G., and Norris, G., 1977, Discrimination and correlation of highest Jurassic and lowest Cretaceous terrestrial palynofloras in north-west Europe: Palynology, v. 1, p. 79-94.

Filatoff, J., 1975, Jurassic palynology of the Perth Basin, Western Australia: Palaeontographica, Abt. B 154, p. 1-113.

Fisher, M.J., 1972, The Triassic palynofloral succession in England: Geoscience and Man, v. 4, p. 101-109.

Fisher, M.J. and Bujak, J., 1975, Upper Triassic palynofloras from Arctic Canada: Geoscience and Man, v. 11, p. 87-94.

Gitmez, G.U. and Sarjeant, W.A.S., 1972, Dinoflagellate cysts and acritarchs from the Kimmeridgian (Upper Jurassic) of England, Scotland, and France: Bull. British Museum Natural History (Geol.), v. 21, p. 171-257.

Harland, R., Morbey, S.J., and Sarjeant W.A.S., 1975, A revision of the Triassic to lowest Jurassic dinoflagellate *Rhaetogonyalax*: Palaeontology, v. 18, p. 847-864.

Harris, W.K. 1978, Palynology of cores from deep sea drilling sites 327, 328, and 330, South Atlantic Ocean: Init. Rept. Deep Sea Drill. Proj. 36, p. 761-815.

Hedlund, R.W. and Beju, D., 1978, Stratigraphic palynology of selected Mesozoic samples DS DP Hole 327A and Site 330: Init. Rept. Deep Sea Drilling Proj. 36, p. 817-831.

Ioannides, N.S., Stavrinos, G.N. and Downie, C., 1977, Kimmeridgian microplankton from Clavel's Hard, Dorset, England: Micropaleontology, v. 22, no. 4, p. 443-478.

Jansonius, J., 1962, Palynology of Permian and Triassic sediments, Peace River area, Western Canada: Palaeontographica Abt. B, 110, p. 35-98.

Lam, K. and Porter, R., 1977, The distribution of palynomorphs in the Jurassic rocks of the Brora Outlier, N.E. Scotland: Jour. Geol. Soc. London, v. 134, p. 45-55.

Lund, J.J., 1977, Rhaetic to lower Liassic palynology of the onshore southeastern North Sea Basin: Geol. Survey of Denmark II Series No. 109, p. 1-128.

Morbey, J.J., 1975, The palynostratigraphy of the Rhaetian Stage, Upper Triassic in the Kendelbachgraben, Austria: Palaeontographica, Abt. B 152, p. 1-75.

Norris, G., 1978, Phylogeny and a revised supra-generic classification for Triassic-Quaternary organic walled dinoflagellate cysts (Pyrrhophyta). Part I. Cyst terminology and assessment of prevoius classifications: N. Jb. Geol. Pal., Abh., v. 155, p. 300-317.

Norris, G., and Dörhöfer, G., 1977, Upper Mesozoic dinoflagellate biogeography: North American Paleont. Convention II, Abstracts.

Norris, G. and Hedlund, R.W., 1972, Transapical sutures in dinoflagellate cysts: Geoscience and Man, v. 2, p. 49-56.

Sarjeant, W.A.S., 1962, Microplankton from the Ampthill Clay of Melton, south Yorkshire: Palaeontology, v. 5, p. 478-497.

——————————, 1963, Fossil dinoflagellates from Upper Triassic sediments: Nature, V. 199, p. 353-354.

——————————, 1968, Microplankton from the upper Callovian and lower Oxfordian of Normandy: Rev. Micropaleontol., v. 10, p. 221-242.

————————, 1974, Fossil and living dinoflagellates: Academic Press, London.

Schurrman, W.M.L., 1977, Aspects of Late Triassic Palynology 2. Palynology of the "Gries et Schiste a *avicula contorta*" and "argiles de Levallois" (Rhaetian) of Northeastern France and Southern Luxemberg: Rev. Paleobot. Palyn., v. 23, p. 159-253.

Shaw, A.B., 1964, Time in Stratigraphy: McGraw-Hill Book Company, New York, p. 1-365.

Tan, J.T., and Hills, L. V., 1978. Oxfordian-Kimmeridgian dinoflagellate assemblage, Ringnes Formation, Arctic Canada: Geol. Survey Canada, Paper 78-1C, p. 63-73.

Wall, D., 1965, Microplankton, pollen and spores from the Lower Jurassic in Britain: Micropaleontology, v. 11, p. 151-190.

Waterhouse, J.B. and Bonham-Carter, G., 1975, Global distribution and character of Permian biomes based on brachiopod assemblages: Canadian Jour. Earth Sci., v. 12, p. 1085-1146.

Wiggins, V.D., 1976, Upper Triassic-Lower Jurassic dinoflagellates: Amer. Assoc. Strat. Palyn. 9th Annual Meeting, Halifax, 1976, Abstracts: p. 29.

Wille, W. and Gocht, H. 1979, Dinoflagellaten aus dem Lias Sudwestdeutschlands: N.Jb. Geol. Palaont. Abh., v. 158, no. 2, p. 221-258.

The Continental Crust and Its Mineral Deposits, edited by D.W. Strangway, Geological Association of Canada Special Paper 20

A REASSESSMENT OF THE FIT OF PANGAEA COMPONENTS AND THE TIME OF THEIR INITIAL BREAKUP

A. Hallam
Department of Geological Sciences, University of Birmingham, P.O. Box 363, Birmingham, B15 2TT, England

ABSTRACT

Recent geophysical and geological research on the continental margins has demonstrated that the 1 km contour used in various computer reconstructions does not correspond closely with the true geological edge of the continents, which in many places must occur below the ocean at least 2 to 4 km deep. This and other considerations lead to a slight revision of the classic Bullard fit for the Atlantic continents. The principal key to reconstruction of Gondwanaland is the position of Madagascar. Recent palaeomagnetic and oceanographic work strongly supports the northerly fit originally proposed by du Toit. Recognition that the Falkland Plateau is subsided continental crust provides an additional constraint on reconstruction.

With regard to the dating of breakup of the Pangaea components, a distinction is drawn between initial taphrogenic activity (rifting and volcanism) and the creation of new ocean floor by spreading. The earliest phase of true ocean opening in this latter sense took place in the southern North Atlantic with the movement of Africa away from North America; this happened at the earliest in the mid-Jurassic and possibly as late as early late Jurassic. East and West Gondwanaland began to split up no earlier than early Cretaceous (Valanginian). By late Cretaceous (Santonian) times all the major Gondwana components were separated by ocean except Australia-East Anarctica and South America-West Antarctica, between which initial spreading commenced in the Eocene and Oligocene respectively. Total separation of North America and Eurasia was accomplished in the late Eocene with the opening of the Greenland Sea.

RÉSUMÉ

La recherche récente en géophysique et en géologie sur les marges continentales a démontré que le contour de 1 km utilisé dans différentes reconstructions par ordinateur ne correspond pas très bien au véritable rebord géologique des continents, lequel, en plusieurs endroits, se retrouve sous l'océan à des profondeurs d'au moins 2 à 4 km. Cette considération et d'autres conduisent à une légère révision dans l'ajustement classique par Bullard des continents atlantiques. La clé principale pour cette reconstruction du Gondwana est la position de Madagascar. Les travaux récents en paléomagnétisme et en océanographie supportent fortement l'ajustement au nord proposé originellement par du Toit. La reconnaissance du fait que le plateau de Falkland est une croûte continentale qui s'est affaissée fournit des contraintes additionnelles à la reconstruction.

Pour ce qui est de la datation de la rupture des composantes de la Pangée, on fait une distinction entre l'activité taphrogénique initiale (effondrement et volcanisme) et la création de nouveaux fonds océaniques par expansion. La phase la plus ancienne d'ouverture réelle d'un océan dans ce sens s'est produite dans le sud de l'Atlantique Nord quand l'Afrique s'est éloignée de l'Amérique du Nord; ceci s'est produit au plus tôt au Jurassique moyen et peut-être aussi tard qu'au début du Jurassique supérieur. L'est et l'ouest du Gondwana commencèrent à se séparer avant le début du Crétacé (Valanginien). A la fin du Crétacé (Santonien), toutes les composantes principales du Gondwana étaient séparées par l'océan à l'exception des ponts de l'Australie à l'Antarctique de l'Est et de l'Amérique du Sud à l'Antarctique de l'Ouest; l'expansion initiale a débuté à l'Eocène et à l'Oligocène respectivement pour ces régions. La séparation totale entre l'Amérique du Nord et l'Eurasie a été complétée à la fin de l'Eocène avec l'ouverture de la mer du Groënland.

INTRODUCTION

The most direct method of fitting together the components of Wegener's super-continent Pangaea is the purely geometric one, at a geologically reasonable bathymetric contour. The widespread acceptance of the 1 km isobath least squares computer fits of Bullard *et al.* (1965) and Smith and Hallam (1970), respectively for the continents bordering the Atlantic and Indian Oceans, testifies to the general success of this method. Nevertheless the matter is not quite as straightforward as once seemed possible, and many points of detail have remained in dispute.

Firstly, there have been a number of misfits, the most serious of which are the overlap of South and Central America, where pre-Mesozoic continental rocks occur, and the geologically implausible projection of the West Antarctic peninsula east of Patagonia, overlapping the Falkland Plateau, which is now known to be subsided continental crust. Secondly, purely geometric fits are not always unambiguous and need to be constrained by matching geological features between the opposed continental areas. Orogenic belts extending at a high angle to the continental margins, are the best controls, yielding fairly precise and geographically unambiguous results. This method cannot be applied, however, in the case of Madagascar, which is a key piece of the Gondwanaland jigsaw puzzle. Geological arguments have been used to support both a northerly position against East Africa and a southerly position against Mozambique and South Africa; the controversy has only recently been resolved by latitudinal matching based on paleomagnetic data.

A further difficulty is that the 1 km isobath does not correspond closely to the true edge of continental crust; the computer fits cited are too tight. Recent oceanographic work suggests that this geologically critical boundary generally lies between 2 and 4 km, and locally may be even lower, with extensive sectors of thinned and subsided continental crust lying beneath deep water (Talwani and Eldholm, 1972; Jansa and Wade, 1975; Pautot *et al.*, 1976; Barker and Griffiths, 1977; Haworth, 1977; Powell *et al.*, in press).

There are two other approaches to checking the accuracy of geometric fits. A general control is provided by palaeomagnetic pole determinations for different continents for times when they are believed to have been united. More detailed control can be achieved by oceanographic research, by matching fracture zones and by tracing the record of magnetic anomalies created by sea-floor spreading. This latter method can give accurate results back to anomaly 34 (around 80 Ma ago). Further back in the Cretaceous the anomaly record is much poorer because of a so-called palaeomagnetic quiet interval, but a good record reappears in the early Cretaceous and later Jurassic. Back beyond 155 Ma, however, there was another quiet interval, and older plate movements must be inferred by extrapolation on the assumption of a given spreading rate.

With regard to the timing of continental breakup, an important distinction must be drawn between initial intracontinental taphrogenic activity and the creation of true ocean floor by spreading; in other words, between rifting and drifting. Many millions of years may have elapsed between the two processes. Intracontinental tension across lines of subsequent ocean opening is characterized by faulting which produces horst and graben structures, often associated with alkaline or subalkaline volcanism, and by more general subsidence due to the thinning of continental crust, which may lead to the incursion of shallow sea.

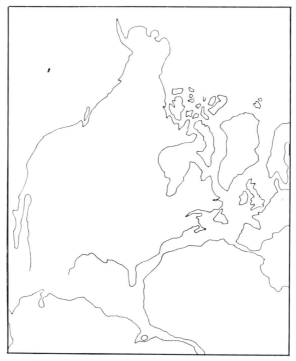

Figure 1. The Bullard fit of the Central and North Atlantic continents. Simplified from Bullard *et al.*, 1965.

Failure to distinguish between rifting and drifting has in the past led to much confusion about the timing of continental separation. Evidence of normal faulting and basaltic volcanism led Dietz and Holden (1970) to infer that the Atlantic and Indian Oceans began to form as early as the Triassic, and many workers, as recently as Dingle (1973), have considered the shallow marine deposits in East Africa, Madagascar and Western Australia as evidence that the Indian Ocean was already in existence in the Jurassic. Furthermore, the basal Jurassic salt deposits on the North Atlantic margins have been cited, by analogy with the Red Sea, as indicating that oceanic crust was present at that time (Kinsman, 1975). However, as Evans (1978) has pointed out, such salts are known to overlie continental deposits, though it is conceivable that they could extend further offshore and eventually come to lie on oceanic crust. Nowhere in the Atlantic has this been proven by drilling, and the Red Sea analogy is questionable, as will be discussed later. It is quite possible that some of these salts formed in desiccated basins below sea level and were later flooded by the sea. Even the replacement of salts by marine deposits in the Atlantic probably relates to the early Jurassic and mid-Cretaceous eustatic rise of sea level rather than to sea-floor spreading within the rift zone.

Independent evidence for dating the formation of oceanic straits comes from deep ocean boreholes and from fossils in the separated continents, because oceanic separation leads to genetic isolation of terrestrial and shelf organisms and hence to morphological divergence. These biogeographic aspects are most recently reviewed by Hallam (in press).

In many if not most cases it can be argued that the lines of ocean opening were controlled by (1) narrow zones of crustal weakness marked by late Precambrian and

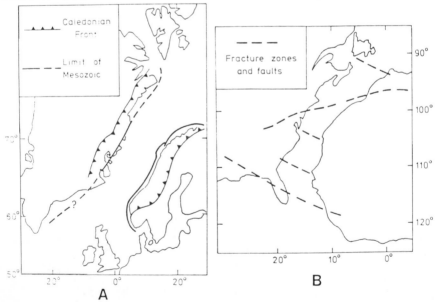

Figure 2. A. The fit between Greenland and Scandinavia, simplified from Talwani and Eldholm, 1977.
B. The fit between Africa and North America, simplified from Le Pichon et al., 1977.

Phanerozoic orogenic belts and/or by (2) sites of long-continued subsidence indicated by thick sequences of shallow marine and continental sediments ranging in age into the late Palaeozoic. This is perhaps most convincingly demonstrated in the northern part of the North Atlantic embracing the margins of Greenland and Norway, where a substantially continuous sedimentary sequence, which ranges back to the Permian, occurs within the Caledonian orogenic belt (Fig. 2A). That sector of the Atlantic between Africa and North America is bordered by Hercynian fold belts which roughly parallel the coastline. Le Pichon *et al.* (1977) have stressed the importance of late Palaeozoic tectonic structures in determining the pattern of Mesozoic rifting.

In the case of Gondwanaland, Cox (1978) maintains that both the South Atlantic and Indian Ocean rifts show a broad dependence on basement structures. Thus the zone of extension marked by substantial Karroo volcanics is largely taken up along the Zambesi and Limpopo zones where younger Precambrian orogenic zones intersect the Archaean craton. Both the East African and West Australian margins follow zones of more or less continuous sedimentation dating back to the late Palaeozoic.

In the ensuing account the North Atlantic and Gondwana continents will be dealt with separately. There will be no attempt to review all the reconstructions that have been proposed, but the cited references provide an adequate source for older literature.

THE NORTH ATLANTIC CONTINENTS

The only fit for the southern part of the North Atlantic that is an improvement on the classic Bullard fit (Fig. 1) is that of Le Pichon *et al.* (1977), which takes into account the new information on subsided continental crust and uses the 3 km isobath between Africa and North America and the 2 km isobath for the younger rifts further north. Consequently there is more separation between the present coastlines (Fig. 2B). One improvement is the absence of overlap between Africa and the Iberian peninsula, which has plagued previous reconstructions. On the other hand, Sclater, *et al.* (1977) object that the reconstruction fails to close Portugal to the Grand Banks. This may be due to the existence in the intervening sector of subsided continental crust below 2 km water depth. In this connection it is noteworthy that Pautot *et. al.* (1976) discovered Hercynian granite in the Western Approaches of the English Channel at depths of 4 km and inferred its probable presence at even greater depths.

Le Pichon *et al.* (1977) follow other geophysicists in treating the so-called Quiet Magnetic Zones between Africa and North America as oceanic crust generated during a magnetic quiet interval prior to 155 Ma, though Hayes and Rabinowitz (1975) acknowledge that the more landward part of the zone could in fact be subsided continental crust. In consequence the overlap of South and Central America, which is so glaring in the Bullard reconstruction, is only slightly alleviated. It effectively disappears if most or all of the Quiet Magnetic Zone is treated as continental crust, as I suggested tentatively on geological grounds a few years ago (Hallam, 1971). This subject will be returned to when the opening history of the Atlantic is discussed.

For the northern part of the North Atlantic, the recent reconstruction by Talwani and Eldholm (1977), based on detailed geophysical analysis of the Norwegian and Greenland Seas, differs slightly from the Bullard reconstruction in that the European sector from Svalbard to Scandinavia is shifted somewhat southwards in relation to

Greenland (Fig. 2A). The Vøring Plateau off Norway is continental crust that has subsided to about 2 km depth (Talwani and Eldholm, 1972). It has become apparent that the trailing continental margins throughout the North Atlantic region have been considerably stretched and thinned, and Le Pichon et al. (1977) speculate on the possibility of hot creep in the lower continental crust.

The opening history of the North Atlantic is evidently a complicated one. Following the pioneer work of Pitman and Talwani (1972), several distinct phases have been generally accepted. The earliest sector to open was that between Africa and North America, but disagreement persists about the exact time when this began. Because the magnetic anomaly record stops at 155 Ma, the older history must be inferred by extrapolation back through time, or from geological and palaeontological evidence, and none of these data have yet been accepted as decisive.

Pitman and Talwani (1972) extrapolated back to the continental edge as defined by the Bullard fit, on the assumption of a rate of spreading unchanged from the 1 cm/year rate established from the anomaly record, and thereby arrived at a figure of 180 Ma. Sclater et al. (1977), using substantially the same data and also assuming a constant rate of spreading, inferred an opening data of 165 Ma. Bearing in mind what was stated earlier about the 1 km isobath being too shallow to mark the transition from continental to oceanic crust, these figures should be regarded as approximating an upper age limit. Luyendyk and Bunce (1973), on the other hand, inferred a much older data in the Triassic by using spreading data in conjunction with evidence of taphrogeny from the continental margin, and assumed an early phase of slow spreading at about 0.6 cm/year. Such an inference is, of course, lacking in any support from ocean-floor data.

Indeed, the oldest deep-ocean sediments overlying apparent oceanic basaltic crust are no older than Oxfordian and occur in the North American Basin east of Cape Hatteras (Hollister et al., 1974) and the Cape Verde Basin (Lancelot et al., 1977). There is thus a time gap of at least 50 million years between the earliest evidence of tensional activity and some geophysical extrapolations, and firm proof of the divergence of plates as signified by the generation of new oceanic crust.

It is therefore desirable to bring other evidence to bear on the problem. Jansa and Wade (1975) have argued on grounds of sedimentary facies distribution, with regressive clastic deposits spreading east and west from a region between the present coastlines, that final separation of the African from the North American plate and initiation of sea-floor spreading took place in early Bajocian times (about 170 Ma, a figure close to that of Sclater et al., 1977). Palaeobiogeographic evidence from ammonites and bivalves seems to indicate, however, a lack of persistent free marine communication across the central part of the Atlantic until the end of the middle or the start of the late Jurassic (Hallam, 1977). This date is close to that of the oldest proven sediments deposited on oceanic crust, and raises the possibility that most or all of the Quiet Magnetic Zone could be subsided continental crust. A comparatively late date of opening is also favoured by Galton (1977) to account for the close similarity of dinosaurs in the late Jurassic of the western United States and East Africa.

There is no easy way to resolve this apparent conflict of evidence, but the solution may lie in a more thorough analysis of Central American tectonics. Indeed, by using a combination of geological and palaeomagnetic evidence, it may be possible to demonstrate translations and/or rotations of continental blocks which would eliminate the problem of continental overlap. Such movements have been demonstrated further north in the western Cordillera (Jones *et al.*, 1977). Coney's (1979) analysis of the opening of the Central Atlantic and Caribbean in the early to mid Jurassic utilizes a megashear. That part of Mexico and Central America southwest of the fault moved with South America initially. Significantly, the geometry invoked allows a land bridge to exist between North and South America long after the initial opening of the Central Atlantic, perhaps even to the end of the Jurassic. If Coney's analysis proves acceptable, one can readily account for the palaeobiogeographic data while accepting the oceanic character of the Quiet Magnetic Zone.

The younger opening history can be dated more precisely because of the good magnetic anomaly record. The next sector to open was that between Newfoundland and Ireland. The current best estimate on ocean-floor magnetic anomaly data indicates a mid-Cretaceous commencement some 90 to 95 Ma ago, when a line opening extended from the Azores-Gibraltar Ridge northwards into the Rockall Trough. Between this time and 80 Ma, the Rockall Trough opened, Newfoundland and Ireland separated and Spain rotated in an anticlockwise direction, opening up the Bay of Biscay. Just before 80 Ma, the line of opening jumped to the west of the Rockall Trough and initiated the Charlie Gibbs Fracture Zone. This new pattern persisted until the start of the Tertiary, when the Norwegian Sea began to form. Spreading extended into the Labrador Sea in the late Cretaceous (Kristoffersen, 1978).

The youngest phase of opening has been studied in detail by Talwani and Eldholm (1977). The Norwegian Sea began opening in the Palaeocene (60 to 63 Ma). From this time until the end of the Eocene the Labrador Sea was also opening and the motion of Greenland was north-westwards relative to Eurasia; Greenland slid past the Barents Shelf and Svalbard in transcurrent fashion. A land connection persisted between Svalbard and Greenland until the end of the Eocene (38 Ma), after which the Greenland Sea began to form. Since this time the opening up of the Norwegian-Greenland Sea can be simply described in terms of the separation of the Eurasian and North American plates.

THE GONDWANA CONTINENTS

On grounds of both geometry and a good magnetic anomaly record, the fits between South America and Africa, and Australia and East Antarctica, are comparatively straightforward and amply supported by geological matching (Smith and Hallam, 1970). Although disagreement on points of detail may remain, these fits can be regarded as essentially uncontroversial. The relationships between east and west Gondwanaland have proved more difficult to determine, the primary problems being the position of Madagascar with respect to Africa and India and the location of the West Antarctic Peninsula.

Geological arguments have been put forward for Madagascar having moved southwards from a position against Tanzania and Kenya (du Toit, 1937; Smith and Hallam, 1970), northeastwards from South Africa (Flores, 1970) or away from the African coast at about the same latitude as today (Kent, 1973). None of this evidence has yet proved conclusive, and since there is no magnetic anomaly record to resolve the dispute, the best available test appears to be a determination of palaeolatitude by rock magnetism. The results published from Madagascar rocks of different ages by

McElhinny *et al*. (1976) unambiguously support the northerly position of the island favoured by du Toit and Smith and Hallam (Fig. 3A).

A serious difficulty remains with the Smith and Hallam reconstruction. The West Antarctic Peninsula overlaps the Falkland Plateau, which is now known to be subsided continental crust down to a depth of at least 3.2 km and probably more than 5 km (Barker *et al*., 1977). The solution to this problem favoured by Powell *et al*. (in press) is to shift India southwards with respect to Madagascar, thereby bringing the southern limit of Jurassic sediments in the two countries more into line (Fig. 3B). An important consequence of this is a duplication of the Mesozoic-Cenozoic orogenic belts of Patagonia and West Antarctica. Such a duplication is even more pronounced in the reconstruction of Barron *et al*. (1978; Fig. 3C), which is very similar to that of Tarling (1972), and does not seem to be geologically plausible (Dalziel, in press). Furthermore, in favouring a southerly position for Madagascar, the Barron reconstruction ignores the palaeomagnetic results of McElhinny *et al*. (1976). Another unacceptable feature is the separation of India from Madagascar, whose Jurassic sequences and faunas show very strong resemblances, suggesting a former adjacent position (Smith and Hallam, 1970). Katz and Premoli (1979) also argue for a close juxtaposition of India and Madagascar on the basis of matching Precambrian lineaments.

Other recent reconstructions (Barker and Griffiths, 1977; Norton and Sclater, 1979) take into consideration oceanographic evidence of fracture zones and magnetic anomalies. Although uncertainties persist because of the extrapolation necessary prior

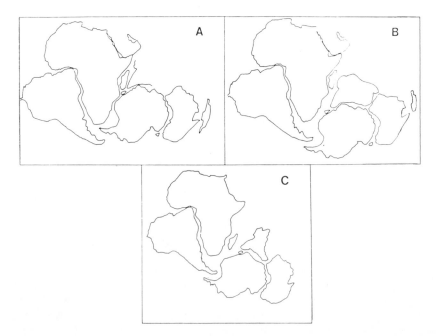

Figure 3. Various proposed fits for the Gondwana continents. A: Smith and Hallam, 1970. B: Powell *et al*., in press. C: Barron *et al*., 1978.

to anomaly 34, both groups of workers agree in essentials with the Smith and Hallam reconstruction.norton and Sclater have produced a comprehensive analysis which may well stand as the definitive work on the subject; hence their reconstruction is accepted here (Fig. 4). The most notable, though minor, differences from the Smith and Hallam reconstruction are a slight southward shift of India relative to Madagascar, with Patagonia and West Antarctica being brought more into line, and a slight eastward shift of Antarctica with respect to Australia. The positioning of India relative to Madagascar in the Norton and Sclater reconstruction is supported by the matching of Precambrian lineaments (Katz and Premoli, 1979). More general support for the Smith-Hallam/Norton-Sclater reconstruction of the Gondwana continents comes from a recent review of palaeomagnetic data by Embleton and Valencio (1977), who are also able to resolve the hitherto apparently anomalous results for the Mesozoic of Australia.

It is generally agreed that the position of the West Antarctic peninsula relative to East Antarctica remains the most uncertain feature, and one or more parts have probably been separated by spreading and/or displaced by transcurrent movements. Thus the reconstructions of Barker and Griffiths (1977, models A and B) and Cox (1978) differ from each other and from the other reconstructions reviewed here. The kink that occurs in most reconstructions in the fold belt in Patagonia and in its continuation in West Antarctica may be an inherited early Mesozoic feature (Dalziel, in press).

Although many workers have dated the commencement of breakup of Gondwanaland as early to mid-Jurassic or even late Triassic (on the bases of the eruption of the Karroo volcanoes and the entry of seawater into the Indian Ocean region), the divergence of polar wandering paths determined from paleomagnetism clearly indicates that active disintegration did not begin until well into the Cretaceous (McElhinny, 1973). This is supported both by paleobiogeography, with many indications that widespread continental connections persisted into this period (Hallam, in press), and by sea-floor spreading data (Norton and Sclater, 1979).

The oldest magnetic anomalies that have been recognized (125 to 130 m.y.) occur in the Cape Basin off the west coast of South Africa (Larsen and Ladd, 1973) and off southwestern Australia (Markl, 1978). A clear pattern of magnetic anomalies indicating spreading along the whole length of the South Atlantic is only present beginning 80 Ma ago (Le Pichon and Hayes, 1971). Not until this time was there a free ocean connection with the North Atlantic, when the transverse Rio Grande-Walvis Ridge sank below 1 km depth (Van Andel et al., 1977).

Some uncertainty remains about the creation of the South Atlantic in the early and mid-Cretaceous. Cande and Rabinowitz (1978) assume rigid plate tectonics and argue for a commencement of spreading between Brazil and Angola at the same time as further south, but at a much slower rate because of the narrower oceanic gap. In the absence of a clear anomaly sequence prior to anomaly 34 there is no proof of this, and paleobiogeographic data seem to demand the persistence of a land link well into the Cretaceous (Buffetaut and Taquet, 1979; Hallam, in press). The Aptian of the Angolan and Brazillian margins is characterized by thick evaporite sequences extending well offshore, deposited on continental crust that probably subsided below sea level as a graben before an inrush of sea in the Albian (Evans, 1978). Even this latter event probably preceded continental separation and creation of ocean floor by millions of years, as in the case of Lower Jurassic marine sediments overlying thick salt deposits on the margins of the southern part of the North Atlantic.

Sea-floor spreading data indicate that India began to separate from Australia-Antarctica in the Valanginian, and from Madagascar in the late Cretaceous, as suggested by a lava sequence on the eastern side of Madagascar. Subsequently India began a rapid northward movement toward Eurasia. Both Barker and Griffiths (1977) and Norton and Sclater (1979) agree that Madagascar must have separated from Africa in the mid-Cretaceous.

Total disintegration of the Gondwana components did not take place until the Tertiary. Sea-floor spreading data indicate that Australia began to separate from Antarctica in the Eocene (53 Ma; Weissel *et al*., 1977) (New Zealand had broken away from Australia-Antarctica much earlier, about 80 Ma). The separation of South America from West Antarctica occurred when the Drake Passage opened in the Oligocene, allowing the creation of a circum-polar ocean current system (Barker and Burrell, 1977).

The youngest rifting event in the creation of new ocean floor occurred in the Red Sea and Gulf of Aden. There has been controversy about how much of the floor of the

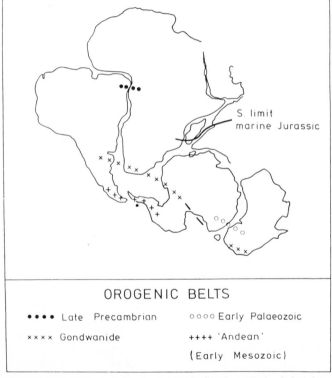

OROGENIC BELTS

•••• Late Precambrian   ∘∘∘∘ Early Palaeozoic

×××× Gondwanide   ++++ 'Andean'

                               (Early Mesozoic)

Figure 4. Proposed reconstruction of Gondwanaland before late Mesozoic breakup, with some geological tie-lines.

Red Sea is genuine oceanic as opposed to thinned continental crust. Although uncertainties remain, geological and geophysical evidence now seems to favour the view that the zone of true oceanic crust is no wider than 70 km and did not start to form until the Pliocene (3.5 Ma), after a long precursory phase of rifting and volcanism (Le Pichon and Francheteau, 1978). The alternative view, that the Red Sea began to open up as early as the Oligocene, implies that the thick Miocene salt was laid down on oceanic crust, as Kinsman (1975) maintained.

CONCLUDING REMARKS

It follows from the above discussion that continental fits cannot yet be made as precisely as earlier work suggested. Further extensive geological and geophysical work on the continental margins is required to determine the best bathymetric estimate of continent-ocean transition. This can be expected to vary considerably in different sectors, and may well be at a deeper level for regions of older ocean. When this has finally been achieved it should be possible to correct for the stretching and thinning that has taken place in the passive continental margins and thereby make a more accurate restoration of original relative position. Uncertainty will probably persist for a long time about the precise original geographic location of West Antarctica and the Central American region.

REFERENCES

Barker, P.F. and Burrell, J., 1977, The opening of Drake Passage: Marine Geol., v.25, p. 15-34.

Barker, P.F., Dalziel, I.W.D. *et al.*, 1977, Initial Reports of the Deep Sea Drilling Project, v.35: Washington, U.S. Govt. Printing Office.

Barker, P.F. and Griffiths, D.H., 1977, Towards a more certain reconstruction of Gondwanaland: Phil. Trans. Royal Soc. London, v. B 279, p. 143-159.

Barron, E., Harrison, C.G.A. and Hay, W.W., 1978, A revised reconstruction of the southern continents: EDS, v. 59, p. 436-449.

Bullard, E.C., Everett, J.E. and Smith, A.G., 1965, The fit of the continents around the Atlantic: Phil. Trans. Royal Soc. London, v. A 258, P. 41-51.

Buffetaut, E. and Taquet, P., 1979, An early Cretaceous terrestrial crocodilian and the opening of the South Atlantic: Nature, v. 280, p. 486-487.

Cande, S.C. and Rabinowitz, P.D., 1978, Mesozoic seafloor spreading bordering conjugate continental margins of Angola and Brazil: Proc. 10th offshore Technol. Conf., Houston, Texas, p. 1869-1872.

Coney, P.J., 1979, Mesozoic – Cenozoic Cordilleran plate tectonics: Geol. Soc. Amer. Mem. 152, p. 33-50.

Cox, K.G., 1978, Flood basalts, subduction and the break-up of Gondwanaland: Nature, v. 274, p. 47-49.

Dalziel, I.W.D., in press, The early (pre-Middle Jurassic) history of the Scotia Arc region: a review and progress report: in Craddock C. ed., SCAR (IUGS 3rd Sympos.) Antarctic Geol. Geophys., Madison, Wisc., 1977).

Dietz, R.S. and Holden, J.C., 1970, Reconstruction of Pangaea: breakup and dispersion of continents, Permian to present: Jour. Geophys. Research, v. 75, p. 4939-4956.

Dingle, R.V., 1973, Mesozoic palaeogeography of the southern Cape, South Africa: Palaeogeog., Palaeoclimatol., Palaeoecol., v. 13, p. 203-213.

Du Toit, A.L., 1937, Our wandering continents: Edinburgh, Oliver and Boyd, 336 p.

Embleton, B.J.J. and Valencio, D.A., 1977, Palaeomagnetism and the reconstruction of Gondwanaland: Tectonophysics, v. 40, p. 1-12.

Evans, R., 1978, Origin and significance of evaporites in basins around Atlantic margin: Amer. Assoc. Petrol. Geol. Bull., v: 62, p. 223-234.

Flores, G., 1970, Suggested origin of the Mozambique Channel: Trans. Geol. Soc. South Africa, v. 73, p. 1-16.

Galton, P., 1977, The Upper Jurassic ornithopod dinosaur *Dryosaurus* and a Laurasia – Gondwanaland connection: in West R.M., ed., Paleontology and Plate Tectonics, Milwaukee Publ. Mus. Spec. Publ. Biol. Geol. no. 2, p. 41-54.

Hallam, A., 1971, Mesozoic geology and the opening of the North Atlantic: Jour. Geol., v. 79, p. 129-157.

Hallam, A., 1977, Biogeographic evidence bearing on the creation of Atlantic seaways in the Jurassic: in West R.M., ed., Paleontology and Plate Tectonics, Milkwaukee Publ. Mus. Spec. Publ. Biol. Geol. no. 2, p. 23-34.

Hallam, A., in press, The relative importance of plate movements, eustasy and climate in controlling major biogeographic changes since the early Mesozoic: In Rosen D.E., ed., Vicariance Biogeography: congruence of earth history with plant and animal distributions: New York, Columbia Univ. Press.

Haworth, R.T., 1977, The continental crust north-east of Newfoundland and its ancestral relationship to the Charlie fracture zone: Nature, v. 266, p. 266-249.

Hayes, D.E. and Rabinowitz, P.D., 1975, Mesozoic magnetic lineations and the magnetic quiet zone of northwest Africa: Earth Planet. Sci. Letters, v. 28, p. 105-115.

Hollister, C.D. Ewing, J.I. et al. 1976, Initial reports of the Deep Sea Drilling Project, v. 11: Washington, U.S. Govt. Printing Office.

Jansa, L.F. and Wade, J.A., 1975. Geology of the continental margin off Nova Scotia and Newfoundland: Geol. Survey Canada Paper 740-30, v. 2, p. 51-105.

Jones, D. L., Silberling, N.J. and Hillhouse, J., 1977, Wrangellia – a displaced terrane in northwestern North America: Can. J. Earth Sci., v. 14, p. 2565-2577.

Katz, M.B. and Premoli, C., 1979, India and Madagascar in Gondwanland: a fit based on matching Precambrian lineaments: Nature, v. 279, p. 312-315.

Kent, P.E., 1973, East African evidence of the palaeoposition of Madagascar: in Tarling, D.H., and Runcorn, S.K., eds., Implications of continental drift to the earth sciences, London, Academic Press, v. 2, p. 873-878.

Kinsman, D.J. 1975, Salt floors to geosynclines: Nature, v. 255, p. 375-378.

Kristoffersen, Y., 1978, Sea-floor spreading and the early opening of the North Atlantic: Earth Planet. Sci. Letters, v. 38, p. 273-290.

Lancelot, Y., Seibolt, E. et al., 1977, Initial reports of the Deep Sea Drilling Project, v. 38, p. 273-290.

Lancelot, Y., Seibold, E. et al., 1977, Initial reports of the Deep Sea Drilling Project, v. 41: Washington, U.S. Govt. Printing Office.

Larsen, R.L. and Ladd, J.W., 1973, Evidence for the opening of the South Atlantic in the early Cretaceous: Nature, v. 246, p. 209-212.

Le Pichon, X. and Hayes, D.E., 1971, Marginal offsets, fracture zones and the early opening up of the South Atlantic: Jour. Geophys. Res., v. 76, p. 6283-6308.

Le Pichon, X., Sibuet, J. C. and Francheteau, J., 1977, The fit of the continents around the North Atlantic Ocean: Tectonophys., v. 38, p. 169-209.

Le Pichon, X. and Francheteau, J., 1978. A plate-tectonic analysis of the Red Sea – Gulf of Aden area: Tectonophys., v. 46, p. 369-406.

Luyendyk, B.P. and Bunce, E.T., 1973, Geophysical study of the north-west African Margin off Morocco: Deep-Sea Res., v. 20, p. 537-550.

McElhinny, M.W., 1973, Palaeomagnetism and Plate Tectonics: Cambridge Univ. Press, 358 pp.

McElhinny, M.W., Embleton, B.J.J., Daly, L. and Pozzi, J.-P., 1976, Paleomagnetic evidence for the location of Madagascar in Gondwanaland: Geology, v. 4, p. 455-457.

Markl, R.G., 1978, Further evidence for the early Cretaceous breakup of Gondwanaland off southwestern Australia: Mar. Geol., v. 26, p. 41-48.

Norton, I.O. and Sclater, J.G., 1979, A model for the evolution of the Indian Ocean and the breakup of Gondwanaland: Jour. Geophys. Research, v. 84, p. 6803-6830.

Pautot, G., Renard, V., Auffret, G. and Pastouret, L., 1976, A granite cliff deep in the North Atlantic: Nature, v. 263, p. 669-672.

Pitman, W.C., and Talwani, M., 1972, Sea floor spreading in the North Atlantic: Geol. Soc. Amer. Bull., v. 83, p. 619-646.

Powell, C. McA., Johnson, B.D. and Veevers, J.J., in press, A revised fit of East and West Gondwanaland: Tectonophysics.

Sclater, J.G., Hellinger, S. and Tapscott, C., 1977, The paleobathymetry of the Atlantic Ocean from the Jurassic to the present: Jour. Geol., v. 85, p. 509-522.

Smith, A.G. and Hallam, A., 1970, The fit of the southern continents: Nature, v. 225, p. 139-144.

Talwani, M. and Eldholm, O., 1972, Continental margin off Norway: a geophysical study: Geol. Soc. Amer. Bull., v. 83, p. 3775-3606.

Talwani, M. and Eldholm, O., 1977, Evolution of the Norwegian-Greenland Sea: Geol. Soc. Amer. Bull., v. 88, p. 969-999.

Tarling, D.H., 1972, Another Gondwanaland: Nature, v. 238, p. 92-93.

Van Andel, T.J.H., Thiede, J., Sclater, J.S. and Hay, W.W., 1977, Depositional history of the South Atlantic Ocean during the last 125 million years: Jour. Geol., v. 85, p. 651-698.

Weissel, J.K., Hayes, D.E. and Herron, E., 1977, Plate tectonics synthesis: the displacements between Australia, New Zealand and Antarctica since the late Cretaceous: Marine Geol., v. 25, p. 231-277.

The Continental Crust and Its Mineral Deposits, edited by D.W. Strangway,
Geological Association of Canada Special Paper 20

A CAMBRO-PERMIAN PANGAEIC MODEL CONSISTENT WITH LITHOFACIES AND BIOGEOGRAPHIC DATA

A. J. Boucot
Department of Geology,
Oregon State University,
Corvallis, Oregon, 97331

Jane Gray
Department of Biology,
University of Oregon,
Eugene, Oregon, 97403

ABSTRACT

Paleozoic lithofacies and biogeographic data are more consistent with, and more readily explained by, a pangaeic reconstruction integrated with a reasonable surface-current circulation pattern, than they are with available non-pangaeic hypotheses. As put forward here, the pangaeic hypothesis requires that the supercontinent be more or less centered on the South Pole during the Cambrian and that it move slowly northward throughout the Paleozoic with the Siberian portion as the leading edge reaching high northern latitudes by the Permian. Several changes in global climatic gradient are also required to explain the lithofacies and biogeographic data: low gradients in the Early Cambrian and Late Devonian-earlier Carboniferous; high gradients from the Middle Cambrian through Middle Devonian; very high gradients in the latest Ordovician, and the later Carboniferous through Permian.

A single pangaeic landmass is most consistent with the following:

1) The extensive mixing of faunas from different biogeographic realms in southeast Kazakhstan from the Cambrian through the Middle Devonian.

2) The presence in the Andean region throughout the Paleozoic of a single, essentially stable, north-south boundary that separates warm-water faunas from cold-water faunas.

3) The steady southerly movement of the east-west, warm-cold water boundaries in the Americas (southern North America to south of the Amazon from the Cambrian through the Permian) and in

the western Old World (parts of northern Europe in the Cambrian to North Africa by the Permian).

4) The distribution through time of known Paleozoic bauxites, laterites, shallow-water marine evaporites, redbeds and other lithologic evidences of climatic conditions. The distribution of the important Paleozoic marine phosphorites, such as the Phosphoria Formation of the North American Permian, is also consistent with the pangaeic hypothesis.

5) The boundaries between inferred equatorial regions of coal formation and preservation that suggest humid conditions, and middle latitude regions with evaporites that suggest arid climatic conditions.

6) The high-level similarities of Paleozoic lithofacies and biofacies in the circum-Mediterranean region.

RESUMÉ

Les données de lithofaciès et de biogéographie du Paléozoïque s'accordent mieux et peuvent s'expliquer plus facilement par une reconstruction pangéique combinée à un patron adéquat de circulation des courants de surface, qu'elles ne le sont par les hypothèses qui ignorent la Pangée. Comme on l'explique ici, l'hypothèse pangéique suppose que le supercontient était plus ou moins centré sur le Pôle Sud au Cambrien et qu'il s'est déplacé lentement vers le nord au cours du Paléozoïque avec la portion sibérienne comme partie avant pour atteindre des latitudes nord élevées au cours du Permien. On suppose aussi plusieurs changements dans le gradient climatique global pour expliquer les données de lithofaciès et de biogéographie: des gradients faibles au début du Cambrien et de la fin du Dévonien au début du Carbonifère; des gradients élevés du Cambrien moyen au Dévonien moyen; des gradients très élevés à la toute fin de l'Ordovicien et de la fin du Carbonifère jusqu'au Permien.

Une masse terrestre pangéique unique est plus susceptible d'expliquer les faits suivants:

1) Le mélange à grande échelle de faunes provenant de milieux biogéographiques différents dans le sud-est du Kazakhstan à partir du Cambrien jusqu'au Dévonien moyen.

2) La présence dans la région andéenne au cours du Paléozoïque d'une limite nord-sud unique, essentiellement stable, qui sépare les faunes d'eau chaude des faunes d'eau froide.

3) Le mouvement continu vers le sud des limites est-ouest, eau chaude-eau froide dans les Amériques (du sud de l'Amérique du Nord jusqu'au sud de l'Amazone, à partir du Cambrien jusqu'au Permien) et dans la partie ouest de l'Ancien Monde (portion de l'Europe du Nord au Cambrien jusqu'en Afrique du Nord au Permien).

4) La distribution dans le temps des dépôts paléozoïques connus de bauxites, latérites, évaporites marines d'eau peu profonde, de lits rouges et d'autres preuves lithologiques des conditions climatiques. La distribution des phosphorites marines importantes du Paléozoïque, telle la formation de Phosphoria du Permien d'Amérique du Nord s'accorde aussi avec l'hypothèse pangéique.

5) Les limites entre les régions équatoriales reconstruites de formation et de préservation du charbon, ce qui indique des conditions humides, et les régions de latitude moyenne où les évaporites indiquent des conditions climatiques arides.

6) Les ressemblances très marquées de lithofaciès et biofaciès de la région entourant la Méditerranée.

INTRODUCTION

This paper is essentially an "appendix" to an earlier paper by Boucot and Gray (1979), which attempted a pangaeic reconstruction of the Paleozoic. Our original pangaeic reconstruction was based primarily on biogeographic data and the distribution of carbonate and non-carbonate rocks. In the present paper we have gathered together additional lithologic data having potential value for climatic interpretation.

Some, such as the redbeds, bauxites, and kaolins and other products of tropical weathering, support our earlier conclusions. Others, such as the phosphorites, provide information that permits a variety of interpretations. We also consider the possible climatic significance of Paleozoic marine evaporites, coal deposits and black shale facies. We regard all the additional lithologic data as consistent, rather than compelling, evidence for the pangaeic hypothesis.

We use the term "pangaeic" even though we do not show North America directly connected with any other continent. This is a matter of convenience; it also reflects the present state of ignorance regarding the changing dimensions through time of the Appalachian-Ouachita and Caledonian geosynclinal systems (or the Iapetus Ocean of some authors), as well as the nature of the fundament, if any, underlying the water body or bodies involved. It is also important to realize that there is good evidence from Paleozoic stratigraphy, lithofacies, and paleogeography for termination of the Uralian Geosyncline in the Kazakhstan region during the Paleozoic (Boucot, 1969). The Uralian Geosyncline can best be regarded as a feature which widened to the north away from southeastern Kazakhstan and which partially split the Early Paleozoic (Silurian and Devonian) platform of the Eurasian region into the Russian and Siberian platforms adjoining the Angara region. The available geologic evidence does not support interpretations which treat the area of southeastern Kazakhstan, and the Russian and European platforms as "plates". The habit of automatically treating almost all geosyncline belts, tectonic belts, "sutures", ophiolite belts, and other features as plate boundaries does not reflect careful evaluation of *all* evidence, and adds little to our understanding of pre-Mesozoic paleogeography. In studies of the so-called "Rhaeic Ocean", for example, some geologists automatically assumed a 1:1 correlation between geosynclines, tectonic belts, and ophiolite belts with former subduction zones and accompanying oceans, and have chosen to ignore contrary evidence. This hypothetical Late Paleozoic Ocean is actually located on the site of a former Lower Paleozoic platform sedimentary sequence including such units as the Silurian black graptolitic shales, which occur almost continuously from the central Sahara into southern Scandinavia (Berry and Boucot, 1967; Jaeger, 1976). In other words, some geosynclines, such as the Hercynian Geosyncline in much of Europe, as well as the Rocky Mountain Geosyncline in North America, occur on the sites of former platform regions. The geologic evidence for a major change in character from platform to geosyncline does not imply the presence of a former ocean.

Van Andel (1979) has summarized many of the concepts necessary for plate tectonic syntheses. As he notes, the oceanic geophysical and geological data supporting plate tectonics are compelling for the post-Paleozoic. The lack of such data for the pre-Mesozoic, however, contributes to the present clash of opinions regarding Paleozoic continental configurations(s). Any attempt to objectively synthesize the welter of published paleogeographies will quickly reveal that there is no generally held opinion. The inconsistencies both during the same time interval and between time intervals are about equal to the number of publications on the topic.

We view these conflicting paleogeographies as a necessary stage leading to a more fundamental synthesis that will be consistent with most of the known physical and biological data. We have interjected a pangaeic interpretation for the Paleozoic as an alternative to the available syntheses, and as encouragement to those inclined towards a broader view than the usual time-interval-by-time-interval or biotic-group-by-biotic-group approaches.

In our recent synthesis (Boucot and Gray, 1979), we paid particular attention to the available global biogeographic data for the Paleozoic: data on macrobenthos such as corals, brachiopods, trilobites, echinoderms, and molluscs, together with more limited information from pelagic organisms such as graptolites, and from conodonts, chitinozoans, some foraminiferans, and plants. We also used the data of lithofacies, in particular the contrast provided by carbonate facies rocks thought to be characteristic of tropical to warm temperate regions of the past, and the widespread terrigenous facies rocks thought to represent temperate to cold climatic conditions.

The agreement between the biotic and sedimentologic data is high. In that synthesis we provided a surface-current circulation pattern, period-by-period, which is consistent with both the biogeographic and lithofacies information, although there is some lithologic information which we did not consider. The present paper attempts to integrate such information into our previous synthesis, to see whether it fits the pattern of surface-current water circulation proposed with the pangaeic reconstruction.

There are a number of assumptions necessary to a global paleogeographic pangaeic synthesis. First, we assume that global climatic gradients have changed through time. Earlier we (Boucot and Gray, 1979), reviewed the Paleozoic evidence that supports this assumption, noting that there was a much higher global climatic gradient at the end of the Ordovician than either before or after it, until the Late Carboniferous-Early Permian interval. Elsewhere (Boucot and Gray, 1978) we have emphasized the well known, very low climatic gradient in the Triassic. Second, we assume a steady drift of the entire pangaeic supercontinent from a Cambrian position centered on the South Pole (Fig. 1) to a Permian position straddling the equator but reaching into high northern and southern latitudes (Fig. 6), rather than a movement of the Earth's axis of rotation. Third, and most important, we assume that present biologic and physical processes also operated during the Paleozoic. For example, we assume that the larvae of some marine benthos have always been able to delay metamorphosis for great lengths of time while drifting and being dispersed widely by currents (see Scheltema, 1979). These larvae tend to represent the more abundant species rather than the endemic, rarer species. We also assume that sedimentary rocks provide the same environmental connotations through time. Products of tropical weathering, such as bauxites, laterites, and kaolin, carry the same climatic implications when they are found in the Paleozoic; marine evaporites and coal deposits in the Paleozoic indicate regions of excessive evaporation and humidity, respectively.

We have used the work of McKelvey (1967) and Sheldon (1964a, b) in considering the possible significance of sedimentary phosphate deposits. Following their conclusions, we assume that the phenomena responsible for such deposits were absent at high latitudes in the past, as they are at present. However, we note that the distribution of most of the Paleozoic phosphorite deposits may be explained in a variety of ways; in most cases their presence merely suggests, rather than proves, the presence of any one oceanographic situation.

Finally, we repeat (see Boucot and Gray, 1979) that there are several pieces of lithofacies data which make a Paleozoic pangaeic interpretation particularly attractive. Prominent among these are the striking similarities between the Lower Paleozoic lithofacies of the circum-Mediterranean region and the Atlantic facies of South America and Africa. Also important in considering the relations of Europe to northern Asia are the Silurian-Devonian lithic similarities in the Uralian, southeast Kazakhstan, and Siberian Platform regions (Boucot, 1969). Many kaleidoscopic Paleozoic reconstructions treat the Siberian Platform and China as continental units entirely independent of, and well removed from, the remainder of Europe and Asia during most of the pre-Carboniferous; we do not consider the evidence convincing for such a view.

GLOBAL CLIMATIC GRADIENTS

Earlier, we (Boucot and Gray, 1979) mentioned the presence of a globally low climatic gradient during the Early Cambrian (preceded by a similar climatic gradient in the later Precambrian), based on the widespread distribution of Lower Cambrian carbonate rocks. These carbonate rocks are notable in such regions as North Africa, where post-Lower Cambrian rocks are characterized by neither carbonates nor warm-water faunas, but by Atlantic Realm biotas which moved in after the Early Cambrian.

We suggested (Boucot and Gray, 1979, p. 477) that there was a very low global climatic gradient from the Late Devonian through the Early Carboniferous. The evidence for this included the appearance of the brachiopod *Tropidoleptus* and goniatites, both considered to be warm-water stenotherms, near the top of the marine sequence in beds of later Middle Devonian and possibly early Upper Devonian age, within the cool or cold Malvinokaffric Realm, plus the absence of any evidence for pronounced biogeographic, latitudinally correlated differentiation of land plants or marine invertebrates during the Early Carboniferous. The Malvinokaffric Realm goniatites could, of course, have floated *post mortem* into a cold region, but in the case of sessile benthic *Tropidoleptus*, this would have been unlikely. There is no positive evidence that a Gondwana Realm was present throughout the Early Carboniferous, despite excellent evidence for its presence during Late Carboniferous to Permian, and Silurian to Middle Devonian times. (During the Silurian-Devonian, ''Malvinokaffric Realm'' is used in place of the more or less geographically equivalent ''Gondwana Realm'' of the later intervals.)

McPherson (1979) has presented additional evidence compatible with an unusually low climatic gradient during the Late Devonian-Early Carboniferous interval. He reports paleosols in southern Victoria Land that are characterized by calcretes of the type occurring in warm, monsoonal regions today. Similar calcretes also occur in the Old Red Sandstone (Friend and Williams, 1978), where they are dated as Frasnian age by means of *Groenlandaspis* (Ritchie, 1975). Ritchie (1975) notes that this freshwater fish has a very cosmopolitan distribution – Antarctica, Australia, Greenland, and northwestern Europe (Ireland, Britain) – which is consistent with a low climatic gradient. Boucot (1975) has commented earlier on the unusually cosmopolitan distribution of Upper Devonian marine organisms; to find a similar distribution for at least one non-marine vertebrate is of more than casual interest. It is important to note that the Early Devonian of Antarctica is definitely of the ''cool'' climate Malvinokaffric Realm type, entirely lacking Old Red Sandstone-type sedimentation, carbonates, evaporites, and related marine fauna.

Crowell (1978) has summarized evidence for Gondwana region glaciations of the

Permo-Carboniferous. He makes a strong case for glaciation beginning no earlier than the close of the Early Carboniferous (Crowell, 1978, p. 1357), which would have ended the globally equable climate of the Late Devonian-Early Carboniferous. Crowell also spells out the difficulties in evaluating the many potentially interacting factors responsible for glacial climate, as well as for climate in general.

PALEOMAGNETIC EVIDENCE DURING THE PALEOZOIC

We have earlier (Boucot and Gray, 1979, p. 465) suggested that paleomagnetic interpretation of pre-Carboniferous Paleozoic geography is based on limited evidence. Heckel and Witzke (1979) also question the compatibility of magnetic and lithic data from Siberia and North America during the Devonian. The two very different geographies provided by Morel and Irving (1978) for the older Paleozoic, both based on paleomagnetic evidence, make one skeptical about having to choose any one reconstruction as the "correct" one. At best the paleomagnetic evidence provides no information about longitude. Much of our present debate about pre-Carboniferous paleogeography concerns which interpretation – kaleidoscopic or pangaeic – better fits the paleomagnetic evidence. Although paleomagnetics has great potential value for guiding one's thinking about the changing paleogeographies of the past, perhaps it is too early for the non-specialist to trust its conclusions for the pre-Carboniferous.

PALEOCLIMATIC AND BIOGEOGRAPHIC IMPLICATIONS OF SEDIMENTARY ROCKS

In the Appendix we have listed varied sedimentary rock types relevant to climatic and biogeographic interpretations not considered by Boucot and Gray (1979). In order to enlarge our earlier pangaeic interpretation, we will consider the significance of these rocks in a period-by-period summary.

Redbeds

Redbeds are excellent indicators of a climatic regime which results in the production of abundant hematite as well as orange and yellow hydrated iron oxide compounds. Today such compounds are not forming to any extent in cold regions, except through reworking of older rocks, but they are common in warm regions. Redbeds are known as far back as the Precambrian in association with other indicators of warm climate such as abundant carbonate rocks. The data provided in the Appendix and in the period-by-period summary are not intended as an exhaustive review of redbed occurrences, but merely as an indication of their distribution through time, as well as a record of their association with other warm-climate indicators.

We use the term redbeds to include red rocks deposited under non-marine conditions, such as the famous Old Red Sandstone of the Devonian, as well as marine redbeds. Marine redbeds aid climatic interpretation in that they commonly include iron compounds (oxidized in the non-marine environment under warm conditions) which have not been reduced after deposition in the marine environment (see Ziegler and McKerrow, 1975; Walker, 1967, 1975). Those marine redbeds high in manganese minerals, chiefly carbonates, have additional climatic significance for they indicate non-marine weathering capable of producing laterite-type, tropical manganese deposits (Pavlides, 1962). One should also note that redbeds occur in desert regions, monsoonal regions, and in everwet regions, in contrast with evaporites, which never occur in very humid regions, and with laterites, bauxites, and kaolins, which are seldom generated in desert regions.

Laterite and Related Soils

Lateritic soils are seldom recognized in the Paleozoic, but a few occur in regions which on other grounds appear to have been climatically "warm" (Young, 1976); there is no evidence of lateritic weathering in "cold" or "cool" regions of the Paleozoic. The extensive literature of soil science indicates many types of lateritic weathering in many climatic regimes, but it is clear that these are strictly warm-climate soils.

Kaolin, Boehmite, and Diaspore Deposits

Kaolin deposits are products of warm-temperate to tropical weathering in a variety of climatic regimes. John Hower (written commun., 1979) points out that the occurrence of kaolin is dependent upon the type of substrate rock undergoing weathering. Not all aluminous bedrock gives rise to a kaolinized end product if the original composition is not compatible; illites and other clay mineral types are also possible products under warm-climate weathering regimes. Diaspore and boehmite deposits form through the long-term alteration of kaolin deposits.

Bauxite, Diaspore, and Emery Deposits

Bauxites and their regional metamorphic equivalents, diaspore and emery, are exclusively products of tropical and subtropical weathering (see Bárdossy, 1973, 1977, 1979 and references). Their significance as indicators of warm environments, from the Precambrian to the Cenozoic, is unquestioned, but their scarcity in the pre-Mesozoic makes them of little more than supplementary interest in the present context. The known Paleozoic deposits all occur in regions that on other grounds, such as the presence of abundant carbonate rocks, represent warm climates. No bauxites of Paleozoic age are known from the Atlantic, Malvinokaffric or Gondwana Realms.

Coals

Modern views make it clear that coals cannot be employed as indicators of tropical to warm-temperate environments. Kräusel (1964) has emphasized that coal deposits indicate only the presence of a humid climate; they do not necessarily correlate with a warm climate, as many of the modern as well as ancient peats and coals were and are being deposited in cold-climate conditions. Schopf (1972) also supports this position, although with somewhat different examples. Smith and Eriksson (1979) describe a depositional model for cold-temperate coals.

The Paleozoic record suggests that the Permo-Carboniferous Gondwanic coals of South Africa, Antarctica, Peninsular India and the Sydney Basin represent cool to cold environments, whereas the Carboniferous coals of eastern North America, Europe and Asia represent warm-temperate to subtropical or tropical conditions. Thus it is impractical to view coals as anything more than indicators of high humidity combined with conditions suitable for the preservation of organic materials, such as freedom from destruction by bacterial activity and oxidation. Such conditions are in fact less common in tropical regions than elsewhere.

At least for the Late Devonian (Heckel and Witzke, 1979) and Pennsylvanian (Wanless, 1966; Heckel, 1977), it can be shown that the humid conditions necessary for coal formation appear in the equatorial region and on either side of the dry belts bordering it. In regions where a pattern of coal and evaporite belts can be detected, latitudinal relations can be reconstructed, but where scattered occurrences of coal are unaccompanied by other indicators, climatic interpretation is difficult and unreliable. In addition, there is no reason to think that widths of climatic belts have remained constant through time. Fluctuations have been the rule. When additional topographic complications are added, with their climatic effects, it is clear that the presence or absence of coal deposits is difficult to interpret.

Evaporites

Meyerhoff (1970) has compiled the global distribution, in time and space, of marine evaporites (see also Zharkov, 1974). From this work it is clear that marine evaporites of Paleozoic age are unknown in biogeographic realms considered to have been cold or cool. This includes the Cambrian Atlantic Realm, the Ordovician Atlantic Realm (Paleotethys, Malvinokaffric, Mediterranean of some authors), the Malvinokaffric Realm of the Silurian and Devonian, and the Permo-Carboniferous Gondwana Realm. In other words, the Southern Hemisphere Realms, which by varied lithologic and biologic criteria can be shown to have been cold or cool, lacked conditions favourable to the formation of evaporites. Conversely, evaporites are well known and widely distributed in warmer, northern Realms during the Paleozoic.

The mutually exclusive occurrence of evaporites and coals emphasizes climatic differences inherent in their production. For example, areas favourable to the production and preservation of coal in the Permo-Carboniferous, i.e., consistently high humidity regions, lack evaporites. The prominent Mississippian age evaporites of the Maritime Provinces of eastern Canada contrast with the prominent Pennsylvanian coal deposits of the same region. In turn, the prominent Northern Hemisphere Pennsylvanian coal deposits provide a contrast with the prominent Permian evaporites of the same region. Coals of the Gondwana Realm are not associated with marine evaporites of any sort.

Another class of evidence is provided by Heckel and Witzke (1979), who indicate that in the Devonian, areas subject to evaporite deposition moved in a manner consistent with the Pangaeic movements discussed earlier, i.e., the evaporite "belt" moved from a northwest to a south-central position in North America.

Black Shales

A "black shale" facies is widespread during certain time intervals on many of the major Paleozoic platforms. It is characterized by finely laminated, non-bioturbated strata containing a fauna which is largely, if not entirely, of the pelagic or epipelagic type and can be interpreted as representing a depositional environment in the Benthic Assemblage 5-6 range or even deeper, i.e. equivalent to shelf-margin depths. Benthic Assemblage 5-6 corresponds to the shelf margin region (Boucot, 1975), 4-5 to the subphotic zone through outer shelf region, 3 to the subtidal photic zone, while 1 and 2 correspond, respectively, to high and low intertidal zones. The possibility that these Paleozoic platform black shale facies represent anoxic epicontinental waters, at least in the Benthic Assemblage 5-6 range, could provide additional evidence about Paleozoic surface-water circulation and water mass characteristics. Were true oceanic deposits recognized in the Paleozoic, it might be determined that the platform black shales represent the upper parts of an anoxic ocean of the type suggested by some for certain post-Paleozoic intervals. This would carry further implications for surface-water circulation patterns. It is still uncertain whether Paleozoic geosynclinal radiolarian-rich cherts, sometimes associated with black shales, are truly oceanic or merely deep-water marginal deposits.

While black shales deposited below depths equivalent to the continental shelf are widespread on some of the platforms, they are not widely enough distributed in time or space to support a pangaeic reconstruction except for the Silurian-Early Devonian of the European-North African region.

Phosphorites

Sheldon (1964a, b) and McKelvey (1967) have summarized the potential significance of sedimentary phosphorites in the geologic record. They indicate that most phosphorites were formed during the later Cenozoic at latitudes extending from about 10 to 40 degrees but centering at about 32 degrees, and that similar latitudinal restriction is reasonable for the older deposits. They also indicate that "divergence upwelling" on the western sides of continents combined with trade wind circulation explains some of these deposits, whereas "dynamic upwelling" on the eastern sides of other continents, sometimes in combination with specific geographic configurations, explains many of the others. However, in the Paleozoic there is another category of phosphorite deposits not discussed by these authors: phosphorites on platform interiors well away from platform margins. For these deposits it is necessary to invoke sufficiently deep water, possibly the equivalent of continental-margin depths, to provide the oxygen minimum conditions required for phosphorite formation, and to provide a surface-current circulation from geosynclinal or open ocean regions capable of supplying the necessary phosphorus-rich water.

Paleozoic phosphorite deposits can thus be sorted into platform and platform margin-geosynclinal types. Platform margin-geosynclinal deposits can be due either to oceanic divergence upwelling or to dynamic upwelling, depending on continental configuration. Therefore, the presence of Paleozoic phosphorites cannot be used to discriminate between the dynamic upwelling and divergence upwelling modes of formation. Paleozoic phosphorites are best treated as circumstantial evidence in agreement with several hypotheses of former continental configurations and surface-water circulation patterns. They do not afford independent evidence confirming any one hypothesis.

The common association of marine phosphorites with outer shelf to upper bathyal, "geosynclinal" suites of bedded cherts (lydite type cherts), black shales, and manganiferous carbonate nodules indicates a relatively deep-water origin. This environment can be situated on a platform, as in some of the Mississippian examples (see Lane, 1978) as well as on a slope leading down to a true oceanic environment. Some of these non-platform phosphorite associates are interlayered with typical platform carbonate sequences. But these interlayered sequences are almost always near or adjacent to platform margins, implying that the phosphorite strata represent an intertongue from the "geosynclinal" or oceanic continental margins rather than a real platform occurrence. Thus, phosphorites are not an invariably reliable indication of divergence upwelling on the mid-latitude, western margins of a true ocean. Phosphorites associated with anoxic black shales may be deposited on a continental platform well removed from the oceanic margins although ultimately connected with the oceanic reservoir as a source of phosphorus (Heckel, 1977; Lane, 1978). Caution is in order when trying to interpret the environmental significance of pre-Carboniferous phosphorites where our understanding of paleogeography is unclear.

PERIOD-BY-PERIOD SUMMARY

Early Cambrian

The minor Lower Cambrian phosphorites of eastern Newfoundland, southern New Brunswick, and Bornholm may be due to either dynamic upwelling or divergence upwelling. The major Lower Cambrian phosphorites of the Chinese Platform — Mongolia, Vietnam, China, and parts of the Soviet Union — are not amenable to the divergence upwelling interpretation if one employs a pangaeic restoration. These Asiatic phosphorites could, however, be viewed as the result of some pattern of dynamic upwelling allied to the northerly bordering Angara Geosyncline as a source of phosphate-rich water. Phosphorites from Kazakhstan in the Angara Geosyncline can be ascribed to either dynamic upwelling or to divergence upwelling depending on the continental reconstruction one adopts. The phosphorites of the Angara Geosyncline intertongue to the north on the flank of the Siberian Platform with carbonates and manganiferous carbonates, and they interbed with a phosphate-rich sequence of rocks to the south, on the flank of the Chinese Platform. Phosphorites of the geosynclinal areas intertongue with terrigenous and volcanic-rich strata. The South Australian phosphorite deposits may have some relation to the nearby Tasman Geosyncline in a dynamic upwelling mode.

Chinese Platform rocks include far more dark-coloured mudstone and argillaceous, non-carbonate-rich rocks than many Lower Cambrian platforms. Such rocks have been categorized by Berry and Boucot (1970) as Platform Mudstones. The Platform Mudstones commonly indicate significantly greater water depth than one normally finds on platforms. In terms of lithofacies and depth, as well as biota, the Chinese deposits might be compared to the Chattanooga Shale of the latest Devonian-Mississippian, as well as to the Silurian graptolite shales of Europe and North Africa. The Sinian and post-Lower Cambrian Early Paleozoic strata of the Chinese Platform are of the Platform Carbonate type, except for the Lower Silurian Platform Mudstones.

R. Heath (written commun., 1979) points out that the generation of widespread phosphorites requires a suitable oxygen minimum. All that can be suggested for the Chinese Platform is that it might have been covered to a great enough depth, possibly about equivalent to shelf margin, for a transplatform current regime to have maintained the necessary oxygen minimum. Other platforms. such as the North American during the Late Devonian-Mississippian Chattanooga Shale interval, have been situated at such depths and have had phosphorites associated with them.

Ziegler and McKerrow (1975) have discussed the distribution of Lower Cambrian red marine sediments from a number of areas in Britain and the Northern Appalachians, which are consistent with the distribution in the same areas of carbonate rocks. Lower Cambrian manganiferous sediments are also known in Britain and the Northern Appalachians.

The major evaporites of the Cambrian and Ordovician may be placed in a southern hemisphere dry belt if one employs a pangaeic reconstruction. Lower Cambrian evaporites (Meyerhoff, 1970) are entirely consistent with the presence of a mid-latitude arid belt in the Southern Hemisphere.

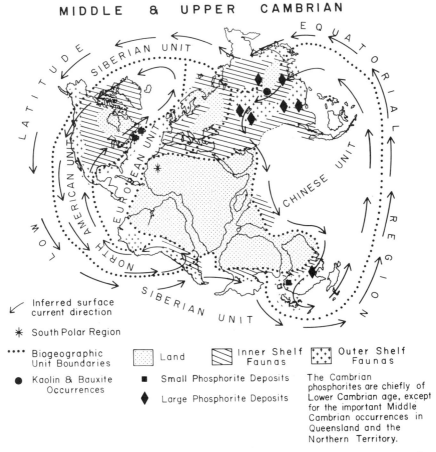

Figure 1. Pangaeic reconstruction for Middle and Late Cambrian time. All maps (Figs. 1-6) show South Polar ("star") and "equatorial regions" whose placement is considered to be generalized. We have not shown other latitudinal parallels due to lack of precise criteria for positioning them. The figures are *not* to be viewed as maps with precise coordinates, nor are they always geographically precise with regard to continent size and latitudinal placement. The figures are intended only as diagrams for consideration in evaluating the reasonableness of a Paleozoic pangaeic hypotheses and the consistency of the biogeographic and various climatic criteria which we have discussed.

ORDOVICIAN

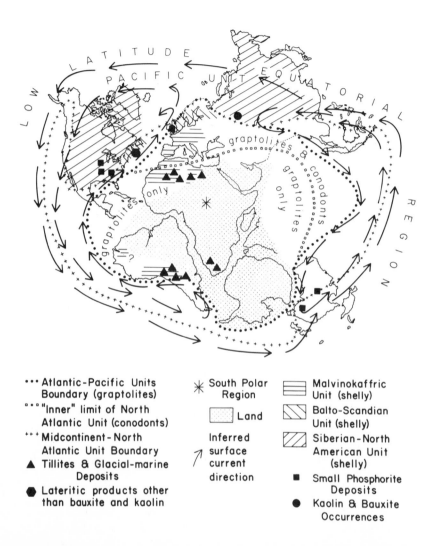

··· Atlantic-Pacific Units
    Boundary (graptolites)
°°° "Inner" limit of North
    Atlantic Unit (conodonts)
⁺⁺⁺ Midcontinent-North
    Atlantic Unit Boundary
▲ Tillites & Glacial-marine
    Deposits
● Lateritic products other
    than bauxite and kaolin

✳ South Polar
    Region
▢ Land

    Inferred
╱ surface
    current
    direction

▤ Malvinokaffric
    Unit (shelly)
◲ Balto-Scandian
    Unit (shelly)
▨ Siberian-North
    American Unit
    (shelly)
■ Small Phosphorite
    Deposits
● Kaolin & Bauxite
    Occurrences

Figure 2. Pangaeic reconstruction for the Ordovician. Recent information (Rong, 1979) indicates that at least the southern half of China belongs to the Malvinokaffric Unit (with a *Hirnantia* -type fauna) in the Ashgill. Thus the boundary of the Siberian-North American Unit is well to the north of central China during that interval. However, by Silurian time the Siberian-North American Unit boundary (the Uralian-Cordilleran Region of the Upper Silurian) must move to the south to account for the warm-water Chinese faunas of the Silurian and Devonian. As a result of this new information it should be noted that cold-water faunas cover an area during the Ashgill exceeding that of the Gondwanan biota during the Permo-Carboniferous.

UPPER SILURIAN

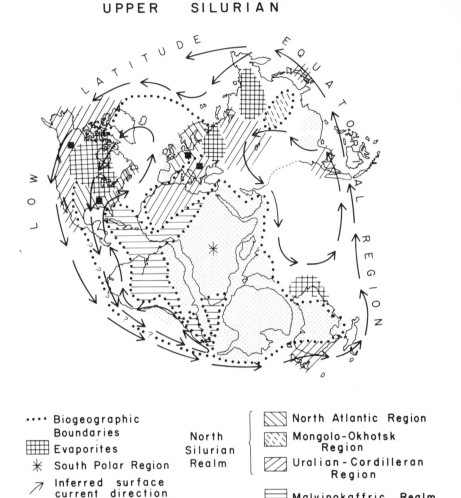

•••• Biogeographic Boundaries		⊠ North Atlantic Region
⊞ Evaporites	North Silurian Realm	⊠ Mongolo-Okhotsk Region
✳ South Polar Region		⊘ Uralian-Cordilleran Region
↗ Inferred surface current direction		
▦ Land		⊟ Malvinokaffric Realm
■ Small Phosphorite Deposits		

Figure 3. Pangaeic reconstruction for the Late Silurian. We have indicated the position of Late Silurian evaporites in what can be construed as a latitudinally oriented dry belt occurring well to the south of the equatorial region. Cock's (1979) recent description of an Upper Silurian Uralian-Cordilleran Region brachiopod fauna from Iran helps to further test the biogeographic units shown here.

Middle and Late Cambrian

Figure 1 for the Middle and Late Cambrian indicates the major Middle Cambrian phosphorites of Queensland and the adjacent Northern Territory in a mid-latitude position on the western side of the Gondwana region; this is consistent with the hypothesis of divergence upwelling. The Upper Cambrian phosphorite of Australia (Victoria) can be viewed in the same manner. The Queensland phosphorites represent outer shelf depth deposits, as indicated by their associated Outer Detrital Belt trilobite faunas (Palmer, 1978).

The Cambrian bauxites of the Sayan Mountains (USSR) are compatible with the tropical to subtropical climatic conditions indicated by the associated carbonate rocks. Cambrian redbeds, both marine and non-marine, are restricted to regions rich in carbonate rocks, as are marine, manganese-rich strata; no redbeds occur in the Atlantic Realm during the Middle and Late Cambrian. The Cambro-Ordovician, possibly Upper Cambrian emery deposit (Massachusetts) is consistent with the restoration shown in Figure 1. Upper Cambrian black shales (such as the alum shales) present in parts of Northern Europe are too restricted in occurrence to be of any value for our purposes. Evaporites are minimal during this time interval.

Ordovician

Figure 2 indicates latest Ordovician tillites and glacial-marine deposits in the Malvinokaffric Realm of the Southern Hemisphere, which is also poor or lacking in carbonate rocks in most places. It should be noted that cold-climate indicators, both biologic and physical, are more widespread at the end of the Ordovician, i.e. there was a great expansion of cold-climate area at the end of the period.

Bárdossy (1973, 1977) reports the presence of Ordovician bauxite in Kazakhstan, while Dewey (1974) has described the lateritic deposits of possible latest Ordovician or slightly earlier age from Nova Scotia. Both of these occurrences are consistent with the distribution of both carbonate and non-carbonate lithofacies as well as the biogeography of the Ordovician.

The minor phosphorites of eastern, central, and western North America, Britain, and central and eastern Australia are additional evidence for more than one mechanism providing phosphate-rich water to both the platform interiors and to the geosynclinal regions. For example, Middle Ordovician phosphorite from Wales may represent either divergence upwelling on the western side of a continent, or dynamic upwelling with a narrow seaway related to the Caledonian-Appalachian geosynclinal complex. The Lower Ordovician phosphorite of Australia (Victoria) (Hill, 1976) with its associated shale and chert, occurs in a graptolite-rich Benthic Assemblage 6 environment where either upwelling mechanism, could have operated.

Ordovician redbeds, both marine and non-marine, are restricted to regions rich in carbonate rocks; the same is true for marine, manganese-rich strata.

Marine evaporites of the Ordovician (Meyerhoff, 1970) are consistent with the existence of a mid-latitude Southern Hemisphere arid belt.

During the Ordovician, the black shale facies is not widespread over any of the major platforms, although it is common in many of the geosynclinal regions (the graptolitic facies), just as the allied agnostid-olenid black shale facies is abundant in the geosynclines of the Cambrian.

Silurian

The Silurian (Fig. 3) is marked by a significant change in climate. First, there are no important Lower Silurian marine evaporites, which in the Late Ordovician indicate a low-latitude, potentially arid belt. The major Silurian evaporites (which are Upper Silurian where their age can be demonstrated) can all be placed in a Southern Hemisphere dry belt distinctly south of the equatorial region (Fig. 3). Second, in contrast to the high climatic gradient of the latest Ordovician, only a moderate climatic gradient is present by the Late Silurian.

There are no important phosphorites known from the Silurian. The minor platform phosphorites of Podolia, central Germany, northwestern Canada and Arkansas are consistent with dynamic upwelling and current transport. The abundant marine and non-marine redbeds are restricted to carbonate-rich regions, as are the marine, manganiferous strata. Lateritic kaolins and bauxites are unknown in the Silurian. The absence of widespread, diversified land plants in the Silurian precludes the development of coals.

During the Silurian, black shale (the graptolitic facies) is abundant for the first time on a major platform (the European-North African). The similarity in lithofacies and fauna both in the black shale and in the non-black shale facies of the Ordovician is one of the major reasons for considering Europe and North Africa as a single platform during the Early Paleozoic. The Lower Silurian of the Chinese Platform is also of the black shale, graptolitic facies. During parts of the Llandovery there are extensive tongues of graptolitic, black shale facies present on the Siberian-Kolyma Platform, which formed a single lithofacies-tectonic entity during the Early Paleozoic.

Devonian

Figure 4 for the later Early Devonian indicates no major phosphorite deposits. Minor phosphorites recognized in such places as Pennsylvania, the Pyrenees, and Australia (Victoria) can be accounted for by dynamic upwelling from proximate geosynclinal regions which would have provided phosphate-rich waters, and by current transport across platforms situated at about shelf-margin depth, which would have provided a suitable oxygen minimum environment.

The presence of significant Middle and Upper Devonian bauxites extending north to south in the Urals as well as in Kazakhstan (Bárdossy, 1977), together with lateritic clays between the Late Silurian and Middle Devonian of Ohio, are climatically consistent with the presence of carbonate rocks in the same regions. The widespread Old Red Sandstone facies of the extra-Malvinokaffric regions also conforms to the warm climatic conditions indicated by physical and biogeographic data. On Figure 4 we have also shown the positions of a low-latitude coal belt and a mid-latitude Southern Hemisphere evaporite belt postulated by Heckel and Witzke (1979) for the Late Devonian in North America. The boundary drawn by them (the C/E boundary of Fig. 4) amounts to a low-latitude humid belt and a mid-latitude arid belt. It is probable that there was no extensive high northern-latitude land area at that time.

During the Early Devonian, the black shale, graptolitic facies covered much of North Africa and southern and central Europe — an extension of the Silurian environment — but extensive black shale is absent on the other major platforms of the Early

LATER   LOWER   DEVONIAN

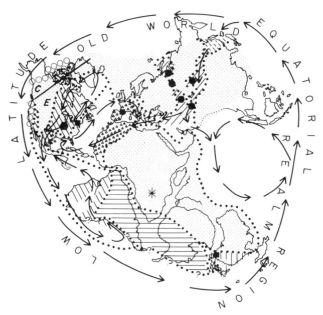

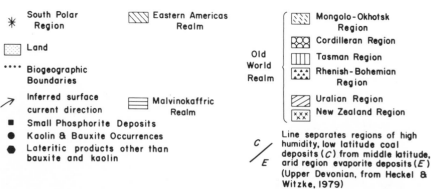

✳	South Polar Region	▨ Eastern Americas Realm	▧ Mongolo-Okhotsk Region

*   ✳ South Polar Region
*   ▦ Land
*   •••• Biogeographic Boundaries
*   ↗ Inferred surface current direction
*   ■ Small Phosphorite Deposits
*   ● Kaolin & Bauxite Occurrences
*   ◕ Lateritic products other than bauxite and kaolin

*   ▨ Eastern Americas Realm
*   ▤ Malvinokaffric Realm

Old World Realm

*   ▧ Mongolo-Okhotsk Region
*   ▩ Cordilleran Region
*   ▥ Tasman Region
*   ⦂ Rhenish-Bohemian Region
*   ▨ Uralian Region
*   ⊠ New Zealand Region

C ╱ E   Line separates regions of high humidity, low latitude coal deposits (C) from middle latitude, arid region evaporite deposits (E) (Upper Devonian, from Heckel & Witzke, 1979)

Figure 4. Pangaeic reconstruction for the later Early Devonian. The equatorial belt shown offshore from western South America is drawn unnaturally close in order to fit the diagram into a reasonably sized space.

Devonian. Beginning in the later Early Devonian in eastern North America, there is a Benthic Assemblage 5-6 black mudstone to shale environment (Esopus Shale and parts of the Shriver Formation) which gradually expands until by Famennian times it covers much of the continent (older parts of the Chattanooga Shale). Similarly widespread black shale facies area absent from the major platforms. The styliolinid black shale facies of the Devonian is widespread within the geosynclinal regions, but almost absent from platforms outside of North America.

Early Carboniferous

Figure 5 shows that a significant Northern Hemisphere landmass is present for the first time in the Early Carboniferous. There are abundant kaolinitic clay and bauxite resulting from lateritic weathering, which is consistent with the abundance of carbonate rocks and the absence of any really cold-climate rocks or biota. The Angara Floral Unit and the Taimyr-Alaskan Unit (based on Foraminifera) are best viewed as warm-temperate rather than as temperate or colder equivalents, despite their northerly positions.

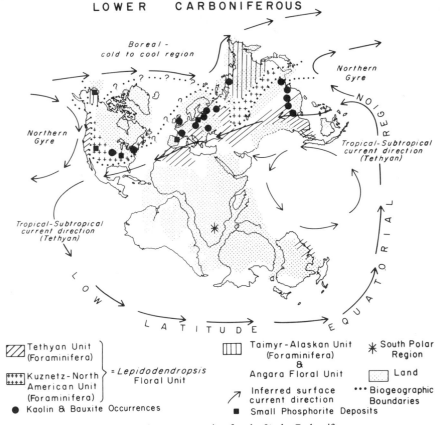

Figure 5. Pangaeic reconstruction for the Early Carboniferous.

The phosphorites of northern Alaska and the Rocky Mountains may be interpreted as the results of oceanic divergence upwelling on the western border of North America, and those of Southern France as the result of dynamic upwelling in a geosynclinal region. Lane (1978) has pointed out that in the starved basin "Kulm" facies of eastern North America (upper Chattanooga and its equivalents), as well as in northern Europe and the western U. S., there is an association of black shales and phosphatic sediments representing deep-water conditions offshelf from carbonate platform sediments, although still belonging to the platforms. The Chattanooga Shale phosphates require a geosynclinal or oceanic source to the south, the Appalachian-Ouachita Geosyncline, which is inferred to have been located at low, equatorial latitudes. Dynamic upwelling into the platform interior is reasonable as a cause for these deposits. The Mississippian phosphates in the Cordilleran region of western North America, although well interior of the inferred oceanic margin, are consistent with divergence upwelling of phosphate-rich waters onto the continental platform. The same may be true for the phosphates of northern Alaska. The European phosphates may be associated with the transport of phosphate-rich waters across the platform from a low-latitude geosynclinal region.

The abundance of redbeds, marine and non-marine, in the Early Carboniferous is in agreement with the other climatic indicators. The evaporites of the Early Carboniferous (Meyerhoff, 1970) in such areas as the Northern Appalachians, and the coals on the Russian Platform suggest low-latitude humid conditions and mid-latitude Northern Hemisphere arid conditions. The coal-evaporite conditions coincide with the foraminiferal biogeographic data, viz. Tethyan Unit for the coals and Kuznetz-North American Unit for the evaporites.

Late Carboniferous-Early Permian

Figure 6 for the late Carboniferous-Early Permian indicates a highly differentiated climatic regime. The Gondwana Realm is characterized by the presence of many glacial centers, although it was emphasized by Crowell (1978) that these are not all contemporaneous. Possible glacial-marine beds of Upper Permian age are known only in northern Siberia, where they occur with the Boreal Realm marine fauna and Angara Floral Unit. The presence of warm Tethyan Realm marine faunas seaward of cooler climate faunas is in agreement with the gyral circulation patterns, as well as with the oceanic divergence upwelling responsible for the important phosphorites of western North America. The Coal/Evaporite boundary deduced by Heckel (1977) for the Late Carboniferous in North America is supported by the mid-latitude arid conditions in the Northern Hemisphere and lower latitude humid conditions. The important Permian evaporites of eastern and western Europe are far enough north to correlate latitudinally with those known from similar positions in North America.

The minor phosphorite present in Ireland may be due to dynamic upwelling, and the Alberta phosphates to divergence upwelling at mid-latitudes on the western side of Pangaea. Minor South African phosphates are of the platform type, occurring in fairly shallow water, and are probably composed of phosphorous derived by dynamic upwelling.

The bauxites and kaolins of China imply a climate favouring lateritic weathering, as do the abundant carbonate rocks. The widespread redbeds of the Late

UPPER CARBONIFEROUS-LOWER PERMIAN

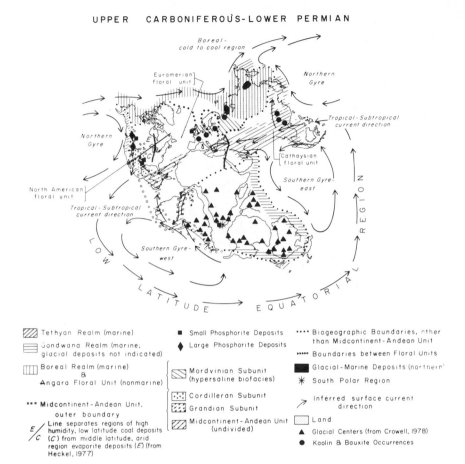

Tethyan Realm (marine)

Gondwana Realm (marine; glacial deposits not indicated)

Boreal Realm (marine) & Angara Floral Unit (nonmarine)

••• Midcontinent-Andean Unit, outer boundary

E/C Line separates regions of high humidity, low latitude coal deposits (C) from middle latitude, arid region evaporite deposits (E) (from Heckel, 1977)

■ Small Phosphorite Deposits

♦ Large Phosphorite Deposits

Mordvinian Subunit (hypersaline biofacies)

Cordilleran Subunit

Grandian Subunit

Midcontinent-Andean Unit (undivided)

•••• Biogeographic Boundaries, other than Midcontinent-Andean Unit

►►►► Boundaries between Floral Units

Glacial-Marine Deposits (northern)

✳ South Polar Region

↗ Inferred surface current direction

Land

▲ Glacial Centers (from Crowell, 1978)

● Kaolin & Bauxite Occurrences

Figure 6. Pangaeic reconstruction for the interval late Carboniferous Early Permian. The area of the landmasses which includes much of eastern North America and Europe as well as parts of North Africa in the North Atlantic region, is not drawn to scale. We have used the same base map for all six figures for comparative purposes, but we see no reason to conclude that the relations of the various blocks remained constant during the Cambrian through Permian interval. Thus, the land area cited above might actually have been far smaller than indicated if the continental blocks were in somewhat different positions. Note that Timor, Tanimbar, Seram, and eastern Sulawesi have been left attached to southeastern Asia rather than being partly, together with the Sahul Shelf, left with Australia (see Audley-Charles, 1978, for a preferred interpretation following in Teichert's pioneering footsteps, which leaves them attached to Australia-New Guinea in the Permian).

Carboniferous-Early Permian correlate with the carbonate—non-carbonate rock distribution pattern. Abundant redbeds in the Late Permian of the former Gondwana Realm of the Southern Hemisphere herald the very low climatic gradient characteristic of the Early Triassic. Redbeds are notable by their absence during the Late Carboniferous-Early Permian in the Gondwana Realm, and by their appearance there in the Late Permian.

Neither the Late Carboniferous nor the Permian are characterized anywhere by widespread platform black shale facies, although Heckel (1977) has described intermittantly present black shales over much of the North American Midcontinent.

ACKNOWLEDGEMENTS

We are grateful to Dr. Ross Heath, College of Oceanography, Oregon State University, Corvallis, and Dr. R.P. Sheldon, U.S. Geological Survey, for giving us some useful thoughts on the origins of marine phosphorites. We are indebted to Dr. John Hower, University of Illinois, Urbana, for comments regarding the origin of kaolin, and to Dr. M. Kuzvart, Prague, for advice concerning the distribution of kaolins. Dr. G. Bárdossy, Budapest, provided information on the distribution of ancient bauxites. Dr. Norman Hatch, U.S. Geological Survey, Denver, and Dr. John Rosenfeld, University of California, Los Angeles, both assisted with information concerning the age of the Chester, Massachusetts emery deposit. Dr. Phillip Heckel and Mr. Brian Witzke, University of Iowa, Iowa City, provided us with a helpful review of an early draft of the manuscript. Dr. Jørn Thiede, University of Oslo, Norway, provided a number of critical, useful comments on the manuscript, as did Dr. Brian Chatterton, University of Alberta, Edmonton, Canada.

REFERENCES

Audley-Charles, M.G., 1978, The Indonesian and Philippine Archipelagos: in Moullade, M., and Nairn, A. E. M., eds., The Phanerozoic geology of the world II, The Mesozoic, A: Elsevier, p. 165-207.

Bárdossy, G., 1973, Bauxite formation and plate tectonics: Acad. Scient. Hungaricae Acta Geol., v. 17, p. 141-154.

————————, 1977, Karsztbauxitok: Budapest, Akademiai Kiado, 398 p.

————————, 1979, Growing significance of bauxites: Episodes, p. 22-25.

Bárdossy, G., and Fontboté J.M., 1977, Observations on the age and origin of the reported bauxite at Portilla de Luna, Spain: Econ. Geol., v. 72, p. 1355-1358.

Barnes, V.E., 1954, Phosphorite in Eastern Llano Uplift of Central Texas: University of Texas, Bureau of Economic Geology, Report of Investigations No. 23, 9 p.

Berry, W.B.N. and Boucot, A.J., 1967, Continental stability – A Silurian point of view: Journal of Geophysical Research, v. 72, p. 2254-2256.

————————, eds., 1970, Correlation of the North American Silurian rocks: Geol. Soc. America Spec. Paper 102, 289 p.

————————, eds., 1974, Correlation of the Silurian rocks of the British Isles: Geol. Soc. America Spec. Paper 154, 154 p.

Bidaut, H., 1953, Note préliminaire sur un mode de formation possible des phosphates Dinantiens des Pyrénées: International Geological Congress, 19th, Alger 1953, Comptes Rendus, Sect. XI Origine des gisements de phosphates de chaux, p. 185-190.

Blisset, A.H. and Callen, R.A., 1967, Myponga phosphate deposit: Mineral Resources Review, Dept. Mines S. Australia, No. 127, p. 93-102.

Boucot, A.J., 1969, The Soviet Silurian: Recent impressions: Geol. Soc. America Bull. v. 80, p. 1155-1162.

_____, 1975, Evolution and extinction rate controls: Amsterdam, Elsevier, 427 pp.

Boucot, A.J. and Gray, J., 1978, Comment on 'Catastrophe theory: Application to the Permian mass extinction': Geology, v. 6, p. 646-647.

_____, 1979, Epilogue: A Paleozoic Pangaea? in Gray, J. and Boucot, A.J., eds., Historical biogeography, plate tectonics, and the changing environment: Corvallis, Oregon State University Press, p. 465-482.

Bracewell, S., 1962, Bauxite, alumina and aluminium: London, Overseas Geol. Surveys, Mineral Resources Division, 235 p.

British Sulphur Corporation, 1964, A world survey of phosphate deposits: London, The British Sulphur Corp., Ltd., Second Edition, 206 p.

_____, 1971, World survey of phosphate deposits: London, The British Sulphur Corp., Ltd., Third Edition, 180 p.

Brown, C.E., 1966, Phosphate deposits in the basal beds of the Maquoketa Shale near Dubuque, Iowa: in Geol. Survey Research 1966, U.S. Geol. Survey Prof. Paper 550-B, p. B152-B158.

Bushinski, G.I., 1964, On shallow water origin of phosphorite sediments: in van Straaten, L.M.J.U., ed., Deltaic and shallow marine deposits: Developments in sedimentology, v. 1: Amsterdam, Elsevier, p. 62-70.

_____, 1969, Old phosphorites of Asia and their genesis: (Academy of Sciences of the USSR, Geological Institute) Jerusalem, Israel Program for Scientific Translations, 266 p.

Callen, R.A., 1969a, Sedimentary phosphate exploration. Pt. 2, The Cambrian and Proterozoic of Fleurieu Peninsula, the Mount Lofty Ranges, and Yorke Peninsula: Mineral Resources Review, Dept. Mines S. Australia, No. 130, p. 80-94.

_____, 1969b, Sedimentary phosphate exploration. Pt. 1, The Cambrian of the Flinders Ranges: Mineral Resources Review, Dept. Mines S. Australia, No. 130, p. 9-34.

Carter, W.D., 1969, The W.L. Newman Phosphate Mine, Juniata County, Pennsylvania: Pennsylvania Geol. Survey, 4th series, Information Circular 64, 16 p.

Cayeux, L., 1939, Les phosphates de chaux sedimentaires de France: Service Carte Geologique de France Topographie Souterraines, v. 1, 312 p.

Charrin, V., 1963, Vers la concessibilite des phosphates de chaux metropolitains: Chimie et Industries, v. 90, p. 444-449.

Choubert, G., 1959, Coup d'oeil sur la fin du Precambrien et le debut du Cambrien dans le Sud marocain: Notes et Mem. Serv. Geol. Maroc, v. 17, p. 7-34.

Cocks, L.R.M., 1979, A silicified brachiopod fauna from the Silurian of Iran: British Museum Natural History (Geol.) Bull., v. 32, p. 25-42.

Conant, L.C. and Swanson, V.E., 1961, Chattanooga Shale and related rocks of central Tennessee and nearby areas: U.S. Geol. Survey Prof. Paper 357, 91 p.

Cook, P.J., 1967, Winnowing – an important process in the concentration of the Stairway Sandstone (Ordovician) phosphorites of central Australia: Jour. Sediment. Petrology, v. 37, p. 818-828.

_____, 1972a, Petrology and geochemistry of the phosphate deposits of Northwest Queensland, Australia: Econ. Geol., v. 67, p. 1193-1213.

_____, 1972b, Sedimentological studies on the Stairway Sandstone of central Australia: Australia, Bureau of Mineral Resources, Geol. and Geophys. Bull. 95, 73 p.

_____, 1976, Georgina Basin phosphatic province, Queensland and Northern Territory-regional geology: in Knight, C.L., ed., Economic geology of Australia and Papua New Guinea. 4. Industrial minerals and rocks: Australasian Instit. Mining and Metallurgy Monograph Series No. 8, p. 245-250.

Crowell, J.C., 1978, Gondwanan glaciation, cyclothems, continental positioning, and climate change: American Jour. Sci., v. 278, p. 1345-1372.

Dale, T.N., 1915, The Cambrian manganese deposits of Conception and Trinity Bays, Newfoundland: Proc. American Philosophical Soc., v. 54, p. 371-456.

Davies, D.C., 1875, The phosphorite deposits of North Wales: Quarterly Jour. Geol. Soc. London, v. 31, p. 357-367.

Demina, V.N., Gulyaev, G.P., and Kolokoltsev, V.G., 1970, O Paleozoiskikh gibbsitovikh boksitakh Timana: Akademia Nauk SSSR, Geologiya Rudnikh Mestorozhdenii, v. 2, p. 65-72.

Dennison, J.M., 1963a, Geology of the Keyser Quadrangle: West Virginia Geol. and Econ. Survey, Geologic Map GM-1.

_____, 1963b, Geology of the Keyser Quadrangle: Privately printed, University of Illinois, 32 p.

Dewey, J.F., 1974, Bears Brook Volcanic Group: Dunn Point Formation, McGillivray Brook Formation, Malignant Cove Formation: in Boucot, A.J., et al., 1974, Geology of the Arisaig area, Antigonish County, Nova Scotia: Geol. Soc. America, Spec. Paper 139, pp. 89-91.

Dmitriev, F.L., Kurochka, V.P., Maevskaya, N.D., Pushkina, S.A., and Frolov, G.N., 1975, Davsonit v nizhnekamennoogolnikh otlozheniyakh Pripyatskoi Vpadini v Belorussii: in Bushinski, G.I., ed., Problemi geniza boksitov: Akademia Nauk SSSR, Nauchnii sovet po rudoobrazovannoo, Izdatel'stvo "Nauka", p. 291-302.

Font-Altaba, M. and Closas, J., 1960, A bauxite deposit in the Paleozoic of Léon, Spain: Econ. Geol., v. 55, p. 1285-1290.

Franke, W., Eder, W., and Engel, W., 1975, Sedimentology of Lower Carboniferous shelf-margin (Velbert Anticline, Rheinisches Schiefergebirge, W. Germany): Neues Jahrbuch für Geologie und Paläontologie, Abh., v. 150, p. 314-353.

Freyer, G., and Tröger, K., 1959, Über Phosphoritknollen im vogtländisch-ostthüringischen Silur: Geologie, v. 8, p. 168-188.

Friend, P.F. and Williams, B.P.J., eds., 1978, A field guide to selected outcrop areas of the Devonian of Scotland, the Welsh Borderland and South Wales: International Symposium on the Devonian System (P.A.D.S. 78), Paleontol. Assoc., 106 p.

Geyer, A.R., Smith, R.C., II, and Barnes, J.H., 1970, Mineral collecting in Pennsylvania: Pennsylvania Geol. Survey, 4th series, General Geology Report 33, p. 201-202.

Gorzynski, Z., 1968, Carboniferous bauxites and argillites: International Geological Congress, 23rd, Prague 1968, v. 14, Proceedings Symposium I, Genesis of the kaolin deposits, p. 115-124.

Harder, E.C., 1949, Stratigraphy and origin of bauxite deposits: Geol. Soc. America Bull., v. 60, p. 887-908.

Heckel, P.H., 1977, Origin of phosphatic black shale facies in Pennsylvanian cyclothems of Mid-continent North America: Amer. Assoc. Petroleum Geol. Bull., v. 61, p. 1045-1068.

Heckel, P.H. and Witzke, B.J., 1979, Devonian world palaeogeography determined from distribution of carbonates and related lithic palaeoclimactic indicators: Palaeontol. Assoc. Spec. Papers in Palaeontology, no. 23, p. 99-124.

Herbert, C., 1976, Phosphate-New South Wales: in Knight, C.L., ed., Economic geology of Australia and Papua New Guinea. 4. Industrial minerals and rocks: Australasian Instit. Mining and Metallurgy Monograph Series No. 8, p. 278-280.

Hill, M., 1976, Phosphate-Victoria: in Knight, C.L., ed., Economic geology of Australia and Papua New Guinea. 4. Industrial minerals and rocks: Australasian Instit. Mining and Metallurgy Monograph Series No. 8, p. 280-282.

Il'in, A.V., 1977, The Arasan manganese ore deposit: Lithology and Mineral Resources, v. 12, p. 196-202.

Inners, J.D., 1975, The stratigraphy and paleontology of the Onesthequaw Stage in Pennsylvania and adjacent states: Ph.D. Dissertation, University of Massachusetts, Amherst.

Jaeger, H., 1976, Das Silur und Unterdevon vom thuringischen Typ in Sardinien und seine regionalgeologische Bedeutung: Nova Acta Leopoldina, Neuefolge, No. 224, Bd. 45, Franz-Kossmat-Symp., p. 263-299.

Johns, R.K., 1976, Phosphate-South Australia: in Knight, C.L., ed., Economic geology of Australia and Papua New Guinea. 4. Industrial minerals and rocks: Australasian Instit. Mining and Metallurgy Monograph Series No. 8, p. 282-285.

Keyser, F. de, and Cook, P.J., 1972, Geology of the Middle Cambrian phosphorites and associated sediments of northwestern Queensland: Australia Bureau of Mineral Resources, Geol. and Geophys. Bull. 138, 79 p.

Kharin, G.S., 1969, O zakonomernostyakh razmeshteniya i usloviyakh obrazonvaniya Devonskikh boksitov Salairia: Akademia Nauk SSSR, Sibirskoe Otdelenie, Institut Geologii i Geofiziki, v. 9, p. 39-44.

Kireev, F.A., 1975, Lateritnie kori bibetrivaniya severoonezhskogo boksitonosnogo rayona: in Bushinski, G.I., ed., Problemi genezisa boksitov: Akademia Nauk SSSR, Nauchnii sovet po rudoobrazovannoo, Izdatel'stvo "Nauka", p. 112-125.

Kochnev, E.A., Titova, A.P., Gentshke, O.L., and Chastnikova, L.C., 1975, Genezis i otsenka perspektiv Paleozoiskoi boksitonosnosti zapadnogo Uzbekistana: in Bushinski, G.I.,ed., Problemi genezisa boksitov: Akademia Nauk SSSR, Nauchnii sovet po rudoobrazovannoo, Izdatel'stvo "Nauka", p. 225-234.

Kolokoltsev, V.G., 1975, Paleogeograficheske Osobennosti Formirovaniya Boksitov Kedra-Vapovskikh Mestorozhsenii Timana: in Bushinski, G.I., ed., Problemi genezisa boksitov: Akademia Nauk SSSR, Nauchnii sovet po rudoobrazovannoo, Izdatel'stvo "Nauka", p. 135-156.

Kozlowski, R., 1931, Fosforyty w utwoeach kambryjskich Sandomierza: Service Géologique de Pologne Bull., v. VI, p. 752-756.

Kräusel, R., 1964, Introduction to the palaeoclimatic significance of coal: in Nairn, A.E.M., ed., Problems of palaeoclimatology: Wiley-Interscience, p. 53-57.

Kuzvart, M., ed., and Konta, J., 1968, Kaolin and laterite weathering crusts in Europe: Acta Univ. Carol., Geologica, Nos.1/2, p. 1-19.

Lane, H.R., 1978, The Burlington Shelf (Mississippian, north-central United States): Geologica et Palaeontologica, v. 12, p. 165-176.

Logan, A. and Hills, L.V., eds., 1973, The Permian and Triassic Systems and their mutual boundary: Canadian Soc. Petroleum Geol. Memoir 2, 766 p.

Loughnan, F.C., 1975, Laterites and flint clays in the Early Permian of the Sydney Basin, Australia, and their palaeoclimatic implications: Jour. Sediment. Petrol., v. 45, p. 591-598.

MacRae, J. and McGugan, A., 1977, Permian stratigraphy and sedimentology - southwestern Alberta and southeastern British Columbia: Canadian Petrol. Geol. Bull., v. 25, p. 752-766.

Mansfield, G.R., 1927, Geography, geology, and mineral resources of part of southeastern Idaho: U.S. Geol. Survey Prof. Paper 152, 453 p.

Matthew, W.D., 1893, On phosphate nodules from the Cambrian of southern New Brunswick: Trans. New York Academy of Science, v. 12, series 1, p. 108-120.

McGugan, A. and Rapson-McGugan, J.E., 1972, The Permian of the southeastern Cordillera: Internat. Geol. Congress, 24th, Montreal 1972, Guide to Excursion A16, 27 p.

McKee, E.D. and Crosby, E.J., coordinators, 1975, Paleotectonic investigations of the Pennsylvanian System in the United States: U.S. Geol. Survey Prof. Paper 853, Pt. I, 349 p., Pt. II, 192 p.

McKee, E.D., Oriel, S.S., and others, 1967, Paleotectonic investigations of the Permian System in the United States: U.S. Geol. Survey Prof. Paper 515, 271 p.

McKelvey, V.E., 1967, Phosphate deposits: U.S. Geol. Survey Bull. 1252-D, p. D1-D21.

McKelvey, V.E., Williams, J.S., Sheldon, R.P., Cressman, E.R., Cheney, T.M., And Swanson, R.W., 1959, The Phosphoria, Park City, and Shedhorn Formations in the western phosphate field: U.S. Geol. Survey Prof. Paper 313-A, p. A1-A47.

McPherson, J.G., 1979, Calcrete (caliche) palaeosols in fluvial redbeds of the Aztec Siltstone (Upper Devonian), southern Victoria Land, Antarctica: Sediment. Geol., v. 22, p. 267-285.

Meyerhoff, A.A., 1970, Continental drift: Implications of paleomagnetic studies, meteorology, physical oceanography, and climatology: Jour. Geol., v. 78, p. 1-51.

Miser, H.D., 1922, Deposits of manganese ore in the Batesville District, Arkansas: U.S. Geol. Survey Bull. 734, 273 p.

Morel, P., and Irving, E., 1978, Tentative paleocontinental maps for the Early Phanerozoic and Proterozoic: Jour. Geol., v. 86, p. 535-561.

Nagorskii, M.P., 1958, Genezis Devonskikh boksitov Salairskogo Kryazha: in Boksiti ikh Mineralogiya i Genezisa: Akademia Nauk SSSR, Otdelenie, Geologii-Geograficheskikh Nauk, Izdatel'stvo Nauk SSSR, p. 306-318.

Nikiforova, O.I., and Obut, A.M., 1965, Siluriiskaya Sistema: Stratigrafiya SSSR, v. X, Izdatel'stvo "Nedra", 529 p.

O'Brien, M.V., 1953, Phosphatic horizons in the Upper Carboniferous of Ireland: Internat. Geol. Congress, 19th, Alger 1953, Comptes Rendus, Sect. XI, Origine des Gisements de phosphates de chaux, p. 135-143.

Oswald, D.H., ed., 1967, International symposium on the Devonian System, Calgary 1967: Calgary, Alberta Soc. Petrol. Geol., v. 1, 1055 p.; v. 2, 1377 p.

Ovtracht, A., 1967, Le Devonien du Domaine Nord-Pyreneen Oriental: in Oswald, D.H., ed., International Symposium on the Devonian System, Calgary 1967: Calgary, Alberta Soc. Petrol. Geol., v. 2, p. 27-35.

Palmer, A.R., 1978, Cambrian lithofacies and biofacies patterns in western United States – A model for the Georgina Basin: in Cook, P.J., and Shergold, J.H., eds., Proterozoic-Cambrian phosphorites: Project 156, UNESCO-IUGS, p. 30.

Paproth, E., Stoppel, D., and Conil, R., 1976, Revision micropaleontologique des sites Dinantiens de Zippenhaus et de Cromford (Allemagne): Société Belge Géologie, Paléontologie, Hydrologie Bull., v. 82, p. 51-139.

Patterson, S.H., 1967, Bauxite reserves and potential aluminum resources of the World: U.S. Geol. Survey Bull. 1228, 176 p.

Patton, W.W., Jr. and Matzko, J.J., 1959, Phosphate deposits in northern Alaska: U.S. Geol. Survey Prof. Paper 302-A, 17 p.

Pavlides, L., 1962, Geology and Manganese deposits of the Maple and Hovey Mountains area, Aroostook County, Maine: United States Geol. Survey Prof. Paper 362, 116 p.

Pietzner, H. and Werner, H., 1964, Die Kaolinitsynthese aus Steinkohlenaschen und ihre Bedeutung für die Enstehung der Tonsteine des Ruhrkarbons: Congrès International de Stratigraphie et de Géologie du Carbonifère, 5th, Paris 1963, Compte Rendu, v. 2, p. 667-678.

Poole, F.G., and Sandberg, C.A., 1977, Mississippian paleogeography and tectonics of the western United States: in Stewart, J.H., Stevens, C.H., and Fritsche, A.E., eds., Paleozoic paleogeography of the western United States: Los Angeles Pacific Section of the Society of Economic Paleontologists and Mineralogists, Pacific Coast Paleogeography Symposium 1, p. 67-85.

Poulsen, C., 1960, The Palaeozoic of Bornholm: Internat. Geol. Congress, 21st, Norden 1960, Guide to Excursions A 46 and C 41, 14 p.

Regnéll, G. and Hede, J.E., 1960, The Lower Palaeozoic of Scania. The Silurian of Gotland: Internat. Geol. Congress, 21st, Norden 1960, Guide to Excursions A22 and C17 (Guidebook d), 87 p.

Ritchie, A., 1975, *Groenlandaspis* in Antarctica, Australia and Europe: Nature, v. 254, p. 569-573.

Rong, Jia-yu, 1979, The *Hirnantia* fauna of China with comments on the Ordovician-Silurian boundary: Acta Stratigraphica Sinica, v. 3, p. 1-29.

Runnels, R.T., 1949, Preliminary report on phosphate-bearing shales in eastern Kansas: Kansas Geol. Survey Bull. 82, 1949 Reports of Studies, pt. 2, p. 37-48.

Sapozhnikov, D.G., ed., 1971, Platformennie boksit SSSR: Akademia Nauk SSSR, Institut Petrografi, Mineralogii i Geochimi, Izdatel'stvo "Nevka", 383 p.

Scheltema, R.S., 1979, Dispersal of pelagic larvae and the zoogeography of Tertiary marine benthic gastropods: *in* Gray, J., and Boucot, A.J., eds., Historical biogeography, plate tectonics, and the changing environment: Corvallis, Oregon State University Press, p. 391-397.

Schmid, H.S. de, 1917, A reconnaissance for phosphate in the Rocky Mountains; and for graphite near Cranbrook, B. C.: Summ. Rept. Mines Branch Canada for 1916, No. 454, p. 22-35.

Schopf, J.M., 1972, Coal, climate and global tectonics: *in* Tarling, D.H., and Runcorn, S.K., eds., Implications of continental drift to the Earth Sciences: New York, Academic Press, p. 609-622.

Sheldon, R. P., 1964a, Exploration for phosphorite in Turkey - a case history: Econ. Geol., v. 59, p. 1159-1175.

—————————, 1964b, Paleolatitudinal and paleogeographic distribution of phosphorite: United States Survey Prof. Paper 501-C, p. C106-C113.

Smith, G. le B. and Eriksson, K.A., 1979, A fluvioglacial and glaciolacustrine deltaic depositional model for Permo-Carboniferous coals of the northeastern Karoo Basin, South Africa: Palaeogeog., Palaeoclimatol., Palaeoecol., v. 27, p. 67-84.

Smith, R.W. and Whitlach, G.I., 1940, The phosphate resources of Tennessee: Tennessee Division of Geol. Bull. 48, 444 p.

Snelgrove, A.K., 1938, Mines and mineral resources of Newfoundland: Geol. Survey Newfoundland Information Circular No. 4, p. 119-120.

Spence, H.S., 1920, Phosphate in Canada: Mines Branch, Canada, No. 396, 156 p.

Störr, M., Kuzvart, M., and Neuzil, J., 1978, Age and genesis of the weathering crust of the Bohemian Massif: Schriftenreihe Geol. Wissenschaftliche, v. 11, p. 265-281.

Summerson, C.H., 1959, Evidence of weathering at the Silurian-Devonian contact in central Ohio: Jour. Sediment. Petrol., v. 29, p. 425-429.

Thorslund, P. and Jaanusson, V., 1960, The Cambrian, Ordovician, and Silurian in Västergötland, Närke, Dalarna, and Jämtland, central Sweden: Internat. Geol. Congress, 21st, Norden 1960, Guide to Excursions A23 and C18 (Guidebook e), 51 p.

Trubina, K.N., 1958a, Drevnyaya Lateritnaya kora Bivetrivaniya v rayone Severo-Onezhskikh Mestorozhdenii boksitov: *in* Strakhov, N.M., and Bushinski, G.I., eds., Boksiti, ikh Mineralogiya i Genezis: Akademia Nauk SSSR, Otdelenie Geologo-Geograficheskikh Nauk, Izdatel-stvo Akad. Nauk SSSR, p. 319-334.

—————————, 1958b, Boksitonosnie otlozheniya podmoskovnogo basseina: *in* Strakhov, N.M., and Bushinski, G.I., eds., Boksiti ikh Mineralogiya i Genezis: Akademia Nauk SSSR, Otdelenie Geologo-Geograficheskikh Nauk, Izdatel'stvo Akad. Nauk SSSR, p. 335-346.

van Andel, T.H., 1979, An eclectic overview of plate tectonics, paleogeography, and paleoceanography: *in* Gray. J., and Boucot, A.J., eds., Historical biogeography, plate tectonics, and the changing environment: Corvallis, Oregon State University Press, p. 9-25.

Vetter, P., 1964, Bassin de Decazeville: *in* Voyage d'étude No. 4: Congrès International de Stratigraphie et de Géologie du Carbonifère, 5th, Paris 1963, Compte Rendu, v. 1, p. 63-80.

Walker, T.R., 1967, Formation of red beds in modern and ancient deserts: Geol. Soc. America Bull., v. 78, p. 353-368.

_____, 1975, Red beds in the western interior of the United States: in McKee, L.D., Crosby, E.J., et al., Paleotectonic investigations of the Pennsylvanian System in the United States: U.S. Geol. Survey Prof. Paper 853, p. 49-56.

Wanless, H.R., 1966, Local and regional factors in Pennsylvanian cyclic sedimentation: Kansas Geol. Survey Bull. 169, v. 2, p. 593-606.

Waugh, B., 1973, The distribution and formation of Permian-Triassic red beds: in Logan, A., and Hills, L.V., eds., The Permian and Triassic Systems and their mutual boundary: Canadian Soc. Petrol. Geol. Memoir 2, p. 678-693.

Woodrow, D.L., Fletcher, F.W., and Ahrnsbrak, W.F., 1973, Paleogeography and paleoclimate at the deposition sites of the Devonian Catskill and Old Red facies: Geol. Soc. America Bull., v. 84, p. 3051-3064.

Young, A., 1976, Tropical soils and soil survey: Cambridge University Press, 467 p.

Zaitsev, N.S. and Il'in A.V., 1970, Khubsugulskii Fosforitonosni Bassenii (MNR): Akademia Nauk SSSR, Doklady, Geol., v. 192, p. 391-394.

Zharkov, M.A., 1974, Paleozoic salt-bearing formations of the World: Acad. Sci.USSR, Siberian Branch, Trans. Inst. Geol. Geophysics, Publishing House "Nedra", 391 p.

Ziegler, A.M., and McKerrow, W.S., 1975, Silurian marine red beds: American Jour. Sci., v. 275, p. 31-56.

APPENDIX

Redbeds

Lower Cambrian

Newfoundland and Wales (Dale, 1915). Marine manganiferous beds (carbonate and oxide); associated fauna indicates that these were deposited in a shelf-margin environment.
Northern Appalachians (Ziegler and McKerrow, 1975). Marine redbeds.
Britain (Ziegler and McKerrow, 1975). Marine redbeds.
Northern Mongolia (Il'in, 1977). Marine manganiferous deposits associated with phosphorites.
Soviet Asia (Bushinski, 1964, 1969). Marine manganiferous, chiefly carbonate, beds.
North Africa (Choubert, 1959). Red sandstones.

Ordovician

New York, Ontario, and adjacent areas. Upper Ordovician non-marine Queenston Beds.
Northern Appalachians. Marine redbeds throughout the Ordovician, including such things as the Sillery Formation.
Central Appalachians. Upper Ordovician non-marine and nearshore marine redbeds of the Sequatchie and Juniata Formations. Many marine redbeds in both the Middle and Upper Ordovician.
Southern Appalachians. Middle and Upper Ordovician marine redbeds.
Arkansas (Miser, 1922). Marine manganiferous beds in the Upper Ordovician Fernvale Formation.

Silurian

Britain, Scandinavia, Baltic Region (Ziegler and McKerrow, 1975; Ziegler et al., in Berry and Boucot, 1974). Marine redbeds; non-marine redbeds in Scotland, southern Sweden, southern Norway.

Northern Appalachians (Pavlides, 1962). Manganiferous marine redbeds in northern Maine, adjacent New Brunswick.

Eastern North America (Berry and Boucot, 1970). Non-marine redbeds including units such as the Vernon-Camillus, Bloomsburg, Medina, Kenogami Formations.

Midcontinent (Miser, 1922). Marine manganiferous nodules in the Lower Silurian Cason Shale, Arkansas; some beds are iron rich as well.

Siberian Platform (Nikiforova and Obut, 1965). Non-marine redbeds.

Devonian

Extra-Malvinokaffric regions of the World (see papers in Oswald, 1967). Non-marine Old Red Sandstone facies in eastern North America, arctic North America, northern Europe (east and west), northern Asia, Australia; marine redbeds are also widespread. Manganiferous nodules with lydites from an Upper Devonian marine sequence of the Pyrenees (Ovtracht, 1967).

Carboniferous

United States (McKee and Crosby, 1975). Non-marine Pennsylvanian redbeds, widespread.

Central Appalachians. Mississippian non-marine redbeds, widespread.

Permian

Global (see papers in Logan and Hills, 1973). Marine and non-marine redbeds.

United States (McKee *et al.*, 1967). Marine and non-marine redbeds.

South America (Waugh, 1973). Non-marine redbeds, Upper Permian.

Laterite and Related Soils

Ordovician-Silurian

Nova Scotia (Dewey, 1974). Above unfossiliferous deposits of possible Upper Ordovician age and immediately beneath fossiliferous, marine beds of Lower Silurian, early Llandovery age. Lateritic beds of monsoonal-type weathering.

Silurian-Devonian

Ohio (Summerson, 1959). Lateritic type paleosols, at boundary between uppermost Silurian or lowermost Devonian and the Middle Devonian.

Devonian

Eastern North America (Woodrow *et al.*, 1973). Paleosols, characteristic of monsoonal weathering from Old Red Sandstones Facies.

Britain (Friend and Williams, 1978). Paleosols, characteristic of monsoonal weathering from Old Red Sandstone Facies.

Victoria Land, Antarctica (McPherson, 1979). Upper Devonian paleosol characterized by calcrete of monsoonal weathering type in an Old Red Sandstone environment.

Kaolin, Which May Include Boehmite and Diaspore

Upper Cambrian

North America (Holser and Pykonen have submitted the following data on a sample of white clay collected from immediately beneath the angular unconformity between the Upper Cambrian and the Precambrian Baraboo Quartzite in the stone quarry at Baraboo Falls, Wisconsin) Their data is consistent with the presence of a tropical to subtropical climate that was characterized by intervals of high humidity.

"David Pykonen and I have examined the material by X-ray diffraction of hydrated and glycolated samples, and by differential thermal analysis. The XRD data were compared with standard patterns in the X-ray Powder Diffraction File and in Brown (1961) and Malloy and Kerr (1961); the DTA data were compared with standards in Kauffman and Dilling (1950) and Todor (1976). The main component is clearly kaolinite, with a rather sharp XRD pattern starting with a peak at 7.2A, and endothermic DTA peak at 550-600C. Subordinate amounts of illite (slightly disordered muscovite) give a weaker but well-defined XRD pattern beginning with a characteristic peak at 10.0A, and a strong endothermic DTA peak between 100 and 250C. A weak endothermic DTA peak at 350C suggests that some gibbsite is present; the strongest XRD line for gibbsite at 4.82A may be one observed in the pattern at 4.90A, although there is a coincident line in the typical illite pattern at 4.9A.

All XRD peaks are accounted for by the above mineral constituents, and all DTA peaks except an unexplained *exothermic* peak at 650C. It would be difficult to exclude the presence of some hydrated halloysite, whose XRD and DTA peaks overlap those of kaolinite and illite, but the sharpness of the observed XRD pattern is in contrast to the usual broad lines of halloysite. A scanning electron micrograph would show up any halloysite by its characteristic tubular shape.

References

Brown, G., (ed) (1961) *X-ray Identification and Crystal Structures of Clay Minerals.* Mineralogical Society, London, 544 p.
Kauffman, A.J., Jr. and Dilling E.D. (1950) Differential thermal curves of certain hydrous and anhydrous minerals, with a description of the apparatus used. *Econ. Geol. 45,* 222-244.
Malloy, M.W., and P.F. Kerr (1961) Diffractometer patterns of A.P.I. reference clay minerals. *Amer. Mineralog. 46,* 582-605.
Todor, D.N. (1976) *Thermal Analysis of Minerals.* Abacus Press, Tunbridge Wells, Kent, England, 256 p."

Upper Carboniferous and Permian

Europe (Störr *et al.,* 1978; Kuzvart and Konta, 1968; Pietzner and Werner, 1964). Widespread.
Eastern North America (Patterson, 1967; Keller *in* McKee and Crosby, 1975). High alumina, commonly high kaolin underclays sometimes with associated boehmite and diaspore, Pennsylvanian of Pennsylvania, Ohio, Maryland, Kentucky, Missouri.
Manchuria (Patterson, 1967). Kaolin deposits, high alumina shales, between Ordovician and mid-Carboniferous strata; also high alumina beds in the Permian.
Australia (Loughnan, 1975). Kaolinite-boehmite mixtures and lateritic paleosols lying above Lower Permian basalts. These evidences of very warm conditions in the Early Permian occur in a sequence that includes undoubted tillites, as well as *Glossopteris* flora. These kaolin-bauxite mixtures, and lateritic paleosols are entirely consistent with the presence of a marked warm interglacial episode. We are indebted to Franklyn van Houten for pointing out this significant occurrence.

Bauxite, Diaspore, and Emery Deposits

Bárdossy (1973) provides an excellent summary of bauxite occurrences in the interval Cambrian through Cretaceous, with maps for the Cambrian, Devonian, Carboniferous, Permian.

Cambrian

Chester, Massachusetts (Norman Hatch, written commun., 1979). Emery deposits, within the Cambro-Ordovician bracket and possibly of Upper Cambrian age.
Southern Siberia (Patterson, 1967; Sapozhnikov, 1971). Bauxites, Sayan Region.

Devonian

Spain (Font-Altaba and Colsas, 1960; Bárdossy and Fontboté, 1977). Bauxite originally thought to be of Lower Devonian age is now known to be of Tertiary age.
Northern and Central Urals (Harder, 1949; Patterson, 1967; Sapozhnikov, 1971). Eifel age bauxites.
Central Iran (Paul Sartenaer, person commun., 1979, from Nasser Valleh). Bauxite in limestone, Yazd area, 150 km SE of Isphahan, Eifel-Givet age.
Siberia (Patterson, 1967; Nagorskii, 1958; Kharin, 1969). Eifel age bauxites, Salair Region.
China (Harder, 1949). Bauxites of possible Devonian age.

Carboniferous and Permian

Scotland (Bracewell, 1962). Mid-Carboniferous bauxites, Ayrshire.
France (Vetter, 1964). Upper Carboniferous bauxite, Massif Central.
Poland (Gorzynski, 1968). Bauxite associated with diaspore-rich beds; overlain by Westphalian age beds, underlain by Precambrian rocks.

Russia
Leningrad Region (Patterson, 1967). Bauxite, between Devonian and Carboniferous.
Onega Region (Trubina, 1958a; Kireev, 1975). Early Carboniferous bauxites, associated kaolins, and hematiferous, lateritic deposits.
Moscow Basin Region (Trubina, 1958b). Early Carboniferous bauxites and halloysites.
Southern Timan Region (Demina, et al., 1970; Kolokoltsev, 1975). Kaolinized and bauxitic materials between the Late Devonian and Carboniferous.
Belorussia and Pripiat Regions (Dmitriev et al., 1975). Carboniferous dawsonite and bauxite.
Soviet-Turkestan (Kochnev et al., 1975). Carboniferous bauxite, Tien Shan Region.

China
Yunan (Patterson, 1967). Bauxites between Devonian and Carboniferous.
Kweichow (Patterson, 1967). Bauxites between Ordovician and Early Carboniferous.
Szechuan (Patterson, 1967). Permian bauxites.
Shantung (Patterson, 1967). Late Palaeozoic, i.e. post-Devonian bauxites, associated with coal deposits.

Phosphorites

Cambrian

Eastern Newfoundland (Snelgrove, 1938). Lower Cambrian nodular phosphate above manganiferous carbonates, Avalon Peninsula.

Southern New Brunswick (Matthew, 1893). Lower Cambrian nodular phosphate.

Southern Sweden (British Sulphur Corporation, 1964, p. 199). Cambrian phosphorite, Dalarna.

Scandinavia (Thorslund and Jaanusson, 1960; Regnéll and Hede, 1960; Poulsen, 1960). Lower and Middle Cambrian phosphoritic sandstones and phosphate nodules in sandstone; also local aggregations of linguloids, orbiculoid, and other inarticulate brachiopod shells.

Poland (Kozlowski, 1931). Middle Cambrian phosphoritic sandstone and phosphatic nodules.

Soviet Asia, China and Viet-Nam (Bushinski, 1964, 1969). Lower (mainly) and Middle Cambrian phosphorites. Soviet occurrences from central Kazakhstan easterly into the Maritime Provinces are closely associated with the Cambro-Ordovician Angara Geosyncline.

Mongolia (Zaitsev and Il'in 1970; Il'in 1977). Cambrian, geosynclinal phosphorites associated with manganiferous strata.

Australia

Victoria (British Sulphur Corporation, 1971, p. 171). Upper Cambrian phosphorite.

South Australia (Blisett and Callen, 1967; Callen, 1969a, b; Johns, 1976). Lower Cambrian platform phosphorites.

Northwest Queensland and adjacent Northern Territory (De Keyser and Cook, 1972; Cook, 1972a, 1976). Major Middle Cambrian phosphorites.

Ordovician

Tennessee, Kentucky, Alabama (Smith and Whitlach, 1940; British Sulphur Corporation, 1971, p. 25-29). Middle and Upper Ordovician limestones with phosphatic horizons.

Arkansas (British Sulphur Corporation, 1971, p. 24).

Iowa (Brown, 1966). Basal phosphorite in Upper Ordovician Maquoketa Formation.

Idaho (Mansfield, 1927). Thin layers of Middle Ordovician age in Swan Peak Quartzite.

North Wales (Davies, 1875). Thin phosphorite in geosynclinal Caradoc age beds.

Australia (Cook, 1967, 1972b; Herbert, 1976; Hill, 1976). Middle Ordovician phosphorite in platform rocks, Amadeus Basin; phosphorite in a shale-chert-sandstone, shelf-margin deposit, New South Wales; minor deep-water Lower and Upper Ordovician phosphorites, Victoria.

Bolivia (Harry Tourtelot, written commun., 1979). Minor phosphatic sandstone rich in lingulid brachiopods is the basis of earlier reports (British Sulphur Corporation, 1971, p. 65) of phosphorite in the Bolivian Ordovician. These are insignificant occurrences according to the unpublished report prepared by Tourtelot (written commun., 1979).

Silurian

Germany (Freyer and Tröger, 1959). Phosphorite nodules in a Platform Mudstone sequence, Vogtland.

Arkansas (Miser, 1922). Phosphorite in Lower Silurian dark-colored shale of Cason Formation, associated with small nodules of manganese carbonates.

Podolia (Cayeux, 1939). Phosphorite nodules in dark-colored shales.

Northwestern Canada (B. Chatterton, written comm., 1979) reports "There are minor Silurian phosphorite deposits of Middle Silurian age developed in the western Mackenzie Mountains (in tongues of the Road River Formation — e.g. near Grizzly Bear Lake)."

Devonian

Pennsylvania (Dennison, 1963a, b; Carter, 1969; Geyer *et al.*, 1970; Inners, 1975). Phosphatic nodules from the Helderberg age (Lower Devonian) Mandata Shale; Lower Devonian Needmore Shale, phosphatic nodules.

Germany. Phosphorites of Devonian age have been reported in the older literature from the Lahn-Dill Synclinoria (viz. British Sulphur Corporation, 1971, p. 136), but modern studies (W. Haas, written comm., 1979) indicate that these phosphorites actually occur in the overlying Early Carboniferous Liegende-Alaun-Schiefer at many places.

Australia (Hill, 1976). Lower Devonian, phosphatic shale and chert, Victoria.

Early Carboniferous

North American Platform (Lane, 1978; Poole and Sandberg, 1977; Conant and Swanson, 1961; Smith and Whitlach, 1940). Phosphate sediments, eastern North America, "Külm" facies, and in black shales, western North America. Some of the phosphorite in basal Hardin Sandstone Member of the Chattanooga Shale, Tennessee, may represent reworked Ordovician phosphorite.

Northern Alaska (Patton and Matzko, 1959). Phosphorite in Alapah Limestone Member of the Lisburne Group.

Poland (Cayeux, 1939). Külm facies phosphatic nodules.

Germany (Paproth *et al.*, 1976; Franke *et al.*, 1975). Külm facies phosphatic sediments.

France (Bidaut, 1953; Cayeux, 1939). Dinantian in Pyrenees and Montagne-Noire.

Late Carboniferous

Eastern Kansas (Runnels, 1949; Heckel, 1977). Phosphatic shale from a cyclothemic sequence.

Texas (Barnes, 1954). Phosphorite at base of Pennsylvanian Marble Falls Limestone, central Texas, and above the Mississippian age Barnett Shale.

Alberta Rockies (de Schmid, 1917; Spence, 1920). Phosphate rock in the Rocky Mountain Quartzite.

Ireland (O'Brien, 1953). Phosphorites in Külm facies, County Clare.

Permian

Western United States (McKelvey *et al.*, 1959; British Sulphur Corporation, 1971, p. 8-21). Rocky Mountain phosphate field.

Alberta (McGugan and Rapson-McGugan, 1972; MacRae and McGugan, 1977). Ishbel Group phosphorites, western and southwestern Alberta. Ishbel Group may be regarded as the northern extension of the Phosphoria Formation.

South Africa (British Sulphur Corporation, 1971, p. 85). Phosphatic nodules associated with manganiferous shales in the Ecca and Dwyka near Durban.

The Continental Crust and Its Mineral Deposits, edited by D.W. Strangway,
Geological Association of Canada Special Paper 20

STRUCTURAL TELESCOPING ACROSS THE APPALACHIAN OROGEN AND THE MINIMUM WIDTH OF THE IAPETUS OCEAN

Harold Williams
Department of Geology,
Memorial University of Newfoundland,
St. John's, Newfoundland A1B 3X5

ABSTRACT

The three major orogenic episodes recognized in the Appalachian Orogen, viz. the Taconic (Ordovician), the Acadian (Devonian), and the Alleghanian (Permian-Carboniferous), all involve compression and shortening across the orogen. Only the Taconic Orogeny can be related directly to the closure of Iapetus and the destruction of its continental margins.

The widest preserved remnant of Iapetus ophiolite suites overlain by volcanic arc sequences, occurs in the interior part of the Appalachian Orogen in northeast Newfoundland. Its preservation is controlled by the geometry of opposing continental margins. This remnant has a present width of 150 km, but palinspastic restoration of its steeply dipping, imbricated ophiolite suites implies a minimum width of 180 km before Alleghanian deformation, 500 km before Acadian deformation and 1000 km before Taconic deformation. Transported ophiolite suites like the Bay of Islands Complex, and related plutonic complexes that originated in the oceanic tract require an additional 100 km in the initial width of Iapetus. Indirect considerations such as the geometry of modern subduction zones and the duration of island arc volcanism imply a still wider Iapetus.

An analysis of structural telescoping across the western margin of Iapetus (ancient continental margin of eastern North America), combined with palinspastic restoration of coeval sedimentary sequences within the Humber Arm Allochthon imply a continental slope/rise of at least 200 km in width.

The 10,000 km length of the Appalachian-Caledonides Orogen, which includes equally continuous tectonic-stratigraphic facies belts related to ancient continental margins, demands a major, wide Iapetus; modern continental margins of comparable dimensions border major oceans.

RÉSUMÉ

Les trois épisodes orogéniques majeurs reconnus dans l'orogène des Appalaches, c'est-à-dire l'épisode taconique (Ordovicien), acadien (Dévonien) et alléghanien (Permien-Carbonifère), impliquent tous une compression et un racourcissement perpendiculaire à l'orogène. Seulement l'orogénèse taconique peut se rattacher directement à la fermeture de l'Iapetus et à la destruction de ses bordures continentales.

Le plus grand vestige de l'Iapetus qui a été préservé, les suites ophiolitiques surmontées de séquences d'arcs volcaniques, se rencontre dans la partie intérieure de l'orogène des Appalaches au nord-est de Terre-Neuve. Sa préservation est contrôlée par la géométrie des bordures continentales opposées. Ce vestige a une largeur actuelle de 150 km mais la restauration palinspastique de ses suites ophiolitiques imbriquées à fort pendage implique une largeur minimum de 180 km avant la déformation alléghanienne, de 500 km avant la déformation acadienne et de 1000 km avant la déformation taconique. Les suites ophiolitiques transportées comme le complexe de Bay of Islands et les complexes plutoniques associés qui ont comme origine un segment océanique impliquent un 100 km additionnel dans la largeur de l'Iapetus. Les considérations indirectes, comme la géométrie des zones modernes de subduction et la durée du volcanisme d'arc insulaire impliquent un Iapetus encore plus large.

Une analyse du télescopage structural normal à la bordure ouest de l'Iapetus (ancienne bordure continentale de l'est de l'Amérique du Nord) combinée avec une restauration palinspastique des séquences sédimentaires de même époque à l'intérieur de l'allochthone de Humber Arm impliquent un complexe pente/glacis continental d'au moins 200 km de largeur.

La longueur de 10 000 km de l'orogène appalachien-calédonien qui comprend également les zones continues de faciès tectonostratigraphiques reliés aux anciennes bordures continentales laisse supposer un océan Iapetus très grand et très large; les marges continentales modernes de dimensions comparables se situent en bordure d'océans majeurs.

INTRODUCTION

Since the advent of plate tectonics, models for the development of the Appalachian Orogen follow the suggeston of Wilson (1966) and involve the generation and destruction of a late Precambrian-Early Paleozoic Iapetus Ocean (e.g., Dewey 1969; Stevens, 1970; Bird and Dewey, 1970; Harland and Gayer, 1972; Williams *et al.* 1972; St. Julien and Hubert, 1975.) Realization that the on-land ophiolite suite of rock units represents ancient oceanic crust and mantle (e.g. Stevens, 1970; Church and Stevens, 1970, 1971; Williams, 1971; Dewey and Bird, 1971) allows the identification and delineation of vestiges of Iapetus wherever such rocks are recognized in the Appalachian Orogen. In the absence of ophiolites, the position of the Iapetus suture within the orogen can be deduced by stratigraphic and structural analyses of rocks that evolved at its continental margins (Williams and Stevens, 1974; Kennedy, 1976; Williams, 1978a).

Rocks of the Appalachian Orogen range in age from Late Precambrian to Early Mesozoic, but only rocks of Middle Ordovician and earlier age are important in defining the margins and vestiges of Iapetus. These older rocks show sharp contrasts in thickness, facies and structural style in different parts of the orogen, and this observation has led to several stratigraphic-tectonic zonations of the northern parts of the system (Williams *et al.*, 1972, 1974; Williams, 1976, 1978b, 1979). In recent syntheses, the Appalachian Orogen is divided into 5 zones, which from west to east are the Humber, Dunnage, Gander, Avalon and Meguma Zones (Williams, 1976, 1978a, 1978b, 1979, Schenk, 1978). This zonation (Fig. 1) fits well with plate models for the

Appalachian Orogen: rocks of the Humber Zone record the development and destruction of the western margin of Iapetus or the ancient Atlantic-type margin of eastern North America; the Dunnage Zone represents vestiges of Iapetus with island arc sequences and local melanges built upon oceanic crust; and the Gander Zone records the development and destruction of a continental margin that lay to the east of Iapetus. (More easterly tectonic-stratigraphic zones of the Northern Appalachians, viz. the Avalon and Meguma Zones, are outside the area of concern in this analysis.)

Thus, vestiges of Iapetus (Dunnage Zone) can be traced in the Northern Appalachians from Newfoundland through northern New Brunswick and the Quebec Eastern Townships (Fig. 1). Farther south the oceanic tract is completely destroyed, but transported vestiges occur in Maryland (Morgan, 1977), and small mafic-ultramafic bodies, possibly of ophiolitic affinity, can be traced to the southern extremity of the Appalachian Orogen (Stevens et al. 1974; Williams and Talkington, 1977; Williams, 1978a).

The western margin of Iapetus (Humber Zone) can be identified by stratigraphic analyses along the full length of the Appalachian Orogen from Newfoundland to Alabama (Rodgers, 1968; Williams and Stevens, 1974; Rankin, 1975). It was initiated in the late Precambrian and evolved as a stable margin up to the time of its Middle Ordovician destruction.

The eastern margin of Iapetus (Gander Zone) is more difficult to identify, possibly because it evolved in places as an Andean-type margin (Strong et al., 1974; Pajari et al., 1977; Williams and Doolan, 1978). It has been traced throughout the Northern

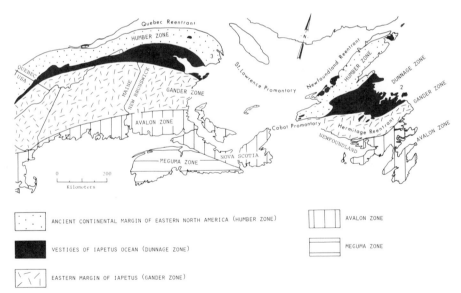

Figure 1. Tectonic elements and zonal subdivision of the Northern Appalachians. Ophiolite complexes in the Humber and Gander Zones represent highly allochthonous Iapetus remnants. Numbers refer to localities mentioned in text: 1, Fredericton Trough; 2, New World Island; 3, Gaspe Peninsula.

Appalachians from Newfoundland to Long Island Sound, New York, but in the Southern Appalachians its stratigraphic record appears to be entirely obliterated. The three major orogenic events of the Appalachian orogenic cycle are Taconic (Ordovician), Acadian (Devonian) and Alleghanian (Permian-Carboniferous). All contributed to shortening and structural telescoping of the Iapetus tract and its continental margins.

Purpose and Scope

The purpose of this paper is to attempt a palinspastic reconstruction of the Cambrian-Ordovician Iapetus Ocean using structural and stratigraphic criteria. Some elements essential to this analysis are deeply buried and others altogether missing. Existing remnants of Iapetus have been shortened by repeated folding and faulting, and some have been severed from their roots, displaced over hundreds of kilometres, and stacked as piles of nappes. A reconstruction of this sort can be done best in places where in situ remnants of the ancient oceanic tract are preserved and where the ancient continental margins of the ocean can be identified in the stratigraphic record. This approach also involves consideration of allochthonous rocks that originated in the oceanic tract. Estimates of initial width and original relationships are based therefore on present surface geometry.

The first consideration is to determine the minimum width of Iapetus as given by its present broadest remnant (Dunnage Zone) in the Appalachian Orogen. To this measured width, the total effects of repeated structural shortening and telescoping are added. Allowance is made also for an area within the oceanic tract that is equal to the widths of transported ophiolitic sequences now found outside the oceanic Dunnage Zone. The width of the Iapetus ocean that was destroyed by subduction is estimated from the geometry of modern subduction zones, rates of modern subduction and the duration of Paleozoic island arc volcanism as preserved in the stratigraphic record.

This simple but powerful analysis demands a major Iapetus Ocean of dimensions similar to those invoked by paleomagnetic data (Briden *et al.*, 1973; McElhinny, 1973; Morel and Irving, 1978) and on paleontologic grounds (Cowie, 1971; Williams, 1972; Skevington, 1974; Spjeldnaes, 1978).

The Lifespan of Iapetus and the Wilson Cycle

The lifespan of Iapetus is not clearly defined, for both the times of its initiation and eventual destruction are equivocal. Certainly it was in existence during the Cambrian and Early Ordovician, for at least 100 Ma.

The time of initiation of Iapetus is the same as the age of mafic dykes and volcanic rocks that cut and unconformably overlie Grenvillian basement along its western margin. The volcanic rocks are associated with thick clastic sequences, and all are interpreted as the products of rifting and the initiation of the Iapetus cycle (Williams and Stevens, 1969, 1974; Strong and Williams, 1972; Rankin, 1976). Isotopic dates of volcanic rocks and dikes range from about 800 Ma (Rankin *et al.*, 1969; Pringle *et al.*, 1971) to 600 Ma (Stukas and Reynolds, 1974a; Doig, 1970) or younger (Clifford, 1968). The initiation of rifting is defined stratigraphically as Early Cambrian or older, for the volcanic rocks and associated clastic sediments are overlain by fossiliferous Lower Cambrian rocks (Williams and Stevens, 1969).

Ophiolite suites, interpreted as the floor of Iapetus, are overlain stratigraphically by fossiliferous Lower Ordovician sequences in Notre Dame Bay, Newfoundland (Upadhyay *et al*., 1971; Kean and Strong, 1975) and by Lower to Middle Ordovician rocks in New Brunswick (Ruitenberg *et al*., 1977; Pajari *et al*., 1977) and Quebec (St. Julien and Hubert, 1975). Where dated isotopically, the Newfoundland ophiolites are late Cambrian to early Ordovician, about 500 Ma old (Mattinson, 1975, 1976; Stukas and Reynolds, 1974b).

The main period of island arc volcanism associated with an open Iapetus occurred during the Early and early Middle Ordovician, for volcanic products of that age directly overlie the ophiolite suite conformably and gradationally (Upadhyay *et al*., 1971). Locally, in Notre Dame Bay, Newfoundland, this period of volcanism commenced earlier, for volcanic rocks of island arc affinity are cut by granite dated isotopically at 510 Ma (Williams *et al*., 1976). Still older volcanic rocks, dated paleontologically as Middle Cambrian (Kay and Eldredge, 1968), occur as blocks in the Dunnage Melange.

During Middle Ordovician, a cessation of volcanism in the oceanic tract is indicated by the deposition of a thin black shale sequence across the island arc volcanic rocks and also across melanges such as the Dunnage in Northeast Newfoundland (Dean, 1978; Hibbard and Williams, 1979). This period of black shale deposition coincided with, or immediately followed, the destruction of the margins of Iapetus and the obduction and emplacement of ophiolite complexes such as the Bay of Islands Complex (Williams and Stevens, 1974; Dean, 1978; Nelson, 1979; Casey and Kidd, 1979). This period of orogenesis is attributed to the classical Taconic Orogeny, and it appears that the oceanic tract was all but destroyed at this time. Intense polyphase deformation and metamorphism affected rocks at the margins of Iapetus in the Humber and Gander Zones, but deformation was less intense across the oceanic Dunnage Zone, where ophiolite complexes are well preserved locally.

Volcanism continued in the Northern Appalachians throughout the Silurian, Devonian and well into the Carboniferous Period. However, volcanic rocks formed during these periods of activity are in most places terrestrial, or they are associated with terrestrial sedimentary rocks. As well, they occur across the entire orogen, showing no preference for localization at the former site of the Iapetus oceanic tract. Similarly, deformation, intrusion and metamorphism associated with the Devonian Acadian Orogeny and Carboniferous Alleghanian Orogeny show no spatial relationship to a Cambrian-Ordovician Iapetus. A sub-Silurian unconformity that is recognized across the deformed continental margins and locally within the oceanic tract of Iapetus, coupled with the lack of any convincing stratigraphic evidence for the existence of a Silurian or younger ocean and contemporaneous continental margins, creates serious difficulties in attempts to relate these later phases of metamorphism, deformation and plutonism to subduction and continental collision (Williams, 1979).

A Silurian Iapetus is permissible only where the base of Silurian stratigraphic sections is unexposed. Thus the Fredericton Trough in New Brunswick has been suggested as the site of a Silurian ocean (McKerrow and Ziegler, 1971), although it is well south of the Cambrian-Ordovician site of Iapetus. Another mid-Paleozoic ocean has been proposed still farther southeast between the Avalon and Meguma Zones of Nova Scotia (Keppie, 1977), and Silurian olistostromes of New World Island, Newfoundland, have been interpreted as oceanic trench-fill deposits (McKerrow and Cocks, 1977, 1978).

There is certainly nothing in the Silurian, Devonian or Carboniferous records to allow an Iapetus restoration of the type attempted here for the Cambrian-Ordovician period. Clearly the Ordovician vestiges of Iapetus were shortened and telescoped by Acadian and Alleghanian orogenic episodes, but these events also affected other parts of the orogen, regardless of their prehistory as either oceanic or continental.

DEFINITION OF THE WIDEST VESTIGE OF IAPETUS

The broadest area in the internal part of the Appalachian Orogen that is floored by rocks of the ophiolite suite is taken as the widest vestige of Iapetus. Such an area provides a point of commencement for a palinspastic restoration, for the ancient Iapetus must have been far wider than its present widest remnant.

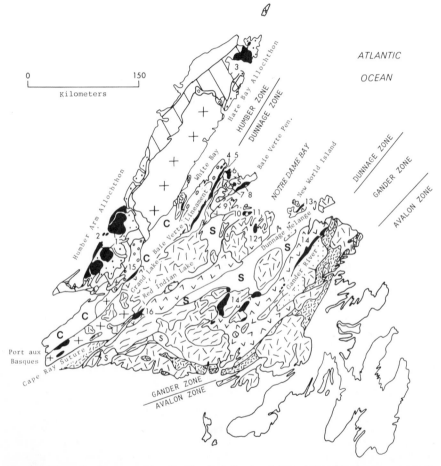

Figure 2. Tectonic elements of insular Newfoundland and the relationships between tectonic-stratigraphic zones and the vestiges and margins of Iapetus. Numbers refer to names of ophiolite complexes and features mentioned in text: 1, Bay of Islands Complex; 2, Little Port

The widest remnant of Iapetus in the Appalachian Orogen is the Northeast Newfoundland Dunnage Zone, where a Lower Ordovician volcanic terrane is floored by local ophiolite occurrences over a present width of 150 km (Figs. 1 and 2). From west to east this terrane contains such ophiolite suites as the Birchy Complex (Williams *et al.*, 1977), Advocate Complex (Kennedy, 1975; Williams *et al.*, 1977), Point Rousse Complex (Norman and Strong, 1975; Williams *et al.*, 1977; Kidd *et al.*, 1978), Nippers Harbour and Betts Cove Complexes (Upadhyay *et al.*, 1971; DeGrace *et al.*, 1976), South Lake Complex (Dean, 1978), Herring Neck Group (Williams and Payne, 1975; Jacobi, pers, commun. 1979) and the Gander River Ultramafic Belt (Kean, 1974; Stevens *et al.*, 1974; Blackwood, 1978). As well, there are local occurrences of sheeted dykes and ophiolitic volcanic rocks at the Springdale Peninsula (Smitheringale, 1972) and Pilleys Island (Strong, 1972) and gabbro at Brighton Island (Stukas and Reynolds,

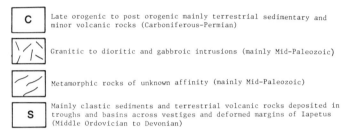

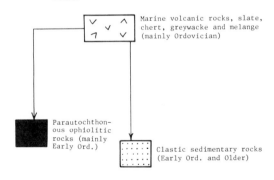

Complex; 3, St. Anthony Complex; 4, Birchy Complex; 5, Advocate Complex; 6, Point Rousse Complex; 7, Nippers Harbour Complex; 8, Betts Cove Complex; 9, Springdale Peninsula; 10, Pilleys Island; 11, Brighton Gabbro; 12, South Lake Complex; 13, Herring Neck Group; 14, Gander River Ultramafic Belt; 15, Glover Island Complex; 1), Annieopsquotch Complex.

1974b). Newly defined ophiolite occurrences, the Annieopsquotch Complex (Herd and Dunning, 1979) at Lloyds River and the Glover Island Complex (Knapp *et al.*, 1979) of Grand Lake, occur southwestward along strike of the coastal exposures (Fig. 2).

The Dunnage Melange (Kay, 1976; Hibbard and Williams, 1979), which occurs within the ophiolitic Dunnage Zone, is also interpreted as overlying oceanic crust, for intrusions localized in the melange contain abundant mafic and ultramafic inclusions. A similar melange that occurs 30 km to the east, and is possibly continuous with the Dunnage in subsurface, contains ultramafic blocks further implying an ophiolitic basement (Pajari and Currie, 1978).

Important questions concerning the evolution of the Northeast Newfoundland ophiolitic Dunnage Zone are: a) whether the zone represents the vestige of a major ocean, a marginal ocean, or several small ocean basins, b) whether its volcanic arc sequences and melanges are parts of a single complex or represent a composite terrane or orogenic collage, and c) where subduction zones are located and what are their polarities. Although essential to plate tectonic models, these questions are disregarded in the following analysis, and the width of Iapetus is estimated from the size of its preserved remnants, regardless of its composite of singular nature.

Rocks of the Dunnage Zone are much less deformed and metamorphosed than those of nearby parts of the Humber and Gander Zones. Local preservation of the widest vestige of Iapetus in Newfoundland appears to relate to the form of adjacent continental margins preserved in the Humber and Gander Zones (Thomas, 1977; Williams, 1979). Thus the Dunnage Zone oceanic tract is widest and best preserved at the matching Newfoundland and Hermitage Re-entrants (Fig. 1), whereas at the matching St. Lawrence and Cabot Promontories, it has been completely destroyed.

SHORTENING AND TELESCOPING OF IAPETUS DURING THE APPALACHIAN OROGENIC CYCLE

The three major orogenic events that are a part of the Appalachian orogenic cycle, viz. the Alleghanian, Acadian and Taconic Orogenies (Rodgers, 1967), all contributed to shortening and structural telescoping of the presently exposed ophiolitic rocks of the Dunnage Zone in northeast Newfoundland. The effects of Alleghanian deformation are minor, those of Acadian Orogeny much more substantial, and those of Taconic Orogeny the most important.

The Newfoundland Dunnage Zone is an ideal area for palinspastic restoration of Iapetus, for apart from its well-preserved Ordovician rocks and structures, it contains Silurian and Devonian rocks across the entire width of the zone. Thus one can clearly judge the shortening and telescoping effects of Acadian Orogeny as evident in Silurian-Devonian rocks and separate these effects from earlier shortening associated with Taconic Orogeny. Similarly, local areas of deformed Carboniferous rocks in the Grand Lake area, and undeformed Carboniferous rocks at Red Indian Lake, allow an estimate of the effects of Alleghanian deformation on older structures.

In the following analysis the effects of each episode of deformation on the 150-km wide vestige of Iapetus in northeast Newfoundland are treated separately, thus allowing estimates of the minimum width of Iapetus as the effects of each are removed.

Alleghanian Orogeny

The effects of the Alleghanian Orogeny at the northern extremity of the Appalachian Orogen are minimal, compared to its effects in parts of mainland Canada and the Southern Appalachians. Deformed Carboniferous rocks occur in a narrow northeast-trending zone in Newfoundland from Port aux Basques to White Bay (Fig. 2). Folding varies from open to tight with steeply dipping beds, and locally narrow zones of older basement rocks occur between steeply dipping Carboniferous strata, suggesting thrust imbrication. Most of this deformation, as recognized in Carboniferous rocks, affects a zone immediately west of the Iapetus Dunnage Zone, except at Grand Lake, where the ophiolitic Glover Island Complex is cut by Carboniferous faults.

Isotopic dates of about 300 Ma, obtained from deformed intrusions at the Baie Verte Peninsula, suggest Carboniferous magmatism and deformation there (Bell and Blenkinsop, 1975a; DeGrace et al., 1976), but this cannot be tested stratigraphically. Similarly, Carboniferous isotopic ages are common among deformed intrusions within the Gander Zone along the eastern margin of the Dunnage Zone (Bell and Blenkinsop, 1975b; Bell et al., 1977, 1979). Locally within central parts of the Dunnage Zone, e.g. at Red Indian Lake, plant-bearing Carboniferous sandstones are undeformed.

Telescoping across the Dunnage Zone as a result of Carboniferous shortening is difficult to estimate but it is judged, conservatively, to be only a small fraction of the present 150 km width. Thus the width of the Dunnage Zone was probably no more than 180 km at the onset of Alleghanian Orogeny.

Acadian Orogeny

Acadian Orogeny has been described as the climactic event in the Northern Appalachians (Neale et al., 1961). It affected wide areas of the orogen, and in eastern Canada its effects are recognized across all of the tectonic-stratigraphic zones. Silurian and Devonian volcanic rocks and intrusions occur well outside the areas of Ordovician igneous activity in the Canadian Appalachians, and Acadian deformation affected all of the Taconic deformed zone and broad areas to the east (Keppie et al., in prep.).

Acadian deformation is expressed mainly by open to tight upright folds with steep axial plane cleavages, and thrust faults are important locally (Dean and Strong, 1977; Williams et al., 1977). Acadian regional metamorphism reaches amphibolite facies in some places, but across the Dunnage Zone, regional metamorphic facies is greenschist or lower.

Most of the Ordovician ophiolitic and arc volcanic rocks across the Newfoundland Dunnage Zone are steeply dipping, vertical, or overturned, mainly as a result of Acadian deformation. This conclusion is based on the fact that nearby Silurian rocks display the same structural style. The vestiges of Iapetus across the entire Dunnage Zone are therefore much shortened and telescoped by Acadian compression. The structural style of the Dunnage Zone as reflected mainly in Silurian rocks is portrayed schematically in Figure 3.

Across the Baie Verte Lineament at the west margin of the Dunnage Zone, Acadian shortening apparently reduced an oceanic domain originally 30 to 50 km wide to a steep structural belt now a mere 1 to 5 km wide (Kidd, 1977); this represents a

Humber Zone | Dunnage Zone | Gander Zone

STYLE AND EFFECTS OF ALLEGHANIAN OROGENY
CARBONIFEROUS TO PERMIAN
REDUCED OCEANIC TRACT TO PRESENT WIDTH OF 150 KM

STYLE AND EFFECTS OF ACADIAN OROGENY
SILURIAN TO DEVONIAN
REDUCED OCEANIC TRACT TO MINIMUM WIDTH OF 180 KM

STYLE AND EFFECTS OF TACONIC OROGENY
MIDDLE TO LATE ORDOVICIAN
REDUCED OCEANIC TRACT TO MINIMUM WIDTH OF 500 KM

CONVERGING IAPETUS WITH SUBDUCTION
LATE CAMBRIAN TO LOWER ORDOVICIAN
MINIMUM WIDTH 1000 KM

SPREADING IAPETUS BEFORE SUBDUCTION
LATE PRECAMBRIAN TO CAMBRIAN
ESTIMATED WIDTH 1000-3000 KM

Figure 3. Destruction of Iapetus and structural telescoping across northeast Newfoundland during the Appalachian orogenic cycle.

minimum shortening of 6:1. Acadian shortening elsewhere is interpreted as much less, judging from the style of upright folds in Silurian rocks with steeply to moderately dipping limbs.

Eastward-directed thrusts at the Baie Verte Lineament affect Silurian-Devonian rocks of the Mic Mac Lake Group (Kidd, 1977; Williams *et al.*, 1977), and some of the major steep faults of Notre Dame Bay are now interpreted as southeast-directed thrusts which were subsequently folded, e.g. the Lukes Arm, Lobster Cove and Chanceport Faults (Dean and Strong, 1977). These and related faults that repeat Ordovician-Silurian stratigraphic sections at New World Island (Kay, 1976) could themselves account for several tens of kilometres of shortening across the Dunnage Zone.

Acadian shortening across the entire Dunnage Zone is estimated conservatively between 2:1 and 3:1. This implies that the Dunnage Zone, which had a minimum width of 180 km before Alleghanian deformation, was at least 500 km wide before Acadian deformation. This compares favourably with widths of present marginal ocean basins, like the Sea of Japan.

Taconic Orogeny

Taconic Orogeny led to the Ordovician destruction of Iapetus and to a complete and permanent change in depositional environments across the ocean basin and its continental margins. The effects of this orogeny are best demonstrated by local unconformities in the Humber Zone and along the margins of the Dunnage Zone; they are also obvious in the stratigraphic record of the Humber Zone and by the presence of large transported terranes that originated within the Iapetus ocean and at its western margin.

On the western margin of Iapetus in the *Humber Zone*, the first intimations of Taconic Orogeny and destruction of the continental margin of eastern North America are a pre-Middle Ordovician disconformity in a shallow water carbonate sequence, and the appearance of black shales followed by siltstones and greywackes above the carbonates. Easterly-derived flysch wedges containing sparse ophiolitic detritus are first recorded in the Lower Ordovician (Arenigian) parts of transported sequences in western Newfoundland; the flysch wedges transgressed westward where they are of early Middle Ordovician (Llanvirnian) age above the carbonate bank sequence. Transported Cambrian-Ordovician clastic sequences comprise the lowest structural slices above the flysch wedges and are succeeded structurally by a variety of volcanic, igneous, and metamorphic rocks in higher structural slices. The ophiolite suites of the Bay of Islands and St. Anthony Complexes form the highest structural slices and represent transported oceanic crust and mantle derived from the Dunnage Zone (Fig. 2).

Ophiolitic blocks in olistostromal melange beneath the lowest sedimentary slices suggest that the slices were assembled before final emplacement (Stevens and Williams, 1973). A local unconformity between transported sedimentary rocks of the Humber Arm Allochthon and overlying Middle Ordovician (Caradocian) limestones of the Long Point Group (Rodgers, 1965; Schillereff and Williams, 1979) confirms the Middle Ordovician emplacement of the allochthon.

Sedimentary rocks that constitute the lower structural slices of the Humber Arm and Hare Bay Allochthons are interpreted as continental slope/rise deposits (Stevens, 1970; Williams, 1975). The superposition of several separate slices of coeval sedimentary rocks within a present width of 50 km, coupled with the shortening effects of transport and post-emplacement deformation, all suggest an initial width of about 200 km for the slope/rise at the western margin of Iapetus. Thus the western margin of Iapetus, represented now by a narrow, 25 km zone of polydeformed and metamorphosed rocks, once rivalled the wide stable margin of the modern Atlantic.

On the eastern margin of Iapetus in the *Gander Zone*, pre-Caradocian deformation is recorded by an unconformity between the Davidsville Group (west) and deformed mafic-ultramafic complexes of probable ophiolitic affinity along the Gander River (Kennedy and McGonigal, 1972; Stevens *et al.*, 1974; Blackwood, 1978). Farther east, the structural style in continental margin clastic rocks is one of recumbent folding, polydeformation and increasing metamorphism away from the ophiolitic rocks. This style is rather analagous to that in easterly parts of the Humber Zone, implying a wide eastern margin of Iapetus comparable to that in the west.

Within the Newfoundland *Dunnage Zone*, ophiolitic rocks of the Nippers Harbour Complex and nearby Lower Ordovician volcanic rocks gradationally above the Betts Cove Complex are overlain unconformably by rocks interpreted as Silurian (Schroeter, 1971; Neale *et al.*, 1975). Similar unconformities are recognized along the Baie Verte Lineament (Kidd, 1977) and beneath dated Silurian rocks of White Bay (Williams, 1977). Thus the margins of Iapetus in Newfoundland were affected by Taconic deformation. Similarly, at Gaspé Peninsula in Quebec, deformed rocks of the Humber Zone continental margin and Dunnage Zone oceanic tract are crossed acutely by Silurian and Devonian rocks of the Gaspé Synclinorium, leaving absolutely no doubt of the Taconic destruction of Iapetus in this region (Williams, 1978a).

The initiation of intrusion and deformation within the Dunnage Zone oceanic domain is more difficult to ascertain. The trondjhemitic Twillingate pluton has been dated isotopically at 510 Ma (Williams *et al.*, 1976) and possible correlatives at Baie Verte Peninsula and Mansfield Cove near Springdale Peninsula are dated at 500 and 594 Ma respectively (R.D. Dallmeyer, pers. commun. 1979; Bostock, 1978). The Twillingate pluton cuts mafic to silicic volcanic rocks of island arc affinity and both are locally mylonitic. The deformed rocks are in turn cut by undeformed mafic dikes with $^{40}Ar/^{39}Ar$ ages between 470 and 440 Ma. These relationships suggest intrusion and local intense deformation commencing in late Cambrian to early Ordovician.

The Dunnage Melange of Northeast Newfoundland is the same age as nearby volcanic arc sequences. It has been interpreted either as a trench-fill related to subduction (Kay, 1976), or a major olistostrome deposited in a back-arc basin (Hibbard and Williams, 1979). Its matrix is dated as uppermost Cambrian-lowermost Ordovician (Tremadocian), although some of its blocks are slightly younger (Arenigian). Regardless of origin, its presence implies a major disturbance within the Dunnage Zone during this early stage of development.

The Birchy and Advocate Complexes of the Baie Verte Lineament (Fig. 2) include several overlapping structural slices of ophiolitic rocks. Similarly the Point Rousse Complex is represented in several structural blocks. All of these ophiolitic complexes at the Baie Verte Peninsula, including the Nippers Harbour and Betts Cove Complexes, are steeply dipping to overturned, but the facing direction of their stratigraphic units is almost everywhere from west to east. This structural style at the Baie Verte Peninsula suggests a Taconic shortening by thrust imbrication of at least 5:1. Thus, the present 40 km width of the steeply dipping ophiolite suites between the Baie Verte Lineament and Betts Cove once represented at least 200 km of ocean floor. Rocks of the Dunnage Zone do not appear to be imbricated by Taconic Orogeny in this same way throughout, but even a modest estimate of 2:1 overall shortening indicates a pre-Taconic Iapetus of 1000 km width.

A 1000 km width for the Dunnage Zone is not unreasonable when one considers the variety of its contrasting Ordovician rocks. For example, the Dunnage Melange forms a steep belt exceeding 5 km in maximum width. It is judged therefore to be extremely thick. If an olistostrome, the deposit must have had lateral dimensions of many tens of kilometres. On the other hand, if the melange represents a subduction phenomenon, its mere presence implies the destruction of many tens or a few hundred kilometres of oceanic crust.

Extreme shortening across the Dunnage Zone of northeast Newfoundland is likely, because of the thickness and nature of its present crust, compared to that of the bounding Humber and Gander Zones. Across the ophiolitic terrane of the Newfoundland Dunnage Zone the present crust is approximately 45 km thick, compared to 30 km in nearby zones (Dainty et al., 1966; Sheridan and Drake, 1968). Furthermore, the additional thickness is attributed to a dense basal layer that could represent imbricated ancient oceanic crust.

PALINSPASTIC RESTORATION OF HIGHLY ALLOCHTHONOUS OPHIOLITE SUITES

Another palinspastic exercise that bears upon the width of Iapetus involves the disassembly of Taconic allochthons and the restoration of their ophiolitic assemblages from their present positions to their places of origin. In western Newfoundland, the Humber Arm and Hare Bay Allochthons (Williams, 1975) contain a sampling of rocks that originated within Iapetus.

Ophiolitic suites such as the Bay of Islands Complex in the Humber Arm Allochthon are interpreted universally now as a sampling of oceanic crust and mantle. Likewise the Little Port Complex of the Humber Arm Allochthon is interpreted as ophiolitic (Karson and Dewey, 1978). The evidence that these complexes originated within the Iapetus Ocean is virtually unassailable. Consequently, in addition to the foregoing considerations bearing on the minimum width of Iapetus, the oceanic tract must be further widened to accommodate that part which is represented within Taconic allochthons.

The widest massif of the Bay of Islands Complex is 25 km. However its rock units are represented now in an upright fold with moderately dipping limbs. Furthermore, the base of this upright structure is truncated by the structural surface of final emplacement and its top is an erosional peneplain. The unfolded restored width of the Bay of Islands Complex is closer therefore to 50 km.

The Bay of Islands Complex has a dynamothermal aureole at its stratigraphic base, interpreted as the result of obduction of hot mantle material. The aureole, now frozen into the folded ophiolite slice, was produced from volcanic rocks and gabbros that are at least in part of ophiolitic affinity. Unstacking of the ophiolite from its aureole therefore requires an initial oceanic width of about 100 km.

The Little Port Complex, now lying west of the Bay of Islands Complex, consists of several blocks and imbricated slices. Its present width of about 10 km represents an initial width of at least 30 km.

Similarly, the St. Anthony Complex of the Hare Bay Allochthon is 30 km wide, and a pronounced magnetic anomaly offshore suggests the presence of a steep imbricated ophiolite belt (Haworth *et al.*, 1976). This local occurrence, complete with aureole of Bay of Islands type, demands a similar effective width of about 100 km.

Restoration of the Humber Zone continental margin indicates that transported ophiolite suites such as the Bay of Islands Complex travelled a minimum distance of 400 km from their place of formation to their present position.

INDIRECT CONSIDERATIONS AND THE WIDTH OF IAPETUS

The duration of ocean opening that preceded Cambrian-Ordovician island arc volcanism is difficult to assess, but the oldest possible ages for the commencement of opening (about 800 Ma) provide ample time for the formation of a 2000 to 4000 km-wide ocean at a conservative spreading rate of 2 cm/yr.

There is no apparent correlation between the amount of preserved ophiolite and the original width of the oceanic tract, for in places such as the Cape Ray Suture in southwest Newfoundland (Brown, 1973), the Dunnage Zone has been completely destroyed (Fig. 2). However, a relationship is to be expected between the duration of island arc volcanism and the amount of oceanic crust that is subducted to sustain such volcanism.

Modern island arcs are built above a sinking slab that reaches depths of at least 100 to 150 km (Isacks and Barazangi, 1977). Gentle dips require the subduction of 250 to 350 km of ocean floor. Furthermore, at a depth of 350 km, the sinking slab is 500 to 1000 km in length.

These estimates of the width of destroyed oceanic crust, based on the geometry of modern subduction zones, are supported by the thickness and duration of volcanic arc deposition in northeast Newfoundland. There, volcanic arc sequences of 10 to 15 km thick were built up over a minimum period of 50 m.y. This period is judged from the minimum age of the oldest volcanic arc rocks that are cut by plutons dated at 510 Ma, and the youngest volcanic arc sequences dated stratigraphically as Middle Ordovician (about 450 to 460 Ma). To sustain arc volcanism over this period, even at a conservative subduction rate of 2 cm/yr, requires the destruction of 1000 km of oceanic crust of course, 1000 km of subducted oceanic crust does not imply 1000 km of ocean closing, for the subducted crust may be continually replaced by crustal generation at an active ridge. However, if active spreading ceased before subduction was initiated, or if it ceased at some time during the 50 Ma subduction episode, then significant ocean closure is demanded. It is also possible that subduction led to the opening of marginal ocean basins, which is suggested by the common association of ophiolitic suites and island arc volcanic rocks in the Dunnage Zone. Thus local spreading may have occurred in some places, while in others the main Iapetus Ocean was being destroyed.

An Iapetus of 2000 km width is reasonable if one considers the total length of the Appalachian-Caledonides chain and therefore the length of the Iapetus Ocean and its continental margins. The Appalachian Orogen extends 3500 km southwestward from Newfoundland along the Atlantic seaboard to Alabama in the southeastern United States. From there it is continuous in subsurface with the Ouachita System in Arkansas and extends to southern Mexico. On the east side of the present Atlantic, the British Caledonides have natural continuations northward through eastern Greenland and Scandinavia, and beyond to Svalbard. Therefore, before the opening of the present Atlantic, the Appalachian-Caledonides Orogen had a continuous length of 10,000 km, with extremities undefined. It seems unlikely that such a continuous orogen, which included equally continuous tectonic-stratigraphic facies belts related to ancient continental margins and an ocean basin, could have been produced by the opening of a narrow rift, a marginal ocean, or even a narrow major ocean. The scale of the phenomenon seems more in keeping with a major wide Iapetus, just as modern facies belts at present continental margins, of comparable dimensions to the length of the Appalachian-Caledonides Orogen, border major oceans.

An Iapetus wider than 2000 km is in agreement with Early Paleozoic faunal contrasts on opposite sides of the Appalachian Orogen (Cowie, 1971; Williams, 1972; Skevington, 1974; Spjeldnaes, 1978). Furthermore, a closing and destruction of Iapetus is suggested by the more cosmopolitan faunal assemblages that first appear in the Middle Ordovician and continue throughout younger periods (Spjeldnaes, 1978). Paleomagnetic evidence, although presently crude, is also in agreement with these estimates (Briden et al., 1973; McElhinny, 1973; Morel and Irving, 1978).

SUMMARY

Structural and stratigraphic palinspastic restoration of Iapetus, combined with indirect considerations, leads to an estimated minimum width of Iapetus of about 2000 km. The palinspastic restoration alone, based on the present surface geometry of the well-exposed northeast Newfoundland cross section, implies a minimum width of 1000 km. An additional 1000 km is plausible using indirect evidence such as the geometry of modern island arcs, the duration of island arc volcanism in northeast Newfoundland and the scale of the Appalachian-Caledonides mountain chain.

An Iapetus of at least 1000 km in width as based on palinspastic restoration was reduced to 500 km or slightly less by Taconic thrusting, folding and obduction of oceanic crust across the ancient continental margin of eastern North America. Iapetus, sensu stricto, was destroyed at this time. Subsequent compression during Acadian Orogeny reduced its post-Taconic width to a mere 180 km, which was further shortened to its present 150 km width by local Alleghanian movements.

ACKNOWLEDGEMENTS

I wish to thank T.J. Calon, J.D. Smewing, D.F. Strong and P.F. Williams who kindly reviewed the manuscript and suggested several improvements. I wish also to acknowledge financial support of my own field work, travel and map compilations through grants by the Natural Science and Engineering Research Council, the Killam Program of the Canada Council and the Department of Energy, Mines and Resources.

REFERENCES

Bell, K. and Blenkinsop, J., 1975a, Preliminary report on radiometric (Rb/Sr) age-dating, Burlington Peninsula, Newfoundland: Newfoundland Department of Mines and Energy, Mineral Development Division, unpubl. report, 9p.

_____, 1975b, The geochronology of eastern Newfoundland: Nature, v. 254, p. 410-411.

Bell, K., Blenkinsop, J. and Strong, D. F., 1977, The geochronology of some granitic bodies from eastern Newfoundland and its bearing on Appalachian Evolution: Canadian Jour. Earth Sci. v. 14, p. 456-476.

Bell, K., Blenkinsop, J., Berger, A. R. and Jayasinghe, N. R., 1979, Newport granite: its age, geological setting, and implications for the geology of northeastern Newfoundland: Canadian Jour. Earth Sci., v. 16, p. 264-269.

Bird, J. M. and Dewey, J. F., 1970, Lithosphere plate-continental margin tectonics and the evolution of the Appalachian Orogen: Geol. Soc. America Bull., v. 81, p. 1031-1060.

Blackwood, R.F., 1978, Northeastern Gander Zone, Newfoundland: in Gibbons, R. V., ed., Report of Activities for 1977: Newfoundland Department of Mines and Energy, Mineral Development Division, Report 78-1, p. 72-79.

Bostock, H. H., 1978, The Roberts Arm Group, Newfoundland: geological notes on a Middle or Upper Ordovician arc environment: Geol. Survey Canada Paper 78-15.

Briden, C. C., Morris, W. A. and Piper, J.D.A., 1973, Paleomagnetic studies in the British Caledonides – VI, Regional and global implications: Geophys. Jour. Royal Astron. Soc. v. 34, p. 107-134.

Brown, P.A., 1973, Possible cryptic suture in southwest Newfoundland: Nature Physical Science, v. 245, No. 140, p. 9-10.

Casey, J.F. and Kidd, W.S.F., 1979, Erosional unconformity above the Bay of Islands ophiolite complex and the para-allochthonous nature of overlying sedimentary rocks: Geol. Soc. America, Abst. with Programs, v. 11, No. 1, p. 6.

Church, W.R. and Stevens, R.K., 1970, Mantle peridotite and Early Paleozoic ophiolite complexes of the Newfoundland Appalachians (abst.): Program and Abstracts, International Symposium on Mechanical Properties and Processes in the Mantle, June 24 - July 3, Flagstaff, Arizona.

_____, 1971, Early Paleozoic ophiolite complexes of the Newfoundland Appalachians as mantle - ocean crust sequences: Jour. Geophys. Research, v. 76, No. 5, p. 1460-1466.

Clifford, P.M., 1968, Flood basalts, dike swarms and sub-crustal flow: Canadian Jour. Earth Sci., v. 5, No. 1, p. 93-96.

Cowie, J.W., 1971, Lower Cambrian faunal provinces in space and time: *in* Middlemiss, F. A., Rawson, P. F. and Newall, G., eds., Faunal provinces in space and time: Geol. Jour., Spec. Issue No. 4, p. 31-46.

Dainty, A.M., Keen, C.E., Keen, M.J. and Blanchard, J.E., 1966, Review of geophysical evidence on crust and upper mantle structure on the Eastern Seaboard of Canada: American Geophys. Union, Monograph 10, p. 349-369.

Dean, P.L., 1978, Volcanic stratigraphy and metallogeny of Notre Dame Bay, Newfoundland: Memorial University of Newfoundland, Geology Report No. 7, 205p.

Dean, P.L. and Strong, D.F., 1977, Folded thrust faults in Notre Dame Bay, Central Newfoundland: American Jour. Sci., v. 277, p. 97-108.

DeGrace, J.R., Kean, B.F., Hsu, E., and Green, T., 1976, Geology of the Nippers Harbour map-area (2E/13), Newfoundland: Newfoundland Department of Mines and Energy, Mineral Development Division, Report 76-3, 73p.

Dewey, J.F., 1969, Evolution of the Appalachian/Caledonian Orogen: Nature, v. 222, No. 5189, p. 124-129.

Dewey, J.F. and Bird, J.M., 1971, Origin and emplacement of the ophiolite suite: Appalachian ophiolites in Newfoundland: Jour. Geophys. Research, v. 76, p. 3179-3208.

Doig, Ronald, 1970, An alkaline rock province linking Europe and North America: Canadian Jour. Earth Sci., v. 7, p. 22-28.

Harland, W.B. and Gayer, R.A., 1972, The Arctic Caledonides and earlier oceans: Geological Magazine, v. 109, p. 289-314.

Haworth, R.T., Poole, W.H., Grant, A.C. and Sanford, B.V., 1976, Marine geoscience survey, northeast of Newfoundland: Geol. Survey Canada Report 76-1A, p. 7-15.

Herd, R.K. and Dunning, G.R., 1979, Geology of Puddle Pond map area, Southwestern Newfoundland: Geol. Survey Canada Paper 79-1A, p. 305-310.

Hibbard, J.P. and Williams, Harold, 1979, The regional setting of the Dunnage Melange in the Newfoundland and Appalachians: American Jour. Sci., v. 279, p. 993-1021.

Isacks, B.L. and Barazangi, M., 1977, Geometry of Benioff Zones: Lateral segmentation and downwards bending of the subducted lithosphere: in Talwani, M. and Pitman, W.C. III, eds.: Island Arcs, Deep Sea Trenches and Back-arc Basins: American Geophys. Union, Maurice Ewing Series 1, p. 99-114.

Karson, J. and Dewey, J.F., 1978, Coastal complex, western Newfoundland: An early Ordovician oceanic fracture zone: Geol. Soc. America Bull., v. 89, p. 1037-1049.

Kay, Marshall, 1976, Dunnage Melange and subduction of the Protacadic Ocean, Northeast Newfoundland: Geol. Soc. America Spec. Paper 175, 49p.

Kay, Marshall and Eldredge, Niles, 1968, Cambrian Trilobites in central Newfoundland volcanic belt: Geol. Magazine, v. 105, No. 4, p. 372-377.

Kean, B.F., 1974, Notes on the geology of the Great Bend and Pipestone Pond ultramafic bodies: Newfoundland Department of Mines and Energy, Mineral Development Division, Report of Activities, p. 33-42.

Kean, B.F. and Strong, D.F., 1975, Geochemical evolution of an Ordovician island arc of the Central Newfoundland Appalachians: American Jour. Sci., v. 275, p. 97-118.

Kennedy, M.J., 1975, Repetitive orogeny in the northeastern Appalachians – new plate models based upon Newfoundland examples: Tectonophysics, v. 28, 39-87.

Kennedy, M.J., 1976, Southeastern margin of the northeastern Appalachians: Late Precambrian orogeny on a continental margin: Geol. Soc. America Bull., v. 87, p. 1317-1325.

Kennedy, M.J. and McGonigal, M., 1972, The Gander Lake and Davidsville groups of northeastern Newfoundland: new data and geotectonic implications: Canadian Jour. Earth Sci., v. 9, p. 452-459.

Keppie, J.D., 1977, Tectonics of southern Nova Scotia: Nova Scotia Dept. Mines Paper 77-1, 34p.

Keppie, J.D., Bursnall, J.T., Ruitenberg, A.A., St. Julien, P., and Williams, Harold, in prep., Time of deformation map of the Canadian Appalachians.

Kidd, W.S.F., 1977, The Baie Verte Lineament, Newfoundland: ophiolite complex floor and mafic volcanic fill of a small Ordovician marginal basin: in Talwani, M. and Pitman, W.C., eds., Island arc, Deep Sea Trenches and Back-arc basins: American Geophys. Union, Maurice Ewing Series 1, p. 407-418.

Kidd, W.S.F., Dewey, J.F. and Bird, J.M., 1978, The Ming's Bight ophiolite complex, Newfoundland: Appalachian oceanic crust and mantle: Canadian Jour. Earth Sci., v. 15, p. 781-804.

Knapp, D., Kennedy, D. and Martineau, Y., 1979, Stratigraphy, structure and regional correlation of rocks at Grand Lake, Western Newfoundland: Geol. Survey Canada, Paper 79-1A, p. 317-325.

Mattinson, J.M., 1975, Early Paleozoic ophiolitic complexes of Newfoundland: isotopic ages of zircons: Geology, v.3, p. 181-183.

—————————, 1976, Ages of zircons from the Bay of Islands ophiolite complex, western Newfoundland: Geology, v. 4, p. 393-394.

McElhinny, M.W., 1973, Paleomagnetism and plate tectonics: Cambridge University Press, 358p.

McKerrow, W.S. and Cocks, L.R.M., 1977, The location of the Iapetus Ocean suture in Newfoundland: Canadian Jour. Earth Sci., v. 14, p. 488-499.

—————————, 1978, A lower Paleozoic trench-fill sequence, New World Island, Newfoundland: Geol. Soc. America Bull., v. 89, p. 1121-1132.

McKerrow, W.S. and Ziegler, A.M., 1971, The Lower Silurian paleogeography of New Brunswick and adjacent areas: Jour. Geol., v. 79, p. 635-646.

Morel, P. and Irving, E., 1978, Tentative paleocontinental reconstructions for the early Phanerozoic and Proterozoic: Jour. Geol., v. 86, p. 535-561.

Morgan, B.A., 1977, The Baltimore Complex, Maryland, Pennsylvania and Virginia: in Coleman, R.G. and Irwin, W.P., eds., North American Ophiolites: Oregon Department of Geology and Mineral Industries Bulletin 95, p. 41-49.

Neale, E.R.W., Beland, J., Potter, R.R. and Poole, W.H., 1961, A preliminary tectonic map of the Canadian Appalachian region based on age of folding: Canadian Instit. Mining Metallurgy Trans., v. 64, p. 405-412.

Neale, E.R.W., Kean, B.F. and Upadhyay, H.D., 1975, Post-ophiolite unconformity, Tilt Cove-Betts Cove area, Newfoundland: Canadian Jour. Earth Sci., v. 12, p. 880-886.

Nelson, K.D., 1979, Stratigraphy in the Badger Bay-Seal Bay area of Western Notre Dame Bay, and its relationship to Ordovician tectonics in Western Newfoundland: Geol. Soc. America, Abst. with Programs, v. 11, no. 1, p. 47.

Norman, R.E. and Strong, D.F., 1975, The geology and geochemistry of ophiolitic rocks exposed at Ming's Bight, Newfoundland: Canadian Jour. Earth Sci., v. 12, p. 777-797.

Pajari, G.E. and Currie, K.L., 1978, The Gander Lake and Davidsville Groups of northeastern Newfoundland: a re-examination: Canadian Jour. Earth Sci., v. 15, p. 708-714.

Pajari, G.E., Jr., Rast, N., and Stringer, P., 1977, Paleozoic volcanicity along the Bathurst-Dalhousie Geotraverse, New Brunswick, and its relations to structure: in Baragar, W.R.A., ed., Volcanic Regimes in Canada: Geol. Assoc. Canada, Spec. Paper No. 16, p. 111-124.

Pringle, I.R., Miller, J.A. and Warrell, D.M., 1971, Radiometric age determinations from the Long Range Mountains, Newfoundland: Canadian Jour. Earth Sci., v. 8, p. 1325-1330.

Rankin, D.W., 1975, The continental margin of Eastern North America in the Southern Appalachians: the opening and closing of the Proto-Atlantic Ocean: American Jour. Sci., v. 275-A, p. 298-336.

Rankin, D.W., 1976, Appalachian salients and recesses: Late Precambrian continental breakup and the opening of the Iapetus Ocean: Jour. Geophys. Research, v. 81, p. 5606-5619.

Rankin, D.W., Stern, T.W., Reed, J.C. Jr., and Newell, M.F., 1969, Zircon ages of felsic volcanic rocks in the Upper Precambrian of the Blue Ridge, Appalachian Mountains: Science, v. 166, p. 741-744.

Rodgers, John, 1965, Long Point and Clam Bank Formations, Western Newfoundland: Geol. Assoc. Canada Proceedings, v. 16, p. 83-94.

—————————, 1967, Chronology of tectonic movements in the Appalachian region of eastern North America: American Jour. Sci., v. 265, p. 408-427.

—————————, 1968, The eastern edge of the North American continent during the Cambrian and Early Ordovician: in Zen, E-an, et al., eds., Studies of Appalachian Geology: Northern and Maritime: John Wiley and Sons, New York, p. 141-150.

Ruitenberg, A.A., Fyffe, L.R. and McCutcheon, S.R., 1977, Evolution of pre-Carboniferous tectonostratigraphic zones in the New Brunswick Appalachians: Geosci. Canada, v. 4, p. 171-181.

St. Julien, P. and Hubert, C., 1975, Evolution of the Taconian orogen in the Quebec Appalachians: American Jour. Sci., v. 275A, p. 337-362.

Schenk, P.E., 1978, Synthesis of the Canadian Appalachians: Geol. Survey Canada Paper 78-13, p. 111-136.

Schillereff, H.S. and Williams, Harold, 1979, Geology of Stephenville map area, Newfoundland: Geol. Survey Canada Paper 79-1A, p. 327-332.

Schroeder, T.M., 1971, Geology of the Nippers Harbour area, Newfoundland: M.Sc. Thesis, University of Western Ontario, 88p.

Sheridan, R.E. and Drake, C.L., 1968, Seaward extension of the Canadian Appalachians: Canadian Jour. Earth Sci., v. 5, p. 337-373.

Skevington, D., 1974, Controls influencing the composition and distribution of Ordovician graptolite faunal provinces: in Rickards, R.B., Jackson, D.E. and Hughes, C.P. eds., Graptolite studies in honour of O.M.B. Bulman, 59-73: Spec. Paper Palaeontology No. 13.

Smitheringale, W.G., 1972, Low potash Lushs Bight tholeiites: ancient oceanic crust in Newfoundland: Canadian Jour. Earth Sci., v. 9, p. 574-588.

Spjeldnaes, N., 1978, Faunal provinces and the Proto-Atlantic: in Bowes, D.R. and Leake, B.E., eds.: Crustal evolution in northwestern Britain and adjacent regions: Geol. Jour. Spec. Issue No. 10, p. 139-150, Seal House Press, Liverpool, U.K.

Stevens, R.K., 1970, Cambro-Ordovician flysch sedimentation and tectonics in west Newfoundland and their possible bearing on a Proto-Atlantic Ocean: in Lajoie, J., ed., Flysch Sedimentology in North America: Geol. Assoc. Canada Spec. Paper No. 7, p. 165-177.

Stevens, R.K., and Williams, Harold, 1973, The emplacement of the Humber Arm Allochthon, western Newfoundland: Geol. Soc. America, Program, NE Section Annual Meeting, Allentown, Pennsylvania, v. 5, No. 2, p. 222.

Stevens, R.K., Strong, D.F. and Kean, B.F., 1974, Do some Eastern Appalachian ultramafic rocks represent mantle diapirs produced above a subduction zone?: Geology, v. 2, p. 175-178.

Strong, D.F., 1972, Sheeted diabases of Central Newfoundland: new evidence for Ordovician seafloor spreading: Nature, v. 235, p. 102-104.

Strong, D.F. and Williams, Harold, 1972, Early Paleozoic flood basalts of northwest Newfoundland: their petrology and tectonic significance. Geol. Assoc. Canada Proceedings, v. 24, No. 2, p. 43-54.

Strong, D.F., Dickson, W.D., O'Driscoll, D.F., Kean, B.F. and Stevens, R.K., 1974, Geochemical evidence for eastward Appalachian subduction in Newfoundland: Nature, v. 248, p. 37-39.

Stukas, V. and Reynolds, P.H., 1974a, $^{40}Ar/^{39}Ar$ dating of the Long Range dikes, Newfoundland: Earth and Planetary Sci. Letters, v. 22, p. 256-266.

——————————, 1974b, $^{40}Ar/^{39}Ar$ dating of the Brighton Gabbro Complex, Lushs Bight terrane, Newfoundland: Canadian Jour. Earth Sci., v. 11, p. 1485-1488.

Thomas, W.A., 1977, Evolution of Appalachian-Ouachita salients and recesses from reentrants and promontories in the continental margin: American Jour. Sci., v. 277, p. 1233-1278.

Upadhyay, H.D., Dewey, J.F. and Neale, E.R.W., 1971, The Betts Cove ophiolite complex, Newfoundland: Appalachian oceanic crust and mantle: Geological Association of Canada Proceedings, v. 24, p. 27-34.

Williams, A., 1972, Distribution of brachiopod assemblages in relation to Ordovician palaeogeography: in Hughes, N.F., ed., Organisms and Continents through time: Spec. Paper Palaeontology No. 12, p. 241-269.

Williams, Harold, 1971, Mafic-ultramafic complexes in Western Newfoundland Appalachians and the evidence for their transportation: A review and interim report: Geol. Assoc. Canada Proceedings, A Newfoundland Decade, v. 24, p. 9-25.

—————————, 1975, Structural succession, nomenclature and interpretation of transported rocks in Western Newfoundland: Canadian Jour. Earth Sci., v. 12, p. 1874-1894.

—————————, 1976, Tectonic stratigraphic subdivision of the Appalachian Orogen: Geol. Soc. America, Abst. with Programs, v. 8, No. 2, p. 300.

—————————, 1977, The Coney Head Complex: Another Taconic Allochthon in West Newfoundland: American Jour. Sci., v. 277, p. 1279-1295.

—————————, (compiler), 1978a, Tectonic-Lithofacies map of the Appalachian Orogen: Memorial University of Newfoundland, St. John's, Newfoundland, Map No. 1.

—————————, 1978b, Geological development of the northern Appalachians: its bearing on the evolution of the British Isles: in Bowes, D.R. and Leake, B.E., eds., Crustal Evolution in northwestern Britain and adjacent regions: Geol. Jour. Special Issue No. 10, Seal House Press, Liverpool, U.K., p. 1-22.

—————————, 1979, Appalachian Orogen in Canada: Canadian Jour. Earth Sci., Tuzo Wilson Volume, v. 16, p. 792-807.

Williams, Harold and Doolan, B.L., 1978, Margins and vestiges of Iapetus: Geol. Soc. America – Geol. Assoc. Canada, Abst. with Programs, v. 10, No. 7, p. 517.

Williams, Harold and Payne, J.G., 1975, The Twillingate Granite and nearby volcanic groups: an island arc complex in northeast Newfoundland: Canadian Jour. Earth Sci., v. 12, p. 982-995.

Williams, Harold and Stevens, R.K., 1969, Geology of Belle Isle - Northern extremity of the deformed Appalachian miogeosynclinal belt: Canadian Jour. Earth Sci., v. 6, p. 1145-1157.

—————————, 1974, The ancient continental margin of eastern North America: in Burk, C.A. and Drake, C.L., eds., The Geology of Continental Margins: Springer-Verlag, New York, p. 781-796.

Williams, Harold and Talkington, R.W., 1977, Distribution and tectonic setting of ophiolites and ophiolitic melanges in the Appalachian Orogen: in Coleman, R.G. and Irwin, W.P., eds., North American Ophiolites: Oregon Dept. Geology and Mineral Industries Bull. 95, p. 1-11.

Williams, Harold, Dallmeyer, R.D. and Wanless, R.K., 1976, Geochronology of the Twillingate Granite and Herring Neck Group, Notre Dame Bay, Newfoundland: Canadian Jour. Earth Sci., v. 13, p. 1591-1601.

Williams, Harold, Hibbard, J.P. and Bursnall, J.T., 1977, Geologic setting of asbestos-bearing ultramafic rocks along the Baie Verte Lineament, Newfoundland: Geol. Survey Canada, Paper 77-1A, p. 351-360.

Williams, Harold, Kennedy, M.J. and Neale, E.R.W., 1972, The Appalachian structural province: in Prince, R.A. and Douglas, R.J.W., eds., Variations in Tectonic Styles in Canada: Geol. Assoc. Canada, Spec. Paper 11, p. 181-261.

Williams, H., Kennedy, M.J. and Neale, E.R.W., 1974, The northeastward termination of the Appalachian Orogen: in Nairn, A.E.M. and Stehli, F.G., eds., The Ocean Basins and Margins, v. 2: Plenum Press, New York, p. 79-123.

Wilson, J.T., 1966, Did the Atlantic close and then re-open?: Nature, v. 11, No. 5050, p. 676-681.

The Continental Crust and Its Mineral Deposits, edited by D.W. Strangway,
Geological Association of Canada Special Paper 20

NEW PALEOMAGNETIC EVIDENCE FOR DISPLACED TERRANES IN BRITISH COLUMBIA

E. Irving
Earth Physics Branch,
Energy, Mines & Resources, 601 Booth Street,
Ottawa, Ontario K1A 0E4

J.W.H. Monger
Geological Survey of Canada
Vancouver, B.C. V6B 1R8

R.W. Yole
Department of Geology,
Carleton University,
Ottawa, Ontario K1S 5B6

ABSTRACT

Geological evidence and new paleomagnetic results suggest that the evolution of the western Canadian Cordillera has been dominated by two major tectonic blocks of exotic origin. The outer, western block (sometimes called Wrangellia) consists of Vancouver Island, the Queen Charlotte Islands and parts of southern Alaska. The inner Stikine terrane comprises much of central and northwestern British Columbia. These terranes are separated from one another, and from the rest of North America, by narrow remnants of former oceanic crust. The western oceanic remnants between Wrangellia and the Stikine terrane have a minimum age of Middle Jurassic. The eastern remnants between the Stikine terrane and cratonic North America have a minimum age of Late Triassic. In the Late Triassic, the Stikine terrane was situated about 13° south of its present position, and at least part of Wrangellia was situated over 20° south of its present position relative to cratonic North America. Wrangellia moved northwards during the Jurassic, and geological evidence indicates that it became attached to the Stikine terrane in the Late Jurassic or Early Cretaceous. The amalgamated terranes of Wrangellia and the Stikine terrane apparently remained at about 13° south of their present position until the Late Cretaceous or early Tertiary, and were in this position when the emplacement of the plutons of the Coast Ranges of British Columbia commenced. During the latest Cretaceous or early Tertiary the amalgamated terranes, now containing the Coast Plutonic Complexes, moved northwards to arrive at their present position relative to cratonic North America in pre-Oligocene time. This motion occurred along major transcurrent faults, such as the Pinchi Fault and the Denali-Shakwak system. We suggest that Wrangellia and the Stikine terrane were formerly attached to oceanic plates, and that the northward motions that we have observed paleomagnetically reflect the general northward motion of the floor of the Pacific Ocean relative to North America since the Triassic. The fault systems bounding the Stikine terrane to the east might therefore be the Late Cretaceous and early Tertiary analogues of the present-day San Andreas transform fault system.

RÉSUMÉ

L'évidence géologique et les résultats des nouvelles études paléomagnétiques suggèrent que l'évolution de la partie ouest de la Cordillère canadienne a été dominée par deux grands blocs tectoniques d'origine exotique. Le bloc occidental extérieur (désigné quelquefois sous le terme de Wrangellia) comprend l'île de Vancouver, les îles Reine-Charlotte et certaines parties du sud de l'Alaska. La région intérieure de Stikine comprend la plus grande partie du centre et du nord-ouest de la Colombie-Britannique. Ces régions sont séparées l'une de l'autre et du reste de l'Amérique du Nord par des vestiges étroits d'une ancienne croûte océanique. Les vestiges océaniques occidentaux entre les terrains de Wrangellia et de Stikine ont un âge minimum du Jurassique moyen. Les vestiges orientaux entre les terrains de Stikine et le craton nord-américain ont un âge minimum de la fin du Trias. A la fin du Trias, les terrains de Stikine étaient situés environ 13° au sud de leur position actuelle et au moins une partie de Wrangellia était située située plus de 25° au sud de sa position actuelle par rapport au craton nord-américain. La région de Wrangellia s'est déplacée vers le nord au cours du Jurassique et l'évidence géologique indique qu'elle s'est rattachée aux terrains de Stikine à la fin du Jurassique ou au début du Crétacé. Les terrains amalgamés de Wrangellia et de Stikine sont apparemment demeurés à environ 13° au sud de leur position actuelle jusqu'à la fin du Crétacé ou au début du Tertiaire et ils étaient dans cette position lorsqu'a débuté la mise en place des plutons de la chaîne Côtière de Colombie-Britannique. A la toute fin du Crétacé ou au début du Tertiaire, les terrains amalgamés comprenant maintenant les complexes plutoniques Côtiers, se sont déplacés vers le nord pour arriver à leur position actuelle par rapport au craton de l'Amérique du Nord avant l'Oligocène. Ce mouvement s'est produit le long de failles transversales majeures comme la faille de Pinchi et le système Denali-Shakwak. On suggère que les terrains de Wrangellia et de Stikine étaient auparavant rattachés à des plaques océaniques et que les mouvements vers le nord que révèle le paléomagnétisme reflètent le mouvement général vers le nord du fond de l'océan Pacifique par rapport à l'Amérique du Nord depuis le Trias. Les systèmes de failles qui limitent les terrains de Stikine à l'est peuvent ainsi être les analogues de la fin du Crétacé et du début du Tertiaire du système actuel de failles transformantes de San Andreas.

INTRODUCTION

The western, volcano-sedimentary "eugeosyncline" of the North American Cordillera is a collage of terranes of varied origins that were added to the ancient western continental margin at different times. These exotic terranes are now amalgamated to form that part of the Cordillera which lies west of the line WW of Figure 1. WW marks the western limit of Precambrian and Paleozoic pericratonic sequences of the Cordillera. This boundary is also delineated by a transition from crustal rocks with Sr^{87}/Sr^{86} initial ratios of greater than 0.706 in the east (indicating continental origin) to those with ratios less than 0.704 in the west (oceanic or island arc origin) (Petö and Armstrong, 1976). Several of these exotic terranes have now been defined on the basis of gross stratigraphy, volcanic chemistry, faunal assemblages and/or paleomagnetic records that are anomalous with respect to those of cratonic North America (Irving and Yole, 1972; Monger, 1977; Muller, 1977; Hillhouse, 1977; Jones *et al.*, 1977). In this article, we summarize our recently obtained paleomagnetic results for two of these terranes in the western Cordillera of Canada. Detailed accounts of our investigations are given in Monger and Irving (1980) and Yole and Irving (1980).

One of the consequences of plate tectonics is that terranes marginal to the major plates will, from time-to-time, be transferred from one plate to another because plate boundaries do not always retain the same configuration. Professor J.T. Wilson was one of the first to invoke tectonic transfer of this sort, when he suggested that during the Mesozoic opening of the North Atlantic fragments of Europe and Africa remained attached to North America, because the line of opening did not coincide with the line along which the ancestral Atlantic closed during the Paleozoic (Wilson, 1966). Terranes can also be transferred across transcurrent junctures between plates, Baja California being the best known modern example. It is evidence for this type of tectonic transfer in the Canadian Cordillera which we shall now summarize.

Several possible exotic terranes have been identified in Alaska and Canada (Fig. 1), and we shall briefly describe these in sequence from west to east. Paleomagnetic data from the Alaska Peninsula and adjacent areas in southern Alaska (Packer and Stone, 1974; Stone and Packer, 1977, 1979) indicate that it is a displaced terrane composed of Jurassic and younger rocks (SAJ). A region sometimes referred to as Wrangellia (Muller, 1977, Jones *et al.*, 1977), with distinctive late Paleozoic and Triassic strata, lies east of SAJ, extending through southeastern Alaska to the Queen Charlotte Islands and Vancouver Island, and possibly southward to the Hell's Canyon area of Oregon and Washington. South and east of Wrangellia in southeastern Alaska, an assemblage of Precambrian, Paleozoic and Mesozoic sedimentary, volcanic and plutonic rocks has been designated the Alexander terrane (Berg *et al.*, 1972). East of Wrangellia there is the Coast Plutonic Complex of British Columbia (CPC of Fig. 1). Magnetic inclinations observed in these bodies (for example in the Howe Sound Pluton (Symons, 1973) and in the Stevens Pass granodiorite situated further south in Washington (Beck and Noson, 1972)) are systematically shallower than would be expected from observation on cratonic North America, indicating that they were emplaced 10 or 20° south of their present position relative to cratonic North America. Although these data suffer from the absence of accurate corrections for possible post-emplacement tilt and individual results may be subject to errors from this source, the effect is systematic over 400 km and the displacement is probably real (summary in Irving 1979, Fig. 3). To the east of the Coast Plutonic Complex the Stikine terrane occupies a large area in northwestern British Columbia, and is characterized by a distinctive sequence of late Paleozoic and early Mesozoic sedimentary and volcanic rocks, some of which may extend into southeastern Alaska (Monger, 1977; Monger and Price, 1979). The easternmost of the presumed exotic terranes is characterized by the Cache Creek tectonostratigraphic assemblage (Monger, 1977; Monger and Price, 1979), which now forms a discontinuous belt extending from southern British Columbia to the Yukon, and which is marked by horizontal lines in Figure 1. Sedimentary and volcanic rocks of the Cache Creek Group, of Mississippian to Triassic age, are interpreted as an oceanic

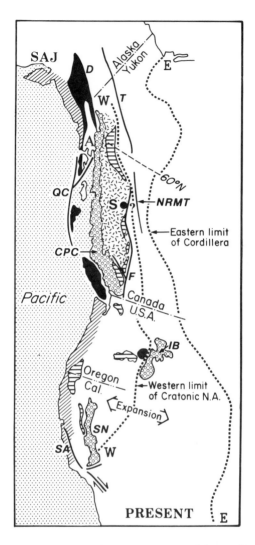

Figure 1. Sketch-map of some of the major tectonic elements of the cordillera. *EE* is the easterly limit of Cordilleran deformation. *WW* is the westerly limit of the North American craton. The ornament is as follows: major plutons, small rings (*CPC* Coast Plutonic Complex, *IB* Idaho batholith, *SN* Sierra Nevada batholith); Wrangellia is black; Alexander terrane (*A*) is left blank; Stikine terrane has an ornament of random lines; mid-Jurassic and older oceanic crust and rocks deposited on it an ornament of horizontal lines; late Mesozoic and Tertiary oceanic crust with rocks deposited on it an ornament of diagonal lines. Some major faults are marked: *D* Denali, *F* Fraser, *NRMT* northern Rocky Mountain Trench, *QC* Queen Charlotte Island-Chatham Strait System, *SA* San Andreas, *T* Tintina. The Pinchi Fault is not labelled but it is immediately east of *S* which is the centre of the sampling area in the Stikine block. *SAJ* is the Late Jurassic displaced terrane of southern Alaska (Packer and Stone, 1974).

subduction complex (Monger and Price, 1979). East of the Cache Creek belt, a volcano-sedimentary complex, possibly including elements from both marginal basins and magmatic arc-subduction complexes, separates the presumed exotic terranes from pericratonic sequences of the eastern Cordillera (Monger and Price, 1979). Between the probable ancient cratonic margin (*WW*) and the eastern limit of Cordilleran deformation (*EE* of Figure 1), the structural history can, at least in part, be related to the accretion of the western exotic terranes.

The paleomagnetic results summarized below show that during Cretaceous and/or later times Wrangellia, the Stikine terrane and the Cache Creek belt moved northward relative to North America to reach their present positions. Geological evidence and paleomagnetic data from parts of Wrangellia suggest an earlier, pre-Cretaceous history of significant motions with respect to North America.

THE PALEOMAGNETIC METHOD

At a sampling locality the magnetic inclination (I) relative to the paleohorizontal provides a measure of the paleolatitude λ_0 since $\tan I = 2 \tan \lambda$. This can be compared with the expected paleolatitude (λe) calculated from paleomagnetic results of coeval rocks from the adjacent craton. If λ_0 and λe differ, then it is probable that displacements in a latitudinal sense have occurred. The paleomagnetic method provides a quantitative measure of latitudinal displacements only, but, because of the fortunate circumstance that during the Mesozoic and Cenozoic the paleomagnetic declinations are generally northerly and the western margin of North America had roughly a north-south orientation, the presence (or absence) of longitudinal displacements can be inferred from the presence (or absence) of belts of rock of oceanic origin that are the remains of subducted or obducted oceans. This geometrical feature, which can be readily understood by reference to the map sequence given by Irving (1979), makes the western Cordillera of North America a particularly suitable place to study displaced terranes, because the paleomagnetic data provides estimates of latitudinal movements, and the presence or absence of large longitudinal movements can be estimated geologically.

An important problem in determining the paleolatitude of small tectonic blocks is that of deciding whether the calculated paleolatitude is north or south of the paleo-equator. Unless the polarity of the geomagnetic field is known for *exactly* the *same* interval of time represented by the rocks sampled, a single observation of inclination will not determine whether the block was situated in the northern or southern hemisphere. Nor can declinations be used because very large tectonic rotations are known to occur (see below). The incompleteness of the record of known reversals and the limits to the accuracy of geological correlations mean that this problem can only be solved when the frequency of reversals is less than about 1 per 10^7 years. However once a time sequence of paleolatitudes has been observed, and continuity with the present geographical latitude established, the sign of paleolatitude can be determined, but this record will always be very difficult to obtain in a single small tectonic block because the rock sequences representing the required time span are seldom present. In practice, there are three conditions under which the sign of the paleolatitude of exotic blocks can be effectively established: 1) When one of the two possible polarity options leads to paleogeographically improbable consequences then the other option is more likely to be correct; 2) when several paleopoles in a time sequence are observed in the same block, that polarity option which provides a continuous drift trajectory (that is, provides for a continuous historical development) is very probably the correct one; and 3) when the geomagnetic field did not reverse, or did so infrequently during the interval of time concerned, the paleogeographic north-seeking direction can be identified and the sign of the paleolatitude be determined. As we shall see, all three conditions apply to our results from the Stikine block and the problem can be satisfactorily solved. However, none of the conditions are applicable to our results from Vancouver Island, so that it is not yet possible to say with certainty whether its paleolatitude was north or south, but some tentative constraints may be found by considering recent data from Alaska (see below).

Of paramount importance in studies of paleolatitudes is the accurate determination of the paleohorizontal plane. In an intrusive body such accuracy is difficult or impossible to obtain unless 1) vertical contact relations can be observed at each sampling location (a relationship often difficult to establish in large batholiths in orogenic belts where contacts are commonly faulted); 2) there is some observable internal feature such as gravity-controlled layering in cumulates; or 3) there is evidence of contact-controlled current flow. We have therefore confined our sampling to places where the paleohorizontal can be accurately determined from sedimentary bedding, from attitudes in volcanic rocks, from layering in intrusive rocks, or from intrusive relationships in sills. All samples from the Stikine terrane and about one third of those from Vancouver Island have been obtained from well-exposed sequences above the treeline (1500 to 2500 m).

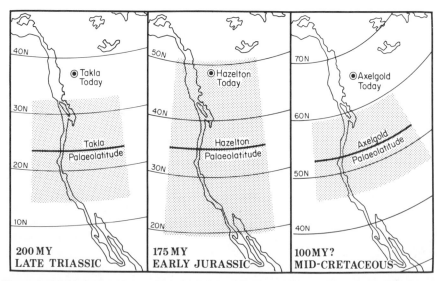

Figure 2. Paleolatitudes observed for the Stikine terrane compared with those observed from cratonic North America. Base maps from Irving (1979). The shaded error zones ($P = 0.05$) combine the statistical errors in the base maps and in the paleolatitudes observed from the Stikine terrane. From Irving and Monger (1980).

RESULTS FROM THE STIKINE TERRANE

The oldest rocks studied from the Stikine terrane are the Late Triassic Takla Group (late Karnian to early Norian). This group consists of up to 5000 m of mainly dark grey-green and locally red pillow basalt, massive basalt, and subaerial flows with associated volcaniclastics. Samples were obtained from two localities (14 sites, 84 samples) from mainly subaerial flows whose attitudes are clearly visible. The mean inclination is 41°, yielding a paleolatitude of 23°. This is compared in Figure 2 with Upper Triassic paleolatitudes calculated from cratonic North America. The Takla paleolatitudes are significantly farther south than would be expected if the sampling localities had not moved relative to North America.

Lower Jurassic rocks of the Hazelton Group were sampled at three localities (15 sites, 85 samples) spread over 100 km. The Hazelton Group consists of marine and non-marine calc-alkaline volcanics and intercalated sediments. It ranges in age from late Sinemurian to early Callovian, and extends across the eastern and southern parts of the Stikine terrane. The rocks sampled are subaerial red basalt and red tuffs. Their mean inclination is 54° and their corresponding mean paleolatitude is 35°, which is 14° further south than expected, but the difference is statistically marginal. The large error shown in Figure 2 is caused by the considerable uncertainty in the Early Jurassic field direction for cratonic North America as it is presently known. It should be noted that in estimating these errors a very conservative weighting procedure has been used so that the errors are almost certainly overestimated (Irving and Monger, 1980).

The third rock unit studied is the Axelgold intrusion. It is a layered anorthositic gabbro which is intrusive into the Cache Creek Group. K-Ar dates indicate an age of about 100 Ma in the mid-Cretaceous (see discussion section). In the section sampled (13 sites, 67 samples) the intrusion comprises alternating layers of anorthosite and anorthositic gabbro with dips varying from 10 to 15°. According to Irvine (1975) the layering was produced by sedimentation of crystals from magmatic currents, and can be assumed to have been approximately horizontal. The mean inclination of 69° after correction for tilt corresponds to a paleolatitude of 52°.

The characteristic magnetizations of the Takla, Hazelton and Axelgold rocks are very stable. The magnetizations have been perturbed in places by lightning strikes and by viscous components along the present field, but these effects can be removed by partial demagnetization. No complex overprints were observed.

The results from these Late Triassic, Early Jurassic and mid-Cretaceous rock units are remarkably consistent. They yield mean paleolatitudes for the eastern part of the Stikine terrane and adjacent rocks of the Cache Creek Group that are close to the latitude of the California-Oregon boundary for these times; that is, until mid-Cretaceous time the Stikine terrane, together with the Cache Creek Group, was located about 1300 km south of its present position relative to North America. Readers will notice that in Figure 2 we have chosen the northern latitude option for the Stikine terrane. The southerly option would place it in the latitude of Peru in the Triassic, northern Chile in the Jurassic, and Patagonia in the Cretaceous – first a progressively southward displacement, followed by a northward displacement of over 10,000 km since the mid-Cretaceous, an improbable sequence of events. The southerly option would also require that the polarity of the Axelgold intrusion be reversed; that is unusual for rocks of this age, which are almost invariably normally magnetized. The three conditions listed above apply to the northerly option for the Stikine terrane, whereas none apply to the southerly position.

Figure 2 only records the paleolatitudes derived from the observations of inclination. The variations of declination, which record the relative rotations among localities, are much more complicated. Within the Stikine terrane itself the rotations seem to be predominantly anticlockwise relative to the North America craton. They are 43° and 62° anticlockwise for the two localities at which the Takla Group was sampled, and 51° and 103° anticlockwise for two of the localities of the Hazelton Group but 14° clockwise for the third. Although the Stikine terrane has been a coherent geological entity, as indicated by continuity of strata and excellent agreement among paleolatitude estimates, it has evidently not behaved as an internally unified plate. On its eastern margin the Axelgold intrusion shows 63° of clockwise rotation. Clockwise rotations are ubiquitous in the cordillera west of the limit EE of Figure 1 (Beck, 1976; Irving, 1979, Fig. 3); the large anticlockwise rotations that we have observed in the Stikine terrane are exceptional. It is of course geometrically possible that the Stikine rotations are, in fact, much larger clockwise rotations (360° minus the above values) but we consider this improbable. These non-uniform rotations may be related to deformation within the Stikine terrane that occurred during its motion northwards. However, most of the rotations are in the wrong sense for ubiquitous dextral shear such as would be expected on the ''ball-bearing'' model of Beck (1976).

These conclusions are in apparent conflict with those of Symons (1974) based on his results from the Jurassic Topley intrusions of the Stikine terrane. He argues that the area ''has not been tectonically rotated or translated relative to stable North America''. We believe that his results do not necessarily support such a conclusion. The magnetizations of the Topley rocks are evidently complex (Symons, 1973, Fig. 3). The more stable magnetizations observed by Symons are consistent with our results from the slightly older Hazelton rocks; the less stable magnetizations are parallel to the early Tertiary geomagnetic field and are therefore probably secondary magnetizations acquired during the early Tertiary igneous activity and mineralization that affected the Topley area (Irving and Monger, 1980).

RESULTS FROM VANCOUVER ISLAND

We now summarize results from the Karmutsen Formation that augment our earlier preliminary studies of that formation (Irving and Yole, 1972). The Karmutsen Formation consists of up to 6000 m of submarine pillow basalts, massive lavas and diabase sills, and it is of mainly Late Triassic age. It extends over 500 km from Vancouver Island to the Queen Charlotte Islands. Our new results are from Vancouver Island (38 sites, 225 samples, 2 specimens from each).

As our original study showed, the magnetization of the Karmutsen is complex (Irving and Yole, 1972). Two families of directions are present, referred to as X and Y. The X magnetizations, which we found at 28 sites, characterize the Karmutsen throughout its great thickness. They have high unblocking fields (ranging up to 100 millitesla), high unblocking temperatures (between 500 and 700°C), and consistent directions through the vertical thickness of the formation. Moreover, similar values of

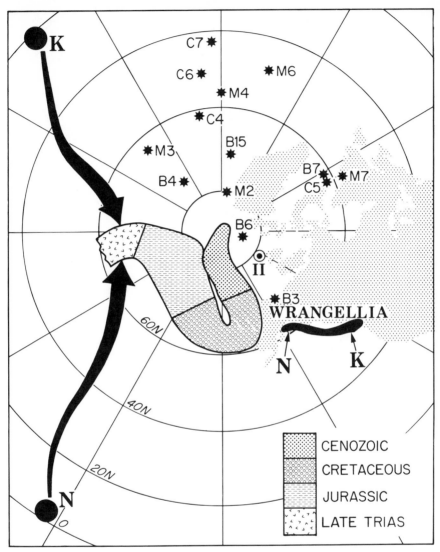

Figure 3. The pole positions from Wrangellia compared with the path of apparent polar wandering for North America. The poles N and K are calculated from the characteristic magnetization of the Nikolai Greenstone (Hillhouse, 1977) and the Karmutsen Formation (X magnetization). The arrows point to the pole position that these rocks would have given if Wrangellia had not moved relative to North America. The stars are poles calculated from the Karmutsen Y magnetization at individual sites. II is the pole for the Island Intrusions (Symons, 1971). From Yole and Irving (1980).

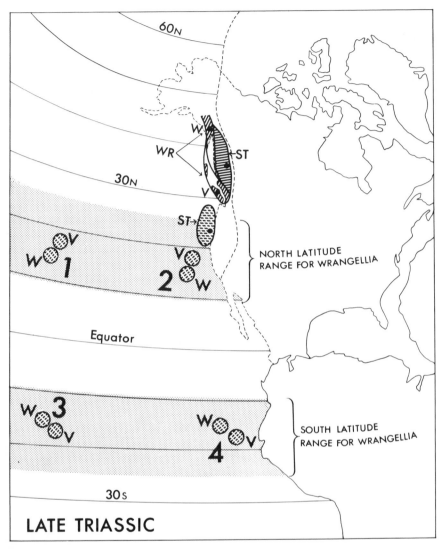

Figure 4. Paleolatitudes and reconstruction of Wrangellia *WR* (*W* = Wrangell Mountains, *V* = Vancouver Island) and the Stikine terrane (*ST*) in the Late Triassic. Present positions shown in solid bars. Postulated past positions shown by simplified shapes with dashed bars. For Wrangellia the two shaded zones are the possible northern and southern paleolatitudes. The width of shaded zone indicates the error in the paleolatitude estimates and the zone itself the longitude indeterminancy. 1, 2, 3 and 4 are therefore four possible positions for Wrangellia; in all cases the Wrangell block is shown outboard of Vancouver Island which may not necessarily have been the case. For the Stikine terrane *ST* only the northern option is given and a position close to North America is preferred for reasons given in the text. From Yole and Irving (1980).

inclination have been found by Hillhouse (1977) in the coeval Nikolai Greenstones of southern Alaska, demonstrating lateral consistency over 1500 km. Siliceous sediments baked at the contact with a diabase sill yield the same direction. The vertical and horizontal consistency, the partial contact test, and the high magnetic stability indicate that X is the original magnetization acquired in the Late Triassic. The X magnetization yields a pole position near the present position of Arabia, which is distant from the Late Triassic pole position from cratonic North America. During the Late Triassic Vancouver Island was evidently not in its present position relative to North America.

The mean inclination of the X magnetization is 33°, corresponding to a paleolatitude of 18°. As shown in Figure 4 this is 13° or 44° further south than the present position of Vancouver Island relative to North America, depending upon whether the paleolatitude is taken to be north or south of the paleoequator. Possible positions relative to the Late Triassic assembly of the continents are represented by north and south latitudinal belts (shaded) which are centred at either 18°N, comparable to the southwestern United States, or at 18°S latitude, comparable to northwestern South America at that time. Positions 1, 2, 3 and 4 of Figure 4 show alternative positions for Wrangellia (Vancouver Island, the Queen Charlotte Islands and the Wrangell mountains of Alaska) determined from our results and those of Hillhouse (1977). The belts indicate the indeterminacy in longitude and the ambiguity in the sign of the paleolatitude. Irving and Yole (1972) favoured the southerly option. Muller (1977) favours the northern option, deriving Wrangellia from off-shore "California". A similar problem arises for southern Alaska (SAJ of Fig. 1), and, because some of the rocks (the Nabesna sequence) that have been studied by Packer and Stone (1974) may overlie Wrangellian terrane, the recent discussion of Stone (1979) is relevant here. Stone has cogently argued that the southerly latitude option is to be preferred because it is in good accord with the plate tectonic reconstructions of the Pacific Ocean given by Hilde *et al*. (1976) based on the independent evidence of magnetic anomalies, and because the inferred motion of SAJ is then continuously northward. If the northerly option is taken there is a sharp change in direction of SAJ motion in the Cretaceous from south-north to north-south (Stone and Packer, 1977, 1979). Although such a change cannot be discounted, the continuity argument seems a more reasonable one to apply; it would also provide indirect support for the southerly option for Wrangellia (Fig. 4). The matter will not be decided until a time sequence of paleolatitude has been observed from Wrangellia.

The Y magnetizations have more scattered directions. The pole positions from the 14 sites at which Y magnetizations have been observed and plotted in Figure 3. Their remanent coercive forces and blocking temperatures are much more variable and often low, and their properties generally more heterogeneous. For these and other reasons (Irving and Yole, 1972; Yole and Irving 1980) we regard the Y magnetizations as overprints acquired after formation, either during deep burial, or at the time of emplacement of the Island Intrusions, a series of large granitoid batholiths that intrude the Karmutsen. The poles for the Y magnetizations do not generally fall close to the path of apparent polar wandering for the North American craton, indicating that Vancouver Island was not situated in the present position relative to North America when they were acquired. Many of the Y magnetizations are consistent with a more southerly position for Vancouver Island than at present.

The southerly position for Vancouver Island (Wrangellia) during the late Triassic to mid-Cretaceous is in conflict with the interpretation by Symons (1971) of his paleomagnetic results from the Jurassic Island Intrusions. Symons's pole (II of Fig. 3), which he regards as representative of the primary magnetization, indicates a more northerly position. However, the unblocking fields and the unblocking temperatures of these granitoid rocks are low (Yole and Irving 1980), and, in the absence of a contact test, we would argue that the magnetization is likely to be of secondary origin and could have been acquired long after intrusion. Symons's data are therefore not necessarily inconsistent with our own, because they may reflect later events in the Cretaceous or Cenozoic.

DISCUSSION

The manner in which the Stikine terrane and Wrangellia may have been assembled and accreted to North America is sketched in Figure 5. This scheme is critically dependent on the accuracy of the ages of the rocks studied and the ages of their magnetizations. The Takla and Hazelton Groups and the Karmutsen Formation are well-dated paleontologically and their characteristic magnetizations are very stable and probably reflect the geomagnetic field direction at the time they were formed (Yole and Irving, 1980; Monger and Irving, 1980). The paleomagnetic results from the Axelgold intrusion are considered satisfactory, but the geological constraints on its age are wide (it cuts post-Oxfordian structures); intrusions of this type are notoriously difficult to date radiometrically. Concordant biotite and hornblende ages range from 108 to 98 Ma (mid-Cretaceous), but since one hornblende yielded 155 Ma (Irvine, 1975) a Late Jurassic or earlier Cretaceous age is also possible. We accept a mid-Cretaceous age for the Axelgold intrusion, because it is the most probable on the basis of presently available radiometric work, and because it shows only normal polarities which are characteristic of the field in the interval 110 to 80 Ma, whereas the Late Jurassic and Early Cretaceous are characterized by frequent reversals (Larson and Pitman, 1972). We stress the uncertain nature of this method of age determination, and note that many of the details of our scheme of Figure 5 are dependent on it. However our general conclusion, that the Stikine terrane and Vancouver Island were once situated much further south, it is not affected by this uncertainty.

Between the Late Triassic and mid-Cretaceous the Stikine terrane apparently lay off what is now the northwestern United States. During the Late Triassic the Wrangell Mountains and Vancouver Island lay either close by a little to the south, or a long way further south if the southern hemisphere option for their paleolatitudes is taken (positions 3 or 4 of Fig. 4). Wrangellia and the Stikine terrane had apparently come together by the mid-Cretaceous, because the Coast Plutonic Complex, mainly emplaced during the mid- to Late Cretaceous and Early Tertiary (Monger and Price, 1979), effectively sealed their contact in British Columbia. In northwestern British Columbia rocks of the Stikine terrane can be traced into the eastern margin of the Coast Plutonic Complex, and in southwestern British Columbia rocks of Wrangellia can be traced into its western margin. The Coast Plutonic Complex comprises a series of granitoid batholiths that now form the core of the Coast Ranges of British Columbia as a continuous entity. This southerly position is consistent with the southward displacement observed paleomagnetically in the Coast Plutonic Complex and the Stevens Pass

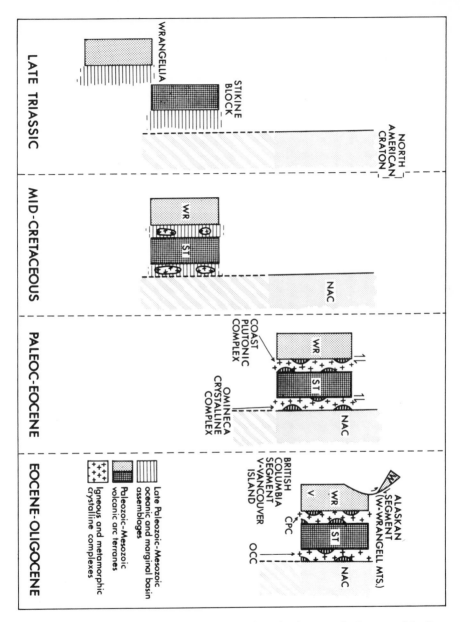

Figure 5. Diagrammatic representation of the assembly of major tectonic elements of the Canadian Cordillera. Explanation in text.

granodiorite (see Introduction), and as we have already seen, some of the overprints observed in the Karmutsen are also consistent with a more southerly situation. The combined terranes moved northwards relative to North America in the Late Cretaceous or early Tertiary. As there is no evidence of post-Jurassic oceanic terrane east of the Stikine terrane, we presume that the motion occurred along strike-slip faults. The motion could have been distributed among several large fault systems of which the Pinchi, Finlay and Fraser Faults and the northern Rocky Mountain trench are possible candidates. Figure 5 also shows, in diagrammatic fashion, the later dismemberment of Wrangellia in the Tertiary, possibly along the Denali, Shakwak, Chatham Strait fault system between about 50 and 35 Ma ago as documented by Lanphere (1978).

IMPLICATIONS FOR METALLOGENESIS

The model suggested above for the evolution of the western Cordillera has obvious implications for metallogenesis and mineral exploration. General relationships between tectonic setting and mineral deposits have long been known and exploited (McCartney and Potter, 1962), and recently such relationships have received intensive study (Sutherland Brown *et al.*, 1971; Hodder and Hollister, 1972; James, 1971; Mitchell and Garson, 1972; Sawkins, 1972; Rona, 1973; Sillitoe, 1972a, b; Touray, 1973). Studies by Godwin (1975) and Griffiths (1977) have attempted to establish metallogenic trends in the western Cordillera of Canada related to presumed subduction zones. The identification of such ancient zones, and of other kinds of boundaries that might have separated the tectonostratigraphic elements ("blocks", "belts", "terranes", "complexes" etc.) of the Cordilleran mosaic is still in an embryonic stage. However, it is clear from previous evidence and from our new results that some of these elements underwent large latitudinal displacements with respect to one another and to the North American craton during the Mesozoic and Tertiary. Our results emphasize the importance of large strike-slip displacements which have significant implications for metallogenesis. More precise definition of these tectonostratigraphic elements, and of the times of their amalgamation with one another and with the North American craton, is essential, not only for tectonic synthesis, but as an aid to exploration for metals.

Earth Physics Branch Contribution No. 817

REFERENCES

Beck, M.E., 1976, Discordant paleomagnetic pole positions as evidence of regional shear in western Cordillera of North America: American Jour. Sci., v. 276, p. 694-712.

Beck, M.E. and Noson, L., 1972, Anomalous palaeolatitude in Cretaceous granitic rocks: Nature, v. 235, p. 11-13.

Berg, H.C., Jones, D.L. and Richter, D.H., 1972, Gravina-Nutzotin belt – tectonic significance of an upper Mesozoic sedimentary and volcanic sequence in southeastern Alaska: U.S. Geol. Survey, Prof. Paper 800-D, p. D1-D24.

Godwin, C.I., 1975, Imbricate subduction zones and their relationship with Upper Cretaceous to Tertiary porphyry deposits in the Canadian Cordillera: Canadian Jour. Earth Sciences, v. 12, p. 1362-1378.

Griffiths, J.R., 1977, Mesozoic-early Cenozoic volcanism, plutonism and mineralization in southern British Columbia: a plate-tectonic synthesis: Canadian Jour. Earth Sciences, v. 14, p. 1611-1624.

Hilde, T.W.C., Uyeda, S. and Kroenke, K., 1976, Tectonic history of the western Pacific: in Drake, C.L. ed., Geodynamics Progress and Prospects: p. 1-15.

Hillhouse, J.W., 1977, Paleomagnetism of the Triassic Nikolai Greenstone, McCarthy Quadrangle, Alaska: Canadian Jour. Earth Sci., v. 14, p. 2578-2592.

Hodder, R.W. and Hollister, V.F., 1972, Structural features of porphyry copper deposits and the tectonic evolution of continents: Canadian Instit. Mining and Metallurgy, Bull., v. 65, p. 41-45.

Irvine, T.N., 1975, Axelgold layered gabbro intrusion McConnel Creek map-area, British Columbia: Geol. Survey Canada, Paper 75-1, Part B, p. 81-88.

Irving, E., 1979, Paleopoles and paleolatitudes of North America and speculations about displaced terrains: Canadian Jour. Earth Sci., v. 16, p. 669-694.

Irving, E. and Yole, R.W., 1972, Paleomagnetism and the kinematic history of mafic and ultramafic rocks in fold mountain belts: Earth Physics Branch Publ. 42, no. 3, p. 87-95.

James, D.E., 1971, Plate tectonic model for the evolution of the Central Andes: Geol. Soc. America Bull., v. 82, p. 3325-3346.

Jones, D.L., Silberling, N.J. and Hillhouse, J.W., 1977, Wrangellia – a displaced terrane in northwestern North America: Canadian Jour. Earth Sci., v. 14, p. 2565-2577.

Lanphere, M.L., 1978, Displacement history of the Denali Fault System, Alaska and Canada: Canadian Jour. Earth Sci., v. 15, p. 817-822.

Larson, R.L. and Pitman, W.C., 1975, World-wide correlation of Mesozoic magnetic anomalies and its implications: Geol. Soc. America Bull., v. 83, p. 3645-3662.

McCartney, W.D. and Potter, R.R., 1962, Mineralization as related to structural deformation, igneous activity and sedimentation in folded geosynclines: Canadian Mining Jour., v. 83, p. 83-87.

Mitchell, A.H.G. and Garson, M.S., 1972, Relationship of porphyry copper and circum-Pacific tin deposits to paleo-Benioff zones: Trans. Instit. Mining and Metallurgy, v. 81, p. B10-B25.

Monger, J.W.H., 1977, Upper Paleozoic rocks of the western Canadian Cordillera and their bearing on Cordilleran evolution: Canadian Jour. Earth Sci., v. 14, p. 1832-1859.

Monger, J.W.H. and Irving, E., 1980 Northward displacement of north-central British Columbia. Nature, v. 285, p. 289-294.

Monger, J.W.H. and Price, R.A., 1979, Geodynamic evolution of the Canadian Cordillera – progress and problems: Canadian Jour. Earth Sci., v. 16, p. 770-791.

Muller, J.E., 1977, Evolution of the Pacific Margin, Vancouver Island, and adjacent regions: Canadian Jour. Earth Sci., v. 14, p. 2062-2085.

Packer, D.R. and Stone, D.B., 1974, Paleomagnetism of Jurassic rocks from southern Alaska, and the tectonic implications: Canadian Jour. Earth Sci., v. 11, p. 976-997.

Petö, P. and Armstrong, R.L., 1976, Strontium isotope study of the composite batholith between Princeton and Okanogan Lake: Canadian Jour. Earth Sci., v. 13, p. 1577-1683.

Rona, P.A., 1973. Plate tectonics and mineral resources: Sci. American, v. 229, p. 86-95.

Sawkins, F.J., 1972, Sulfide ore deposits in relation to plate tectonics: Jour. Geol., v. 80, p. 377-396.

Sillitoe, R.H., 1972a. A plate tectonic model for the origin of porphyry copper deposits: Econ. Geol., v. 67, p. 184-197.

_____, 1972b. Relation of metal provinces in western America to subduction of oceanic lithosphere: Geol. Soc. America Bull., v. 83, p. 813-818.

456   IRVING *ET AL.*

Stone, D.B., 1979, Paleomagnetism and paleogeographic reconstruction of southern Alaska: in Sisson, A., ed., Proceedings 6th annual Geological Society of Alaska Symposium Anchorage, Alaska: p. K1-K17.

Stone, D.B. and Packer, D.R., 1977, Tectonic Implications of Alaska Paleomagnetic Data: Tectonophysics, v. 37, p. 183-201.

—————————, 1979, Paleomagnetic data from the Alaska Peninsula: Geol. Soc. America Bull., v. 90, p. 545-560.

Sutherland Brown, A., Cathro, R.J., Panteleyev, A. and Ney, C.S., 1971, Metallogeny of the Canadian Cordillera: Canadian Instit. Mining and Metallurgy Bull., v. 64, p. 37-61.

Symons, D.T.A., 1971, Paleomagnetism of Jurassic Intrusions, B.C.: Geol. Survey Canada Paper 70-63, p. 1-17.

—————————, 1973, Concordant Cretaceous palaeolatitudes from felsic plutons in the Canadian Cordillera: Nature, v. 241, p. 59-61.

—————————, 1974, Paleomagnetic results from the Jurassic Topley intrusions near Endako, British Columbia: Canadian Jour. Earth Sci., v. 10, p. 1099-1108.

Touray, J.C., 1973, Un modèle en métallogenique: La Recherche, no. 34, mai, p. 488-489.

Wilson, J.T., 1966, Did the Atlantic close and then re-open?: Nature, v. 211, p. 676-681.

Yole, R.W. and Irving, E., in prep., Displacement of Vancouver Island, paleomagnetic evidence from the Karmutsen Formation.

The Continental Crust and Its Mineral Deposits, Edited by D. W. Strangway,
Geological Association of Canada Special Paper 20

A PLATE TECTONIC CONTEST IN ARCTIC CANADA

J. Wm. Kerr

Geological Survey of Canada,
Institute of Sedimentary and Petroleum Geology,
Calgary, Alberta T2L 2A7

ABSTRACT

The Canadian Arctic Rift System connects the North Atlantic Ocean with the Arctic Ocean through North America. It is a dormant branch of the larger rift system that includes the Mid-Atlantic Ridge and the Nansen-Gakkel Ridge of the eastern Arctic Ocean. The branch loops into North America, surrounds Greenland and loops back again.

The Canadian Arctic Rift System was formed by two plate tectonic episodes that originated on opposite sides of the North American Plate and were propagated toward each other. Both were strongly controlled by pre-existing structures, which either guided the propagating faults or impeded their growth.

The Boreal Rifting Episode began first, with extensional structures propagated southeast-ward from the western Arctic Ocean into the continent. This partly fragmented the Canadian Arctic Islands, but its advance was halted by a structural obstacle. The younger Eurekan Deformation was propagated northwestward from the Mid-Atlantic Ridge, rotating Greenland and Canada apart.

In latest Cretaceous to mid-Tertiary time the two plate tectonic episodes were active simultaneously, indenting the continent from different sides. The Eurekan Deformation was the more efficient. It progressively interfered with, neutralized, and finally overpowered the other.

Extension faults of the Eurekan Deformation broke northwestward through the Arctic Islands in mid-Tertiary time to connect Baffin Bay with the older western Arctic Ocean. When the structural connection between the two oceans was finally achieved, the Canadian Arctic Rift System became dormant.

RÉSUMÈ

Le système de rift de l'Arctique canadien met en connexion l'océan Atlantique Nord et l'océan Arctique à travers l'Amérique du Nord. C'est une branche inactive du système de rift

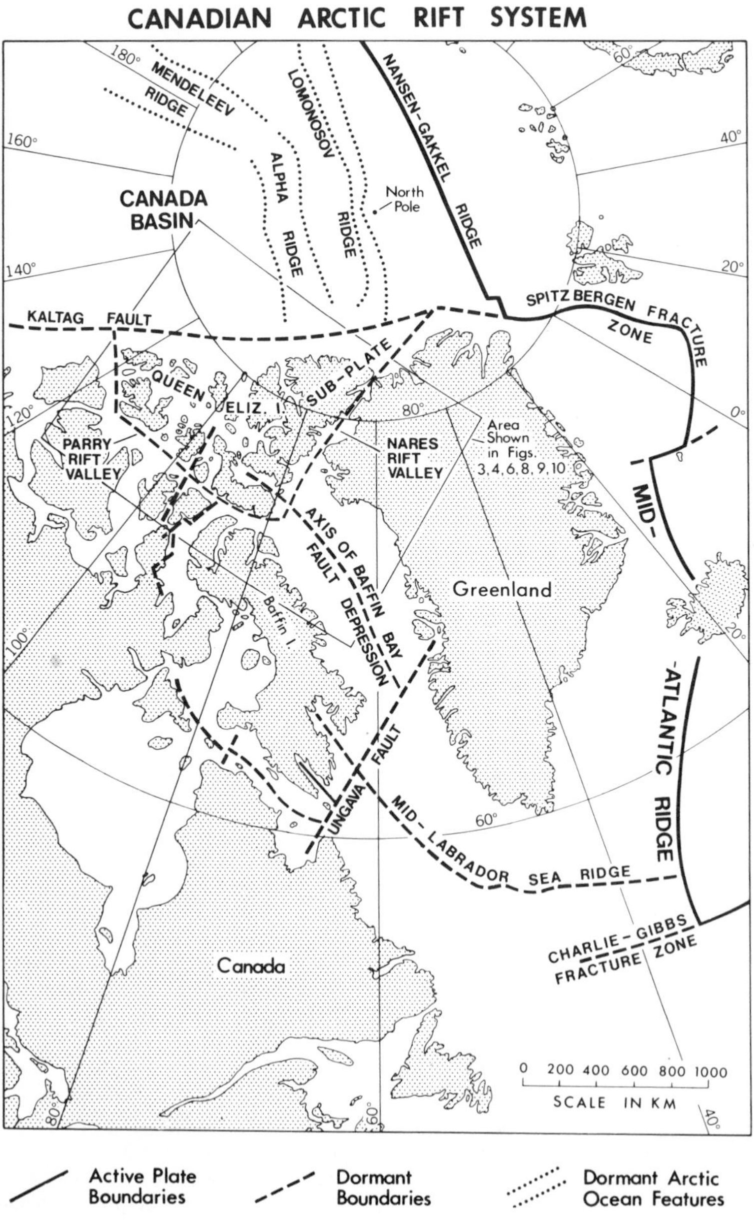

CANADIAN ARCTIC RIFT SYSTEM

Figure 1. The Canadian Rift System, a dormant feature formed by plate movements. The system includes the Mid-Labrador Sea Ridge, parts of the Kaltag Fault, and the structures shown here that lie between those two features. The Queen Elizabeth Islands are the triangular island group lying within the Queen Elizabeth Islands Sub-plate.

plus imposant qui comprend la crête médio-atlantique et la crête Nansen-Gakkel dans l'est de l'océan Arctique. L'embranchement fait une boucle en Amérique du Nord, contourne le Groënland et forme une autre boucle.

Le système de rift de l'Arctique canadien a été formé par deux épisodes de tectonique des plaques qui ont leur origine sur les côtés opposés de la plaque de l'Amérique du Nord et se sont propagés l'un vers l'autre. Les deux épisodes étaient fortement contrôlés par des structure préexistantes qui ont soit guidé la propagation des failles soit empêché leur croissance.

L'épisode d'effondrement boréal a d'abord débuté, avec des structures d'extension se propageant au sud-est de l'océan Arctique occidental dans le continent. Cet événement a partiellement fragmenté les îles de l'Arctique mais son avance a été entravée par un obstacle structural. La déformation d'Eureka, plus récente, s'est propagée au nord-ouest à partir de la crête médio-atlantique et a fait subir une rotation au Groënland pour le séparer du Canada.

De la fin du Crétacé au milieu du Tertiaire, les deux épisodes de tectonique des plaques agissaient simultanément, fissurant le continent sur différents fronts. La déformation d'Eureka a été plus efficace. Elle a progressivement interféré avec l'autre, l'a neutralisée pour finalement la maîtriser.

Les failles d'extension de la déformation d'Eureka se sont propagées au nord-ouest à travers les îles de l'Arctique au milieu du Tertiaire pour rattacher la baie de Baffin à l'océan Arctique occidental plus ancien. Quant le lien structural entre les deux océans a finalement été établi, le système de rift de l'Arctique devint inactif au Canada.

INTRODUCTION

The Canadian Arctic Rift System (Kerr, 1973, 1979a) is an extensional plate structure more than 4500 km long (Fig. 1), which extends through the North American Continent to connect the North Atlantic Ocean with the Arctic Ocean. It developed gradually from Mississippian to mid-Tertiary time, and then became dormant, apparently rather suddenly.

This rift system has received little attention in plate tectonic studies because it is dormant. This may be a mistake. After all, if we can determine why the plate tectonic process stopped in this structure, then we also might obtain unique insight into why the process began.

A close examination of the history of the Canadian Arctic Rift System may shed light upon some basic problems of plate tectonics: 1) Why do plates have their particular jagged shapes? 2) How are plates deformed during plate separation? 3) Does pre-existing structure of the continent affect deformation within continental parts of the plate? 4) Can two plate tectonic events interfere with each other?

BEFORE PLATE BREAKUP

Plate breakup of the Canadian Arctic was preceded by a long constructional phase (Fig. 2), during which a large protocontinent developed. This phase, which has been described in detail earlier (Kerr, 1979a), involved the formation of a Precambrian crystalline basement or Shield (Sequence 1), and deposition upon it of large sedimentary basins (Sequences 2 and 3). The constructional phase ended in Late Devonian or early Mississippian time, at the end of the Ellesmerian Orogeny (Fig. 2). That event involved the magmatic phase of the Pearya Geanticline, when plutonic rocks were intruded. This orogeny deformed Sequences 1 to 3 and welded them to the continental crust.

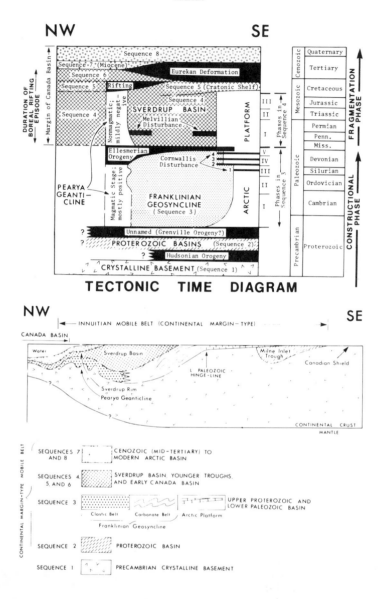

Figure 2. Sequences in the Canadian Arctic and their present relationships. Tectonic events (solid black pattern) produced widespread unconformities. The constructional phase (above) ended with the Ellesmerian Orogeny, when the upper boundary of Sequence 3 (cross-section) was exposed and being eroded. The fragmentation phase involved two plate tectonic events. The Boreal Rifting Episode began first and affected the northwest region, advancing southeastward. The Eurekan Deformation began later, originating in the southeast and advancing northwestward.

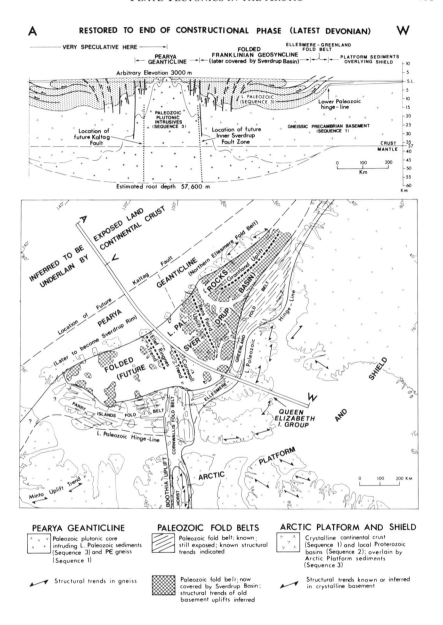

Figure 3. Structural configuration of the Canadian Arctic in latest Devonian time, at the end of the Ellesmerian Orogeny, which coincided with the end of the constructional phase. The entire region probably was a continental crust, above sea level, and being eroded. The present islands shown here had not yet formed; their outlines are shown for reference only. For location *see* Figure 1. These structures guided later geologic events.

At the end of the constructional phase the entire area shown by Figure 3 probably contained continental crust. South and east of the lower Paleozoic hinge-line (Fig. 3), where the Canadian Shield and Arctic Platform are now exposed, the crust may have had a normal continental thickness (37 km). This was made up largely of the original gneissic Precambrian basement (Sequence 1), overlain by local Proterozoic basins (Sequence 2, not shown), and a widespread thin sheet of lower Paleozoic Arctic Platform sediments (Sequence 3). That region had been a stable cratonic area throughout the development of Sequence 3.1

Northwest of the hinge-line was the folded Franklinian Geosyncline, containing more than 12,000 metres of latest Proterozoic and lower Paleozoic sediments (Sequence 3). The geosyncline had been deformed into several major fold belts (Fig. 3). The Parry Islands Fold Belt and Ellesmere-Greenland Fold Belt were relatively surficial décollement structures, formed by south to southeastward overriding of the Ellesmerian Orogeny (Fig. 2). The northern Ellesmere Fold Belt had deep-seated, more intense deformation, and was part of the Pearya Geanticline. The Cornwallis Fold Belt, part of the Boothia Uplift, was basement-controlled. It was formed by the Cornwallis Disturbance, which was partly contemporaneous with and related to the Ellesmerian Orogeny (Kerr, 1977). A large region now obscured by the Sverdrup Basin probably also was a Paleozoic geosynclinal fold belt, with both longitudinal and transverse fold trends that resulted from basement uplifts.

At the end of the constructional phase (Fig. 3), the Pearya Geanticline was an exposed linear tectonic axis that trended northeast through the Canadian Arctic Islands (Trettin *et al.*, 1972; Kerr, 1979a). In its core are late Precambrian (Grenvillian) gneissic rocks (Sinha and Frisch, 1976), which may be remnants of a Shield (Sequence 1). The geanticline was deformed episodically in early Paleozoic time, culminating in the Ellesmerian Orogeny with plutonic intrusion, folding, uplift, and erosion. It apparently was a mountain range of great relief at that time, with the plutonic rocks presumably widening at depth to a rigid core and a thickened root zone (Fig. 3, cross-section). The Pearya Geanticline is exposed today only in the northeast, as the Northern Ellesmere Fold Belt. It is inferred that the geanticline continued farther southwest at the end of the constructional phase, but later became covered by rocks of the Sverdrup Rim (Fig. 3, map).

The structure that existed in the region of the present Arctic Ocean at the end of the constructional phase is very speculative. The ocean probably did not exist (Kerr, 1979a). It is inferred that there was exposed land underlain by continental crust northwest of the Pearya Geanticline, possibly representing a mirror image of the folded geosyncline to the southeast (Fig. 3, cross-section). An alternative hypothesis (Trettin and Balkwill, 1979) suggests that a spreading ocean existed on the site of the Canada Basin, with subduction along its southeastern margin producing the Ellesmerian Orogeny in the adjacent Canadian Arctic Archipelago.

The crust that existed at the end of the constructional phase (Fig. 3) was the substructure beside which and within which the ocean basins and their branches were to form in the younger fragmentation phase (plate breakup). The internal structure of that crust in large measure controlled the shapes and histories of the plates.

At the end of the constructional phase most of the crustal column in the Canadian Arctic consisted of rigid crystalline rock (Fig. 3, cross-section). Folded sediments formed only a thin upper level of this crust. That level comprised local basins of Sequence 2 (not shown), and widespread rocks of Sequence 3. The sedimentary level was a great deal less competent than underlying crystalline rocks.

The differences existing in the crust (Fig. 3, cross-section) caused a great variation in its overall competence. The total crust probably was most competent in the region of the Pearya Geanticline; there the crust was very thick and was composed largely of isotropic plutonic intrusives. In the southeast, in the Arctic Platform and Shield, the crust may have been somewhat less competent, being thinner there and composed largely of gneisses whose foliation provided anisotropy. The least competent crust was in between, where there was a thick sedimentary column and presumably a thinned crystalline layer. The strong structural trends present in the crystalline layers of the crust provided an obstacle to certain kinds of deformation and an aid to others.

PLATE BREAKUP HISTORY

The fragmentation phase of the Canadian Arctic included two plate tectonic events (Fig. 2), the Boreal Rifting Episode, and the Eurekan Deformation. These events brought about the evolution of ocean basins and their branches within the continent (Kerr, 1979a). The fragmentation phase extended from about mid-Mississippian time to the present, and coincided with deposition of Sequences 4 through 8.

Boreal Rifting Episode

The Queen Elizabeth Islands Sub-plate (Fig. 1) contains within it the Queen Elizabeth Islands Group. Much of the fragmentation history of the Canadian Arctic discussed below centres on the evolution of that sub-plate, beginning with an unbroken continent (Fig. 3), and progressing to present geography (Fig. 1).

The initial fragmentation of the Canadian Arctic (Fig. 4) was part of the Boreal Rifting Episode (Kerr, 1979a). This began in Mississippian time shortly after the Ellesmerian Orogeny, and apparently resulted from spread on the Alpha Ridge.

Prior to the Boreal Rifting Episode, the entire area shown by Figure 4 was part of a largely exposed protocontinent that had been produced by the earlier constructional phase (Fig. 3). The Sverdrup Basin and possibly the Canada Basin developed unconformably upon part of this protocontinent as rifting and subsiding basins.

The Boreal Rifting Episode (Kerr, 1979a) occurred intermittently during the deposition of Sequences 4, 5, and 6 (Fig. 2). Rifting apparently originated with the Alpha Ridge as the active spreading centre, and first affected the Canada Basin. Side effects of this rifting reached some distance southeastward into the continent, causing the Sverdrup Basin to form by crustal fracturing. The Boreal Rifting Episode did not break fully through the continent to the southeast, and therefore was aborted (Figs. 4 and 5). Several major features that were formed during the Boreal Rifting Episode are discussed below.

Alpha Ridge

The Alpha Ridge (Fig. 1) is part of a shallow feature trending across the Arctic

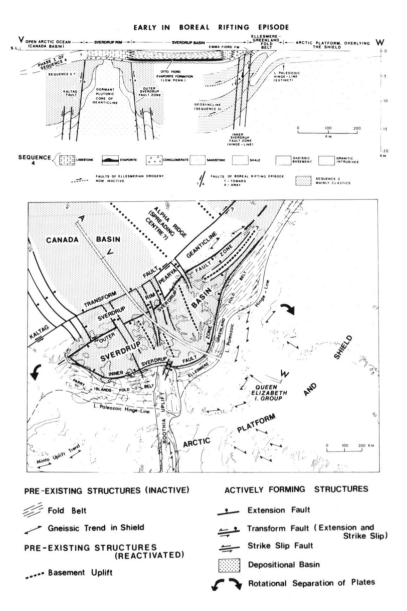

Figure 4. Initial fragmentation of the Canadian Arctic to form early stages of the Canada Basin and the Sverdrup Basin. This represents late Paleozoic time (Carboniferous to Permian) in an early stage of the Boreal Rifting Episode, when the lower part of Sequence 4 was being deposited (Fig. 2). The islands had not yet formed (cf. Fig. 1). The structures in the map and cross-section were controlled by the older structures shown in Figure 3.

Ocean, connecting the continental shelves of Canada and the U.S.S.R. Suggestions for its structure and origin include: a continental fragment (King *et al.*, 1966), an extinct sea-floor spreading centre (Hall, 1973; Ostenso and Wold, 1971), and a former subduction zone (Herron *et al.*, 1974). DeLaurier (1978) concluded that the Alpha Ridge has never been a spreading centre. His evidence that it has not been a spreading centre since Cretaceous time is that Late Cretaceous (Maastrichtian) fossils have been recovered from it (Clark, 1975); his evidence that it does not represent an older spreading centre that became extinct in pre-Tertiary time is its high relief, which, he theorized, could not have persisted to the present. The author (Kerr, 1979a) considers that the Alpha Ridge *is* an extinct spreading centre whose activity extended from about mid-Mississippian to latest Cretaceous time and then ceased. The present high relief may be the result of a submerged continental remnant beneath the Alpha Ridge, as suggested for other spreading centres in the world (Kerr, 1967b).

Sverdrup Basin and Sverdrup Rim

The Sverdrup Basin (Balkwill, 1978; Kerr, 1979a), is a successor basin that developed unconformably on the Franklinian Geosyncline during the Boreal Rifting Episode (Fig. 2). Sedimentation was controlled by two major hinge-lines, the Inner Sverdrup Fault Zone on the southeast, and the Outer Sverdrup Fault Zone on the northwest (Fig. 4). Both fault zones were controlled by older structures.

The Sverdrup Rim (Meneley *et al.*, 1975; Balkwill, 1978; Kerr, 1979a) contains rocks equivalent to those of the Sverdrup Basin but they are thinner. Subsidence of the Sverdrup Rim was less than the Sverdrup Basin, presumably because the rim was supported by the massive granitic core and root zone of the Pearya Geanticline which resisted great subsidence (Fig. 4, cross-section). This was the non-magmatic or mildly negative stage of the Pearya Geanticline, when slight subsidence occurred and thin sediments were deposited upon it to form the Sverdrup Rim (Fig. 2).

Rocks of Sequence 4 in the Sverdrup Basin thinned northwestward onto the Sverdrup Rim. The sequence may thicken again from there farther northwestward into the sedimentary wedge reported along the margin of the Canada Basin by Sobczak and Weber (1973). The Outer Sverdrup Fault Zone separates the Sverdrup Basin from the Sverdrup Rim, and is a major hinge-line.

The Sverdrup Basin consists of rocks of Sequences 4 and 5 (Fig. 2). Sequence 4 is by far the thickest and contains three phases. The sequence began with extreme crustal fracturing and merged into broader downwarping (Balkwill, 1978; Kerr, 1979a).

Phase I of Sequence 4 (Fig. 4, cross-section) began with mid-Mississippian non-marine conglomerates (Emma Fiord Formation), deposited in small downfaulted structural basins on the northern and southern margins of the Sverdrup Basin (Kerr and Trettin, 1962; Kerr, 1976). These probably occur throughout the Sverdrup Basin and represent crustal fracturing that marked the beginning of the Boreal Rifting Episode (Fig. 2). Fracturing accelerated and by Pennsylvanian time the entire Sverdrup Basin was severely faulted and subsided more widely. An upper Mississippian to lower Pennsylvanian basal conglomerate was deposited throughout the basin and overstepped southeast and northwest onto older rocks (Fig. 4, cross-section). In Phase I the Sverdrup Basin was an overall low created by active faulting, with internal horsts and anticlines providing coarse clastic sediments to adjacent grabens and synclines.

As the crustal fracturing event continued (Fig. 4, map), the Sverdrup Rim was breached further (Meneley *et al.*, 1975; Balkwill, 1978). Faulting in the Sverdrup Basin was gradually replaced by more general subsidence in Pennsylvanian and Permian time. The basin became a broad sag between the craton and the Sverdrup Rim, and fine-grained marine deposition became widespread. This sagging probably was controlled by reactivated deep-seated extension faults, but displacement no longer extended fully through the sedimentary column. Pennsylvanian and Permian marine faunas of the Sverdrup Basin have Asian affinity (Nassichuk, 1975), and probably migrated from Asia via the Canada Basin and the grabens that cut through the Sverdrup Rim. Deep marine rocks were deposited along the axis of the Sverdrup Basin, primarily evaporites and shales; mainly limestone was deposited along the margins and on the Sverdrup Rim. The Melvillian Disturbance (Fig. 2) produced an unconformity between Pennsylvanian and Permian rocks (Thorsteinsson and Tozer, 1970). It apparently was a pulse of stepped-up activity in the long-lasting Boreal Rifting Episode (Kerr, 1979a).

Phase II of Sequence 4 (Lower to mid-Upper Triassic) includes very thick siltstone and shale in the axial part of the Sverdrup Basin and thinner sand on the margin, as this was a time of widespread marine deposition. The hinge-line continued to persist along the southeast near the present southeastern limit of exposure (Fig. 4). The Sverdrup Rim continued to be relatively higher than the Sverdrup Basin, but received some sediments. The lack of carbonate and evaporites suggests that the Sverdrup Rim had been breached further and no longer had a major impounding effect on the Sverdrup Basin. Facies patterns indicate a continuation of the earlier tectonic regime, with great extension or crustal fracturing occurring beneath the Canada Basin, which was the main marine depocentre, and lesser extension beneath the Sverdrup Basin (Balkwill, 1978).

Phase III of Sequence 4 (mid-Upper Triassic to Lower Cretaceous) was a time when widespread terrigenous clastic rocks accumulated during low relief. The craton southeast of the Sverdrup Basin was a sedimentary source, so non-marine sands predominate along the margin of the basin, with shales along the axis. Non-marine sand units prograded northward fully across the Sverdrup Basin during times of complete marine withdrawal. During this phase, low to moderate tectonic activity continued in the Sverdrup Basin, with relative subsidence and some uplift, but no strong faulting.

During Sequence 4 the Sverdrup Basin was a small basin lying on the continent, adjacent to a larger developing oceanic basin, the Canada Basin (Fig. 4). The events that affected the Sverdrup Basin probably were secondary events, marginal to larger scale rifting in the Canada Basin. The Canada Basin was most likely in existence as a marine basin by late Paleozoic time, for Nassichuk (1975) and Nassichuk and Davies (1980) report that it was the source of marine faunas of the Sverdrup Basin. From faunal evidence (Tozer, 1960) and facies patterns (Balkwill, 1978), the Canada Basin certainly appears to have been in existence by Triassic time.

Sequence 5 marked a fundamental change in sedimentary and tectonic patterns in the Canadian Arctic, beginning in Early Cretaceous (Valanginian) time (Fig. 2).

The oldest unit of this sequence is the Isachsen Formation, a widespread non-marine unit that was deposited largely conformably in the Sverdrup Basin, but also over-stepped unconformably southeastward far beyond the established margin of Sverdrup Basin, and onto the cratonic shelf. This initial blanket of non-marine sediments of the Isachsen Formation was succeeded by alternating marine and non-marine units as wider subsidence followed. The overstepping by Sequence 5 was part of a wide-spread transgression that covered much of the North American Continent (Williams and Stelck, 1975) as well as western Greenland (Henderson *et al.*, 1976). There first was a basin-wide marine regression and exposure, followed by deltaic alluviation of sands, mainly from the southeast, covering the entire Sverdrup Basin. Tectonism associated with the Isachsen Formation included faulting, and apparently emanated from northwest of the Sverdrup Basin (Roy, 1974; Rahmani, 1977; Balkwill, 1978). This probably was an enlarged pulse of the Boreal Rifting Episode, reflecting an increase in spreading on the Alpha Ridge. Faulting was slightly more extensive than it had been in late Paleozoic time. An area southwest of the main Sverdrup Basin was affected, suggesting that the Canada Basin had been enlarged toward the southwest. The extension associated with this pulse also diminished rapidly southeastward at the southeastern margin of the Sverdrup Basin, indicating that the Shield to the southeast was still an effective obstacle to the advance of the Boreal Rifting Episode.

The Inner Sverdrup Fault Zone is remarkably parallel to the older, lower Paleozoic hinge-line farther southeast (Fig. 4). The present southeastern limit of exposure of the Sverdrup Basin is approximately at the inner fault zone, but overlaps it slightly, indicating that the fault zone also controlled later preservation. The Outer Sverdrup Fault Zone Fault extends along the Pearya Geanticline, presumably guided by its dormant core.

The Sverdrup basin formed initially in late Paleozoic time (Fig. 4), when block faulting fractured the crust, creating high relief and depositing syntectonic conglom-erates as the basal unit of the basin. The faults at the southeastern margin died out upward into rocks of the Sverdrup Basin (Kerr, 1976). Extension in the Sverdrup Basin decreased southeastward abruptly at the inner fault zone. The older Ellesmere-Greenland Fold Belt was affected slightly as was the northern end of the Boothia Uplift (Fig. 4), but there were no known effects on the Arctic Platform or Shield.

The Sverdrup Basin apparently was formed by extension that originated farther northwest in the Canada Basin, and may have been tied to that basin by northwest-trending extension faults that cut through the Pearya Geanticline. An origin for the Sverdrup Basin by crustal extension is supported by geophysical observations (Forsyth *et al.*, 1979; L.W. Sobczak, pers. commun., 1979), that the crust thins generally from the margin of the basin toward its centre.

The Outer Sverdrup Fault Zone is a major hinge-line between the Sverdrup Rim and the Sverdrup Basin (Fig. 4). It includes major deep faults reported by L. W. Sobczak (pers. commun., 1979) that occur north of Hassel Sound and east of Ellef Ringnes Island. They trend northeast and are downdropped toward the southeast, affect Paleozoic and Precambrian rocks, and die out upward within the Sverdrup Basin. This fault zone probably began to form in mid-Mississippian time (Kerr, 1979a), continuing into Pennsylvanian time, with the resulting high relief producing conglomerates.

Canada Basin

The Canada Basin (Fig. 1) is a large, deep part of the Arctic Ocean, lying northwest of the Queen Elizabeth Islands and southwest of the Alpha Ridge. It has depths greater than 3500 m over a wide area. Various suggestions on its origin were summarized earlier (Kerr, 1979a).

Little is known about the margin of the Canada Basin adjacent to the Canadian Arctic Islands, because the rocks there have not been drilled in the offshore area. A seaward thickening continental terrace wedge was reported there by Sobczak and Weber (1973). Kerr suggested (1979a) that the Canada Basin began to form in mid-Mississippian time, contemporaneous with the block faulting that initiated the Sverdrup Basin (Fig. 4). In the Canada Basin there apparently was greater extension than in the Sverdrup Basin. These two basins are separated by the Sverdrup Rim, which is cored by the Pearya Geanticline (Fig. 4, cross-section).

An extension of the Kaltag Fault (Fig. 4) follows the edge of the continental shelf along the southeast margin of the Canada Basin (Norris, 1974; Yorath and Norris, 1975). It is nearly parallel to the Pearya Geanticline, and perpendicular to the Alpha Ridge. It is a strike-slip fault in northern Alaska and the Yukon, where it is considered to be right lateral. Kerr (1979a) suggested that in the Canada Basin region, the Kaltag Fault is a transform fault, with left lateral displacement adjacent to the Arctic Islands as well as downdropping on its northwest side.

Mechanism of Boreal Rifting Episode

The mechanism of the Boreal Rifting Episode is shown diagrammatically in Figure 5. Relationships suggest that the Alpha Ridge was a spreading centre, with the Kaltag Transform Fault transverse to it and controlled by the Pearya Geanticline. Great extension and downfaulting of the Canada Basin was transformed across this fault into lesser extension and downfaulting of the Sverdrup Rim and Sverdrup Basin. The abrupt reduction in the amount of extension from the Canada Basin to the Sverdrup Basin may have resulted from the presence of the Pearya Geanticline. The rigidity of the geanticline and its dormant linear plutonic core caused the geanticline to continue as a high relative to the subsiding Sverdrup Basin and Canada Basin (Fig. 4, cross-section). This was the non-magmatic or mildly negative stage of the Pearya Geanticline (Fig. 2).

The Pearya Geanticline thus was a transverse linear structure that impeded the southeastward advance of extensional structures, localized the Kaltag Fault, and thereby controlled the continental margin. Nevertheless, substantial extension took place within the geanticline; however, this was concentrated in deep and narrow transverse grabens (Fig. 5). Some extension continued through the geanticline to the southeast, where deep subsidence created the Sverdrup Basin above the former Franklinian Geosyncline (Fig. 4, cross-section). It is inferred that the Sverdrup Basin subsided more readily than did the Sverdrup Rim because the crust beneath it was thinner and less competent. That column, formed of a thin crystalline layer overlain by sediments, was a crust in which more widely distributed extension faults and subsidence could occur. It appears that the faults of the Boreal Rifting Episode were propagated southeastward through the trend of the Pearya Geanticline. Once through, they met certain northwest-trending uplifts that they were able to follow along more readily (cf. Figs. 3 and 4).

A second transverse linear structure farther southeastward within the continent also impeded the southeastward continuation of extensional structures of the Boreal Rifting Episode (cf. Figs. 3, 4 and 5). This trend localized the southeastern margin of the Sverdrup Basin at the Inner Sverdrup Fault Zone. Southeast of this fault zone the crystalline crust is overlain by a fold belt of mainly carbonate rocks. There the

MECHANISM OF BOREAL RIFTING EPISODE

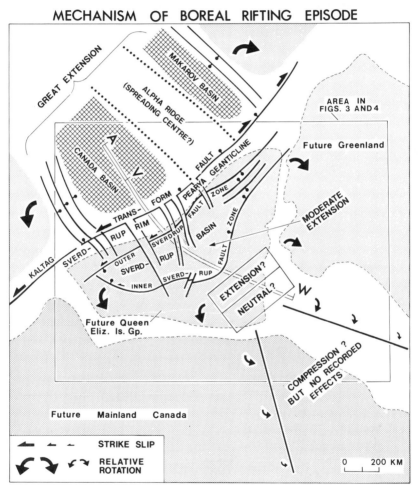

Figure 5. Early in the Boreal Rifting Episode (late Paleozoic time, cf. Fig. 4). Activity apparently originated with the Alpha Ridge as a spreading centre, and with a tendency to be propagated southeastward. Propagation was hindered by the Pearya Geanticline, resulting in formation of the Kaltag Transform Fault along its northwest margin. Extension decreased from northwest to southeast, abruptly across the Pearya Geanticline, gradually across the Sverdrup Basin, and abruptly again across the Inner Sverdrup Fault Zone.

continental crust apparently was extended little by the Boreal Rifting Episode. The structure that ultimately controlled the Inner Sverdrup Fault Zone was the crystalline crust, which thickened gradually southeastward, achieving its normal thickness at the then extinct lower Paleozoic hinge-line (Fig. 4, cross-section). The competence of the total crustal column presumably increased gradually southeastward within the region of the deformed Franklinian Geosyncline (Fig. 3). The Inner Sverdrup Fault Zone formed where the competence of the total crustal column increased sufficiently that a zone formed (the Inner Sverdrup Fault Zone), across which extension decreased rapidly southeastward. Extension died out rapidly into neutrality southeast of the Inner Sverdrup Fault Zone. Theoretically, compression may have occurred at this time still farther to the southeast (Fig. 5); however, no compressional deformation of that age is known within the stable crystalline continental crust.

The Canadian Arctic was the site of a great rotational deformation during the Boreal Rifting Episode (Fig. 5). This is indicated by the southeastward reduction in extension. The Alpha Ridge appears to have been a spreading centre that fractured a protocontinent, and the rotation emanated from the ridge. Blocks on either side of the spreading centre rotated apart, forming the Canada Basin on one side and Makarov Basin on the other. The Alpha Ridge and the continental blocks southwest of the Canada Basin also were rotating apart relatively, the result being propagation of faults southeastward into the Sverdrup Basin.

The structural trends in the crystalline rocks of the Canadian Arctic, both the Pearya Geanticline and the Shield, controlled, impeded, and variously halted the advance of extensional structures of the Boreal Rifting Episode (cf. Figs. 3, 4 and 5). Structures in the crystalline basement rocks of the Shield (Sequence 1) controlled the location of the lower Paleozoic hinge-line (Fig. 3), which in turn controlled the location of the Inner Sverdrup Fault Zone (Fig. 4). The trend of the intrusive cored Pearya Geanticline (Fig. 3) localized the Kaltag Fault along its northwest margin. It also localized the northwest margin of the Sverdrup Basin along the Outer Sverdrup Fault Zone. The part of the Pearya Geanticline that was high during initial fragmentation (Fig. 3) has remained high to the present day and forms the mountains of northern Ellesmere and Axel Heiberg Islands. This may have persisted as a high because of the proximity of the Alpha Ridge. The part of the geanticline farther southwest collapsed during the Boreal Rifting Episode to form the moderately negative Sverdrup Rim above it (Fig. 4).

The Boreal Rifting Episode fractured and subsided the Canada Basin during deposition of Sequences 4 and 5 (Fig. 2). A lesser amount of extension reached into the continent where the Sverdrup Basin was formed on continental crust. This process apparently continued intermittently from mid-Mississippian time to Late Cretaceous time. The boundaries separating zones of greater and lesser extension (Fig. 5) remained relatively fixed during this interval. The southeastern boundary of great extension, the Kaltag Fault, was very persistent. The Boreal Rifting Episode had the tendency to penetrate southeast from the Alpha Ridge through the North American Continent, and southeast from the Canada Basin into the Sverdrup Basin. The southeastward advance of this rifting episode was aborted, apparently because of the obstacle created by structural trends within the northern part of the North American Continent. The southeastern limit of its effects is slightly southeast of the Inner

Sverdrup Fault Zone. It did not reach the lower Paleozoic hinge-line, nor the Arctic Platform and Shield beyond it, where a thick crystalline layer formed most or all of the continental crust (Fig. 4). The total thickness in the southeast may have been near normal for a continental crust (about 37 km), for that is the approximate thickness today (Sander and Overton, 1965; Sobczak and Weber, 1973).

There is no indication today of a root zone under the present Sverdrup Rim, similar to that which may have existed there at the end of the constructional phase (L.W. Sobczak, pers. commun., 1979). The root zone inferred there at the end of the constructional phase (Fig. 3) presumably was eliminated later, during the process of subsidence of the Sverdrup Rim (Fig. 4).

Eureka Sound Formation

Sequence 6 is the Eureka Sound Formation, a complex unit intimately involved with tectonic activity. It was deposited while the two plate tectonic episodes central to this study were affecting the North American Arctic (Kerr, 1979a): the Boreal Rifting Episode and a partly younger event, the Eurekan Deformation (Fig. 2).

During deposition of the lower part of Sequence 6 (Fig. 6, early; Fig. 7, lower part), increased activity of the Boreal Rifting Episode fractured the northwest part of the continent severely. The Pearya Geanticline and Sverdrup Rim were raised and eroded. The Sverdrup Basin was fractured by arches within it that were uplifted and eroded, and that basin no longer existed as a single sedimentary basin. Parry Channel apparently opened up from the west as a submarine rift valley, achieving a major breakthrough of the Sverdrup Rim and underlying Pearya Geanticline. Eastward advance of this rift valley was halted by the crosswise trend of Precambrian rocks of the Boothia Uplift, and the Southeast Bathurst Fault Zone developed at its termination (cf. Figs. 3 and 6). This faulting and fracturing from the northwest apparently occurred during Late Cretaceous (Campanian-Maastrichtian) to Early Tertiary (Paleocene) time, while the lower part of Sequence 6 was being deposited (Fig. 6, early, and Fig. 7, lower part). The Boreal Rifting Episode may also have resulted in continued deposition of a thick and widespread succession of Sequence 6 in the continental terrace wedge along the margin of the active Canada Basin.

Deposition of the upper part of Sequence 6 was closely associated with the Eurekan Deformation.

Eurekan Deformation

The Eurekan Deformation (Kerr, 1979a) was a plate tectonic episode that formed much of the southeastern part of the Canadian Arctic Rift System (Fig. 1) between Late Cretaceous (Campanian/Maastrichtian) and mid-Tertiary (Miocene or possibly later) time. It was contemporaneous with the deposition of Sequences 6 and 7 (Fig. 2). The Eurekan Deformation includes two related and complementary phenomena, the Eurekan Rifting Episode (extension or rifting) and the Eurekan Orogeny (compression).

The Eurekan Rifting Episode (Kerr, 1977, 1979a) formed the Mid-Labrador Sea Ridge, the Baffin Bay Fault Depression, and most of the fault-controlled channels in the southeastern part of the Queen Elizabeth Islands Sub-plate (Fig. 1). The episode

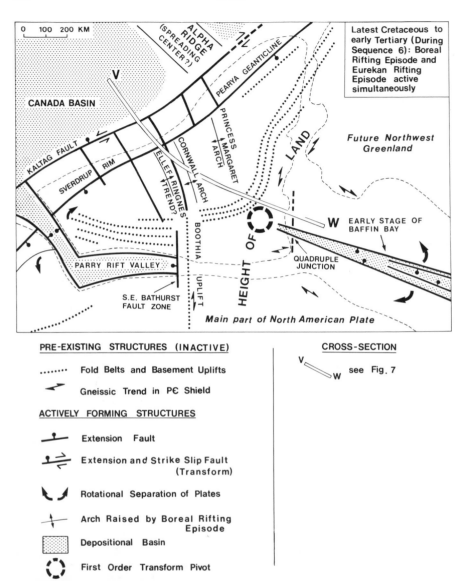

Figure 6. Simultaneous development of rifting episodes during Sequence 6 (after Kerr, 1979a, Fig. 10). At first, in Late Cretaceous time, the Boreal Rifting Episode was advancing southeast, fracturing the crust, and causing uplift and erosion of the Pearya Geanticline, Sverdrup Rim, and arches within the Sverdrup Basin (Fig. 7, early). Later, the younger Eurekan Rifting Episode began to advance northwest and enlarged rapidly. It neutralized the Boreal Rifting Episode, causing collapse of former highs, the Sverdrup Rim, and intrabasin arches, with encroachment of Tertiary rocks onto each (Fig. 7, late).

began in the southeast in the area of Labrador Sea, and advanced northwestward to Baffin Bay and the Arctic Islands. In its latest phases it connected with faults in the northwest that had been initiated by the Boreal Rifting Episode.

A Contest of Plate Tectonic Episodes

The Eurekan Deformation began to affect the Canadian Arctic in Late Cretaceous time, while the Boreal Rifting Episode was still active (Fig. 2). Regional considerations (Kerr, 1979a) and magnetic striping (Srivastava, 1978) suggest that Baffin Bay was in an early stage, beginning to be fragmented by Late Cretaceous (Maastrichtian) to Early Tertiary (Paleocene) time (Fig. 6, early; Fig. 7, lower part).

The Eurekan Deformation was propagated northwest, so the first structures to

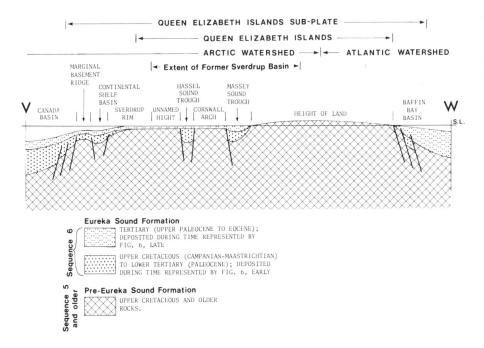

Figure 7. Generalized cross-section from the Canada Basin to the Baffin Bay Basin reconstructed to about Eocene time (from Kerr, 1979a, Fig. 15), showing Sequence 6 of the Innuitian Mobile Belt (see Fig. 2). The two basins had been actively forming as rifted oceanic basins, with the Queen Elizabeth Islands Sub-plate being a continental transition between. Two events are shown here. The lower part of Sequence 6 was deposited mainly in the northwest, as substantial activity of the Boreal Rifting Episode continued (Fig. 6, early). The upper part of Sequence 6 was deposited more widely, when the Boreal Rifting Episode had nearly expired, but the Eurekan Rifting Episode was in an early stage and accelerating (Fig. 6, late). The time of the reconstruction shown above is just prior to the climactic phase of the Eurekan Orogeny (Fig. 8), when Sequence 6 and older rocks were deformed by extension southeast of the height of land, and by compression in the northwest.

reach Baffin Bay may have been extension faults producing a narrow rift valley (Fig. 6, early). The Boreal Rifting Episode was well developed and continued to be active at this time; however, its effects seemed unable to advance southeast of the Sverdrup Basin and also had been halted in Parry Rift Valley. Thus, two plate tectonic episodes were occurring simultaneously on opposite sides of the continent, and were advancing toward each other. Each had produced indentation and partial fragmentation of that continent. One had produced the Canada Basin and Sverdrup Basin, but its southeast advance seemed stalled. The other was in an early stage and just beginning to form Baffin Bay; however, this was advancing rapidly northwest, and its structural influence was increasing.

Faults of the Eurekan Rifting Episode could advance northwest easily in Baffin Bay because they followed structural trends in the crystalline basement (Fig. 6). These faults could not easily advance northwest beyond Baffin Bay, however, because of the obstruction by crosswise structural trends in the basement of the present southeastern Queen Elizabeth Islands Group (cf. Figs. 3 and 6). As a result of this obstruction a first-order transform pivot and a quadruple junction began to form.

The Eurekan Deformation apparently accelerated markedly in late Paleocene time (Fig. 6, late; Fig. 7, upper part). The effective domain of the Boreal Rifting Episode simultaneously receded to the northwest and its influence became less prominent. The effects of the Boreal Rifting Episode were neutralized in the Sverdrup Basin, perhaps due to interference by the enlarged Eurekan Rifting Episode. The former fragmentation of the Sverdrup Basin was replaced by widespread collapse, including the formerly uplifted structures: Sverdrup Rim, and arches within the Sverdrup Basin such as Cornwall Arch (Fig. 6, late; Fig. 7, upper part of Sequence 6). Collapse resulted in encroachment of the upper part of Sequence 6 (Upper Paleocene to Eocene) onto the former highs. The Pearya Geanticline, however, remained high in the northeast near Ellesmere Island, and was a source for Lower Tertiary strata that became finer and prograde southeastward (Bustin, 1977). The persistence of the high in the northeast may have been related to the proximity of the spreading centre occupying the Alpha Ridge to the northwest. Paleocene-Eocene sedimentation presumably continued farther northwest in the Canada Basin and the adjacent continental shelf, but if tectonic activity there was diminished the strata may have been thinner than contemporaneous rocks of Baffin Bay (Fig. 7, upper part of Sequence 6).

By Paleocene time (Fig. 6, late; Fig. 7, late) the main sub-plate boundaries within the Canadian Arctic Rift System were beginning to take on their present form. The rifting episodes had formed two oceanic basins, separated by a broad height of land where little rifting or crustal extension had occurred. This height of land was an isthmus connecting the main part of North America through the Queen Elizabeth Islands to the future northwest Greenland.

As the Eurekan Deformation progressed wider fragmentation affected the Baffin Bay region. Thick Paleocene-Eocene sediments accumulated in the Eclipse Trough during tectonism (Miall et al., 1980), and a similar column may occur in Lancaster Aulacogen (Daae and Rutgers, 1975; Kerr, 1979b). Both of these were rift-controlled, deeply subsiding basins that continued to develop as the Eurekan Deformation advanced (Fig. 8). They apparently were tied to greater rifting in Baffin Bay, where even thicker sediments may have accumulated.

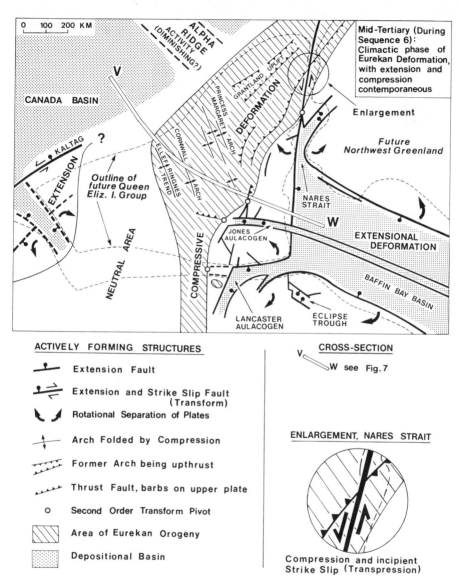

Figure 8. Climactic phase of the Eurekan Deformation (after Kerr, 1979a, Fig. 20). This represents mid-Tertiary time (between middle Eocene and early Miocene), when extension in the southeast (Eurekan Rifting Episode) caused compressional deformation farther northwest (Eurekan Orogeny). Farther west, in a large neutral area, extension that earlier emanated from the Canada Basin was neutralized by the Eurekan Deformation. In the extreme west, the Banks Island area, extension of the Boreal Rifting Episode continued uninterrupted, presumably because this was too far west to be affected by the Eurekan Deformation. This followed shortly after the situation depicted in Figure 6.

In mid-Tertiary time (between middle Eocene and early Miocene) the Eurekan Deformation reached its climactic phase (Fig. 8). It caused extensional deformation in the southeast (the Eurekan Rifting Episode), affecting mainly Baffin Bay and its branches; it also caused complementary compressive deformation in the northwest (the Eurekan Orogeny), affecting mainly the Sverdrup Basin.

The Eurekan Deformation was strongly controlled by the pre-existing structures (cf. Figs. 3, 6 and 8). Extension faults propagating northwest apparently were stopped by structures within the crystalline basement in the southeastern Queen Elizabeth Islands Group (Fig. 8). As spread and collapse continued in the Baffin Bay Basin the quadruple junction in northwest Baffin Bay grew, as fault bounded blocks rotated away from each other to enlarge and deepen the intervening marine channels. A main branch of the quadruple junction was deflected westward into Lancaster Aulacogen, where it was stopped and deflected further southward by the Boothia Uplift (cf. Fig. 3). Another branch was deflected northward into Nares Strait. A third branch of this junction penetrated the Queen Elizabeth Islands Group and formed the Jones Aulacogen in Jones Sound. The height of land that had formed earlier between the two rifted domains (Fig. 6) was being broken down on its southeastern side as parts collapsed into the sea (Fig. 8), and its summit presumably was migrating northwest. At this time (Fig. 8) the islands in the eastern Arctic were beginning to take on the shapes they have today (Fig. 1). The general outlines of most of the islands were produced by faults.

In the large region of extensional deformation in the southeast several plates were rotating away from each other. The relative rotational separation of adjacent blocks is shown in paired arrows located within land masses (Fig. 8). The lateral movement apart of land masses indicated by each pair of arrows probably was not great. The intervening deep seaways probably formed largely by foundering of continental crust, and to a lesser degree by lateral separation. Since the pairs of arrows indicate relative movement, a single arrow of one pair cannot be related directly to a single arrow of another pair.

Nares Strait was active in the climactic phase of the Eurekan Deformation (Fig. 8). Kerr suggested (1967a, 1979a) that the strait contains a complex submarine rift valley. In the south it had minor rotational opening and major foundering. Farther north it had compression and minor strike-slip faulting. Such displacement probably involved a pivot, with extension in the south and transpression in the north, in the sense of Harland (1971, 1973). Because of the absence of great strike-slip in Nares Strait, Kerr (1967b, 1979a) suggested that Baffin Bay formed by major foundering and minor lateral separation. Most authors favour great strike-slip displacement in Nares Strait and great lateral separation in Baffin Bay (Keen et al., 1974; Keen and Hyndman, 1979; Jackson et al., 1979).

The Eurekan Rifting Episode continued to separate and rotate apart the blocks on either side of Baffin Bay, but the extension faults could not advance northwest. Consequently, a pincer arrangement developed (Fig. 8), with compressive deformation of the Eurekan Orogeny occurring in the northwest. This compressive orogeny affected central and eastern parts of the Sverdrup Basin that earlier had been neutralized. Regionally, the area of extensional deformation was transformed into an apically opposed area of compressive deformation through a large intermediate zone

(Fig. 8). Each area was made up of blocks that moved relative to adjacent blocks, apart in the extensional area and toward one another in the compressive area. Several transform pivots that developed between the areas of extensional and compressive deformation were points about which rotation was concentrated, transforming extensional deformation in one block into compressive deformation in another. All levels of the column in the Sverdrup Basin were affected by the Eurekan Orogeny (Fig. 8), including the hitherto underformed upper part of the Eureka Sound Formation (Fig. 7, upper part).

It appears that the Eurekan Deformation was not strong enough to cause compressive deformation throughout the entire Queen Elizabeth Islands Group (Fig. 8). In a large area the Boreal Rifting Episode was neutralized. Uppermost Cretaceous (Maastrichtian) and younger rocks are not folded along the Arctic Coastal Plain on

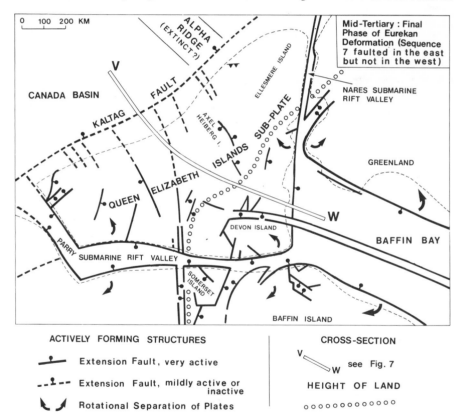

Figure 9. Final phase of the Eurekan Deformation and severing of the continent in early Miocene or later time (after Kerr, 1979a, Fig. 21). Faults were able to break northward through the obstructions in Nares Submarine Rift Valley and westward through obstructions in Parry Submarine Rift Valley. For the first time a structural connection from the Atlantic Ocean to the Arctic Ocean was made through the North American Continent. Compare with Figure 10.

northern Ellef Ringnes Island (Meneley *et al.*, 1975), suggesting that this was a neutral area of essentially continuous sedimentation throughout the Eurekan Orogeny. The folding that occurred farther southeast in the Sverdrup Basin is good evidence that the Eurekan Orogeny originated in the southeast rather than the northwest; it also supports the suggestion that the Boreal Rifting Episode did not affect the Sverdrup Basin after that basin collapsed in about Late Paleocene time (Fig. 7, upper part).

During the climactic phase of the Eurekan Deformation, extensional deformation continued farther west (Fig. 8), where there was differential uplift and faulting from Maastrichtian to Eocene time, without any interruptions by compressional deformation (Miall, 1975). This presumably was part of the Boreal Rifting Episode and was a side effect of extensional faulting that continued in part of the Canada Basin. The process that produced the Boreal Rifting Episode in the Canada Basin probably continued (i.e., activity of the Alpha Ridge), but that deformation was completely neutralized except in the west, where extension prevailed through the entire time span of the Eurekan Deformation.

Compressive deformation of the Eurekan Orogeny continued until as late as middle Eocene time, for rocks of that age are deformed on Axel Heiberg Island and overlain by lower Miocene rocks (Balkwill and Bustin, 1975). These are respectively Sequences 6 and 7 (Fig. 2). This phase of deformation may have continued into Oligocene time, for compressive movement of that age probably occurred in the Lake Hazen Fault Zone of northern Ellesmere Island (Miall, 1979).

Severing of the Continent by the Eurekan Rifting Episode

The final severing of the North American Continent in the Arctic (Fig. 9) resulted from the final pulse of the Eurekan Rifting Episode, which occurred in early Miocene or later time (Kerr, 1979a). For the first time the Queen Elizabeth Islands Sub-plate became completely surrounded by fault zones. This was achieved as faults in the southeast broke through the height of land to connect with faults that had formed earlier in the northwest by the Boreal Rifting Episode (cf. Figs. 8 and 9).

The nature of the severing and the activity of various faults is inferred from the distribution and setting of Miocene rocks (Sequence 7). Nearly all of the major faults associated with plate breakup in the Arctic occur at sea only (Fig. 9). Few of the faults that occur on land have associated Miocene rocks. Thus, the final phase is rather speculative. Miocene rocks are known in three regions and have a different setting in each (Fig. 10).

In the long Arctic Coastal Plain (Fig. 10), Miocene rocks are the northwest-dipping, unconsolidated sands of the Beaufort Formation (Thorsteinsson and Tozer, 1970; Hills and Matthews, 1974). On northwest Ellef Ringes Island that formation is undeformed and conformable with Upper Cretaceous rocks (Meneley *et al.*, 1975), as part of the northwest-dipping continental terrace wedge. That area apparently subsided without tectonic interruption from Late Cretaceous to mid-Miocene or later time. On western Banks Island (Fig. 10) Miocene rocks also are part of a conformable northwest-dipping continental terrace wedge. In the offshore area there may be a concordant sequence extending from Oligocene to present-day rocks (Miall, 1975). Although rifting deformation apparently continued until Eocene time in the Banks

Island area, it has been dormant since then. Thus, the Arctic Coastal Plain and the continental terrace wedge to the northwest apparently were not affected by the final pulse of deformation that severed the continent in Miocene time.

On Axel Heiberg Island (Figs. 9 and 10) lower Miocene conglomerates have been subjected to both penecontemporaneous and post-depositional faulting (Balkwill and Bustin, 1975). This is the youngest dated extensional faulting in the Canadian Arctic Islands; it is presumed to be of Miocene age, and is assigned to the Eurekan Rifting Episode.

On northern Ellesmere Island (Fig. 10) conglomerates presumed to be equivalent to the Miocene Beaufort Formation are cut by thrust faults directed to the northwest (Wilson, 1976). This is the youngest compressive deformation in the Canadian Arctic Islands and is assigned to the Eurekan Orogeny.

The structural setting of Miocene rocks (Figs. 9 and 10) suggests that this last pulse of faulting (Miocene or younger) was a continuation of the Eurekan Deformation, originating in the southeast and propagated northwest. There was extension in the southeast (Axel Heiberg Island), compression in the northeast (northern Elles-

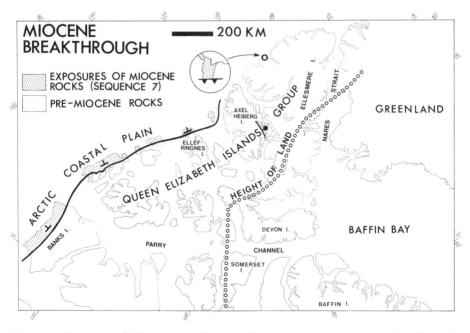

Figure 10. Exposures of Miocene rocks (Sequence 7) on present geography, showing effects of the final pulse of the Eurekan Deformation in early Miocene or later time (cf. Fig. 9). This is evidence that the Miocene pulse, which advanced northwest from Baffin Bay, was propagated mainly through Nares Strait, and to a lesser degree through Parry Channel. There was extensional deformation on Axel Heiberg Island, and compressional deformation on northern Ellesmere Island, but the Arctic Coastal Plain was not affected. The height of land is still expressed in topography and bathymetry.

mere Island), and stability in the northwest and west. This pulse apparently severed the continental crust in the region of the height of land in two places, by means of Parry Submarine Rift Valley and Nares Submarine Rift Valley (Figs. 9 and 10). This completed the outline of the Queen Elizabeth Islands Sub-plate, and for the first time made the structural connection, by rifted zones, between Baffin Bay and the Arctic Ocean. It apparently was accomplished by Nares Submarine Rift Valley more than by Parry Submarine Rift Valley (Fig. 9).

Assuming the final pulse of faulting affecting the region was due to the Eurekan Deformation, many areas would have been affected. Structures in the marine areas probably were rejuvenated, including Baffin Bay, Nares Submarine Rift Valley, central to eastern parts of Parry Submarine Rift Valley, and other channels within the Canadian Arctic Islands. This, of course, is speculative. In the main waterways there may be a Miocene succession (Sequence 7) equivalent to the Beaufort Formation, bounded above and below by stratigraphic breaks that represent pulses in the Eurekan Rifting Episode (see Kerr, 1979b). The Eurekan Deformation ceased activity in mid-Tertiary time (early Miocene or later), after extensional deformation severed the continent and reached the Arctic Ocean (Figs. 9 and 10). Apparently there has been no strong tectonism since then in the study area.

It appears that the breakthrough of the final pulse was rather intense in Nares Submarine Rift Valley, because associated faulting extended a large distance from that structure (Fig. 10). This suggests that the obstacle within the height of land in that area was rather strong, and that the rotational couple producing extension and complementary compression was substantial. In contrast the final formation of Parry Submarine Rift Valley by breaking through the height of land there (Boothia Uplift; cf. Fig. 6), may have occurred with little violence. This is suggested because of the lack of disruption of the Beaufort Formation in the Arctic Coastal Plain. This lack of disruption by the final pulse (Figs. 9 and 10) suggests that the faults which broke westward through the Boothia Uplift in that pulse simply connected with pre-existing faults in the west end of the rift valley that had formed earlier by the Boreal Rifting Episode (cf. Fig. 6) but had been dormant for some time (Fig. 8). The eastern ends of the older faults would have been rejuvenated in the final pulse (Fig. 10), but the western ends, where the undisturbed Beauford Formation occurs, might have been inactive.

DORMANT RIFT SYSTEMS

The Canadian Arctic Rift System is now dormant (Fig. 1); it became dormant when the Eurekan Deformation ceased after its final Miocene or younger pulse (Fig. 9). The system may have been tectonically stable or partly in extension since then, i.e., after about 20 Ma ago. There may have been minor continued activity at sea where the major zones of weakness of the rift system exist, but these are not expressed on land. There is little seismic activity along the system now (Basham et al., 1977). The patterns of earthquakes are not characteristic of plate margins, but rather are an expression of the readjustment of existing structures to a regional stress field.

This paper suggests that the Alpha Ridge was an active spreading centre until Late Cretaceous or Early Tertiary time (Figs. 4 and 6). Spread then diminished, being neutralized by the compressional effects of the Eurekan Rifting Episode (Fig. 8).

From then on the ridge ceased being a spreading centre in the usual sense of its flanks moving apart relative to each other. Farther west, however, where the Eurekan Orogeny had no effect, the movement of rock away from the Alpha Ridge *relative to the Canadian Arctic Islands* continued on the Kaltag Fault (Fig. 8), for extensional deformation continued on Banks Island until mid-Tertiary (Eocene) time (Miall, 1975). All spreading activity on the Alpha Ridge apparently ended with the final Miocene or post-Miocene breakthrough from Baffin Bay to the Arctic Ocean (Figs. 9 and 10). Maastrichtian (latest Cretaceous) fossils dredged from the Alpha Ridge suggest that its flanks have not moved apart since that time (Clark, 1975).

The Canadian Arctic Rift System and Alpha Ridge are dormant structures within the North American Plate (Fig. 1). Since they became dormant, the entire region of northern Canada, Greenland, and that part of the Arctic Ocean west of the Nansen-Gakkel Ridge has travelled as part of the North American Plate. Seismic activity within that region may be localized along dormant sub-plate boundaries and may result from the overall rotation of that plate.

The Queen Elizabeth Islands Sub-plate is circumscribed by dormant major faults that produced its general outline (Figs. 1 and 9). The sub-plate was internally broken by lesser faults, also dormant, that produced the major outlines of the islands of the Queen Elizabeth Islands Group. The present detailed outlines of most islands in the Canadian Arctic resulted from erosional modification of fault bounded blocks. Clear examples are Baffin, Somerset, and Devon Islands (cf. Figs. 9 and 10).

The major height of land that extended northeastward through the Canadian Arctic in mid-Tertiary time (Figs. 6 and 7) still exists, and is expressed by topography and bathymetry (Fig. 10). It is an area of mainly continental crust, isolated between oceanic basins that were formed by different rifting episodes (Fig. 9). The Canadian Arctic is thus a structural and physiographic transition between the Atlantic Ocean in the southeast and the Arctic Ocean in the northwest (Kerr, 1979a). In a structural sense, the marginal parts of the transition are most oceanic (Labrador Sea, Baffin Bay, and the Canada Basin), while the central part of the transition is most continental (the height of land in the central Canadian Arctic Islands).

BASEMENT CONTROL OF PLATE BOUNDARIES

The formation of both large and small structures in the Canadian Arctic by the Boreal and Eurekan plate tectonic events was controlled to a remarkable degree by pre-existing structures. The locations of the Kaltag Fault and the Outer Sverdrup Fault Zone apparently were controlled by the Pearya Geanticline (cf. Figs. 3 and 4). The Sverdrup Rim may reflect an inactive crystalline core beneath the Pearya Geanticline. The location of the Inner Sverdrup Fault Zone was controlled by a combination of older basement structure and lower Paleozoic fold belts. The northwest-trending faults within the Sverdrup Basin (Fig. 4) were guided by the trends of older basement uplifts (Fig. 3).

In the Eurekan Deformation, Baffin Bay opened along trends in the Precambrian crystalline basement that presumably guided and facilitated extension faults (cf. Figs. 3 and 6). The northwestward advance of Baffin Bay was hindered by strong transverse-trending basement structures that curved eastward and then northeast-ward in the southeastern part of the Queen Elizabeth Islands Sub-plate (cf. Figs. 1, 3 and 8). This broad obstacle distributed extension in the southeast and compressive deformation in the northwest (Fig. 8).

A branch of Baffin Bay that was deflected westward formed Lancaster Aulacogen, which was also guided by structural trends (Fig. 8). Lancaster Aulacogen in turn was halted and deflected southwestward by the north-trending gneisses comprising the core of the Boothia Uplift (cf. Fig. 6). Simultaneously, another branch of Baffin Bay was deflected northward into Nares Submarine Rift Valley (Fig. 8).

The present northeast-trending height of land in the Canadian Arctic (Fig. 10) was similarly controlled by structure. Its northwest side began to develop in an early stage of the Boreal Rifting Episode, and was well developed by Cretaceous-Tertiary time (Fig. 6, early; Fig. 7, early). At that time the height of land was part of a largely unbroken continental region of thick crystalline crust lying southeast of the fractured Sverdrup Basin (cf. Figs. 4 and 6). The southeastern margin of the height of land was formed later, by faulting that produced Baffin Bay and its branches (Figs. 8 and 9). The bay and its branches in turn had been controlled by basement structures that had deflected the quadruple junction (Fig. 6). The southeastern extension of the height of land was controlled by basement structure of the Boothia Uplift. In subsequent events the height of land in turn controlled the location of compressive deformation (cf. Figs. 6 and 8). It also was a source area for rocks of Sequence 6 (Fig. 7), a summit between watersheds of the Atlantic and Arctic Ocean Basins, with sediments shed northwestward and southeastward by river systems controlled by structure.

SUMMARY

This section will address the four basic problems outlined in the introduction.

1) The lands and seas in the Canadian Arctic owe their shapes to plate tectonic processes, specifically to plate divergence. The shapes were controlled by the interplay between plate tectonic stresses and pre-existing structures. The plate boundaries here are zones of extension faults that were propagated into the continent from the oceanic areas. The general location of major plate boundaries may be determined by subcrustal forces that initiate plate breakup; however, the shapes of intermediate (e.g., the Greenland Sub-plate) to small-scale tectonic structures (local faults) were controlled primarily by structural trends in the crystalline Precambrian basement.

The pre-existing structure that influenced plate boundaries in the Canadian Arctic most strongly is the Pearya Geanticline (Fig. 3). It consisted of Precambrian crystalline gneissic basement and overlying sediments that were intruded by Devonian plutonic rocks to form a rigid structural axis. At the onset of the fragmentation phase this axis presumably was a linear mountain belt with a thickened and rigid granitic core and root zone. It localized the northwest margin of Arctic Canada, in that the Kaltag Fault was forced to follow along its northwest margin (Fig. 4).

2) During plate separation in the Arctic, plates were deformed in two ways: by marginal breakup and by internal compression. The marginal parts of plates (zones at least several hundreds of kilometres wide) were affected by extension faults, to form rift valleys, aulacogens, and major fault zones. These margins also broke up into sub-plates such as the Somerset Island and Devon Island Sub-plates (Fig. 9). Those sub-plates are bounded by fault zones, but also were greatly affected internally by faults that are side effects of plate breakup and separation.

In the Queen Elizabeth Islands Sub-plate there was a substantial breakup, with largely downfaulting and only a small amount of lateral separation (Fig. 9). This pattern may characterize the margins of divergent plates in early stages of plate breakup.

Deformation appears to diminish cratonward within the continent. It also appears to diminish centrally within each sub-plate, and is greatest within those sub-plates on the initiating or oceanic side of the continent. The interior parts of a plate or sub-plate can be deformed compressionally, by either folds, or thrust and reverse faults. The zones of compressional deformation are apically opposed to the zones of extensional deformation; a pair of each type forms a couple, with extension being the cause of the couple and compression the effect (Fig. 8).

3) Deformation within plates and sub-plates is strongly and closely guided by pre-existing structures. The greatest control is exerted by the structure of the crystalline rocks, which form the continental crust in the southeast, and the root zone of the tectonic axis in the northwest. Structures within sedimentary fold belts exert a minor control on younger plate tectonic deformation.

Where pre-existing gneissic structure forms a zone that coincides with a preferred direction of extensional fault propagation, the fault propagates readily along it. The joint direction perpendicular to the gneissic trend provides a second, less preferred direction of propagation for local faults.

Gneissic trends that are oblique to the preferred direction of fault propagation are obstacles; the advance of rift zones is impeded or halted and unusual structures may result. Because separation in the seaway behind such an obstacle continues although faulting cannot advance, rotation concentrates about a pivot near the obstacle, on the oceanic side. Compressive deformation occurs on the far, or continental side, of the obstacle, apically opposed to the extensional deformation on the oceanic side. The pivot thus transforms one type of deformation into another across the obstacle; this can occur between structures of various sizes.

4) Two plate tectonic events affected the Canadian Arctic during late Paleozoic to mid-Cenozoic time, each being propagated toward the other from oceanic areas. Because they were separated by a substantial breadth of continent, the events at first operated independently, with no observable effects upon each other. As they advanced farther and farther toward each other, however, the isthmus of continent between them narrowed, and their effects began to overlap. A contest for supremacy developed: only one stress pattern could exist in a given area, but there were two competing influences. The result was a merging, with final dominance of one. The tectonic episode that was most efficient for the prevailing circumstances won out, first by neutralizing and finally by overwhelming the effects of the other.

Both tectonic episodes ceased activity shortly after their active zones merged. This cessation may have been the direct result of merging. Alternatively, and more likely, it was the result of an additional, external factor.

ACKNOWLEDGEMENTS

The author is grateful to J. Dixon, W. W. Nassichuk, E.R.W. Neale, D. K. Norris, P. Ohlendorf, L. W. Sobczak and R. Thorsteinsson for information and helpful advice.

REFERENCES

Balkwill, H.R., 1978, Evolution of Sverdrup Basin, Arctic Canada: American Assoc. Petroleum Geol. Bull., v. 62, p. 1004-1028.

Balkwill, H.R. and Bustin, R.M., 1975, Eureka Sound Formation at Flat Sound, Axel Heiberg Island, and Chronology of the Eurekan Orogeny: Geol. Survey Canada Paper 75-1B, p. 205-207.

Basham, P.W., Forsyth, D.A. and Wetmiller, R.J., 1977, The Seismicity of Northern Canada: Canadian Jour. Earth Sci., v. 14, p. 1646-1667.

Burke, K, 1976, Development of Grabens Associated with the Initial Ruptures of the Atlantic Ocean: in Bott, M.H.P., ed., Sedimentary Basins of Continental Margins and Cratons, Tectonophysics, v. 36 (1-3), p. 93-112.

Bustin, R.M., 1977, The Eureka Sound and Beaufort Formations, Axel Heiberg and west-central Ellesmere Islands, District of Franklin: M.Sc. Thesis, University of Calgary, 208 p.

Clark, D.L., 1975, Geological History of the Arctic Ocean Basin: in Yorath, C.J., Parker, E.R. and Glass, D.J., eds., Canada's Continental Margins and Offshore Petroleum Exploration: Canadian Soc. Petroleum Geol. Memoir 4, p. 589-611.

Daae, H.D. and Rutgers, A.T.C., 1975, Geological history of the Northwest Passage: Canadian Petroleum Geol. Bull., v. 23, p. 84-108.

DeLaurier, J.M., 1978, The Alpha Ridge is not a spreading centre: in Sweeney, J.F., ed., Arctic Geophysical Review: Publications of the Earth Phys. Branch, Ottawa, v. 45, p. 35-50.

Forsyth, D.A., Mair, J.A. and Fraser, I., 1979, Crustal Structure of the central Sverdrup Basin: Canadian Jour. Earth Sci., v. 16, p. 1581-1598.

Hall, J.K., 1973, Geophysical evidence for ancient sea-floor spreading from Alpha Cordillera and Mendeleyev Ridge: in Pitcher, M.G., ed., Arctic Geology, American Assoc. Petroleum Geol. Memoir 19, p. 542-561.

Harland, W.B., 1971, Tectonic transpression in Caledonian Spitzbergen: Geol. Magazine, v. 108, p. 27-41.

_____, 1973, Tectonic Evolution of the Barents Shelf and Related Plates: in Pitcher, M.G., ed., Arctic Geology: American Assoc. Petroleum Geol. Memoir 19, p. 559-608.

Henderson, G., Rosenkrantz, A. and Schiener, E.J., 1976, Cretaceous-Tertiary sedimentary rocks of West Greenland: in Escher, A. and Watt, W.S., eds., Geology of Greenland: The Geological Survey of Greenland, p. 340-363.

Herron, E.M., Dewey, J.F. and Pitman, W.C., III, 1974, Plate tectonic model for the evolution of the Arctic: Geology, v. 2, p.377-380.

Hills, L.V. and Matthews, J.V., Jr., 1974, A preliminary list of fossil plants from the Beaufort Formation, Meighen Island, District of Franklin: in Report of Activities, Part B: Geol. Survey Canada Paper 74-1B, p. 224-226.

Jackson, H.R., Keen, C.E., Falconer, R.K.H. and Appleton, K.P., 1979, New Geophysical Evidence for sea floor spreading in central Baffin Bay: Canadian Jour. Earth Sci., v. 16, p. 2122-2135.

Keen, C.E. and Hyndman, R.D., 1979, Geophysical Review of the continental margins of eastern and western Canada: Canadian Jour. Earth Sci., v. 16, p. 712-747.

Keen, C.E., Keen, M.J., Ross, D.I. and Lack, M., 1974, Baffin Bay, Small Ocean Basin formed by sea-floor spreading: American Assoc. Petroleum Geol. Bull., v. 58, p. 1089-1108.

Kerr, J.Wm. 1967a, Nares Submarine Rift Valley and the relative rotation of North Greenland: Bulletin of Canadian Petroleum Geology, v. 15, p. 483-520.

_____, 1967b, A submerged continental remnant beneath the Labrador Sea: Earth and Planetary Sci. Letters, v. 2, p. 283-289.

_____, 1973, Canadian Arctic Rift System – A Summary (Abst.): in Pitcher, M.G., ed., Arctic Geology: American Assoc. Petroleum Geol. Memoir 19, p. 587.

_____, 1976, Geology of Outstanding Arctic Aerial Photographs. 3. Margin of Sverdrup Basin, Lyall River, Devon Island: Bulletin of Canadian Petroleum Geology, v. 24, p. 139-153.
dian Jour. Earth Sci., v. 14, p. 1374-1401
_____, 1979a, Evolution of the Canadian Arctic Islands – A Transition between the Atlantic and Arctic Oceans, Geological Survey of Canada, Open File Report No. 618: to be published in Nairn, A.E.M., Stehli, F.G. and Churkin, M., Jr., The Ocean Basins and Margins, Vol. 5, The Arctic Ocean: Plenum Publishing Corp., New York,
_____, 1979b, Structural Framework of Lancaster Aulacogen, Arctic Canada: Geological Survey of Canada, Open File Report No. 619: to be published as Geol. Survey Canada Bull. 319.

Kerr, J.Wm. and Trettin, H.P., 1962 Mississippian rocks and the mid-Paleozoic earthmovements in the Canadian Arctic Archipelago: Alberta Soc. Petroleum Geol. Jour., v. 10, p. 247-256.

King, E.R., Zietz, I. and Alldredge, L.R., 1966, Magnetic data on the structure of the central Arctic Region: Geol. Soc. America Bull., v. 77, p. 619-646.

Meneley, R.A., Henao, D. and Merritt, R.K., 1975, The northwest margin of the Sverdrup Basin: in Yorath, C.J., Parker, E.R. and Glass, D.J., eds., Canada's Continental Margins and Offshore Petroleum Exploration: Canadian Soc. Petroleum Geol. Memoir 4, p. 531-544.

Miall, A.D., 1975, Post-Paleozoic geology of Banks, Prince Patrick and Eglinton Islands, Arctic Canada: in Yorath, C.J., Parker, E.R., and Glass, D.J., eds., Canada's Continental Margins and Offshore Petroleum Exploration: Canadian Soc. Petroleum Geol. Memoir 4, p. 557-587.

_____, 1979, Tertiary Fluvial Sediments in the Lake Hazen Intermontane Basin, Ellesmere Island, Arctic Canada: Geol. Survey Canada Paper 79-9.

Miall, A.D., Balkwill, H.R. and Hopkins, W.S., Jr., 1980, Cretaceous and Tertiary sediments of Eclipse Trough, Bylot Island area, Arctic Canada, and their regional setting: Geol. Survey Canada Paper 79-23.

Nassichuk, W.W., 1975, Carboniferous ammonoids and stratigraphy in the Canadian Arctic Archipelago: Geol. Survey Canada Bull. 237.

Nassichuk, W.W. and Davies, G.R., 1980, Stratigraphy, Biochronology, and Sedimentology of the Otto Fiord Formation – a major Mississippian-Pennsylvanian evaporite of subaqueous origin in the Canadian Arctic Archipelago: Geol. Survey Canada Bull. 286.

Norris, D.K., 1974, Structural Geometry and Geological History of the northern Canadian Cordillera: in Wren, A.E. and Cruz, R.B., eds., Proceedings of the 1973 National Convention: Canadian Soc. Explor. Geophys., p. 18-45.

Ostenso, N.A. and Wold, R.J., 1971, Aeromagnetic survey of the Arctic Ocean: techniques and interpretations: Marine Geophysical Research, v. 1, p. 178-219.

Rahmani, R.A., 1977, Fault Control on Sedimentation of Isachsen Formation in Sverdrup Basin: Geol. Survey Canada Paper 77-1B, p. 157-161.

Roy, K.J., 1974, Transport Directions in the Isachsen Formation (Lower Cretaceous), Sverdrup Islands, District of Franklin: Geol. Survey Canada Paper 74-1, pt. A, p. 351-353.

Sander, G.W. and Overton, A., 1965, Deep seismic refraction investigation in the Canadian Arctic Archipelago: Geophysics, v. 30, p. 87-96.

Sinha, A.K. and Frisch, T.O., 1976, Whole Rock Rb/Sr and Zircon U/Pb ages of metamorphic rocks from northern Ellesmere Island, Canadian Arctic Archipelago. II. The Cape Columbia Complex: Canadian Jour. Earth Sci., v. 13, p. 774-780.

Sobczak, L.W. and Weber, J.R., 1973, Crustal Structure of Queen Elizabeth Islands and polar continental margin, Canada: in Pitcher, M.G., ed., Arctic Geology, American Assoc. Petroleum Geol. Memoir 19, p. 517-525.

Srivastava, S.P., 1978, Evolution of the Labrador Sea and its bearing on the early evolution of the North Atlantic: Geophys. Jour. Royal Astron. Soc., v. 52, p. 313-357.

Thorsteinsson, R. and Tozer, E.T., 1970, Geology of the Arctic Archipelago: in Douglas, R.J.W., ed., Geology and Economic Minerals of Canada, Chapter X, Economic Geology Report No. 1, Fifth Edition: Geol. Survey Canada, p. 548-590.

Tozer, E.T., 1960, Summary account of Mesozoic and Tertiary stratigraphy, Canadian Arctic Archipelago: Geol. Survey Canada Paper 60-5.

Trettin, H.P. and Balkwill, H.R., 1979, Contributions to the Tectonic History of the Innuitian Province: Canadian Jour. Earth Sci., v. 16, p. 748-769.

Trettin, H.P., Frisch, T.O., Sobczak, L.W., Weber, J.R., Niblett, E.R., Law, L.K., De-Laurier, J.M. and Whitham, K., 1972, The Innuitian Province: in Price, R.A. and Douglas, R.J.W., eds., Variations in Tectonic Styles in Canada: Geol. Assoc. Canada Spec. Paper 11, p. 83-179.

Williams, G.D. and Stelck, C.R., 1975, Speculations on the Cretaceous Paleogeography of North America: in Caldwell, W.G.E., ed., The Cretaceous System of the Western Interior of North America: Geol. Assoc. Canada Spec. Paper 13, p. 1-29.

Wilson, D.G., 1976, Studies of Mesozoic Stratigraphy, Tanquary Fiord to Yelverton, Ellesmere Island, District of Franklin: Geol. Survey Canada Paper 76-1A, p. 449-451.

Yorath, C.J. and Norris, D.K., 1975, The tectonic development of the southern Beaufort Sea and its relationship to the origin of the Arctic Ocean Basin: in Yorath, C.J., Parker, E.R. and Glass, D.J., eds., Canada's Continental Margins and Offshore Petroleum Exploration: Canadian Soc. Petroleum Geol. Memoir 4, p. 589-611.

The Continental Crust and Its Mineral Deposits, edited by D.W. Strangway,
Geological Association of Canada Special Paper 20

THE GRENVILLE PROVINCE:
A PALEOMAGNETIC CASE-STUDY OF
PRECAMBRIAN CONTINENTAL DRIFT

David J. Dunlop, Derek York and Glenn W. Berger[1]
Geophysics Laboratory, University of Toronto,
Toronto, Ontario M5S 1A7

Kenneth L. Buchan[2]
Geological Survey of Canada, 601 Booth Street,
Ottawa, Ontario K1A 0E8

J. Mark Stirling[3]
Geophysics Laboratory, University of Toronto,
Toronto, Ontario M5S 1A7

ABSTRACT

The major problem in establishing apparent polar wander paths (APWP's) for Precambrian orogens is assigning ages to ancient pole positions. Magnetic and isotopic systems are "frozen" or "blocked" at different temperatures and therefore, if the orogen cooled slowly, at significantly different times. Our approach is to use regional cooling curves derived by Berger et al. (1979) and Berger and York (1980) from $^{40}Ar/^{39}Ar$ step-heating data. The paleopoles are dated by comparing their magnetic blocking temperatures to the cooling curve.

The Haliburton intrusions of the Grenville Province cooled from 650°C to 200°C between 1000 and 800 Ma ago. The two magnetizations preserved by the Haliburton intrusions, with blocking temperatures of 520 to 650°C and <250°C, record the paleofield at around 980 and 820 Ma respectively. The corresponding paleomagnetic poles provide well-dated tie points for a Grenville Loop in the North American APWP between 1050 and 800 Ma, and rule out the possibility of a collision between "Grenvillia" and the rest of the Canadian Shield in this time interval.

An earlier episode of divergence followed by collision may be recorded by a 1200-Ma hornblende $^{40}Ar/^{39}Ar$ date for the Thanet gabbro (regional grade: low amphibolite) and novel

[1]Now at Department of Physics, Simon Fraser University, Burnaby, B.C. V5A 1S6
[2]Now at Department of Physics, Memorial University of Newfoundland, St. John's, Newfoundland A1B 3X7
[3]Now at School of Business, University of Manitoba, Winnipeg, Manitoba R3T 2N2

magnetizations preserved by the Thanet and Cordova gabbros. These magnetizations, if primary, are inconsistent with the presently known North American APWP. They imply the existence of a small ocean between "Grenvillia" and "Interior Laurentia", closing after 1200 Ma by ≈90° rotation about a pole northeast of Newfoundland. Although it would provide a mechanism for the Grenvillian orogeny, this scenario is at present speculative and requires cautious testing.

RÉSUMÉ

Le problème principal dans l'établissement des trajets apparents de migration polaire pour les orogènes du Précambrien réside dans l'assignation d'âges à d'anciennes positions des pôles. Les systèmes magnétiques ou isotopiques sont "gelés" ou "bloqués" à différentes températures et, par conséquent, si l'orogène s'est refroidi lentement, à des époques significativement différentes. Notre approche a été d'utiliser les courbes régionales de refroidissement dérivées par Berger *et al.* (1979) et Berger et York (1979) à partir de données ^{40}Ar/^{39}Ar de chauffage par paliers. On date les paléopôles en comparant leurs températures de blocage magnétique à la courbe de refroidissement.

Les intrusions de Haliburton de la province de Grenville ont refroidi de 650° à 200°C entre 100 et 800 Ma. Les deux aimantations préservées par les intrusions de Haliburton, avec des températures de blocage de 520 à 650°C et de moins de 250°C, sont les témoins du champ magnétique qui existait il y a 980 et 820 Ma respectivement. Les pôles paléomagnétiques correspondants fournissent des points d'attache bien datés pour la boucle de Grenville du trajet apparent de migration polaire pour l'Amérique du Nord entre 1050 et 800 Ma et excluent la possibilité d'une collision entre le Grenville et le reste du bouclier canadien dans cet intervalle de temps.

Un épisode précoce de divergence suivi d'une collision a pour témoin une datation de 1200 Ma par ^{40}Ar/^{39}Ar sur la hornblende du gabbro de Thanet (degré régional: début des amphibolites) et de nouvelles aimantations préservées par les gabbros de Thanet et de Cordova. Ces aimantations, si elles sont primaires, ne s'accordent pas avec le trajet apparent de migration du pôle pour l'Amérique du Nord. Elles impliquent l'existence d'un petit océan entre les régions de "Grenvillia" et de "Laurentia intérieure", lequel s'est refermé après 1200 Ma par une rotation d'environ 90° autour d'un pôle situé au nord-est de Terre-Neuve. Bien qu'il fournisse un mécanisme pour l'orogénèse grenvillienne, ce scénario est spéculatif pour le moment et requiert une vérification minutieuse.

INTRODUCTION

The Grenville structural province of the Canadian Shield is a region of generally high-grade metamorphism with a long and complex thermal history. As a consequence, the dual problems of metamorphic paleomagnetism – determining a reliable paleomagnetic pole position and deciding which of a host of possible whole-rock and mineral isotopic ages best dates the paleopole – are particularly severe in the Grenville.

The paleomagnetist must resolve a vector sum of two, or occasionally more, components of natural remanent magnetization (NRM). These may be partial thermal remanent magnetizations (pTRM's) acquired at a variety of blocking temperatures (T_B's) and times during slow cooling. Included in this category is partial thermal overprinting (during a reheating event in a compound cooling history) of the primary TRM acquired during initial cooling. NRM components may also include chemical remanent magnetization (CRM) acquired at lower temperature, hence later than implied by the apparent T_B given by a thermal demagnetization experiment.

The geochronologist faces similar problems, since radiogenic systems vary greatly in closure temperature (T_C) to isotopic migration. For example, in the Haliburton-Hastings region of southern Ontario, which will be the focus of this paper, U/Pb zircon ages of 1300 to 1200 Ma (Silver and Lumbers, 1966) "see through" the Grenvillian orogeny and date an earlier cooling. K/Ar whole-rock and mineral ages tend to record 800 to 1000 Ma, with younger ages corresponding to minerals with lower T_C's.

In papers by Berger *et al.* (1979) and Berger and York (1980), T_C's of hornblende, biotite and plagioclase to argon diffusion are calibrated and, in turn, used to date paleopoles with similar T_B's. This method of *thermochronometry* is based on two facts. First, the $^{40}Ar/^{39}Ar$ step heating method generates both an age and a value of T_C for each mineral. A regional cooling curve is generated by the $^{40}Ar/^{39}Ar$ data alone, with no input from magnetic or other results. Second, stepwise thermal demagnetization and $^{40}Ar/^{39}Ar$ step heating are perfectly parallel experiments, since magnetization blocking and argon diffusion are both simple thermal activation processes (York, 1978).

Apart from a kinetic correction of laboratory T_B's for slow cooling in nature (Pullaiah *et al.*, 1975), magnetic T_B's and isotopic T_C's are analogous quantities. If a particular NRM component is a pTRM, its (corrected) T_B allows its age to be read from the isotopically determined cooling curve. If the NRM component is a CRM, the true T_B may be much lower than the apparent T_B, and the age read from the cooling curve is an upper limit.

THE "GRENVILLE PROBLEM"

Figure 1 illustrates what has come to be known in Precambrian paleomagnetism as the "Grenville Problem". Paleomagnetic poles from the Grenville Province and its sister orogen, the Sveconorwegian Province of the Baltic Shield, fall into two groups. B poles, the more northerly ones, are concordant with the late Precambrian apparent polar wander path (APWP) for Laurentia (proto-North America). A poles, however, are discordant. They are unlike Precambrian paleopoles of any age elsewhere.

One possible explanation is that A poles record divergence between two protocontinents, "Grenvillia" and "Interior Laurentia", during the late Precambrian. This possibility is all the more intriguing because continental collision would provide a mechanism for the crustal thickening and subsequent uplift and deep erosion that occurred in the Grenvillian orogeny (Dewey and Burke, 1973).

To test this two-plate hypothesis, we require a "window" or area of locally lower metamorphic grade where we can be sure of seeing through the orogeny magnetically and isotopically. Otherwise, we cannot easily decide if A poles in general are precollisonal. In southern Ontario's Haliburton-Hastings Highlands and adjoining Hastings Basin (Fig. 2), with its well-documented isograds (Lumbers, 1964), we have such a local window.

HIGH-GRADE TERRANE: THE HALIBURTON INTRUSIONS

The first example of multicomponent NRM in Grenville rocks (Buchan and Dunlop, 1973, 1976) came from the Haliburton intrusions (Glamorgan gabbro-anorthosite complex; Bark Lake and Dudmon diorites). The higher-T_B A component

of NRM agreed closely in direction with A directions from the single-component
Allard Lake anorthosite (Hargraves and Burt, 1967) and the double-component Morin
anorthosite (Irving *et al.*, 1974a). The lower-T_B B component agreed in a more general
way with a number of Grenville B results, including some believed to be late-stage if
not actually post-orogenic (Murthy, 1971; Park and Irving, 1972).

At the time of the Haliburton study, there were no thermochronometric data
available for individual Grenville units, although *regional* whole-rock K/Ar thermo-
chrons had been proposed by Harper (1967). However, the converging Grenvillian and
Interior Laurentian APWP's (Fig. 3) envisaged by Irving *et al.* (1974b), Stewart and
Irving (1974) and Buchan and Dunlop (1976) were not without time constraints. The
Interior Laurentian APWP segment (Track 3A of Irving and McGlynn, 1976, or part of
the Logan Loop of Robertson and Fahrig, 1971) was largely based on Keweenawan
lavas and rapidly chilled dikes of low metamorphic grade. Rb/Sr ages of 1150 Ma and
about 1050 Ma could be assigned respectively to the north end of the track and to its
juncture with the Grenville Track.

Thus, for a collision to be recorded by the A-B sequence of Grenville poles, two
conditions would have to hold.

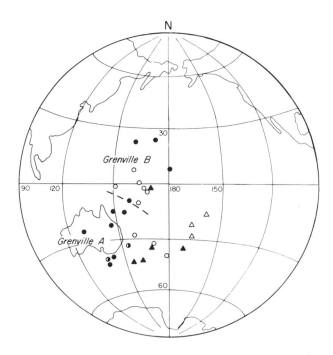

Figure 1. Grenville A and B paleopoles. Circles and triangles represent poles from the Grenville
and Sveconorwegian Provinces respectively, after closing the Atlantic. Open, closed and half-
closed symbols denote reversed, normal and mixed-polarity poles in that order. For identification
of individual poles, see Buchan (1978, Fig. 13).

1) Absolute ages: B poles would have to be about 1050 Ma old, the age of the proposed juncture.

2) Relative ages: A poles would have to be older than B poles for *convergence* to be recorded. This is the sequence implied by the relative T_B's of A and B NRM's if both are pTRM's.

Figure 4 illustrates the continental convergence corresponding to the converging APWP's of Figure 3. The reconstructions are non-unique in paleolongitude, but rotations and paleolatitudes of the protocontinents are well-determined. The Grenville Front is shown as the plate boundary solely for convenience in drawing. The actual suture zone between the now-united plates must lie well south of the Front, since metamorphosed equivalents of pre-orogenic formations in the Superior and Southern Provinces can be traced as far as 150 km into the Grenville (Wynne-Edwards, 1972, p. 285-7).

However, with the advent of thermochronometry (Berger *et al.*, 1979), the two-plate scenario of Figures 3 and 4 must be abandoned. $^{40}Ar/^{39}Ar$ step heating dates for hornblendes, biotites and feldspars from all three Haliburton intrusions define a simple cooling history (see Fig. 8, Bark Lake cooling curve). The A NRM component has (corrected) T_B's similar to the hornblende T_C's, and dates at 980 Ma. The B NRM has T_B's like the T_C's of the K-feldspar and some of the plagioclases, and dates at about 820 Ma. These are maximum ages, based on the assumption that A and B components are pTRM's, not CRM's.

Although the Haliburton A pole is older than the B pole (assuming pTRM's), both are younger than 1050 Ma, the time of the proposed juncture of Grenvillian and Interior Laurentian APWP's (Fig. 3). Neither component sees through the Grenvillian orogeny. The Haliburton A and B poles – and probably the Grenville A-B pole sequence in general – postdate by at least 150 Ma any collision that may have occurred between Grenvillia and Interior Laurentia.

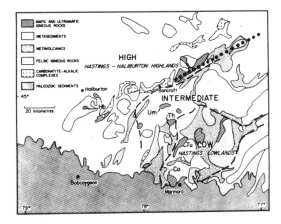

Figure 2. Geology and metamorphic zonation of the Hastings area (after Lumbers, 1964 and Buchan, 1978). Dashed lines separate regions of high (amphibolite to granulite), intermediate (low amphibolite) and low (greenschist) metamorphic grade. The symbols Hb, Th, Um, Tu and Co identify the Haliburton intrusions and the Thanet, Umfraville, Tudor and Cordova gabbros.

The Haliburton A and B poles do not define a convergent juncture between the APWP's of colliding plates, but they do provide two well-dated tie points for a Grenville Loop (Fahrig *et al.*, 1974; Stewart and Irving, 1974; McWilliams and Dunlop, 1975, 1978) in the APWP of a single Laurentian plate over the interval 1050 to 820 Ma (Fig. 5). The paleomagnetic record of this time interval outside the Grenville remains practically non-existent (see, however, Elston and Grommé, 1974; Halls, 1975; Watts, 1979).

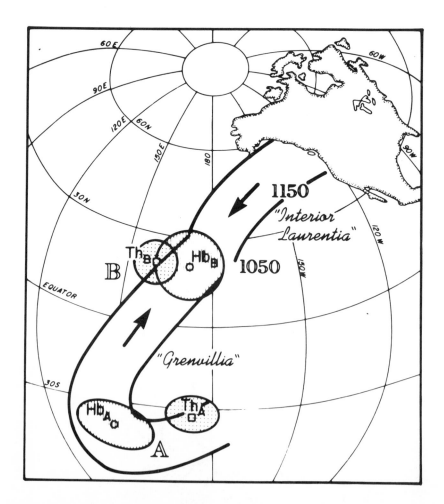

Figure 3. Hypothetical convergent APWP's for Grenvillian and Interior Laurentian protocontinents for the time interval 1150 to 1050 Ma. Hb_A, Hb_B, Th_A, Th_B are paleopoles for A and B NRM components from the Haliburton intrusions and the Thanet gabbro. A and B are the regions of Grenville A and B poles. Base map after Morris and Roy (1977). Dating of Hb_A and Hb_B (see text) rules out this convergent two-plate model.

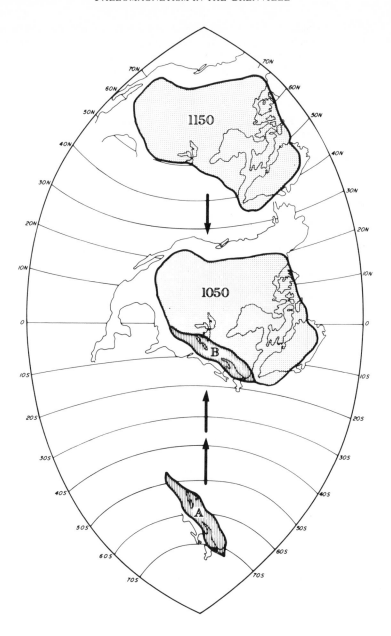

Figure 4. The hypothetical continental collision implied by the convergent APWP's of Figure 3. A and B stand for the ages of Grenville (more specifically Haliburton) A and B poles. Since these both turn out to be younger than 1000 Ma, the collision pictured is impossible.

Figure 6 illustrates the drift of the Laurentian protocontinent from 1150 to 820 Ma implied by the APWP of Figure 5. It incorporates three latitudinal shifts for the Hastings area, each about 50° towards or away from the equator and occupying a time span of roughly 100 Ma. The minimum rate of drift (not allowing for any paleolongitude shift) is 5 to 6 cm/yr, which is quite consistent with present-day plate velocities.

INTERMEDIATE-GRADE TERRANE:
THE THANET AND UMFRAVILLE GABBROS

With the failure of magnetic and isotopic recorders in high-grade rocks to penetrate the Grenville orogeny, we turn to intermediate-grade terrane (regionally low amphibolite facies). The Umfraville and Thanet gabbros (Fig. 2) were examined in a pioneering study by Hood (1958). Detailed recent work by Symons (1978) and Palmer *et al.* (1979) has refined Hood's equatorial B pole. The Umfraville body was intruded at 1180 ± 20 Ma (Symons, 1978). The 911 Ma K/Ar isochron age of Palmer *et al.* (1979) shows, however, that the NRM, with T_B's of 500 to 600°C, is a much younger metamorphic overprint. A precise age for the Umfraville pole must await $^{40}Ar/^{39}Ar$ dating.

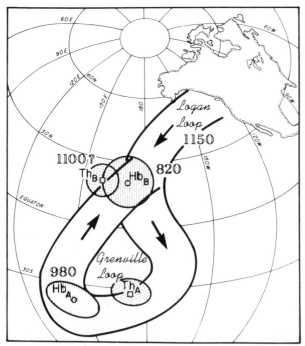

Figure 5. The Laurentian APWP between 1150 and 820 Ma. Symbols as in Figure 3. Hb_A (980 Ma) and Hb_B (820 Ma) are tie-points for the Grenville Loop. To judge by the positions of the Thanet poles on the APWP, Th_A should be 1050 to 1000 Ma old and Th_B should be either ≈ 1100 Ma or ≈ 800 Ma old, depending upon whether it falls on the old or the young end of the Grenville Loop. The actual thermochronometric ages of Th_A and Th_B are discussed in the text and shown in Fig. 8.

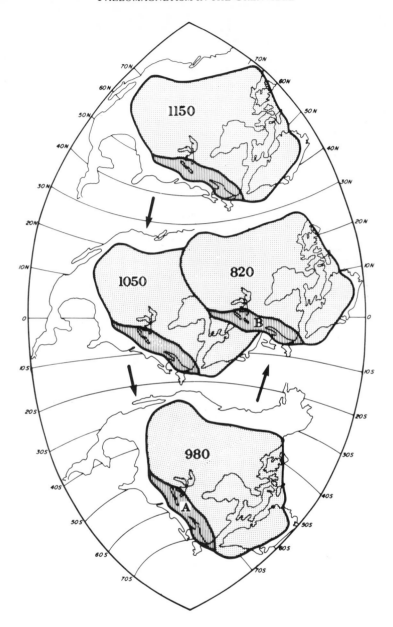

Figure 6. Drift of the Laurentian protocontinent between 1150 and 820 Ma implied by the APWP of Fig. 5. Drift rate is ≥5 cm/yr. Paleolongitude is indeterminate; the relative longitudes shown are arbitrary.

DUNLOP ET AL.

The Thanet gabbro (Buchan, 1978) yields multicomponent NRM's. Both A and B directions are different from Hood's early result. In contrast to the Haliburton study, T_B's of the Thanet B NRM are always above those of the A_1 and A_2 components (Fig. 7). If all components are pTRM's, the B component is older than either A component. Based on the positions of the Thanet paleopoles on the now-calibrated Grenville and Logan Loops (Fig. 5), one can put forward the testable hypothesis that the Thanet A_1 and A_2 NRM's are about 1000 Ma old and the Thanet B NRM about 1100 Ma old.

Let us examine the thermochronometric data (Fig. 8). The hornblende T_C's exceed the probable orogenic reheating temperature of 500 to 550°C of the Thanet body. They see through the Grenvillian metamorphism to an intrusion age of about 1200 Ma. Biotite and plagioclase ages were reset following the peak of burial reheating. A compound cooling history (curve b) is thus preferred by Berger and York (1980).

The thermochronometric data are consistent with the hypothesized ages of about 1100 Ma for the Thanet B NRM and about 1000 Ma for the A NRM, provided the peak temperature attained in the orogeny was slightly higher than 500 to 550°C, and provided the A_2 component is of chemical rather than thermal origin. A peak temperature of 575°C is not incompatible with the regional low-amphibolite grade of metamorphism. Such a temperature, reached slightly earlier than 1100 Ma according to curve (c) of Figure 8, would serve to thermally remagnetize the Thanet B NRM without affecting the 1200-Ma age recorded by the hornblende clock.

The A_2 component has blocking temperatures adjacent to or slightly overlapping those of the B component, yet its direction diverges about 90° from the B direction. The A_2 NRM is likely a post-1100 Ma CRM, acquired contemporaneously with the A_1 component whose direction is similar. Assuming the A_1 component is a thermal overprint, the estimated acquisition time of both A magnetizations is between 1050 and 1000 Ma (Fig. 8).

By this model, the Thanet paleopoles fit reasonably well on the time-calibrated Grenville and Logan Loops of Figure 5. The B pole was recorded as a pTRM overprint at 1150 to 1100 Ma and the A_2 and A_1 poles as CRM and pTRM overprints respectively at 1050 to 1000 Ma.

An alternative simple model can explain the data. The B component may be a late (~800 Ma) CRM and the Thanet B pole simply another member of the general Grenville B population.

A third interpretation invoking independent Grenvillian and Interior Laurentian plates is also possible. Let us suppose that curve (b) of Figure 8 is correct and that the Thanet B component *is* a 1200 Ma pTRM that sees through the Grenvillian metamorphism. Its pole position is then inconsistent with the Laurentian APWP, being rotated an angular distance of 35° from the 1250 to 1200 Ma Laurentian tie point (Mackenzie poles, Irving and McGlynn, 1976, Track 3B, Fig. 6) at 0°, 170°W.

The corresponding tectonic picture (Fig. 9) is one of independent Grenvillian and Interior Laurentian plates, rotated about 35° with respect to each other, at 1250 to 1200 Ma. The intervening small ocean closes via small rotation (accompanying major northward drift of both plates) between 1200 and 1150 Ma. In this model, Thanet B and A poles record the Grenvillian orogeny, or at least its culmination, as a collisional event prior to 1150 Ma.

LOW-GRADE TERRANE: THE CORDOVA AND TUDOR GABBROS

It is critical to determine whether the Thanet B component sees through the Grenvillian metamorphism or was reset. In the centre of the Hastings Basin (Fig. 2), where the metamorphic grade is as low as upper greenschist, a pre-Grenvillian magnetization is more likely to have survived than in the intermediate-grade rocks of the Thanet area.

The Tudor gabbro (Hood, 1958; Palmer and Carmichael, 1973) gives a single paleopole lying northwest of Grenville B poles. Its K/Ar age of 680 Ma (Hayatsu and Palmer, 1975) is much younger than any other Grenville poles we have considered.

The Cordova gabbro, however, yields Grenville A and B NRM components and, in addition, the novel C magnetization shown in Figure 10. The C direction is D = 195°E, I = +18° (α_{95} = 13.5°, k = 17.1, N = 8 samples) with corresponding paleopole

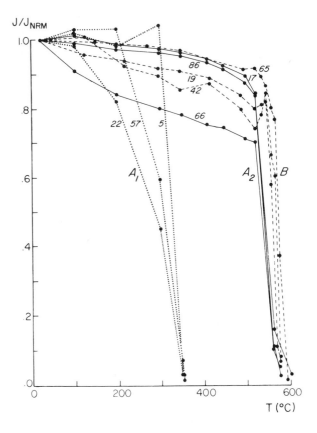

Figure 7. Decrease of magnetization intensity during stepwise thermal demagnetization of Thanet gabbro samples exhibiting A_1, A_2 and B components of NRM or a superposition of A_2 and B. (After Buchan, 1978).

at either 97°W, 35°S or 83°E, 35°N. Thirteen samples from 8 sites have NRM vectors that upon alternating-field demagnetization swing toward or away from or stabilize at this C direction.

The only similar magnetic directions elsewhere in the North American Precambrian are about 1700 Ma old. The Cordova gabbro is certainly not this old. It intrudes the Belmont volcanics, whose probable age is about 1300 Ma. The Cordova C paleopole therefore cannot be incorporated into the Laurentian APWP as presently known.

The Cordova C pole does, on the other hand, fit the interpretation of the Thanet B pole as a 1200 Ma pTRM advanced in the last section, provided the C component is somewhat older than 1200 Ma. The plate reconstruction implied by the Cordova C pole (Fig. 9) predicts a larger ocean (about 90° rotation about the same pole of opening as for the Thanet B paleopole) between Grenvillia and Interior Laurentia prior to 1200 Ma.

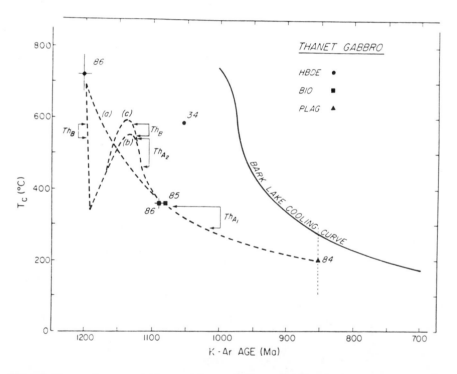

Figure 8. Thermochronometric data and three possible cooling histories for the Thanet gabbro. Compound cooling history (b) is preferred because the metamorphic grade indicates that reheating did not exceed 500 to 550°C. The (corrected) T_B ranges shown for Thanet B, A_2 and A_1 NRM components (cf. Fig. 7) imply respective ages of ≈1200, ≈1100 and 1050 to 1000 Ma if all components are pTRM's. The cooling curve for the Bark Lake diorite (one of the Haliburton intrusions) is shown for comparison. (After Berger and York, 1980).

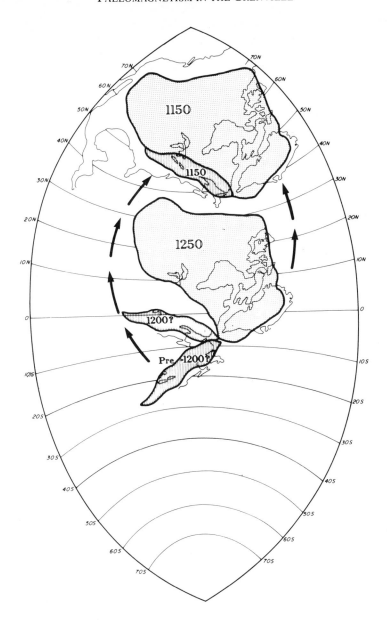

Figure 9. Continental reconstructions implied by assuming that the Thanet gabbro B and Cordova gabbro C magnetizations see through the Grenvillian orogeny. The 1250-Ma position of Interior Laurentia is given by Mackenzie poles (Irving and McGlynn, 1976, Fig. 6). The positions of Grenvillia at 1200 Ma and pre-1200 Ma are determined by the Thanet B-pole and the Cordova C-pole respectively. Although the position of Grenvillia is not recorded at 1150 Ma, collision of the two plates (the Grenvillian orogeny) is arbitrarily assumed to be complete by this time.

The data of Figure 10 are preliminary, and thermal work remains to be done. The hypothesis of separate Grenvillian and Interior Laurentian plates around 1200 Ma is likewise tentative and largely untested. Only ^{40}Ar/^{39}Ar thermochronometry and thermal demagnetization of the Cordova samples will indicate whether the Cordova C NRM is primary and pre-orogenic or the result of post-800 Ma remagnetization.

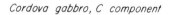

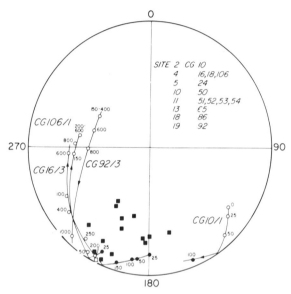

Figure 10. Preliminary C-component directional data for the Cordova gabbro. Circles denote successive positions of the NRM vector at increasing levels of alternating-field demagnetization (labels give fields in Oersteds); squares depict stable directions after optimal demagnetization. Open, closed symbols are upward, downward directions respectively. Equal-area projection. (Data from J.M. Stirling, work in progress.)

CONCLUSIONS

Amphibolite and granulite-facies rocks of the Grenville Province have been reheated during burial to temperatures sufficient to reset the bulk of their primary magnetization. U/Pb and Rb/Sr dating of such overprinted magnetizations is unsatisfactory. Only K/Ar mineral ages are reset at temperatures comparable to the blocking temperatures for magnetic overprinting.

With the advent of ^{40}Ar/^{39}Ar thermochronometry (Berger *et al.*, 1979), which amounts to cross-calibrating magnetic blocking temperatures and isotopic closure temperatures, paleomagnetic poles from slowly cooled orogens can now be dated with acceptable precision. The only proviso is that the magnetizations must have been acquired thermally. A chemical overprint may be much younger than its thermochronometric age.

Thermochronometry has demonstrated that:

1) A and B NRM components in the Haliburton intrusions (Buchan and Dunlop, 1973, 1976) are orogenic overprints with ages of 980 and 820 Ma respectively (Berger *et al.*, 1979);

2) the corresponding Haliburton A and B paleopoles are tie-points for a Grenville Loop in the Laurentian APWP between 1050 and 820 Ma;

3) Grenville A and B poles in general do not record a collision between Grenvillian and Interior Laurentian protocontinents.

The Thanet gabbro from the Hastings Basin, a region of exceptionally low metamorphic grade for the Grenville, has preserved a hornblende K/Ar age of 1200 Ma, believed to date initial cooling of the intrusion. If the B NRM component of the Thanet gabbro and the C NRM component of the Cordova gabbro are likewise interpreted as primary, one can "look through" the Grenvillian orogeny via these NRM components and see divergent Grenvillian and Interior Laurentian plates at and before 1200 Ma. In this interpretation, the intervening small ocean closed between about 1250 and 1150 Ma by ≈90° relative rotation of the plates about a pole of closing northeast of Newfoundland, accounting for the Grenvillian orogeny as a collisional event. This highly speculative and controversial scenario awaits testing by thermochronometry. A non-collisional, one-plate model is equally probable on the present evidence.

ACKNOWLEDGEMENTS

Lunch-hour debates with M.E. Bailey, C.J. Hale and L.D. Schutts on the Grenville and other paleomagnetic topics have been an unfailing source of fresh ideas and useful criticism. Dr. G.W. Berger's work on thermochronometric dating of paleomagnetic poles was made possible by a Negotiated Development Grant from the National Research Council of Canada. Other aspects of our research were supported by National Research Council Operating Grants and Department of Energy, Mines and Resources Research Agreements.

REFERENCES

Berger, G.W. and York, D., 1980, Looking through the Grenvillian metamorphic veil with [40]Ar/[39]Ar dating of the Thanet gabbro, southern Ontario: Canadian Jour. Earth Sci., in press.

Berger, G.W., York, D. and Dunlop, D.J., 1979, Calibration of Grenvillian palaeopoles by [40]Ar/[39]Ar dating: Nature, v. 277, p. 46-47.

Buchan, K.L., 1978, Magnetic overprinting in the Thanet gabbro complex, Ontario: Canadian Jour. Earth Sci., v. 15, p. 1407-1421.

Buchan, K.L. and Dunlop, D.J., 1973, Magnetisation episodes and tectonics of the Grenville Province: Nature Phys. Sci., v. 246, p. 28-30.

_____, 1976, Paleomagnetism of the Haliburton intrusions: superimposed magnetizations, metamorphism and tectonics in the late Precambrian: Jour. Geophys. Res., v. 81, p. 2951-2967.

Dewey, J.F. and Burke, K.C.A., 1973, Tibetan, Variscan and Precambrian basement reactivation: products of continental collision: Jour. Geol., v. 81, p. 683-692.

Elston, D.P. and Grommé, C.S., 1974, Precambrian polar wandering from Unkar Group and Nankoweap Formation, eastern Grand Canyon, Arizona: *in* Geology of northern Arizona, part 1: Regional Studies: Geol. Soc. America, p. 97-119.

Fahrig, W.F., Christie, K.W. and Schwarz, E.J., 1974, Paleomagnetism of the Mealy Mountain anorthosite suite and of the Shabogamo gabbro, Labrador, Canada: Canadian Jour. Earth Sci., v. 11, p. 18-29.

Halls, H.C., 1975, Shock-induced remanent magnetisation in late Precambrian rocks from Lake Superior: Nature, v. 255, p. 692-695.

Hargraves, R.B. and Burt, D.W., 1967, Paleomagnetism of the Allard Lake anorthosite suite: Canadian Jour. Earth Sci., v. 4, p. 357-369.

Harper, C.T., 1967, On the interpretation of potassium argon ages from Precambrian shields and Phanerozoic orogens: Earth Planetary Sci. Letters, v. 3, p. 128-132.

Hayatsu, A. and Palmer, H.C., 1975, K-Ar isochron study of the Tudor gabbro, Grenville Province, Ontario: Earth Planetary Sci. Letters, v. 25, p. 208-212.

Hood, P.J., 1958, Paleomagnetic studies of some Precambrian rocks in Ontario: Ph.D. Thesis, University of Toronto, 200 p.

Irving, E., Emslie, R.F. and Ueno, H., 1974b, Upper Proterozoic poles from Laurentia and the history of the Grenville structural province: Jour. Geophys. Res., v. 79, p. 5491-5502.

Irving, E. and McGlynn, J.C., 1976, Proterozoic magnetostratigraphy and the tectonic evolution of Laurentia: Phil. Trans. Royal Soc. London, v. A280, p. 433-468.

Irving, E., Park, J.K. and Emslie, R.F., 1974a, Paleomagnetism of the Morin complex: Jour. Geophys. Res., v. 79, p. 5482-5490.

Lumbers, S.B., 1964, Preliminary report on the relationship of mineral deposits to intrusive rocks and metamorphism in part of the Grenville Province of southeastern Ontario: Ontario Dept. Mines Prelim. Rept. 1964-4, 33 p.

McWilliams, M.O. and Dunlop, D.J., 1975, Precambrian paleomagnetism: magnetizations reset by the Grenville orogeny: Science, v. 190, p. 269-272.

——————————, 1978, Grenville paleomagnetism and tectonics: Canadian Jour. Earth Sci., v. 15, p. 687-695.

Morris, W.A. and Roy, J.L., 1977, Discovery of the Hadrynian polar track and further study of the Grenville problem: Nature, v. 266, p. 689-692.

Murthy, G.S., 1971, The paleomagnetism of diabase dikes from the Grenville Province: Canadian Jour. Earth Sci., v. 8, p. 802-812.

Palmer, H.C. and Carmichael, C.M., 1973, Paleomagnetism of some Grenville Province rocks: Canadian Jour. Earth Sci., v. 10, p. 1175-1190.

Palmer, H.C., Hayatsu, A., Waboso, C.E. and Pullan, S., 1979, A paleomagnetic and K-Ar study of the Umfraville gabbro, Ontario: Canadian Jour. Earth Sci., v. 16, p. 459-471.

Park, J.K. and Irving, E., 1972, Magnetism of dikes of the Frontenac Axis: Canadian Jour. Earth Sci., v. 9, p. 763-765.

Pullaiah, G., Irving, E., Buchan, K.L. and Dunlop, D.J., 1975, Magnetization changes caused by burial and uplift: Earth Planetary Sci. Letters, v. 28, p. 133-143.

Roberston, W.A. and Fahrig, W.F., 1971, The great Logan paleomagnetic loop: the polar wander path from Canadian Shield rocks during the Neohelikian era: Canadian Jour. Earth Sci., v. 8, p. 1355-1372.

Silver, L.T. and Lumbers, S.B., 1966, Geochronological studies in the Bancroft Madoc area of the Grenville Province, Ontario, Canada (abst.): Geol. Soc. America Abst. with Programs, v. 87, p. 156.

Stewart, A.D. and Irving, E., 1974, Paleomagnetism of Precambrian sedimentary rocks from N.W. Scotland and apparent polar wandering path for Laurentia: Geophys. Jour. Royal Astron. Soc., v. 37, p. 51-72.

Symons, D.T.A., 1978, Paleomagnetism of the 1180 Ma Grenvillian Umfraville gabbro, Ontario: Canadian Jour. Earth Sci., v. 15, p. 956-962.

Watts, D.R., 1979, The paleomagnetism of the Keweenawan sediments: the Eileen and Middle River sections and the Fond du Lac formation, (abst.): Eos (Trans. American Geophys. Union), v. 60, p. 241.

Wynne-Edwards, H.R., 1972, The Grenville Province: in Price, R.A., and Douglas, R.J.W., eds., Variations in Tectonic Styles in Canada: Geol. Assoc. Canada, Spec. Paper 11, 688 p.

York, D., 1978, A formula describing both magnetic and isotopic blocking temperatures: Earth Planetary Sci. Letters, v. 39, p. 89-93.

The Continental Crust and Its Mineral Deposits, edited by D.W. Strangway,
Geological Association of Canada Special Paper 20

VOLCANISM ON EARTH THROUGH TIME

Kevin Burke and W.S.F. Kidd
Department of Geological Sciences,
State University of New York at Albany
Albany, New York 12222

ABSTRACT

Volcanism is widespread on Earth and apparently always has been. In this respect the
Earth contrasts with the Moon, where volcanism stopped about 3 Ga ago, and with Mars, where
it may have stopped 1 Ga ago. Although the terrestrial volcanism that occurred before the oldest
preserved rocks formed was probably similar to later volcanism, the very earliest volcanic
activity on Earth could have been like that on the Moon, if the Earth acquired its water late
during the high impact flux.

Volcanism plays a vital part in all three stages of lithospheric evolution active on Earth
today. Basalt forms at divergent plate boundaries; tholeiitic and calc-alkaline rocks, most
characteristically andesite, dominate where arcs form above subduction zones; highly potassic
volcanics are associated with active continental collision, crustal thickening, and fractionation.
These three processes appear to have operated at plate margins throughout most of the Earth's
history.

Ultramafic komatiites, forming perhaps 5% of basaltic piles and indicating the existence of
ultramafic melts, are peculiar to the Archean, and are presumably due to the greater heat
generation of the early Earth, although their precise significance is ambiguous. Non-plate
margin (hot-spot) alkaline and tholeiitic volcanism is also recorded in old rocks. Remnants of
flood basalts and associated sills and dyke swarms formed by rifting episodes are clearly
displayed in the Canadian shield up to 2.5 Ga ago, and in Greenland back to 3.6 Ga.

Sea level variations during the Phanerozoic have been interpreted as indicating variations
in spreading ridge volume and hence (on an Earth of roughly constant volume) of plate creation
and destruction rates. Precambrian sea level fluctuations cannot be interpreted in the same way
because the sediment record is incomplete and because timing has not been adequately resolved. Estimates of episodicity in plate activity in the Precambrian depend mainly on the
occurrence of peaks in age abundance. This method is highly misleading because the preserved
areas of particular ages are much too small to be representative of the world at those times. As

far back as the geological record, goes, volcanism has been continuously active on Earth and has occurred mainly at plate boundaries. Phanerozoic history indicates some episodicity, and the Precambrian was probably similar.

RÉSUMÉ

Le volcanisme est très répandu sur toute la Terre et apparemment il l'a toujours été. Sous cet aspect, la Terre contraste avec la Lune où le volcanisme s'est arrêté il y a 3 Ga et avec Mars où il a dû cesser il y a 1 Ga. Bien que le volcanisme terrestre antérieur aux plus anciennes roches préservées ait été probablement semblable au volcanisme plus tardif, l'activité volcanique la plus ancienne sur terre a pu être semblable au volcanisme lunaire si la Terre a acquis son eau plus tard durant le flux élevé d'impacts.

Le volcanisme joue un rôle vital dans les trois stades d'évolution de la lithosphère comme on l'observe sur la Terre aujourd'hui. Le basalte se forme aux limites de plaques divergentes; les roches tholéiitiques et calco-alcalines, en particulier l'andésite, dominent là où des arcs se forment audessus des zones de subduction; on retrouve les volcaniques très potassiques en association avec les zones de collision continentales actives, l'épaississement de la croûte et son fractionnement. Ces trois processus semblent avoir agi aux limites des plaques durant la plus grande partie de l'histoire de la Terre.

Les komatiites ultramafiques, qui forment peut-être 5% des empilements basaltiques et indiquent l'existence de magmas ultramafiques, sont caractéristiques de l'Archéen et on présume qu'elles sont dues à la plus grande production de chaleur dans la Terre primitive; toutefois, leur signification précise est encore ambiguë. Le volcanisme alcalin ou tholéiitique en dehors des bordures de plaques (points chauds) se retrouve aussi dans les roches anciennes. Des vestiges de coulées de basalte de même que les sills et les essaims de dykes associés qui se sont formés durant des épisodes d'effondrement sont clairement exposés dans le bouclier canadien jusqu'à il y a 2,5 Ga et jusqu'à 3,6 Ga au Gröenland.

On interprète les variations du niveau marin au cours du Phanérozoïque comme indiquant des variations dans le volume des crêtes en expansion et par conséquent (sur une Terre de volume à peu près constant) comme un indice des taux de création et de destruction de plaques. On ne peut interpréter les fluctuations de niveau marin au Précambrien de la même façon parce que le registre sédimentaire est incomplet et parce que la chronologie est inadéquate. Les estimations de l'épisodicité de l'activité des plaques au Précambrien dépendent surtout de la présence de pics d'abondance d'âges. Cette méthode porte beaucoup à confusion parce que les régions où les roches de certains âges sont préservées sont beaucoup trop petites pour être représentatives du monde à ces époques. Aussi loin que le registre géologique s'étende, le volcanisme a été continuellement actif sur Terre et s'est produit surtout aux bordures des plaques. L'histoire phanérozoïque indique une certaine épisodicité et c'était probablement semblable au Précambrien.

INTRODUCTION: VOLCANISM ON EARTH COMPARED WITH THAT ON OTHER TERRESTRIAL PLANETS

Active volcanism is widespread on Earth and is recorded in rocks of all ages. Strikingly, the oldest preserved rocks on Earth (at Isua in West Greenland) appear to be volcanic breccias. With the exception of a very small volume produced by partial melting following impact (e.g., basaltic rocks preserved in plutonic facies at Sudbury and the Bushveld, and suevites at numerous localities) terrestrial volcanic rocks appear to be mainly the products of partial melting generated fundamentally by the decay of radioactive heat-generating nuclides.

In this respect the Earth contrasts with some other terrestrial planets and satellites. On the Moon, volcanism ended about 3 Ga ago, and much, if not all, of lunar volcanism was related to impacts about 4 Ga ago or earlier. By 3 Ga ago, radioactive heat generation within the Moon seems to have declined to a rate sufficiently low to have been entirely removed through the surface by conduction. This suggests to us that the Moon's heat-generating nuclides were concentrated close to the surface by that time.

On Mars, earlier volcanism appears to have been impact-dominated. However, the spectacular later volcanism of the Tharsis area and some older similar features more likely resulted from partial melting through radioactive decay of heat-generating nuclides. Martian volcanism seems to have ceased, but there is some uncertainty as to exactly how long ago this happened. We do not yet know about volcanism on Venus, but extremely violent volcanic activity is in progress on Io (Morabito et al., 1979). This volcanism, perhaps unique in the solar system, appears to be the result of tidal forces (Peale et al., 1979). Tidal forces do not appear to be major generators of volcanic heat elsewhere in the solar system, although tidal influences triggering eruptions have been recognized on the Earth (e.g., Mauk and Johnston, 1973).

SECULAR VARIATION OF TERRESTRIAL VOLCANISM

This short review deals only with volcanism, but plutonic igneous rocks are associated with all volcanic rocks, and most plutonic rocks exposed or emplaced at high crustal levels at one time underlay volcanic rocks. In a few, probably exceptional, areas shallow intrusives do not appear to be linked to volcanic rocks. Examples exist in the young, high-level granites of the Himalayas (Gansser, 1964) where the absence of volcanic superstructures may be due to intrusion in a dominantly compressional environment, or to rapid and effective erosion following uplift. This contrasts with conditions on the Tibetan plateau to the north, where compression is now less dominant and young volcanism has been widespread.

Estimates of the variation of heat generation in the Earth through radioactive decay with time are shown in Figure 1. Small additions through such processes as tidal dissipation are significant through all of geologic time, and much larger contributions to heat generation were made before 4 Ga by core formation and impact. Like Turcotte and Burke (1978), we assume that heat escapes from the Earth roughly as soon as it is generated, and generally within 0.5 Ga.

Volcanism is, and apparently always, has been, one of the dominant processes by which heat is removed from the Earth's interior. At present, heat leaves the Earth in three main ways: by conduction, by the eruption and rapid cooling of igneous rocks, and by the longer-term process of the cooling of ocean floor as it ages, subsides, and moves across the Earth's surface. We have shown elsewhere (Burke and Kidd, 1978) that because some of the ancient continental crust is and was of normal thickness and its base did not suffer widespread melting, the continental conductive thermal gradient about 2.5 Ga. ago in the lower crust was not significantly greater than it is now. Since Figure 1 indicates that two or three times as much heat was being generated at that time, this greater terrestrial heat must have been removed by the other two processes: igneous rock emplacement and ocean-floor aging. Ocean-floor aging is linked to divergent plate boundaries; most other igneous activity today also occurs at plate boundaries.

We and our colleagues (Burke and Dewey, 1972; Dewey and Burke, 1973; Burke *et al.*, 1976; Burke *et al.*, 1977) have discussed at length elsewhere why we consider most igneous activity to have been plate-boundary related throughout recorded geologic history (about the last 3.8 Ga). Others have interpreted the early lithosphere as not broken into rigid plates and consequently most tectonic and igneous activity as unrelated to plate margins. It is unnecessary to repeat the arguments for and against these interpretations. Let it suffice to say that the rocks and structures formed in early times so closely resemble later ones, that any radically different process operating then must have been capable of simulating the results of present plate tectonics. It is mainly for this reason that we prefer to regard the ancient world as broken into plates. There are two minor but significant differences between early and present-day plate tectonics. First, the size of suture-bounded areas in Archean terranes (notably the Superior Province) is generally smaller than in later terranes. This is consistent with the idea that greater lengths of plate boundary and/or faster moving plates were needed to dissipate the extra Archean heat (Burke and Kidd, 1978). Secondly, Archean terranes contain most known examples of a unique class of volcanic rocks, the ultramafic komatiites, which form a small proportion of Archean basaltic piles. Archean komatiites are presumably related to the greater heat generation of the early Earth, and their restricted occurrence has been used by various workers as a datum point in analyzing the thermal history of the Earth. Clearly, this is qualitatively

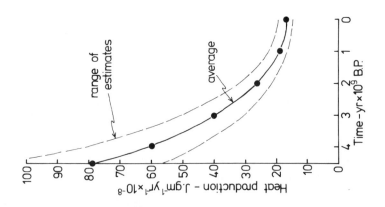

Figure 1. Heat production in the Earth from decay of radioactive isotopes of U, Th, and K through geological time (after Lee, 1967).

correct (the Earth was hotter), but we know too little about the details of the origin, environment, and mechanism of eruption of the Archean komatiites to make quantitative inferences about the significance of these peculiar rocks. In summary, we conclude from the composition, distribution, and structure of volcanic rocks occurring among the oldest of terrestrial rocks that most volcanic activity from 3.8 Ga on was, as now, related to plate margins.

VOLCANISM AT PLATE MARGINS: DIVERGENT MARGINS

There are three types of plate boundaries (Wilson, 1965) and it is appropriate to consider volcanic activity associated with each of them. Tuzo Wilson has long advocated the integration of solid-earth sciences, pointing out that geology, geophysics, and geochemistry are not realistically separable, and are strongest when results from all fields are considered together. There is no better example of this integration than the study of divergent plate-margin volcanism (which takes place at oceanic spreading ridges, and except in anomalous places like Iceland and the Afar, occurs two or more kilometres below sea level). Geophysical methods have dominated the study of oceanic spreading ridge volcanism and have been complemented by geochemical study of dredged pillow-lavas and limited submersible reconnaissance. The realization that ophiolite sequences in mountain belts represent small samples of ocean floor and preserve material produced at divergent plate boundaries has permitted integrated studies of the type favoured by Wilson, which have done much to reveal the essential features of divergent plate-boundary magmatism.

Figure 2 shows the results of one study of this kind (based on Dewey and Kidd, 1977). Basalt rising from partially melted mantle occupies a magma chamber roughly triangular in cross-section; volcanic material is erupted from this chamber to form the pillow lavas of the ocean floor,the underlying sheeted dykes, and the cumulate and non-cumulate plutonic rocks that overlie the depleted mantle.

The recognition of discrete major volcanoes within the axial valley of the FAMOUS area (Ballard and van Andel, 1977; Ramberg and van Andel, 1977) and across-strike compositional variations in basalt (Bryan and Moore, 1977) has led to interpretations of slowly spreading boundaries having along strike diversity and episodic structural development. This kind of interpretation is supported by near-bottom magnetic anomaly studies (e.g., Macdonald, 1977), in contrast to earlier, continuously evolving models, which leaned heavily on magnetic anomaly patterns mapped at or above the sea surface. The small-scale topographic and magnetic structures occurring at oblique angles to regional spreading directions (e.g., Macdonald and Holcombe, 1978) are typical of the kind of complexity being recognized in these detailed studies. The study of small variations in basalt composition, particularly in trace-element distributions (White and Bryan, 1977), has led to the realization that mid-ocean ridge basalts cannot be interpreted as simple products of either fractional crystallization or partial melting, but that a complex interaction of these processes, complicated by episodic magma injection into evolving magma chambers (O'Hara, 1977), is more likely to have taken place.

The differences between fast-spreading ridges (with no axial rift) and slow-spreading ridges (with rifts) have been illuminated by seismic refraction studies that show much stronger evidence of active magma chambers at fast-spreading ridges

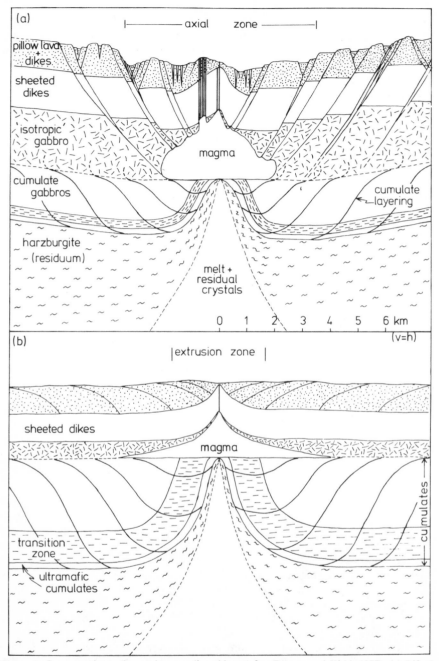

Figure 2. Cross-sections of oceanic spreading ridges (after Dewey and Kidd, 1977). (a) Rifted, slow-spreading ridge. (b) Non-rifted, fast-spreading rise crest.

than at slowly spreading riges (Orcutt *et al.,*1975; Fowler, 1977). It has even been suggested that magma bodies may only be episodically present (Tapponnier and Francheteau, 1978) at slowly spreading ridges, but the geochemical data do not support this interpretation (Walker *et al.,* 1980). Thermal calculations (Sleep, 1975) suggesting episodic magma chambers along slowly spreading ridges are therefore unlikely to be correct. Clearly, there is some minimum spreading rate below which the cooling effect of circulating seawater will overwhelm the heat supplied by new magma, but this does not seem to have happened along most of the present ridge system.

The hot-spot (i.e., non-plate margin-type) volcanism that occurs in places like the Afar, Iceland, and the Azores will be considered in a later section because we interpret it not so much as an anomalous type of divergent plate-boundary volcanism but as a normal part of the wide spectrum of hot-spot volcanism. Divergent plate boundaries have developed across the sites of these hot-spots.

All but a tiny proportion of oceanic crust has been subducted- that is, permanently removed from the Earth's surface at least in any recognizably original form. Of the tiny sample that has escaped this fate, much is badly shredded and dismembered, particularly the material preserved in steeply-inclined zones and /or at deeper structural levels, even in Mesozoic and Tertiary orogenic belts. The general observation that recognizable ophiolite complexes become scarcer in older orogenic belts is explained by the greater uplift and erosion that have generally occurred there, compared with younger belts. Other reasons for their reported scarcity may include the past unfamiliarity of workers in older belts with the utility of the ophiolite concept, the possibility that major continental collision and suturing was generally more effective and intense further back in time, and the effect of the smaller length of older orogenic belts remaining exposed or preserved compared with younger ones. Well-described ophiolite complexes of Paleozoic age are known from the Appalachians and Caledonides (Bird*et al.,* 1971, 1978; Dewey and Bird, 1971; Church and Stevens, 1971; Sturt *et al.,* 1979), and less detailed descriptions of some of the obviously widespread Paleozoic complexes in the Urals and central Asian orogenic belts are available (for example Abdulin *et al.,* 1974, Makarychev and Shtreys, 1973).

Definite examples of pre-Paleozoic ophiolite complexes have been identified only in Pan-African orogenic belts in Morocco (LeBlanc, 1976) and Saudi Arabia (reported in Brown, 1978), and in the older Baikal orogenic belt in the U.S.S.R. (Klitin and Pavlova, 1974). Many other dismembered pieces surely remain to be identified in Proterozoic orogenic belts, since the well-studied belts, for example the Labrador-Cape Smith-Nelson (Wilson, 1968b) and the Coronation orogenic belts (Hoffman, 1980), so clearly developed through rifting and later collision (Wilson cycles); dated dykes reveal that in both cases rifting began about 2.15 Ga ago. The suggestion (Burke *et al.,* 1976) that oceanic crustal samples (dismembered ophiolite complexes) should exist in Archean greenstone belts has not, to our knowledge, been confirmed. Because there is an extensive greenstone-belt terrane (the Birrimian of West Africa) of similar age to the Coronation orogen, it seems to us very likely that dismembered ophiolites will eventually be recognized in Archean Greenstone belts. Since the Birrimian greestones were formed at the same time that ocean opening and closing (which must have involved sea-floor spreading and subduction) is recorded in the Coronation and other orogens, samples of the oceanic crust of that time may be preserved in the Birrimian terrane. Since this terrane does not differ in any significant way from older Archean greenstone terranes, the inference that oceanic crust may be preserved in Archean terranes seems to us unexceptionable. Extensive areas of mafic volcanics of tholeiitic compositions appropriate to the ocean floor exist in the Archean greenstone belts, but equally widespread plutonic equivalents are generally lacking. Moores (1973) suggested that older Pre-cambrian oceanic crust contained more anorthosite than younger material; this has yet to be verified. It is possible that the extensive tholeiitic submarine lavas of the greenstone belts include, besides lava generated as ocean floor and island arcs, large thicknesses generated by intraplate flood-basalt type magmatism; this is discussed below in the section on intraplate volcanism.

VOLCANISM AT PLATE MARGINS: TRANSFORM MARGINS

Very much less volcanism occurs at transform boundaries than at convergent and divergent plate boundaries and all, or nearly all, of what does occur there is associated with local extensional areas, commonly termed "pull-aparts" "Leaky transforms" are most probably the same features, less well defined because they are submarine. The young alkalic basalts dredged along the St. Paul and Romanche fracture zones are the only rocks presently known that are likely to have been erupted on oceanic transform/fracture zones away from the places where spreading ridge axes abut such zones. Even in these two cases, the tectonic details of their occurrence and the local abundance of the volcanics (or lack of it) are poorly known. It is also hard to prove that these volcanics are not related to areas of "hot-spot" activity, with relatively little associated volcanism, like the Jos Plateau and Air regions of the African plate. These areas are clearly identifiable since they are not submarine, lie within a plate, and are obviously associated with a structurally and topographically defined uplift. The fact that there is an uplift at St. Paul's Rocks perhaps favours the idea that these volcanics are "hot-spot"-related rather than related to secondary extension across the transform/fracture zone.

Pull-apart basins along large strike-slip faults on land (e.g., the Salton Sea, the Dead Sea rift, near the western end of the Altyntagh Fault, and small basins along the North Anatolian fault) reveal that such pull-aparts are not common, that they are usually small relative to the extent of the transform system along which they occur, and that the volume of volcanics directly associated with the pull-apart is also rather minor. It may be that larger volumes of magma occur at depth and that the surface expression is not representative of the amount of magmatism in these pull-aparts; the geothermal activity in the Salton Sea area may be evidence of this.

It has been proposed that other volcanic rocks occurring in some places in the general region of large continental transform fault systems (e.g., the young Arabian basalts, the Hsing-An basalts of China, and some of the young volcanics in the vicinity of the San Andreas Fault) are in a less direct way related to the transform-zone tectonics. Since present evidence for a strong connection is unconvincing, we treat such volcanics under intraplate volcanism.

Perhaps because of the small volume of transform-related volcanics, and their susceptibility to later tectonic disruption, no well-documented examples are known from older, inactive transform zones. For example, well-studied pull-apart basins formed on an extensive, large-displacement Carboniferous strike-slip fault zone (probably an old transform system) through the Canadian Maritime Provinces (Belt, 1969) do not contain any known examples of contemporaneous volcanics. The exposed portion of the possibly correlative (Wilson, 1962) Great Glen Fault system does not show any obvious pull-apart structures. A much older example (around 1.8 Ga) of a preserved piece of large-displacement strike-slip fault, the McDonald Fault, also has no known volcanics associated with its movement, despite having very thick accumulations of coarse clastic sediments (P.F. Hoffman, pers. commun.) which probably were preserved by pull-apart tectonics.

VOLCANISM AT PLATE MARGINS: CONVERGENT MARGINS

In present-day arc systems andesitic volcanism dominates, although overall about 20% is basaltic (Ewart, 1976). The volcanic zone always lies above a subducting slab of oceanic lithosphere, generally where the latter is between 100 and 150 km deep. It is therefore inferred that the descending lithosphere is involved in production of arc magmas, although exactly how and where these magmas are produced is unresolved. While great variability exists in the detailed features of volcanic arcs, the most prominent contrast in volcanism is that silicic, large-volume ignimbrites with high Na and K content are concentrated in, if not confined to, areas underlain by continental-type lithosphere, even though, as in New Zealand and southern Alaska, this may consist of (geologically) recently accreted material. Some involvement of the lithosphere underlying these Andean-type volcanic arcs in the generation of their magmas seems likely.

Arcs built on oceanic crust tend to contain proportionately more mafic and less silicic volcanics, although precise estimates are difficult to make since most of these arcs are submerged, and preservation is biased against pyroclastic and clastic materials. Burke et al. (1976) pointed out that the collision and accretion of this kind of arc, together with remnants of the intervening marginal basin floors and fill, will produce greenstone-granodiorite belt geology like that of the Archean Superior Province and similar terranes elsewhere. These include at least one that is of post-Archean age (the Birrimian of West Africa - 1.8 Ga), and thus contemporary with orogenic belts clearly formed from the operation of the Wilson cycle. The present southwestern Pacific contains areas that we envisage as close analogues to those that formed the greenstone terranes. Andesites in present arcs do not have any significant geochemical differences from andesites in Archean greenstone belts. As the overwhelming bulk of andesites is made at convergent margins today, possibly in response to the liberation of water from hydrated phases in the subducted slab at significant depths in the mantle, we see no good reason to suppose that they were made in any grossly different way in the past. Greenstone-type terranes are not unique to the Archean. We pointed out above (see also Burke and Dewey, 1972) the example of the Birrimian, and smaller analogues can be found in Paleozoic and Mesozoic orogenic belts (Burke et al., 1976). More extensive analogues probably exist, awaiting proper description, in the Paleozoic orogenic area of Central Asia (Burke et al., 1978a). The area of older orogens preserved intact from the tectonic effects of younger orogenies decreases with age. It is entirely possible that the sample we have of Archean

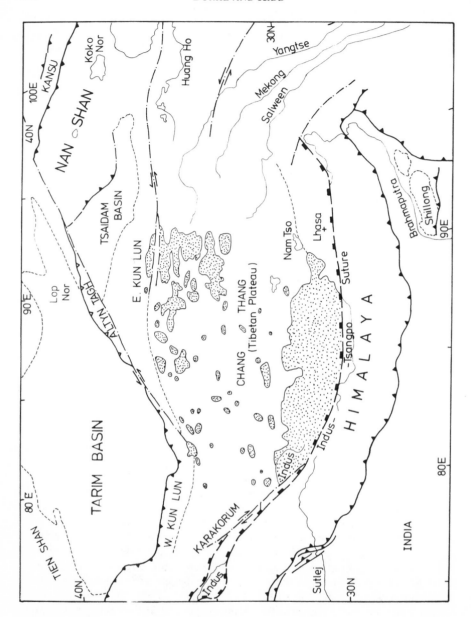

Figure 3. Tectonic sketch map of the Tibetan Plateau and surrounding regions, emphasizing the Neogene-Recent volcanic rocks (stippled). Lines with black triangles – active thrust boundaries; line with open triangles – inactive thrust boundary; dot-dash lines – active transcurrent faults.

orogenic terranes is not representative of their original proportions, due to the accidents of the siting of later rifts and, hence, collisions (Wilson cycles). Assuming, however, that the greater abundance of greenstone belts in the Archean reflects a real secular tectonic change, then it is reasonable to suppose that it is connected with the Earth's greater heat production in the past, which, as suggested above, resulted in faster plate motion and greater length of plate boundary. Such properties would most probably have led to more arc generation in a given time, thus accounting for the greater abundance of greenstone-type terranes in the past without excluding them from post-Archean terranes.

Plate convergence eventually leads to continental collisions. These, in recent examples (Fig. 3), give rise to volcanics very similar to those found in Andean arcs (Burke et al., 1974; Kidd, 1975; Sengör and Kidd, 1979). Older examples of such calc-alkaline, K-rich volcanics and/or their subjacent post-kinematic granites are well known in collisional orogens up to about 2 Ga, e.g. the granites and other plutons of the Appalachian-Caledonian belt (Dewey and Kidd, 1974); the post-kinematic granites and local rhyolites of the Pan African; the rhyolites of the St. Francois Mountains (Bickford and Mose, 1975), related to the Elsonian orogeny which we suggest is collision-induced; and the volcanics of the Bear Province (Hoffman et al., in prep.), related to collision about 1.8 Ga ago at the end of the Wilson cycle on the site of the Coronation orogen. Pieces of continental crust older than 1.8 Ga are either too small or too deeply uplifted and eroded to distinguish collisional from arc-related magmatic products with any confidence, but we see no reason to suppose their absence.

INTRAPLATE TYPE VOLCANISM: HOT SPOTS AND FLOOD BASALTS

Young volcanism not related to plate boundaries is widespread (Burke and Kidd, 1975). Despite the great variety in its expression, we feel that there are sufficient common elements among different instances of non-plate margin volcanism that it is inappropriate not to treat them in a single category. Many areas of active intraplate volcanism are well-exposed in Africa and have been well studied; the association in them of volcanics and a structurally and topographically high area is well established (Burke and Whiteman, 1973; Thiessen et al.,1979). The volcanics of Dakar, near sea level, are not an exception, since subsurface data (Spengler and Deteil, 1966) reveal an underlying youthful structural elevation. There is a complete spectrum of intraplate volcanic areas in Africa, from uplifts without any volcanism (e.g., Fouta Djallon), through minor volcanism of alkaline type (e.g. Air), to more abundant alkaline volcanism (e.g., Tibesti), to areas of voluminous tholeiitic flood basalts together with comparatively minor alkaline volcanics (e.g., Ethiopia). In all cases a structural uplift is associated with the volcanic area; the size of this uplift varies, although most tend to be 100 to 200 km across (Burke and Whiteman, 1973). There seems to be a tendency for the areas with larger volumes of volcanics to have larger diameters of uplifts, although it becomes possible to resolve subsidiary uplifts within the larger ones (e.g., Ahaggar; Black and Girod, 1970). Thus it is not wholly clear whether the larger uplift structures are merely groups of smaller ones, or whether the smaller diameter uplifts in them are secondary to the larger, but generally lower amplitude uplifts.

This spectrum of intraplate volcanism can be seen less well displayed on other continents, but it is more significant that it is also developed in the oceans, and that the alkaline and tholeiitic volcanics in oceanic hot spots are essentially indistinguishable from those in continental hot spots. Again, the African plate provides many of the best examples. Oceanic intraplate volcanic areas also show a variation in volume from the largest, like Hawaii, with abundant tholeiite and minor alkaline volcanics, to relatively small edifices, like Ascension, that are mainly alkaline volcanics. Intraplate volcanic areas of very small volumes, like some of those on the African continent, are harder to detect in the oceans, and uplifts without accompanying volcanism harder still. Nevertheless, careful study has detected them in one area (Menard, 1973).

Because alkaline magmas are erupted in places within present island arcs and collisional orogens, it is not possible to detect "intraplate" type or "hot-spot" type activity unambiguously within such zones of convergence even though it may well occur. However, it is possible to detect such volcanism along divergent plate boundaries because of the alkaline character of some of the magmas, the excess volume of magma, and the associated structural uplift. The occurrence of "hot-spot" type volcanism at discrete and long-lasting sites along divergent plate boundaries is, we suggest, one of the most significant properties of their distribution, and helps considerably in winnowing the many hypotheses put forward for their origin. In particular, since the lithosphere is thin to virtually non-existent at spreading ridge crests and is created progressively away from them, it is unlikely that either the crack propagation hypotheses of Turcotte and Oxburgh (1973) and Oxburgh and Turcotte (1974), or the dense anchor-asperity hypothesis of Shaw and Jackson (1973) is correct. An "active mantle" hypothesis for hot-spot origin, such as proposed by Wilson (1963), is more satisfactory.

A similar spectrum of size and volume of magmas can be seen in those hot spots located on divergent plate boundaries. The central and north Atlantic contains the best examples. These range from the anomalously elevated area at 45°N on the ridge, where there seems to be relatively little excess volcanism, through the Azores, with mostly alkaline volcanism constructing small islands above sea level, to the extreme of Iceland, with its voluminous excess tholeiitic volcanism and, in relative terms, minor alkaline magmatism. Anomalous (with respect to spreading-ridge basalt) geochemical signatures, particularly $^{87}Sr/^{86}Sr$ (White et al., 1976) correlate exactly with discrete, anomalously elevated ridge crest areas in Iceland, 45°N, the Azores, and near 34°N (Colorado Seamount). The distinctive geochemistry of the "excess" magmas argues strongly for a separate (deeper) source for them (Schilling, 1973) compared with normal spreading-ridge basalts. It is because the latter are usually produced in a fairly constant amount through a wide range of spreading rates that the "excess" character of the hot-spot magmatism along ridge axes can be easily detected.

The connection between flood basalts and the largest hot spots is clearly shown by Iceland. The island itself consists of flood basalts, and the hot-spot tracks that lead away from Iceland go to large areas of flood basalts in East Greenland and the northern British Isles (Fig. 4). These basalts were erupted during the initial rifting that lead to successful sea-floor spreading in the northern Atlantic beginning about 60

Ma ago. The relics of associated central alkaline volcanoes of that age are well known from Scotland. A younger but similar situation is seen in the Afar (Fig. 3), where the Ethiopian Traps erupted before and up to the opening of the Red Sea and Gulf of Aden. If the Pacific plate were not moving so rapidly with respect to the source of the Hawaiian magmas, the accumulation of igneous material at present rates of production would clearly rival Iceland, and although the Hawaiian Emperor hot-spot trace (Fig. 4) does not lead back to a site of rifting and a large flood basalt pile, it is

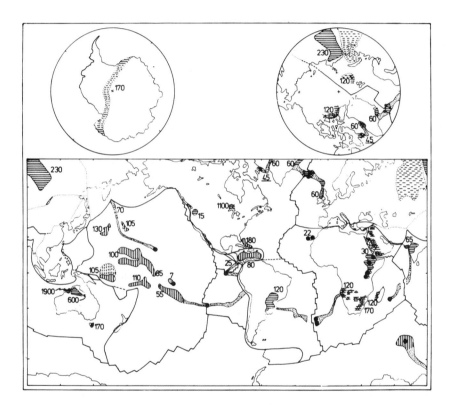

Figure 4. World map with plate boundaries showing larger active hot spots (black circles); their tracks, i.e., volcanic accumulations left behind hot spots on moving plates (stipple); large areas of flood basalts and associated sills (horizontal ruling, dashed where basalts mostly subsurface); submarine equivalent of flood basalts, mostly sill complexes (vertical ruling – dashed where tentative). Approximate ages of flood basalts and sills given in millions of years, as are the ages of the oldest portions of some hot-spot tracks. Number underlined is age of youngest part of Davis Strait hot-spot track.

appropriately grouped with other flood basalt-producing objects. Several other large hot spots do have tracks leading back to rifting sites and accumulations of flood basalts (Fig. 4), in particular Tristan/Gough to the Kaokoveld and Parana basalts, which were erupted just before and up to the opening of the South Atlantic 120 Ma ago; and Reunion to the Deccan Traps, which were erupted about 65 Ma ago just before rifting in the Gulf of Khambat and removal of the Seychelles from India (McKenzie and Sclater, 1971). The Galapagos hot spot, although its tracks do not lead back to preserved flood basalts, is sited on a spreading ridge that started about 25 Ga ago by rifting across older oceanic crust generated at the East Pacific Rise (Hey, 1977).It may not be a coincidence that the Galapagos hot spot, as shown by its area and the volume of magmatic products in its tracks, is a relatively large one, like others that have been listed as associated with flood basalt production and initial ocean opening. Ocean opening in the Labrador Sea-Baffin Bay was accompanied by the flood basalt volcanism (about 60 Ma) that is now preserved on Disko Island, and Cape Dyer of Baffin Island. The hot spot that produced this volcanism made the shallow sill to Davis Strait as a track, but is now obviously extinct. Perhaps it is also not a coincidence that spreading ceased in the Labrador Sea-Baffin Bay at the same time or shortly after the hot spot died. This case illustrates the point emphasized by Vogt (1972), that the volumes of magma generated in hot-spot sites vary with time, although, unlike him, we do not think there is strong evidence for synchroneity in this variation among many hot spots. Thus, the amount of magma produced by the Hawaiian hot spot, judged by the volume of the track, was very small at the time of and shortly after the bend formed in the Hawaiian-Emperor chain. Its volume may have been no more impressive at that time than one of the smaller present ocean island hot spots, like St. Helena. The point we wish to make here is that voluminous flood basalt-producing hot-spots are the same kind of object as the smaller ones, and that one can change into the other, and back again, with time. They may also die out, in terms of their volcanic expression, temporarily or permanently. They vary a great deal in the length of time during which large quantities of tholeiitic magma are produced, from the short burst of the Columbia River basalts (Baksi and Watkins, 1973) and Deccan Traps (Wellman and McElhinny, 1970) to the more extended histories of Iceland and Hawaii.

Large flood tholeiite events in the oceans seem, in some instances, to produce huge sill complexes, probably because of the limited abilities of even mafic lava to travel underwater before chilling. Such sill complexes (Fig. 4) have been identified underlying large portions of the Caribbean (Burke et al., 1978b) and the submarine plateaus of the western Pacific (Winterer, 1976). The Mid-Pacific Mountains, one of the latter, may perhaps be traced to the hot spots of Easter Island and/or Pitcairn Island through the Tuamotu-Line Islands track. Sill complexes are, of course, also important components of may continental flood basalt events, particularly the early early Jurassic Karoo and Ferrar dolerites of South Africa and the Antarctic, and the mid-Cretaceous Isachsen diabase of the Canadian Arctic (Fig. 4).

It is of note that all major flood basalt events (where at least some of the extensive lavas and/or sills are still preserved) were connected with extensive rifting and, in most cases, with successful opening of an ocean. Two large and well known ones not associated with successful ocean opening are the Columbia River basalts and the Siberian Traps.

Well-preserved remnants of older flood basalt and/or sill complexes are seen back to 2.15 Ga. Most remnants are found within old rifts and aulacogens because the accidents of erosion through geological time have claimed any more extensive basalts lying relatively higher outside them. The most prominent flood basalt remnants of Paleozoic or older age lying outside old rifts and of any significant extent are the 600 Ma-old Antrim basalts of N.W. Australia, the 1100 Ma-old Keweenawan lavas and sills north of Lake Superior, and the 1900 Ma-old Kimberly Plateau basalts and sills, also in N.W. Australia. Within old rifts there are many more examples; prominent ones in North America are the rifts of about 1100 Ma, which include the Coppermine River (N.W. Territories), part of which may be outside the rift defined by Burke and Dewey (1973), Seal Lake (Labrador), Gardar (S. Greenland), Keweenawan and its extension in the mid-continent rift and gravity high, and various aulacogens along the Cordilleran margin of the North American craton, including one exposed in part in the Grand Canyon. Another prominent rifting episode is that shown by the extensive tholeiitic sill complex (about 2.15 Ga old) that invades what we interpret as the initial rifting facies clastics and volcanics (arkoses, sandstones and basalts, etc. of the Seward and part of Attikamagen Formations) preserved in the fold and thrust belt of the Labrador Trough (Baragar, 1967).

Associated with this 2.15 Ga rifting and ocean opening event is an extensive dyke swarm (Fahrig and Wanless, 1963; Stevenson, 1968) that runs (in Archaen basement) subparallel with the northern end of the Labrador Trough and then turns to run obliquely along the Cape Smith belt, the continuation of the Labrador Trough in northern Ungava. This dyke swarm is the last relict of a flood basalt event; a very similar although smaller example is seen close to and subparallel with the western margin of the Appalachians in northern Newfoundland (Williams, 1967), where it (and nearby small relicts of flood basalts) was associated with the early Cambrian or slightly older opening of the ocean that Tuzo Wilson called the Proto-Atlantic (1966), now termed the Appalachian Ocean or Iapetus. Dyke swarms as relicts of flood basalt and rifting events are common. The MacKenzie dykes, which extend a great distance NNW across the western part of the Canadian shield are of the same age as the Keweenawan rifting and flood basalt event. There are no well-preserved segments of continental rifted margins older than 2.15 Ga, the opening age of the Labrador Trough and Coronation oceans. However, there are examples of extensive tholeiitic dyke swarms, from which we infer that rifting events like those recorded in younger rocks took place. The Ameralik dykes in West Greenland (McGregor, 1973), which are at least 3.0 Ga and perhaps as much as 3.6 Ga old (Pankhurst et al., 1973) are evidence of the oldest rifting event directly recorded. The nature of the older rocks they cut, containing calc-alkaline volcanics, is to us indirect evidence of subduction, sea-floor spreading and yet older rifting.

One feature of low-metamorphic grade Archean volcanic terranes is that they do not – with very minor exceptions which could be younger than Archean (Cooke and Moorhouse, 1969) – contain alkaline volcanic and plutonic rocks like those that have been emplaced at present hot-spot sites and are preserved sparingly in rocks up to about 2.5 Ga old. Judging by the largest present hot spots, perhaps most of them were large in the Archean and erupted huge quantities of flood basalt-type tholeiite. Lavas and sills generated in this way, if preserved from subduction, might be represented within the extensive basalts in greenstone belts. Alternatively, efficient subduction of Archean oceanic lithosphere may have removed essentially all the hot-spot material and may have preferentially preserved island arc edifices, which even today contain only rare occurrences of alkalic volcanics. Added to the small area of preserved Archean rocks, this small chance of preservation of alkalic volcanics may explain why they are so rare in such terranes.

FLUCTUATION IN VOLCANIC ACTIVITY?

An important question is whether, and if so by how much, the amount of volcanism on Earth has fluctuated with time. This question is fraught with difficulties. Older rocks are generally preserved in smaller proportions relative to younger ones, so that the occurrence of volcanic rocks per unit time could be expected to be less. Long ago, however, since heat generation was greater, volcanism may also gave been greater, and these effects may partly cancel. Because oceanic rocks are destroyed by subduction and obduction, the record is necessarily incomplete; generally only arc and continental rocks are preserved. In spite of these difficulties, some workers have discerned episodicity in igneous activity, especially in the Precambrian. The crude technique, popular twenty years ago, of plotting histograms of isotopic ages has fallen into disuse with the recognition of the complexity of the record. A particularly significant observation is that the proportions of continental area representative of a particular time interval and available for study are so small. For example, the Superior Province, which is the largest Archean area on Earth, representing about half the total area of Archean rocks not obscured by later events, amounts to only 1% of continental area. A more sophisticated approach considering the isotopic compositions of Nd, Sr, and Rb has been used by several authors (e.g., McCulloch and Wasserberg, 1978) to show that many Precambrian continental rocks became isolated from the main mantle resevoir during an interval of about 200 Ma ago roughly 2.7 to 2.5 Ga ago. This is a most interesting and unexpected result.

Turcotte and Burke (1978) used an indirect method to estimate volcanic fluctuation with time during the Phanerozoic. Realizing that sea level responds to the volume of the mid-ocean ridges, they inferred that the times when the continents were most flooded were times when the ridges were most active. By crudely calculating the proportions of heat escaping through conduction and ocean-floor aging, they were able to estimate that nearly twice as much heat was escaping the Earth through the cooling of aging ocean floor during the late Cretaceous episode of continental flooding as is escaping in this way now. In order to keep the Earth at roughly the same volume, they inferred that plate consumption also peaked at this time, which is consistent with the familiar high concentration of circum-Pacific batholithic emplacement during the Late Cretaceous and the Late Cretaceous concentration of emplacements of Tethyan ophiolites.

CONCLUSIONS

It has been impossible to cover all aspects of volcanism through time in such a short review, but it seems appropriate to point out that the history of volcanic activity on Earth as summarized here is best viewed as Tuzo Wilson (1968a) first proposed, in the context of cycles of ocean opening and closing. Not only can rift, plate-margin, and collisional volcanism be well accommodated within the framework of "Wilson Cycles" (Dewey and Burke, 1974), but so can the role of hot spot (intraplate type) volcanism, especially of the Hawaiian kind (Wilson, 1963).

REFERENCES

Abdulin, A.A., Bespalov, V.P., and Shelepova, T.N., 1974, Ophiolite belts of Kazakhstan, Doklady Akad. Nauk. SSSR., v. 218, p. 159-162 (20-22 in translation).

Baksi, A.K. and Watkins, N.D., 1973, Volcanic production rates: comparison of oceanic ridges, islands and the Columbia River basalts: Science, v. 180, p. 493-496.

Ballard, R.D. and van Andel, T.H., 1977, Morphology and tectonics of the inner rift valley at 36°50'N on the Mid-Atlantic Ridge: Geol. Soc. America Bull., v. 88, p. 507-530.

Baragar, W.R.A., 1967, Wakuach Lake Map area, Quebec Labrador (230): Geol. Survey Canada, Mem. 344.

Belt, E.S., 1969, Newfoundland Carboniferous stratigraphy and its relation to the Maritimes and Ireland: in Kay, M., ed., North Atlantic Geology and Continental Drift, Amer. Assoc. Petrol. Geol. Mem. 12., p, 734-753.

Bickford, M.E. and Mose, D.G., 1975, Geochronology of Precambrian rocks in the St. Francois Mountains, southeastern Missouri: Geol. Soc. America Spec. Paper 165, 48 p.

Bird, J.M., Dewey, J.F., and Kidd, W.S.F., 1971, Proto-Atlantic ocean crust and mantle: Appalachian/Caledonian ophiolites: Nature Phys. Sci., v. 231, p. 28-31.

———————, 1978, The Mings Bight ophiolite complex, Newfoundland: Appalachian oceanic crust and mantle: Canadian Jour. Earth Sci., v. 15, p. 781-804.

Black, R., and Girod, M., 1970, Late Palaeozoic to recent igneous activity in West Africa and its relationship to basement structure: in African Magmatism and Tectonics: Clifford, T.N. and Gass, I.G., eds., Edinburgh, Oliver and Boyd, p. 185-210.

Brown, G.F., 1978, Arabian-Nubian shield discussed on the spot: Geotimes, v. 23, p. 24.

Bryan, W.B., and Moore, J.G., 1977, Compositional variations of young basalts in the Mid-Atlantic Ridge rift valley near 36°49'N: Geol. Soc. America Bull., v. 88, p. 556-570.

Burke K.C.A. and Dewey, J.F., 1972, Orogeny in Africa: in Dessauvagie T.F. and Whiteman, A.J., eds., African Geology Univ. Ibadan Press, p. 583-608.

———————, 1973, Plume-generated triple junctions: key indicators in applying plate-tectonics to old rocks; Jour. Geol. v. 81, p. 406-433.

Burke, K.C.A., and Kidd, W.S.F., 1975, Earth, heat flow in: in McGraw-Hill Yearbook of Science and Technology, McGraw-Hill, New York, p. 165-169.

———————, 1978, Were Archean continental geothermal gradients much steeper than those of today?: Nature, v. 272, p. 240-241.

Burke, K.C.A. and Whiteman, A.J., 1973, Uplift, rifting and the breakup of Africa: in Tarling, D.H. and Runcorn, S.K., eds., Implications of Continental Drift to the Earth Sciences, v. 2: Academic Press, London, p. 735-755.

Burke, K.C.A., Dewey, J.F. and Kidd, W.S.F., 1974, The Tibetan Plateau: its significance for tectonics and petrology: Geol. Soc. America Abstracts with Program 6, v. p. 1027-1028.

———————, 1976, Dominance of horizontal movements, arc and microcontinental collisions during the later permobile regime: in Windley, B.F., ed., The Early History of the Earth: Wiley, London, p. 113-129.

———————, 1977, World distribution of sutures – the sites of former oceans: Tectonophysics, v. 40, p. 69-99.

Burke, K.C.A., and Delano, L., Dewey, J.F., Edelstein, A., Kidd, W.S.F., Nelson, K.D., Sengör, A.M.C. and Stroup, J., 1978a, Rifts and Sutures of the World, Contract Rept.

NAS5-24094: Geophysics Branch ESA Division Goddard Space Flight Center, Greenbelt, MD, 238 p.

Burke, K., Fox, P.J., and Sengör A,M.C., 1978b, Buoyant ocean floor and the evolution of the Caribbean: Jour. Geophys. Res. v. 83, p. 3949-3954.

Church, W.R. and Stevens, R.K., 1971, Early Palaeozoic ophiolite complexes of the Newfoundland Appalachians as mantle-oceanic crust sequences: Jour. Geophys. Res., v. 76, p. 1460-1466.

Cooke, D.L. and Moorhouse, W.W., 1969, Timiskaming volcanism in the Kirkland Lake area, Ontario, Canada: Canadian Jour. Earth Sci., v. 6, p. 117-132.

Dewey, J.F. and Bird, J.M., 1971, Origin and emplacement of the ophiolite suite: Appalachian ophiolites in Newfoundland: Jour. Geophys. Res., v. 76, p. 3179-3206.

Dewey, J.F. and Burke, K.C.A., 1973, Tibetan, Variscan and Precambrian basement reactivation: products of continental collision: Jour. Geol., v. 81, p. 683-692.

——————————, 1974, Hot spots and continental break-up: implications for collisional orogeny: Geology, v. 2, p. 57-60.

Dewey, J.F. and Kidd, W.S.F., 1974, Continental collisions in the Appalachian-Caledonian orogenic belt: variations related to complete and incomplete suturing: Geology, v. 2, p. 543-546.

——————————, 1977, Geometry of plate accretion: Geol. Soc, Amer. Bull., v. 88, p. 960-968.

Ewart, A., 1976, Mineralogy and chemistry of modern orogenic lavas – some statistics and implications: Earth Planet. Sci. Letters, v. 31, p. 417-432.

Fahrig, W.F. and Wanless, R.K., 1963, Age and significance of diabase dyke-swarms of the Canadian shield: Nature, v. 200, p. 934-937.

Fowler, C.M.R., 1977, Crustal structure of the mid-Atlantic ridge crest at 37°N: Geophys. Jour. Royal Astron. Soc., v. 47, p. 459-491.

Gansser, A., 1964, The Geology of the Himalayas: New York, Wiley-Interscience, 289 p.

Hey, R., 1977, Tectonic evolution of the Cocos-Nazca spreading center: Geol. Soc. America Bull., v.88, p. 1404-1420.

Hoffman, P., 1980, Precambrian Wilson Cycle: in Strangway, D.W., ed., The Continental Crust of the Earth and Its Mineral Deposits: Geol. Assoc. Canada Spec. Paper 20.

Kidd, W.S.F., 1975, Widespread late Neogene and Quaternary calc-alkaline volcanism on the Tibetan Plateau: EOS, Trans. American Geophys. Union, v. 56, p. 453.

Klitin, K.A. and Pavlova, T.G., 1974, Ophiolite complex of the Baikal fold zone: Doklady Akad. Nauk SSR, v. 215, p. 413-416, (p. 33-36 in translation).

LeBlanc, M., 1976, Oceanic crust at Bou Azzer: Nature, v. 261, p. 34-35.

Lee, W.H.K., 1967, Thermal history of the Earth: Unpub. Ph.D. Thesis, University of California, Los Angeles.

Macdonald, K.C., 1977, Near-bottom magnetic anomalies, asymmetric spreading, oblique spreading, and tectonics of the Mid-Atlantic Ridge near 37°N: Geol. Soc. America Bull., v. 88, p. 541-555.

Macdonald, K.C. and Holcombe, T.L., 1978, Inversion of magnetic anomalies and sea-floor spreading in the Cayman Trough; Earth Planet. Sci. Letters, v. 40, p. 407-414.

McCulloch, M.T. and Wasserburg, G.T., 1978, Sm-Nd and Rb-Sr chronology of continental crust formation; Science, v. 200, p. 1003-1011.

McGregor, V.R., 1973, The early pre-Cambrian gneisses of the Godthaab district, West Greenland: Phil. Trans. Royal Soc. London, v. A 273, p. 343-358.

McKenzie, D. and Sclater, T.G., 1971, The evolution of the Indian Ocean since the late Cretaceous: Geophys. Jour. Royal Astron. Soc., v. 24, p. 437-528.

Makarychev, G.I. and Shtreys, N.A., 1973, Tectonic position of ophiolites of the southern Tien Shan., Doklady Akad. Nauk. SSSR, v. 210, p. 1164-1166 (P. 92-93 in translation).

Mauk, F.T. and Johnston, M.J.S., 1973, On the triggering of volcanic eruptions by Earth tides: Jour. Geophys. Res., v. 78, p. 3356-3362.

Menard, H.W., 1973, Epeirogeny and plate tectonics: EOS, Trans. American Geophys. Union, v. 54, p. 1244-1255.

Moores, E.M., 1973, Plate tectonic significance of Alpine peridotite types: in Tarling, D.H. and Runcorn, S.K., eds. Implications of Continental Drift to the Earth Sciences: v. 2: Academic Press, London, p. 963-975.

Morabito, L.A., Synnot, S.P., Kupferman, P.N. and Collins, S.A. 1979, Discovery of currently active extraterrestrial volcanism: Science, v. 204, p. 972.

O'Hara, M.J., 1977, Geochemical evolution during fractional crystallization of a periodically refilled magma chamber: Nature, v. 266, p. 503-507.

Orcutt, J., Kennett, B., Dorman, L. and Prothero, W., 1975, a low-velocity zone underlying a fast-spreading rise crest: Nature, v. 256, p. 475-476.

Oxburgh, E.R. and Turcotte, D.L., 1974, Membrane tectonics and the east African rift: Earth. Planet. Sci. Letters, v. 22, p. 133-140.

Pankhurst, R.J., Moorbath, S. and McGregor, V.R., 1973, Late event in the geological evolution of the Godthaab district, West Greenland: Nature Phys. Sci., v. 243, p. 24-26.

Peale, S.J., Cassen, P. and Reynolds, R.T., 1979, Melting of Io by tidal dissipation: Science, v. 203, p. 892-894.

Ramberg, I.B. and van Andel, T.H., 1977, Morphology and tectonic evolution of the rift valley at 36°30'N, Mid-Atlantic Ridge: Geol. Soc. America Bull., v. 88, p. 577-586.

Schilling, J.G., 1973, Iceland mantle plume: Nature, v. 246, p. 141-143.

Sengör, A.M.C. and Kidd, W.S.F., 1979, Post-collisional tectonics of the Turkish-Iranian plateau and a comparison with Tibet: Tectonophysics, v. 55, p. 361-376.

Shaw, H.R. and Jackson, E.D., 1973, Linear island chains in the Pacific: result of thermal plumes or gravitational anchors?: Jour. Geophys. Res., v. 78, p. 8634-8652.

Sleep, N.H., 1975, Formation of oceanic crust: some thermal constraints: Jour. Geophys. Res., v. 80, p. 4037-4042.

Spengler, A. de E. and Delteil, J.R., 1966, Le bassin secondaire-tertiare de Cote d'Ivoire: in Reyre, Du, ed., Bassins sedimentaires du littoral africain 1e partie: Assoc. Serv. Geol. Afr. Paris, p. 99-113.

Stevenson, I.M., 1968, A geological reconnaissance of Leaf River map area, New Quebec and Northwest Territories: Geol. Survey Canada Mem. 356.

Sturt, B.A., Thon, A. and Furnes, H., 1979, The Karmoy ophiolite, southwest Norway: Geology, v. 7, p. 316-320.

Tapponnier, P. and Francheteau, J., 1978, Necking of the lithosphere and the mechanics of slowly accreting plate boundaries: Jour. Geophys. Res., v. 83, p. 3955-3970.

Thiessen, R., Burke, K. and Kidd, W., 1979, African hotspots and their relation to the underlying mantle: Geology, v. 7, p. 263-266.

Turcotte, D.L. and Burke, K.C.A., 1978, Global sea-level changes and the thermal structure of the Earth: Earth Planet Sci. Letters, v. 41, p. 341-346.

Turcotte, D.L. and Oxburgh, E.R., 1973, Mid-plate tectonics: Nature, v. 244, p. 337-339.

Vogt, P.R., 1972, Evidence for global synchronism in mantle plume convection and possible significance for geology: Nature, v. 240, p. 338-342.

Walker, D., Shibata, T. and DeLong, S.E., 1980, Abyssal tholeiites from the Oceanographer Fracture Zone: Contrib. Mineral. Petrol., in press.

Wellman, P. and McElhinny, M.W., 1970, K-Ar age of the Deccan Traps, India: Nature, v. 227, p. 595-596.

White, W.M. and Bryan, W.B., 1977, Sr-isotope, K, Rb, Cs, Sr, Ba and rare-earth geochemistry of basalts from the FAMOUS area: Geol. Soc. America Bull., v. 88, p. 571-576.

White, W.M., Schilling, J.G., and Hart, S.R., 1976, Evidence for the Azores mantle plume from strontium isotope geochemistry of the central North Atlantic: Nature, v. 263, p. 659-663.

Williams, H., 1967, Geological Map of Newfoundland 1:1m: Geol. Survey Canada Map 1231A.

Wilson, J.T., 1962, Cabot Fault, an Appalachian equivalent of the San Andreas and Great Glen Faults and some implications for continental displacement: Nature, v. 195, p. 135-138.

——————————, 1963, A possible origin of the Hawaiian Islands : Canadian Jour. Phys., v. 41, p. 863-870.

——————————, 1965, A new class of faults and their bearing on continental drift: Nature, v. 207, p. 343-348.

——————————, 1966, Did the Atlantic close and then reopen?: Nature, v. 211, p. 676-681.

——————————, 1968a, Static or mobile Earth: the current scientific revolution: Proc. Amer. Phil. Soc., v. 112, p. 309-320.

——————————, 1968b, Comparison of the Hudson Bay arc with some other features: in Beals, C.S., and Shemstone, D.A., eds., Science, History and Hudson Bay: Dept. Energy, Mines, and Resources, Ottawa.

Winterer, E.L., 1976, Anomalies in the tectonic evolution of the Pacific: in The geophysics of the Pacific Ocean Basin and its margin: Amer. Geophys. Union Mono. 19, p. 269-278.

The Continental Crust and Its Mineral Deposits, edited by D.W. Strangway,
Geological Association of Canada Special Paper 20

WOPMAY OROGEN: A WILSON CYCLE OF EARLY PROTEROZOIC AGE IN THE NORTHWEST OF THE CANADIAN SHIELD

P. F. Hoffman
Precambrian Division, Geological Survey of Canada,
588 Booth St., Ottawa, Ontario K1A 0E4

ABSTRACT

The 2.1 to 1.8 Ga Wopmay Orogen is interpreted as a complete Wilson Cycle by means of direct and detailed comparison with Cenozoic tectonic environments. At 2.1 Ga, a continental plate came to rest over hot spots located at present Great Slave Lake and Coronation Gulf. A system of rifts developed after intrusion of alkaline-peralkaline complexes and mafic dyke swarms, the meridional Wopmay rift joining the hot spots and another rift extending into the present East Arm of Great Slave Lake. East-west extension led to continental breakup along the Wopmay rift, where thick bimodal volcanics and arkosic sediments subsided and a west-facing continental terrace and rise were deposited. West-dipping subduction of oceanic lithosphere led to collisional accretion of an arc-bearing microcontinent. During the collision the west-dipping thrust complex of continental margin rocks stripped from the subducting plate was intruded by the syntectonic Hepburn and Wentzel Batholiths, which are folded tabular bodies with extensive "hot-side-up" and "hot-side-down" metamorphic aureoles. After the collision, an east-dipping subduction zone developed west of the accreted microcontinent. A major arc-parallel transcurrent fault, reflecting a dextral component of oblique subduction, accompanied renewed calc-alkaline magmatism in the microcontinent and truncated the earlier collision zone. A remote terminal collision to the west is inferred from a 1000 km wide swath of conjugate transcurrent faults indicating east-west compression. At this time, the more rigid Slave Province indented and cracked the northwestern Churchill Province, triggering the 1.8 Ga lower Dubawnt alkaline volcanism. It is concluded that plate tectonics was operative in the early Proterozoic but magmatism was more intense than today in certain environments and the release of mafic tholeiites from subducting plates in trenches seems to be a Proterozoic speciality.

RÉSUMÉ

On interprète l'orogène de Wopmay, datant de 2,1 à 1,8 Ga, comme un cycle de Wilson complet au moyen de comparaisons directes et détaillées avec les milieux tectoniques du Cénozoïque. Il y a 2,1 Ga, une plaque continentale vint reposer sur des points chauds localisés près de la position actuelle du Grand Lac de l'Esclave et du golfe Coronation. Un système de fosses s'est formé après l'intrusion de complexes alcalins-peralcalins et d'essaims de dykes mafiques, la fosse méridionale de Wopmay rejoignant les points chauds et une autre fosse qui s'étendait dans l'actuel bras est du Grand Lac de l'Esclave. L'extension est-ouest a mené à une rupture continentale le long du fossé de Wopmay où d'épais sédiments bimodaux volcaniques et arkosiques se sont affaissés alors qu'une terrasse continentale orientée vers l'ouest et un glacis se formaient. La subduction avec pendage à l'ouest de la lithosphère océanique a provoqué l'accrétion par collision d'un microcontinent surmonté d'un arc. Le complexe de chevauchement avec pendage à l'ouest des roches continentales de bordure arrachées à une plaque en subduction a été envahi par les batholites syntectoniques de Hepburn et de Wentzel, lesquels sont des masses tabulaires plissées avec des auréoles métamorphiques très marquées aux faces chaudes du haut et du bas. Après la collision, une zone de subduction avec pendage à l'est s'est développée à l'ouest du microcontinent d'accrétion. Une faille transversale majeure, parallèle à l'arc, témoin d'une composante à droite de subduction oblique, a accompagné le retour du magmatisme calco-alcalin dans le microcontinent et a tronqué la zone de collision préexistante. On suppose une collision terminale loin à l'ouest à partir d'un faisceau de failles conjuguées transversales de 1000 km de longueur indiquant une compression est-ouest. A ce moment, la province de l'Esclave, plus rigide, a brisé et fracturé la province de Churchill au nord-ouest, pour devenir responsable de la phase initiale du volcanisme alcalin de Dubawnt il y a 1,8 Ga. On en conclut que la tectonique des plaques était active au début du Protérozoïque mais que le magmatisme était plus intense qu'aujourd'hui dans certains milieux et que la production des tholéiites mafiques à partir de plaques en subduction dans les fossés semble être une caractéristique du Protérozoïque.

INTRODUCTION

There is still no agreement as to whether there were mobile lithospheric plates (plate tectonics) in the Precambrian (Kerr, 1978). Cogent theoretical arguments have been made in favour of plate tectonics (Burke *et al.*, 1976) but most Precambrian geologists remain reluctant to apply the concept of the Wilson Cycle (the "life cycle of ocean basins" of Wilson, 1968) to Precambrian orogenic belts.

Wopmay Orogen (McGlynn, 1970) is an exceptionally well-exposed 1.8 to 2.1 Ga orogenic belt flanking the Archean Slave Province of the Canadian Shield (Fig. 1). The zonation of the orogen is best developed at the north end (Fig. 2), where there is an unusually complete sedimentary succession in the foreland (Zones 1 and 2). Study of this succession led to recognition of the "Coronation Geosyncline" (Hoffman *et al.*, 1970; Fraser *et al.*, 1972; Hoffman, 1973) now an obsolete term but one which, at the time, helped draw attention to the similarities of the orogen to Phanerozoic mountain belts. Since 1973, the northern 200 km of Zone 4 and 100 km of Zone 3 have been mapped in detail, as have the two related basins in the foreland, Kilohigok Basin (Campbell and Cecile, 1976a, 1976b, 1979) and Athapuscow Aulacogen (Hoffman *et al.*, 1977; Hoffman, 1977). It is now possible to make more specific comparisons with Cenozoic tectonic regimes, about which knowledge has also increased rapidly in recent years.

When this comparative approach is applied to the evolution of the orogen, it is hard to resist the conclusion that there was a complete Wilson Cycle, even though rocks of strictly oceanic origin are absent. It is a special pleasure to present this interpretation because J. Tuzo Wilson himself participated in some of the pioneer mapping of the orogen (Lord, 1942).

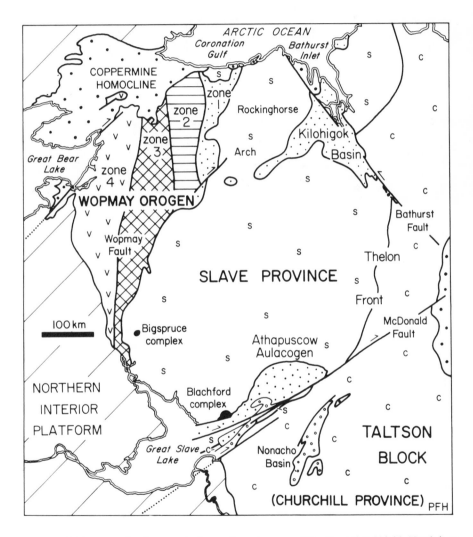

Figure 1. Major tectonic elements in the northwest corner of the Canadian Shield. North is to the top of the page.

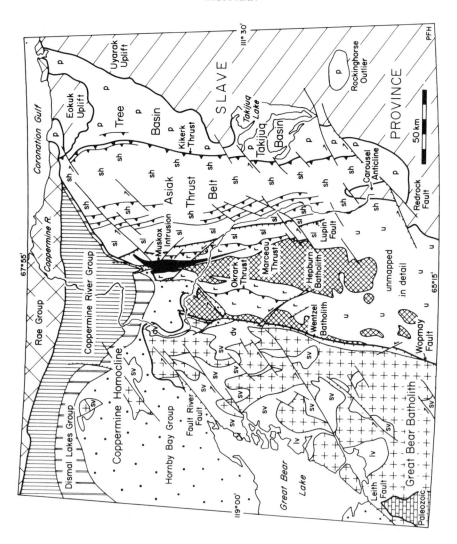

Figure 2. Generalized geology of the north end of Wopmay Orogen. Kikerk Thrust is the boundary between Zone 1 and 2, Lupin Fault between Zone 2 and 3, and Wopmay Fault between Zone 3 and 4. The Asiak Thrust Belt is only partly mapped in detail and there are many thrust faults (with barbs) not indicated here. Lithotectonic units: lv – LaBine Group, sv – Sloan Group, dv – Dumas Subgroup, r – pre-orogenic rift and continental rise facies plus orogenic deposits, sl – pre-orogenic continental slope facies plus orogenic deposits, sh – pre-orogenic continental shelf facies plus orogenic deposits, p – platform facies plus orogenic deposits, u – pre-orogenic rift, rise and slope facies undivided.

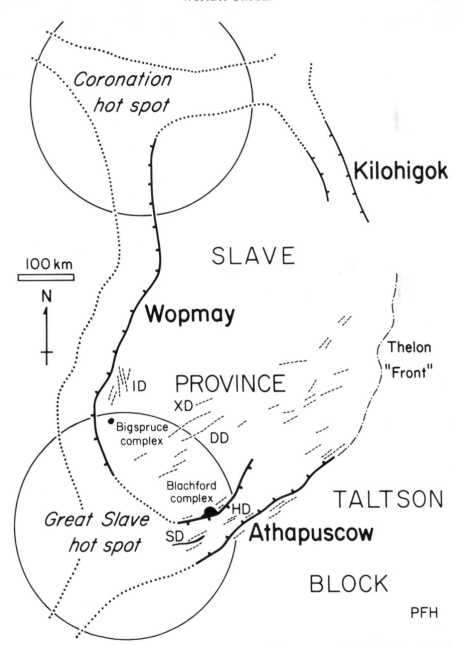

Figure 3. Hot spot model for the origin of the Wopmay, Athapuscow and Kilohigok rifts. Dotted rift margins are uncontrolled. Fine dashed lines are mafic dykes. Igneous units: ID – Indin Dykes, XD – "X" Dykes, DD – Dogrib Dykes, HD – Hearne Dykes, SD – Simpson Islands (alkaline) Dyke.

RIFTING, CONTINENTAL BREAKUP, AND DEVELOPMENT OF
AN ATLANTIC-TYPE CONTINENTAL MARGIN

The model presented here for the early stages of the Wopmay Wilson Cycle is not new but much of the supporting evidence is, especially that concerning magmatic activity during rifting. The comparison is based on the Cenozoic evolution of the classic Afro-Arabian rift system (Cloos, 1939; Gass, 1970; LeBas, 1971; Gass et al., 1978). It is postulated that about 2.1 Ga ago a relatively thick plate of continental lithosphere came to rest over a pair of deep mantle hot spots (Burke and Dewey, 1973; Dewey and Burke, 1974) located at present Great Slave Lake and Coronation Gulf (Fig. 3). The distance between the hot spots, about 800 km, exactly equals the average spacing of Neogene hot spots in Africa (Thiessen et al., 1979). It is believed that triple rift junctions evolved from each hot spot (Hoffman et al., 1974). A north-south trending rift connected the two hot spots and evolved to become the Wopmay Orogen. Two other rifts extended eastward from the hot spots into what would become the Athapuscow Aulacogen and Kilohigok Basin. The Athapuscow rift developed along the Thelon Front (Fig. 1), an older boundary separating Archean granite-greenstone belts of the Slave Province (McGlynn and Henderson, 1970) from probably Archean high-grade (granulite in part) gneisses, subsequently retrograded (Fraser et al., 1978), of the Taltson Block (new name), which extends southeastward for 400 km to the Cree Lake mobile zone of Saskatchewan (Lewry et al., 1978). Rifts are also believed to have extended westward from the hot spots but these are strictly hypothetical. The lithospheric plate, having been weakened by the hot spots and related rifts, responded to east-west tension of unspecified origin by breaking along the Wopmay and presumably also along the western rifts, creating a new and expanding ocean basin (Fig. 4). Following break-up, subsidence along the Wopmay continental margin led to the development of an Atlantic-type continental terrace (shelf-slope) and rise. In contrast, the Athapuscow and Kilohigok rifts failed to produce ocean basins, and instead became deeply-subsiding transverse troughs, along which sediment was channeled toward the continental margin. Some of the supporting evidence for this model comes from 1) igneous intrusions in the Western Slave Province, 2) rift volcanism and sedimentation, and 3) the continental terrace and rise.

Former hot spots and related rifts can be recognized on old eroded cratons by the presence of alkaline-peralkaline intrusive complexes and swarms of diabase dykes. Although knowledge of the western Slave Province is sketchy, three alkaline-peralkaline complexes have recently been described, all 2.1 ± 0.1 Ga in age. The most alkaline one is at Bigspruce Lake (Fig. 3) and contains nepheline syenites and carbonatite (Martineau and Lambert, 1974). The largest complex is at Blachford Lake and comprises an older, mildly alkaline suite and a younger peralkaline suite of plutons (Davidson, 1978). The third complex is a vertically differentiated dyke, composed mainly of potash-rich gabbro, which is exposed in a basement horst near the west end of Athapuscow Aulacogen (Badham, 1979b; Burwash and Cavell, 1980). The Hearne Dykes (new name) parallel the aulacogen for at least 240 km and comprise the largest dyke swarm. Although undated radiometrically, these dykes are

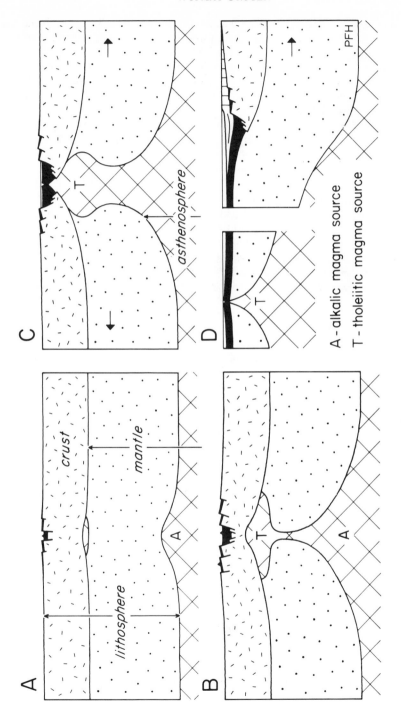

Figure 4. Lithosphere-asthenosphere evolution and magmatism during rifting (A-B), continental break-up (C), and subsidence of an Atlantic-type continental margin (D).

known to cut the Blachford complex (Davidson, 1978) and the alkaline gabbro dyke (Reinhardt, 1972), but are overlain unconformably by early rift sediments of the aulacogen (Hoffman *et al.*, 1977). East of Wopmay Orogen are the 2.1 Ga Indin Dykes and the less well dated Dogrib and "X" Dykes (McGlynn and Irvine, 1975). None of these igneous intrusions was considered when the hot spot model was originally proposed (Hoffman *et al.*, 1974) and therefore they constitute independent confirmation of the model.

There is also important new evidence for early rifting in Wopmay Orogen and Athapuscow Aulacogen. The westernmost thrust slices in Zone 3 of the orogen contains an 8 to 10 km thick succession called the Akaitcho Group, which underlies the continental rise facies of the Odjick Formation (Fig. 5) and contains major bimodal (basalt-rhyolite) volcanic rocks (Easton and Hoffman, 1980). The older part of this succession, the base of which is not exposed, consists of arkosic turbidites intruded by huge anastomosing sills of rhyolite porphyry. The younger part consists of intercalated basalt, rhyolite, and pelitic sediments, intruded by gabbro sills. The uppermost part may be correlative with basalt flows (Vaillant Formation) erupted directly onto Archean rocks at the base of the continental terrace succession (Fig. 5). All these basalts are tholeiitic, suggesting that by this time rifting was well advanced (Fig. 4). Alkaline basalt does occur in the Union Island Group, the oldest autochthonous rift succession in Athapuscow Aulacogen, occupying a narrow (15 km) strip along the Thelon Front. This succession is composed mainly of dolomite and very black shale, possibly lacustrine, in which there are two localized basalt units (Hoffman *et al.*, 1977). The older and more marginal basalt is alkaline, whereas the younger and more medial one is tholeiitic (Goff and Scarfe, 1976). Trace elements of both are typical of continental rift volcanism in the Cenozoic. There is another, more impressive, rift succession in the aulacogen, the Wilson Island Group. More than 6 km thick, this succession begins with voluminous bimodal lavas and granite-pebble conglomerate, followed by alluvial and deltaic clastics with westward paleocurrents (Hoffman *et al.*, 1977). However, this succession is entirely allochthonous and, until its age is determined radiometrically, it is impossible to know whether it is related to Wopmay Orogen or to an older (2.1 to 2.5 Ga) rifting event.

The pre-orogenic sedimentary history of Wopmay Orogen (Fig. 5) is comparable to that of Atlantic-type continental margins (Hoffman, 1973), although the rift succession underlies the inner continental rise rather than the outer shelf as in the North American Atlantic margin (Sheridan, 1974). Subsidence along such continental margins post-dates the onset of sea-floor spreading and results primarily from thermal contraction of the lithosphere (Sleep, 1971; Watts and Ryan, 1976; Turcotte and Ahern, 1977; Keen, 1979), complemented by sediment loading and possibly other factors (Bott, 1979). Examination of modern continental margins indicates that it takes about 50 million years to reach a stage of maturity comparable to that of the Wopmay continental margin. With a modest half-spreading rate of 2.5 cm per year, an ocean basin 2500 km wide would be produced in that time. The effects of greater radioactive heat production in the early Proterozoic would be to increase the rate of sea-floor spreading (Burke *et al.*, 1976) and decrease the rate of subsidence at the continental margin, thus increasing the estimated width of the ocean. What became of this ocean? That is related to the destruction of the continental margin.

WEST-DIPPING SUBDUCTION LEADING TO
CONTINENT-MICROCONTINENT COLLISION

Most previous attempts to account for the Wopmay orogeny (Hoffman, 1973; Hoffman and McGlynn, 1977; Badham, 1978b) have invoked Cordilleran models, presupposing east-dipping lithospheric subduction and west-dipping back-arc thrusting, without continental collision. Recent mapping, however, makes a comparison with Cenozoic collision zones, as favoured from the outset by Dewey and Burke (1973), much more promising (Hoffman, 1979). The collision model presented here involves west-dipping subduction of oceanic lithosphere, ending with collision of the Wopmay continental margin with a microcontinent (Fig. 6). The reasons for specifying a microcontinent as the accreted element are discussed later. At this point, the fundamental tectonic zonation of the orogen (Fig. 1) should be outlined.

Zone 1 comprises the autochthonous sediments of Takijuq and Tree Basins (Fig. 2). In both basins, thin (600m) equivalents of shelf sediments to the west are overlain by a relatively thin (800m), distal flysch (Fig. 5). In Takijuq Basin, the flysch is overlain by calcareous laminites, possibly evaporitic at the top, above which is a red

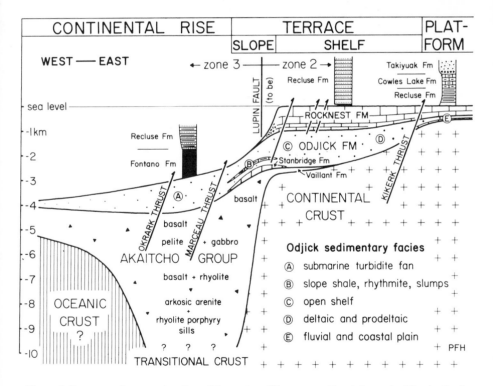

Figure 5. Pre-orogenic reconstruction of the mature Wopmay continental margin. The depth of the continental rise is drawn the same as in modern oceans. Subsequent orogenic deposits are shown in the columnar sections.

sandy molasse. Folding within these basins is related mainly to Laramide-type basement uplifts, the compressional axes of which strike northwest-southeast, oblique to
that of the orogen as a whole.

Zone 2 is allochthonous, having moved eastward relative to the autochthon on a
myriad of thrust faults, the most important of which is the Kikerk Thrust (new name)
at the east boundary of the zone. The thick (2-4 km) shelf sequence is overlain
sharply by a very thin (80 m) hemipelagic shale, which passes abruptly upwards into
a thick-bedded, sandy, feldspathic flysch, at least 1.5 km thick, in which paleocurrents are precisely axial to the tectonic folds and are from the north.

Zone 3 is a continuation of Zone 2 and is divided into three subzones by two
major east-vergent thrust faults, Okrark Thrust on the west and Marceau Thrust on
the east. The oldest rocks are the early rift deposits (Akaitcho Group) of the western
and central subzones. In the central subzone, these deposits are overlain by the
continental rise facies of the Odjick Formation, which is of a continental slope facies
in the eastern subzone. In both the central and eastern subzones, the Odjick Formation is overlain by about 1.5 km of hemipelagic shale, in which there are spectacular

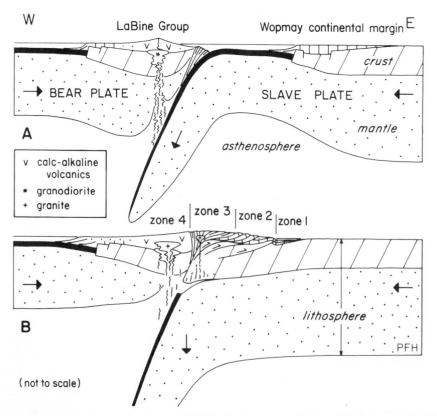

Figure 6. First collision in Wopmay Orogen, resulting in accretion of the Great Bear arc-bearing
microcontinent.

sedimentary breccias related to the Lupin Fault, a major west-side-down growth fault at the east boundary of the zone. The shale grades upward into flysch that is similar to but more distal than that in Zone 2. The thrusts formed relatively early in the structural sequence and were followed by very large, upright, kink-type folds. Syn- to post-tectonic plutons of the Hepburn Batholith in the central subzone, and Wentzel Batholith in the western subzone, began to be emplaced during tightening of the late kink folds. The early plutons are large tabular bodies of granite, but younger ones become progressively smaller, more discordant, and more basic in composition (Hoffman et al., 1980). Metamorphic isograds, ranging from biotite to sillimanite + orthoclase, are related to the early tabular granites. They occur both in "hot-side-up" and "hot-side-down" configurations, and indicate emplacement depths of 10 to 12 km (St-Onge and Hoffman, 1980). The two batholiths have most but not all of the general S-type characteristics (Chappell and White, 1974).

Zone 4 consists of little-metamorphosed, broadly folded, enormously thick calc-alkaline volcanic rocks, mostly ignimbrites, and intercalated continental sediments (Hoffman and McGlynn, 1977; Hoffman, 1978). They are intruded by large, tabular, epizonal, I-type plutons, mainly granitic in composition, some of which are co-magmatic with the volcanics and some younger. The folds trend northwest-southeast and are asymmetrical, so that deeper stratigraphic levels are exposed to the southwest, and higher levels to the northeast. An undated granitic-metamorphic basement complex is locally exposed on the west side of the zone (McGlynn, 1976, 1979).

The boundary between Zone 4 and Zone 3 is a fundamental discontinuity, the Wopmay Fault. This longitudinal transcurrent fault was apparently active during volcanism and marked the east boundary of the subsiding volcano-plutonic depression of Zone 4. The Wentzel Batholith is truncated by the fault, and the presumed westward extension of Zone 3 has probably been displaced northward and covered by volcanic rocks and sediments of Zone 4. Locally, the youngest of these strata onlap the Wentzel Batholith unconformably, showing that it was unroofed at that time. A few of the post-volcanic plutons of Zone 4 also occur east of the fault and a couple actually straddle it. Thus, the fault must have stopped moving before the end of plutonism.

Cenozoic collisional orogens are comparable both in the general sequence of tectonic events (Dewey, 1977; Roeder, 1979) and in the overall scale of their zonation (Fig. 7) to Wopmay Orogen. The first major event leading to the destruction of the Wopmay continental margin was the foundering of the continental terrace. This is indicated by the abrupt superposition of hemipelagic shale on the peritidal Rocknest shelf facies (Fig. 5) and was associated with the Lupin growth fault. This situation is comparable to the ongoing collision in the Banda Arc, where the north Australian continental terrace, broken by numerous growth faults, is being pulled into the Timor Trough by sinking of the north-dipping subducted slab of oceanic lithosphere beneath the Banda volcanic arc (Von der Borch, 1979). Homologous growth faults (Buchanan and Johnson, 1968) are well developed in the Ouachita (late Paleozoic) collisional orogen (Graham et al., 1975). The hemipelagic shale passes upward into feldspathic flysch of granitic-metamorphic provenance (Jeletsky, 1974). The flysch was deposited by turbidity currents that flowed southward along the trench or trough axis, apparently from an uplifted area of more advanced collision to the north. The events considered so far occurred just as predicted by the collision model. A more subtle

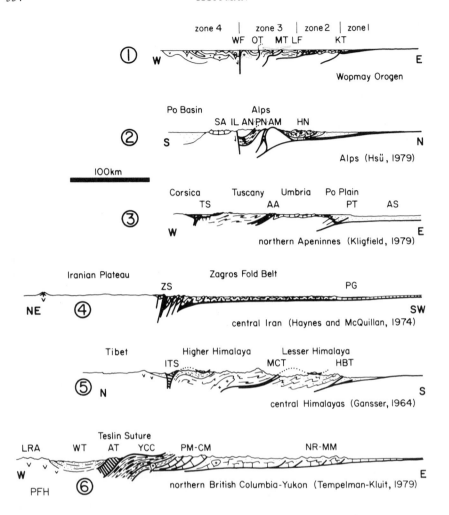

Figure 7. Comparison of diverse collisional orogens. All are fundamentally similar in origin and owe their differences principally to variable post-collision histories. 1) Wopmay Orogen: WF – Wopmay Fault, OT – Okrark Thrust, MT – Marceau Thrust, LF – Lupin Fault, KT – Kikerk Thrust. 2) Alps: SA – Southern Alps, IL – Insubric (Peri-Adriatic) Line, AN – Austro-Alpine nappes, PN – Pennine nappes, AM – Aar Massif, HN – Helvetic nappes. 3) Apennines: TS – Tyrrhenian Sea, AA – Alpi Apuane shear zone, PT – Po thrust, AS – Adriatic Sea. 4) Zagros: ZS – Zagros suture, PG – Persian Gulf. 5) Himalayas: ITS – Indus-Tsangpo suture, MCT – Main Central Thrust, HBT – Himalayan Boundary Thrust. 6) Canadian Cordillera: LRA – Lewes River arc, WT – Whitehorse Trough, AT – Anvil Range ophiolitic complex, YCC – Yukon cataclastic complex, PM-CM – Pelly and Cassiar Mountains, NR-MM – Northern Rockies and Mackenzie Mountains.

prediction, not yet confirmed, is that there should be an hiatus at or near the top of the Rocknest shelf, caused by emergence of the shelf as it passed over the outer swell of the subduction zone, a feature dictated by the flexural rigidity of the lithosphere (Molnar *et al.*, 1977; Caldwell *et al.*, 1977).

The next major event in Wopmay Orogen may be unparalleled in the Cenozoic. This was the intrusion of swarms of thick, weakly differentiated gabbro sills, apparently just before the onset of folding and thrusting. Stratigraphically, the sills are concentrated in the hemipelagic shale, but many intrude the underlying terrace sediments and the overlying flysch. Geographically, the greatest concentration of sills is just up-slope from the Lupin Fault, but the two features are not precisely contemporaneous. Fault movement occurred during hemipelagic shale deposition, presumably on the upper slope of the trench or trough, whereas the gabbros intrude flysch and could not, therefore, have been intruded until after the sediments passed below the trench or trough axis. Homologous mafic magmatism, generally tholeiitic and including extrusive equivalents, is a major component of the early Proterozoic Circum-Ungava Geosyncline (Dimroth *et al.*, 1970), and is significant in the late Proterozoic Damara orogen (Kröner, 1977) of Namibia and in the Dalradian of the early paleozoic Caledonides in Ireland and Scotland (Pitcher and Berger, 1972, p. 43-50). Possible Mesozoic (Roeder and Mull, 1978) and Cenozoic (Roeder, 1978) examples are rare, and only one example in an active trench (Kulm *et al.*, 1973) is known to me. If the collision model is correct, since the early Proterozoic there has been a decrease in the capacity of subducting lithosphere to release magma in trenches. This has obvious implications for Archean greenstone belts.

Deformation of the Wopmay continental margin deposits involved east-west shortening, vertical thickening, and eastward translation relative to the Slave craton. There is no empirical evidence for either gravity gliding (Lemoine, 1973) or gravitational spreading (Price, 1973) at the structural level now exposed. The two major thrust faults and related folds in Zone 3 appear to be the oldest compressional structures in the orogen. They predate the high-temperature metamorphism (St-Onge and Hoffman, 1980) and both thrust sheets were later kinked to form paired anticlines and synclines, with the synclines in the more external position. Neither they nor the thrusts in the western part of Zone 2 involve basement. On the other hand, Kikerk Thrust, far into the foreland, appears to be both younger and more deeply rooted. It cuts the calcareous laminites above the flysch (Fraser, 1974) and probably post-dates the molasse in Takijuq Basin. The basement exposed in Carousel Anticline (Fig. 2) is presumed to have moved with the Kikerk Thrust sheet, although this is uncertain because Zone 2 has not yet been systematically mapped. Overall, the pattern of deformation is very similar to that in Cenozoic collisions. Folds and thrusts, without basement involvement, are formed in flysch detached from the subducting plate before collision, as in the Makran (White, 1979), and in terrace sediments underlying flysch during the early stages of collision, as in the Zagros (Haynes and McQuillan, 1974; Colman-Sadd, 1978). Roeder (1973, 1979) and Dewey (1977) describe how flat thrusts become sigmoidally kinked when plate convergence continues after collision. Continued convergence also leads to crustal underthrusting in the foreland, as in the Himalayas (LeFort, 1975; Warsi and Molnar, 1977; Bird, 1978), and perhaps also in the Alps (Hsü, 1979) and the Apennines (Kligfield, 1979). Kikerk Thrust may be analogous to the Himalayan Boundary Thrust but there is much less crustal stacking in Wopmay Orogen than in the Himalayas, which is exceptional in this respect among Cenozoic collision orogens.

The syn- to post-tectonic plutons in Zone 3 of Wopmay Orogen are "Hercynotype" (Pitcher, 1979) in character and setting. Similar but less abundant plutons occur in those Cenozoic collision orogens sufficiently uplifted to reveal them, such as the Alps (Hsü, 1979) and the Apennines (Kligfield, 1979). Kikerk Thrust may be analogous to the Himalayan Boundary Thrust but there is much less crustal stacking 1976; Roeder, 1978), but there are various explanations for the required heating in an environment said to be cold in classical plate tectonics. The explanations include: 1) crustal thickening (Dewey and Burke, 1973), 2) shear heating along crustal underthrusts (Andrieux *et al.*, 1977; Carmignani *et al.*, 1978), 3) delamination of the subducting lithosphere, causing injection of hot asthenosphere along the base of the crust (Bird, 1978), and 4) breakaway of the subducted slab, causing general upwelling of asthenosphere in the collision zone (Bird *et al.*, 1975).

The extensive area of "hot-side-up" metamorphism in Wopmay Orogen is yet another point of comparison with the Himalayas (Pecher, 1975; LeFort, 1976; Thakur, 1977), where there is disagreement as to its significance. In Wopmay Orogen the "hot-side-up" metamorphism does not result from structural inversion. St-Onge and Hoffman (1980) relate it to the floor of Hepburn Batholith, now largely eroded away. They note that the anatectic melt must have originated at depths exceeding the 10 km at which the batholith was emplaced (St-Onge and Carmichael, 1979), because of the presence of garnet in most of the plutons (Green and Ringwood, 1968; Brown and Fyfe, 1970; Green, 1976).

The continental calc-alkaline volcanic rocks and post-tectonic I-type plutons of Zone 4 in Wopmay Orogen are comparable to magmatism affecting the overriding plate in Cenozoic collisions. Ideally, as in the Turkish-Iranian Plateau (Sengör and Kidd, 1979), this occurs in two stages. The first is the well-known Andean calc-alkaline activity that occurs before collision in bands where actively subducting lithosphere is 100 to 150 km beneath the surface. This type of arc magmatism may be absent, as in the Alps, where the width of oceanic lithosphere consumed is less than about 500 km (Dewey *et al.*, 1973). The second stage occurs after collision, where there is no longer any lithospheric subduction, and most of the magmas are generated by deep crustal anatexis (Dewey and Burke, 1973; Toksöz and Bird, 1977). Although not well studied, relatively silica- and potash-rich calc-alkaline rocks, commonly ignimbritic, and minor alkaline basalts appear to be a typical association (Kidd, 1975; Sengör and Kidd, 1979).

A similar two-fold division may be present in Wopmay Orogen, although the available geochemistry and geochronology is strictly of a reconnaissance nature. The volcanic succession in Zone 4 is divided into two groups separated by a disconformity, an older LaBine Group and a younger Sloan Group. Most of the lower volcanics of the LaBine Group are tholeiitic basalt (McGlynn, 1976, 1979; Wilson, 1979), variably up to 1.5 km thick, on top of which 2 km of rhyolite ignimbrite was erupted concurrently with the subsidence of major calderas. The LaBine Group ends with repeated andesite-dacite-rhyolite cycles, 5 km thick in total, which can be related to resurgent doming in the calderas and emplacement of shallow quartz monzodiorite-granodiorite-granite plutons (Hilderbrand, 1979, and pers. commun.) The LaBine Group is exposed only on the west side of Zone 4, making it impossible to determine possible arc polarity from regional variation in the potash/silica ratio. In the central and eastern parts of the zone, it is buried by the enormously thick and nearly uninterrupted dacite-rhyodacite-rhyolite ignimbrites of the Sloan Group. The highest part of this succession (Dumas Subgroup), which is preserved only along the Wopmay Fault, also contains weakly alkaline (potassic) basalts (Easton, pers. commun.). The Sloan Group is intruded by large tabular granodiorite-granite plutons, probably coeval with the ignimbrites, and by distinctly younger syenogranites and small diorites (Hoffman and McGlynn, 1977) a few of which scatter east of Wopmay Fault into Zone 3.

Badham (1973) has argued for an Andean environment (but with east-dipping subduction) for parts of the LaBine Group, whereas the upper Sloan Group was deposited on the unroofed Wentzel Batholith of Zone 3 and therefore both it and the late plutons must post-date the collision, in agreement with the two-stage model. However, Badham (1979a) challenges the collision model, noting that the lack of deep erosion in Zone 4 precludes a high plateau like Tibet. But Tibet owes its extreme (5 km) elevation primarily to the unusually long history of subduction beneath it (Toksöz and Bird, 1977), and other overriding plates have highly variable elevations after collision. The Turkish-Iranian Plateau averages only 1.1 km above sea level and the Pannonian Basin, located on the overriding plate of the Oligocene collision in the Carpathians (Burchfiel, 1980), had typical post-collision volcanism (calc-alkaline andesite and rhyolite ignimbrite in the Miocene-Pliocene and basalt in the Pliocene-Pleistocene) and has subsided up to 5 km below sea level (Stegena *et al.*, 1975; Horvath and Stegena, 1977). Therefore, the lack of deep erosion in Zone 4 of Wopmay Orogen does not rule out collision.

Although the overall zonation of Wopmay Orogen is consistent with a simple collision model, problems remain. Where, for example, is the suture? Where are the oceanic rocks that should be found in the highest thrusts of the subducting plate? Where is the fore-arc sedimentary basin that should be near the front of the overriding plate? What is the significance of the Wopmay Fault? These and other problems lead to speculation that more than two plates may be involved and that what is exposed of Wopmay Orogen may be only part of a much larger compound orogenic system.

THE ROLE OF TRANSCURRENT FAULTING IN THE OROGEN AND THE POSSIBILITY OF A REMOTE SECOND COLLISION

Most previous tectonic models for Wopmay Orogen assumed an east-dipping, not west-dipping, subduction zone beneath Zone 4. Perhaps this view should not be totally abandoned. There are several features that can be most readily explained if the initial collision was followed by the generation of a second subduction zone, dipping to the east and located on the west side of the accreted microcontinent (Fig. 8). Generation of the second subduction zone would follow the breakaway and descent of the earlier west-dipping slab. The new subduction zone could eventually lead to closure of the ocean west of the accreted microcontinent, resulting in another, perhaps terminal, continental collision. This collision zone would be cryptically located west of Zone 4, but perhaps its distant effects can be seen in the Shield. Among the features that can be accounted for in the three-plate model are: 1) calc-alkaline intrusions in Athapuscow Aulacogen, 2) the Wopmay Fault, and 3) late conjugate transcurrent faulting of supraregional extent.

The simple collision model fails to account for a string, 250 km long, of prominent calc-alkaline laccoliths in Athapuscow Aulacogen (Hoffman *et al.*, 1977; Badham, 1978a). The vast majority of these intrusions were emplaced at the base of a thick evaporite collapse breccia, which underlies a molasse that is correlative with that in Zone 1 of Wopmay Orogen. The intrusions are coeval with or younger than the molasse, which was probably deposited during the unroofing of Zone 3 in the orogen. The problem posed by these intrusions is that they are located east of the orogen and therefore on the subducting plate, whereas calc-alkaline rocks generally occur on overriding plates. There seems nothing in the evolution of the aulacogen

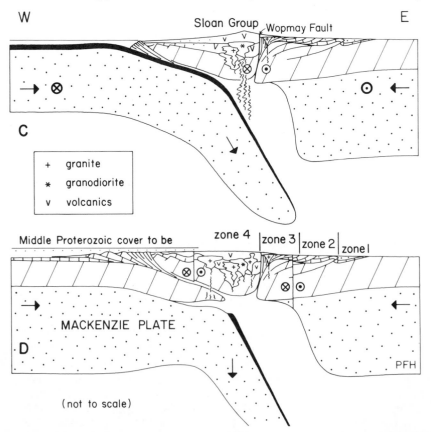

Figure 8. Renewed subduction west of the accreted microcontinent, leading to terminal continental collision. The collision suture is possibly marked by the Keith Arm gravity high (see Fig. 9).

itself that would have triggered these intrusions. Furthermore, there is a distinct change in composition of the intrusions from diorite in the western half of the aulacogen to quartz monzonite (IUGS definition) in the eastern half, a change suggestive of an east-dipping subduction zone. But why have very few of these intrusions been found outside the aulacogen? Perhaps it is because they achieve their broad (up to 25 km) laccolithic form only at high stratigraphic levels, and those few that have been found in the basement rocks exposed outside the aulacogen are very small (Davidson, 1978). Alternatively, if the lithosphere beneath the aulacogen was still relatively thin (Fig. 4), the asthenospheric wedge above the subducting slab, where calc-alkaline magmas are generated (Anderson et al., 1978), would be thicker than elsewhere in the foreland. This could be critical if the dip of the subducting slab was shallow, as indeed it must have been if the entire string of laccoliths was generated at slab depths of 100 to 150 km.

In the simple collision model, the boundary between Zones 3 and 4 should be an intercontinental suture. The boundary is in fact the Wopmay Fault, which runs the entire exposed length of the orogen (Fig. 1). The singularity and relative straightness of this fault are comparable to the kind of longitudinal transcurrent faults that have been scything the western Cordilleran Orogen since the Mesozoic (Monger and Price, 1979) and which are fundamentally related to oblique convergence (or the special case of pure strike-slip) across an arc-trench system (Fitch, 1972). As mentioned previously, the Wopmay Fault was active during Sloan Group volcanism but had ceased moving before the late syenogranite and diorite plutons. Movement included a major dip-slip component, west-side-down, and probably a dextral transcurrent component. The sense of transcurrent motion is suggested by the oblique northwest-southeast trend of folds (Wilcox et al., 1973) in Zone 4, an obliquity that cannot be entirely attributed to rotation during the later period of conjugate transcurrent faulting (Freund, 1970) still to be discussed. The fault severely truncates Wentzel Batholith and the regional structure of Zone 3. This and the fact that outliers of the Sloan Group lie unconformably on Wentzel Batholith and high-grade metamorphics of Zone 3, proving that they were already unroofed, suggests that fault movement post-dates the collision. This can be easily accommodated in the three-plate model by assuming oblique covergence across the east-dipping subduction zone, with the oceanic plate moving northeastward relative to the continental plate (Fig. 8). In this model, the western extension of Zone 3 and the suture itself have been carried northward by movement on the Wopmay Fault and buried by the Sloan Group. The Sloan Group could be the product of east-dipping subduction rather than of strictly post-collisional activity as in the simple collision model. Perhaps it also covers the fore-arc sedimentary basin of the original west-dipping subduction zone. Thus, in the three-plate model, the Wopmay Fault is interpreted as a transcurrent fault related to oblique east-dipping subduction that post-dates the first collision. The fault truncates and has displaced the suture zone of the first collision, which is perhaps now covered by the younger Sloan Group volcanics. The extraordinarily profuse and complex magmatism of Zone 4 is related to its unique location above two successive and oppositely dipping subduction zones, one or both of which led to continental collision.

What happened to the ocean west of the accreted microcontinent? There is evidence in the northwest of the Shield and beyond of another east-west compressional event that could reflect a terminal collision to the west. In Wopmay Orogen, there is an impressive system of conjugate transcurrent faults (Fig. 2), northwest left-slip and northeast right-slip, that post-dates even the late syenogranites and diorites. This is the last compressional event in the orogen and all younger structures, including dip-slip reactivation of the conjugate transcurrent faults, indicate episodic east-west extension during the middle Proterozoic (Hoffman, 1980). The system of conjugate transcurrent faults is comparable to that developed in central Europe in response to Alpine compression (Ahorner, 1970; Illies, 1974; Sengör, 1976) and in southeastern Iran during Cenozoic convergence, probably collisional, between the Lut and Helmand cratonic blocks (Freund, 1970). In western Canada, the conjugate transcurrent faults may extend far beyond Wopmay Orogen (Fig. 9). The major left-slip (140 km) faulting in Kilohigok Basin (Campbell and Cecile, 1979) and right-slip (more than 95 km) faulting in Athapuscow Aulacogen (Reinhardt, 1969; Thomas, et al., 1976; Hoffman et al., 1977) appear to be of the same age. The general absence of such faults elsewhere in the Slave Province (McGlynn, 1977) suggests that it behaved as a rigid cratonic block that forcibly indented the Churchill Province (Gibb, 1978). This indentation may have cracked the Churchill lithosphere, giving rise to the 1.8 Ga alkaline volcanism of the Dubawnt Group and related west-northwest trending lamprophyre dykes (LeCheminant et al., 1979b; LeCheminant et al., 1979a; Blake, in press). West of the Shield, several of the larger transcurrent faults can be traced in

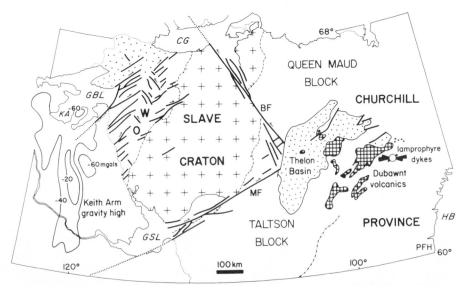

Figure 9. Conjugate transcurrent faults (heavy lines) caused by east-west compression related to a terminal continental collision suture possibly located along the Keith Arm gravity high. The relatively rigid Slave craton has indented and cracked the Churchill Province lithosphere, giving rise to the Dubawnt volcanics and lamprophyres. Dyke orientation data courtesy of A.N. LeCheminant. Gravity data from Earth Physics Branch (1974). Faults: BF – Bathurst Fault, MF – McDonald Fault. Locations: CG – Coronation Gulf, GBL – Great Bear Lake, KA – Keith Arm, GSL – Great Slave Lake, HB – Hudson Bay.

the subsurface to the Cordilleran mountain front (Burwash *et al.*, 1964; Balkwill, 1971) and their subsequent reactivation may have locally controlled sedimentation in the Cordilleran foreland (Sikabonyi and Rodgers, 1959; Lis and Price, 1976). The dominance of northeast over northwest transcurrent faults, especially west of the Slave Province, requires an overall component of right-lateral shear (Arthaud and Matte, 1977). This is consistent with the obliquity of east-dipping subduction leading to the proposed collision. The total extent of the conjugate transcurrent faults is unknown, but must be at least 1000 km wide, and could reach as far south as the Colorado Lineament (Warner, 1978).

How are these comparative models for the evolution of Wopmay Orogen to be tested? There is a parable, attributed to Rudolf Trümpy, that an evolutionary model for the Alps was devised without the use of fossils and that the model worked well . . . until the first fossil was found! The situation in Wopmay Orogen in regard to geochronology is somewhat similar. Except for placing the beginning and the end of the proposed Wilson Cycle at about 2.1 and 1.8 Ga, the evolutionary model presented here was arrived at virtually without reference to geochronology. Geochronology could be decisive, for example, in determining to what extent magmatism in Zone 4 pre-dates or post-dates the syn-tectonic granites of Zone 3. The orogen will be a severe challenge to geochronology because, unless Precambrian tectonics operated in slow-motion compared with the Cenozoic, differences of 10 to 20 Ma will be critical. The comprehensive geochronological investigation of the orogen initiated in 1979 by W.R. Van Schmus (University of Kansas) should provide a definitive test for the comparative tectonic approach of this paper.

DISCUSSION

In plate tectonics, the subduction of oceanic lithosphere is fundamental to orogeny. Although uncertainties remain, it appears that the principal driving force is the downward pull of relatively dense subducting slabs (Richer, 1977). In the earliest stages of orogeny involving the Wopmay continental margin, there is a clear indication that subduction was the operating mechanism. It is known that shallow-water carbonate sedimentation is capable of keeping pace with very rapid rates of subsidence, rates in excess of 1 km per million years (Schlager, 1979). Therefore, the foundering of the Rocknest carbonate shelf (Fig. 5), the first event in the destruction of the continental margin, must have been very abrupt. The foundering was accompanied by deposition of a very thin hemipelagic shale, so sediment loading can be ruled out as a causal factor. As the shelf foundered, the Lupin growth fault developed as a result of east-west tension at the upper continental slope. These facts indicate that the foundering was caused by a sudden and profound downward pull from the west. Descent into a west-dipping subduction zone seems to be the only appropriate and sufficient mechanism.

An observation that is commonly cited as evidence against plate tectonics in the Precambrian (and years ago the same argument was used in the Phanerozoic) is the presence of sialic basement beneath the orogen. No such basement is exposed in Zone 3 at the north end of Wopmay Orogen, but it does occur in the eastern part of the zone to the south (Frith *et al.*, 1974; Frith *et al.*, 1977). However, this alone does not imply that the sediments of Zone 3 were originally deposited on sialic basement.

There has probably been at least 50 km of tectonic shortening, without basement involvement, in the sediments of Zone 2. Therefore, if thick sialic basement originally extended as far as the western edge of Zone 2, it must have subsequently underthrusted Zone 3 for at least 50 km. It is not surprising that sialic basement exists beneath Zone 3, but it was tectonically emplaced and cannot be used as evidence that the orogen was originally ensialic. In this context, it is worth noting that almost the entire mass of the Himalayas consists of sialic basement and ensialic sediments (Gansser, 1964), despite the fact that more than 5000 km of oceanic lithosphere was consumed between India and Eurasia (Powell, 1979).

The reversal in polarity of a subduction zone following accretion of a microcontinental arc is not unique to Wopmay Orogen. A virtually identical model has been proposed for the Grampian Orogeny (early Ordovician) of the Scottish Caledonides (Mitchell, 1978). In the northern Canadian Cordillera, a Mesozoic reversal from west- to east-dipping subduction following accretion of the Stikine Block has been documented by Tempelman-Kluit (1979). And in the riot of collisions along the southern flank of Eurasia, the older north Tethyan suture resulted from predominantly south-dipping subduction (Sengör, 1979), whereas the younger and still active south Tethyan subduction zone is north-dipping.

Are there any differences between early Proterozoic and Phanerozoic tectonics? It is dangerous to generalize from only one orogen, but magmatism, in a variety of tectonic settings, seems to have been somewhat more prolific in Wopmay Orogen than in the Phanerozoic. The early rift volcanics of the Akaitcho Group are unusually thick, 8 to 10 km including intercalated sediments, although there are few examples in the Phanerozoic where this facies is so well preserved as in Wopmay Orogen. As noted earlier, mafic tholeiites derived from subducting plates in trenches seem to be very common in the Proterozoic but become progressively rarer in the Phanerozoic. The syntectonic batholiths generated during the first Wopmay collision (Zone 3) are perhaps more voluminous and span a greater compositional range, granite to pyroxenite, than is typical of the Phanerozoic. The felsic calc-alkaline rocks of the overriding plate (Zone 4) are extraordinarily voluminous, perhaps uniquely so, and are of a general type that is abundant on several shields in the mid-Proterozoic (1.3 to 1.9 Ga). It is tempting to relate the seemingly more prolific magmatism to a warmer, more radioactive lithosphere in the early Proterozoic. If this were true, the lithosphere should also have been thinner. However, this is not borne out by evidence from intracratonic alkaline magmatism. The 2.2 Ga nepheline syenites and carbonatites of the Bigspruce Lake complex and the 1.8 Ga trachytes and lamprophyres of the Dubawnt Group suggest respective depths of 100 to 150 km and 150 to 200 km to the base of the lithosphere at the time of magma generation (Gass et al., 1978). At that time, the respective cratons were not more than 0.4 Ga and 0.8 Ga old, indicating a rate of thickening of cratonic lithosphere (Jordan, 1978, 1979) not much different from that in the last billion years of Earth history.

Why is it that most geologists working in other early Proterozoic orogens (e.g., Dimroth, 1972; Sims, 1976; Gee, 1979) have been reluctant to apply the Wilson Cycle? I suspect it is because, until recently, the nature of Cenozoic collisions was not well understood. It is encouraging to note that this understanding is now being widely applied to the late Proterozoic Pan-African orogens (e.g., Al-Shanti and Mitchell, 1975; Caby, 1978; Porada, 1979; Pedreira, 1979).

ACKNOWLEDGMENTS

It is a pleasure to acknowledge the outstanding efforts of my senior mapping assistants over the years, I.R. Bell, M.P. Cecile, R.M. Easton, R.S. Hildebrand, M.R. St-Onge and R. Tirrul. I have also had valuable discussions with F.H.A. Campbell, A. Davidson, A.N. LeCheminant and J.C. McGlynn. I wish to record my gratitude to Marie and David Eby, and to Lillian and John McGlynn for their generosity and kindness to me while I was working on this paper. The manuscript was critically read by D.M. Carmichael, A. Davidson, J.C. McGlynn and two anonymous reviewers.

REFERENCES

Ahorner, L., 1970, Seismo-tectonic relations between the graben zones of the Upper and Lower Rhine Valley: in Illies, J.H. and Mueller, S., eds., Graben Problems: E. Schweizerbart'sche Verlagsbuchhandlung, Stuttgart, p. 155-166.

Al-Shanti, A.M.S. and Mithcell, A.H.G., 1975, Late Precambrian subduction and collision in the Al Amar-Idsas region, Arabian Shield, Kingdom of Saudi Arabia: Tectonophysics, v. 31, p. T41-T47.

Anderson, R.N., DeLong, S.E., and Schwarz, 1978, Thermal model for subduction with dehydration in the downgoing slab: Jour. Geol. v. 86, p. 731-739.

Andrieux, J., Brunel, M. and Hamet, J., 1977, Metamorphism, granitization and relations with the Main Central Thrust in central Nepal: $^{87}Rb/^{87}Sr$ age determinations and discussion: in Himalaya: Sciences de la Terre: Colloques Internationaux du Centre de la Recherche Scientifique, No. 268, Paris, p. 31-39.

Arthaud, F. and Matte, P., 1977, Late Paleozoic strike-slip faulting in southern Europe and northern Africa: result of a right-lateral shear zone between the Appalachians and the Urals: Geol. Soc. America Bull., v. 88, p. 1305-1320.

Badham, J.P.N., 1973, Calc-alkaline volcanism and plutonism from the Great Bear batholith, N.W.T.: Canadian Jour. Earth Sci. v. 10, p. 1319-1328.

_____, 1978a; Magnetite-apatite-amphibole-uranium and silver-arsenide mineralizations in lower Proterozoic igneous rocks, East Arm, Great Slave Lake, Canada: Econ. Geol., v. 73, p. 1474-1491.

_____, 1978b, Has there been an oceanic margin to western North America since Archean time?: Geology, v. 6, p. 621-625.

_____, 1979a, Reply on "Has there been an oceanic margin to western North America since Archean time?": Geology, v. 7, p. 227.

_____, 1979b, Geology and petrochemistry of lower Aphebian (2.4-2.0 Ga) alkaline plutonic and hypabyssal rocks in the East Arm of Great Slave Lake, Northwest Territories: Canadian Jour. Earth Sci., v. 16, p. 60-72.

Balkwill, H.R., 1971, Reconnaissance geology, southern Great Bear Plain, District of Mackenzie: Geol. Survey Canada Paper 71-11, 47 p. (complete with Map 5-1971).

Bird, P., 1978, Initiation of intracontinental subduction in the Himalaya: Jour. Geophys. Research, v. 83, p. 4975-4987.

Bird, P., Toksöz, M.N., and Sleep, N.H., 1975, Thermal and mechanical models of continent-continent covergence zones: Jour. Geophys. Research, v. 80, p. 4405-4416.

Blake, D.H., 1980, Volcanic rocks of the Paleohelikian Dubawnt Group in the Baker Lake-Angikuni Lake areas, District of Keewatin, N.W.T.: Geol. Survey Canada Bull. 309., in press.

Bott, M.H.P., 1979, Subsidence mechanisms at passive continental margins: in Watkins, J.S., Montadert, L. and Dickerson, P.W., eds., Geological and Geophysical Investigations of

Continental Margins: American Assoc. Petroleum Geol. Memoir 29, p. 3-9.

Brown, G.C. and Fyfe, W.S., 1970, The production of granitic melts during ultrametamorphism: Contrib. Mineral. Petrol., v. 28, p. 310-318.

Buchanan, R.S. and Johnson, F.K., 1968, Bonanza gas field – a model for Arkoma Basin growth faulting: in Cline, L.M., ed., Geology of the western Arkoma Basin and Ouachita Mountains, Oklahoma: Oklahoma City Geol. Soc. Guidebook, p. 75-85.

Burchfiel, B.C., 1980, East European Alpine system and the Carpathian orocline as an example of collision tectonics: Tectonophysics, in press.

Burke, K. and Dewey, J.F., 1973, Plume-generated triple junctions: key indicators in applying plate tectonics to old rocks: Jour. Geol., v. 86, p. 406-433.

Burke, K., Dewey, J.F. and Kidd, W.S.F., 1976, Dominance of horizontal movements, arc and microcontinental collisions during the later permobile regime: in Windley, B.F., ed., The Early History of the Earth: New York, John Wiley and Sons, p. 113-130.

Burwash, R.A. and Cavell, P.A., 1980, Uranium-thorium enrichment in alkaline olivine basalt magma, Simpson Islands Dyke, Northwest Territories, Canada: Contrib. Mineral. Petrol. in press.

Burwash, R.A., Baadsgaard, H., Peterman, Z.E. and Hunt, G.H., 1964, Precambrian: in McCrossan, R.G. and Glaister, R.P., eds., Geological History of Western Canada: Alberta Soc. Petroleum Geol., Calgary, p. 14-19.

Caby, R., 1978, Paleogynamique d'une marge passive et d'une marge active au Precambrien superieur: leur collision dans la chaine panafricaine du Mali: Bulletin de la Societe geologique de la France, v. 20, p. 857-861.

Caldwell, J.G., Turcotte, D.L., Haxby, W.F., and Karig, D.E., 1977, Thin elastic plate analysis of outer rises: in Talwani, M. and Pitman III, W.C., eds., Island Arcs, Deep Sea Trenches and Back-Arc Basins: American Geophys. Union, Washington, p. 467.

Campbell, F.H.A. and Cecile, M.P., 1976a, Geology of the Kilohigok Basin, Bathurst Inlet, N.W.T.: in Report of Activities, Part A: Geol. Survey Canada Paper 76-1A, p. 369-377.

_____, 1976b, Geology of the Kilohigok Basin, District of Mackenzie: Geol. Survey Canada, Open File Map 342 (1:500,000 scale).

_____, 1979, The northeastern margin of the Aphebian Kilohigok Basin, Melville Sound, Victoria Island, District of Franklin: in Current Research, Part A: Geol. Survey Canada Paper 79-1A, p. 91-94.

Carmignani, L., Giglia, G. and Kligfield, R., 1978, Structural evolution of the Apuane Alps: an example of continental margin deformation in the northern Apennines, Italy: Jour. Geol., v. 86, p. 487-504.

Chappell, B.W. and White, A.J.R., 1974, Two contrasting granite types: Pacific Geol., v. 8, p. 173-174.

Cloos, H., 1939, Hebung-Spaltung-Vulcanismus: Geologische Rundschau, v. 30, p. 405-527.

Colman-Sadd, S.P., 1978, Fold development in Zagros simply folded belt, southwest Iran: American Assoc. Petroleum Geol. Bull., v. 62, p. 984-1003.

Davidson, A., 1978, The Blachford Lake Intrusive Suite: An Aphebian alkaline plutonic complex in the Slave Province, Northwest Territories: in Current Research, Part A: Geol. Survey Canada Paper 78-1A, p. 119-127.

Dewey, J.F., 1977, Suture zone complexities: a review: Tectonophysics, v. 40, p. 53-67.

Dewey, J.F. and Burke, K.C.A., 1973, Tibetan, Variscan, and Precambrian basement reactivation: products of continental collision: Jour. Geol., v. 81, p. 683-692.

_____, 1974, Hot spots and continental break-up: implications for collisional orogeny: Geology, v. 2, p. 57-60.

Dewey, J.F., Pitman III, W.C., Ryan, W.B.F. and Bonin, J., 1973, Plate tectonics and the evolution of the Alpine system: Geol. Soc. America Bull., v. 84, p. 3137-3180.

Dimroth, E., 1972, The Labrador geosyncline revisited: American Jour. Sci., v. 272, p. 487-506.

Dimroth, E., Baragar, W.R.A., Bergeron, R. and Jackson, G.D., 1970, The filling of the Circum-Ungava Geosyncline: in Baer, A.J., ed., Symposium on Basins and Geosynclines of the Canadian Shield: Geol. Survey Canada Paper 70-40, p. 45-142.

Earth Physics Branch, 1974, Bouguer Anomaly Map of Canada: Gravity Map Series No. 74-1, Department of Energy Mines and Resources, Ottawa (scale 1:5,000,000).

Easton, R.M. and Hoffman, P.F., 1980, Stratigraphy, structure and geochemistry of the Akaitcho Group (early Proterozoic), north-central Wopmay Orogen, Hepburn Lake map-area (86J), District of Mackenzie: in Current Research, Part B: Geol. Survey Canada Paper 80-1B, in press.

Fitch, T.J., 1972, Plate convergence, transcurrent faults and internal deformation adjacent to southeast Asia and the western Pacific: Jour. Geophys. Research, v. 77, p. 4432-4460.

Fraser, J.A., 1974, The Epworth Group Rocknest Lake area, District of Mackenzie: Geol. Survey Canada Paper 73-39, 23 p., complete with Map 1384A.

Fraser, J.A., Heywood, W.W., and Mazurski, M.A., 1978, Metamorphic map of the Canadian Shield: Geol. Survey Canada Map 1475A.

Fraser, J.A., Hoffman, P.F., Irvine, T.N., and Mursky, G., 1972, The Bear Province: in Price, R.A. and Douglas, R.J.W., eds., Variations in Tectonic Styles in Canada: Geol. Assoc. Canada Spec. Paper 11, p. 453-504.

Freund, R., 1970, Rotation of strike-slip faults in Sistan, southeast Iran: Jour. Geol., v. 78, p. 188-200.

Frith, R., Frith, R.A., and Doig, R., 1977, the geochronology of the granitic rocks along the Bear-Slave structural province boundary, northwest Canadian Shield: Canadian Jour. Earth Sci., v. 14, p. 1356-1373.

Frith, R.A., Frith, R., Helmstaedt, H., Hill, J. and Leatherbarrow, R., 1974, Geology of the Indin Lake area, District of Mackenzie: in Report of Activities, Part A: Geol. Survey Canada Paper 74-1A, p. 165-172.

Gansser, A., 1964, Geology of the Himalayas: Interscience Publishers, New York, John Wiley and Sons, 289 p.

Gass, I.G., 1970, The evolution of volcanism in the junction area of the Red Sea, Gulf of Aden and Ethiopian rifts: Phil. Trans. Royal Soc. London, Series A, v. 267, p. 369-381.

Gass, I.G., Chapman, D.S., Pollack, H.N. and Thorpe, R.S., 1978, Geological and geophysical parameters of mid-plate volcanism: Phil. Trans. Royal Soc. London, Series A, v. 288, p. 581-597.

Gee, R.D., 1979, Structure and tectonic style of the Western Australian Shield: Tectonophysics, v. 58, p. 327-369.

Gibb, R.A., 1978, Slave-Churchill collision tectonics: Nature, v. 271, p. 50-52.

Goff, S.P. and Scarfe, C.M., 1976, Volcanological evidence for the origin of the East Arm, Great Slave Lake, N.W.T.: Program with Abstracts, Geol. Assoc. Canada, v. 1, p. 47.

Graham, S.A., Dickinson, W.R. and Ingersoll, R.V., 1975, Himalayan-Bengal model for flysch dispersal in the Appalachian-Ouachita system: Geol. Soc. America Bull., v. 86, p. 273-286.

Green, T.H., 1976, Experimental generation of cordierite- or garnet-bearing granitic liquids from a pelitic composition: Geology, v. p. 85-88.

Green, T.H. and Ringwood, A.E., 1968, Origin of garnet phenocrysts in calc-alkaline rocks: Contrib. Mineral. Petrol., v. 18, p. 163-174.

Hamet, J. and Allegre, C.J., 1976, Rb-Sr systematics in granite from central Nepal (Manaslu): significance of the Oligocene age and high $^{87}SR/^{86}Sr$ ratio in Himalayan orogeny: Geology, v. 4, p. 470-472.

Haynes, S.J. and McQillan, H., 1974, Evolution of the Zagros suture zone, southern Iran: Geol. Soc. America Bull., v. 85, p. 739-744.

Hildebrand, R.S., 1979, Litho-stratigraphic map of the northeast $^{1}/_{2}$ of Great Bear Lake: Department of Indian and Northern Affairs, Yellowknife, N.W.T.: Econ. Geol. Series Map 1979-3.

Hoffman, P.F., 1973, Evolution of an early Proterozoic continental margin: the Coronation geosyncline and associated aulacogens of the northwestern Canadian shield: Phil. Trans. Royal Soc. London, Series A, v. 273, p. 547-581.

Hoffman, P.F., 1977, Geology of the East Arm of Great Slave Lake, District of Mackenzie: Geol. Survey Canada Open File Map 475 (1:50,000 scale).

Hoffman, P.F., 1978, Geology of the Sloan River map-area (86K), District of Mackenzie: Geol. Survey Canada Open File Map 535 (1:125,000 scale).

Hoffman, P.F., 1979, Comment on "Has there been an oceanic margin to western North America since Archean time?": Geology, v. 7, p. 226.

Hoffman, P.F., 1980, Conjugate transcurrent faults in north-central Wopmay Orogen (early Proterozoic) and their dip-slip reactivation during post-orogenic extension, Hepburn Lake map-area (86J), District of Mackenzie: in Current Research, Part A: Geol. Survey Canada Paper 80-1A, p. 183-185.

Hoffman, P.F. and McGlynn, J.C., 1977, Great Bear Batholith: a volcano-plutonic depression: in Baragar, W.R.A., Coleman, L.C. and Hall, J.M., eds., Volcanic Regimes in Canada: Geol. Assoc. Canada Spec. Paper 16, p. 170-192.

Hoffman, P.F., Bell, I.R., Hildebrand, R.S. and Thorstad, L., 1977, Geology of the Athapuscow Aulacogen, East Arm of Great Slave Lake, District of Mackenzie: in Report of Activities, Part A: Geol. Survey Canada Paper 77-1A, p. 117-129.

Hoffman, P.F., Dewey, J.F. and Burke, K., 1974, Aulacogens and their genetic relation to geosynclines, with a Proterozoic example from Great Slave Lake, Canada: in Dott, R.H., Jr. and Shaver, R.H., eds., Modern and Ancient Geosynclinal Sedimentation: Soc. Econ. Paleontol. Mineral. Spec. Publ. 19, p. 38-55.

Hoffman, P.F., Fraser, J.A. and McGlynn, J.C., 1970, The Coronation Geosyncline of Aphebian age, District of Mackenzie: in Baer, A. J., ed., Symposium on Basins and Geosynclines of the Canadian Shield: Geol. Survey Canada Paper 70-40, p. 200-212.

Hoffman, P.F., St-Onge, M.R., Easton, R.M., Grotzinger, J. and Schulze, D.E., 1980, Syntectonic plutonism in north-central Wopmay Orogen (early Proterozoic), Hepburn Lake map-area (86J), District of Mackenzie: in Current Research, Part A: Geol. Survey Canada Paper 80-1A, p.171-177.

Horvath, F., and Stegena, L., 1977, The Pannonian basin: a Mediterranean interarc basin: in Biju-Duval, B. and Montadert, L., eds., Structural History of the Mediterranean Basins: Editions Technip, Paris, p. 333-340.

Hsü, K.J., 1979, Thin-skinned plate tectonics during Neo-Alpine orogenesis: American Jour. Sci., v. 279, p. 353-366.

Illies, J.H., 1974, Taphrogenesis and plate tectonics: in Illies, J.H. and Fuchs, K., eds., Approaches to Taphrogenesis: E. Schweizerbart'sche Verlagsbuchhandlung, Stuttgart, p. 433-460.

Jeletzky, O.L., 1974, Unroofing of the Hepburn Batholith: a petrographic study of the Aphebian Recluse Formation, Epworth Group, Northwest Territories:Unpublished B.S. Thesis, Carleton University, Ottawa, 73 p.

Jordan, T.H., 1978, Composition and development of the continental tectosphere: Nature, v. 274, p. 544-548.

Jordan, T.H., 1979, The deep structure of the continents: Sci. American, v. 240, p. 92-107.

Keen, C.E., 1979, Thermal history and subsidence of rifted continental margins – evidence from wells on the Nova Scotian and Labrador shelves: Canadian Jour. Earth Sci., v. 16 p. 505-522.

Kerr, R.A., 1978, Precambrian tectonics: Is the present the key to the past?: Science, v. 199, p. 282-285, 330.

Kidd, W.S.F., 1975, Widespread late Neogene and Quaternary calc-alka-line vulcanism on the Tibetan Plateau: EOS, Trans. American Geophys. Union, v. 56, p. 453.

Kligfield, R., 1979, The northern Apennines as a collisional orogen: American Jour. Science, v. 279, p. 676-691.

Kröner, A., 1977, Precambrian mobile belts of southern and eastern Africa – ancient sutures or sites of ensialic mobility? A case for crustal evolution towards plate tectonics: Tectonophysics, v. 40, p. 101-135.

Kulm, L.D., Scheidegger, K.F., Prince, R.A., Drymond, J., Moore, T.C. Jr. and Hussong, D.M., 1973, Tholeiitic basalt ridge in the Peru Trench: Geology, v. 1, p. 11-14.

LeBas, M.J., 1971, Per-alkaline volcanism, crustal swelling, and rifting: Nature, v. 230, p. 85-87.

LeCheminant, A.N., Lambert, M.B., Miller, A.R. and Booth, G.W., 1979a, Geological studies: Tebesjuak Lake map area, District of Keewatin: in Current Research, Part A: Geol. Survey Canada Paper 79-1A, p. 179-186.

LeCheminant, A.N., Leatherbarrow, R.W. and Miller, A.R., 1979b, Thirty Mile Lake map area, District of Keewatin: in Current Research, Part B: Geol. Survey Canada Paper 79-1B, p. 319-327.

LeFort, P., 1975, Himalayas: the collided range. Present knowledge of the continental arc: American Jour. Science, v. 275-A, p. 1-44.

LeFort, P., 1976, A thermal model of intracontinental subduction. Explanation of the Himalayan inverted metamorphism: Proc. National Academy of Lincei, v. 21, p. 209-213.

Lemoine, M., 1973, About gravity gliding tectonics in the western Alps: in DeJong, K.A. and Scholten, R., eds., Gravity and Tectonics: New York, John Wiley and Sons, p. 201-216.

Lewry, J.F., Sibbald, T.I.I. and Rees, C.J., 1978, Metamorphic patterns and their relation to tectonism and plutonism in the Churchill Province in northern Saskatchewan: in Fraser, J.A. and Heywood, W.W., eds., Metamorphism in the Canadian Shield: Geol. Survey Canada Paper 78-10, p. 139-154.

Lis, M.G. and Price, R.A., 1976, Large-scale block faulting during deposition of the Windermere Supergroup (Hadrynian) in southeastern British Columbia: in Report of Activities, Part A: Geol. Survey Canada Paper 76-1A, p. 135-136.

Lord, C.S., 1942, Snare River and Ingray Lake Map-areas, Northwest Territories: Geol. Survey Canada Memoir 235, 55 p.

Martineau, M.P. and Lambert, R.St.J., 1974, The Bigspruce Lake nepheline-syenite/ carbonatite complex, N.W.T.: Program with Abstracts, Geol. Assoc. Canada, 1974, p. 59.

McGlynn, J.C., 1970, Bear Province: in Douglas, R.J.W., ed., Geology and Economic Minerals of Canada: Geol. Survey Canada Economic Geology Report No. 1, p. 77-84.

McGlynn, J.C., 1976, Geology of the Calder River (86F) and Leith Peninsula (86E) map-areas, District of Mackenzie: in Report of Activities, Part A: Geol. Survey Canada Paper 76-1A, p. 359-361.

McGlynn, J.C., 1977, Geology of Bear-Slave structural provinces, District of Mackenzie: Geol. Survey Canada Open File Map 445.

McGlynn, J.C., 1979, Geology of the Precambrian rocks of the Riviere Grandin and in part of the Marian River map areas, District of Mackenzie: in Current Research, Part A: Geol. Survey Canada Paper 79-1A, p. 127-131.

McGlynn, J.C. and Henderson, J.B., 1970, Archean volcanism and sedimentation in the Slave Structural Province: in Baer, A.J., ed., Symposium on Basins and Geosynclines in the Canadian Shield: Geol. Survey Canada Paper 70-40, p. 31-44.

McGlynn, J.C. and Irvine, E., 1975, Paleomagnetism of early Aphebian diabase dykes from the Slave Structural Province, Canada: Tectonophysics, v. 26, p. 23-38.

Mitchell, A.H.G., 1978, The Grampian orogeny in Scotland: arc-continent collision and polarity reversal: Jour. Geol., v. 86, p. 643-646.

Molnar, P., Chen, W.P., Fitch, T.J., Tapponnier, P., Warsi, W.E.K. and Wu, F.T., 1977, Structure and tectonics of the Himalaya: a brief summary of relevant geophysical observations: in Himalaya: Science de la Terre: Colloques Internationaux du Centre National de la Recherche Scientifique, No. 268, Paris, p. 269-282.

Monger, J.W.H. and Price, R.A., 1979, Geodynamic evolution of the Canadian Cordillera – progress and problems: Canadian Jour. Earth Sci., v. 16, p. 770-791.

Pecher, A., 1975, The main Central Thrust of the Nepal Himalaya and the related metamorphism in the Modi-Khola cross-section (Annapurna Range): Himalayan Geol., v. 5, p. 115-131.

Pedreira, A.J., 1979, Possible evidence of a Precambrian continental collision in the Rio Pardo basin of eastern Brazil: Geology, v. 7, p. 445-448.

Pitcher, W.S., 1979,The environmental control of granitic emplacement in orogenic belts: in Abstracts with Programs, Geol. Soc. America, v. 11, p. 496.

Pitcher, W.S. and Berger, A.R., 1972, The Geology of Donegal: A Study of Granite Emplacement and Unroofing: New York, John Wiley and Sons, 435 p.

Porada, H., 1979, The Damara-Ribeira orogen of the Pan-African-Brasiliano cycle in Namibia (Southwest Africa) and Brazil as interpreted in terms of continental collision: Tectonophysics, v. 57, p. 237-265.

Powell, C.McA., 1979, A speculative tectonic history of Pakistan and surroundings: some constraints from the Indian Ocean: in Farah, A. and DeJong, K.A., eds., Geodynamics of Pakistan: Geol. Survey Pakistan, Quetta, p. 5-24.

Price, R.A., 1973, Large-scale gravitational flow of supracrustal rocks, southern Canadian Rockies: in DeJong, K.A. and Scholten, R., eds., Gravity and Tectonics: New York, John Wiley and Sons, p. 491-502.

Reinhardt, E.W., 1969, Geology of the Precambrian rocks of Thubun Lakes map-area in relationship to the McDonald Fault system, District of Mackenzie: Geol. Survey Canada Paper 69-21, 29 p.

Reinhardt, E.W., 1972, Occurrences of exotic breccias in the Petitot Islands (85H/10) and Wilson Island (85H/15) map-areas, East Arm of Great Slave Lake, District of Mackenzie: Geol. Survey Canada Paper 72-25.

Richter, F.M., 1977, On the driving mechanism of plate tectonics: Tectonophysics, v. 38, p. 61-88.

Roeder, D., 1973, Subduction and orogeny: Jour. Geophys. Research, v. 78, p. 5005-5024.

Roeder, D., 1978, Three central Mediterranean orogens: a geodynamic synthesis: in Closs, H., Roeder, D., and Schmidt, K., eds., Alps, Apennines, Hellenides: IUCG Scientific Report 38, Schweizerbart'sche, Stuttgart, p. 589-620.

Roeder, D., 1979, Continental collisions: Reviews of Geophysics and Space Physics, v. 17, p. 1098-1109.

Roeder, D. and Mull, C.G., 1978, Tectonics of Brooks Range ophiolites, Alaska: American Assoc. Petroleum Geol. Bull. v. 62, p. 1696-1702.

Schlager, W., 1979, Drowning of carbonate platforms: in Abstracts with Programs, Geol. Soc. America, v. 11, p. 511-512.

Sengör, A.M.C., 1976, Collision of irregular continental margins: implications for foreland deformation of Alpine-type orogens: Geology, v. 4, p.779-782.

Sengör, A.M.C., 1979, The suture zone of Permo-Triassic Tethys: extent and tectonic history: in Abstracts with Programs, Geol. Soc. America, v. 11, p. 513-514.

Sengör, A.M.C. and Kidd, W.S.F., 1979, Post-collisional tectonics of the Turkish-Iranian Plateau and a comparison with Tibet: Tectonophysics, v. 55, p. 361-376.

Sheridan, R.E., 1974, Atlantic continental margin of North America: in Burk, C.A. and Drake, C.L., eds., The Geology of Continental Margins: New York, Springer-Verlag, p. 391-408.

Sikabonyi, L.A. and Rodgers, W.J., 1959, Paleozoic tectonics and sedimentation in the northern half of the west Canadian basin: Alberta Soc. Petroleum Geol. Jour., v. 7, p. 193-216.

Sims, P.K., 1976, Precambrian tectonics and mineral deposits, Lake Superior region: Econ. Geol., v. 71, p. 1092-1118.

Sleep, N.H., 1971, Thermal effects of the formation of Atlantic continental margins by continental break up: Geophys. Jour. Royal Astron. Soc., v. 24, p. 325-350.

Stegena, L., Geczy, B. and Horvath, F., 1975, Late Cenozoic evolution of the Pannonian basin: Tectonophysics, v. 26, p. 71-90.

St-Onge, M.R. and Carmichael, D.M., 1979, Metamorphic conditions, northern Wopmay Orogen, N.W.T.: in Program with Abstracts, Geol. Assoc. Canada, v. 4, p. 81.

St-Onge, M.R. and Hoffman, P.F., 1980, "Hot-side-up" and "hot-side-down" metamorphic isograds in north-central Wopmay Orogen, Hepburn Lake map-area (86J), District of Mackenzie: in Current Research, Par A: Geol. Survey Canada Paper 80-1A, p. 179-182.

Tempelman-Kluit, D.J., 1979, Transported cataclasite, ophiolite and granodiorite in Yukon: evidence of arc-continent collision: Geol. Survey Canada Paper 79-14, 27 p.

Thakur, V.C., 1977, Divergent isograds of metamorphism in some parts of Higher Himalayan zone: in Himalaya: Science de la Terre: Colloques Internationaux du Centre National de la Recherche Scientifique, No. 268, Paris, p. 433-441.

Thiessen, R., Burke, K. and Kidd, W.S.F., 1979, African hotspots and their relation to the underlying mantle: Geology, v. 7, p. 263-266.

Thomas, M.D., Gibb, R.A. and Quince, J.R., 1976, New evidence from offset aeromagnetic anomalies for transcurrent faulting associated with the Bathurst and McDonald faults, Northwest Territories: Canadian Jour. Earth Sci., v. 13, p. 1244-1250.

Toksöz, M.N. and Bird, P., 1977, Formation and evolution of marginal basins and continental plateaus: in Talwani, M. and Pitman III, W.C., eds., Island Arcs, Deep Sea Trenches and Back-Arc Basins: American Geophys. Union, Washington, p. 379-394.

Turcotte, D.L. and Ahern, J.L., 1977, On the thermal and subsidence history of sedimentary basins: Jour. Geophys. Research, v. 82, p. 3762-3766.

Von der Borch, C.C., 1979, Continent-island arc collision in the Banda arc: Tectonophysics, v. 54, p. 169-193.

Warner, L.A., 1978, The Colorado Lineament: a middle Precambrian wrench fault system: Geol. Soc. America Bull., v. 89, p. 161-171.

Warsi, W.E.K. and Molnar, P., 1977, Gravity anomalies and plate tectonics in the Himalayas: in Himalaya: Sciences de la Terre: Colloques Internationaux du Centre de la Recherche Scientifique, No. 268, Paris, p. 463-478.

Watts, A.B. and Ryan, W.B.F., 1976, Flexure of the lithosphere and continental margin basins: Tectonophysics, v. 36, p. 25-44.

White, R.S., 1979, Deformation of the Makran continental margin: in Farah, A. and DeJong, K.A., eds., Geodynamics of Pakistan: Geol. Survey Pakistan, Quetta, p. 295-304.

Wilcox, R.E., Harding, T.P. and Seely, R., 1973, Basic wrench tectonics: American Assoc. Petroleum Geol. Bull., v. 57, p. 74-96.

Wilson, A., 1979, Petrology and geochemistry of the upper Hottah Lake sequence, Hottah Lake, District of Mackenzie: Unpublished B.Sc. Thesis, McMaster University, Hamilton, 88 p.

Wilson, J.T., 1968, Static or mobile earth: the current scientific revolution: Proc. American Phil. Soc. v. 112, p. 309-320.

THE GLOBAL VIEW

The Continental Crust and Its Mineral Deposits, edited by D. W. Strangway,
Geological Association of Canada Special Paper 20

EPISODICITY, SEQUENCE, AND STYLE AT CONVERGENT PLATE BOUNDARIES

John F. Dewey
Department of Geological Sciences,
State University of New York at Albany,
Albany, New York 12222

ABSTRACT

Arcs may be classified as extensional, neutral or compressional. Extensional arcs such as
the Marianas are intra-oceanic, have plentiful, mainly basaltic magmatism, few earthquakes in
the overriding plate, back-arc basin plate accretion, a thin basaltic crust, low relief, small or
absent sedimentary dispersal fans, an ophiolitic fore-arc basement, deep trenches, and steep
Benioff Zones. Compressional arcs such as the Peruvian Andes lie on continental margins, have
little, mainly silicic magmatism, many large earthquakes in the overriding plate, back-arc thrusting, a thick continental crust, high relief, thick sedimentary dispersal fans, continental fore-arc
basements with inner trench wall tectonic erosion, shallow trenches, and flat Benioff Zones.
Neutral arcs such as the central American arc have exactly intermediate characteristics, usually
with subduction-accretion prisms and, commonly, arc-parallel transforms. These variations
may be due to the relationship between the following vectors of plate motion in an inert or
slowly moving asthenosphere reference frame: roll-back of the subduction hinge or trench line
due to vertical sinking of the underrriding plate, and the motion of the overriding plate –
towards or away from the trench line. Where the overriding plate retreats from the trench line,
or where it advances more slowly than roll-back, the arc is extensional. Where the overriding
plate advances at the same rate as roll-back, the arc is neutral. Where the overriding plate
advances faster than roll-back, the arc is compressional. The subduction of fracture zones and
other boundaries between lithosphere of different buoyancy causes segmentation of underriding
and overriding plates. The progressively changing age of lithosphere being subducted produces
episodic variation between extensional and compressional arcs and, hence, fundamental sequential changes in tectonic style.

RÉSUMÉ

On peut classifier les arcs insulaires comme des arcs en extension, neutres ou en compression. Les arcs en extension, comme ceux des îles Mariannes, ont les caractéristiques suivantes: ils sont intra-océaniques, iasails ont du magmatisme abondante de composition surtout basaltique, peu de séismevemens dans la plaque chevauchante, une accrétion de plaque dans le bassin derrière l'arc, une croûte basaltique mince, un relief faible, des cônes de dispersion sédimentaires petits ou absents, un socle ophiolitique en avant de l'arc, des fossés profonds et des zones de Benioff raides. Les arcs en compression, comme les Andes péruviennes, reposent sur les marges continentales, ont du magmatisme peu abondant surtout silicique, plusieurs séismes forts dans la plaque chevauchante, du chevauchement en arrière de l'arc, une croûte continentale épaisse, un relief très marqué, des cônes sédimentaires de dispersion épais, un socle continental en avant de l'arc avec érosion tectonique du mur interne, des fossés peu profonds et des zones de Benioff plates. Les arcs neutres comme l'arc de l'Amérique centrale ont des caractéristiques exactement intermédiaires, habituellement avec des prismes de subduction-accrétion et, souvent, avec des failles transformantes parallèles à l'arc. Ces variations peuvent être dues à la relation entre les vecteurs de mouvement de plaques suivants dans un cadre de références formé par une asthénosphère inerte ou se déplaçant lentement: le recul de la charnière de subduction ou de la ligne du fossé causé par l'enfoncement vertical de la plaque sous-jacente et le mouvement de la plaque chevauchante vers la ligne du fossé ou en s'en éloignant. Là où la plaque chevauchante recule par rapport à la ligne du fossé, ou là où elle avance plus lentement que le recul de la charnière, l'arc est en extension. Lorsque la plaque chevauchante avance au même taux que recule la charnière, l'arc est neutre. Lorsque la plaque chevauchante avance plus rapidement que recule la charnière, l'arc est en compression. La subduction des zones de fractures et des autres limites entre des lithosphères de différentes densités provoque la segmentation de la plaque chevauchée et de la plaque chevauchante. Le changement progressif d'âge de la lithosphère subissant la subduction produit une variation' épisodique entre les arcs en extension et en compression et, de là, des changements séquentiels fondamentaux dans le style tectonique.

INTRODUCTION

Great stratigraphic and structural complexity characterizes the beginning of the Wilson Cycle (early continental rifting) and immense structural complexity typifies the end of the Wilson Cycle (continental collision). The intervening protracted history of subduction also has an extraordinarily complex and subtle range of geometric and kinematic relationships. Continental and continent/arc collision, transpressional segments of transforms, and subduction-accretion, although complicated, produce distinctive and reasonably well-understood deformations. It is less clear why the tectonics and petrology of non-collisional convergent boundaries vary so greatly in space and time. Trench, arc-trench-gap/arc, back arc systems above a subducting oceanic slab, hereinafter called arcs for brevity and simplicity, vary tremendously in their broad-scale characteristics. The two end members are the central Peruvian segment of the Andes and the Marianas, between which exist a vast array of arcs of intermediate characteristics (Fig. 1). The contrasts between the central Peruvian Andes and the Marianas may be summarized as follows.

The Marianas (Fig. 2C) are characterized by a deep open trench (to 11 km) that subducts Jurassic crust, extensional faulting on the outer trench wall, a fairly steep Benioff Zone (Fig. 3), widespread intra-arc extension, back-arc basin spreading in the Marianas Trough, more earthquakes in the underriding than the overriding plate,

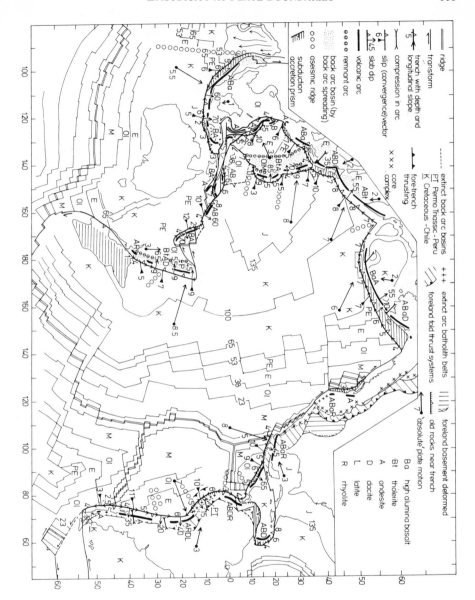

Figure 1. Tectonic map of the Pacific Ocean and its marginal arcs and orogens. Age of oceanic crust: P – Pliocene and younger, M – Miocene; Ol – Oligocene, E – Eocene, PE – Paleocene, K – Cretaceous, J – Jurassic. Depth of trenches given in kilometres; slip vectors and 'absolute' plate motion given in cm/yr. 'Absolute' plate motion signifies plate motion in a hot-spot reference frame (vectors from Chase, 1978).

a rather thin mafic/intermediate volcanic/plutonic crust, and tholeiitic/andesitic volcanism. In consequence of the dominant arc extension and mafic/intermediate volcanism, the arc has a subdued geomorphological expression, a relatively quiet tectonic and eruptive style of high-temperature, low viscosity lavas with low-lying mainly submerged cones and shields with fringing reefs, and poorly developed mafic/intermediate volcaniclastic sedimentary dispersal fans. The arc-trench-gap basement consists of ophiolite with a small outer subduction-accretion prism.

In contrast, the central Peruvian segment of the Andean subduction system (Fig. 2A) has a shallower trench (to 6 km) that subducts Eocene crust, thrust faulting on the outer trench wall (Prince and Kulm, 1975), major thrust faulting in the underriding Nazca Plate up to 200 km from the trench, a Benioff Zone with very shallow dip down to about 200 km and a steeper, deeper portion below a seismic gap (Fig. 3), widespread intra-arc compression, back-arc thrusting over a foreland trough, more and higher energy earthquakes in the overriding than the underriding plate, no volcanism, and thick (70 km) continental crust gradually tapering trenchward to less than 10 km at the Arequipa Massif where Precambrian granulites occur almost at the

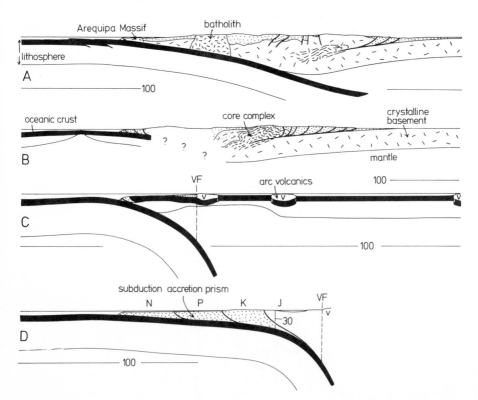

Figure 2. True scale sections across the Pacific margin (scale in kilometres). A. Central Peru. B. Western Canada. C. Marianas. D. Alaska (age of subduction-accretion prism: J – Jurassic, K – Cretaceous, P – Paleogene, N – Neogene).

inner trench wall (Dalmayrac *et al.*, 1977; Shackleton *et al.*, 1979). Where volcanic activity occurs along strike in southern Peru, it is andesitic/dacitic/latitic and rhyolitic. In consequence of the dominant compression and, where it occurs, the intermediate/silicic volcanism, the arc has a spectacular geomorphic expression with very high recent uplift rates, a somewhat violent tectonic and eruptive style with low-temperature high-viscosity lavas, and thick extensive sedimentary dispersal fans, mainly in the Peruvian/Bolivian foreland trough, consisting of silicic volcanic/plutonic and metamorphic crystalline detritus. The arc-trench-gap basement consists of Precambrian crystalline rocks. The rear-arc portion, at least, of the Peruvian segment appears to be similar to the Cretaceous/early Tertiary Canadian Cordillera with progressive listric thrusting across an a-type subduction zone (defined by Bally, 1975, as a subduction zone that consumes continental lithosphere).

A further variant that may occur within arcs of all types is arc-parallel strike-slip or transform faulting (Fitch, 1972), frequently along the volcanic axis, as in Sumatra, presumably because the arc lithosphere is hottest, thinnest, and weakest along that line. The Andaman Sea back-arc basin appears to be a pull-apart offset segment in the Sumatra/Shan transform system (Fig. 1).

A contrast of particular importance is that the Mariana-type arc has an ophiolitic fore-arc basement above a steep Benioff Zone (N-type of Fig. 3), whereas the Peruvian type has a thin continental crystalline basement above a shallow-dipping Benioff Zone to about 200 km (O-type of Fig. 3). As pointed out by Barazangi and Isacks (1976), a wedge of asthenosphere occurs in the Mariana-type at shallow levels between the overriding and underriding plate, and the shear interface between the two plates has a cross-sectional length of less than 100 km. In contrast, the cross-sectional length of the shear interface between the two plates in central Peru is at least 400 km, and there is no shallow-level asthenosphere wedge. Larger earthquakes characteristically occur less often where the shear interface is long (Kelleher *et al.*, 1974).

A third type of fore-arc basement is the tapering subduction-accretion prism (Fig. 1) above a very flat upper Benioff Zone (Figs. 2D, 3). Such prisms are widest where sediment supply to the trench is greatest; they can be seen in major positions of the subduction zones at Sumatra, Alaska, Cascade, Japan, Shikoku, Central America, and Barbados. The tectonic and petrologic characteristics of this type of arc are intermediate between the Mariana and Peruvian types, that is, with neither rear-arc extension nor thrusting but occasionally with arc-parallel transforms.

In summary, the geological expression of subduction varies from one extreme of rather quiescent upper plate extension with mainly mafic volcanism, and an ophiolitic fore-arc basement above a low-velocity zone wedge and steep, deep Benioff Zone, to the other extreme of more violent upper plate compression with more silicic volcanism, and a continental fore-arc basement above a gentle dipping Benioff Zone with no asthenosphere wedge beneath the arc. Upon the spectrum between these two extremes may be superposed arc-parallel transforms and subduction-accretion prisms.

Lateral variation between these extremes may be gradual, but more typically is abrupt, with a sharp lateral segmentation of both overriding and underriding plates (Barazangi and Isacks, 1976). Lateral segmentation is particularly clear along the

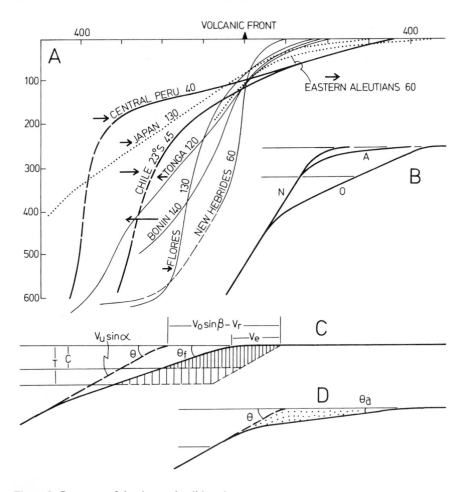

Figure 3. Geometry of the downgoing lithosphere.

A. True scale sections across Pacific subduction zones showing the geometry of associated Benioff Zones (distance and depth in kilometres). Arrows indicate advance (to the right) or retreat (to the left) of overriding plate. Number on slab is approximate age in millions of years of the oceanic crust being subducted at the trench.

B. Three suggested forms of Benioff Zone. N — normal, where overriding plate does not advance oceanward across position of downgoing lithosphere. A — modified by flat dip at shallow level (above about 30 km) below subduction-accretion prism. O — dip, down as much as 300 km, flattened by oceanward advance of overriding plate across downgoing lithosphere.

C. Geometric aspects of Benioff Zone flattened by overriding plate advance. The flattened slab dip is expressed by

$$\theta_f = \tan^{-1} \frac{V_u \sin \alpha \cdot \sin \theta}{V_u \sin \alpha \cdot \cos \theta + V_o \sin \beta - V_r - V_e}$$

where θ is the dip of the unmodified Benioff Zone, V_u the velocity at which the descending slab is inserted along its length into the asthenosphere, V_o the velocity of the overriding plate with

Andean subduction system, where the central Peruvian and central Chilean non-volcanic zones coincide with shallower-dipping Benioff Zones and the northern and central Chilean volcanic zones with steeper dips (Fig. 1). A further correlation is between the Andean non-volcanic zones and seamount chain subduction in the case of central Peru and central Chile, with a trench/trench/ridge triple junction in southern Chile and a trench/trench/transform triple junction in Costa Rica where a ridge segment has just been subducted. The volcanic gap in the Tonga-Kermadec arc is similarly associated with seamount chain subduction, suggesting, as noted by Kelleher and McCann (1977), the probable importance of the subduction of various kinds of buoyant or upstanding zones in arc tectonics.

In the following sections a model is developed to account for the major variations in arc tectonics outlined above.

REFERENCE FRAMES AND PREVIOUS THEORIES

The striking differences in arc tectonics cannot be explained by variations in convergence rate or direction between underriding and overriding plates (relative plate motion). The Peruvian, Japanese, and Tonga arcs, despite their fundamental differences, all have roughly head-on convergence at a rate of about 10 cm/yr. Convergence rate is clearly unimportant but, as Fitch (1972) argued and as will be further developed below, the angle of convergence is critical in the development of intra-arc transforms.

Molnar and Atwater (1978) drew attention to the fact that the east Pacific trenches subduct thin lithosphere capped by Cenozoic oceanic crust from 0 to about 50 Ma old, whereas the west Pacific trenches subduct thick lithosphere capped mainly by Mesozoic oceanic crust older than about 100 Ma. They argued that cold, thick, old lithosphere sinks (vertical component) into the asthenosphere at a subduction zone faster than hot, thin, young lithosphere, causing a fast 'roll-back' or oceanward retreat of the subduction hinge in the former case and a slow or non-existent 'roll-back' in the latter. They concluded that if the roll-back is faster than the convergence rate, back-arc extension results and if roll-back is slower than convergence rate, back-arc compression results.

A further group of arguments depends less upon relative plate motion than so-called 'absolute' motion, better termed plate motion relative to a non-plate reference frame. Such reference frames have been developed by no-net rotation of the lithosphere (Minster et al., 1974), plate boundary velocity minimization criteria (Kaula, 1975), balancing systems of forces on plates (Gordon et al., 1978), using a hot-spot frame to define an inert or slowly moving asthenosphere or mean mesosphere frame (Minster and Jordan, 1978), and using subducting slabs as a reference

respect to the subduction hinge or trench line, V_r the velocity at which the subduction hinge retreats oceanward, V_e the velocity of inner trench wall tectonic erosion, and α the angle between the trench and the direction of motion of the underriding plate (α and β shown in Fig. 4). Vertical line ornament indicates cross-sectional area of crust (narrow-spacing) and subcrustal lithosphere (wide-spacing) tectonically eroded or shaved during flattening of slab dip from Θ to Θ_f.

D. Shallow subduction dip (Θ_a) produced by subduction-accretion.

frame (Kaula, 1975; Jordan, 1975). The rationale for the last method is that because it is difficult to displace dipping slabs rapidly and laterally in the asthenosphere (Havemann, 1972), the slab plays an anchoring role (Tullis, 1972). Hager and O'Connell (1978, 1979) have developed a constant viscosity mantle flow model in which sub-plate flow is driven by plate motion, which in turn is driven mainly by lithosphere thickness variation with a contribution from subducting slabs. This model predicts the dips of subduction zones rather well and provides a further possible reference frame in the geometry of asthenosphere flow rather than a frame of material particles.

Wilson and Burke (1972) suggested that back-arc extension and back-arc compression result, respectively, from the oceanic plate advancing on a stationary (with respect to the deep mantle) continental plate and a continental plate advancing on a stationary oceanic plate. Hyndman (1972), Chase (1978), and Uyeda and Kanamori (1979) argued that the motion of the overriding plate, with respect to a trench line fixed or nearly fixed by the subducting slab in an asthenosphere reference frame, determines arc tectonics and the retreat causes back-arc spreading. Wu (1978) argued that spreading behind the Marianas arc is due to the anchoring of the trench line after the leading edge of the Pacific Plate encountered the mesosphere, followed by the continued westward motion of the Philippine Plate.

Models of plate motion in some form of inert or slowly moving asthenosphere reference frame (Chase, 1978; Gordon et al., 1978; Hager and O'Connell, 1979; Minster et al., 1974; Minster and Jordan, 1978, 1979) have the following four features in common. First, plates that carry mostly oceanic crust move faster than those carrying mainly continental crust. Secondly, the fastest-moving plates have the greatest proportion of subducting edges to area. Thirdly, velocities increase towards the equator. Fourthly, and most critical for the ensuing analysis, the overriding plate has a component of motion towards the trench line in the east Pacific and away from the trench line in the west and southwest Pacific (Fig. 1).

THE MODEL

The model proposed here combines the overriding plate motion models of Hyndman (1972) and Chase (1978), and the subduction-hinge, roll-back model of Molnar and Atwater (1978), attaching little or no importance to the motion of the underriding plate or to the overriding/underriding relative plate motion except for its possible induction of intra-arc strike-slip. Figure 4 illustrates the essentials of the model, in which the reference frame for the three primary plate motion vectors (V_o, V_u, V_g, defined below) is an inert or slowly moving asthenosphere into which the descending lithosphere is ingested by a combination of vertical sinking (V_g) and lengthwise insertion (V_u). The slab is considered to act as an anchor (Tullis, 1972), unable to move its position laterally except in an oceanward direction (by roll-back due to sinking). This is supported by the absence of 'overturned' Benioff Zones. Thus, the subduction hinge defines the trench line, which can move only oceanward ($V_r = v_g \cot \theta$), quickly for rapidly sinking, thick, old lithosphere and slowly for slowly sinking, thin, young lithosphere. The underriding plate may move obliquely through the subduction hinge, the component normal to the trench line being $V_u \sin \alpha$. The vector (V_m) along which the slab moves into the asthenosphere is a function of V_u and

V_g (Fig. 4C). The overriding plate may retreat or advance with respect to the subduction hinge, depending upon the relationship between V_o (motion in the asthenosphere reference frame) and V_r. Where V_o has a component ($V_o \sin \beta$) away from the trench, a gap must open between the overriding and underriding plates. Observation indicates that such gaps do not develop at the trench line but along the volcanic axis within the arc (where the lithosphere is thin, hot, and weak). This means that, under such conditions, some coupling exists between the frontal portion of the arc (F) and the retreating subduction hinge (Fig. 4A). Where V_o, in the asthenosphere frame of

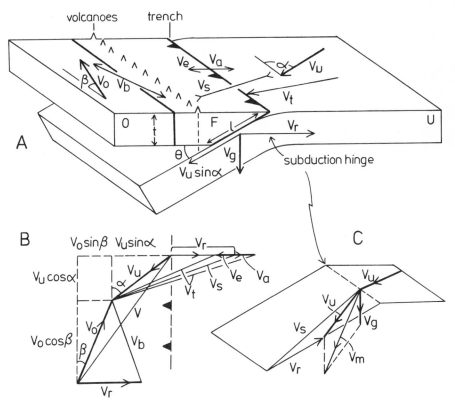

Figure 4. Block diagram (A) and plan (B) of vectors and parameters used to describe plate motion at a subduction zone. Reference frame for V_o, V_u, V_g and V_r is an inert asthenosphere or hot-spot frame. V_o – velocity of overriding plate (O), V_u – velocity of underriding plate (U), V_g – velocity of sinking of subducting slab due to negative buoyancy, V_r – velocity of oceanward retreat of subduction hinge ($V_r = V_g \cot \Theta$), V_b – slip vector between overriding plate (O) and frontal sliver (F), V – slip vector on interface between V and F, V_t – slip vector at trench, V_e – subduction erosion vector, V_a – subduction accretion vector, V – convergence slip vector between O and U. Maximum intra-arc strike-slip rate is given by $V_o \cos \beta + V_u \cos \alpha$.
C. Subduction rate (V_m) into asthenosphere as a function of slab insertion (V_u) and slab sinking (V_g) rates.

reference, has a component ($V_o \sin \beta$) towards the trench, a gap between the overriding plate (O) and the frontal portion of the arc (F) opens where $V_r > V_o \sin \beta$, that is, where the subduction hinge retreats faster than the oceanward movement of O. The relative motion between O and F defines the back-arc or intra-arc opening or extension velocity and direction (V_b). Where the overriding plate moves oceanward at a rate such that $V_o \sin \beta = V_r$, the arc may be classified as tectonically neutral, except that intra-arc transforms can develop where V_o and/or V_u are oblique to the trench line. Where the overriding plate advances oceanwards faster than the subduction hinge retreats ($V_o \sin \beta > V_r$), it overrides the subduction hinge, flattening the upper part of the downgoing slab (Fig. 3C) and compressing the frontal portion of the overriding plate. Tectonic erosion or the shaving of slivers from the leading edge of O (Smith, 1976) results from slab flattening, initially involving mainly lithospheric mantle, but more and more crust as the flattening proceeds.

We may thus define three basic kinds of arc: 1) *extensional* (Mariana-type, Fig. 2C), where the overriding plate retreats from the trench or where the subduction hinge retreats faster than the overriding plate advances, 2) *neutral* (Alaskan-style, Fig. 2D and perhaps Japan/Kuriles), where the overriding plate advances at the same rate as the subduction hinge retreats, and 3) *compressional* (Peruvian-type, Fig. 2A), where the overriding plate advances faster than the subduction hinge retreats. Foretrench-bulge gravity highs (Bodine and Watts, 1979) are particularly well developed where overriding plates advance oceanward (Uyeda and Kanamori, 1979), suggesting that buckling, rather than bending, resulting from slab downward flexure is the dominant factor in their development. All three types of arc may have associated intra-arc transforms, depending upon how the frontal portion of the arc is coupled to the overriding and underriding plates, the maximum combined rate of slip on all transforms in the arc being $V_o \cos \beta + V_u \cos \alpha$ (Fig. 4B). The extent to which F is strike-slip coupled to U determines V_s, the direction and rate of slip on the interface between U and F (Fig. 4B). The slip direction and rate at the trench is $V_t = V_s$ except where increased by the velocity of subduction-accretion (V_a) or reduced by the velocity of surface tectonic erosion (V_e), that is, the velocity at which material is subducted from the base of the inner trench wall. The subduction velocity *sensu stricto* – the rate at which the underriding plate vanishes beneath the leading edge of the overriding plates – is the component of V_t normal to the trench. Thus we see that the convergence rate (OvU = V), the interface slip rate (UvF = V_s), the slip rate at the trench (V_t), and the subduction rate are generally different vectors and should not be confused. Of particular importance is that the tectonics of the arc is controlled largely by V_o and V_r, the convergence vector V and the slab insertion vector V_u being of minor importance. Buoyant and upstanding zones such as seamounts, seamount chains, plateaus, and small microcontinents clearly play an important role in generating local complexities at subduction zones (Kelleher and McCann, 1977), but their effects appear to be secondary and superposed upon the primary character of arcs generated by plate motion. The following section examines five combinations of V_o, V_u, V_r, V_a and V_e, their derivative vectors V, V_b, V_s and V_t, and the resultant tectonic, geomorphological, and petrologic expressions on the leading edge of the overriding plate predicted as arc geology.

PREDICTED ARC GEOLOGY

In the five models developed below, V_u and V_r are constants.

1) Overriding slab (O) retreats obliquely and fore-arc block (F) moves ocean-ward normal to the trench line (Fig. 5). Here, a back-arc basin opens obliquely in the V_b direction and the arc is under dominant extension. Extension in this Mariana or Tonga-type arc allows the widespread rise of basaltic partial melts together with andesite. Low-lying basaltic/andesitic shields and cones with fringing reefs give rise to only minor volcaniclastic dispersal fans and, consequently, subduction-accretion will be associated only rarely with this kind of intra-oceanic arc and only where the trench has a suitable longitudinal slope from a clastic source (e.g., Barbados accretion prism). The deep-seated arc geology is likely to consist of a complicated ophiolite-complex foundation, whose complexity may be related to nucleation of the subduction zone on an oceanic fracture zone (Karson and Dewey, 1978). The ophiolite foundation will be injected by 'permissive' diorite and gabbro bodies and capped

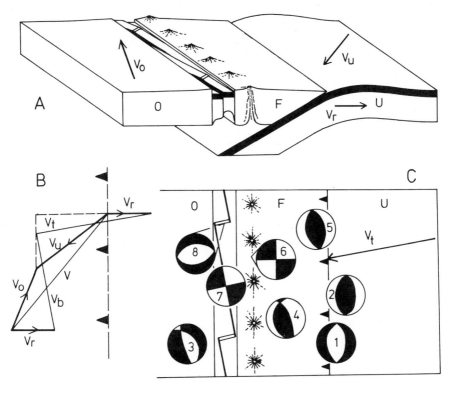

Figure 5. Block diagram (A), vector plan (B), and tectonic plan (C) of a subduction system where the overriding plate (O) retreats obliquely (V_o) from the subduction hinge, which retreats oceanward (V_r). Vector nomenclature explained in Fig. 4 caption. Lower hemisphere stereographic projections of nodal planes associated with faulting expected in various parts of the subduction system; black-compressional first motions. Full discussion in text.

by an andesite/basalt carapace. First motion plots (Fig. 5C) illustrate some predicted variations in strain type and orientation. Plots 1 and 2 refer, respectively, to extension and compression above and below the neutral surface in the subduction-accretion hinge. Plot 3 indicates down-dip compression within the subducting slab. Plot 4 is generated by slip (V$_s$) on the U/F interface, 5 is a thrust solution near the inner trench wall, 6 refers to minor right-lateral arc-parallel strike-slip faulting, and 7 and 8 are, respectively, extension and strike-slip motion on the back-arc ridge/transform boundary.

2) Overriding slab (O) advances obliquely such that V$_o$ sin β < V$_r$ and fore-arc block (F) moves oceanward normal to the trench line (Fig. 6). A tectonic situation similar to 1 develops, but with the addition of a larger component of O/F strike-slip motion, which lends greater obliquity to the back-arc opening vector (V$_b$).

3) Overriding slab (O) advances obliquely such that V$_o$ sin β = V$_r$; V$_o$ cos β and part of V$_o$ cos α are taken up as strike-slip between O and F (V$_b$) (Fig. 7). Such an arc is neither strongly extended nor compressed and may be called a neutral arc with no strike-slip faulting (Alaska/Aleutian) or major intra-arc transforms (Sumatra). Intra-arc strain patterns are thus dominated by strike-slip-related events (Fig. 7C) such as offsetting ridge segments (9) and compressional offsets (10). The Andaman Sea ap-

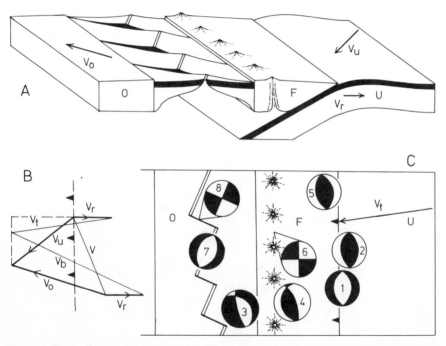

Figure 6. Block diagram (A), vector plan (B), and tectonic plan (C) of a subduction system where the overriding plate (O) advances obliquely oceanward (V$_o$) at a rate, normal to the trench line, insufficient to keep up with the oceanward retreat of the subduction hinge (V$_r$). Vector notation and ornament as in Figs. 4 and 5. Full discussion in text.

pears to be such an intra-arc basin generated at a ridge offset of the Sumatra/Shan transform system (Fig. 1). Gabbro, diorite, and minor more silicic plutons are injected in a strike-slip regime, probably in offset pull-apart segments in particular, and thus intrusive styles and plutonic fabrics will vary from isotropic, permissive, gap-filling to smeared out, with ductile to cataclastic shear zone fabrics. Subduction-accretion prisms are more likely to be associated with long-lived neutral arcs, because extensional arcs migrate oceanward to become isolated from continental sediment sources and, as will be argued below, compressional arcs commonly incur tectonic erosion at the inner trench wall.

4) Overriding slab (O) advances such that $V_o \sin \beta > V_r$, F advances oceanward normal to the trench, and $V_o \cos \beta$ is taken up as strike-slip within the arc (V_b) (Fig. 8). This is a hypothetical situation in which the frontal portion of O does not collapse under compression, but behaves as a fairly rigid beam overriding and flattening the dip on the downgoing plate. In such an arc, extensional cracks (Fig. 8C, TG), parallel with the surface projection of V_s, may nucleate magmatism. Magmatism will be otherwise inhibited in such an arc, except in pull-apart segments on transforms, because cracks in other orientations will be closed under compression.

5) Overriding slab (O) advances such that $V_o \sin \beta > V_r$ and the frontal portion of O is almost coupled, in a trench-parallel strike-slip sense, to U (Fig. 9). In this arc

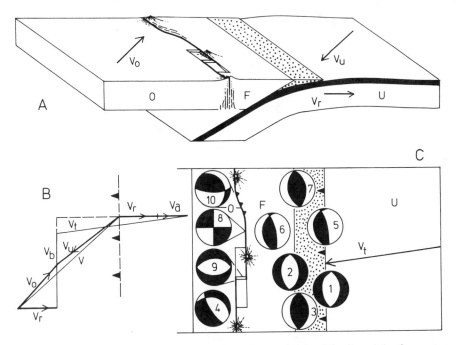

Figure 7. Block diagram (A), vector plan (B), and tectonic plan (C) of a subduction system where the overriding plate (O) advances obliquely oceanward (V_o) at a rate, normal to the trench line, that exactly keeps pace with the oceanward retreat of the subduction hinge (V_r). Vector notation and ornament as in Figs. 4 and 5. Full discussion in text.

there is no discrete frontal or fore-arc block; rather, a wide frontal zone of O above a flattened down-going slab takes up a large portion of the relative motion between O and the subduction hinge by transform and distributed strike-slip faulting parallel with the arc, and by a-type subduction combined with a general shortening strain (Fig. 9B). Little magmatism, except on cracks parallel with V_s or in pull-aparts on transforms, will occur. The long shallow O/U interface will prevent the eruption of basalts over most of the width of the arc until an asthenosphere wedge appears a long way from the trench.

    This describes a Peruvian-type arc, where compression due to over-riding of the subduction hinge is transmitted to the underriding plate and oceanic thrusting (Hus-

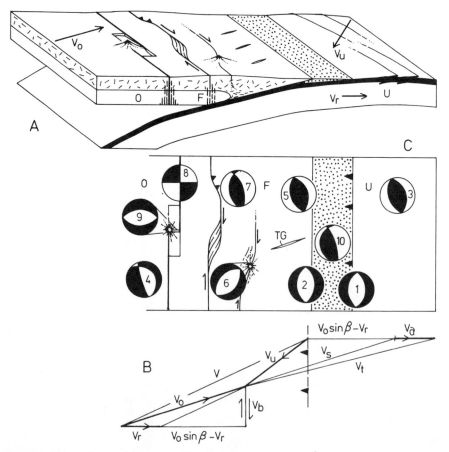

Figure 8. Block diagram (A), vector plan (B), and tectonic plan (C) of a subduction system where the overriding plate (O) advances obliquely oceanward (V_o) at a rate, normal to the trench line, greater than the rate of oceanward retreat of the subduction hinge (V_r) and where the maximum strike-slip component of V_o is taken up by a transform fault between O and F. Vector notation and ornament as in Figs. 4 and 5. Full discussion in text.

song *et al.*, 1975) ensues (Fig. 9A, T). Shortening and thickening (James, 1971) of the overriding continental crust may be accomplished by a combination of general, pervasive collapse, a-type subduction, and the underplating of slivers sliced off and carried down from the frontal part of the overriding plate by tectonic erosion; all of these mechanisms may be episodic. The effect of crustal thickening is to produce minimum melting potassic/silicic partial melts and rapid isostatic uplift. The silicic liquids will generally be unable to reach the surface in a regime of crustal collapse and shortening except in occasional gashes and transform pull-aparts, and therefore are likely to have forceful diapiric and lateral injection styles with complex and variable planar and linear fabrics in plutons. Cordilleran-type core complexes (Davis and Coney, 1979) are a likely result of a-type subduction, crustal thickening, collapse (Fig. 2A, B; Fig. 9A) where core complex metamorphics may be extruded continentward as nappes with a frontal fold and listric thrust system above an undeformed basement, which is being swallowed in an a-type subduction zone. Occasionally, as in Wyoming, the foreland basement is heavily involved in the frontal thrust

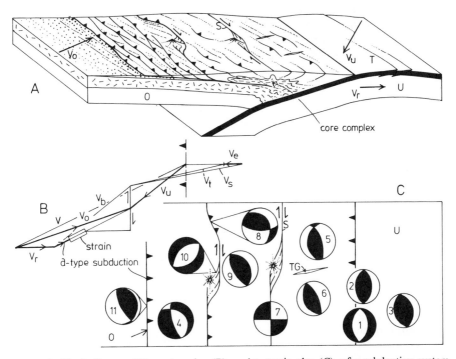

Figure 9. Block diagram (A), vector plan (B), and tectonic plan (C), of a subduction system where the overriding plate (O) advances obliquely oceanward (V_o) at a rate, normal to the trench line, greater than the rate of oceanward retreat of the subduction hinge (V_r) and where relative displacement between O and the inner wall of the trench is distributed between an a-type subduction zone, strain, and strike-slip displacement. Vector notation and ornament as in Figs. 4 and 5. Full discussion in text.

belt (Smithson *et al.*, 1978), probably due to pre-existing crustal heterogeneities. Sedimentary dispersal fans from such arcs consist mainly of silicic volcanic and plutonic and metamorphic detritus, and are mainly deposited as thick sequences in foreland troughs adjacent to a-type subduction zones. Most of the sediment transported oceanward either flows along the trench or is carried down the subduction zone to be added to the component shaved from the continental edge of the trench.

ARC SEGMENTATION

Circum-Pacific arcs exhibit clear along-strike variation between extensional, neutral, and compressional types. This variation can be either gradual or abrupt, and affects the overriding as well as the underriding slab (Fig. 1; Barazangi and Isacks, 1976). Abrupt segmentation is particularly well displayed in the downgoing slab under the Marianas, where the slab is broken into 'fingers' with different, though steep, dips and under the arcs of the east Pacific. Under the Andes, the slab is broken into at least five distinct, differently dipping segments with corresponding segmentation of the overriding plate (Fig. 1). There are two obvious reasons for segmentation of the downgoing slab. First, there is likely to be segmentation at a trench/trench/transform triple junction (the transform being between two over-riding slabs) of the central American-type. In central America, the left-lateral Guatemala transform separates the Mexican arc from the central American arc. The Mexican arc subducts Miocene and younger crust along a subduction zone dipping at 45°, while the central American arc subducts Oligocene crust along a Benioff Zone dipping at 65°. The Mexican arc moves oceanward faster than the central American arc, and flattens the underriding plate. The volcanic chain is correspondingly much further from the trench in the Mexican arc, where a wide arc-trench gap is clearly underlain by continental basement extending close to the trench, and where tectonic erosion is likely occurring. In contrast, the narrow central American arc-trench gap consists of a subduction-accretion prism. Secondly, segmentation is likely to be associated with fracture zone subduction. Fracture zones are contacts between oceanic crust of different ages in a lithosphere of different thickness. Hence, on subduction, the sinking velocity (V_g) of the older, colder, thicker lithosphere will be greater than that of the adjacent younger, hotter, thinner lithosphere and will give rise to a steeper dipping subducting slab. Similarly, roll-back velocity (V_r) will be higher for the older lithosphere than for the younger lithosphere. This disparity across the subducting fracture zone could, with an appropriate V_o/V_r combination, produce an along-strike change from an extensional to a neutral or compressional arc (Fig. 10A).

This suggests that back-arc basin spreading may terminate along strike where there is a strong age contrast in the subducting lithosphere. Such an age contrast between Jurassic and Oligocene ocean floor exists at the southern end of the Marianas (Fig. 1), approximately coinciding with the right-lateral transform southern termination of the frontal arc block of the Marianas behind which back-arc plate accretion is occurring in the Marianas Trough (Karig *et al.*, 1978). In the Marianas, the Philippine Plate moves away from the trench line and roll-back should be relatively fast. In the Yap-Palau arcs, V_o is almost parallel with the trench line and roll-back should be very slow. An age contrast in the subducting lithosphere also exists across the northern end of the Palau Ridge southeast of Kyushu. The Oligocene/Miocene crust of the Parece Vela Basin is subducted beneath the compressional Shikoku arc where volcanism is scant. The neutral Okinawa arc to the southwest subducts Paleocene crust of the Philippine Basin. Frontal arc pinning by buoyant, difficult-to-subduct, or obstructive lithosphere may prevent the normal oceanward retreat of subduction hinges. For example, the northern end of the Marianas may be pinned by the subduction of the Marcus-Necker Rise, with the consequent obstruction of back-arc opening at that position in the Bonin/Marianas arc.

Figure 10B schematically illustrates the effect of an overriding plate advancing towards a trench where the subduction hinge is rolling back at a laterally varying velocity, which is related to the variation, from old to young, of the oceanic floor being subducted. This is a very complicated condition in which the frontal portion of the arc must either be broken into a large number of individual blocks or suffer lateral stretching, because V_b constantly varies in direction and rate along the arc, creating a

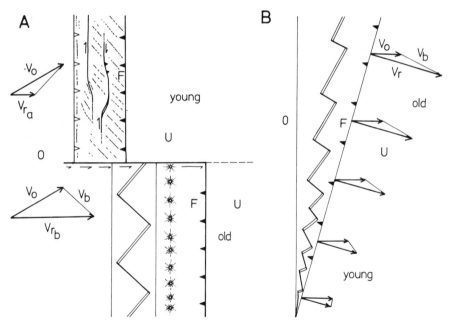

Figure 10. A. Schematic plan of the possible lateral tectonic contrast in a subduction system where the overriding plate advances obliquely oceanward (V_o) and a fracture zone separates young and old oceanic lithosphere, the young lithosphere having a subduction hinge that retreats slowly oceanward (V_{ra}) and the old having a hinge that retreats rapidly (V_{rb}). Vector notation and symbols as in Figs. 4-9. Full discussion in text.

B. Schematic plan of a subduction system where the overriding plate advances oceanward (V_o) and the subduction hinge retreats at a rate (V_r) that varies continuously along the trench as a result of the continuous change in age of the subducting lithosphere. Vector notation and ornament as in Fig. 4-10. Full discussion in text.

hinging effect such that the variation does not occur about a single O/F rotation pole. The active Lau back-arc basin behind the Tonga-Kermadec arc (Sclater *et al.*, 1972) narrows southwards to a hinge in the Taupo-Rotorua volcanic field of the North Island of New Zealand, probably because of the pinning effect of the Campbell Plateau subduction-accretion prism.

An important effect of underriding plate segmentation across fracture zones is that, due to the intersection of the trench line with ocean floor of different ages and depths, the longitudinal slope of trenches is interrupted. Thus, longitudinal turbidite flow in trenches may be strongly controlled by fracture zone walls, either enhanced (where flow drops to an older, deeper floor) or blocked (where flow meets a fracture wall that bounds a younger, higher floor). This will produce rather complicated stratigraphical/structural relationships in subduction-accretion prisms because one segment of a prism may grow rapidly while an adjacent segment may accrete very slowly, producing sharp offsets in the inner trench wall.

ARC VARIATION WITH TIME

During some part of the Wilson Cycle between continental separation and continental collision, not only must the subduction of oceanic lithosphere occur, but the lithosphere that is being subducted in one or more arcs must progressively change its age and thickness. In the subduction history of any major ocean, the variation of subducting lithosphere with time is likely to be spatially and temporally complicated. In the present Pacific subduction zones, lithosphere of widely different ages is being subducted with the consequent variations noted by Molnar and Atwater (1978). Any individual subduction zone will consume lithosphere that gradually changes its age, thickness, subductability (V_g), and roll-back velocity (V_r). The Andean arc north of the Chile Ridge triple junction subducts progressively younger lithosphere, and, to the south, progressively older (Fig. 1). The Kurile and Japan arcs subduct progressively older ocean floor while the Marianas and Kermadecs consume progressively younger ocean floor. A change in the age pattern of ocean floor subduction will occur when a ridge is subducted, where a trench/trench/ridge triple junction moves along an arc, or where the inflection of a magnetic bight is subducted under certain geometric relations between the slip vector and the inflexion. Thus, if V_r, dependent on V_g, in turn dependent on the age of the ocean floor, is an important factor in determining, together with V_o, the tectonic characteristics of an arc, the progressive change in age of lithosphere subducted should produce progressive variations in arc tectonics. In the Peruvian Andean arc, extension occurred in Permo-Triassic times (Noble *et al.*, 1978) with basalt/comendite magmatism, and again in late Cretaceous/Paleocene time when the giant permissive Peruvian batholith was intruded (Pitcher, 1972). In the southern Chilean Andes, a back-arc basin opened and closed during the Mesozoic (Dalziel *et al.*, 1973). Thus, episodic phases of extension and compression could result from either a progressively or a suddenly changing roll-back velocity combined with changes in the motion of the overriding plate. Figure 11 illustrates a possible early sequence of events once subduction has begun following an Atlantic-type phase of expansion. The early subduction of old crust (A), where O retreats or where $V_o \sin \beta < V_r$, produces an extensional arc with the opening of back-arc basins above a steep subduction zone. As younger crust is subducted (B) and $V_o \sin \beta > V_r$, compression closes the back-arc basin and causes ophiolite obduction. Subsequently, as very young crust is subducted, the arc becomes Andean. Conversely, such a sequence may be common in reverse order prior to termination of the Wilson Cycle by continental collision. This has fundamental implications for the geological history of orogenic belts, in which the important stages in a Wilson Cycle might be:

1) Rifting, and alkaline and basaltic magmatism associated with continental stretching, mainly permissive silicic iron-enriched to alkaline plutonism;

2) growth of Atlantic-type margins;

3) subduction, opening of back-arc basins in extensional arcs dominated by basaltic volcanism, few or no clastic wedges at this 'orogenic' stage, permissive batholith styles;

4) closing of back-arc basins, ophiolite obduction, compressional arcs, uplift, beginning of clastic wedges, much strike-slip faulting parallel with arc;

5) increasingly silicic but decreasing volcanism, development of metamorphic core complexes, forceful intrusion, complex polyphase deformation, extrusion of nappes and foreland fold thrust systems across a-type subduction zones;

6) gradual transition back to more neutral and extensional arcs; and

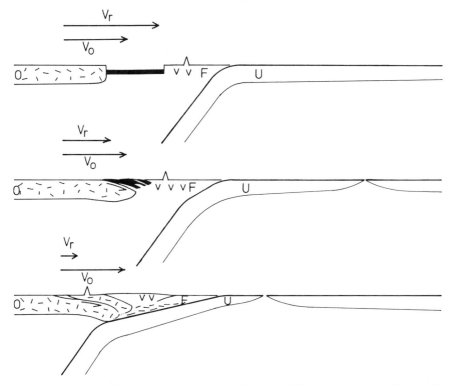

Figure 11. Sequence of subduction-related tectonic regimes likely during the middle part of a Wilson Cycle. Vector notation as in Fig. 4. Full discussion in text.

7) continental collision with all its structural complexity. It is emphasized that this is an extremely generalized "history;" many variations will be induced by multiple arcs and ridges, local as well as global plate motion re-organizations, and the kinds of natural spatial variations observed today in the Pacific. However, the model outlined in this paper seems capable of accounting for a great deal of the extraordinary variation in the geology of arcs in space and time, in particular episodic arc "orogeny" and changes in arc kinematics.

ACKNOWLEDGEMENTS

I am grateful to many friends and colleagues with whom, over many years, I have had stimulating discussions about orogeny and subduction, particularly Clark Burchfiel, Bill Dickinson, Ron Oxburgh, Dietrich Roeder, Robert Shackleton, Alan Smith and John Sutton. From these discussions I have derived many of the ideas expressed in this paper, which was written during a leave of absence at the University of Calgary, to whose faculty, staff and students in the Department of Geology I am deeply indebted for their hospitality and assistance.

REFERENCES

Bally, A.W., 1975, A geodynamic scenario for hydrocarbon occurrences: World Petroleum Congr., Tokyo 1975, paper PD-1, p. 33-34.

Barazangi, M. and Isacks, B.L., 1976, Spatial distribution of earthquakes and subduction of the Nazca plate beneath South America: Geology, v. 4, p. 686-692.

Bodine, J.H. and Watts, A.B., 1979, On lithospheric flexure seaward of the Bonin and Mariana Trenches: Earth Planet. Sci. Letters, v. 43, p. 132-148.

Chase, C.G., 1978. Extension behind island arcs and motions relative to hot spots: Jour. Geophys. Research, v. 83, p. 5385-5387.

Dalmayrac, B., Lancelot, J.R., and Leyreloup, A., 1977, Two-billion-year granulites in the late Precambrian metamorphic basement along the southern Peruvian Coast: Science, v. 198, p. 49-52.

Dalziel, I.W.D., de Wit, M.J. and Palmer, K.F., 1973, A fossil marginal basin in the southern Andes: Nature, v. 250, p. 291-294.

Davis, G.H. and Coney, P.J., 1979, Geologic development of the Cordilleran metamorphic core complexes: Geology, v. 7, p. 120-124.

Fitch, T.J., 1972, Plate convergence, transcurrent faults and internal deformation adjacent to southeast Asia and the western Pacific: Jour. Geophys. Research, v. 77, p. 4432-4460.

Gordon, R.G., Cox, A. and Harter, C.E., 1978, Absolute motion of an individual plate estimated from its ridge and trench boundaries: Nature, v. 274, p. 752-755.

Hager, B.H. and O'Connell, R.J., 1978, Subduction zone dip angles and flow driven by plate motion: Tectonophysics, v. 50, p. 111-133.

Hager, B.H. and O'Connell, R.J., 1979, Kinematic models of large-scale flow in the Earth's mantle: Jour. Geophys. Res., v. 84, p. 1031-1048.

Havemann, H., 1972, Displacement of dipping slabs of lithosphere plates: Earth Planet. Sci. Letters, v. 17, p. 129-134.

Hussong, D.M., Odegard, M.E. and Wipperman, L.K., 1975, Compressional faulting of the oceanic crust prior to subduction in the Peru-Chile trench: Geology, v. 3, p. 601-604.

Hyndman, R.D., 1972, Plate motions relative to the deep mantle and the development of subduction zones: Nature, v. 238, p. 263-265.

James, D.E., 1971, Andean crustal and upper mantle structure: Jour. Geophys. Research, v. 76, p. 3246-3272.

Jordan, T.H., 1975, The present-day motions of the Caribbean plate: Jour. Geophys. Research, v. 80, p. 4433-4439.

Karig, D.E., Anderson, R.N. and Bibee, L.D., 1978, Characteristics of back-arc spreading in the Mariana Trough: Jour. Geophys. Research, v. 83, p. 1213-1226.

Karson, J. and Dewey, J.F., 1978, The coastal complex, western Newfoundland: an early Ordovician fracture zone: Geol. Soc. America Bull., v. 89, p. 1037-1049.

Kaula, W.M., 1975, Absolute plate motions by boundary velocity minimizations: Jour. Geophys. Research, v. 80, p. 244-248.

Kelleher, J. and McCann, W., 1977, Bathymetric highs and the development of convergent plate boundaries: in Talwani, M. and Pitman, W.C. III, eds., Island Arcs and Deep Sea Trenches: American Geophys. Union, p. 115-122.

Kelleher, J., Savino, J., Rowlett, H. and McCann, W., 1974, Why and where great thrust earthquakes occur along island arcs: Jour. Geophys. Research, v. 79, p. 4889-4899.

Minster, J.B. and Jordan, T.H., 1978, Present-day plate motions: Jour. Geophys. Research, v. 83, p. 5331-5354.

Minster, J.B., Jordan, T.H., Molnar, P. and Haines, E., 1974, Numerical modelling of instantaneous plate tectonics: Geophys. Jour. Royal Astron. Soc., v. 36, p. 541-576.

Molnar, P. and Atwater, T., 1978, Interarc spreading and cordilleran tectonics as alternates related to the age of subducted oceanic lithosphere: Earth Planet. Sci. Letters, v. 41, p. 330-340.

Noble, D.C., Silberman, M.L., Megard, F. and Bowman, H.R., 1978, Comendite and basalt in the Mitu Group, Peru: evidence for Permian-Triassic lithospheric extension in the central Andes: Jour. Research, v. 6, p. 453-457.

Pitcher, W.S., 1972, The Coastal Batholith of Peru: some structural aspects: Proc. XXIV Int. Geol. Congr. Montreal, Section 2, p. 156-163.

Prince, R.A. and Kulm, L.D., 1975, Crustal rupture and the initiation of imbricate thrusting in the Peru-Chile Trench: Geol. Soc. America Bull., v. 86, p. 1639-1653.

Sclater, J.G., Hawkins, J.W., Mammerickx, J. and Chase, C.G., 1972, Crustal extension between the Tonga and Lau Ridges: petrologic and geophysical evidence: Geol. Soc. Amer. Bull., v. 83, p. 505-518.

Shackleton, R.M., Ries, A.C., Coward, M.P. and Cobbold, P.R., 1979, Structure, metamorphism and geochronology of the Arequipa Massif of coastal Peru: Geol. Soc. London Jour., v. 136, p. 195-214.

Smith, A.G., 1976, Plate tectonics and orogeny: a review: Tectonophysics, v. 33, p. 215-285.

Smithson, S.B., Brewer, J., Kaufman, S. and Oliver, J., 1978, Nature of the Wind River thrust, Wyoming, from COCORP deep-reflection data and from gravity data: Geology, v. 6, p. 648-652.

Tullis, T.E., 1972, Evidence that lithospheric slabs act as anchors (abst.): Trans. American Geophys. Union, v. 53, p. 522.

Uyeda, S. and Kanamori, H., 1979, Back-arc opening and the mode of subduction: Jour. Geophys. Res., v. 84, p. 1049-1061.

Wilson, J.T. and Burke, K., 1972, Two types of mountain building: Nature, v. 239, p. 448-449.

Wu, F.T., 1978, Benioff Zones, absolute motion and interarc basin: Jour. Phys. Earth, v. 26, p. 39-54.

The Continental Crust and Its Mineral Deposits, edited by D.W. Strangway,
Geological Association of Canada Special Paper 20

THE THERMAL HISTORY OF THE EARTH

Dan McKenzie
Bullard Laboratories
Department of Earth Sciences
Madingley Rise, Madingley Road,
Cambridge CB3 0EZ
Nigel Weiss
Department of Applied Mathematics and Theoretical Physics,
Silver Street, Cambridge CB3 9EW

ABSTRACT

The large convective heat flux implied by plate tectonic models of island arcs suggests that the thermal history of the Earth is controlled by convection rather than conduction, and that the variation of viscosity with temperature is important. A detailed numerical study of steady state convection in fluids with temperature dependent viscosity provides empirical relationships between the heat flux and temperature. The expressions also give accurate descriptions of the time dependent behaviour, and show that the surface heat flux adjusts to the rate of heat generation in a time that is short compared with the half lives of the radioactive elements in the mantle. The results imply that the Earth has been cooling ever since it was formed, and provide a simple explanation for the origin of ultrabasic Precambrian lavas.

RÉSUMÉ

Les gigantesques flux thermiques convectifs qu'impliquent les modèles de tectonique des plaques pour les arcs insulaires suggèrent que l'histoire thermique de la Terre est contrôlée par la convection plutôt que par la conduction et que la variation de viscosité avec la température est importante. Une étude numérique détaillée de la convection en régime permanent dans les fluides dont la viscosité dépend de la température fournit des relations empiriques entre le flux thermique et la température. Les expressions donnent aussi des descriptions précises de son comportement en fonction du temps et montrent que le flux thermique de surface s'ajuste au taux de génération de chaleur dans un temps relativement court par comparaison avec les demi-vies des éléments radioactifs dans le manteau. Ces résultats impliquent que la Terre s'est refroidie depuis sa formation et ils fournissent une explication valable pour l'origine des laves ultrabasiques du Précambrien.

INTRODUCTION

The thermal history of the Earth has preoccupied geophysicists for more than 100 years, though our approach to the problem is now very different from that of Kelvin's in the nineteenth century. The general acceptance of plate tectonics started with the confirmation of Wilson's (1965) ideas about transform faults, and quickly led to models in which heat was principally transported by convection, not conduction. This change avoided the unrealistically high temperatures which all simple thermal models of the mantle had produced.

Much present interest in the Earth's thermal history is related to a general revival of interest in the Precambrian, especially the Archaean. A more general question to which the thermal history is also relevant is the problem of the driving mechanism. It is now clear that the tectonic deformation of Mars has been much less extensive than that of Earth. Even Venus, which has a higher surface temperature than the Earth, appears to be covered with large craters with central peaks (Malin and Saunders, 1977; Campbell et al., 1979). The only bodies in the solar system other than the Earth which show major surface activity are Io and Europa. It will be of great importance to test our present views about mantle convection when more information becomes available about their internal structure, but in the next ten years it is probable that we will learn more about the various regimes of convection in a silicate body by studying the Precambrian on Earth than by looking elsewhere. In terms of understanding the dynamics of planetary interiors the results of the exploration of the solar system have been disappointing.

There never has been any doubt that large scale deformation has occurred throughout the recorded history of the Earth. The principal difficulty in saying more is the direct consequence of this deformation. No Precambrian sea floor is known anywhere, though there are some ophiolite bodies of this age which may be relicts of ocean basins. Large parts of the continents are older than 0.6Ga (1Ga = 10^9 years) but few of the oldest of these regions have escaped later deformation. Therefore much of our knowledge of Earth processes during the period older than -2Ga has to be reconstructed carefully from metamorphosed fragments, a procedure not always free from controversy.

The approach taken in this paper and in our earlier attempt to investigate this subject (McKenzie and Weiss, 1975) is to attempt to construct a simple parameterized convective thermal model, the consequences of which can be then compared with what geological observations are available. There is no doubt that the models discussed here and in McKenzie and Weiss (1975) are great simplifications, but we hope that disagreements between their results and the observations will lead to an eventual improvement in our understanding.

Certain features of the early Precambrian are not in doubt. The heat generated by radioactive decay must have been considerably greater than it is now. How much greater depends on the Potassium-Uranium ratio, and that most generally accepted is the model proposed by Wasserburg et al. (1964). We use this model for all our calculations. The principal problem in doing so is that of knowing the distribution of K, U and Th between the crust and mantle. The best way to deal with this problem is to use the convective models discussed here to calculate the isotope distributions (O'Nions et al. 1979) and the age distribution of the continental crust. In the discussion below we do not allow for differentiation of the crust from the mantle, and therefore probably underestimate the change in the heat generation rate with time.

The principal observation about the early Precambrian which we use is the existence of magnesium-rich lavas (Mg O > 20%) which must have been erupted at temperatures of at least 1600°C. Such lavas are not now being produced. This observation was used in McKenzie and Weiss (1975) to argue that the mantle temperature in the Archaean must have been 200 to 300°C greater than it is now, and hence to calculate the plate thickness and surface temperature gradient. These calculations were speculative because at the time little was known about the influence of temperature dependent viscosity on convective heat transport, or about the accuracy of the simplified equation governing the time dependent behaviour. These questions have now been examined (Hewitt, McKenzie and Weiss, in prep.; Daly 1978), though problems with resolution have limited the investigation to Rayleigh numbers less than 10^6 and aspect ratios of little more than one. There are considerable difficulties still to be overcome before these results can be extrapolated with confidence. This work has, however, strongly confirmed the accuracy of the simplified equation governing the variation of temperature with time (see next section), though it has not supported Tozer's (1972) suggestion concerning the parameterization of variable viscosity.

A more controversial question is the depth to which the convective flow associated with plate tectonics extends. We have argued in the past that the focal mechanisms of earthquakes and the geometry of the sinking slabs suggest that there is a barrier to convection at the depth of about 700 km. This argument is controversial (Davies, 1977; O'Connell, 1977), but recent work by Peltier and Wu (pers. commun.) on the gravity field in the Canadian Artic shows that a lower mantle viscosity of about 10^{24} poise is required to account for the correlation between the gravity and the shape of the Pleistocene ice cap. Furthermore O'Nions et al. (1979) have argued that the observed depletion of the upper mantle in large ion lithophile elements is greater than would be expected if the continents had been produced from the entire mantle. They suggest that only the upper mantle above 700 km has been the source region, and that the lower mantle has not been involved. These geochemical observations therefore suggest that there has been little transport of material between the upper and lower mantle, implying a considerably greater viscosity than 10^{24}. If this geochemical model is correct it provides a source for the undepleted mantle material observed in hot spot vulcanism. Since the upper mantle is heated from below as well as from within, jets of hot fluid will detach from the lower boundary, and may entrain a small volume of undepleted lower mantle material in their cores. This behaviour depends only on the geometry of the heat sources, and occurs for both constant and temperature dependent viscosity. Since the volume of magma erupted by hot spots is small, only a small amount of mixing between the reservoirs is required during the history of the Earth. Therefore this model of upper mantle convection can account for the existence and geochemistry of hot spots, which Wilson (1963) proposed to explain oceanic island vulcanism. His observation that the magma source remains relatively stationary as the plate moves is consistent with Richter and McKenzie's (1978) view that the plate motions are decoupled from the convective circulation in the upper mantle.

Though none of these new observations support whole mantle convection we nonetheless use such a model to discuss the thermal history. A one layer model is easier to investigate than a two layer one, and provides a reference for more complicated model calculations. Furthermore until estimates are available for the viscosity and thermal conductivity of the lower mantle it is not clear how efficiently it will transport heat.

CONVECTIVE MODELS

The most obvious approach to mantle convection is to attempt to include all the physical effects which are known to be important in the governing equations, and then to solve these equations by numerical methods. However this approach is impractical for a variety of reasons. The resulting equations are so complicated that programming errors are probable and hard to detect. Because the finite differences schemes involve many effects, numerical errors and instabilities are difficult to recognize. Since the parameters governing mantle behaviour are poorly known, the convective behaviour must be explored by varying each independently. If the number of parameters is large the number of numerical experiments which are necessary becomes prohibitive. For these reasons we investigated the convective behaviour when only one or two parameters are allowed to vary, and thus we were often not able to reach the parameter range of geophysical interest. We obtained empirical expressions from the numerical results which allow interpolation, and with less confidence, extrapolation.

Probably the two most important effects in the thermal history of the Earth are the variation of viscosity with temperature and the extent to which the flow is maintained by internal heating. Tozer (1972) argued that the strong dependence of viscosity on temperature caused the mean viscosity to adjust to the rate of heat generation. If the heat supply exceeds the convective heat transport the temperature increases, lowering the viscosity and hence increasing the convective transport. Since the viscosity in the mantle varies rapidly with temperature, the temperature changes required during the Earth's evolution are small, and the rate of heat generation should differ little from the heat flux through the surface. We investigated whether this argument was valid by carrying out extensive numerical calculations which will be described fully elsewhere (Hewitt, McKenzie and Weiss, in prep.). We used

$$v(T) = v_0 \exp(-CT) \qquad (1)$$

to describe the temperature dependence of the viscosity, where C and v_0 are constants. We fixed the Rayleigh number

$$R_0 = g \alpha L^4 (F + HL)/kv_0\kappa \qquad (2)$$

where g is the acceleration due to gravity, α the thermal expansion coefficient, L the depth of the layer, F the heat flux through the base of the layer, H the rate of internal heat generation, k the conductivity and κ the diffusivity of the fluid. For each R_0 we solved the full two dimensional time dependent convective equations for a given value of C until we reached a steady state. We continued to increase C until the resolution of the 48 x 48 mesh became insufficient. Such experiments were performed for convection driven entirely by heating from below, by half the heat supplied from below and half from within, and by all the heat supplied from within. Our aim was to find some definition of the mean viscosity and the Rayleigh number which would remove the effect of the variation of viscosity with temperature. The most successful choice was the mean viscosity used by Parmentier *et al.* (1976)

$$\bar{v} = \int_V v(T) \, \dot{e}^2 \, dx \, dz / \int_V \dot{e}^2 \, dx \, dz \tag{3}$$

where $\dot{e}$ is the strain rate. The corresponding Rayleigh number R_2 is given by

$$R_2 = R_0 v_0 / \bar{v} \tag{4}$$

Figure 1 shows the way in which the average velocity

$$<\bar{u}'> = \int_V |u'| \, dx \, dz / \int_V dx \, dz \tag{5}$$

where u' is the dimensionless velocity, depends on R_2 and $\bar{v}$ when the flow is driven by heating from below. The variable viscosity values differ little from corresponding values obtained with constant viscosity at the same R_2. This definition of R_2 does not, however, remove the influence of $\bar{v}$ on the mean temperature (Fig. 2). The definition suggested by Booker (1976) was less successful than (3) in removing the dependence of $\bar{v}$. Because of our failure to find such a definition we chose to describe the dependence of any variable X on R_2 and $\bar{v}$ by empirical laws

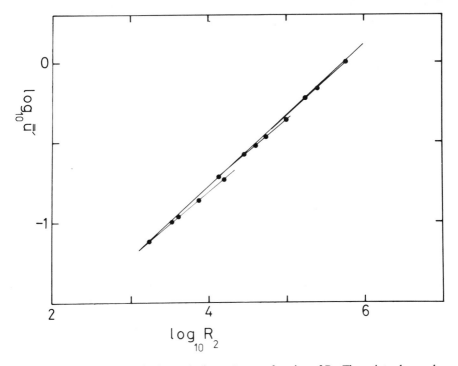

Figure 1. The mean dimensionless velocity $<u'>$ as a function of R_2. The points show values obtained from the numerical solutions, the lines are empirical fits (6). The long upper straight line is for constant viscosity, the shorter lines show the behaviour as C is increased at constant R_0.

$$\log_{10}X = A \log_{10} R_2 + B \log_{10}\bar{v} + C \tag{6}$$

and to determine the constants A, B and C by fitting planes by least squares to results of the full steady convective calculations. The continuous lines in Figures 1 and 2 are obtained from these empirical expressions. The procedure is satisfactory over the range where numerical results are available, but unfortunately R_2 for whole mantle convection with a viscosity of 10^{22} poise is more than 10^8, or two orders of magnitude beyond the resolution of a 48 x 48 mesh. Extrapolation of expressions such as (6) in this way is not very reliable. An alternative method of extrapolation is to require the boundary layer thickness δ and the temperature difference between the surface and the interior of the fluid at large Rayleigh numbers to depend only on the heat flux through the surface and to be independent of the layer depth. This argument leads to

$$\delta \propto R_0^{-0.25} \tag{7}$$

for the constant viscosity case (McKenzie et al., 1974). This expression agrees well with

$$\delta \propto R_0^{-0.24} \tag{8}$$

obtained from numerical experiments when the heat is supplied from below but not with

$$\delta \propto R_0^{-0.17} \tag{9}$$

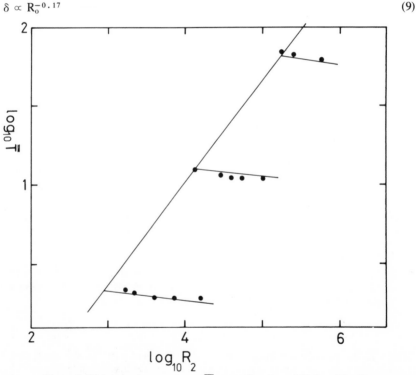

Figure 2. The mean temperature $<\overline{T}>$ as a function of R_2 (see Fig. 1).

when the heat is supplied from within. Equation (9) shows that the boundary thickness increases with L, and therefore becomes less stable, when the aspect ratio was one. Kulacki and Emara (1977) and Kulacki and Nagle (1975) have carried out a number of laboratory experiments with rigid upper and lower boundaries and internal heating to determine the relationship between the Nusselt number and the Rayleigh number for water with a Prandtl number of less than seven. They define the Nusselt number as

$$N = Q_pL/kA\overline{\Delta T} \tag{10}$$

where Q_p is the power input, A the area of the layer and $\overline{\Delta T}$ the mean temperature difference between the top and the bottom of the layer. Kulacki and Emara find $N \propto R_o^{0.227}$ whereas our numerical experiments with fixed aspect ratio and free boundaries give $N \propto R_o^{0.154}$ when N is defined in terms of $<\overline{T}>$. It is possible that the difference between the numerical and laboratory experiments results from the difference in Prandtl number and boundary conditions. It is, however, more likely to result from fixing the aspect ratio in the numerical experiments, which suppresses a variety of possible instabilities (McKenzie et al., 1974). Numerical experiments at large Rayleigh numbers with both constant and variable viscosity and internal heating show that the upper boundary layer is unstable, forming a cold sheet which changes the aspect ratio from 1 to 0.5 if the experiment is carried out in a box of aspect ratio 1, or to time dependent behaviour if the aspect ratio is 2 or larger. This instability reduces $<\overline{T}>$, and hence increases the Nusselt number. Preliminary laboratory experiments at large Rayleigh numbers and a Prandtl number of about 200 (Carrigan, pers. commun.) also show that the aspect ratio of the time dependent three dimensional flow is considerably smaller than 1. It is likely that the same type of circulation occurred in Kulacki and Emara's (1977) experiment, and that this change is aspect ratio increases the Nusselt number. Though no similar experiments have yet been carried out at a large Prandtl number it is likely that they will show that $N \propto R_o^{0.25}$ is a better approximation than $N \propto R_o^{0.15}$. We therefore required $<\overline{T}> \propto R_o^{0.75}$ and $\delta \propto R_o^{-0.25}$ and used the constant viscosity cases both with and without internal heating to estimate constants of proportionality. Expressions obtained in this way are denoted as boundary layer relations, those obtained from (6) as empirical.

The boundary layer expressions derived by McKenzie et al. (1974) are valid only for constant viscosity, and it is not obvious how they can best be extended to include temperature dependence. We chose to use (1) to obtain $\overline{v} = v (<\overline{T}>)$ and used $R = R_2$ where R_2 is defined by (4). Hence the viscosity and R_2 vary rapidly with $<\overline{T}>$. The same procedure was used by McKenzie and Weiss (1975) and is based on Tozer's (1972) argument.

The discussion above is concerned with descriptions of steady state flow, but thermal history calculations require descriptions of the time dependent behaviour. McKenzie and Weiss (1975) argued that the rate of change in the mean temperature was given by the difference between the rate of heat loss E and F + HL

$$\rho C_pL \ \frac{d<\overline{T}>}{dt} = (F + HL) - E \tag{11}$$

where ρ is the mean density and C_p the specific heat. From the steady state calculations we can obtain $E = E(<\overline{T}>, \overline{v})$ and hence use (11) to follow the variation of $<\overline{T}>$ and E with time as F + HL changes. Recently a number of other authors (Daly, 1978; Sharpe and Peltier, 1978; Stevenson and Turner, 1979) have used the same approach, though only Daly compared the parameterized behaviour with that of the full convective calculations. To test whether (11) provides an accurate description of the time dependent behaviour we compared the surface heat flux (Fig. 3) and mean temperature (Fig. 4) obtained from the full numerical calculation with that obtained from (11) when the flux was suddenly changed by a factor of two and the viscosity was temperature dependent. Both the heat flux and the temperature obtained from the empirical calcuation agree excellently with those obtained from the finite difference results. The constant viscosity boundary layer expressions give the correct flux but too high a temperature, whereas the temperature dependent boundary layer expressions allow the flux out of the layer to adjust too rapidly and give too low a temperature. These results suggest that the boundary layer calculations will lie on either side of the true temperature, and show that Tozer's (1972) argument overestimates the effect of temperature dependent viscosity.

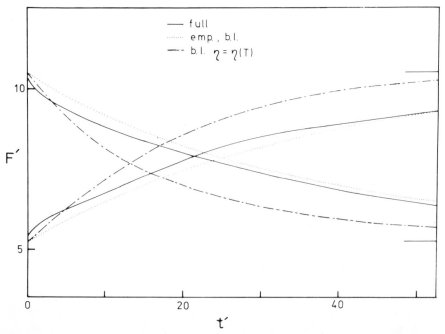

Figure 3. The dimensionless surface heat flux for a convecting layer driven only by internal heating when the heat generation is decreased by a factor of 2 (decreasing curves), run to steady state and then increased by a factor of two (increasing curves). $R_0 = 2.4 \times 10^3$, initially, $= 1.2 \times 10^3$ after the flux is decreased, $C = 1.4$ throughout. The continuous curves show the flux for the numerical calculation, the dotted line shows the flux for the empirical expressions and constant viscosity boundary layer expressions using (10), which cannot be distinguished on this plot, and the dot-dashed line shows the flux for the variable viscosity boundary layer expressions and (10).

MANTLE CONVECTION

We can now use the expressions obtained in the last section to investigate the Earth's thermal history. Because the expressions were obtained for a plane layer we use a layer depth of 2000 km, to give approximately the same surface area to volume ratio as that of the mantle. Most geochemists now believe that the K/U ratio is approximately 10^4, rather than the Chondritic value (see O'Nions $et\ al.$, 1979), and therefore we use Wasserburg $et\ al.$'s (1964) estimates of the abundance of the heat producing elements. The calculations described below neglect the depletion of the mantle with time.

The variation of the surface heat flux with time depends on whether the heat is supplied from below (Fig. 5) or from within (Fig. 6). The difference between the constant viscosity boundary layer curves is small, but that between the empirical expressions with constant viscosity of 2×10^{21} stokes is considerable. As explained in the last section the empirical results from internal heating are unlikely to be correct because the Rayleigh number is greater than 10^8 and boundary layer instabilities will occur. The boundary layer curve with variable viscosity was obtained using

$$v = D\Theta \exp (E/k\Theta) \tag{12}$$

where Θ is the absolute temperature, $E = 4.25$ ev the activation energy and k is Boltzmann's constant. D is a constant chosen to give a viscosity of 2×10^{21} stokes at 1350°C. The relationship between Θ and $<\overline{T}>$ is given by

$$\Theta = <\overline{T}> + T_0 \tag{13}$$

where T_0 is a constant. To obtain T_0, the temperature difference across the mechanical

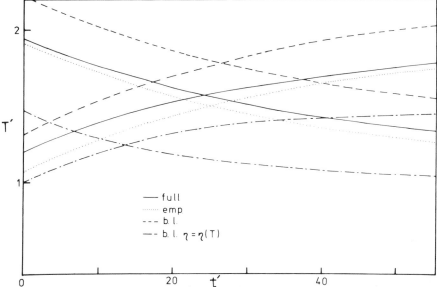

Figure 4. The mean temperature for the four cases in Figure 3. Though the empirical and constant viscosity boundary layer (b.1.) cases produce the same flux, their temperatures are different.

boundary layer, we fixed the total temperature difference between the surface and the convecting region to be 1350°C. We then found the steady state value of $<\overline{T}>$ for the present day radioactive heat flux and $v = 2 \times 10^{21}$ stokes. Equation (13) then gives T_0 by subtraction. Once T_0 has been obtained in this way we can obtain $\Theta(t)$ and compare the values with geological estimates of the temperature. Though the absolute value of $\Theta(t)$ depends on T_0, and hence on the estimate of the convecting region, temperature variations with time do not. All calculations were started at $t = 0$ with the surface heat flux equal to the heat generation rate. This initial condition implies that the Earth was formed hot and has since been cooling. There are a variety of mechanisms which could have produced initial high temperatures. In most models of the Earth's accumulation (Safronov, 1978; Kaula, 1979; Sharpe and Peltier, 1978) some of the kinetic energy heats the planetary interior. Core formation, heating by short lived radioactive elements and tidal dissipation due to the proximity of the Moon may all heat the interior substantially. In contrast a cold origin requires all these effects to be small, since there is no known way of removing heat from the interior except by convection, which implies high temperatures.

Figure 5 shows the surface heat flux as a function of time for two examples with different power law relationships between the Rayleigh and Nusselt numbers when all

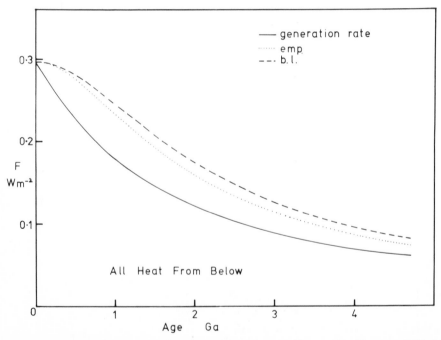

Figure 5. Surface heat flux as a function of time (broken lines) and heat generation rate (solid line) for a layer 2000 km thick containing K, U and Th in concentrations given by Wasserburg *et al.* (1964), and using heat generation rates and time constants given by Wetherill (1966). All heat was supplied from below. In both cases the viscosity is kept constant.

the heat is supplied from below. These calculations show that the difference between the heat generation rate and the heat loss is not large even when the viscosity is constant. Daly (1978) used a single exponential rate of decay for the heat generation and obtained larger differences between the rate of heat loss and generation. In our calculations there are four different time constants. Initially, the time dependence is controlled by the decay of K, and the difference between heat production and heat loss increases rapidly. After 2Ga the decay of U^{238} and Th become more important than that of K, and the ratio of the heat flux to heat generation does not continue to increase.

Figure 6 shows similar calculations, but with the heat supplied within the fluid. The empirical curve with constant viscosity is the only one to show a large difference between heat loss and heat generation and, as explained in the last section, this behaviour is probably the result of fixing the aspect ratio at 1. The curve obtained using Tozer's argument, labelled boundary layer $\eta = \eta(T)$ and corresponding to the expressions used by McKenzie and Weiss (1975), is the only curve in either Figure 5 or 6 in which the viscosity depends on the temperature. The numerical experiments suggest that this calculation considerably underestimates the difference between the heat flux and heat generation (Fig. 3). Though there is some uncertainty in the internally heated calculations, the curves in Figures 5 and 6, obtained using a constant viscosity suggest that the surface heat flux is no more than about 30 per cent greater than the heat generation rate. Though comparison of the full numerical calculations with calculations similar to those of McKenzie and Weiss (1975) shows that Tozer's argument overestimates the effect of the temperature dependence of viscosity, any decrease of

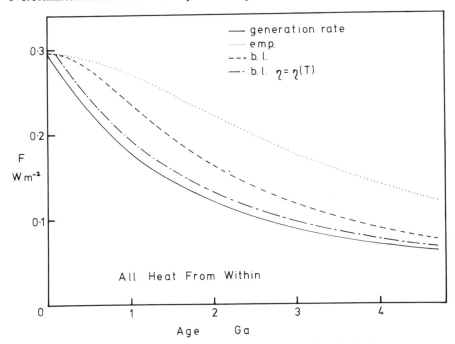

Figure 6. As for Figure 5, but with the heat generation uniformly distributed.

viscosity with temperature must reduce the difference between heat flux and heat generation by reducing the change in temperature required to change the Nusselt number. Hence these calculations suggest that whole mantle convection produces heat flux which differs by less than about 30 per cent from the present rate of heat generation.

The mean temperatures obtained from (13) for the two boundary layer models in Figure 6 are shown in Figure 7 together with some estimates of the melting temperature of peridotitic komatiites (Bickle *et al.*, 1977). These lavas, though not the basaltic komatiites, are restricted to the early Precambrian. For this reason the mantle temperatures are believed to have decreased with time. Figure 7 shows that the convective thermal evolution is consistent with the existence of high temperatures in the Archaean,

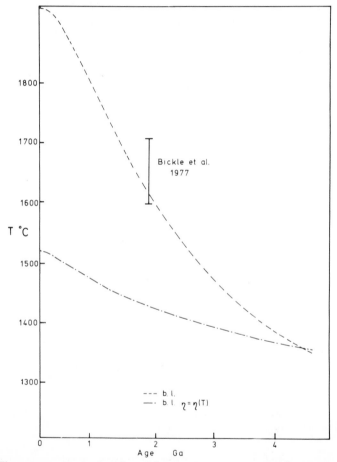

Figure 7. The temperature as a function of time for the two boundary layer models in Figure 6. Both curves were required to pass through 1350°C at 4.55Ga. The vertical bar shows the approximate range of melting temperatures obtained by Bickle *et al.* (1977) for komatiites of age −2.7 Ga.

and with their absence at later times. Though the curve which passes through the estimates of Bickle et al. (1977) represents constant viscosity convection, the uncertainties in the present mantle temperature are large, and smaller increases which should result when the temperature dependence of viscosity can be properly parameterized may well be consistent with the observations. The importance of Figure 7 is that it shows that there is no obvious problem in fitting the determinations of mantle temperature as a function of time with a simple parameterized convective model.

No similar agreement is apparent between observed and calculated geotherms. Mean geotherms at three periods (Fig. 8) show considerable steepening in the past, a result which Bickle (1978) has argued is not consistent with temperature gradients estimated from metamorphic minerals. Though these estimates are not very precise (England, 1979), Bickle (1978) believes that this disagreement requires that most of the heat loss must occur in oceanic areas through plate creation. The difficulty with this argument, and indeed with all arguments which require the lithosphere beneath old oceans and continents to differ in thickness, is that the thickness of both the thermal and mechanical boundary layers in a steady state is controlled by the mantle viscosity, which in turn is controlled by the temperature (Parsons and McKenzie, 1978). Since the komatiites strongly suggest that the mantle temperatures were greater, and hence the viscosity less, in the early Precambrian, both boundary layers should have been thinner in steady state. Exactly the same argument requires old lithosphere to be the

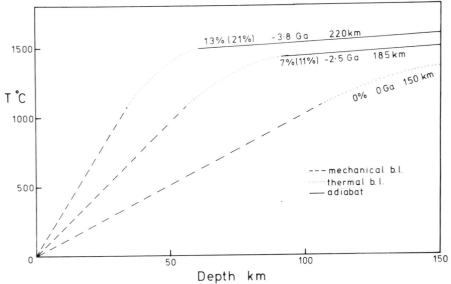

Figure 8. Geotherms calculated from the variable viscosity boundary layer model in Figure 6 for three different times, showing the three convective regions. The mechanical boundary layer forms at ridges and is destroyed at trenches. The thermal boundary layer is involved in the small scale flow. The region below these boundary layers is adiabatic. The percentages on the curves show the percentage of partial melt at the base of the thermal boundary layer using H = 208 cal/gm (outside the brackets) and H = 109 cal/gm (inside the brackets), and a melting point gradient of 4.8°C kb. The depth in km at which partial melting starts is also shown.

same thickness in continental and oceanic areas. The only obvious escape from this conclusion is to assume a different composition of the boundary layer in the mantle beneath continents both now and in the past. Then the viscosity at the same temperature could be higher, and hence the mechanical boundary layer could be thicker. Surface wave velocities across shields and old oceanic plates are slightly different (Forsyth 1977, Brune and Dorman 1963), though there appears to be no significant difference between the rate of heat loss from the mantle beneath shields and old oceans (Sclater *et al.*, 1980). Whether Bickle's argument is correct depends on the accuracy of the pressure estimates obtained from metamorphic minerals. On this question there appears to be no general agreement, despite the wide variety of experiments which have been carried out. Since both the pressures and temperatures involved are easily accessible in the laboratory, and since the field observations are not in dispute, this is an obvious area in which progress in understanding Precambrian thermal regimes can be expected.

CONCLUSIONS

Numerical experiments on convection in fluids whose viscosity varies with temperature have provided empirical equations governing the variation of temperature with Rayleigh number and mean viscosity, but no simple models to describe this behaviour. Neither Tozer's (1972) nor Booker's (1976) models can describe the observations. In contrast the time dependent behaviour is easily modelled with a simple equation proposed by McKenzie and Weiss (1975). This simple equation, together with a variety of laws relating the convective heat flux to the mean temperature can describe the Earth's convective thermal history. Comparisons of numerical calculations of convection in a fluid with temperature dependent viscosity with a parameterization based on Tozer's argument show that the thermal adjustment time is considerably larger than his model implies. All models show that the mean temperature of the mantle has decreased with time, and that the surface heat flux probably differs by less than about 30 per cent from the rate of heat generation. All the models investigated above involve whole mantle convection, which at present seems unlikely, and none take account of the depletion of the radioactive elements with time due to continent formation. However, the results appear consistent with what little is known about the Earth's thermal structure in the Archaean, and may imply that the difference between the thickness of continental and oceanic lithosphere was considerably greater then than it is now. Improvements in these calculations will depend on improving our understanding of the influence of temperature dependent viscosity on the Nusselt number, and on the influence of mantle layering.

ACKNOWLEDGEMENTS

We would like to thank E.G. Nisbet and P.C. England for several discussions on these problems. Research on mantle convection at Cambridge is supported by a grant from the Natural Environmental Research Council.

REFERENCES

Bickle, M.J., 1978, Heat loss from the Earth: A constraint on Archaean tectonics from the relation between geothermal gradients and the rate of plate production: Earth Planetary Sci. Letters, v. 40, p. 301.

Bickle, M.J., Ford, C.E. and Nisbet, E.G., 1977, The petrogenesis of peridotitic komatiites: evidence from high-pressure melting experiments, Earth Planetary Sci. Letters, v. 37, p. 97.

Booker, J.R., 1976, Thermal convection with strongly temperature-dependent viscosity: Jour. Fluid Mech., v. 76, p. 741.

Brune, J. and Dorman, J., 1963, Surface waves and Earth structure in the Canadian Shield: Seismol. Soc. America Bull., v. 53, p. 167.

Campbell, D.B., Burns, B.A. and Boriakoff, V., 1979, Venus: further evidence of impact cratering and tectonic activity from Radar observations: Science, v. 204, p. 1425.

Daly, S.F. 1978, Convection with decaying heat sources and the thermal evolution of the mantle: Ph.D. Thesis, University of Chicago.

Davies, G.F., 1977, Whole mantle convection and plate tectonics: Geophys. Jour. Royal Astronom. Soc., v. 49, p. 459.

England, P.C., 1979, Continental geotherms during the Archaean: Nature, v. 277, p. 556.

Forsyth, D.W., 1977, The evolution of the upper mantle beneath mid-ocean ridges: Tectonophys., v. 38, p. 89.

Kaula, W.M. 1979, Thermal evolution of the Earth and Moon growing by planetesimal impacts: Jour. Geophys. Res., v. 84, p. 999.

Kulacki, F.A. and Nagle, M.E. 1975. Natural convection in horizontal fluid layers with volumetric energy sources: Jour. Heat Transfer, v. 97, p. 204.

McKenzie, D.P., Roberts, J.M. and Weiss, N.O., 1974, Convection in the Earth's mantle: towards a numerical simulation: Jour. Fluid Mech., v. 62, p. 465.

McKenzie, D.P. and Weiss, N.O., 1975, Speculations on the thermal and tectonic history of the Earth: Geophys. Jour. Royal Astronom. Soc., v. 42, p. 131.

Malin, M.C. and Saunders, R.S. 1977, Surface of Venus: Evidence of diverse landforms from radar observations: Science, v. 196, p. 987.

O'Connell, R.J., 1977, On the scale of mantle convection: Tectonophys., v. 38, p. 119.

O'Nions, R.K., Evensen, N.M. and Hamilton P.J. 1979, Geochemical modeling of mantle differentiation and crustal growth: Jour. Geophys. Res., v. 84, p. 6091.

Parmentier, E.M. Turcotte, D.L. and Torrance, K.E., 1976, Studies of finite amplitude non-Newtonian thermal convection with application to convection in the Earth's mantle: Jour. Geophys. Res., v. 81, p. 1839.

Parsons, B. and McKenzie, D., 1978, Mantle convection and the thermal structure of the plates: Jour. Geophys. Res., v. 83, p. 4485.

Richter, F. and McKenzie, D., 1978, Simple plate models of mantle convection: Jour. Geophys., v. 44, p. 441.

Safronov, V.S., 1978, The heating of the Earth during its formation, Icarus, v. 33, p. 3.

Sclater, J.G. Jaupart, C. and Galson, D., 1980, The heat flow through oceanic and continental crust and the heat loss of the Earth. Rev. Geophys. Space Phys., v. 18, p. 269.

Sharpe, H.N. and Peltier, W.R., 1978, Parameterized mantle convection and Earth's thermal history: Geophys. Res. Letters, v. 3, p. 99.

Stevenson, D.J. and Turner, J.S., 1979, Fluid models of mantle convection: in McElhinney, ed., The Earth: Its Origin, Evolution and Structure.

Tozer, D.C., 1972, The present thermal state of the terrestrial planets: Phys. Earth Planet. Int., v. 6, p. 182.

Wasserburg, G.J., MacDonald, G.J.F., Hoyle, F. and Fowler, W.A., 1964, Relative contributions of Uranium, Thorium and Potassium to heat production in the Earth: Science, v. 143, p. 465.

Wetherill, G.W., 1966, Radioactive decay constants and energies: in Clark, S.P., ed., Handbook of Physical Constants: Geol. Soc. America, Memoir 97, p. 513.

Wilson, J.T., 1963, A possible origin of the Hawaiian Islands: Canadian Jour. Phys. v. 41, p. 863.

—————————, 1965, A new class of faults and their bearing on continental drift: Nature, v. 207, p. 343.

The Continental Crust and Its Mineral Deposits, edited by D. W. Strangway,
Geological Association of Canada Special Paper 20

METAMORPHISM AND PLATE CONVERGENCE

Akiho Miyashiro
Department of Geological Sciences,
State University of New York at Albany,
Albany, New York 12222

ABSTRACT

Along convergent plate boundaries where subduction of oceanic lithosphere occurs, a high-pressure type metamorphic belt (glaucophane-schist belt) may form in the subduction zone, and a low-pressure (high-temperature) type metamorphic belt may form in the depth of the associated volcanic arc. The P-T conditions of such a high-pressure type metamorphic belt have been ascribed to cooling by the descending slab. A number of models have been proposed for the belt, based on different views on the identification of a Benioff-zone fault in, as well as on the mechanism of uplift of, the metamorphosed rock complex.

In the low-pressure type metamorphic belt, the high geothermal gradient may be partly due to heat transfer from the underlying shallow low-velocity mantle, but would usually be caused mainly by the intrusion of granitic plutons. Movement of the associated H_2O and CO_2 should play a large part in regional-scale temperature distribution.

The existence of regoinal metamorphism and granite formation due to continental collision has recently been established in the Himalayas. The unusually thick continental crust produced by collision may have undergone partial melting in its basal part to produce granitic magma, and the rise of the magma may have led to regional metamorphism. Bird (1978) has speculated that during collision, the mantle lithosphere of the Indian plate was split from the overlying crust, and plunged deeply into the asthenosphere; thus the Himalayan crust was directly heated by the asthenosphere, which led to regional metamorphism and granite formation.

Low-pressure type regional metamorphism (with andalusite) has been common in all geologic times, whereas high-pressure type regional metamorphism (with glaucophane) began near the beginning of the Paleozoic (possibly in the late Proterozoic). This indicates a change in tectonic processes near the beginning of the Paleozoic.

RÉSUMÉ

Le long des limites de plaques convergentes où il y a subduction de lithosphère océanique, une zone de métamorphisme à haute pression (zone de schiste à glaucophane) peut se former dans la zone de subduction et une zone de métamorphisme à basse pression (haute température) peut se former en profondeur dans l'arc volcanique associé. Les conditions P-T d'une telle zone de métamorphisme à haute pression peuvent être attribuées au refroidissement par la plaque descendante. On a proposé un certain nombre de modèles pour expliquer cette zone en se basant sur différentes hypothèses sur l'identification dans le complexe rocheux métamorphisé d'une faille de zone de Bénioff de même que sur les mécanismes de soulèvement du complexe.

Dans la zone métamorphique à basse pression, le gradient géothermique élevé peut être attribuable en partie au transfert de chaleur du manteau à faible vitesse peu profond, mais il serait normalement surtout causé par l'intrusion de plutons granitiques. Le mouvement de H_2O et de CO_2 associé devrait jouer une part importante dans la distribution à l'échelle régionale des températures.

On a récemment établi l'existence d'un métamorphisme régional et de la formation de granite causés par une collision continentale dans les Himalayas. La croûte continentale anormalement épaisse produite par la collision peut avoir subi une fusion partielle à sa base pour former un magma granitique; la montée du magma a pu être responsable du métamorphisme régional. Bird (1978) a émis l'hypothèse que durant la collision, la lithosphère du manteau de la plaque indienne était séparée de la croûte sus-jacente, et qu'elle a plongé profondément dans l'asthénosphère; ainsi, la croûte des Himalayas a été chauffée directement par l'asthénosphère, ce qui a conduit au métamorphisme régional et à la formation de granite.

Le métamorphisme régional de basse pression (avec de l'andalousite) est commun à toutes les époques géologiques alors que le métamorphisme régional de haute pression (avec le glaucophane) a débuté à peu près au début du Paléozoïque (peut-être même à la fin du Protérozoïque). Ceci indique un changement dans les processus tectoniques à peu près au début du Paléozoïque.

HISTORICAL INTRODUCTION

Early Work

It was already known in the late 19th century that regional metamorphism and granite intrusion were genetically related to the formation of folded mountains. This view originated mainly from field associations of intense deformation with regional metamorphic rocks and granites.

Metamorphic petrology based on the phrase rule began in the early 20th century (e.g., Eskola, 1939). In the first half of the century, however, metamorphic petrology had little or no connection with the tectonic aspects of geologic processes. This situation was caused mainly by two factors: First, at that time the Caledonides of Scotland and Norway were the only orogenic belt extensively surveyed from the petrologic viewpoint. Hence, metamorphic petrologists did not have an appropriate understanding of the global diversity of regional metamorphism. For example, due attention was not paid to the regional metamorphism which produced the glaucophane schist group of rocks. Secondly, petrologically investigated areas were so small that spatial connections between P-T conditions and tectonic features were not recognizable.

In the 1950s, petrologic studies were extended to many wide regions of the world including the circum-Pacific: Japan, New Zeland and California in particular. This allowed a better perspective on the global diversity of regional metamorphism, and led

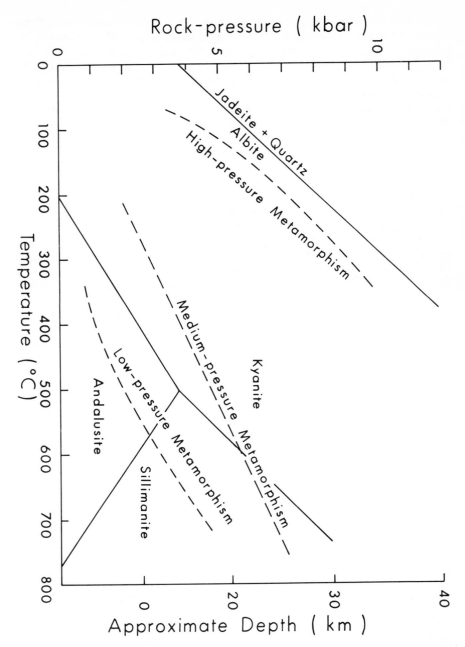

Figure 1. Pressure-temperature relations of the three types of regional metamorphism with different geothermal gradients. The phase relations of Al_2SiO_5 minerals are based on Holdaway (1971), and those of jadeite and quartz on Popp and Gilbert (1972).

to a proposal to classify regional metamorphism into three categories according to geothermal gradient (Miyashiro, 1961): a high-pressure type with a low geothermal gradient ($\leq 10°C/km$), a medium-pressure type with an intermediate geothermal gradient (about $20° - 30°C/km$) and a low-pressure type with a high geothermal gradient ($\geq 40°C/km$). (These values of geothermal gradients have been corrected to recent experimental data on the stability relations of relevant minerals as shown in Fig. 1.) The high-pressure type is characterized by the occurrence of glaucophane and jadeite, the medium-pressure type by the occurrence of kyanite and sillimanite (Barrovian sequence of progressive metamorphic zones) and the low-pressure type by the occurrence of andalusite and sillimanite. In Japan, California and other circum-Pacific regions, a high-pressure type and a low-pressure type metamorphic belt are paired: they run parallel, usually with the high-pressure type belt on the ocean side. Parts of the two belts may belong to the medium-pressure type. The high-pressure type metamorphic belt was regarded as representing an ancient trench zone, where ocean floor had been underthrust beneath continental crust, and the low-pressure type metamorphic belt was believed to represent the belt of granites and volcanic arc (Miyashiro, 1959, 1961, 1967). As the concept of paired metamorphic belt became incorporated into geologic theories based on plate tectonics (e.g., Dewey and Bird, 1970; Hallam, 1973), high-pressure type metamorphism was seen as being genetically related to subduction of a cold oceanic plate.

In the following pages, when high-pressure and low-pressure type metamorphism are mentioned, the word "type" may sometimes be dropped for simplicity of expression, though the addition of the word is desirable to emphasize that the classification is based on geothermal gradients rather than on high or low values of pressure themselves.

More Recent Progress

Paired metamorphic belts occur in circum-Pacific regions such as Chile, California, Japan, Celebes and New Zealand, as shown, for example, in Figure 1 of Miyashiro (1973a) as well as in figures in Ernst (1975, 1977). The contemporaneity of the two belts in a pair has been best established for the Mesozoic pair of western North America (Armstrong and Suppe, 1973), which took more than 100 million years to form.

Although paired belts have been found outside the Pacific, e.g., in Hispaniola (Nagle, 1974) and Jamaica (Draper *et al.*, 1976), in general glaucophane schist belts and paired belts are rare in the Atlantic region. In some cases the older glaucophane-schist belts may have been obliterated by overprinting of later metamorphic events, as in the northern Appalachians (Laird and Albee, 1975). It is also possible that plate motions may have been slower in the Atlantic region than in the Pacific in ancient times just as at present, and slower subduction may not have produced glaucophane schist belts.

In the Alps, the glaucophane-schist belt of the Pennine nappes is not accompanied by a contemporaneous low-pressure metamorphic or granitic belt. Ernst (1973) considered that the Alpine metamorphic belt was formed by the closing of a small ocean basin, and the amount of subduction would have been too small to generate much magma or to cause low-pressure metamorphism. Even though this belt is not paired, however, it is believed to have formed by essentially the same mechanism as the paired belts of the circum-Pacific regions.

However, it does not appear likely that all the regional metamorphic belts of the world were formed by subduction of oceanic lithosphere. Among the Phanerozoic belts, the Hercynides of Europe deviate considerably from the above generalized scheme. They have wide metamorphic terranes of the low-pressure type with no glaucophane-schist belts and no ophiolites. A number of authors have doubted the validity of subduction models for this belt (Zwart, 1967; Ager, 1975; Schermerhorn, 1974). Most Precambrian orogenic belts appear to be more or less similar to the Hercynides. Dewey and Burke (1973) have ascribed the Hercynian and such Precambrian orogenies to continental collisions.

Geologic theories based on plate tectonics have assumed that continental collision is a major cause of orogeny and is probably accompanied by regional metamorphism as exemplified by the Urals and the Appalachians (Wilson, 1968; Hamilton, 1970; Bird and Dewey, 1970). However, since dating of events by magnetic anomalies has not been possible for such Paleozoic orogenic belts, interpretations of time-space relationships have been uncertain. This has made it difficult to distinguish the metamorphism associated with continental collision from that associated with the preceding subduction. On the other hand, in the Himalayas, orogenic events are Cenozoic and therefore easier to decipher. Although around 1970, even the existence of Tertiary metamorphism in the Himalayas was not clear, progress in Himalayan studies since then has established the existence of Tertiary regional metamorphism.

In recent years, very wide Precambrian metamorphic terranes have been petrographically surveyed to give us a better perspective of the nature of regional metamorphism in the remote geologic past.

GENERAL STRUCTURE OF PAIRED METAMORPHIC BELTS

Different pairs of metamorphic belts show considerably different features. For better understanding of the accounts to follow, however, a rather arbitrarily idealized cross-section of paired metamorphic belts is shown in Figure 2.

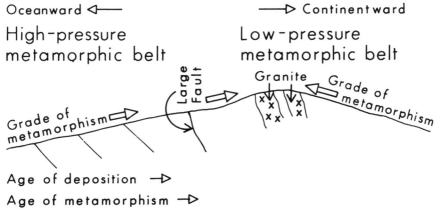

Figure 2. Idealized cross-section of paired metamorphic belts. Arrows indicate the direction of increase in metamorphic grade, in the age of sedimentary deposition and in the age of metamorphic recrystallization. Inclined lines in the high-pressure belt represent bedding.

High-pressure metamorphic belts ususally have oceanward vergence. According to Ernst (1974, 1977), it is a general rule in high-pressure belts that not only metamorphic grade (temperature of metamorphism) but also the age of sedimentary deposition and the age of metamorphic recrystallization tend to increase continentward as shown in Figure 2. The structure is quite asymmetric. On the other hand, low-pressure metamorphic belts show a more symmetric structure, having a thermal axis from which metamorphic grade decreases on both sides. Broadly syn-metamorphic granites occur in a zone along the thermal axis, while post-metamorphic granites (not shown in Fig. 2) may occur in a much wider region.

A large fault exists between the two metamorphic belts of a pair (e.g., the Median Tectonic Line between the Sanbagawa and Ryoke metamorphic belts in southwestern Japan and the Coast Range thrust between the Franciscan and Sierra Nevada metamorphic belts in California). It has usually been assumed to be a thrust, dipping to the continental side, but in many cases the evidence is not clear. As shown in Figure 2, the metamorphic terranes of the two belts may be in direct contact with each other on this fault (e.g., the Sanbagawa-Ryoke pair of southwestern Japan). In other pairs, a virtually unmetamorphosed coeval sedimentary zone, presumably representing an arc-trench gap, occurs between the two belts. This zone is in some pairs on the continental side of the large fault (e.g., in the Franciscan-Sierra Nevada pair), but is in others on the oceanic side (e.g., in the Kamuikotan-Hidaka pair in Hokkaido).

Miyashiro (1972) has suggested that the Sanbagawa-Ryoke pair originally had such an unmetamorphosed sedimentary zone between the two metamorphic belts, that the Median Tectonic Line is not a thrust, but a fault along which the terrane on the oceanic side was uplifted after metamorphism so as to expose the high-pressure type metamorphic rocks at the surface, and that the fault crossed the metamorphic belts at a small angle so that a large lateral movement accompanying the uplift resulted in the disappearance of the unmetamorphosed zone. A large fault, the Morioka-Shirakawa Line, exists along the present-day volcanic front of the Northeast Japan arc, i.e., on the continental side of the subduction zone. This may be a modern example of such faults along which the oceanic side terrane was uplifted after metamorphism.

The Benioff zone is a most remarkable feature of the present-day trenches. It is natural that the large faults between paired metamorphic belts were suspected to be fossil Benioff zones by Miyashiro (1959), even before sea-floor spreading and plate tectonics were formulated. Although Miyashiro later gave up this view for the Median Tectonic Line of Japan, it may be valid in some other cases.

The high-pressure metamorphic terrane of the Franciscan is chaotically disturbed and has no visible sialic basement, while that of the Sanbagawa belt of Japan is not so intensely disturbed and its stratigraphy has been worked out very well. The Pennine-Helvetic high-pressure metamorphic zone of the Alps has exposures of pre-Alpine metamorphic (sialic) basements. Ernst (1970, 1971, 1973) explained the difference between the three high-pressure metamorphic belts by assuming that the original sedimentary piles of the Franciscan were deposited within a Mesozoic trench, while the metamorphic terranes of the Sanbagawa belt and the Pennine-Helvetic zone represents a continental shelf, slope and rise near a trench. Much or most of the metamorphic rocks in the last two regions were derived from much older sediments or basements.

HIGH-PRESSURE TYPE METAMORPHISM IN SUBDUCTION ZONES

Even though we may accept the view that high-pressure metamorphic belts (glaucophane-schist belts) are formed in and around trench zones owing to subduction of cold oceanic lithosphere, there still exist a number of different models. The following is a modification of Miyashiro's classification of such models (1972, p. 648-649), taking more recent discussions into account.

Group 1. In this group of models, the large fault between the two metamorphic belts of a pair is regarded as a fossil Benioff zone. A high-pressure metamorphic complex was originally formed in the foot wall of the Benioff-zone fault (upper surface of the descending slab). After subduction stopped, the metamorphic complex was

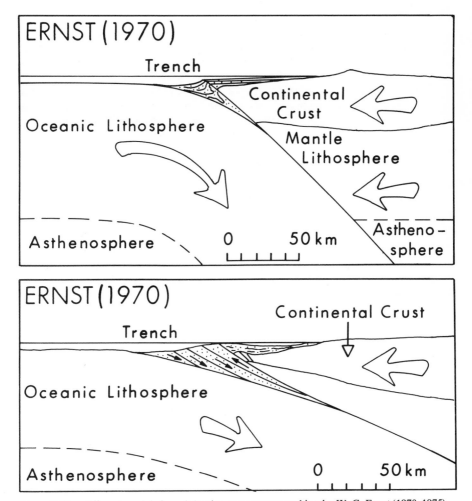

Figure 3. Two different models for subduction-zone metamorphism by W. G. Ernst (1970, 1975).

uplifted to be exposed on the surface by a reverse movement along the same fault. This view is exemplified by the early model of Ernst (1970, 1971), as illustrated in the upper diagram of Figure 3. Here, trench sediments are locked to the descending slab, and so the depth of metamorphism is directly controlled by the amount of descent of the slab. In many cases, however, the former is a few tens of kilometers, while the latter may be hundreds or thousands of kilometers. This group of models may become applicable to such cases by assuming that the sedimentary piles are not so strictly locked. By this assumption, however, the models come to resemble Group 2.

Other authors assumed that the high-pressure metamorphic complex was originally formed in the foot wall of the Benioff-zone fault, and was subsequently scraped off, added to the hanging wall, and then uplifted by upfolding, upward block movement, etc. Scraped-off metamorphic rocks, however, would not show such a regular continentward increase of metamorphic grade as observed.

Group 2. Here, high-pressure metamorphism is assumed to occur mainly in the sedimentary piles overlying a descending slab, i.e., in the hanging wall of a Benioff-zone fault. The resultant metamorphic complex is part of the crust on the continental side of the fault from the beginning, and is subsequently lifted by movement along a newly formed high-angle fault such as the Morioka-Shirakawa Line (Miyashiro, 1972, Fig. 2). Here, subduction tends to produce a stationary distribution of temperature and pressure in the descending slab and the surroundings, and the depth of metamorphism in the sedimentary pile is independent of the total amount of subduction of oceanic lithosphere.

In Group 1, a Benioff-zone fault is situated on the continental-side limit of a high-pressure metamorphic terrane, while in Group 2 it is situated at or near the oceanic-side limit of the same terrane.

Group 3. High-pressure metamorphism may occur either in the foot wall or in the hanging wall of a Benioff-zone fault. In either case, if the position of the subduction zone shifts successively oceanward, the metamorphic complex formed at an earlier time would be included in the hanging wall of a subsequently formed Benioff zone and could be lifted along a newly formed high-angle fault. In the Franciscan terrane, metamorphism involves a number of distinct phases of recrystallization, and the rocks in the western part are generally younger than those in the eastern. This may be the result of successive oceanward shifts of the Benioff zone (Hsü, 1971).

This view may be combined with models of the two earlier groups, as Ernst (1975, 1977) for example did. He assumed that during metamorphism sedimentary piles are subject to successive imbrication and underplating of younger material in their basal parts (Fig. 3, lower diagram). The subduction zone moves oceanward in steps; thus, sediments become younger oceanward. Underplating results in the elevation (buoyant rise) and subsequent exposure of older, once deeply subducted and recrystallized portions of the terrane. Low-grade metamorphic portions of the underplated younger rocks are exposed oceanward, while the coeval high-grade, more deeply subducted sections are generally exposed only after the erosional stripping of the overlying older rocks.

LOW-PRESSURE TYPE METAMORPHISM IN THE BELT
OF VOLCANIC ARC AND GRANITES

Low-pressure type metamorphic belts are usually accompanied by an abundance of granitic masses. This zone may be regarded as representing a deeper part of a volcanic arc. For example, the Mesozoic granitic belt of western North America is closely associated with broadly contemporaneous arc-type volcanic rocks. The intrusion of granitic masses accompanied by circulating aqueous fluids could be a main cause of high temperature and high geothermal gradients (Miyashiro, 1959, 1961, 1967, 1973a, b).

Recent data relevant to tectonic models of such low-pressure type metamorphic belts will now be discussed.

Origin of High Geothermal Gradients

By definition the geothermal gradient of low-pressure metamorphic belts is about 40°C/km or higher. The actual observed geothermal gradients in wide low-pressure metamorphic terranes of young orogenic belts (volcanic arc zones) are between 40° and 175°C/km (Miyashiro, 1973b, p. 140-174), in some cases even higher.

Recent explosion seismological studies in the active arc of Northeast Japan have shown that a low velocity layer of the mantle (with $P_n = 7.5 - 7.7$ km/sec) underlies a 30-km thick crust (Research Group for Explosion Seismology, 1977). A similar structure appears to exist in the Kurile and possibly in other active arcs, as well as in some other high heat-flow regions such as the Basin and Range province of the western United States. Many geophysicists regard the low-velocity layer as representing an asthenosphere. If incipient partial melting of peridotite mantle is occurring there, the temperature should be around 1200°C. Therefore, the average geothermal gradient in the overlying crust should be about 1200°/30 km = 40°C/km. The actual gradient near the surface may be even larger. Thus, the minimum value of geothermal gradient necessary for low-pressure metamorphism may well be produced simply by upward heat transfer through the crust from the underlying shallow asthenosphere.

However, the temperature of 1200°C at the base of crust is much higher than that usually calculated from observed heat flow values. Roy *et al.* (1968), for example, calculated the temperature beneath the Basin and Range province from an observed high heat-flow value and additional assumptions, and obtained a value of about 750°C at the base of a 30-km crust. If we accept this, the average geothermal gradient is only 25°C/km, a value insufficient for low pressure metamorphism.

It is possible that the low seismic velocity of the uppermost mantle does not result from incipient partial melting, but from the presence of rocks other than peridotite. Ringwood (1974), for example, assumed that the magma generated by partial melting of the oceanic crust within a descending slab reacts with the peridotites in the overlying upper mantle wedge to produce pyroxenite masses, which rise through the wedge because of their lower density. During this rise, they undergo partial melting to generate calc-alkalic magmas. If this model is tentatively accepted, the uppermost mantle beneath a volcanic arc may be largely of pyroxenite like composition, giving a lower seismic velocity than peridotite.

Therefore, it is likely that the high geothermal gradients in low-pressure metamorphic belts are usually produced by a combination of simple heat transfer from the underlying mantle with some additional factors, the most important of which would be the upward movement of granitic magmas and associated volatiles. If a granitic mass at, say, 800°C rises to reach a depth of 10 km, the average geothermal gradient above the mass is 80°C/km As shown schematically in Figure 2, low-pressure metamorphic belts usually have a nearly symmetric structure with a median thermal axis (high-temperature axis), where granitic masses, broadly contemporaneous with metamorphism, occur.

Origin of Regional Scale Isograds

If granitic masses are essential in controlling temperature distribution, we come to the question: What is the difference between contact and regional metamorphism? This is a time-honoured problem. It is widely known that two contrasting views have been expressed in relation to the low- to medium-pressure regional metamorphism of the Scottish Caledonides. Barrow (1893) regarded the regional metamorphic terrane as nothing but a large-scale contact aureole around numerous small granitic intrusions. Thus he saw no essential difference between regional and contact metamorphism. On the other hand, Harker (1932, p. 177) emphasized that regional metamorphism is characterized not only by its wide areal extent but also by a regional-scale temperature distribution (progressive mineral zones), independent of the distribution of individual plutonic masses. In other words, regional metamorphism is thought to be distinct from contact (for a review of the two positions, see Miyashiro, 1973b, p. 24-26).

Field observations in other regions suggest an actual situation intermediate between the two extreme views. Syn- or nearly syn-metamorphic granitic rocks tend to be abundant in the highest grade zone, i.e. along the thermal axis. However, isograds do not strictly encircle individual intrusions, but indicate a more regional-scale trend. Such low- to medium-pressure type regional metamorphic terranes presumably grade into contact aureoles with decreasing depth in the crust. For example, Devonian regional metamorphism in the New England Appalachians grades into contact metamorphism towards the northeast (Thompson and Norton, 1968; Hamilton and Myers, 1967, p. 14). The existence of contact aureoles around granites does not in itself contradict the former participation of the same granitic magma in regional-scale heating. Granitic magmas, after participating in regional-scale heating, may continue to move in a declining phase of regional metamorphism, thus leading to the formation of distinct contact aureoles around them.

The regional-scale isograds more or less independent of individual plutons may be largely produced by a flow of hot H_2O related in some way to granitic plutons, or by H_2O and CO_2 rising from the mantle. A large part of such H_2O may be of surficial (not juvenile) origin. Strong evidence for the surficial origin of water in the surroundings of granitic plutons has been obtained from recent studies of oxygen and hydrogen isotopes in rocks, for example, of Mesozoic and Tertiary granitic batholiths in western North America (Taylor, 1977). Large-scale convection of water occurs around granitic intrusions because of their thermal effects, and the flowing water causes regional-scale isotopic exchange reactions. The areas where isotopic exchange occurs are much wider than contact aureoles, and tend to show a regional-scale extension (say, 100 km wide) enclosing a number of small intrusions, just like progressive mineral zones in regional metamorphism. On the other hand, Schuiling and Kreulen (1979) give evidence suggesting the existence of hot CO_2-rich fluids rising from the mantle to cause metamorphism.

Deep Structure of the Arc Zone

There is a great uncertainty in our knowledge concerning the deep structure of island arcs, as exemplified by the above discussion on the nature of the low velocity layer in the uppermost mantle. Hence the formulation of a model for the generation of granitic magmas is difficult.

In early models based on plate tectonics (Oxburgh and Turcotte, 1968; Dickinson, 1970), it was assumed that arc magmas are generated by frictional heating along the upper surface of a descending slab, and that the magmas rise to produce both arc volcanism and granitic intrusion. Some recent models assume a more complicated process such as reactions with material in the upper mantle wedge or the crust, or magma generation in the upper mantle wedge itself. Sr-isotope studies in the 1970s have revealed considerable difficulties in such models (e.g., Kistler and Peterman, 1973; Brooks et al., 1976). This has led some authors to assume that the continental crust is underlain by an ancient mantle lithosphere which may consist of horizontal layers decreasing in age with depth (Brooks et al., 1976; cf. Carter et al., 1978).

METAMORPHISM ASSOCIATED WITH CONTINENTAL COLLISION

The Himalayas are divided into several longitudinal zones showing different rock facies and metamorphism, including the Tibetan, the Higher, and the Lower Himalayas from north to south. Small granitic masses of Tertiary radiometric ages occur in all three named zones. Most of the Higher and a majority of the Lower Himalayas are metamorphosed. They may include metamorphosed Precambrian basements of the Indian subcontinent (partly reactivated) as well as true Tertiary metamorphic rocks formed by continental collision. Though ealrier authors claimed that most or all the metamorphism was Precambrian (e.g., Gansser, 1964), recent workers have preferred Tertiary ages for a large part of the metamorphism (e.g., Frank et al., 1973; Le Fort, 1975). This is corroborated by an Oligocene or younger Rb-Sr isochron age of associated granites (e.g., Hamet and Allegre, 1976).

The major part of Himalayan regional metamorphism appears to be of the medium-pressure type (Barrovian sequence), though some parts belong to the low-pressure type with andalusite, and glaucophane has been claimed to occur at a few places (Ohta, 1973).

The continental collision has generated an unusually thick crust in the Himalayas and Tibet, though the mechanism of its formation is not clear. Partial melting may have occurred in its basal part, leading to the generation of granitic magmas. The Cenozoic granite in the Himalayas and coeval volcanic rocks in Tibet may have resulted from the rise of the magmas.

Peter Bird (1978) has applied the concept of flake tectonics (Oxburgh, 1972) to the Himalayas. It has been assumed that after India collided with Eurasia, the mantle lithosphere of the northern edge of the Indian plate was split from the overlying crustal layer, and plunged deep into the asthenosphere. Thus the Himalayan crust came into direct contact with, and was intensely heated by, the underlying asthenosphere, resulting in regional metamorphism and granite formation.

Many geologists have suggested so-called inverted metamorphism in large parts of the Higher and the Lower Himalayas (e.g., Ray, 1947; Gansser, 1964; Le Fort, 1975). Recent authors have attempted to explain it by syn-metamorphic folding or thrusting (Frank *et al.*, 1973; Le Fort, 1975). The observed evidence for inverted metamorphism is that the strata and thrust in those regions dip generally to the north, while metamorphic grade increases northward to apparently higher horizons. This has been interpreted as indicating that higher-grade metamorphic rocks lie over lower-grade rocks. This assumes that isogradic surfaces are parallel to lithologic and strati-graphic boundaries. However, observations at some places have proved this assumption incorrect (Ohta, 1973, p. 254). It may well be doubtful whether metamorphic grade shows any regular relationship to stratigraphic horizons.

SECULAR CHANGE IN THE TYPE OF REGIONAL METAMORPHISM

We are interested in the relationship between the type of metamorphism and geologic age, because a change with age in the type of regional metamorphism, if any, probably means a change in the nature of tectonic processes. Eskola (1939, p. 368) was the first to suggest a possible relationship between metamorphism and geologic age in this context. He stated that glaucophane and eclogite are very rare in Precambrian basement complexes.

Later, de Roever (1956) discussed the relationship at some length. He claimed that lawsonite is confined to, and glaucophane formed mainly in, Mesozoic and younger metamorphism, while biotite and hornblende formed mainly in Paleozoic and older metamorphism. Here, glaucophane indicates the glaucophane-schist facies, and law-sonite a typical part of the facies. On the other hand, biotite and hornblende suggest temperatures higher than those required for the glaucophane-schist facies. The metamorphism to produce them may be of amphibolite and contiguous facies, as observed in the Hercynides and Caledonides in Europe. Thus, de Roever's claim means that the P-T conditions of metamorphism changed at the boundary between Paleozoic and Mesozoic time.

This view of the glaucophane-schist facies was accepted by later authors, including Ernst (1973) in particular. Ernst added that the jadeite + quartz assemblage and metamorphic aragonite, both characteristic of the typical glaucophane-schist facies, are confined to Mesozoic and later metamorphism. He ascribed the change in metamorphism to the general cooling of the Earth with resultant decrease in geothermal gradient and thickening of plates.

Table 1 of Ernst's (1972) paper, however, showed a considerable number of Paleozoic as well as some possible examples of late Proterozoic glaucophane schists. Subsequent studies identified new areas of Paleozoic glaucophane rocks. The paired metamorphic belts in Chile, having a particularly wide glaucophane-schist terrane, are Paleozoic (Munizaga *et al.*, 1973; Hervé *et al.*, 1974).

On the other hand, there is no evidence for the formation of glaucophane-schist facies rocks in middle Proterozoic and older time. Recently, this fact has been used to support the view that subduction and plate tectonics did not begin until the late Proterozoic (e.g., Duff and Langworthy, 1974; Wynne-Edwards, 1976).

Although de Roever's claim of the rarity of biotite and hornblende in Mesozoic and younger metamorphism may be generally true in Europe, it contradicts observations in other regions of the world. In the circum-Pacific regions such as Japan, Mesozoic low-pressure metamorphic belts are widespread, and biotite is abundant in almost all grades.

Petrographic studies in Precambrian terranes show that most of the world's Precambrian regional metamorphism is of the low-pressure and medium-pressure types. Low-pressure regional metamorphism has been widespread throughout geological time, from the Archaean to the present (e.g., Miyashiro, 1973, p. 110). It is not clear whether such metamorphism was more frequent in Precambrian than in later times. For a more detailed discussion of this problem, see Miyashiro (1973, p. 106-112).

ACKNOWLEDGEMENTS

I am grateful to John M. Allen, Dugald Carmichael, Pat Ohlendorf and Fumiko Shido for their comments on the original manuscript.

REFERENCES

Ager, D.V., 1975, The geological evolution of Europe: Proc. Geol. Assoc. London, v. 86, p. 127-154.

Armstrong, R.L. and Suppe, J., 1973, Potassium-argon geochronology and Mesozoic igneous rocks in Nevada, Utah and southern California: Geol. Soc. America Bull., v. 84, p. 1375-1392.

Barrow, G., 1893, On an intrusion of muscovite-biotite gneiss in the southeastern Highlands of Scotland, and its accompanying metamorphism: Jour. Geol. Soc. London, v. 49, p. 330-358.

Bird, J.M. and Dewey, J.F., 1970, Lithosphere plate-continental margin tectonics and the evolution of the Appalachian orogen: Geol. Soc. America Bull., v. 81, p. 1041-1060.

Bird, P., 1978, Initiation of intracontinental subduction in the Himalaya: Jour. Geophys. Res., v. 83, p. 4975-4987.

Brooks, C., James, D.E. and Hart, S.R., 1976, Ancient lithosphere: its role in young continental volcanism: Science, v. 193, p. 1086-1094.

Carter, S.R., Evensen, N.M., Hamilton, P.J. and O'Nions, R.K., 1978, Neodymium and strontium isotope evidence for crustal contamination of continental volcanics: Science, v. 202, p. 743-747.

de Roever, W.P., 1956, Some differences between post-Paleozoic and older regional metamorphism: Geol. en Mijinb. (N.S.), v. 18, p. 123-127.

Dewey, J.F. and Bird, J.M., 1970, Mountain belts and the new global tectonics: Jour. Geophys. Res., v. 75, p. 2625-2647.

Dewey, J.F. and Burke, K.C.A., 1973, Tibetan, Variscan and Precambrian basement reactivation: products of continental collision: Jour. Geol. v. 81, p. 683-692.

Dickinson, W.R., 1970, Relations of andesites, granites and derivative sandstones to arch-trench tectonics: Rev. Geophys. and Space Phys., v. 8, p. 813-860.

Draper, G., Harding, R.R., Horsfield, W.T., Kemp, A.W. and Tresham, A.E., 1976, Low-grade metamorphic belt in Jamaica and its tectonic implications: Geol. Soc. America Bull., v. 87, p. 1283-1290.

Duff, B.A. and Langworthy, A.P., 1974, Orogenic zones in central Australia: intraplate tectonics?: Nature, v. 249, p. 645-647.

Ernst, W.G., 1970, Tectonic contact between the Franciscan mélange and the Great Valley Sequence—crustal expression of a late Mesozoic Benioff zone: Jour. Geophys. Res., v. 75, p. 886-901.

──────────, 1971, Metamorphic zonations on presumably subducted lithospheric plates from Japan, California and the Alps: Contr. Min. Petrol., v. 34, p. 43-59.

──────────, 1972, Occurrence and mineralogic evolution of blueschist belts with time: Amer. Jour. Sci., v. 272, p. 657-668.

──────────, 1973, Blueschist metamorphism and P-T regimes in active subduction zones: Tectonophysics, v. 17, p. 255-272.

──────────, 1975, Systematics of large-scale tectonics and age progressions in Alpine and circum-Pacific blueschist belts: Tectonophysics, v. 26, p. 229-246.

──────────, 1977, Mineral parageneses and plate tectonic settings of relatively high-pressure metamorphic belts: Fortschr. Miner., v. 54, p. 192-222.

Eskola, P., 1939, Die metamorphen Gesteine: in T.F.W. Barth, C.W. Correns, and P. Eskola, Die Entstehung der Gesteine: Julius Springer, Berlin, p. 263-407.

Frank, W., Hoinkes, G., Miller, C., Purtscheller, F., Richter, W. and Thöni, M., 1973, Relations between metamorphism and orogeny in a typical section of the Indian Himalayas: Tschermaks Min. Petr. Mitt., v. 20, p. 303-332.

Gansser, A., 1964, Geology of the Himalayas: Interscience Publishers, New York, 289 p.

Hallam, A., 1973, A Revolution in the Earth Sciences: Clarendon Press, Oxford, 127 p.

Hamet, J. and Allègre, C.-J., 1976, Rb-Sr systematics in granite from central Nepal (Manaslu): significance of the Oligocene age and high $^{87}Sr/^{86}Sr$ ratio in Himalayan orogeny: Geology, v. 4, p. 470-472.

Hamilton, W., 1970, The Uralides and the motion of the Siberian platforms: Geol. Soc. America Bull., v. 81, p. 2553-2576.

Hamilton, W. and Myers, W.B., 1967, The nature of batholiths: U.S. Geol. Survey Prof. Paper 554C.

Harker, A., 1932, Metamorphism: Methuen, London, 362 p.

Hervé, F., Munizaga, F., Godoy, E. and Aguirre, L., 1974, Late Paleozoic K/Ar ages of blueschists from Pichilemu, central Chile: Earth Planet. Sci. Letters, v. 23, p. 261-264.

Holdaway, M.J., 1971, Stability of andalusite and the aluminum silicate phase diagram: Amer. Jour. Sci., v. 271, p. 97-131.

Hsü, K.J., 1971, Franciscan mélanges as a model for eugeosynclinal sedimentation and underthrusting tectonics: Jour. Geophys. Res., v. 76, p. 1162-1170.

Kistler, R.W. and Peterman, Z.E., 1973, Variations in Sr, Rb, K, Na, and initial Sr^{87}/Sr^{86} in Mesozoic granitic rocks and intruded wall rocks in central California: Geol. Soc. America Bull., v. 84, p. 3489-3512.

Laird, J. and Albee, A. L., 1975, Polymetamorphism and the first occurrence of glaucophane and omphacite in northern Vermont: Geol. Soc. Amer., Abstracts with Programs, v. 7, no. 7, p. 1159.

LeFort, P., 1975, Himalayas: the collided range. Present knowledge of the continental arc: Amer. Jour. Sci., v. 275-A, p. 1-44.

Miyashiro, A., 1959, Abukuma, Ryoke and Sanbagawa metamorphic belts (in Japanese with English abstract): Jour. Geol. Soc. Japan, v. 65, p. 624-637.

──────────, 1961, Evolution of metamorphic belts: Jour. Petrol., v. 2, p. 277-311.

──────────, 1967, Orogeny, regional metamorphism, and magmatism in the Japanese Islands: Medd. fra Dansk Geol. Forening, Kφbenhavn, v. 17, p. 390-446.

──────────, 1972, Metamorphism and related magmatism in plate tectonics: Amer. Jour. Sci., v. 272, p. 629-656.

——————————, 1973a, Paired and unpaired metamorphic belts: Tectonophysics, v. 17, p. 241-254.

——————————, 1973b, Metamorphism and Metamorphic Belts: George Allen and Unwin Ltd., London, 492 p.

Munizaga, F., Aguirre, L. and Hervé, F., 1973, Rb/Sr ages of rocks form the Chilean metamorphic basement: Earth Planet. Sci. Letters, v. 18, p. 89-92.

Nagle, F., 1974, Blueschist, eclogite, paired metamorphic belts and the early tectonic history of Hispaniola: Geol. Soc. Amer. Bull., v. 85, p. 1461-1466.

Ohta, Y., 1973, Geology of the Nepal Himalayas: in Hashimoto, S., Ohta, Y. and Akiba, C., eds., Geology of the Nepal Himalayas: Himalayan Committee of Hokkaido University, Sapporo, Japan, Saikon Publishing Co., p. 235-259.

Oxburgh, E.R., 1972, Flake tectonics and continental collision: Nature, v. 239, p. 202-204.

Oxburgh, E.R. and Turcotte, D.L., 1968, Problems of high heat flow and volcanism associated with zones of descending mantle convective flow: Nature, v. 218, p. 1041-1043.

Popp, R.K. and Gilbert, M.C., 1972, Stability of acmite-jadeite pyroxenes at low pressure: Amer. Mineral, v. 57, p. 1210-1231.

Ray, S., 1947, Zonal metamorphism in the eastern Himalaya and some aspects of local geology: Geol. Mining Metall. Soc. India Quart. Jour., v. 19, p. 117-142.

Research Group for Explosion Seismology, 1977, Regionality of the upper mantle around northeastern Japan as derived from explosion seismic observations and its seismological implications: Tectonophysics, v. 37, p. 117-130.

Ringwood, A.E., 1974, The petrological evolution of island arc systems: Jour. Geol. Soc. London, v. 130, p. 183-204.

Roy, R.F., Blackwell, D.D. and Birch, F., 1968, Heat generation of plutonic rocks and continental heat flow provinces: Earth Planet. Sci. Letters, v. 5, p. 1-12.

Schuiling, R.D. and Kreulen, R., 1979, Are thermal domes heated by CO_2-rich fluids from the mantle?: Earth Planet. Sci. Letters, v. 43, p. 298-302.

Schermerhorn, L.J.G., 1974, Variscan specialists meet in Rennes: Geotimes, v. 19, p. 23-25.

Taylor, H.P., Jr., 1977, Water/rock interactions and the origin of H_2O in granitic batholiths: Jour. Geol. Soc. London, v. 133, p. 509-558.

Thompson, J.B. Jr. and Norton, S.A., 1968, Paleozoic regional metamorphism in New England and adjacent areas: in Zen, E-an, White, W.S., Hadley, J.B. and Thompson, J.B., eds., Studies of Appalachian Geology: Northern and Maritime: Interscience Publishers, New York, p. 319-327.

Wilson, J. Tuzo, 1968, Static or mobile earth: the current scientific revolution: Amer. Phil. Soc. Proc., v. 112, p. 309-320.

Wynne-Edwards, H.R., 1976, Proterozoic ensialic orogenesis: the millipede model of ductile plate tectonics: Amer. Jour. Sci., v. 276, p. 927-953.

Zwart, H.J., 1967, The duality of orogenic belts: Geol. Mijnbouw, v. 46, p. 283-309.

The Continental Crust and Its Mineral Deposits, edited by D.W. Strangway,
Geological Association of Canada Special Paper 20

GLOBAL PLATE MOTION AND MINERAL RESOURCES

Peter A. Rona
Atlantic Oceanographic and Meteorological Laboratories,
National Oceanic and Atmospheric Administration,
15 Rickenbacker Causeway,
Virginia Key,
Miami, Florida 33149

ABSTRACT

Previous studies have related the distribution of mineral and energy resources to plate tectonic settings, providing the exploration geologist with a useful spatial framework. The present study contributes to a framework in both space and time by recognizing that the distribution of mineral and energy resources is coordinated with a globally organized pattern of plate motion within each geologic era and undergoes global reorganization at time intervals of the order of 100×10^6 years during the transition between eras in the geologic time scale. Principles of this coordination between global plate motion and resources that can be applied to guide regional exploration are inferred from consideration of a documented plate reorganization in the early Cenozoic era, with examples based on the distribution of stratabound massive sulphides with volcanic affinities, porphyry copper-molybdenum deposits, and giant oil and gas fields, as follows:

1) The mineral and energy resources formed during a given geologic era exhibit a distinct distribution controlled by tectonic settings created by the global pattern of plate motion prevalent during that era. The globally organized pattern of plate motion favours the formation of certain types of deposits including porphyry copper-molybdenum.

2) A global reorganization of plate motion occurs during the transition between geologic eras. The reorganization involves creation of new tectonic settings, as well as continuation, reorientation, or conversion of pre-existing tectonic settings. The distribution of mineral and energy resources formed in the reorganized tectonic settings changes accordingly. The global reorganization favours the emplacement of ophiolites with potential for stratabound massive sulphides, and possibly the entrapment of hydrocarbons.

3) The lengths of divergent, convergent (subduction, obduction, collision), and transform

plate boundaries change during, and tend to stabilize after, a global reorganization of plate motion. This reapportions the areas of various associated tectonic settings and prospective regions for the formation of mineral and energy resources.

RÉSUMÉ

Des études antérieures ont rattaché la distribution des ressources en minéraux et en énergie à des contextes de tectonique des plaques, ce qui fournit au géologue d'exploration un cadre spatial utile. Cette étude contribue à définir ce cadre dans le temps et dans l'espace en reconnaissant que la distribution des ressources minérales et énergétiques correspond à un patron bien organisé de mouvement des plaques à l'échelle globale pour chaque ère géologique et qu'elle subit une réorganisation globale à des intervalles de temps de l'ordre de 100 Ma durant la transition entre les ères géologiques. Les principes de cette correspondance entre les mouvements des plaques et les ressources qu'on peut appliquer pour guider l'exploration régionale sont tirés d'une considération de la réorganisation documentée des plaques au début du Cénozoïque, avec des exemples basés sur la distribution des sulfures massifs stratifiés d'affinités volcaniques, les dépôts de cuivre-molybdène porphyriques et les immenses champs de pétrole et de gaz, de la manière suivante: 1) Les ressources énergétiques et minérales formées durant une ère géologique donnée montrent une distribution distincte qui est contrôlée par les contextes tectoniques créés par le patron global des mouvements de plaques prévalents durant cette ère. Le patron organisé à l'échelle globale des mouvements de plaques favorise la formation de certains types de dépôts y compris les dépôts porphyriques de cuivre-molybdène. 2) Une réorganisation globale des mouvements de plaques se produit durant la transition entre les ères géologiques. La réorganisation implique la création de nouveaux contextes tectoniques de même que la continuation, la réorientation ou la conversion de contextes tectoniques pré-existants. La distribution des ressources minérales et énergétiques qui se sont formées dans les contextes tectoniques réorganisés change de la même façon. La réorganisation globale favorise la mise en place des ophiolites avec un potentiel pour les sulfures massifs stratifiés et probablement pour le piégeage des hydrocarbures. 3) La longueur des limites de plaques divergentes, convergentes (subduction, obduction, collision) et transformantes changent durant une réorganisation globale du mouvement des plaques et tendent à se réorganiser par après; ceci redistribue les aires des différents contextes tectoniques associés et les régions prometteuses pour la formation des ressources minérales et énergétiques.

INTRODUCTION

The theory of plate tectonics provides a framework for understanding the distribution of mineral and energy resources in space and time. In space, plate motions create tectonic settings that control the formation of these resources. The two major tectonic settings, as recognized in the Wilson cycle, are opening ocean basins, created by divergent plate motions, and closing ocean basins, created by convergent plate motions. Previous studies have related the spatial distribution of mineral and energy resources to major and certain subsidiary tectonic settings. This information is summarized in Table I (e.g., Bally, 1975; Bonatti, 1975; Guild, 1972a; Mitchell and Bell, 1973; Mitchell and Garson, 1976; Rona, 1977; Hutchinson, 1973; Sawkins, 1972; Sillitoe, 1972a, 1972b).

It is more difficult to relate the distribution of mineral and energy resources to the development of tectonic settings through time because of the complexities of plate interactions. In the static Earth model, geotectonic cycles of mountain building followed a predetermined sequence of events based upon a finite number of varia-

TABLE I
TECTONIC SETTINGS AND RESOURCES

Major Tectonic Setting	Subsidiary Tectonic Settings	Energy and Mineral Resources	References
Opening ocean basin	Intraplate continental rifts and hot spots: domal uplifts, incipient rifts, aborted rifts, and aulacogens, Atlantic-type passive continental margins (miogeoclines).	Hydrocarbons, fluorite, Mississippi-Valley-type Pb-Zn, stratabound massive Cu-Fe sulphides, magmatic Cu-Ni, disseminated and vein Sn, magmatic Pt, carbonatite rare earth elements, kimberlite diamonds, certain layered mafic igneous complexes Pt-Cr-Fe-Cu-Ni.	Burke and Dewey, 1973; Dewey and Burke, 1973; Sawkins, 1976
	Sea-floor spreading centres.	Metalliferous sediments (Cu, Fe, Mn, Ni, Zn, Co, Ba), stratiform manganese and iron oxides and iron silicates, Cyprus-type massive Cu-Fe sulphides, podiform chromite, asbestos, Cu-Ni sulphides, platinoids.	Rona, 1978
	Transform fault and fracture zone.	Metalliferous sediments, stratiform manganese oxides and iron silicates, disseminated and vein Cu-Fe-Pb-Zn sulphides, mercury.	Bonatti *et al.*, 1976; McLaughlin *et al.*, 1979
Closing Ocean basin	Subduction zones: oceanic trenches, volcanic island arcs, marginal ocean basins, Andean-type active continental margins.	Hydrocarbons, porphyry Cu-Mo-Ag-Au-Sn, contact metasomatic Cu-Pb-Zn-Ag-Au-Sn-Mo-W, vein Cu-Ag-Au-Sn-W, volcanogenic Kuroko-type massive sulphide Cu-Fe-Pb-Zn-Ag-Ba.	Sawkins, 1972; Sillitoe, 1972a, b; Guild, 1972a, b; Mitchell and Bell, 1973; Mitchell and Garson, 1972; Mitchell, 1976
	Obduction zones: island-arc- or oceanic-ridge-continent collisions.	Ophiolites that may contain mineral resources like those in oceanic lithosphere at sea-floor spreading centres.	Coleman, 1977
	Continent-continent collision zones.	Anatectic lithophile Sn-W-Mo-Bi-U, anorthositic Fe-Ti, certain pegmatitic and morphic gemstones; possible juxtaposition of pre-existing metallogenic provinces; possible entrapment of oil and gas in miogeosynclinal strata by overriding nappes.	Dewey and Burke, 1973; Sillitoe, 1978; Ray and Acharyya, 1976

tions on the theme of sediment accumulation in elongate belts (geosynclines) and subsequent crustal accretion involving relatively small horizontal movement and large vertical uplift. Different tectonic settings and developmental stages of the predetermined geosynclinal cycle were related to the formation in time of hydrocarbon accumulations and specific metal deposits (e.g., Bilibin, 1968). In the dynamic Earth model of plate tectonics, however, geotectonic cycles follow an apparently random, rather than predetermined, sequence of events, based upon an infinite number of variations depending on continually changing plate interactions (Curray, 1975).

Evidence of plate motions during the Mesozoic and Cenozoic eras suggests that although plate motions may appear random within relatively short intervals of time, the motions exhibit a globally organized pattern within each geologic era and undergo global reorganization at time intervals of the order of 100×10^6 years – that is, during the transition between eras in the geologic time scale (Rona and Richardson, 1978). This paper first reviews the spatial relation between plate motions and the distribution of mineral and energy resources, then describes temporal changes in plate motion in the form of a documented global plate reorganization, and finally considers the influence of periodic global reorganization of plate motion on the distribution of mineral and energy resources in space and time.

SPATIAL RELATIONS BETWEEN PLATE MOTIONS AND RESOURCES

The opening of an ocean basin by creation of oceanic lithosphere about a divergent plate boundary is accompanied by a characteristic distribution of mineral and energy resources (Table I). Incipient rifting of a continent, presumably associated with mantle plume activity (Morgan, 1972), creates a complex distribution of subsidiary tectonic settings, including domal uplifts and active and aborted rift zones. "Hot spots" associated with the domal uplifts and other rifts may generate various types of metal deposits (Table I; Sawkins, 1976). The failed arms of three-armed rift systems at triple-point junctions (Burke and Dewey, 1973) generally trend parallel to or at an obtuse angle to the predominant rift zone. They may become aulacogens with long axes transverse to continental margins, which constitute a favourable environment for the accumulation of hydrocarbons (Burke and Wilson, 1976). Continued rifting of continental fragments about a divergent plate boundary may form an evaporite basin, with its long axis parallel to the plate boundary. This also constitutes an environment for the accumulation of hydrocarbons during an early stage in the opening of an ocean basin (Rona, 1969, 1970, 1976).

The establishment of a sea-floor spreading centre in an opening ocean basin creates a tectonic setting for the concentration of hydrothermal mineral deposits. Metalliferous sediments, stratiform metallic oxide encrustations, and massive metallic sulphide bodies may form on or within the basaltic layer (layer 2) of oceanic crust (Rona, 1978). Structural and thermal conditions concentrate these deposits at discrete sites along sea-floor spreading centres. Narrow mineralized strips transverse to the axis of the spreading centres may be generated about the hydrothermal sites both at and between oceanic fracture and transform fault zones; the length of such strips depends on the the persistence of the hydrothermal processes with time and the rate and direction of sea-floor spreading. Magmatic and metamorphic mineral deposits, including podiform chromite and asbestos, that form in ultramafic rocks of oceanic crust and upper mantle may also be distributed in linear zones transverse to the spreading centres (Table I).

The closing of an ocean basin by consumption of oceanic lithosphere at convergent plate boundaries around its perimeter is accompanied by another characteristic distribution of mineral and energy resources. Formation of hydrocarbons is favoured where organic-rich sediments accrete, are preserved, and undergo proper thermal maturation. This can occur: 1) on the landward wall of oceanic trenches above subduction zones parallel to convergent plate boundaries, and 2) in marginal basins between island arcs and continents landward of convergent plate boundaries. These two hydrocarbon settings can be seen at the eastern and western margins, respectively, of the Pacific (Rona and Neuman, 1976a, 1976b). Porphyry, contact metasomatic, and vein-type metal deposits are associated with intrusives that are overlain by co-magmatic extrusives of the calc-alkaline magmatic suite. This configuration develops at structurally controlled sites on Andean-type continental margins and volcanic island arcs above subduction zones parallel to convergent plate boundaries; in addition, Kuroko-type massive sulphide deposits are present on island arcs (Table I; Sillitoe, 1972a, 1972b; Sawkins, 1972; Mitchell and Garson, 1972; Guild 1972a, 1972b; Hutchinson, 1973; Ringwood, 1974; Mitchell, 1976).

Ophiolites that may contain minerals characteristic of oceanic lithosphere are obducted in belts parallel to convergent plate boundaries (Table I; Coleman, 1977). The overthrusting and thickening of continental lithosphere during and following continental collision may generate anatectic lithophile-element deposits (Sillitoe, 1978), iron-titanium deposits in anorthosites (Dewey and Burke, 1973), and certain gemstones associated with pegmatites and zones of high-temperature metamorphism (Sillitoe, 1978); these deposits exist at discrete structurally controlled sites distributed along the continental collision zone. Continental collision may also juxtapose pre-existing metallogenic provinces; for example, collision of the Indian plate with the southeastern margin of the Iran-Afghanistan microcontinent juxtaposed miogeoclinical, ophiolitic, and calc-alkaline metallogenic provinces (Sillitoe, 1978). Nappes in the collision zone may entrap hydrocarbons where present in underlying miogeoclinal strata, as inferred for the Alpine-Himilayan geosyncline (Ray and Acharyya, 1976).

TEMPORAL CHANGES IN GLOBAL PLATE MOTION

The Early Mesozoic era was a time of reorganization of global plate motion; the breakup and dispersion of Pangaea began at that time (Dietz and Holden, 1970). Based on the rate and direction of sea-floor spreading , the incidence of continental collision, and the obduction of ophiolites, Rona and Richardson (1978) deduced a subsequent global reorganization from Mesozoic to Cenozoic patterns of plate motion that occurred primarily during the Eocene epoch (53.5 to 37.5 Ma ago).

Regional changes in plate motion, as synthesized by Rona and Richardson (1978), reveal the following patterns of Early Cenozoic plate reorganization (Figs. 1-4).

1) Major reorientation of those relative plate motions with large N-S components to large E-W components. This pattern is exhibited by changes in the direction of spreading in the Indian Ocean basin (53 to 50 Ma ago) and the North Pacific Ocean

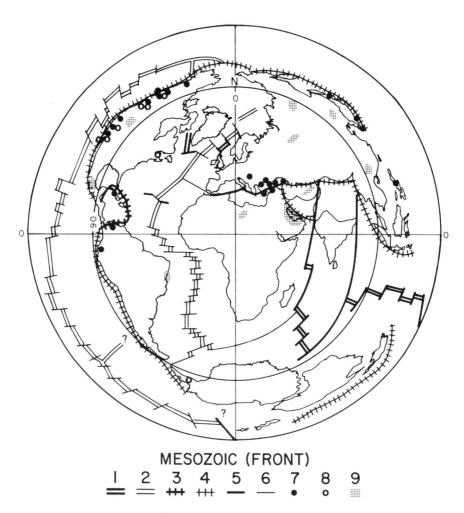

MESOZOIC (FRONT)

| 1 | 2 | 3 | 4 | 5 | 6 | 7 | 8 | 9 |
| = | = | +++ | +++ | — | — | ● | ○ | ▦ |

Figure 1. Mesozoic plate boundaries and certain resources before the Early Cenozoic global reorganization of plate motion; plotted on Lambert equal area projection showing Eocene (50 ±5 Ma) positions of continents (modified from Rona and Richardson, 1978). Front view. (1) Changed divergent boundary; (2) Unchanged divergent boundary; (3) Changed convergent boundary; (4) Unchanged convergent boundary; (5) Changed transform boundary; (6) Unchanged transform boundary; (7) Stratabound massive sulphide deposit with volcanic affinity (Guild, 1972b; Ridge, 1972); (8) Porphyry copper-molybdenum deposit (Hollister, 1973; Argall and Wyllie, 1976; Rowley *et al.*, 1977); (9) Giant oil or gas field (Halbouty *et al.*, 1970; Moody, 1975).

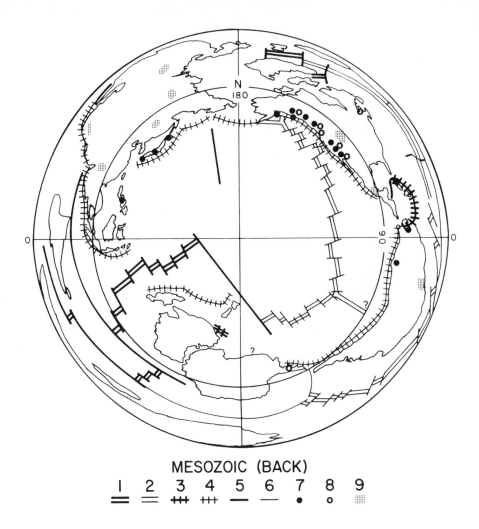

MESOZOIC (BACK)

1	2	3	4	5	6	7	8	9
═	═	+++	+++	—	—	●	○	⠿

Figure 2. Mesozoic plate boundaries and certain resources before the Early Cenozoic global reorganization of plate motion; plotted on Lambert equal area projection showing Eocene (50 ±5 Ma) positions of continents (modified from Rona and Richardson, 1978). Back view. 1-9 as in Figure 1.

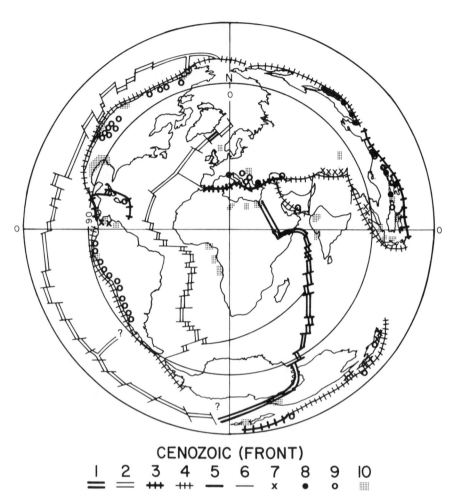

CENOZOIC (FRONT)
```
 I   2   3   4   5   6   7   8   9  IO
 =   =  +++ +++  —   —   x   ●   o  ▦
```

Figure 3. Cenozoic plate boundaries and certain resources after the Early Cenozoic global reorganization of plate motion; plotted on Lambert equal area projection showing Eocene (50 ±5 Ma) positions of continents (modified from Rona and Richardson, 1978). Front view. (1) Changed divergent boundary; (2) Unchanged divergent boundary; (3) Changed convergent boundary; (4) Unchanged convergent boundary; (5) Changed transform boundary; (6) Unchanged transform boundary; (7) Eocene ophiolite; (8) Stratabound massive sulphide deposit with volcanic affinity; (9) Porphyry copper-molybdenum deposit; (10) Giant oil or gas field.

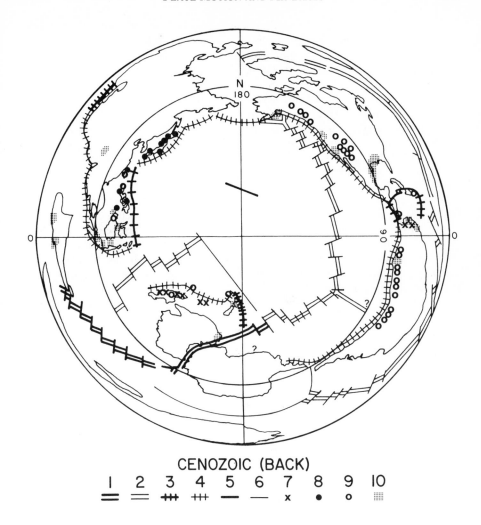

CENOZOIC (BACK)

| 1 | 2 | 3 | 4 | 5 | 6 | 7 | 8 | 9 | 10 |

Figure 4. Cenozoic plate boundaries and certain resources after the Early Cenozoic global reorganization of plate motion; plotted on Lambert equal area projection showing Eocene (50 ±5 Ma) positions of continents (modified from Rona and Richardson, 1978). Back view. 1-10 as in Figure 3.

basin (55 to 50 Ma ago), as well as by a change in the direction of motion of the North Pacific sea floor with respect to the Hawaiian hot spot (43 to 41 Ma ago).

2) Continuation, with only minor reorientation, of those relative plate motions with large E-W components and initiation of new E-W plate motions. This pattern is exhibited by continued E-W sea-floor spreading about the Mid-Atlantic Ridge in the North Atlantic and South Atlantic ocean basins, and by the initiation of predominantly E-W spreading about centres in the Arctic Ocean (63 Ma ago), the northern North Atlantic Ocean basin (56 to 42 Ma ago), and the Red Sea (38 Ma ago).

3) Decrease in the rate of sea-floor spreading concurrent with both major and minor E-W reorientation of sea-floor spreading (patterns 1 and 2). This occurred in the Indian (53 to 40 Ma ago), South Pacific (56 Ma ago), and North Atlantic ocean basins (59 to 50 Ma ago).

4) Change in the nature of motion at certain plate boundaries between convergence-with-compression and strike-slip. Such change occurred antithetically in two different regions in middle to low latitudes. In the Caribbean region, N-S compression and subduction at E-W-trending convergent plate boundaries changed to predominantly E-W strike-slip motion (54 to 38 Ma ago). In the Mediterranean region, predominantly E-W- trending strike-slip motion changed to N-S compression at a convergent plate boundary (53 Ma ago).

5) Conversion of certain transform plate boundaries to convergent plate boundaries as a consequence of major reorientation of plate motions. During the change in direction of sea-floor spreading from large N-S to E-W components in the Indian and North Pacific ocean basins (55 to 41 Ma ago), inferred pre-existing N-S-trending transform faults along the western margin of the Pacific became subduction zones. The opening of marginal basins in the western Pacific was coordinated with changes in the motion of the Pacific plate with respect to the Eurasian plate (Jurdy, 1979).

6) Sequential extinction of one spreading centre and initiation of another with a major reorientation of spreading. Such a change occurred within the same region in middle to low latitudes. The end of E-W spreading in the Tasman Sea (59 to 56 Ma ago) was followed by the beginning of N-S spreading south of Australia (53 to 50 Ma ago).

7) Obduction of ophiolites at certain convergent plate boundaries involved in reorganization of plate motion. This occurred in the Indian Ocean (100 to 43 Ma ago), the South Pacific (52 to 37 Ma ago), the North Pacific (65 to 30 Ma ago), the Caribbean (50 to 30 Ma ago), and the Mediterranean (100 to 37 Ma ago) (Figs. 3, 4).

The nature of the plate interactions which produced the observed patterns of Early Cenozoic global reorganization of plate motion was deduced from changes in length of global plate boundaries (Rona and Richardson, 1978). The reconstructed lengths of convergent and divergent plate boundaries before and after the reorganization were measured and totaled (Table II). For convergent plate boundaries, subduction of oceanic crust, which offers relatively low resistance to plate convergence, was distinguished from collision involving continental or transitional (island arc) crust, which, due to crustal buoyancy, offers relatively high resistance to plate convergence (McKenzie, 1969).

Immediately preceding the Early Cenozoic reorganization of plate motion, the total lengths of convergent and divergent plate boundaries were nearly equal, at

TABLE II
EARLY CENOZOIC PLATE REORGANIZATION

Plate Boundary	Worldwide Total Length* (km)		
	Before	After	Net Change
Convergent: Subduction and Collision	52,000	74,500	+22,500
Convergent: Collision	2,500	19,000	+16,500
Divergent	50,000	50,000	-0-

*Measured with estimated 5% accuracy on 40-cm-diameter globe.

(Modified from Rona and Richardson, 1978)

52,000 and 50,000 km respectively (Table II). After reorganization, the total length of convergent plate boundaries, including both subduction and collision, increased to 74,500 km, while the total length of divergent plate boundaries remained at 50,000 km. Collisional plate boundaries account for 16,500 km of the 22,500 km of convergent boundaries added during the reorganization. Although the global plate reorganization reapportioned the lengths of divergent and convergent plate boundaries, the global rates of lithospheric creation and destruction must have been nearly equal if the Earth's diameter was essentially constant (Chase, 1972).

This increase in E-W-trending collisional plate boundaries during the interval 55 to 40 Ma produced a corresponding increase in resistance, apparently forcing the global plate system to reorganize along lines of less resistance. The reorganization included reorientation of relative plate motions with large N-S components into large E-W components, continuation of pre-existing E-W sea-floor spreading about N-S-trending spreading centres, initiation of new E-W spreading about N-S-trending spreading centres, deceleration of spreading rates, complex antithetic and sequential events limited to certain regions within middle to low latitudes, and obduction of ophiolites. The persistence of about 19,000 km of collisional plate boundaries since 40 Ma has acted to constrain the global pattern of plate motion established by the Early Cenozoic plate reorganization.

GLOBAL REORGANIZATION OF PLATE MOTION AND DISTRIBUTION OF RESOURCES

The changes in plate motion from the Mesozoic to Cenozoic era are coordinated with changes in the distribution of mineral and energy resources. Major features of the pre- and post-reorganizational distribution of known stratabound massive sulphide deposits, porphyry copper-molybdenum deposits, and giant oil and gas fields exemplify these changes (Figs. 1-4).

1) Stratabound massive sulphide deposits with volcanic affinities increased in incidence from Mesozoic to Cenozoic in island arcs of the western North Pacific region, coordinated with continued and new subduction, respectively; they decreased in the Mediterranean region, coordinated with complex changes in microplate interactions within an overall change to convergent motion; they decreased to zero along the western margin of North America, coordinated with a change in direction of relative plate motion; and they also decreased to zero along the northern and southern boundaries of the Caribbean region, coordinated with an overall change from convergent to strike-slip motion. The global emplacement of ophiolites in-

creased during the Early Cenozoic plate reorganization; therefore, the prospective areas for those mineral deposits that may form in oceanic crust and upper mantle increased accordingly (Table I).

2) Porphyry copper-molybdenum deposits increased in incidence from Mesozoic to Cenozoic along the western margin of North America; they initially appeared along the western margin of South America, coordinated with change to a large E-W component of relative plate motion in the Pacific region; and they initially appeared in the western North and South Pacific, eastern and western Caribbean, and Mediterranean regions, coordinated with change to convergent plate motion with subduction.

3) The interpretation of oil and gas fields is complicated by temporal differences between source rock, reservoir rock, and entrapment. Based on the age of the reservoir rock, giant oil and gas fields increased in incidence from Mesozoic to Cenozoic along the western margin of North America and appeared along the western margin of South America, coordinated with change in direction of relative plate motion in the Pacific region; they initially appeared in the western Pacific, coordinated with creation of marginal basins by the formation of volcanic island arcs along new subduction zones, initially appeared along the southern margin of Australia, coordinated with rifting and sea-floor spreading between Australia and Antarctica, increased in incidence in the South Atlantic coordinated with early stages of opening, initially appeared in the Red Sea, coordinated with early opening, and initially appeared along the western margin of India, coordinated with a change from transform to divergent plate motion.

CONCLUSIONS

Coordinating the changes in distribution of certain mineral and energy resources with the Mesozoic to Cenozoic global reorganization of patterns of plate motion has not simply been an exercise in fitting known deposits to past plate boundaries, but reveals principles that can be applied to guide regional exploration (Figs. 1-4).

1) The mineral and energy resources formed during a given geologic era exhibit a distinct distribution controlled by the tectonic settings created by the global pattern of plate motion prevalent during that era. For example, different Mesozoic to Cenozoic distributions of known stratabound massive sulphides with volcanic affinities, porphyry copper-molybdenum deposits, and giant oil and gas fields can be identified and related to such tectonic differences during each of these eras. The broad age-distribution of porphyry copper-molybdenum deposits (Hollister, 1973; Argall and Wyllie, 1976), in particular, suggests that their formation is favoured by the globally organized pattern of plate motion prevalent during a geologic era.

2) A global reorganization of plate motion occurs during the transition between geologic eras. The reorganization involves the creation of new tectonic settings, as well as the continuation, reorientation, or conversion of pre-existing ones. The distribution of mineral and energy resources formed in the reorganized tectonic settings changes accordingly. For example, the changes in distribution from the Mesozoic to the Cenozoic of known stratabound massive sulphide deposits with volcanic affinities, porphyry copper-molybdenum deposits, and giant oil and gas fields are coordinated with the Early Cenozoic reorganization of plate motion. This reorganization

created new N-S-trending sea-floor spreading centres and subduction zones, continued pre-existing N-S-trending sea-floor spreading centres, reoriented E-W-trending sea-floor spreading centres into N-S directions, converted N-S-trending transform faults into subduction zones and E-W-trending subduction zones into continental collision zones, and emplaced ophiolites at sites along the collision zones. The increase in ophiolite emplacement during a global plate reorganization enhances the prospects for stratabound massive sulphides and other deposits that may occur in oceanic crust (Table I). Preliminary evaluation suggests that the generation and entrapment of hydrocarbons may also be favoured by such global plate reorganization between geologic eras (Richardson and Rona, 1980).

3) The lengths of divergent, convergent (subduction, obduction, collision), and transform plate boundaries change during, and tend to stabilize after, a global reorganization of plate motion; this reapportions the areas of various associated tectonic settings and prospective regions for the formation of mineral and energy resources. For example, the near equality in length of divergent and convergent plate boundaries before the Early Cenozoic plate reorganization (Table II) implies a comparable number of areas of associated tectonic settings and prospective regions for resources formed in these tectonic settings (Table I). During the Cenozoic reorganization, the development of subduction zones at new convergent plate boundaries favoured the formation of porphyry copper-molybdenum deposits in the western North and South Pacific regions. Similarly, a relatively high incidence of obduction of prospective ophiolites in the Eocene epoch during this plate reorganization was related to a marked increase in the length of collisional plate boundaries at that time. (Table II). The increased length of collisional plate boundaries after the plate reorganization (Table II) has created an exceptionally large area of collisional tectonic settings with their prospective resources (Table I), while the lengths of other convergent and divergent plate boundaries, with their associated tectonic settings and prospective resources, have remained nearly equal to those prior to reorganization (Table II).

Previous studies relating the distribution of mineral and energy resources to plate tectonic settings have provided a useful spatial framework to guide exploration. The present study contributes a temporal framework by recognizing that the distribution of mineral and energy resources is coordinated with a globally organized pattern of plate motion that lasts for a time interval of the order of 100×10^6 years, corresponding to a geologic era, and that the distribution changes with global reorganization of plate motion between geologic eras. The principles that describe this coordination between global plate motion and resources have been derived from the exceptionally complete record of plate motions preserved in Mesozoic and Cenozoic oceanic lithosphere and continental collision zones, However, if used judiciously, the principles can be applied not only to the Mesozoic and Cenozoic eras, but to delineate tectonic settings and prospective regions for resources formed in prior geologic eras, which are represented by only fragmentary records of plate motions.

ACKNOWLEDGEMENTS

I thank Philip W. Guild of the U. S. Geological Survey for advice on sources of information regarding ages of metal deposits and Lyle B. Fox, Jr., of the National Oceanic and Atmospheric Administration for help in compiling this information.

REFERENCES

Argall, G.O. Jr., and Wyllie, R.J.M., compilers, 1976, World Mining Copper Map, Second Edition: San Francisco, California, Miller Freeman Publications, Inc.

Bally, A.W., 1975, A geodynamic scenario for hydrocarbon occurrences: World Petroleum Congress Proc. 9th, v. 1, paper 3, 12 p.

Bilibin, Y.A., 1968, Metallogenic provinces and metallogenic epochs: Geol. Bull. Queens College, New York, No. 1, 35 p.

Bonatti, E., 1975, Metallogenesis at oceanic spreading centers: in Annual Review Earth Planet. Sci., v. 3: Palo Alto, California, Annual Reviews, Inc., p. 401-431.

Bonatti, E., Guerstein-Honnorez, B.-M., and Honnorez, J., 1976, Copper-iron sulfide mineralizations from the equatorial Mid-Atlantic Ridge: Econ. Geol., v. 71, p. 1515-1525.

Burke, K.C., and Dewey, J.F., 1973, Plume-generated triple junctions: Key indications in applying plate tectonics to old rocks: Jour. Geol., v. 81, p. 406-433.

Burke, K.C., and Wilson, J.T., 1976, Hot spots on the earth's surface: Sci. American, v. 235, No. 2, p. 46-57.

Chase, C.G., 1972, The N plate problem of plate tectonics: Geophys. Jour. Royal Astron. Soc., v. 29, p. 117-122.

Coleman, R.G., 1977, Ophiolites: New York, Springer-Verlag, 229 p.

Curray, J.R., 1975, Marine sediments, geosynclines, and orogeny: in Fischer, A.G., and Judson, S., ed., Petroleum and Global Tectonics: Princeton, New Jersey, Princeton University Press, 322 p., p. 157-222.

Dewey, J.F., and Burke, K.C., 1973, Tibetan, Variscan and Precambrian basement reactivation: Products of continental collision: Jour. Geol., v. 81, p. 683-692.

Dietz, R.S., and Holden, J.C., 1970, Reconstruction of Pangaea: Breakup and dispersion of continents, Permian to present: Jour. Geophys. Research, v. 75, p. 4939-4956.

Guild, P.W., 1972a, Distribution of metallogenic provinces in relation to major earth features: in Petrascheck, W.D., ed., Metallogenic and Geochemical Provinces: New York, Springer, 183 p., p. 10-22.

─────────────, 1972b, Massive sulfides vs. porphyry deposits in their global tectonic settings: Tokyo, Joint Meeting MMIJ-AIME, Print No. G13, 8 p.

Halbouty, M.T., Meyerhoff, A.A., King, R.E., Dott, R.H., Sr., Klemme, H.D., and Shabad, T., 1970, World's giant oil and gas fields, geologic factors affecting their formation, and basin classification: in Halbouty, M.T., ed., Geology of Giant Petroleum Fields: American Assoc. Petrol. Geol. Memoir 14, p. 502-555.

Hollister, V.F., 1973, Regional characteristics of porphyry copper deposits of South America: Soc. Mining Engin. of AIME, Preprint No. 73-1-2

Hutchinson, R.W., 1973, Volcanogenic sulfide deposits and their metallogenic significance: Econ. Geol., v. 68, p. 1223-1246.

Jurdy, D.M., 1979, Relative plate motions and the formation of marginal basins: Jour. Geophys. Research, v. 84, p. 6796-6802.

McKenzie, D.P., 1969, Speculations on the consequences and causes of plate motions: Geophys. Jour., v. 18, p. 1-32.

McLaughlin, R.J., Sorg, D.H., Ohlin, H.N., and Heropoulis, C., 1979, Base- and precious-metal occurrences along the San Andreas Fault, Point Delgada, California: U.S. Geological Survey Open File Rep. 79-584, 11 p.

Mitchell, A.H., 1976, Tectonic settings for emplacement of subduction-related magmas and associated mineral deposits: in Strong, D.F., ed., Metallogeny and Plate Tectonics: Geol. Assoc. Canada Spec. Paper 14, 660 p., p. 3-22.

Mitchell, A.H., and Bell, J.D., 1973, Island-arc evolution and related mineral deposits: Jour. Geol., v. 81, p. 381-405.

Mitchell, A.H., and Garson, M.S., 1972, Relationships of porphyry copper and circum-Pacific tin deposits to palaeo-Benioff zones: Trans. Instit. Mining Metal., Sec. B., v. 81, p. 1310-1325.

————————————, 1976, Mineralization at plate boundaries: Minerals Sci. Eng., v. 8, No. 2, p. 129-169.

Moody, J.D., 1975, Distribution and geological characteristics of giant oil fields: in Fischer, A.G., and Judson, S., eds., Petroleum and Global Tectonics: Princeton, New Jersey, Princeton University Press, 322 p., p. 307-320.

Morgan, W.J., 1972, Deep mantle convection plumes and plate motions: American Assoc. Petroleum Geol. Bull., v. 56, p. 203-213.

Ray, K.K., and Acharyya, S.K., 1976, Concealed Mesozoic-Cenozoic Alpine Himalayan geosyncline and its petroleum possibilities: American Assoc. Petrol. Geol. Bull., v. 60, p. 794-808.

Richardson, E.S., and Rona, P.A., 1980, Global Eocene plate reorganization, implications for petroleum exploration: Suva, Fiji, CCOP/SOPAC Symposium, Petroleum Potential in Island Arcs, Small Ocean Basins, Submerged Margins, and Related Areas, in press.

Ridge, J.D., 1972, Annotated bibliographies of mineral deposits in the western hemisphere: Geol. Soc. America Mem. 131, 681 p.

Ringwood, A.E., 1974, The petrological evolution of island arc systems: Jour. Geol. Soc., v. 130, p. 183-204.

Rona, P.A., 1969, Possible salt domes in the deep Atlantic off northwest Africa: Nature, v. 224, p. 141-143.

————————————, 1970, Comparison of continental margins of eastern North America at Cape Hatteras and northwestern Africa at Cap Blanc: American Assoc. Petrol. Geol. Bull., v. 54, p. 129-157.

————————————, 1976, Salt deposits of the Atlantic: Anais Academia Brasilieria de Ciencias, v. 48 (Supplemento), p. 265-274.

————————————, 1977, Plate tectonics, energy and mineral resources: Basic research leading to payoff: EOS., American Geophys. Union Trans., v. 58, p. 629-639.

————————————, 1978, Criteria for recognition of hydrothermal mineral deposits in ocean crust: Econ. Geol., v. 73, p. 135-160.

Rona, P.A., and Neuman, L.D., 1976a, Plate tectonics and mineral resources of circum-Pacific region: Tulsa, Oklahoma, American Assoc. Petrol. Geol. Memoir 25, p. 48-57.

————————————, 1976b, Energy and mineral resources of the Pacific region in light of plate tectonics: Ocean Management, v. 3, p. 57-78.

Rona, P.A., and Richardson, E.S., 1978, Early Cenozoic plate reorganization: Earth Planet. Sci. Letters, v. 40, p. 1-11.

Rowley, P.D., Williams, P.L., and Schmidt, D.L., 1977, Geology of an Upper Cretaceous copper deposit in the Andean province, Lassiter Coast, Antarctic Peninsula: Washington, U.S. Gov't Printing Office, U.S. Geol. Survey Prof. Paper 984, 36 pp.

Sawkins, F.J., 1972. Sulfide ore deposits in relation to plate tectonics. J. Geol., v. 80, p. 377-397.

Sawkins, F.J., 1976. Metal deposits related to intracontinental hotspot and rifting environments. J. Geol., v. 84, p. 653-671.

Sillitoe, R.H., 1972a. Relation of metal provinces in western America to subduction of oceanic lithosphere. Geol. Soc. Am. Bull., v. 83, p. 813-817.

_____, 1972b. A plate tectonic model for the origin of porphyry copper deposits. Econ. Geol., v. 67, p. 184-197.

_____, 1978. Metallogenic evolution of a collisional mountain belt in Pakistan: a preliminary analysis. J. Geol. Soc. Lond., v. 135, p. 377-387.

ORE DEPOSITS

The Continental Crust and Its Mineral Deposits, edited by D.W. Strangway,
Geological Association of Canada Special Paper 20

URANIUM DEPOSITS THROUGH TIME

Duncan R. Derry
Derry, Michener & Booth,
401 Bay Street,
Toronto, Ontario

ABSTRACT

The distribution of different types of uranium deposits through geological time has been noted by various authors, especially as affecting deposits formed in the Precambrian era. While this is significant, it may have been oversimplified by grouping deposits within geological time limits that are very extensive when compared with Phanerozoic groupings. As more detailed information becomes available, we find that some of the higher-grade deposits have complicated histories involving several stages of concentration separated by long periods of time.

A comparison of the proportion of uranium production and reserves contributed by individual classes (and geological ages) of deposits shows a changing trend in favour of the unconformity class.

Possibly more attention should be paid to metallographic uranium areas and their localizations. Some areas, e.g. Gabon, West Africa and Baker Lake, N.W.T., have more than one type of uranium deposit. A further consideration is that much of the known uranium reserves in the non-Communist world lies within a dozen areas, each ranging between 3000 and 130,000 km². It is possible that such uranium metallographic areas were localized by very ancient fracture systems, now not easily recognized, in the original Earth's crust.

RÉSUMÉ

Plusieurs auteurs ont étudié la distribution des différents types de dépôts d'uranium au cours des temps géologiques, en particulier de ceux qui datent du Précambrien. Bien que cela puisse être significatif, cette distribution a pu être trop simplifiée parce qu'on a regroupé des dépôts dans des limites de temps géologiques qui sont très vastes lorsqu'on les compare aux regroupements du Phanérozoïque. Au fur et à mesure que de l'information plus détaillée devient disponible, on voit que certains des dépôts les plus riches ont des histoires complexes qui impliquent plusieurs stades de concentration séparés par de longues périodes de temps.

626 DERRY

En comparant la part de l'uranium produit et en réserves appartenant à des classes individuelles de dépôts (et à des âges géologiques), on voit une tendances qui favorise la classe discordante.

Peut-être devrait-on porter plus d'attention aux régions d'uranium métallographique et à leurs localisations. Quelques régions, comme le Gabon en Afrique occidentale, et la région de Baker Lake dans les T.N.-O., possèdent plus d'un type de dépôts d'uranium. On considère en outre que la plupart des réserves connues d'uranium dans le monde non-communiste se trouvent dans une douzaine de régions dont les dimensions varient de 3000 à 130 000 km². Il est possible que de telles régions d'uranium métallographique étaient localisées le long d'anciens systèmes de fractures, maintenant difficiles à reconnaître, dans la croûte originale de la Terre.

INTRODUCTION

The time-bound character of uranium deposits has been noted by a number of writers in the last few years. One of the most useful articles on this aspect of uranium mineralization is by David Robertson *et al.* (1978), in the recent number of "Economic Geology" devoted to the geology of uranium deposits in various parts of the world. A figure from this paper is included here (Fig. 1), on which world uranium reserves are plotted against geological time. The figure shows three peaks: the first, between 2800 and 2200 Ma, represents the quartz-pebble conglomerates of Witwatersrand and Elliot Lake; the second, between 2000 and 1500 Ma, includes the unconformity type exemplified by the deposits of Athabasca, Saskatchewan, of South Alligator River in the Northern Territory of Australia and some deposits in Gabon; the third peak, extending from Permian to Tertiary, includes reserves that are mainly modifications of the sandstone type like the deposits of the western U.S., which contribute nearly 50 per cent of current world uranium production.

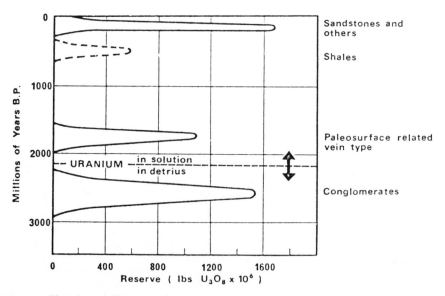

Figure 1. Time Bound Character of Uranium Deposits.

CLASSES OF DEPOSITS

This time/reserve chart will be returned to after a discussion of data presented at an energy resources symposium in Sydney, Australia, in May 1978 (Derry, 1978). Table I gives a breakdown of 1977 uranium production by individual countries into types of deposits. This shows the dominance of the U.S. in production from sandstone type deposits, the dependence in Canada (predominantly) and South Africa (exclusively) on production from the quartz-pebble conglomerate type, and the relatively small but growing part played in production by the unconformity type. The magmatic type (which includes Rossing, the Bancroft area of Ontario, and Mary Kathleen in Australia) also plays a small part in world production to date, but is increasing as Rossing reaches full production and will grow further by the addition of new producers of this type.

Table II gives the same breakdown into classes of individual deposits and groups, but also shows the geochronological age and the approximate grade (in kilograms of U_3O_8 per tonne). The sandstone type deposits mined in the U.S. in 1977 had an average grade of 1.4 kgs/tonne (in 1978 a shade below this); the Elliot Lake deposits had about 1 kg, the Witwatersrand 0.25 kgs (mainly as by-product of gold) and the magmatic type between 0.4 and 1.5 kgs. Compared to this, the unconformity type had a much higher grade, mostly above 2.5 kgs, with some deposits reaching very high figures.

The black shale class has not yet been mined, and relatively little is known about its reserves, but if the price of uranium were profitably higher than the costs of operation, this class might include large resources of the metal. Black shale is probably the least time-bound class, but the majority of known examples lie within the general range of 400 to 45 Ma.

TABLE I
Western World Uranium Production - 1977
(Tonnes U_3O_8)
From Mineral Commodity Summaries 1978 With Some Adjustments

Country	Total by Country	Breakdown Into Types of Deposits			
		Sandstone	Conglom-merate	Unconformity/ Vein	Mag-matic
U.S.A.	13,540	12,870	–	675	–
Australia	900	–	–	–	900
Canada	6,930	–	3,920	2,810	200
France	2,180	1,300	–	880	–
Gabon	1,090	–	–	1,090	–
Niger	1,900	1,900	–	–	–
Republic of South Africa	3,640	–	3,640	–	–
Namibia	1,360	–	–	–	1,360
Other Market Economy Countries	500	500	–	–	–
Total Per Type	32,045	16,570	7,560	5,455	2,460
Per Cent		51.7%	23.6%	17.0%	7.6%

TABLE II
CLASSES OF URANIUM DEPOSITS

	Examples		
Class	Name of Deposit	Age in Million Years	Grade Kilos U_3O_8/Tonne
Sandstone Type	Western U.S.A.	225-45	1.0 to 3.0 Avg. 1.4
	Massif Central, France	280	0.4 to 0.9
	Arlit, etc., Niger	345	2.9
	Beverly (L. Frome), S. Australia	65	2.4
Calcrete Type	Yeelirrie, W. Australia	35	1.5
	Dusa Mareb, Somalia	35 to present	1.6?
Quartz-Pebble Conglomerate Type	Blind River (Elliot Lake), Ontario	2200	1.0
	Witwatersrand, Rep. of S. Africa	2800-2100	0.25
Bituminous Shales and Related Sediments	Ranstad, Sweden	570	0.1 to 0.3
	Chattanooga Shales, U.S.A.	370	0.1 to 0.2
	Phosphate Beds, Florida, etc.	60	0.1 to 0.3
	Black Sea Organic Ooze	Present	0.1
Unconformity-Vein Type	Ranger I, N.T., Australia		3.1
	Ranger II, N. T., Australia		4.0
	Nabarlek, N. T., Australia	pre-1700-1600-	23.0
	Koongarra, N. T., Australia	900-500	5.0
	Jabiluka I, N. T., Australia		2.5
	Jabiluka II, N. T., Australia		4.1
	Rabbit, Lake, Sask., Canada	pre-1800-1350-1050	4.0
	Cluff Lake, Sask., Canada	1050-900	3.0 to 25.0
	Key Lake, Sask., Canada	pre-1800-1350-900	28.0
	Ace Mine, Eldorado, Sask., Canada	1700-1110-270	2.0
	Okla, Gabon	1740	5.0
Magmatic Type	Mary Kathleen, Queensland	1670	1.3
	Bancroft, Ontario, Canada	900	1.0
	Rossing, Namibia	510	0.4
	Ilimaussaq, Greenland	1000	0.5
	Midnite Mine, Washington, U.S.A.	100?	2.8
Volcanic Type	Jadaguda, India	pre-1800	0.6
	Makkovik, Labrador, Canada	1800	1.3 to 8.0
	Baker Lake, N.W.T., Canada	pre-1800-1400	?
Miscellaneous	Roxby Downs, S. Australia	900	0.5+

The volcanogenic class has been added since this table was first prepared. It contains some deposits previously put in a miscellaneous category – the two developed deposits (Kitts and Michelin) in the Makkovik area of Labrador and the Jadaguda deposit in India; probably some deposits in the general area of Baker Lake, N.W.T., are also volcanic in origin. L. W. Curtis read a paper on this type of deposit at the GAC–GSA meetings in Toronto in 1978 and made a good case for a distinct class related directly to volcanic activity. It may be one of the less time-bound classes.

Mention should be made of the relative economic importance of these classes in present and future production. Figure 2 shows the percentage contributions of the four main classes to world uranium production on a yearly basis from 1975 to 1977, indicating only minor changes over this period. But when we compare the average of these three years with the corresponding figures for each class in reserves, i.e., future

ANNUAL WORLD URANIUM PRODUCTION BY DEPOSIT TYPES

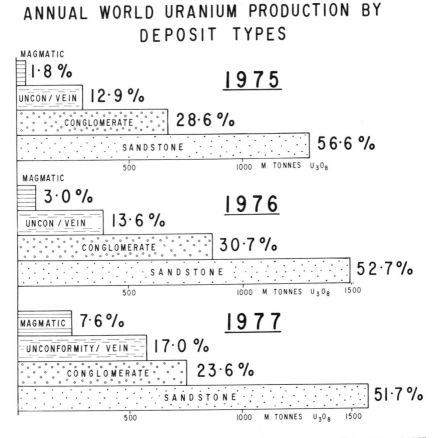

Figure 2. Annual World Uranium Production by Deposit Classes for Years 1975, 1976 and 1977.

production (Fig. 3), a different picture emerges. Moving from the average of world production for 1975-1977 to reserves as of 1977, the sandstone type drops in percentage contribution while the unconformity type increases, and soon may dominate production.

DISCUSSION

A Critical Look at Time-Bound Concepts

The Robertson *et al* (1978) data (Fig. 1) emphasizes the time-bound character of the different types of deposits. Although my figures support this interpretation, I feel that many of us may have overemphasized the time aspect. Let us consider first the quartz-pebble conglomerate type. It is quite true that no significant amounts of uranium have been produced from this type of rock when it is younger than 2200 Ma. I question mechanical concentration as the main agent, but agree that the process of concentration probably depended on atmospheric conditions that may have changed about this time. We must, however, be careful not to group Witwatersrand with Elliot Lake on a time basis too closely. After all, the oldest uraniferous conglomerate at Witwatersrand differs in age (Pretorius, pers. commun.) from the oldest one at Elliot Lake by at least 600 million years — about the total time that has elapsed since the end of Precambrian.

Within the unconformity class, the most economically important deposits occur where middle to upper Proterozoic sediments overlie rocks metamorphosed either by the 2650 Ma orogeny or by the 1700 Ma orogeny. But the more detailed the studies on these intriguing deposits become, the more complex they appear. There seem to have been several stages of remobilization and concentration of the uranium, as seen in two examples:

Key Lake, Saskatchewan (Dahlkamp, 1978)

Assumed Aphebian Source Bed	pre-1800 Ma
Unconformity of Athabasca	1350 Ma
Remobilization (by diabase intrusion?)	1230 Ma
General Remobilization	900 Ma
(some later remobilization 370 and 107 Ma)	

East Alligator River, Northern Territory, Australia (Hegge and Rowntree, 1978)

Cahill Formation (uraniferous)	between 2500-1700 Ma
Kombolgie Unconformity	1600 Ma
Remobilization	900 Ma
Remobilization	500 Ma

These figures show the extreme susceptibility of uranium to remobilization by either supergene or epigene agents.

In addition to the evidence of several phases of mineralization in many uranium "provinces", some areas contain two or more types of deposits. One example is Gabon, where some deposits are unconformity type while others are sandstone type. Another example is the Baker Lake-Thelon Lake area, N.W.T., where there is an unconformity type deposit in the west and deposits of the volcanic type in the east.

Global Distribution of Uranium

One feature that comes up again and again in a study of uranium deposits on a worldwide basis is the spatial relationship of the Proterozoic deposits to Archean granites — not any "old" granite but one that is more uraniferous than the average. It brings to mind the French geologists' separation of Hercynian granites into those that are "fertile" and those that are "sterile". Thus, the problem arises as to why uranium, and some other metals, seem to be concentrated in certain parts of the world and associated with successive granite intrusions in such parts. The outstanding example among other metals is tin. Over half the world's known reserves (and past production) are found within an area on the Malay Peninsula that represents less than half of one per cent of the globe's surface. Another 10 per cent of world tin reserves is in a still smaller area in Bolivia. In both places, the primary tin deposits are directly associated with granitic intrusions of two or more widely-separated periods – between 300 and 50 Ma ago in the Malay Peninsula and between 210 and 23 Ma ago in Bolivia.

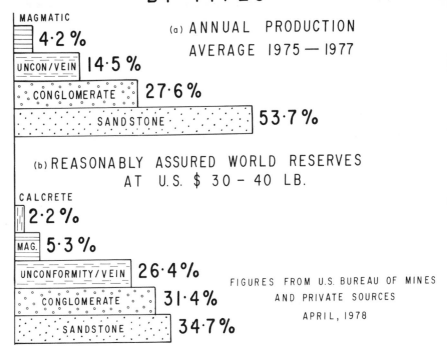

Figure 3. Uranium Percentage Contribution by Classes (a) Annual Production Average 1975-1977 (b) Average of Production Figures above Compared with "Reasonably Assured World Reserves".

In a separate approach to metallographic areas, Routhier (1976) has pointed out that lead-zinc deposits in Spain and France follow a line or trend, unmarked by any surface feature, that transects geological boundaries and orogenies of several periods.

Uranium exploration is still in its early stages, but it is interesting to note that in spite of its very wide distribution as small mineral deposits, over 90 per cent of the known reserves in non-Communist countries are concentrated in a dozen areas, each under 120,000 km^2 (the approximate area of the Colorado Plateau).

We have no satisfactory explanation of these broad metallographic concentrations, although Burwash (1979) has contributed some valuable statistics. Could they be the result of uneven distribution of some metals during the formation of the Earth's crust? Did large scale faulting in the primordial crust produce structures that have been obscured by later sedimentation and orogenies? In a photograph of Europa, one of the moons of Jupiter, taken from Voyager 1 on March 2nd of this year, major fractures, some of which branch, can clearly be seen. Similar escarpments, probably faults which post-date at least some of the meteorite craters, are seen in photos taken from Mariner 10 of the surface of Mercury, and in some of the Viking 1 pictures of the surface of Mars. Similar faulting or major fractures in the original Earth's crust might have resulted in the first redistribution and broad concentration of some metals, perhaps including uranium, in what was initially a homogeneous layer.

Perhaps following Tuzo Wilson, who preceded his contributions to plate tectonics by studies of the fracture patterns of Precambrian terrains, we should search for fracture systems which are invisible at the surface, but might be detected by geophysics.

REFERENCES

Burwash, R.A., 1979, Uranium and Thorium in the Precambrian Basement of Western Canada: II. Petrologic and Tectonic Controls: Canadian Jour. Earth Sci. v. 16, p. 472-483.

Derry, D.R., 1978, Sources of Uranium, Present and Future. Earth Resources Foundation, Sydney, Australia: Australia's Mineral Energy Resources, p. 71-83.

Dahlkamp, F.J., 1978, Geological Appraisal of Key Lake U-Ni Deposits, Northern Saskatchewan: Econ. Geol. v. 73, p. 1430-1449.

Hegge, M.R. and Rowntree, J.C., 1978, Geologic Setting and Concepts on the Origin of Uranium Deposits in the East Alligator River Region, N.T., Australia: Econ. Geol. v. 73, p. 1420-1429.

Robertson, D.S., Tilsley J.E. and Hogg, G.M., 1978, The Time-Bound Character of Uranium Deposits: Econ. Geol. v. 73, p. 1409-1419.

Routhier, P., 1976, A New Approach to Metallogenic Provinces: The Example of Europe: Econ. Geol. v. 71, p. 803-811.

The Continental Crust and Its Mineral Deposits, edited by D.W. Strangway,
Geological Association of Canada Special Paper 20

TECTONIC SETTINGS OF Ni-Cu SULPHIDE ORES: THEIR IMPORTANCE IN GENESIS AND EXPLORATION

A.K. Naldrett and A.J. Macdonald
Department of Geology, University of Toronto, Toronto, Ontario M5S 1A1

ABSTRACT

Magmatic Ni-Cu sulphide ore forms only when immiscible sulphides separate from a silicate magma; the process is thus controlled by the availability of sulphur, not Ni or Cu.

The ores at Sudbury are located at the margin of the Nickel Irruptive, which is thought to have intruded an astrobleme. Irvine (1975) pointed out that silicification of a mafic magma can lower the solubility of sulphur so that the magma becomes saturated and sulphides segregate from it. He suggested that the environment at Sudbury just after impact was conducive to this occurring.

The deposits of the Duluth complex are related to flood basalt magmatism associated with the mid-continental rift. Petrographic and sulphur isotope evidence indicates that much of the sulphur was derived from adjacent country rock slate.

The Noril'sk-Talnakh camp is located at the western edge of the Siberian platform in an area underlain by marine sediments, evaporites, limestones, and lagoonal sediments. Tensional tectonics originating in the Permian have resulted in a zone of crustal thinning, Permian to Tertiary subsidence, and concomitant rapid sedimentation within a zone extending several thousand kilometres from north to south and 1000 kilometres west of Noril'sk. The ore deposits are hosted by feeders to the volcanism that accompanied the initiation of this rifting. Sulphur isotope ratios indicate that the sulphur in the ore has come from the underlying evaporites.

At Sudbury, Duluth and Noril'sk, the reaction of mafic magmas with supracrustal rocks has been a major factor in the formation of the ores. At Duluth and Noril'sk, intracontinental rifting brought mafic magmas into contact with supracrustal sulphur. The Duluth-Noril'sk model has application in exploration. Triple junctions, in particular the aulacogenes that may result, are areas where evaporite formation and mafic magmatism are natural consequences of this tectonic development. Thus, they are likely regions for ore formation.

RÉSUMÉ

Le minerai sulfuré de Ni-Cu d'origine magmatique se forme seulement quand des sulfures immiscibles se séparent d'un magma à silicates; ce processus est aussi contrôlé par la disponibilité du soufre et non du Cu ni du Ni.

A Sudbury, les minerais se concentrent à la bordure de l'irruptif de Nickel dont l'origine pourrait s'expliquer par l'intrusion d'un astroblème. Irvine (1975) a remarqué que la silicification d'un magma mafique peut abaisser la solubilité du soufre de sorte que le magma devient saturé et que les sulfures s'en séparent. Il a suggéré que l'environnement à Sudbury juste après l'impact était favorable à ce type d'occurrence.

Les dépôts du complexe du Duluth se rattachent au magmatisme de laves basaltiques associées à la zone d'effondrement du milieu du continent. L'évidence tirée de la pétrographie et des isotopes de soufre indiquent que la plus grande partie du soufre provenait des ardoises adjacentes.

Le camp de Noril'sk-Talnakh est situé sur la bordure ouest de la plate-forme sibérienne dans une région qui comprend des sédiments marins, des évaporites, des calcaires et des sédiments de lagon. La tectonique de tension dont la source est dans le Permien a formé une zone d'amincissement de la croûte, une subsidence du Permien au Tertiaire et une sédimentation rapide concomitante dans une zone s'étendant sur plusieurs milliers de kilomètres du nord au sud et jusqu'à 1000 km à l'ouest de Noril'sk. Les dépôts de minerai reposent dans des cheminées d'alimentation du volcanisme qui ont accompagné la phase initiale de cet effondrement. Les rapports d'isotopes de soufre indiquent que le soufre dans le minerai provient des évaporites sous-jacentes.

A Sudbury, Duluth et Noril'sk, la réaction des magmas mafiques avec les roches de couverture a été un facteur majeur dans la formation des minerais. A Duluth et Noril'sk, l'effondrement intracontinental a amené les magmas mafiques en contact avec le soufre dans les roches de couverture. Le modèle Duluth-Noril'sk a ses applications en exploration. Les jonctions triples, en particulier les aulacogènes qui peuvent en résulter, sont des régions où la formation d'évaporites et de magmatisme mafique sont les conséquences naturelles de ce développement tectonique. Ainsi, elles sont des régions probables pour la formation de minerais.

INTRODUCTION

Ni-Cu sulphide deposits have long been regarded as magmatic (Vogt, 1918; Wilson, 1953; Souch *et al.*, 1969). This is to say, they result from the formation of droplets of an immiscible sulphide liquid within a mafic or ultramafic silicate magma, the partitioning of Ni, Cu, and other elements including the Pt group into these droplets, and the subsequent concentration of the sulphides, often through gravitational settling. Recently there has been some disagreement as to the importance of magmatic concentration (Barrett *et al.*, 1977), or even as to whether it operates at all (Lusk, 1976; Fleet, 1977; Fleet *et al.*, 1977; Fleet, 1979). One of us (Naldrett, 1979) has discussed this point recently and concluded that magmatic segregation is of overriding importance in the formation of these deposits. We are adopting this viewpoint in the present paper, and refer the reader to the discussion for the supporting evidence.

Figure 1 shows typical Ni and Cu and concentrations in mafic and ultramafic magmas, plotted against the MgO content of the magmas. It is the Ni and Cu which partition into the droplets and ultimately give rise to an orebody. Rajamani and Naldrett (1978) have presented data on the coefficients that govern the partitioning of the metals Ni, Cu, Co, and Fe between sulphide and silicate melts. These indicate that metal concentrations such as those illustrated in Figure 1 are sufficient to account for the tenor of the sulphide ores (Naldrett, 1979).

The abundance of the metals, therefore, is not usually a problem in the formation of a magmatic sulphide ore; the key is the availability of sulphur. Only when supersaturation of sulphide occurs in a silicate magma can the process of partitioning and concentration proceed. In addition, the excess sulphide must be producd at a critcal instant in the emplacement and crystallization of the magma so that the sulphides can settle and concentrate undiluted by the silicates that will also be accumulating as the magma crystallizes.

This paper will draw attention to important features in the tectonic settings of certain major Ni-Cu sulphide camps, show how these features have contributed to sulphide saturation in the magmas in question and, from this, develop conceptual models to aid the selection of favourable regions for Ni-sulphide exploration.

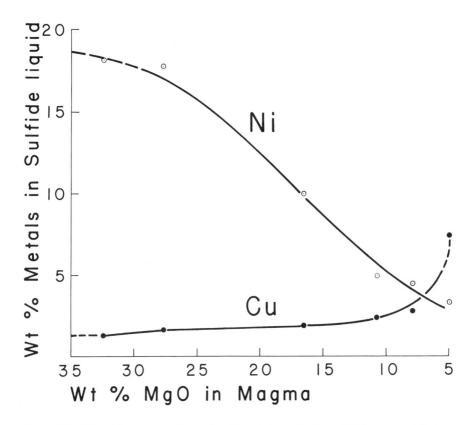

Figure 1. The Ni and Cu contents of typical mafic and ultramafic (komatiitic) magmas plotted as a function of their MgO content (after Rajamani and Naldrett, 1978).

AN OVERVIEW OF Ni-Cu SULPHIDE DEPOSITS

Since Ni-Cu sulphide deposits are intimately associated with mafic and ultramafic igneous magmas, and since this magmatism results directly from specific tectonic processes, a convenient way of regarding these deposits is on the basis of their magmatism and tectonism together – that is, on the basis of their petro-tectonic settings. The settings of the deposits that account for over 95% of Ni-Cu sulphide reserves are:

I. Noritic rocks associated with an astrobleme (the scar resulting from meteorite impact). The Sudbury mining camp in Canada is believed by many to be of this type and to be the sole example.

II. Intrusive equivalents of flood basalts associated with intracontinental rifting. Important examples include the Noril'sk camp of Siberia (Glazkovsky *et al.*, 1977) and the as yet unexploited sulphides at the western edge of the Duluth complex in Minnesota (Bonnichsen, 1972; Mainwaring and Naldrett, 1977).

III. Magmatic activity accompanying the early stages of formation of Precambrian greenstone belts. This comprises two main types:
A. Tholeiitic intrusions such as those hosting the ores of the Pechenga Ni camp of the Kola peninsula, USSR (Gorbunov, 1968) and Lynn Lake, Manitoba, Canada (Ruttan, 1955).
B. Komatiitic lavas and intrusions, especially the more ultramafic variants. Highly ultramafic komatiites (Arndt *et al.*, 1977), extruded at temperatures of 1500 to 1600°C, are restricted to the Archean, where they are hosts to important sulphide deposits, in particular those of the Kambalda and other camps of the Eastern Goldfields of Western Australia (Ross and Hopkins, 1975). Rocks believed to be komatiitic (Peredery, 1979) are hosts to deposits of the Thompson, Manitoba, district. Somewhat less ultramafic komatiites are hosts to rich deposits, as yet unmined, in the northern tip of the Ungava peninsula of Quebec (Barnes, 1979).

IV. Tholeiitic intrusions, generally synchronous with orogenesis in Phanerozoic orogenic belts. These host deposits of lesser importance (eg., the Röna deposit, Norway–Boyd and Mathiesen, 1979).

Figure 2 is taken from data presented by Naldrett (1973). Although these data are incomplete, they do suggest that the amount of nickel (in past production or known reserves) depends upon the settings of the deposits as classified above. The Sudbury district's dominant position in world nickel production is very clear. The second important setting is that of intracontinental rifting. Figures for the reserves of the Noril'sk-Talnakh camp, which falls into this category, are unavailable and therefore may be much larger than shown here. Important zones of sulphide have been outlined along the northern margin of the Duluth complex, which is related to the mid-continental rift system. Drilling suggests that potential reserves in the complex are many times larger than outlined to date (Bonnichsen, 1972). These deposits are low in grade and much richer in Cu than Ni (Cu/Ni ≈ 4). Since the current environmental concerns in Minnesota may prevent their exploitation, these deposists are shown below rather than above the line in Figure 2. Nevertheless, they constitute a major Ni resource.

The third important setting is that of komatiitic volcanism. Controversy surrounds the exact tectonic setting represented by Precambrian greenstone belts and in particular Archean belts. Ultramafic komatiites commonly occur interlayered with basaltic rocks towards the base of cycles of volcanism which become less mafic upwards, in some cases culminating in calc-alkaline andesites, dacites, and rhyolites. Most of the controversy centres on the origin of the calc-alkaline rocks; there is general agreement (Viljoen and Viljoen, 1969; Hunter, 1974; Naldrett and Turner, 1977) that the komatiitic volcanism coincides with an early phase of tensional tectonics. The equivalence or lack of equivalence of ultramafic komatiites with either the alkalic, flood, or oceanic basalts resulting from present-day rifting has, in our opinion, yet to be established. Naldrett and Turner suggested that the conditions leading to komatiitic volcanism also lead to the tapping of sulphide-rich zones within the mantle, accounting for the association. We have nothing to add to these views and, for the remainder of this paper, will concentrate on the astrobleme (Sudbury) and intracontinental rifting settings.

THE ORES OF THE SUDBURY BASIN

Sudbury–the ores, igneous rocks, and geologic setting–has become the subject of an extensive literature. It is not our intention to review it here, but merely to point out the main features, reference some of the most recent papers, and draw attention to some unusual characteristics that may indicate the origin of the deposits. The geology of the Sudbury basin is illustrated in Figure 3. The ore deposits either occur close to the outer margin of the Sudbury Nickel Irruptive or are associated with dyke-like bodies of Irruptive rocks (the offsets) projecting away from the Irruptive. The petrology of the

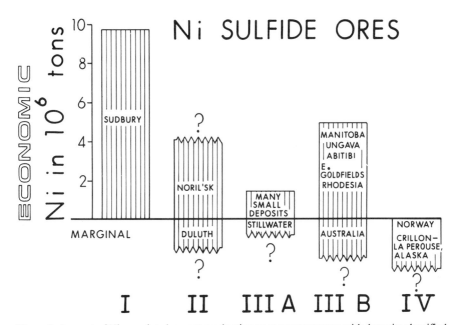

Figure 2. Amount of Ni associated as past production or present reserves with deposits classified according to their petrotectonic setting.

Irruptive has been discussed by many authorities, most recently by Pattison (1979), who concentrates on the rocks closely associated with the ore, the igneous sublayer, and prior to that by Naldrett *et al*. (1970, 1972) and Hewins (1971). The Irruptive has intruded along the contact between the overlying Whitewater series and underlying Archean granites and gneisses (to the north) and Proterozoic greenstones and greywackes (to the south).

Perhaps the most interesting rock formation at Sudbury is the Onaping formation, which forms the base of the Whitewater series. This has been described in detail by Peredery (1972). It is a breccia composed of numerous fragments of country rock, quartzite, granite, gneiss and gabbro, and the devitrified remnants of a volcanic-looking glass set in a matrix of smaller rock fragments and devitrified glassy shards. The base of the Onaping formation, which is in immediate contact with the underlying Irruptive, is marked by irregular-shaped bodies of an igneous rock which is distinct from and intruded by the Irruptive itself and which contains numerous country-rock inclusions.

The Sudbury structure was originally interpreted as a volcanic caldera, the Onaping formation as an ignimbrite which had filled the caldera as it collapsed, and the

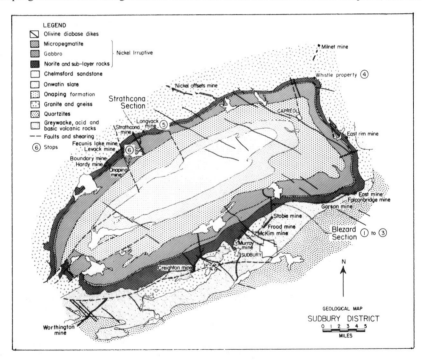

Figure 3. The geology of the Sudbury basin.

Irruptive as an intrusion into the zone of faulting bounding the caldera (Speers, 1957). Two important observations have led many geologists to reinterpret the field evidence. First, Dietz (1964) described the presence of shatter cones in quartzites lying to the south of the present Irruptive margin, and suggested that the structure was a meteorite impact site. Dietz (1972) developed his ideas further and suggested that the ore itself was derived from the meteorite. Second, French (1967) noted rows of inclusions in quartz and feldspar grains in country-rock fragments from the Onaping formation, which, he suggested, were the remnants of lamellae of thetamorphic glass resulting from shock pressures in excess of 100 kb. French interpreted his results to provide strong support for the impact hypothesis.

Following these original observations, the Onaping formation has been compared to the suevite breccias from the Riess structure in Bavaria, which are regarded as impact fall-back breccia (Peredery, 1972). The irregular bodies of igneous rock at the base of this formation have been interpreted as melts generated by the heat of the

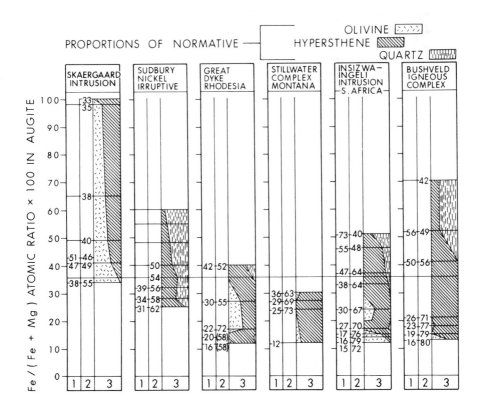

Figure 4. Normative proportions of olivine, hypersthene and quartz in cumulus rocks from a number of well-known layered intrusions compared on the basis of the Fe/(Fe+Mg) ratio of their augite. The same ratio for hypersthene (column 1) and the An content of plagioclase (column 2) are also shown for each sample.

impact; breccia dykes composed of locally derived material have been equated to the pseudo-tachylite veins observed in the vicinity of impact sites (Peredery, 1972); and the triggering of the intrusion of the Irruptive has been attributed to deep crustal fracturing resulting from impact (Guy-Bray, 1972). The impact hypothesis is now widely, although not universally accepted (CF Card and Hutchinson, 1972; Fleet, 1979).

The Sudbury Nickel Irruptive consists of a layer of norite overlying a marginal zone in which the rocks become progressively more quartz-rich towards the basal contact. The norite grades upwards into a gabbro with a gradual decrease in hypersthene content. Cumulus magnetite appears within the gabbro, which itself grades abruptly upwards into a granophyre. The Fe/Mg ratio of both augite and hypersthene increases upwards through the norite and gabbro into the granophyre, and downwards across the marginal zone. The anorthite content of plagioclase increases upwards throughout the norite and gabbro, but remains relatively constant across the marginal zone (Naldrett et al., 1970). In comparison with other large layered intrusions, the Irruptive is unusual in three important aspects: the absence of small-scale (<200 ft.) cyclical units, the absence of visible layering resulting from variation in the proportions of minerals, and the very high quartz content of the rocks.

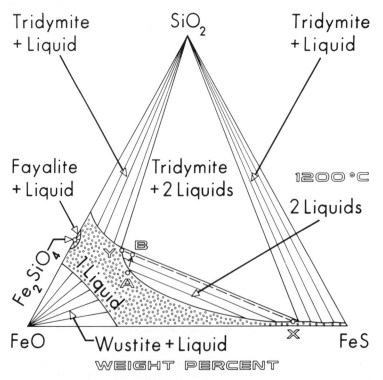

Figure 5. 1200°C isotherm of the FeO-SiO_2-FeS system (drawn from the liquidus diagram of MacLean, 1969) illustrating the effect of adding SiO_2 to a composition represented by point A.

This last point is illustrated in Figure 4, in which the normative proportions of olivine, hypersthene and quartz in cumulus rocks from a number of well known layered intrusions are compared on the basis of the $Fe/(Fe+Mg)$ ratio of their augite. Where data are available, the $Fe/(Fe+Mg)$ ratio of hypersthene and the An content of plagioclase are also shown. The compositions of these minerals vary in a systematic manner with augite, all three compositions serving as indicators of the degree of differentiation of the magma from which they crystallized. The quartz-rich nature of the Sudbury rocks is immediately apparent. The silica content of a magma and hence, to some degree, of the cumulus rocks crystallizing from it, is also an indicator of the degree of its differentiation; it might be argued that the Sudbury magma was more differentiated than the others when it was emplaced. However, if this were true, the compositions of the pyroxenes and plagioclase should also indicate an advanced stage of differentiation, which they do not. A possible explanation for this anomaly is that the Sudbury magma has become highly contaminated with quartz-rich material, resulting in an increase in SiO_2 but no corresponding increase in the Fe/Mg ratio. The presence of numerous partially digested inclusions within the Irruptive (Stevenson and Colgrove, 1968) and the relatively high Sr^{87}/Sr^{86} ratio of 0.706 for the norite (Gibbins and McNutt, 1975) support this viewpoint. The lack of cyclical units and igneous layering also becomes explicable, since an unusually high silica content would result in an unusually viscous magma. This would inhibit convection and the winnowing effect of vertical magma currents which are thought to give rise to the cyclical units and visible layering.

Irvine (1975) has pointed out that silicification of a mafic magma can induce saturation of sulphide, and has suggested that this is responsible for the concentration of sulphide ore at Sudbury. This point is illustrated in Figure 5. Following MacLean (1969), the FeO-SiO_2 side of this diagram can be taken as equivalent to silicate melts and the FeS corner as equivalent to sulphide magmas. A composition represented by point A lies in the field of a homogeneous FeO-SiO_2-FeS liquid. When enough SiO_2 is added to composition A to change it to B, the bulk composition moves out of the single-liquid field into a two-phase field, in this case consisting of a silicate-rich liquid Y, rather poorer in sulphide than A, and a sulphide-rich liquid X. This development of two liquids has been achieved merely by the addition of silica and the input of some heat to promote its assimilation. No crystallization or decrease in temperature has been involved. As Irvine points out, in the natural case of Sudbury, heat induced by impact of a meteorite would assist in the assimilation of rocks fractured by this event. Sulphides would separate without the co-precipitation of silicates, and would thus be free to accumulate as rich orebodies relatively free of dilution by silicate minerals.

Figures 6a to 6e are diagrammatic illustrations of a possible scenario to account for some of the features discussed above. A meteorite struck at Sudbury, blasting out a large crater, 100 km in diameter (Fig. 6a), melting rock at the base of the crater and severely fracturing the underlying crust. Fall-back breccia partially filled the crater. Pressure release induced by the unloading, coupled with diffusion of volatiles towards the zone of low pressure, resulted in melting of the mantle and intrusion of this melt into the zone of fracturing (Fig. 6b). As it moved upwards, the magma encountered an unusually fractured portion of the crust, including, as it rose up to the base of the crater, rocks very hot, even molten, as a result of the impact (Fig. 6c). Most of these rocks were quartzitic or granitic in composition, representative of the target zone of the meteorite. Circumstances were ideal for assimilation; a great deal of siliceous material was ingested by the magma and sulphides were "salted out" to settle into structures around the periphery of the intrusion (Fig. 6d). As Naldrett *et al.* (1970) and Peredery and Naldrett (1975) have pointed out, the final emplacement of the intrusion appears to have occurred as a series of pulses increasingly rich in silica. The latest pulses were relatively poor in sulphide (Duke and Naldrett, 1978), possibly due to the

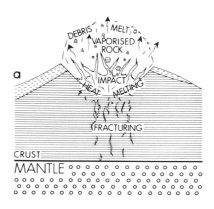

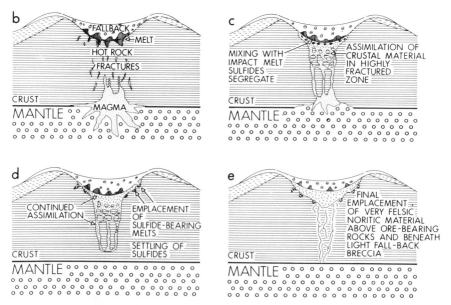

Figure 6. Diagrammatic illustrations of the possible formation of the Sudbury astrobleme (a), partial melting of mantle (b), ascent into the crust and assimilation of fractured, heated and partially molten country rock (c), segregation and settling of sulfide ores around margin of the intrusion (d), and final emplacement of more felsic phases of the Irruptive (e).

loss of a great deal of sulphide during the early stages of silification. As mentioned earlier, the impact hypothesis is not universally accepted, some preferring to believe that Sudbury was a major volcanic centre. The fracturing associated with a series of large volcanic explosions would also provide a setting for assimilation, so the concept of silica-assimilation is not restricted to impact sites.

ORES OF THE DULUTH COMPLEX

The Duluth complex outcrops as an arcuate mass extending about 150 miles northeast from Duluth, Minnesota, to the Canadian border. Detailed descriptions of parts of the complex have been given by Taylor (1964), Weiblen (1965), Phinney (1972), and Bonnichsen (1972). Sims and Morey (1972) have presented a general discussion of the stratigraphic setting, and Weiblen and Morey (1975) an up-to-date summary of the geology of the complex.

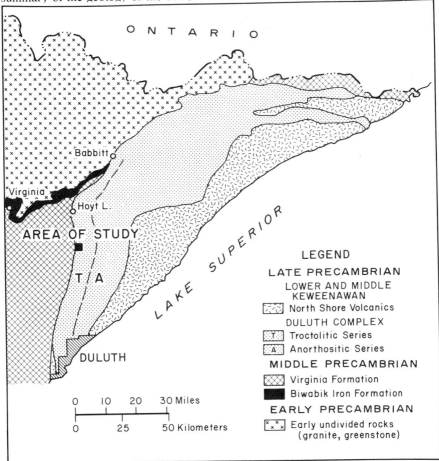

Figure 7. Geologic setting of the Duluth complex (after Mainwaring and Naldrett, 1977).

The complex is closely associated with Keewanawan basalts of the Lake Superior area and was intruded 1.12 Ga ago (Faure *et al.*, 1969) close to the contact of the basalts with older Precambrian rocks which are now exposed to the northwest (Fig. 7). These consist of a complex of Archean felsic intrusions and volcanics, primarily the Giants Range granite, which are overlain to the south unconformably by the south-dipping Biwabic iron formation. This is itself overlain by the Virginia formation, which is composed of black argillites, greywackes, siltstone, graphitic slate, and a sulphide facies iron formation.

The ores of the Duluth complex occur along its western margin, associated with a variety of intrusions which make up the complex, some of which appear to have been feeders for the Keewanawan flood basalts found along the north shore of Lake Superior. Mainwaring and Naldrett (1977) studied one of these, the Water Hen intrusion, together with its associated mineralization. In this intrusion, the sulphides are associated with a zone of dunite and with peridotite layers within overlying troctolite. Although the ores consist of a normal assemblage of pyrrhotite, pentlandite and chalcopyrite occuring interstitial to silicate grains in a characteristic magmatic 'net texture', one unusual characteristic is that in certain zones they contain as much as 10% graphite. Other unusual features of the ore environment include the presence of green (Mg-A1) spinel in some of the troctolites, the occasional occurrence of cordierite, and numerous partially resorbed remnants of what appear to be hornfelsed Virginia slate in the troctolites.

These observations led Mainwaring and Naldrett to question whether the graphite and sulphides might not have been derived from the Virginia formation. The results of their sulphur isotope study are included in Figure 8. Barren sulphides from the Virginia average about $+18\,\delta^{34}S$ per mil, while the Water Hen ore ranges from $+11$ to $+16\,\delta^{34}S$ per mil. These results are consistent with a model in which perhaps as much as 75% of the sulphur was derived from the country rocks. Mainwaring and Naldrett proposed the model illustrated in Figure 9, in which an initial pulse of magma, rich in olivine phenocrysts and contaminated with country rock sulphide, was injected to form the

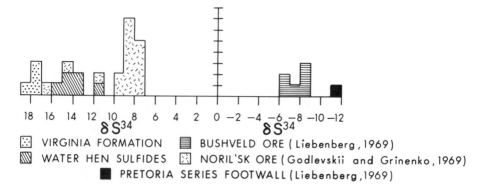

Figure 8. Sulfur isotope ratios from the Water Hen intrusion, Minnesota, the Noril'sk and Bushveld regions (after Mainwaring and Naldrett, 1977).

main, dunite-hosted ore zone which then started to differentiate. Successive influxes of new magma flushed out existing, partially differentiated magma to form the Keewanawan north shore volcanics, and themselves differentiated to give rise to a series of periodite-troctolite-anorthosite cyclical units in which sulphides settled to become enriched in the basal peridotite members of each of these units. Sulphur isotope ratios reported from other locations within the Duluth complex (E.C. Perry, pers. commun.; E.M. Ripley, written commun., 1978) also indicate that much of the sulphur was derived from a crustal source.

As stated previously, the Duluth complex is an intrusive part of the Keewanawan flood basalt province of Lake Superior. The lake itself represents the northern extremity of the mid-continental gravity high which cuts across North America in a northeasterly direction from Kansas to Minnesota and then turns south into central Michigan (Fig. 10). The zone of anomalous gravity is also marked by numerous strong positive and negative magnetic anomalies. Pre-Paleozoic basement rocks intersected in boreholes within the zone include a high proportion of Keewanawan basalts and mafic intrusions. These and the related gravity and magnetic anomalies are expressions of intracontinental rifting which had its greatest development in the Lake Superior area with the extrusion of the Keewanawan flood basalts and the intrusion of the Duluth complex.

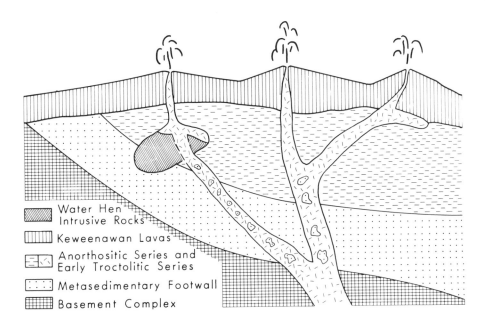

Figure 9. Schematic illustration of the crystallization of the Water Hen intrusion (After Mainwaring and Naldrett, 1977).

ORES OF THE NORIL'SK-TALNAKH AREA

The Noril'sk Cu-Ni deposits are located on the extreme northwestern margin of the Siberian Platform, which has been a stable craton since the end of the Paleozoic (Fig. 11). To the north, the Platform is separated by the Khatanga Trough from a second platform, the Taymyr Peninsula, which has also been stable since the Paleozoic. To the west, a third craton, the East European-Urals Block, is separated from the Siberian Platform by the Yenisei Trough, known geographically as the West Siberian Lowlands. The East European-Urals Block has behaved as a craton since Middle Permian time.

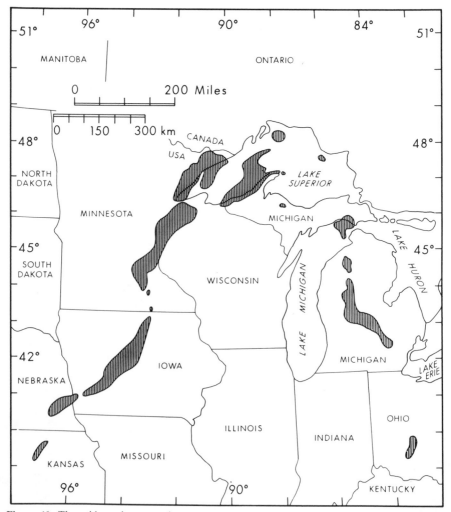

Figure 10. The mid-continent gravity anomaly. Shading indicates areas of positive Bouguer anomaly ()0 milligals).

Within Western Siberia, Lower Paleozoic marine argillaceous sediments are overlain by extensive Devonian evaporite deposits and Lower Carboniferous shallow-water limestones which are in turn unconformably overlain by Middle Carboniferous lagoonal and continental sediments, including coal measures (Smirnov, 1966; Glazkovsky *et al.*, 1977). The emergent sedimentary sequence is covered by an extremely large volume (approximately one million cubic kilometres) of Late Permian and Triassic flood basalt and tuff (Bazunov, 1976; Glazkovsky *et al.*, 1977) known as the Siberian traps. Sill-like tholeiitic intrusions, varying in composition from subalkaline dolerite to gabbro-dolerite (Glazkovsky *et al.*, 1977), were emplaced contemporaneously with, and are feeders to, the extrusive activity. The ore-bearing gabbro-dolerites are differentiated, with picrite and picritic dolerite overlain by more felsic differentiates. The relationships among these sediments, intrusions and extrusives are illustrated in the map of the Noril'sk-Talnakh area (Fig. 12). Individual sills may attain

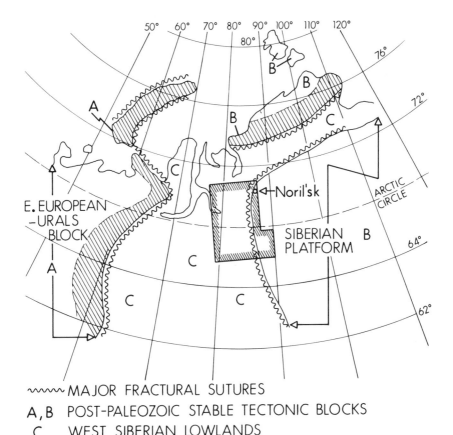

~~~~ MAJOR FRACTURAL SUTURES

A, B POST-PALEOZOIC STABLE TECTONIC BLOCKS

C WEST SIBERIAN LOWLANDS

Figure 11. The main tectonic features of northwestern Siberia (Maksimov and Rudkevich, 1974). The block outlined by shading is the area of Fig. 15.

lengths of 12 km, widths of 2 km and thicknesses of 30 to 350 m. The Cu-Ni-Platinoid mineralization forms relatively persistent ore horizons, occurring as segregations and accumulations of pyrrhotite, pentlandite and chalcopyrite, in the lower portions of the intrusions.

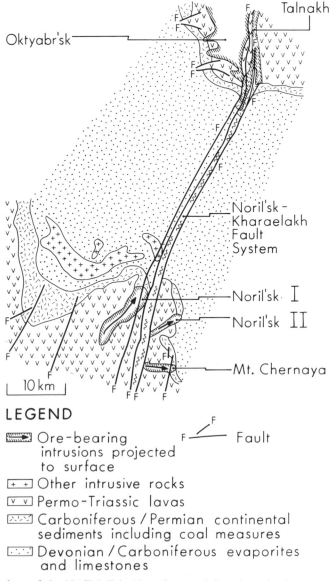

Talnakh

Oktyabr'sk

Noril'sk - Kharaelakh Fault System

Noril'sk I

Noril'sk II

Mt. Chernaya

10 km

LEGEND

▨ Ore-bearing intrusions projected to surface

⊞ Other intrusive rocks

⊻ Permo-Triassic lavas

⊡ Carboniferous / Permian continental sediments including coal measures

⊡ Devonian / Carboniferous evaporites and limestones

F⟋ Fault

Figure 12. Geology of the Noril'sk-Talnakh region (partially schematic) based on maps of Smirnov (1966) and Glazkovsky *et al.* (1977).

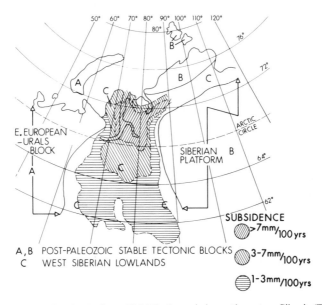

Figure 13. Subsidence during the Early and Middle Jurassic in northwestern Siberia (Tamrazyan, 1971).

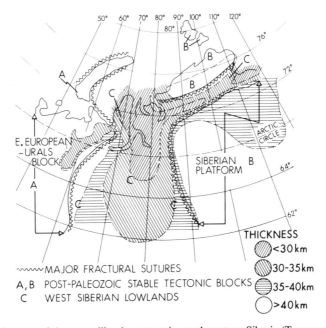

Figure 14. Thickness of the crystalline basement in northwestern Siberia (Tamrazyan, 1971).

The structure of the district is dominated by Late Permian to Triassic block faulting which was coeval with the igneous activity. Individual faults may be over 500 km in length with throws of up to 1000 m. (Smirnov, 1966). The principal faults trend NNE to NE.

Following the extrusion of flood basalt, considerable subsidence took place in the Yenisei and Khatanga Troughs from the Early Triassic to the Mid-Tertiary (Fig. 13), with a resultant sedimentary sequence that attains a thickness in excess of 10 km (Tamrazyan, 1971). The zone of subsidence describes a 'Y'-shaped structure, bounded by block faulting of the type described above (Maksimov and Rudkevich, 1971).

Geophysical measurements of depth to the Mohorovicic Discontinuity (Tamrazyan, 1971) indicate that, despite the considerable thickness of Mesozoic and Early Tertiary sediments in the area, the discontinuity is significantly shallower beneath the Yenisei and Khatanga Troughs than beneath the adjacent platforms (Fig. 14). The coincidence of considerable subsidence with crustal thinning beneath the troughs is indicative of a rifting environment. In its early stages the rifting was accompanied by the widespread intrusions and extrusions already described as occurring on the Siberian platform, and which aeromagmatic surveys indicate underlie broad areas of the lowlands themselves (Fig. 15) (Kulikov et al., 1972).

The mineralized Triassic intrusions represent volcanic conduits radiating outward and upward from intrusive centres and penetrating the sedimentary sequence. Many of the intrusive centres at Noril'sk are associated with a prominent block fault, the

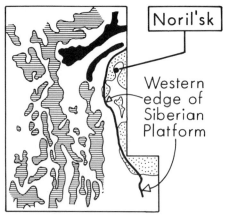

▤ Anomalies indicative of volcanics

■■ Anomalies known to be associated with intrusives

▦ Anomalies associated with known volcanics and intrusives of Siberian Platform

Figure 15. Aeromagnetic anomalies indicative of igneous rocks in a portion of the W. Siberian Lowlands and adjacent Siberian platform (Fig. 12 for location) (Kulikov et al., 1972).

Noril'sk-Kharaelakh fault, which occurs within the Siberian platform but is parallel to the fault system forming the boundary between the platform and the Siberian Lowlands to the west. It is regarded as a deep-seated fault which acted as the route for upwelling magma (Malich and Tuganova, 1976). Both the Mt. Chernaya deposit to the south, and the Talnakh-Oktyabr'sk deposits to the north, are located on this fault. Recently, a further group of gabrro-dolerite hosted Cu-Ni deposits, the North Kharaelakh ore field, with an aerial extent of 1500 km², has been described (Dyuzhikov *et al.*, 1976). These deposits, approximately 375 km north of Noril'sk, also lie close to the northerly extension of the Noril'sk-Kharaelakh fault.

The sulphur isotopic data for the Noril'sk district (see Fig. 8) indicate that the sulphides are anomalously heavy in ³⁴S. These data are inconsistent with mantle derivation of the sulphur, and consistent with its derivation from evaporites.

The geological and geochemical evidence indicates clearly that as the basaltic magmas rose through the sedimentary cover, they assimilated sulphur from the sulphate evaporites that they penetrated (Godlevski and Grinenko, 1963; Kovalenko *et al.*, 1975). The sulphur in calcium sulphate was reduced to sulphide, CaO entered the magma, and iron from the magma reacted with the reduced sulphur so that the end product was droplets of immiscible iron sulphide dispersed through the intrusions. These droplets acted as collectors for Ni, Cu and the Pt group elements which are so enriched in the Noril'sk ores. Tarasov (1970) has pointed to evidence that coal-bearing horizons from the Middle Carboniferous sediments were also assimilated and may have assisted in the reduction of the calcium sulphate.

DISCUSSION

Thus far we have described the geologic setting of the two largest nickel sulphide camps in the world, Sudbury and Noril'sk, and a major nickel resource, the Duluth deposits. In each case we have suggested that interaction with country rocks was responsible for the precipitation of the large amounts of sulphide that form the ores. The possible astrobleme setting for the Sudbury Nickel Irruptive has not, as far as we know, been suggested for other sulphide deposits, and is perhaps not the most useful guide for future exploration. Silicification of mafic magmas is not, of course, tied to meteorite impact, and evidence of extensive assimilation of felsic or siliceous rocks by other intrusions should be interpreted as a favourable omen in evaluating any mafic or ultramafic body as a host for nickel ores.

A much more important lesson emerges from the descriptions of Duluth and Noril'sk. In each case intracontinental rifting has allowed mafic magmas to rise up into the crust, where they have become contaminated by concentrations of crustal sulphur. This model has a wide potential application in exploration. Burke and Wilson (1976) have pointed out that doming of local areas of the Earth's crust often gives rise to Y-shaped rift zones centred on the dome, and that these rifts can be the 'seeds' from which continental rifting and ultimately ocean growth develop. Commonly, two arms of the Y become active rifts, while the third fails to develop, but becomes the locus for continental and lagoonal sedimentation, evaporites, and basic magmatism. The Y-shaped pattern of rifting in Western Siberia is interesting in this context, because none of the rift zones there have developed oceanic crust.

Burke and Wilson point to numerous examples where initial Y-shaped rifts result in the development of oceanic crust along two of the arms, while the third arm fails to develop to the same extent, and shows the characteristics of an aulacogene. One of these is Afar in East Africa. Figure 16, drawn from the data of Baker *et al.* (1972), illustrates the extent of doming in the Afar area by the elevation of Precambrian basement above sea level. Two arms of the rifting cut through the dome as sites of active spreading, and give rise to the Red Sea and the Gulf of Aden; the third (failed) arm connects southward with the East African rift system.

The Afar region is an example of early, extensive flood basalt and ignimbritic volcanism that now largely falls within the zone of rifting. Compared with other flood basalts, this type is unusual in its transitional nature between tholeiitic and alkaline composition and the final fractionation of the magmas into peralkaline felsic rocks (Gass, 1970; Barberi *et al.*, 1972). This igneous activity was followed by continental, largely alkalic, volcanism along the rift margins. Tholeiitic, oceanic-type basalts have developed within the two actively spreading arms and, to a lesser extent, within the Afar triangle (Fig. 17).

The tectonic development of the area has led to periodic inundation by the sea, resulting in thick evaporites (Hutchinson and Engels, 1970). The close association between igneous activity and the evaporites is illustrated in Figure 18, which is a diagrammatic vertical section along line A-B in Figure 16. The Afar region seems ideal for the application of the Noril'sk model to exploration. The geologic setting is particularly attractive in the light of Augustithis' (1978, p. 95) description of halite and sylvite being physically entrapped within feeders to the lavas and later extruded as molten salt at Dallol in the Danihill area.

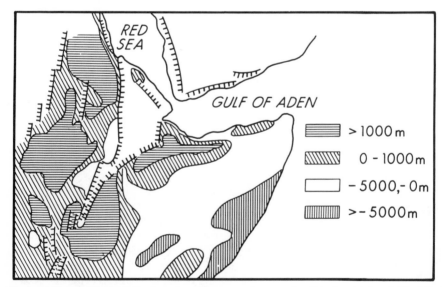

Figure 16. Uplift of Precambrian basement (in metres above sea level) in the vicinity of Afar (after Baker *et al.*, 1972).

CONCLUSIONS

Mineable concentrations of magmatic sulphides are not the normal consequence of mafic or ultramafic magmatism (apart from those sulphides very rich in PGE). An additional factor (or factors) is required to cause sulphides to segregate from a magma and concentrate to a degree that is not achieved by simple closed-system cooling and crystallization of the magma. An analysis of three of the most important nickel camps in the world indicates that assimilation of silica may have been the factor at Sudbury, and assimilation of country rock sulphur the factor at Noril'sk and Duluth. At both Noril'sk and Duluth, intracontinental rifting has allowed magma to rise into the crust, come into contact with concentrations of crustal sulphur and become very rich in sulphide. Major rift zones throughout the world, particularly those in which mafic magmatism and sulphate evaporites occur, are thus prime regions for exploration.

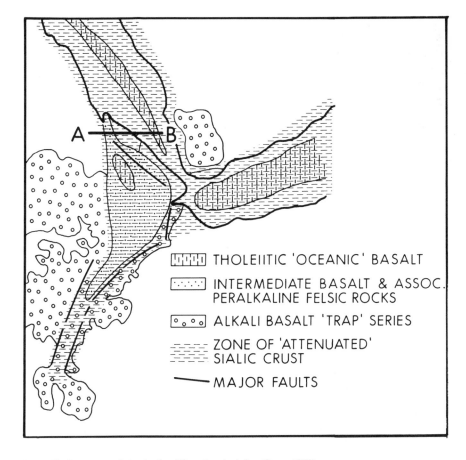

Figure 17. Igneous activity in the Afar triangle (after Gass, 1970).

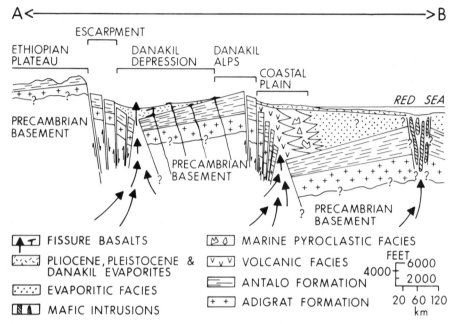

Figure 18. Schematic section illustrating the close association of mafic volcanism and evaporites on the S.W. flank of the Red Sea (after Hutchinson and Engels, 1970).

ACKNOWLEDGEMENTS

Financial support has been received from Natural Sciences and Engineering Research Council Grant No. A4244 to Prof. A. J. Naldrett. In addition, considerable support has been obtained under the N.R.C. Negotiated Development Grant to the Department of Geology, University of Toronto (principal applicant, D. W. Strangway).

REFERENCES

Arndt, N.T., Naldrett, A.J., and Pyke, D.R., 1977, Komatiitic and iron-rich tholeiitic lavas of Munro Township, Northeast Ontario: Jour. Petrol., v. 18, p. 319-369.

Augustithis, S.S., 1978, Atlas of textural patterns of basalts and their genetic significance: Elsevier, Amsterdam, 323 p.

Baker, B.H., Mohr, P.A., and Williams, L.A.J., 1972, Geology of the Eastern rift system of Africa: Geol. Soc. America Spec. Paper 136.

Barberi, F., Tazieff, H., and Varet, J., 1972, Volcanism in the Afar depression: its tectonic and magmatic signficance: Tectonophysics v. 15, p. 19-29.

Barnes, S.J., 1979, Petrology and geochemistry of the Katiniq nickel deposit and related rocks, Ungava, northern Quebec: Unpubl. M.Sc Thesis, University of Toronto, 220 p.

Barrett, F.M., Binns, R.A., Groves, D.I., Marston, R.J., and McQueen, K.G., 1977, Structural history and metamorphic modification of Archean volcanic-type nickel deposits, Yilgarn Block, Western Australia: Econ. Geol., v. 72, p. 1195-1224.

Bazunov, E.A., 1976, Development of the main structures of the Siberian platform: history and dynamics: Tectonophysics, v. 36, p. 289-300.

Bonnichsen, W., 1972, Sulfide minerals in the Duluth complex: in Sims, P.K. and Morey, G.W., eds., Geology of Minnesota: A centennial volume: Minnesota Geol. Survey, St. Paul, Minn., p. 388-393.

Boyd, R. and Mathiesen, C.O., 1979, The nickel mineralization of the Råna mafic intrusion, Nordland, Norway: Canadian Mineral, v. 17, pt. 2, p. 287-298.

Burke, K.C. and Wilson, J.T., 1976, Hot spots on the Earth's surface: Sci. American, v. 235, p. 46-57.

Card, K.D. and Hutchinson, R.W., 1972, The Sudbury structure: Its regional geological setting: in Guy-Bray, J.V., ed., New Developments in Sudbury Geology: Geol. Assoc. Canada Spec. Paper 10, p. 67-78.

Dietz, R.S., 1964, Sudbury structure as an astrobleme: Jour. Geol., v. 72, p. 412-434.

──────────────, 1972, Sudbury astrobleme, splash emplaced sublayer and possible cosmogenic ores: in Guy-Bray, J.V., ed., New Developments in Sudbury Geology: Geol. Assoc. Canada Spec. Paper 10, p. 29-40.

Duke, J.M. and Naldrett, A.J., 1978, A numerical model of the fractionation of olivine and molten sulfide from komatiite magma: Earth Planet. Sci. Letters, v. 39, p. 255-266.

Dyuzhikov, O.A., Fedorenko, V.A., Nestorovskiy, V.S., and Demidovich, V.I., 1976, The new North Kharaelakh ore field and its nickel potential: Doklady, Earth Science Sections, v. 229, p. 123-125.

Faure, G., Chaudhuri, S., and Fenton, M., 1969, Ages of the Duluth gabbro complex and of the Endion sill, Duluth, Minnesota: Jour. Geophys. Res., v. 74, p. 720-725.

Fleet, M.E., 1977, Origin of disseminated copper-nickel sulfide ore at Frood, Sudbury, Ontario: Econ. Geol., v. 72, p. 1449-1456.

──────────────, 1979, A tectonic origin for the Sudbury Shatter Cones (Abst.): Program with Abstracts, v. 4, Geol. Assoc., Canada-Mineral. Assoc. Canada, p. 50.

Fleet, M.E., MacRae, N.D., and Herzberg, C.T., 1977, Partition of nickel between olivine and sulfide: A test for immiscible sulfide liquids: Contrib. Mineral. Petrol., v. 65, p. 191-197.

French, B.M., 1967, Sudbury structure, Ontario: Some petrographic evidence for origin by meteorite impact: Science, v. 156, p. 1094-1098.

Guy-Bray, J.V., 1972, Introduction: in Guy-Bray, J.V., ed., 1972, New Developments in Sudbury Geology: Geol. Assoc. Canada Spec. Paper 10, p. 1-5.

Gass, I., 1970, The evolution of volcanism in the junction area of the Red Sea, Gulf of Aden and Afar triangle: Philos. Trans. Royal Soc. London, Ser. A, v. 267, p. 369-382.

Gibbins, W.A. and McNutt, R.H., 1975, The age of the Sudbury Nickel Irruptive and the Murray granite: Can. Jour. Earth Sci., v. 12, p. 1970-1989.

Glazkovsky, A.A., Gorbunov, G.I., and Sysoev, F.A., 1977, Deposits of nickel: in Smirnov, V.I., ed., Ore Deposits of the USSR; v. II

Godlevski, M.N. and Grinenko, L.N., 1963, Some data on the isotopic composition of sulfur in the sulfides of the Noril'sk deposit: Geochemistry, No. 1, p. 35-41.

Gorbunov, G.I., 1968, Geology and origin of the copper-nickel sulfide ore deposits of Pechenga (Petsamo): Nedra, Moscow, 352 p. (in Russian).

Hewins, R.H., 1971, The petrology of some marginal rocks along the north range of the Sudbury Irruptive: Unpub. Ph.D. Thesis, University of Toronto.

Hunter, D.R., 1974, Crustal development in the Kaapvaal Craton in the Archean: Precambrian Res., v. 1, p. 259-294.

Hutchinson, R.W. and Engels, G.G., 1970, Tectonic significance of regional geology and evaporite lithofacies in northeastern Ethiopia: Philos. Trans. Royal Soc. London, Ser. A, v. 267, p. 313-329.

Irvine, T.N., 1975, Crystallization sequences of the Muskox intrusion and other layered intrusions – 11. Origin of chromitite layers and similar deposits of other magmatic ores: Geochim. Cosmochim. Acta, v. 39, p. 991-1020.

Kovalenko, V.A., Gladyshev, G.D., and Nosik, L.P., 1975, Isotopic composition of sulfide sulfur from desposits of Talnakh ore node in relation to their selenium content: Internatl. Geology Review, v. 17, p. 725-736.

Kulikov, P.K., Belousov, A.P., and Latypov, A.A., 1972, West Siberian Triassic Rift System: Geotectonics (English Edition), No. 6, p. 367-371.

Lusk, J., 1976, A possible volcanic – exhalative origin for lenticular nickel sulfide deposits of volcanic association, with special reference to Western Australia: Canadian Jour. Earth Sci., v. 13, p. 451-458.

MacLean, W.H., 1969, Liquidus phase relations in the $FeS-FeO-Fe_3O_4-SiO_2$ system, and their application in geology: Econ. Geol., v. 64, p. 865-884.

Mainwaring, P.R. and Naldrett, A.J., 1977, Country-rock assimilation and the genesis of Cu-Ni sulfides in the Water Hen intrusion, Duluth Complex, Minnesota: Econ. Geol., v. 72, p. 1269-1284.

Maksimov, Ye. M. and Rudkevich, M. Ya., 1971, Quantitative evaluation of the dynamics of Mesozoic and Cenozoic vertical movements over the West-Siberian plate: Geotectonics, No. 4, p. 245-248.

Malich, N.S. and Tuganova, Ye. V., 1976, Distribution patterns of mineral resources in the Siberian platform cover: Internatl. Geology Review, v. 18, p. 417-424.

Naldrett, A.J., 1973, Nickel sulphide deposits – Their classification and genesis, with special emphasis on deposits of volcanic association: Trans. Canadian Instit. Mining Metallurgy, v. 76, p. 183-201.

―――――――――――, 1979, Ni-Cu sulfide deposits: Magmatic or hydrothermal? A response and a discussion: Econ. Geol. v. 74 (in press).

Naldrett, A.J., Guy-Bray, J.V., Gasparrini, E.L., Podolsky, T., and Rucklidge, J.C., 1970, Cryptic variation and the petrology of the Sudbury Nickel Irruptive: Econ. Geol., v. 65, p. 122-155.

Naldrett, A.J., Hewins, R.H., and Greenman, L., 1972, The main Irruptive and the sub-layer at Sudbury, Ontario: Proc. 24th Int. Geol. Cong., Section 4, 206-214.

Naldrett, A.J. and Turner, A.R., 1977, The geology and petrogenesis of a greenstone belt and related nickel sulfide mineralization at Yakabindie, Western Australia: Precambrian Res., v. 5, p. 43-103.

Pattison, E.F., 1979, The Sudbury sub-layer: Its characteristics and relationships with the main mass of the Sudbury Irruptive: Canadian Mineral. 17, pt. 2, p. 257-274.

Peredery, W.V., 1972, Chemistry of fluidal glasses and melt bodies in the Onaping Formation: in Guy-Bray, J.V., ed., New Developments in Sudbury Geology: Geol. Assoc. Canada Spec. Paper 10, p. 49-59.

―――――――――――, 1979, Relationship of ultramafic amphibolites to metavolcanic rocks and serpentinites in the Thompson belt, Manitoba: Canadian Mineral., v. 17, pt. 2, p. 187-200.

Peredery, W.V. and Naldrett, A.J., 1975, Petrology and the Upper Irruptive rocks, Sudbury, Ontario, Canada. Econ. Geol. 70, 164-175.

Phinney, W.C., 1972. Northwestern part of the Duluth complex; in Sims, P.K. and Morey, G.B., eds., Geology of Minnesota: A centennial volume: Minnesota Geol. Survey, St. Paul, Minn., p. 335-345.

Rajamani, V. and Naldrett, A.J., 1978, Partitioning of Fe, Co, Ni and Cu between sulfide liquid and basaltic melts and the composition of Ni-Cu sulfide deposits: Econ. Geol., v. 73, p. 82-93.

Ross, J.R. and Hopkins, G.M.F., 1975. The nickel sulfide deposits of Kambalda, Western Australia; in Knight, C., ed., Economic Geology of Australia and Papua-New Guinea: Australasian Instit. Mining Metallurgy, Monograph 5, p. 100-121.

Ruttan, G.D., 1955, Geology of Lynn Lake: Canadian Instit. Mining Metallurgy Bull., v. 48, p. 339-348.

Sims, P.K. and Morey, G.B., 1972, Geology of Minnesota: A centennial volume: Minnesota Geol. Survey, St. Paul, Minn., 632 p.

Smirnov, M.F., 1966, The Noril'sk nickeliferous intrusions and their sulfide ores: Nedra Press, Moscow.

Souch, B.E., Podolsky, T., and Geological Staff, 1969, The sulfide ores of Sudbury: Their particular relationship to a distinctive inclusion-bearing facies of the Nickel Irruptive; in Wilson, H.D.B., ed., Magmatic Ore Deposits: Econ. Geol. Monograph 4, p. 252-261.

Speers, E.C., 1957. The age relations and origin of the common Sudbury breccia: Jour. Geol., v. 65, p. 497-514.

Stevenson, J.S. and Colgrove, G.L., 1968, The Sudbury Irruptive: Some petrogenetic concepts based on recent field work: Proc. XXIII Internatl. Geol. Cong., v. 4, p. 27-35.

Tamrazyan, G.P., 1971, Siberian continental drift: Tectonophysics, v. 11, p. 433-460.

Tarasov, A.V., 1970, Structural control of copper and nickel mineralization in Noril'sk I deposit: Internatl. Geology Review, v. 12, p. 933-941.

Taylor, R.B., 1964, Geology of the Duluth gabbro complex near Duluth, Minnesota: Minnesota Geol. Survey Bull., v. 44, 63 p.

Viljoen, M.J. and Viljoen, R.P., 1969. Evidence of the existence of a mobile extrusive peridotitic lava from the Komati formation of the Onverwacht group: Geol. Soc. South Africa, Spec. Publ. 2, p. 87-113.

Vogt, J.H.L., 1918, Die sulphid: Silikatschmelzlosungen: Vid-Selsk. Skr. I.M.N. Kl., no. 1.

Weiblen P.W., 1965, A funnel-shaped gabbro-troctolite intrusion in the Duluth complex, Lake county, Minnesota: Unpub. Ph.D. Thesis, University of Minnesota, Minneapolis, Minn.

Weiblen, P.W. and Morey, G.B., 1975, The Duluth complex – A petrographic and tectonic summary. 36th Ann. Minnesota Mining Symposium: Dept. of Conferences and Continuing Education, Universtiy of Minnesota, Minneapolis, Minn., p. 72-95.

Wilson, H.D.B., 1953, Geology and geochemistry of base metal deposits: Econ. Geol., v. 48, p. 370-407.

The Continental Crust and Its Mineral Deposits, edited by D.W. Strangway,
Geological Association of Canada Special Paper 20, 1980

MASSIVE BASE METAL SULPHIDE DEPOSITS
AS GUIDES TO TECTONIC EVOLUTION

R. W. Hutchinson
Department of Geology,
University of Western Ontario,
London, Ontario N6A 5B7

ABSTRACT

The major family of massive base metal sulphide deposits includes one main group hosted in
volcanic and another in sedimentary rocks. Each group includes differing varieties distinguish-
able by their broad geological characteristics. These varieties span a complete spectrum from
ensimatic to ensialic geological environments and their time distribution suggests an evolutionary
sequence in their development. Some varieties exhibit close metallogenic affiliations with porphyry copper deposits and Mississippi Valley-type lead-zinc ores. Since broad crustal tectonic
processes controlled the space and time distribution of all these deposits, they may serve as
guides to tectonic evolution and paleotectonic processes.

The distribution of the varieties across Phanerozoic orogenic belts aids in reconstructing the
mechanics of ancient plate collisions in complex areas like the early Paleozoic proto-Atlantic of
Newfoundland. The space-time relationships of certain types of deposits suggest broad evolutionary changes in consuming plate margins from a "primitive" Japanese type in Paleozoic time
through a western Pacific type in Mesozoic time to a "mature" Andean type in Tertiary time.
Margins of the Japanese type may have a significant, although obscure, relationship to the
intracratonic-epeirogenic tectonism so characteristic of North America during Paleozoic time.

Precambrian massive sulphide environments suggest that the earliest Archean tectonic
processes operating in greenstone belts were analogous to those of modern accreting plate
boundaries, and later Archean processes to those of early-stage, consuming plate boundaries, but
vertical rather than horizontal motion was dominant. Worldwide Eparchean granitic plutonism
generated a thickened, stable, sialic-continental crust, and Proterozoic tectonic processes were
dominated by rifting of continental crust. Aphebian rifting involved mainly vertical displacements, forming deep, subsiding but cratonic sedimentary basins. A major change from volcanic-
hosted to sedimentary-hosted massive sulphide deposits in mid-Proterozoic time, and a corresponding lack of differentiated, calc-alkaline volcanic sequences, suggest that Helikian rifting

involved lateral separations, with formation of major, rift-bounded sedimentary basins, and possibly generation of the deep ocean basins. This activity preceded the major lateral movements of oceanic and continental plates that characterize subsequent plate tectonics. Evolutionary tectonism along plate boundaries regenerates all the older deposit types in the original sequence but on reduced time and space scales.

RÉSUMÉ

La grande famille des dépôts massifs de sulfures des métaux de base comprend un groupe important qu'on retrouve dans les roches volcaniques et un autre dans les roches sédimentaires. Chaque groupe comprend des variétés identifiables à leurs caractères géologiques généraux. Ces variétés couvrent un éventail complet allant de milieux géologiques ensimatiques à ensialiques et leur distribution dans le temps suggère une séquence évolutive de développement. Certaines variétés montrent des affinités métallogéniques étroites avec les dépôts de cuivre porphyrique et les minerais de plomb-zinc du type Vallée du Mississippi. Puisque des processus tectoniques d'envergure dans la croûte ont contrôlé la distribution dans le temps et dans l'espace de ces dépôts, ils peuvent servir de guides dans l'évolution tectonique et les processus paléotectoniques.

La distribution de ces variétés à travers les zones orogéniques du Phanérozoïque aide à reconstruire la mécanique des anciennes collisions de plaques dans des régions complexes comme le proto-Atlantique du début du Paléozoïque à Terre-Neuve. Les relations spatio-temporelles de certains types de dépôts suggèrent de vastes changements dans l'évolution des bordures de plaques qui se consument à partir d'un type japonais "primitif" au Paléozoïque en passant par un type Pacifique de l'Ouest au Mésozoïque jusqu'à un type andéen de maturité au Tertiaire. Les bordures du type japonais peuvent avoir des relations significatives, bien qu'obscures, avec le tectonisme intracratonique-épéirogénique si caractéristique de l'Amérique du Nord au cours du Paléozpïque.

Les environnements de sulfures massifs du Précambrien suggèrent que les processus tectoniques du tout début de l'Archéen opérant dans les ceintures de roches vertes étaient semblables à ceux des limites modernes de plaques en accrétion et les processus de la fin de l'Archéen semblables à ceux des stades initiaux des bordures de plaques qui se consument, mais le mouvement qui prédominait était vertical plutôt qu'horizontal. Le plutonisme granitique éparchéen à l'échelle du globe a engendré une croûte continentale sialique stable épaissie, alors que les processus tectoniques du Protérozoïque étaient dominés par la rupture et l'effondrement de la croûte continentale. L'effondrement durant l'Aphébien impliquait surtout des déplacements verticaux pour former des bassins sédimentaires s'enfonçant profondément tout en demeurant cratoniques. Au milieu du Protérozoïque, il s'est fait un changement majeur dans les dépôts de sulfures massifs de volcaniques qu'ils étaient, à sédimentaires; une absence correspondante de séquences volcaniques calco-alcalines différenciées suggère que l'effondrement durant l'Hélikien impliquait des séparations latérales avec formation de grands bassins sédimentaires limités par des zones d'effondrement et peut-etre la génération de bassins océaniques profonds. Cette activité a précédé les grands mouvements latéraux de plaques océaniques et continentales qui caractérisent la tectonique des plaques qui a suivi. Le tectonisme évolutif le long des bordures de plaques regénère tous les types anciens de dépôts dans la séquence originale mais sur des échelles réduites dans le temps et l'espace.

INTRODUCTION

Massive base metal sulphide deposits are economically vital to the world's supply of both base and precious metals, but the considerable scientific importance of these ores is perhaps insufficiently appreciated. Massive sulphide deposits occur in rocks of the entire geological column from Archean to Recent and they are found in an extremely

Table 1
MASSIVE BASE METAL SULPHIDE FAMILY
COMMON IDENTIFYING CHARACTERISTICS

Stratigraphically-controlled distribution:
  Stratabound in exhalative volcanogenic group
  Stratiform in exhalative sedimentary group

Massive lens:
  Iron sulphide-rich (py ± po)
  Base metal sulphides (cp, sph, gal)
  Important precious metals (Ag, Au)

Stratigraphically overlying:
  Fe-rich chemical sedimentary rock (I.F.–4 facies, chert, tuff)

Stratigraphically underlying:
  Stringer-disseminated sulphides, Cu-rich
  Altered footwall rock (vent or pipe sometimes recognizable)

Prominent metal-mineral zoning:
  Vertical, within deposits
      Cu(Au) base ⟶ Zn, Pb(Ag) top
      po base ⟶ py top
  Lateral, between deposits
      Cu(Au) proximal ⟶ Zn, Pb(Ag) distal

Fragmental sulphides:
  Mainly in exhalative volcanogenic group

Soft-sediment deformation structures:
  Mainly in exhalative sedimentary group

Widely variable metamorphism ⟶ textural variability

Table 2
MASSIVE BASE METAL SULPHIDE FAMILY

I. EXHALATIVE VOLCANOGENIC GROUP

  HOSTED IN VOLCANIC-DOMINATED, MARINE SUCCESSIONS

  RELATIVELY ENRICHED IN CU AND AU

  STRATABOUND – LENTICULAR

II. EXHALATIVE SEDIMENTARY GROUP

  HOSTED IN SEDIMENT-DOMINATED, MARINE SUCCESSIONS

  RELATIVELY ENRICHED IN PB AND AG

  STRATIFORM – TABULAR ·

wide range of geological environments, from ensimatic to ensialic. The distribution of the different types of massive sulphides in space and time strongly suggests that the differences are the result of tectonic evolution. Consequently these ores may be used as guides to tectonic evolution and to paleotectonic processes. In addition, the massive sulphides have significant metallogenic and genetic links to other important families of ore deposits, notably to porphyry copper deposits and to Mississippi Valley type lead-zinc ores.

MASSIVE BASE METAL SULPHIDE DEPOSITS

The massive base metal sulphides comprise a major family of ore deposits, all members of which share a set of common identifying characteristics. Although not all of these characteristics are present in each deposit, they are well described in the literature (Sangster and Scott, 1976; Sangster, 1972a; Hutchinson, 1973) and are summarized in Table I. They indicate the common exhalative origin of all these ores which were originally deposited as sea-floor chemical precipitates from discharged hydrothermal brines (Hutchinson, 1973).

The massive base metal sulphides can be divided into two main groups (Table II), one hosted mainly in volcanic rocks and called exhalative volcanogenic, and one occurring chiefly in sedimentary rocks, designated exhalative sedimentary (Oftedahl, 1958). These two groups can be further subdivided into a number of different types, although it is emphasized that there are no sharp boundaries between them. Instead there is a natural, continuous spectrum of geologic environments. This subdivision is based on differences in metal content, associated volcanic and sedimentary rock types, age and depositional environment (Tables III and IV) – all of which are controlled by differences in the broad tectonic environment of their subaqueous deposition, which is the main point of this discussion.

The oldest and most "primitive" Zn-Cu-rich type of massive sulphide deposit was laid down in tectonically unstable, deeply subsiding basins that apparently underwent synkinematic deformation. The resulting rock sequences are dominated by thick, well-differentiated, tholeiitic to calc-alkaline, mafic to felsic volcanic rocks (Baragar and Goodwin, 1969; Goodwin, 1968). The associated sedimentary rocks are immature greywackes and volcaniclastics (Hutchinson, 1973).

The second "polymetallic" or "kuroko" Zn-Pb-Cu-rich type was deposited in generally shallower, increasingly cratonic and more highly oxygenated basins, and the associated sedimentary rocks are increasingly epiclastic, craton-derived, and carbonate-sulphate rich. These basins were apparently tensionally controlled during deposition (Uyeda and Nishiwaki, 1980), and the volcanic rocks tend to be of bimodal basaltic-rhyodacitic character. The ores are associated with more explosive, silicic-alkalic volcanism than are those of the primitive type, as evidenced by more abundant fragmental and quartz-porphyritic rocks.

The third type of massive sulphide deposit, "cupreous pyrite", is very different from the two previous ones. It was formed in deep oceanic basins clearly controlled by oceanic crustal rifting and tension. The associated igneous rocks are ultramafic to tholeiitic and the ores occur in basaltic pillow lavas (Hutchinson, 1973). There are only very minor sedimentary rocks, all pelagic. This assemblage is ophiolitic, and the entire succession is ensimatic and oceanic crustal.

Table 3

MASSIVE BASE METAL SULPHIDE FAMILY

	VOLCANIC ROCKS	CLASTIC SEDIMENTARY ROCKS	APPROXIMATE AGE RANGE		EXAMPLES
I. EXHALATIVE VOLCANOGENIC GROUP					
1. PRIMITIVE TYPE ZN-CU: AG-AU	FULLY DIFFERENTIATED SUITES BASALTIC TO RHYODACITIC	VOLCANOCLASTICS GREYWACKES	ARCHEAN EARLY PROTEROZOIC EARLY PHANEROZOIC	>2.5 >1.8 Є-DEV.	NORANDA, KIDD CREEK; JEROME, FLIN FLON, CRANDON WEST SHASTA
2. POLYMETALLIC TYPE PB-ZN-CU: AG-AU	BIMODAL ? SUITES THOLEIITIC BASALTS CALC-ALKALIC LAVAS, PYROCLASTICS	VOLCANOCLASTICS INCREASING CLASTICS MINIMUM CARBONATES	EARLY PROTEROZOIC PHANEROZOIC	>1.8 ORD. MESO. TER.	PRESCOTT, SUDBURY BASIN, MT. ISA NEW BRUNSWICK EAST SHASTA JAPAN
3. CUPREOUS PYRITE TYPE CU: AU	OPHIOLITIC SUITES THOLEIITIC BASALTS	MINOR TO LACKING	PHANEROZOIC	Є-ORD. MESO.	NEWFOUNDLAND CYPRUS, TURKEY, OMAN
4. KIESLAGER TYPE CU-ZN:AU	MAFIC; THOLEIITIC(?) (AMPHIBOLITE)	GREYWACKE, SHALE(?) (BIOT.-AMPHIB. SCHIST)	LATE PROTEROZOIC PALEOZOIC	1.2-.8	MATCHLESS-OTJIHASE S.W. AFRICA; DUCKTOWN, TENN; BESSHI, JAPAN; GOLDSTREAM, B.C.
II. EXHALATIVE SEDIMENTARY GROUP					
5. CLASTIC HOSTED TYPE PB-ZN: AG	MINOR, BASALTIC (GABBROIC-AMPHIBOLITIC INTRUSIVE SHEETS)	ARGILLITE, TURBIDITE	MID PROTEROZOIC PALEOZOIC	1.7-1.0 Є-DEV.	SULLIVAN; BROKEN HILL; McARTHUR RIVER ANVIL, MEGGEN, RAMMELS-BERG, HOWARDS PASS
6. CARBONATE HOSTED TYPE ZN-PB: (AG)	MINOR TO LACKING (TUFFACEOUS BEDS)	CARBONATE LST.-DOL. SANDSTONE, SHALE	LATE PROTEROZOIC PHANEROZOIC	~1.0 MISS.	BALMAT NAVAN, SILVERMINES; TYNAGH, IRELAND

Table 4

MASSIVE BASE METAL SULPHIDE FAMILY

	DEPOSITIONAL ENVIRONMENT	TECTONIC ENVIRONMENT	
		GENERAL CONDITIONS	PLATE TECTONIC SETTING
I. EXHALATIVE VOLCANOGENIC GROUP			
1. PRIMITIVE TYPE ZN-CU: AG-AU	EXTENSIVE, EVOLVING, DEEP TO SHALLOW WATER, THOLEIITIC TO CALC-ALKALINE MARINE VOLCANISM	MAJOR SUBSIDENCE COMPRESSION	SUBDUCTION AT CONSUMING MARGIN, ISLAND ARC
2. POLYMETALLIC TYPE PB-ZN-CU: AG-AU	EXPLOSIVE SHALLOW CALC-ALKALINE-ALKALINE MARINE-CONTINENTAL VOLCANISM	SUBSIDENCE REGIONAL COMPRESSION BUT LOCAL TENSION	BACK-ARC OR POST-ARC SPREADING; CRUSTAL RIFTING AT CONSUMING MARGIN
3. CUPREOUS PYRITE TYPE CU: AU	DEEP THOLEIITIC MARINE VOLCANISM	MINOR SUBSIDENCE TENSION	OCEANIC RIFTING AT ACCRETING MARGIN
4. KIESLAGER TYPE CU-ZN: AU	DEEP MARINE SEDIMENTATION AND THOLEIITIC VOLCANISM	MAJOR SUBSIDENCE COMPRESSION	FORE-ARC TROUGH OR TRENCH
II. EXHALATIVE SEDIMENTARY GROUP			
5. CLASTIC HOSTED TYPE PB-ZN: AG	DEEP MARINE SEDIMENTATION; MINOR THOLEIITIC ACTIVITY	MAJOR SUBSIDENCE TENSION, RIFTING	SEPARATION AT CONTINENTAL RIFT; AULACOGENIC TROUGH OR TRENCH
6. CARBONATE HOSTED TYPE ZN-PB: (AG)	SHALLOW MARINE-SHELF SEDIMENTATION	MINOR SUBSIDENCE, TENSION	SHELF; LOCAL FAULT - CONTROLLED BASIN

The fourth, "kieslager or Besshi", type of deposit is, in general, more highly metamorphosed than the others, although there are specific exceptions. In many respects these deposits are intermediate in geological characteristics between the primitive and the cupreous pyrite types. Like the primitive type, the kieslager type was deposited in unstable subsiding basins with thick sedimentary sequences of greywackes and volcaniclastic rocks and it apparently underwent synkinematic deformation. Like the cupreous pyrite deposits, however, this type is associated with tholeiitic volcanic or plutonic rocks and is characterized by the absence of significant differentiated calc-alkaline or felsic types. The kieslager type exhibits strong sedimentary affinities, like the deposits of the exhalative sedimentary group, to which it thus forms a significant metallogenic link.

The fifth or "clastic-hosted" type of massive sulphide deposit belongs to the exhalative-sedimentary group. It was apparently formed in subsiding and probably cratonic, rift-controlled basins, possibly aulacogenic (Dunnet, 1976), and is associated with thick turbiditic argillite-shale sequences. The minor associated igneous rocks are mainly of tholeiitic and mafic character, including diabase, amphibolite, and basalt.

The sixth and final "carbonate-hosted" type of massive sulphide was formed in shallow cratonic margins, apparently in local, tensional fault-controlled, horst-graben environments with shelf facies carbonate and clastic sedimentary rocks. Thin tuffaceous strata may indicate a little associated volcanism or igneous activity (Schultz, 1961), but this type is totally ensialic in character.

The lack of well-defined boundaries between the volcanic and sedimentary groups is illustrated by the famous Mt. Isa deposit which appears to have some characteristics of both (Mathias and Clark, 1975). The silver-lead-zinc orebodies are of clastic-hosted, exhalative sedimentary nature, but the deeper, silica-shale, copper-rich orebodies have volcanogenic characteristics including a closer association, albeit through faulting, with the Southern and Eastern Creek volcanic rocks. Similarly, no distinct boundaries separate the six types of deposits: the kieslager type is intermediate and gradational in many of its characteristics between the primitive and cupreous pyrite types, and the polymetallic exhalative volcanogenic type shares some characteristics with the carbonate-hosted exhalative sedimentary deposits. There are other similar transitional examples.

AGE

The periods of maximum or major development of each of the six types exhibit a distinct progression through geologic time. Their age ranges and peaks are compared in Figure 1, along with those of porphyry copper and Mississippi Valley type lead-zinc deposits, two other metallogenically related types.

Deposits of the primitive type are the oldest, first appearing in ancient Archean rocks, where they are best developed and most important, as at Noranda, Quebec (Spence and de Rosen Spence, 1975) and Kidd Creek, Ontario (Walker et al., 1975). Deposits of this type occur in many other districts, however: in rocks of early Proterozoic age as at Jerome, Arizona (Anderson and Nash, 1972) and Flin Flon, Manitoba (Sangster, 1972b; Koo and Mossman, 1975), and also in rocks of early Paleozoic age as in the West Shasta district of California (Kinkel et al., 1956).

Deposits of the polymetallic type appear next on the time scale, in early Proterozoic rocks as at Prescott, Arizona (Gilmour and Still, 1968), at Snow Lake, Manitoba (Byers *et al.*, 1965) and, interestingly, in rocks of the Whitewater Group within the Sudbury Basin (Martin, 1957; Card and Hutchinson, 1972). They are perhaps most important, however, in rocks of early Paleozoic age, as in the Bathurst district of New Brunswick, Canada (Smith and Skinner, 1958; Lea and Rancourt, 1958; Stanton, 1959) or in the Tasman Belt in eastern Australia (Burton, 1975; Davis, 1975; Malone *et al.*, 1975). Still other examples are known in Mesozoic and Tertiary successions as at East Shasta, California (Albers and Robertson, 1961) and in the Green Tuff region of Japan (Ishihara, 1974).

The cupreous pyrite type appears late in the geological time scale, only in Phanerozoic rocks. The earliest deposits are perhaps those of pre- or early Ordovician age in Newfoundland (Hutchinson, 1973), but the type is particularly important and best developed in Mesozoic rocks of the Tethyan Belt in southern Europe and the Middle East as in Cyprus, Turkey and Oman (Bear, 1963; Griffits *et al.*, 1972; Carney and Welland, 1974; Gealey, 1977). In terms of tectonics this age is significant because the peak development of these ores coincides with a period of major oceanic crustal generation that accompanied the break-up of Pangaea and the subsequent separation of the Eurasian, African and North American continents.

Deposits of the kieslager type also appear to be relatively late in geologic time. Good examples occur only in rocks of late Precambrian and Paleozoic ages, as in the

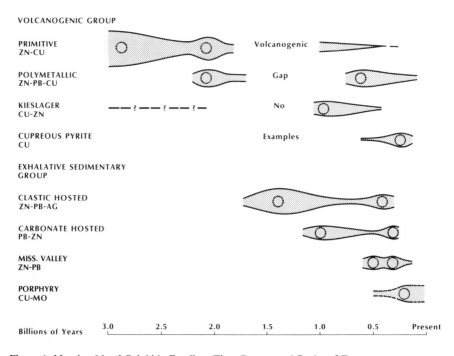

Figure 1. Massive Metal Sulphide Family – Time Ranges and Peaks of Types

Amphibolite Belt of South-West Africa (Goldberg, 1976) and at the Besshi and neighbouring deposits in Japan (Kanehira and Tatsumi, 1970) respectively. The Goldstream (Hoy, 1979) and Hart River deposits are recently discovered Paleozoic examples in British Columbia and the Yukon Territory. Considering their geologic setting however, the absence of kieslager-type deposits in Archean rocks is puzzling (Fig. 1). Archean examples may be present but unrecognized owing to their high metamorphic rank, as at Manitouwadge, Ontario (Brown and Bray, 1960) or Fox Lake-Ruttan-Sherridon, Manitoba (Farley, 1948).

An enigmatic aspect of the time distribution of these four volcanogenic types is the absence of any significant examples in rocks of mid-Proterozoic (Helikian) age. This absence is termed the "volcanogenic gap" (Fig. 1). It is pertinent that this gap is, however, filled by many large and important deposits of the clastic-hosted, exhalative sedimentary type, like those of the Sullivan Mine in southern British Columbia (Morris et al., 1972) and the highly metamorphosed ores of the great Broken Hill lodes in New South Wales (Johnson and Klingner, 1975). This type is also important in mid-to late Paleozoic sequences as in the Selwyn Basin area of the MacKenzie mountains, Yukon Territory, (Carne, 1976) and at the famous old Meggen and Rammelsberg deposits of southern Germany (Gwosdz, 1974; Anger et al., 1966). The tectonic significance of the "volcanogenic gap" and of the prolific development of deposits of the clastic-hosted, exhalative sedimentary type in Helikian time is not understood but merits consideration.

Finally, deposits of the carbonate-hosted, exhalative sedimentary type appear first in late Precambrian rocks of the Grenville structural province in North America as at Balmat and Edwards in New York state (Lea and Dill, 1968) but are especially important in late Paleozoic rocks as at Tynagh, Silvermines and Navan in the Mississippian of Ireland (Pereira, 1967; Derry et al., 1965). Earlier Paleozoic examples in Canada may be the lead-zinc deposits of the Kootenay arc in British Columbia (Sangster, 1970). The geological environment and time distribution of this type is remarkably similar to those of Mississippi Valley type lead-zinc deposits, which suggests an important metallogenic and possibly genetic link between these two types.

Another significant feature is the deep, copper-rich, stringer and disseminated sulphides which occur in altered rocks beneath many massive base metal sulphide deposits (Hutchinson, 1973). This type of mineralization has many geological similarities to, and suggests metallogenic relationships with, porphyry copper deposits (Hutchinson and Hodder, 1972), but the vast majority of porphyry deposits are much younger than the massive sulphides (Fig. 1).

PLATE TECTONIC ENVIRONMENTS

The tectonic environment of formation of the various types of massive sulphides can be reasonably explained in terms of lithospheric plate configurations according to current plate tectonic theory, or Wilson Cycle relationships (Fig. 2). First, it is clear from their geological setting, as well as from recent oceanographic studies (Degens and Ross, 1969; Francheteau et al., 1979) that the ensimatic cupreous pyrite type reflects the tectonic environment of accreting plate margins and forms at spreading centres in mid-oceanic ridges. The remaining three volcanogenic types, however, apparently are generated at different times and/or tectonic positions along consuming plate bound-

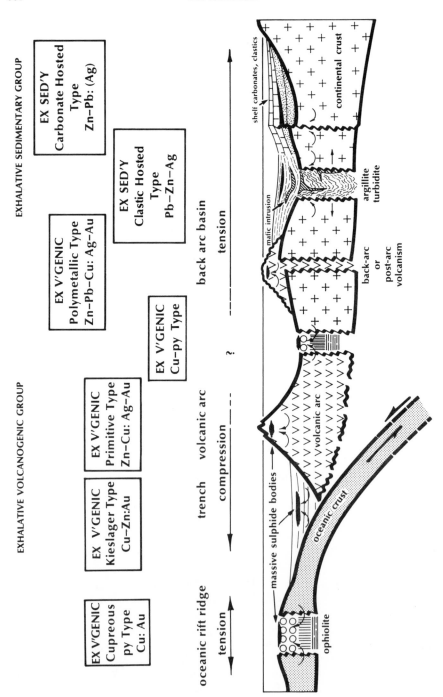

EXHALATIVE SEDIMENTARY GROUP

EX SED'Y
Carbonate Hosted
Type
Zn–Pb: (Ag)

EX SED'Y
Clastic Hosted
Type
Pb–Zn–Ag

EXHALATIVE VOLCANOGENIC GROUP

EX V'GENIC
Polymetallic Type
Zn–Pb–Cu: Ag–Au

EX V'GENIC
Cu–py Type

EX V'GENIC
Primitive Type
Zn–Cu: Ag–Au

EX V'GENIC
Kieslager Type
Cu–Zn:Au

EX V'GENIC
Cupreous
py Type
Cu: Au

back arc basin

tension

volcanic arc

trench

compression

oceanic rift ridge

tension

tension

shelf carbonates, clastics

continental crust

mafic intrusion

argillite
turbidite

back-arc
or
post-arc
volcanism

volcanic arc

oceanic crust

massive sulphide bodies

ophiolite

aries. It is probable that within this tectonic setting the primitive type reflects the differentiated, tholeiitic to calc-alkaline to silicic-alkalic volcanism of early-stage island arc environments. Deposits of the kieslager type are apparently generated in the fore-arc trench environment, where they may understandably share some geological characteristics with both cupreous pyrite and primitive types (Fig. 2), and be subjected, "on average", to higher rank metamorphism than the other types. Deposits of the polymetallic type appear typical of the back-arc basin, or the younger, inner volcanic arc environment. The tectonic environment of the clastic-hosted exhalative sedimentary type is perhaps least understood, but both its geologic environment and current oceanographic studies (Degens and Ross, 1976) suggest that deposits of this type form in continental, possibly aulacogenic rift systems. Finally, deposits of the carbonate-hosted exhalative sedimentary type reflect an ensialic, stable shelf or craton margin.

As shown in Figure 2, not all of these tectonic environments develop coevally; there is a general age progression from deposits on the left of the diagram to deposits on the right. This is particularly true along the consuming plate boundary, where the trench and island arc environments may evolve first and be followed later by evolution of the inner-arc or back-arc basin environments. Thus the combined space- and time-related evolutionary changes in tectonic environments that control the formation of volcanogenic deposits result in progressively younger deposits and an accompanying increase in lead and silver at the expense of copper and gold across a continentward-dipping subduction zone. This compositional change is analogous to the continent-ward increase in the potash content of volcanic rocks noted by Kuno (1966) across volcanic island arcs.

The resulting geological differences among the volcanogenic massive sulphide deposits provide a guide to paleotectonic reconstruction of ancient plate collisions. An example is the Cordilleran margin of North America in Paleozoic-Mesozoic time (Fig. 3). Subduction of Pacific oceanic crust in Devonian time (Fig. 3a) generated the differentiated volcanic rocks and primitive-type Cu-Zn deposits of the West Shasta district (Kinkel et al., 1956) in a volcanic arc above the subducted plate. This was followed in Triassic-Jurassic time (Fig. 3b) by the evolution of a younger, inner-arc or back-arc environment in the East Shasta district (Albers and Robertson, 1961), in which deposits of the polymetallic type were formed. Finally, in Cretaceous time, uplift and emplacement of the Sierra Nevada batholith (Fig. 3c) was accompanied by obduction of Franciscan ophiolitic rocks of the oceanic crustal plate, containing small cupreous pyrite-type deposits (Stinson, 1957), onto the continental margin. This example clearly shows the succession of younger deposits and the increase in lead and silver over copper and gold across the continentward-dipping subduction zone.

The eastern North American Appalachian margin shows the reverse configuration of these deposit types. Here the youngest, Mississippian Zn-Pb-Ag-rich carbonate-hosted sulphide bodies of the Walton Mine in Nova Scotia (Boyle and Jambor, 1966) are farthest east. The polymetallic ores and Silurian-Ordovician felsic

---

Figure 2. Plate Tectonic Relationships: Depositional Environments of Various Massive Base Metal Sulphides

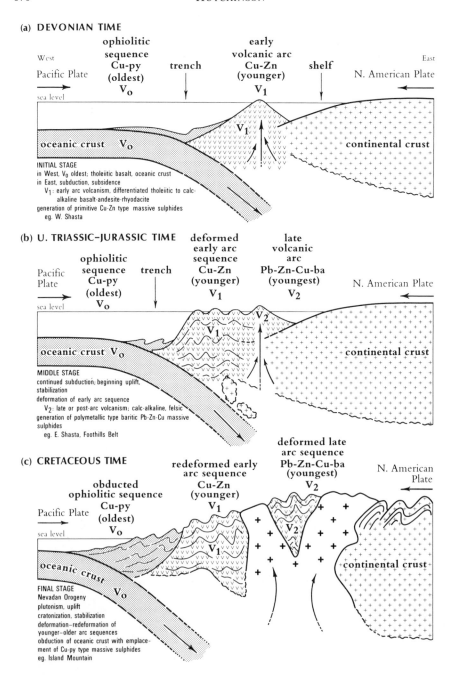

Figure 3. Metallogenic Relationships, North American Pacific Margin

volcanic host rocks like those of Buchans, Newfoundland (Thurlow *et al.*, 1975), northern New Brunswick (Smith and Skinner, 1958; Lea and Rancourt, 1958), coastal Maine and the Piedmont in the southeastern U.S.A. (Gair and Slack, 1978) also lie to the east, or oceanward, whereas the older ophiolitic rocks with deposits of cupreous pyrite type (and asbestos) lie farther west, or continent-ward, as in the ophiolite sheets of western Newfoundland (Duke and Hutchinson, 1974) and the Eastern Townships of Quebec. In some places along the Appalachian belt Cu- and Zn-bearing deposits of the primitive or kieslager types are located between the older and youngest types, west of and in rocks older than those of the felsic volcanic-polymetallic belt. Examples occur at Ming and Rambler Mines, Newfoundland (Heenan and Truman, 1972; Tuach and Kennedy, 1978), Weedon, Solbec and Cupra d'Estrie, Quebec (Sauve *et al.*, 1972) and in northwestern Maine (Mining Engineering, 1978). Father sourth, the Cu-Zn ores of Ducktown, Tennessee (Magee, 1968) also lie in older rocks, west of the polymetallic Piedmont deposits.

In contrast to the Cordilleran belt in California (Fig. 3) this spatial configuration of massive sulphide types strongly suggests an eastward-, or oceanward-dipping subduction zone along the North American Atlantic margin in early Paleozoic time. This may be explained by the paleo-plate configurations of the proto-Atlantic, or Iapetus ocean (Harland and Gayer, 1972), which closed during late Precambrian to early Ordovician time (Fig. 4a). Older ophiolitic rocks containing deposits of the cupreous pyrite type were piled up by obduction to the west by this closure, to form the ophiolite complexes of western Newfoundland and Quebec's Eastern Townships, while an eastward-dipping zone was subducted beneath the European-African continent(s). Deposits of the primitive, or kieslager types were formed in volcanic arc, or fore-arc trench environments above this subduction zone. Continued closure of the ocean in Ordovician-Silurian time (Fig. 4b) generated more felsic volcanic rocks and deposits of the polymetallic type in a slightly younger, back- or inner-arc environment, as at Buchans, and in northern New Brunswick, coastal Maine, and the Appalachian piedmont. Complete closure of the ocean in Devonian time (Fig. 4c) was marked by Acadian orogeny, uplift, and granitic plutonism. Subsequent opening of the modern Atlantic Ocean to the east of the earlier orogenic belt left the present oceanward progression of younger sulphide types and the increase of lead-silver at the expense of copper-gold, providing evidence for an eastward-dipping paleo-subduction zone. A similar eastward-dipping zone is also indicated in Newfoundland by other geologic and geophysical evidence (Church and Stevens, 1971; Haworth, 1978); the metallogenic relationships discussed by those authors are therefore in accord with the interpretation presented here.

It is interesting that the age of two of the types progresses from older in the southern Appalachians to younger in the north. The Cu-Zu-bearing deposits of primitive or kieslager type at Ducktown, Tennessee are found in late Precambrian rocks (Magee, 1968), the comparable Weedon-Solbec-Cupra d'Estrie ores of the Eastern Townships in Cambro-Ordovician rocks (Sauve *et al.*, 1972) and the Ming-Rambler deposits of Newfoundland in early to late Ordovician rocks (Kean *et al.*, in press). Similarly, the polymetallic ores of northern New Brunswick are of Ordovician age (Smith and Skinner, 1958), whereas the comparable deposits of Buchans, Newfoundland are of late Ordovician-early Silurian age (Bell and Blenkinsop, in press). These age relationships suggest that closure of the proto-Atlantic ocean proceeded from

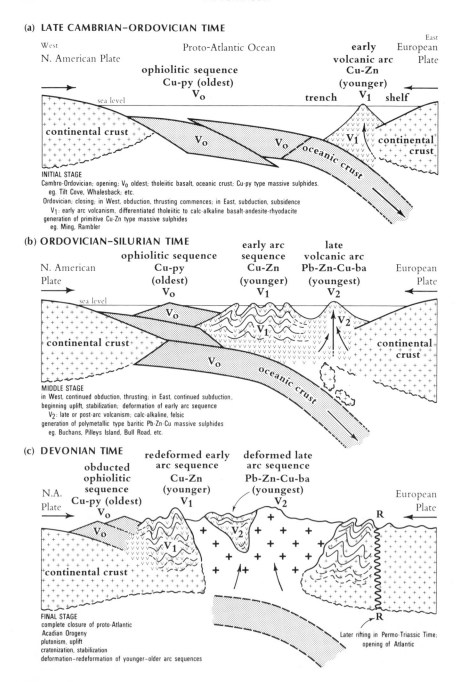

(a) LATE CAMBRIAN–ORDOVICIAN TIME

West
N. American Plate

Proto-Atlantic Ocean
ophiolitic sequence
Cu-py (oldest)
V_0

East
early European
volcanic arc Plate
Cu-Zn
(younger)
trench V_1 shelf

sea level

continental crust V_0 V_0 V_1 continental crust

oceanic crust

INITIAL STAGE
Cambro-Ordovician; opening; V_0 oldest; tholeiitic basalt, oceanic crust; Cu-py type massive sulphides.
eg. Tilt Cove, Whalesback; etc.
Ordovician; closing; in West, obduction, thrusting commences; in East, subduction, subsidence
V_1: early arc volcanism, differentiated tholeiitic to calc-alkaline basalt-andesite-rhyodacite
generation of primitive Cu-Zn type massive sulphides
eg. Ming, Rambler

(b) ORDOVICIAN–SILURIAN TIME

N. American
Plate

ophiolitic sequence
Cu-py
(oldest)
V_0

early arc
sequence
Cu-Zn
(younger)
V_1

late
volcanic arc
Pb-Zn-Cu-ba
(youngest)
V_2

European
Plate

sea level

V_0 V_1 V_2

continental crust V_0 continental crust

oceanic crust

MIDDLE STAGE
in West, continued obduction, thrusting; in East, continued subduction,
beginning uplift, stabilization; deformation of early arc sequence
V_2: late or post-arc volcanism; calc-alkaline, felsic
generation of polymetallic type baritic Pb-Zn-Cu massive sulphides
eg. Buchans, Pilleys Island, Bull Road, etc.

(c) DEVONIAN TIME

N.A.
Plate

obducted
ophiolitic
sequence
Cu-py (oldest)
V_0

redeformed early
arc sequence
Cu-Zn
(younger)
V_1

deformed late
arc sequence
Pb-Zn-Cu-ba
(youngest)
V_2

European
Plate

V_0

V_0 V_1 V_2 R

continental crust

FINAL STAGE
complete closure of proto-Atlantic
Acadian Orogeny
plutonism, uplift
cratonization, stabilization
deformation–redeformation of younger–older arc sequences

R

Later rifting in Permo-Triassic Time;
opening of Atlantic

Figure 4. Metallogenic Relationships, North American Atlantic Margin.

south to north, i.e., in the opposite direction to the opening of the "modern" Atlantic, which appears to have occurred from north to south, judging from the progressive southward "younging" of rift-related evaporite sequences and mafic intrusions (Schenk, 1969; Aymé, 1965; Belmonte et al., 1965).

It is also interesting that farther to the north along this belt in Ireland, Britain and Scandinavia, although not all types are present, the spatial disposition of some of them appears to be the reverse of that in the North American Appalachians. Thus, the older and more copper-rich types, for example Avoca in Ireland, Parys Island in Anglesey, and Skorovas in Norway (Platt, 1977; Thanasuthipitak, 1975; Halls et al., 1977) lie to the southeast of the younger, lead-zinc-silver-rich types (Derry et al., 1965; Pereira and Dixon, 1965; Ford and King, 1965). This suggests a northwesterly-dipping paleo-subduction zone beneath the Caledonides of northwestern Europe, as opposed to the southeasterly-dipping subduction zone of the North American Appalachians. There are other indications of a reversal in dip of the early Paleozoic subduction zone from eastward in the Appalachians to westward in the Caledonides: major thrusting and nappe emplacement involving eastward transport of the allochthons in Norway and Sweden (Sturt and Roberts, 1978; Gee, 1978) is comparable to activity in the Appalachians (Osberg, 1978; Hatcher, 1978), but with westward transport. Thus the interpretation based on spatial relationships of the massive sulphide types is consistent with that derived from structural relationships.

METALLOGENIC-TECTONIC RELATIONSHIPS

The similarities in age and geological environment of massive sulphides of the carbonate-hosted exhalative sedimentary type to the Mississippi Valley-type lead-zinc deposits suggest that the late-stage evolution of inner-arc or back-arc basins in Paleozoic orogenic belts was somehow related to the intra-cratonic, epeirogenic tectonism that was so very important throughout North America during Paleozoic time. This tectonism controlled the prolonged activity of the numerous domes, arches and sedimentary basins within the continent, which in turn influenced emplacement of the Mississippi Valley-type ores. The specific tectonic mechanisms are virtually unknown, but they are relevant to both tectonic geology and mineral exploration, and merit detailed investigation.

The abundance of massive sulphide deposits in the relatively young Tertiary rocks of Japan, combined with the absence there of porphyry copper deposits is similar to the situation found in Paleozoic orogenic belts. Since porphyry copper deposits are elsewhere abundant in Mesozoic and Tertiary orogenic belts, their absence in Japan is particularly puzzling.

Perhaps these time-space relationships may be explained by broad evolutionary changes in the types of consuming plate margins through Phanerozoic time. Possibly margins like that of Japan, with a prominent, off-shore, marine volcanic arc and a broad back-arc basin (Fig. 5a) represent a primitive type that was prevalent in Paleozoic time, underwent evolutionary changes like those discussed above for the North American Cordillera and Appalachians (Figs. 3, 4), and led to extensive generation of all the massive sulphide, as well as Mississippi Valley types, but only rarely to formation of porphyry copper deposits.

Perhaps margins of the West Pacific type, with more restricted back-arc basins,

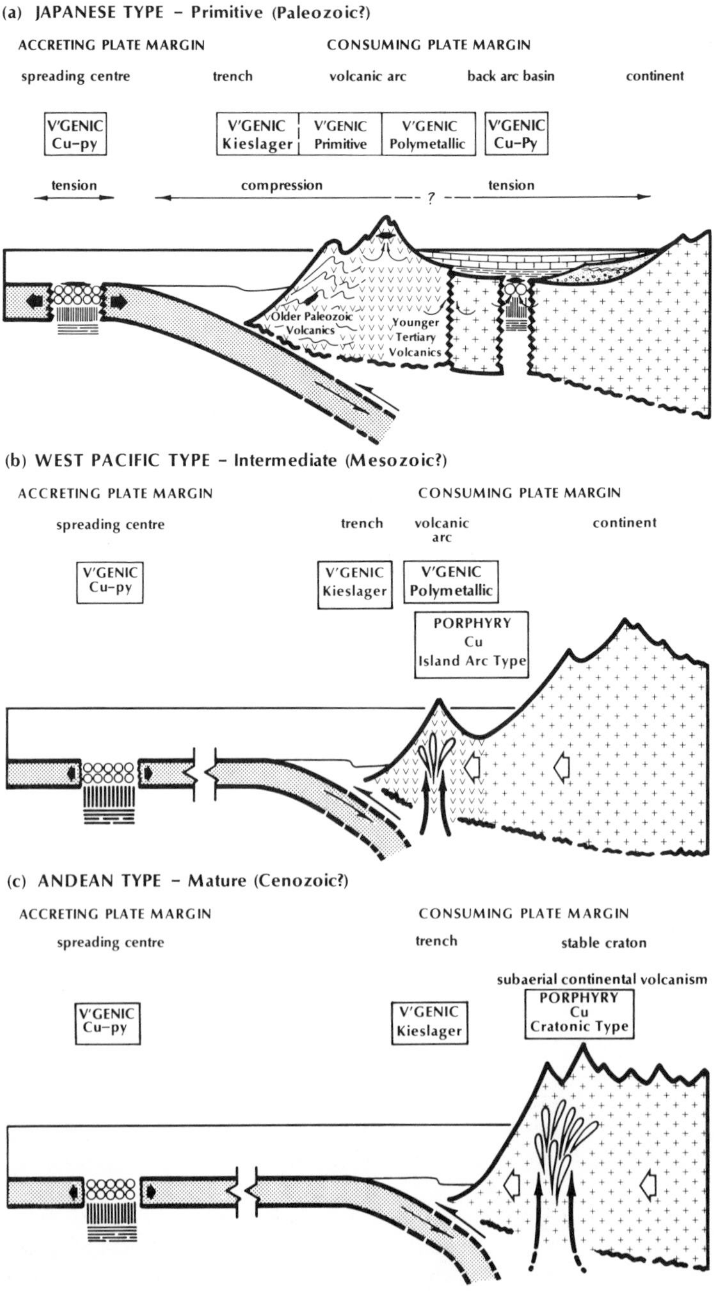

(a) JAPANESE TYPE – Primitive (Paleozoic?)

ACCRETING PLATE MARGIN   CONSUMING PLATE MARGIN

spreading centre   trench   volcanic arc   back arc basin   continent

| V'GENIC Cu-py | | V'GENIC Kieslager | V'GENIC Primitive | V'GENIC Polymetallic | V'GENIC Cu-Py |

tension   compression   – ? – –   tension

Older Paleozoic Volcanics   Younger Tertiary Volcanics

(b) WEST PACIFIC TYPE – Intermediate (Mesozoic?)

ACCRETING PLATE MARGIN   CONSUMING PLATE MARGIN

spreading centre   trench   volcanic arc   continent

| V'GENIC Cu-py | | V'GENIC Kieslager | V'GENIC Polymetallic |

PORPHYRY Cu Island Arc Type

(c) ANDEAN TYPE – Mature (Cenozoic?)

ACCRETING PLATE MARGIN   CONSUMING PLATE MARGIN

spreading centre   trench   stable craton

subaerial continental volcanism

| V'GENIC Cu-py | | V'GENIC Kieslager | PORPHYRY Cu Cratonic Type |

Figure 5. Possible Evolutionary Stages in Consuming Plate Boundaries.

and marine volcanic arcs closer to the overriding cratonic block (Fig. 5b) represent a somewhat later, more mature evolutionary variety of consuming plate boundary that was more widespread in Mesozoic time. This tectonic environment led to the formation of some polymetallic volcanogenic massive sulphides, but more significantly to the evolution of a primitive or "island arc type" of porphyry copper deposit like those of the west Pacific (Gustafson and Titley, 1978), or of the Highland Valley in southern British Columbia (Sutherland Brown, 1976a). It is not within the scope of this paper to describe fully the geologic characteristics of this type of porphyry copper deposit, but for details readers are referred to the "plutonic type" of Sutherland Brown (1976b).

Finally, plate margins of the Andean type (Fig. 5c), in which no back-arc basin is present and in which volcanism is no longer marine but subaerial and continental, may represent the latest and most mature evolutionary variety of consuming plate margin, which became common in Tertiary time. This tectonic environment favoured the generation of a more mature, or "cratonic type" of porphyry copper deposit like those of the Andean belt in South America (Gustafson and Hunt, 1975; Camus, 1975) or of the southwestern United States (Titley and Hicks, 1966). These have been further subdivided into "phallic" and "volcanic" types, which are different from the plutonic type (Sutherland Brown, 1976b). It is interesting and pertinent that evolutionary changes in consuming plate margins, specifically controlled by steep-dipping vs. flat-dipping subduction zones in the case of the Andean and Japanese margins respectively, have been suggested by Uyeda and Nishiwaki (1980) to explain the absence of porphyry copper and the abundance of massive sulphide deposits in Japan.

POSSIBLE PRECAMBRIAN TECTONIC ENVIRONMENTS

While the depositional environments of the various massive sulphide types can be related to Phanerozoic plate tectonic configurations with some certainty, the significance of the important Precambrian massive sulphide types to more ancient tectonic processes is much more obscure. Certain aspects however are important, albeit poorly understood. Although deposits of the ensimatic, cupreous pyrite type are normally destroyed by subduction of oceanic plates and therefore might not be expected in Precambrian rocks, obduction commonly preserved these ores and their distinctive host rocks in Phanerozoic orogenic belts. Their absence in all but the very latest Precambrian rocks suggest that Phanerozoic plate tectonic mechanisms, particularly compressive lateral plate motions, were not important in earlier Precambrian time. Plate tectonic processes are thus considered, like ores and rocks of the ophiolitic, oceanic crustal environment, to be a relatively late tectonic evolutionary development on Earth.

Nevertheless, the important development of the primitive type of massive base metal sulphide deposit in thick, well-differentiated Canadian Archean greenstone belts suggests that Archean tectonic processes were similar in effect, if not in mechanics, to the early island arc tectonic environments of Phanerozoic orogenic belts. Moreover, the Archean greenstone belts of Western Australia and southern Africa contain ultramafic to tholeiitic rocks that, although extrusive and not ophiolite *sensu stricto*, are petrochemically comparable to those of Phanerozoic oceanic crust. Similar rocks are now known to underlie the differentiated volcanic sequences in the Canadian greenstone belts (Jolly, 1977; Arndt *et al.*, 1977).

Considering these relationships from a uniformitarian standpoint, and reasoning by analogy to plate tectonic processes, it is suggested that the earliest Archean tectonism in the greenstone belts resulted in ultramafic-mafic, mantle-derived mag-

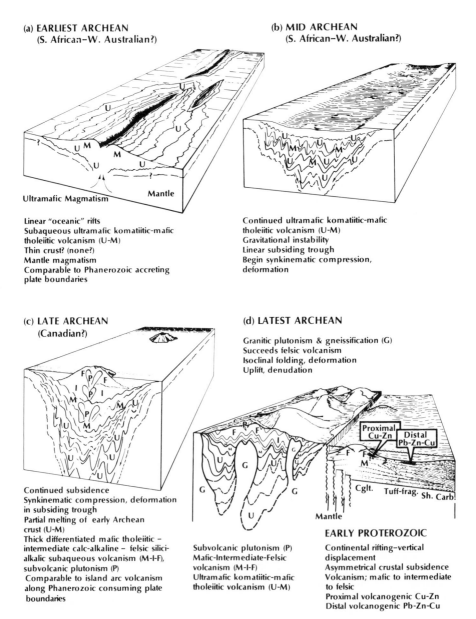

(a) EARLIEST ARCHEAN
(S. African–W. Australian?)

Ultramafic Magmatism   Mantle

Linear "oceanic" rifts
Subaqueous ultramafic komatiitic-mafic
tholeiitic volcanism (U-M)
Thin crust? (none?)
Mantle magmatism
Comparable to Phanerozoic accreting
plate boundaries

(b) MID ARCHEAN
(S. African–W. Australian?)

Continued ultramafic komatiitic-mafic
tholeiitic volcanism (U-M)
Gravitational instability
Linear subsiding trough
Begin synkinematic compression,
deformation

(c) LATE ARCHEAN
(Canadian?)

Continued subsidence
Synkinematic compression, deformation
in subsiding trough
Partial melting of early Archean
crust (U-M)
Thick differentiated mafic tholeiitic –
intermediate calc-alkaline – felsic silici-
alkalic subaqueous volcanism (M-I-F),
subvolcanic plutonism (P)
Comparable to island arc volcanism
along Phanerozoic consuming plate
boundaries

(d) LATEST ARCHEAN

Granitic plutonism & gneissification (G)
Succeeds felsic volcanism
Isoclinal folding, deformation
Uplift, denudation

Proximal Cu-Zn   Distal Pb-Zn-Cu

Cglt. Tuff-frag. Sh. Carb.

Mantle

Subvolcanic plutonism (P)
Mafic-Intermediate-Felsic
volcanism (M-I-F)
Ultramafic komatiitic-mafic
tholeiitic volcanism (U-M)

EARLY PROTEROZOIC

Continental rifting–vertical
displacement
Asymmetrical crustal subsidence
Volcanism; mafic to intermediate
to felsic
Proximal volcanogenic Cu-Zn
Distal volcanogenic Pb-Zn-Cu

Figure 6. Possible Precambrian Tectonic Evolution in Greenstone Belts.

matism. This occurred (Fig. 6a) from fissures or rifts, either in a primitive thin crust, or perhaps, as in the case of modern crustal generation, directly from and on the mantle. Continued activity of this type led (Fig. 6b) to gravitational instability, and to major subsidence or "vertical subduction". As in Phanerozoic subduction zones, reworking of this older, ultramafic-mafic "oceanic" crust by partial melting due to subsidence (Fig. 6c), produced a "second crustal differentiation" with island arc-like volcanism and thick, differentiated volcanic sequences. These sequences contained the earliest primitive type of massive sulphide deposit so typical of the Canadian Archean. Judging from the large size, number, and close spacing of these districts and their deposits, and from the thicknesses and areal extent of their host greenstone belt volcanic sequences, this style of volcanism, and attendant hydrothermal brine discharge, must have occurred on a vast scale not since duplicated on Earth. However, geological and age differences between Canadian Archean sequences on the one hand, and southern African ones on the other (Windley, 1977) suggest that the evolution from early ultramafic to late differentiated volcanism did not proceed everywhere to completion, nor at the same rate.

This interpretation of a thin or even non-existent crust beneath Archean greenstone belt supracrustal sequences is essentially that of the "simatists" as opposed to that of the "sialists" (Baragar and McGlynn, 1978) and it contradicts the evidence for an early, thick sialic crust that arises from the study of ancient Archean gneiss terrains (Hunter, 1970; Moorbath, 1980). This contradiction has been widely discussed by proponents of both views (Viljoen and Viljoen, 1969; Anhaeusser, 1973; Hunter, 1974; Baragar and McGlynn, 1976) and its resolution is fundamental to a proper understanding of Archean geology (Glikson, 1978). Although no solution is proposed here, it is suggested that both views are valid, are soundly based on geological evidence, but arise from studies of very different Archean terrains. If so, it is essential to reconcile the two by integrating them into a broader hypothesis of Archean tectonism.

The culmination of Archean tectonism (Fig. 6d, left) was the granitic plutonism and metasomatism of the Kenoran orogeny. Like Acadian or Nevadan orogeny in the Appalachians and Cordillera, this orogeny uplifted, deformed, and metamorphosed all the earlier rocks. Thus, perhaps for the first time, a thickened sialic continental crust was formed, opening the way to a new evolutionary tectonic style: continental crustal rifting became the dominant tectonic process in Proterozoic time. Its importance is suggested by fossil rift systems (Kumarapeli and Saull, 1966; Kanasewich, 1968), some with associated alkaline complexes, by major yoked basins of early Proterozoic sedimentation like those of the Witwatersrand (Brock and Pretorius, 1964a, 1964b) and by the widespread tholeiitic intrusions of Proterozoic age that formed major sills, dykes and laccoliths, for example the Nipissing Diabase, the Sudbury Irruptive, the Bushveld Complex and many others (Card and Hutchinson, 1972).

It is suggested that initially in Aphebian time (Fig. 6d, right), displacement on the crustal rifts was mainly vertical, perhaps still reflecting the earlier, predominantly vertical Archean tectonism. This led to the reappearance of the older volcanogenic deposits of the immature type and, because of the thickened nature of the continental crust, to the evolutionary appearance of the later deposits of the polymetallic type. The common presence of sulphates in these deposits compared with their rarity in the earlier primitive type suggests that oxygenation of the Earth's hydrosphere in early

Proterozoic time (Garrels *et al.*, 1973) also was a factor in the evolutionary development of the Pb-Ag-rich polymetallic type. In somewhat later, Helikian time lateral displacement and separation became dominant along these continental rift systems, resulting in aulacogens and deep, sediment-filled troughs. The large and important deposits of the clastic-hosted exhalative sedimentary type were formed in this tectonic setting (Dunnet, 1976). Tholeiitic magmatism associated with the rifting formed the diabasic-amphibolitic rocks that are commonly associated with these ores, (Freeze, 1966; Johnson and Klingner, 1975). It is possible that this stage of tectonic evolution marked the first appearance of the Earth's present deep ocean basins, and may have been initiated by a period of widespread global expansion (Glikson, 1979). This might explain the rarity in rocks of this age of compression-generated calc-alkaline andesitic volcanic rocks (Miyashiro, 1975), of obduction-emplaced ophiolite complexes (Glikson, 1979), of exhalative volcanogenic massive sulphide deposits and of eugeosynclinal lode-gold deposits (Hutchinson, 1975). It might also account for the abundance of major mafic intrusions in rocks of the same age (Card and Hutchinson, 1972).

Finally, the latest Precambrian, Hadrynian time may have marked the onset of plate tectonic processes with major lateral movement and interaction of oceanic and continental lithospheric plates. This activity has continued to the present, perhaps with evolutionary changes of its own as discussed here. It is interesting that despite the wide variety of these plate configurations, all of the earlier mineral deposit types reappear in the same time sequence, though on a shorter scale of both time and space, along linear plate boundaries. Perhaps the evolutionary cliché from embryology, "ontogeny recapitulates phylogeny," is in a manner applicable to the history of tectonic processes and metallogenesis.

SUMMARY AND CONCLUSIONS

Massive base metal sulphide deposits can be assigned to two main groups, exhalative volcanogenic and exhalative sedimentary, and can be further subdivided into six distinct types based on their differing geological characteristics. These differences are the result of tectonic evolution through time. Consequently, the various types, along with certain other related ore deposits, constitute another useful guide to paleotectonic evolution and processes. The environments of formation of the six types can be related to various configurations and evolutionary developments along both accreting and consuming plate boundaries in Phanerozoic orogenic belts. The spatial disposition of base metal deposit types across consuming boundaries provides an additional indication of the dip of paleo-subduction zones. The space-time distribution of certain types suggests an evolutionary change in the nature of consuming boundaries, from primitive Japanese type, through west Pacific type to mature Andean type, during Phanerozoic time.

The geologic settings and the space-time distributions of the deposits also suggest that plate tectonic mechanisms were not important in Precambrian time and that Archean greenstone belt tectonism was dominated by subsidence or vertical subduction. Tectonic evolution and metallogeny in the earliest (South African) greenstone belts were otherwise analogous to those of modern accreting plate margins, while the tectonism and resulting metallogeny in later (Canadian) greenstone belts were similar to those of early-stage island arc volcanism at consuming plate margins. Because of the prevalent vertical tectonics, however, the later volcanic environment was superposed directly above the earliest greenstone belts, rather than being laterally separated as in "modern" plate configurations.

Proterozoic tectonism, in turn, was dominated by continental crustal rifting. Initial displacement on these rifts in Aphebian time was mainly vertical, but later Helikean displacements were mainly lateral. The tensional tectonic regime associated with rifting was accompanied by mafic tholeiitic magmatism, but the differentiated calc-alkaline magmatism so characteristic of compressional tectonic regimes was minor or lacking. This may explain the rarity of volcanogenic sulphide deposits associated with calc-alkaline rocks in mid-Proterozoic time. Lateral separation on the continental rifts may have formed the deep ocean basins and eventually initiated late Precambrian-Phanerozoic plate tectonic processes. Various plate configurations, with their own evolutionary developments along plate boundaries, led to repetition in the same sequence, but in a smaller span of both time and space, of all the earlier environments and resulting mineral deposit types.

ACKNOWLEDGEMENTS

Funds to support this study have been supplied by the National Research Council of Canada and this financial aid is gratefully acknowledged. My friends and colleagues G. G. Suffel, R. W. Hodder, R. Kerrich and W. R. Church have provided many helpful suggestions in discussions of the aspects treated above, and have also assisted with bibliographic referencing. Many of the basic field geological relationships on which the arguments are partly based, particularly in Precambrian rocks, were observed during travels while on sabbatical leave in Australia and southern Africa. Assistance in the field was provided by many government, industrial and educational organizations and their geologists too numerous to list here, but to all of whom the writer is deeply grateful. Finally, my graduate students deserve credit for stimulating exchanges of ideas and for their assistance during manuscript preparation.

REFERENCES

Albers, J.P., and Robertson, J.F., 1961, Geology and ore deposits of the East Shasta copper-zinc district, Shasta County, California: United States Geol. Survey Prof. Paper 338, 107 p.

Anderson, C.A., and Nash, J.T., 1972, Geology of the massive sulphide deposits at Jerome, Arizona – a reinterpretation: Econ. Geol., v. 67, no. 7, p. 848-863.

Anger, G., Nielsen, H., Puchelt, H., and Ricke, W., 1966, Sulfur isotopes in the Rammelsberg ore deposit (Germany): Econ. Geol., v. 61, no. 3, p. 511-536.

Anhaeusser, C.R., 1973, The evolution of the early Precambrian crust of southern Africa: Phil. Trans. Royal Soc. London, Ser. A., 273, p. 359-388.

Arndt, N.T., Naldrett, A.J., and Pyke, D.R., 1977, Komatiitic and iron-rich tholeiitic lavas of Munro Township, northeast Ontario: Jour. Petrol., v. 18, part 2, p. 319-369.

Aymé, J.-M., 1965, The Senegal salt basin: in Salt basins around Africa: Instit. Petrol., London, p. 83-90.

Baragar, W.R.A., and Goodwin, A.M., 1969, Andesites and Archean volcanism of the Canadian Shield: in Andesite Congress (Eugene and Bend, Oregon, 1968): Proceedings. Oregon Dept. Geology Mineral Industries Bull. 65, p. 121-142.

Baragar, W.R.A., and McGlynn, J.C., 1976, Early Archean basement in the Canadian Shield: A review of the evidence: Geol. Survey Canada Paper 76-14, 20 p.

——————————, 1978, On the basement of Canadian greenstone belts: discussion: Geosci. Canada, v. 5, p. 13-15.

Bear, L.M., 1963, The mineral resources and mining industry of Cyprus: Cyprus Geol. Survey Dept., Bull. 1, 208 p.

Bell, K., and Blenkinsop, J., in press, A geochronologic study of the Buchans area, Newfoundland: in Swanson, E.A., ed., Buchans Volume, Geol. Assoc. Canada Spec. Paper.

Belmonte, Y., Hirtz, P., and Wenger, R., 1965, The salt basins of the Gabon and Congo (Brazzaville) – a tentative paleogeographic interpretation: in Salt basins around Africa: Instit. Petrol., London, p. 55-74.

Boyle, R.W., and Jambor, J.L., 1966, Mineralogy, geochemistry and origin of the Magnet Cove barite-sulphide deposit, Walton, Nova Scotia: Canadian Instit. Mining and Metallurgy Bull., v. 59, no. 654, p. 1209-1227.

Brock, B.B., and Pretorius, D.A., 1964a, An introduction to the stratigraphy and structure of the Rand gold field: in Haughton, S.H., ed., The geology of some ore deposits in southern Africa: Johannesburg, Geol. Soc. South Africa, v. 1, p. 25-62.

——————————, 1964b, Rand basin sedimentation and tectonics: in Haughton, S.H., ed., The geology of some ore deposits in southern Africa: Johannesburg, Geol. Soc. South Africa, v. 1, 549-600.

Brown, W.L., and Bray, R.C.E., 1960, The geology of the Geco Mine: Canadian Instit. Mining and Metallurgy Bull., v. 53, no. 573, p. 3-11.

Burton, C.C.J., 1975, Rosebery zinc-lead-copper orebody in Knight, C.L., editor, Economic Geology of Australia and Papua New Guinea, 1. Metals: Australasian Instit. Mining and Metallurgy Monograph Series 5, p. 619-625.

Byers, A.R., Kirkland, S.J.T., and Pearson, W.J., 1965, Geology and mineral deposits of the Flin Flon area, Saskatchewan: Saskatchewan Dept. Mineral Resources, Report 62, 95 p.

Camus, F., 1975, Geology of the El Teniente orebody with emphasis on wall-rock alteration: Econ. Geol. v. 70, no. 8, p. 1341-1372.

Card, K.D., and Hutchinson, R.W., 1972, The Sudbury structure: its regional geological setting: in Guy-Bray, J.V., ed., New Developments in Sudbury Geology: Geol. Assoc. Canada Spec. Paper 10, p. 67-78.

Carne, R.C., 1976, Stratabound barite and lead-zinc-barite deposits in eastern Selwyn Basin, Yukon Territory: Dept. Indian and Northern Affairs, Open File Report EGS, 1976-16, 41 p.

Carney, J.N., and Welland, M.J.P., 1974, Geology and mineral resources of the Oman mountains: Instit. Geol. Sciences, Overseas Division, Report No. 27, 49 p.

Church, W.R., and Stevens, R.K., 1971, Early Paleozoic ophiolite complexes of the Newfoundland Appalachians as mantle-ocean crust sequences: Jour. Geophys. Research, v. 76, p. 1460-1466.

Davis, L.W., 1975, Captains Flat lead-zinc orebody: in Knight, C.L., ed., Economic Geology of Australia and Papua New Guinea, 1. Metals: Australasian Instit. Mining and Metallurgy, Monograph Series 5, p. 694-700.

Degens, E.T., and Ross, D.A., eds., 1969, Hot brines and recent heavy metal deposits in the Red Sea: Springer Verlag, New York, 600 p.

Degens, E.T., and Ross, D.A., 1976, Strata-bound metalliferous deposits found in or near active rifts: in Wolf, K.H., ed., Handbook of strata-bound and stratiform ore deposits, v. 4, p. 165-202.

Derry, D.R., Clark, C.G., and Gillat, N., 1965, The Northgate base metal deposit at Tynagh, County Galway, Ireland: Econ. Geol., v. 60, no. 6, p. 1218-1237.

Duke, N.A., and Hutchinson, R.W., 1974, Geological relationships between massive sulphide bodies and ophiolitic volcanic rocks near York Harbour, Newfoundland: Canadian Jour. Earth Sciences, v. 11, p. 53-69.

Dunnet, D., 1976, Some aspects of the Panantarctic cratonic margin in Australia: Phil. Trans. Royal Soc. London, A., v. 280, p. 641-654.

Farley, W.J., 1948, Sherritt Gordon Mine: in Structural Geology of Canadian Ore Deposits; Jubilee Volume: Canadian Instit. Mining and Metallurgy, p. 292-294.

Ford, T.D., and King, R.J., 1965, Layered epigenetic galena-barite deposits in the Golconda Mine, Brassington, Derbyshire,England: Econ. Geol., v. 60, no. 8, p. 1686-1701.

Francheteau, J., et al., 1979, Massive deep sea sulphide ore deposits discovered on the East Pacific Rise: Nature, v. 277, no. 5679, p. 523-528.

Freeze, A.C., 1966, On the origin of the Sullivan orebody, Kimberley, B.C., Tectonic history and mineral deposits of the western Cordillera: Canadian Instit. Mining and Metallurgy, Special Volume No. 8, p. 263-294.

Gair, J.E., and Slack, J.F., 1978, Plate tectonic setting of stratabound massive sulphide deposits of the United States Appalachians, Abst.: in Abstracts with Programs, Geol. Assoc. Canada – Mineral. Assoc. Canada – Geol. Soc. America, p. 405.

Garrels, R.M., Perry, E.A., Jr., and Mackenzie, F.T., 1973, Genesis of Precambrian iron formations and the development of atmospheric oxygen: Econ. Geol., v. 68, no. 7, p. 1173-1179.

Gealey, W.K., 1977, Ophiolite obduction and geologic evolution of the Oman mountains and adjacent areas: Geol. Soc. America, Bull., v. 88, no. 8, p. 1183-1191.

Gee, D.G., 1978, The Swedish Caledonides – a short synthesis: in Caledonian-Appalachian orogen of the north Atlantic region: Geol. Survey Canada Paper 78-13, p. 63-72.

Gilmour, Paul and Still, A.R., 1968, The geology of the Iron King Mine: in Ridge, J.D., ed., Ore Deposits in the United States 1933-1967 (Graton-Sales Vol.) New York: American Instit. Mining and Metallurgy and Petrol. Eng., v. 2, p. 1238-1257.

Glikson, A.Y., 1978, On the basement of Canadian greenstone belts: Geosci. Canada, v. 5, p. 3-12.

_____, 1979, The missing Precambrian crust: Geology, v. 7, p. 449-454.

Goldberg, I., 1976, A preliminary account of the Otjihase deposit, South West Africa: Econ. Geol., v. 71, no. 1, p. 384-390.

Goodwin, A.M., 1968, Evolution of the Canadian Shield: Geol. Assoc. Canada Proceedings, v. 19, p. 1-14.

Griffits, W.R., Albers, J.P., and Oner, Omer, 1972, Massive sulphide copper deposits of the Ergani-Maden area, southeastern Turkey: Econ. Geol., v. 67, no. 6, p. 701-716.

Gustafson, L.B., and Hunt, J.P., 1975, The porphyry copper deposit at El Salvador, Chile: Econ. Geol., v. 70, no. 5, p. 857-912.

Gustafson, L.B., and Titley, S.R., 1978, Porphyry copper deposits of the southwestern Pacific islands and Australia – preface: Econ. Geol., v. 73, no. 5, p. 597-599, and see following papers.

Gwosdz, W., 1974, Die liegunde Schichten der devonischen Pyrit-und Schwerspat-lager von Eisen (Saarland), Meggen und des Rammelsberges: Geologisches Rundschau, v. 63, no. 1, p. 74-92. (English language translation available from Geol. Survey Canada Library, no. 898.)

Halls, C., Reinsbakken, A., Ferriday, I., Haugen, A., and Rankin, A., 1977, Geological setting of the Skorovas orebody within the allochthonous volcanic stratigraphy of the Gjersvik Nappe, central Norway: in Volcanic processes in ore genesis, Geol. Soc. London – Institut. Mining and Metallurgy, Spec. Publ. No. 7, p. 128-151.

Harland, W.B., and Gayer, R.A., 1972, The Arctic Caledonides and earlier oceans: Geological Magazine, v. 109, p. 289-314.

Hatcher, R.D., Jr., 1978, Synthesis of the southern and central Appalachians, U.S.A.: *in* Caledonian-Appalachian orogen of the north Atlantic region: Geol. Survey Canada Paper 78-13, p. 149-158.

Haworth, R.T., 1978, Geophysical evidence for an east-dipping Appalachian subduction zone beneath Newfoundland: Geology, v. 6, p. 522-526.

Heenan, P.R. and Truman, M.P., 1972, The discovery and development of the Ming zone, Consolidated Rambler Mines Ltd., Baie Verte, Newfoundland: Canadian Instit. Mining and Metallurgy Bull., v. 66, no. 729, p. 78-88.

Hoy, T., 1979, Geology of the Goldstream area: British Columbia Ministry Energy Mines and Petrol. Resources, Bull. 71, 49 p.

Hunter, D.F., 1970, The ancient gneiss complex in Swaziland: Trans. Geol. Soc. South Africa, v. 73, p. 107-150.

Hunter, D.F., 1974, Crustal development in the Kaapvaal Craton, 1. The Archean: Precambrian Research, v. 1, p. 259-294.

Hutchinson, R.W., 1973, Volcanogenic sulphide deposits and their metallogenic significance: Econ. Geol. v. 68, no. 8, p. 1223-1246.

——————————, 1975, Lode gold deposits: the case for volcanogenic derivation: Fifth gold and money session, Pacific Northwest Metals and Minerals Conference, Proceedings Volume, Portland, Oregon, p. 64-105.

Hutchinson, R.W., and Hodder, R.W., 1972, Possible tectonic and metallogenic relationships between porphyry copper and massive sulphide deposits: Canadian Instit. Mining and Metallurgy Bull., v. 65, no. 718, p. 34-40.

Ishihara, Shunso, 1974, editor, Geology of kuroko deposits: Mining Geology Special Issue No. 6, Society Mining Geologists of Japan, 435 p.

Johnson, I.R., and Klingner, G.D., 1975, The Broken Hill ore deposit and its environment: *in* Knight, C.L., ed. Economic Geology of Australia and Papua-New Guinea, 1. Metals: Australasian Instit. Mining and Metallurgy Monograph Series 5, p. 476-490.

Jolly, W.T., 1977, Relations between Archean lavas and intrusive bodies of the Abitibi green-stone belt, Ontario-Quebec: *in* Baragar, W.R.A., Coleman, L.C. and Hall, J.M., ed., Volcanic Regimes in Canada: Geological Association of Canada Special Paper 16, p. 311-330.

Kanasewich, E.R., 1968, Precambrian rift: genesis of stratabound ore deposits: Science, v. 161, p. 1002-1005.

Kanehira, K., and Tatsumi, T., 1970, Bedded cupriferous iron sulphide deposits in Japan: a review *in* Tatsumi, T., ed., Volcanism and Ore Genesis: University of Tokyo Press, p. 51-76.

Kean, B.F., Dean, P.L. and Strong, D.F., in press, Regional geology of the central volcanic belt of Newfoundland: *in* Swanson, E.A., ed., Buchans Volume, Geol. Assoc. Canada Spec. Paper.

Kinkel, A.R., Hall, W.E. and Albers, J.P., 1956, Geology and base metal deposits of the West Shasta copper-zinc district, Shasta County, California: United States Geol. Survey Prof. Paper 285, 156 p.

Koo, J., and Mossman, D.J., 1975, Origin and metamorphism of the Flin Flon stratabound Cu-Zn sulphide deposit, Saskatchewan and Manitoba: Econ. Geol., v. 70, no. 1, p. 48-62.

Kumarepeli, P.S., and Saull, V.A., 1966, The St. Lawrence valley system; a North American equivalent of the East African rift valley system: Canadian Jour. Earth Sciences, v. 3, no. 5, p. 639-658.

Kuno, H., 1966, Lateral variation of basalt magma types across continental margins and island arcs: Bull. Volcanology, v. 29, p. 195-222.

Lea, E.R. and Dill, D.B., Jr., 1968, Zinc deposits of Balmat-Edwards district, New York: *in* Ridge, J.D., ed. Ore deposits in the United States 1933-1967 (Graton-Sales Vol.) New York: American Instit. Mining, Metallurgy and Petrol. Eng., v. 1, p. 20-48.

Lea, E.R. and Rancourt, C., 1958, Geology of the Brunswick Mining and Smelting orebodies, Gloucester County, N.B.: Canadian Instit. Mining and Metallurgy Bull., v. 51, no. 551, p. 167-177.

Magee, Maurice, 1968, Geology and ore deposits of the Ducktown district, Tennessee: in Ridge, J.D., ed. Ore deposits in the United States, 1933-1967 (Graton-Sales Vol.) New York: American Instit. Mining, Metallurgy and Petrol. Eng., v. 1, p. 207-241.

Malone, E.J., Olgers, F., Cucchi, F.G., Nicholas, T., and McKay, W.J., 1975, Woodlawn copper-lead-zinc orebody: in Knight, C.L., ed., Economic Geology of Australia and Papua New Guinea, 1. Metals: Australasian Instit. of Mining and Metallurgy Monograph Series 5, p. 701-710.

Martin, W.C., 1957, Errington and Vermilion Lake Mines: in Structural Geology of Canadian Ore Deposits, Congress Volume, Canadian Instit. Mining and Metallurgy, p. 363-376.

Mathias, B.V. and Clark, C.J., 1975, Mount Isa copper and silver-lead-zinc orebodies – Isa and Hilton Mines: in Knight, C.L., ed., Economic Geology of Australia and Papua-New Guinea, 1. Metals: Australasian Institute of Mining and Metallurgy Monograph Series 5, p. 351-372.

Mining Engineering, 1978, A treasure of metals beneath the forest: Mining Engineering, v. 31, p. 481-482.

Miyashiro, A., 1975, Volcanic rock series and tectonic setting: Annual Review of Earth and Planetary Sciences, v. 3, p. 251-270.

Moorbath, S., 1980, The chronology of old rocks: in Strangway, D.W., ed., The Continental Crust and Its Mineral Deposits: Geol. Assoc. Canada Spec. Paper 20, p.

Morris, H.C., et al., 1972, An outline of the geology of the Sullivan Mine, Kimberley, British Columbia in: Major lead-zinc deposits of western Canada, 24th International Geological Congress Guidebook, Field Excursion A-24-C-24, p. 26-34.

Oftedahl, Christoffer, 1958, A theory of exhalative-sedimentary ores: Geologiska Foreningens Stockholm Forhandlingar, v. 80, pt. 1, no. 492, p. 1-19.

Osberg, P.H., 1978, Synthesis of the geology of the northeastern Appalachians, U.S.A.: in Caledonian-Appalachian orogen of the north Atlantic region: Geol. Survey Canada Paper 78-13, p. 137-148.

Pereira, J., 1967, Stratabound Pb-Zn deposits in Ireland and Iran: in Brown, J.S., ed., Genesis of stratiform lead-zinc-barite-fluorite deposits; Econ. Geol. Monograph 3, p. 192-200.

Pereira, J., and Dixon, C.J., 1965, Evolutionary trends in ore deposition: Instit. Mining and Metallurgy, v. 74, p. 505-527.

Platt, J.W., 1977, Volcanogenic mineralization at Avoca, County Wicklow, Ireland, and its regional implications: in Volcanic processes in ore genesis, Special Publication No. 7, Geol. Soc. London – Instit. Mining and Metallurgy, p. 163-170.

Sangster, D.F., 1970, Metallogenesis of some Canadian lead-zinc deposits in carbonate rocks: Proceedings Geol. Assoc. Canada, v. 22, p. 27-36.

Sangster, D.F., 1972a, Precambrian volcanogenic massive sulphide deposits in Canada: a review: Geol. Survey Canada Paper 72-22, 44 p.

Sangster, D.F., 1972b, Isotopic studies of ore-leads in the Hanson Lake – Flin Flon – Snow Lake mineral belt: Canadian Jour. Earth Sci., v. 9, no. 5, p. 500-513.

Sangster, D.F. and Scott, S.D., 1976, Precambrian stratabound massive Cu-Zn-Pb sulphide ores in North America: in Wolf, K.H., ed., Handbook of stratabound and stratiform ore deposits, v. 6, p. 129-211: Elsevier Scientific Publishing Company, Amsterdam.

Sauve, P., Cloutier, J.P., and Genois, G., 1972, Base metal deposits of southeastern Quebec: International Geological Congress Guidebook, Excursion B-07, 24 p.

Schenk, P.E., 1969, Carbonate-sulfate-redbed facies and cyclic sedimentation of the Windsorian stage (middle Carboniferous) Maritime Provinces: Canadian Jour. Earth Sci., v. 6, p. 1037-1066.

Schultz, R.W., 1961, Lower Carboniferous cherty ironstones at Tynagh, Ireland: Econ. Geol., v. 61, no. 2, p. 311-342.

Smith, C.H. and Skinner, R., 1958, Geology of the Bathurst-Newcastle mineral district, New Brunswick: Canadian Instit. Mining and Metallurgy Bull., v. 51, no. 551, p. 150-155.

Spence, C.D. and de Rosen Spence, A.F., 1975, The place of sulphide mineralization in the volcanic sequence at Noranda, Quebec: Econ. Geol., v. 70, no. 1, p. 90-101.

Stanton, R.L., 1959, Mineralogical features and possible mode of emplacement of the Brunswick Mining and Smelting orebodies, Gloucester County, New Brunswick: Canadian Instit. Mining and Metallurgy Bull., v. 52, nom 570, p. 631-643.

Stinson, M.C., 1957, Geology of the Island Mountain copper mine, Trinity County, California: *California: Jour. Mines and Geology*, v. 53, no. 1, p. 9-33.

Sturt, B.A. and Roberts, D., 1978, Caledonides of northern-most Norway (Finnmark): *in* Caledonian-Appalachian orogen of the north Atlantic region: Geol. Survey Canada Paper 78-13, p. 17-24.

Sutherland Brown, A., 1976a, editor, Porphyry deposits of the Canadian Cordillera: Canadian Instit. Mining and Metallurgy Spec. Vol. 15.

Sutherland Brown, A., 1976b, Morphology and classification: *in* Porphyry deposits of the Canadian Cordillera: Canadian Instit. Mining and Metallurgy Spec. Vol. 15, p. 44-51.

Thanasuthipitak, T., 1975 Relationship of mineralization to petrology at Parys Mountain, Anglesey (unpublished Ph.D. thesis, University of Aston, Birmingham, 1974): Abst. in Trans. of Institut. Mining and Metallurgy, Sec. B, v. 84, p. B71.

Thurlow, J.G., Swanson, E.A., and Strong, D.F., 1975, Geology and lithogeochemistry of the Buchans polymetallic sulphide deposits, Newfoundland: Econ. Geol., v. 70 no. 1, p. 130-144.

Titley, S.R., and Hicks, C.L., eds., 1966, Geology of the porphyry copper deposits southwestern North America: University of Arizona Press, Tucson, 287 p.

Tuach, J., and Kennedy, M.J., 1978, The geologic setting of the Ming and other sulphide deposits, Consolidated Rambler Mines, northeast Newfoundland: Econ. Geol. v. 73, no. 2, p. 192-206.

Uyeda, S., and Nishiwaki, C., 1980, Stress field, metallogenesis and mode of subduction: *in* Strangway, D.W., ed., The Continental Crust and Its Mineral Deposits: Geol. Assoc. Canada, Spec. Paper 20, p.

Viljoen, M.J., and Viljoen, R.P., 1969, A reappraisal of granite-greenstone terrains of shield areas using the Barberton model: *in* Upper Mantle Project: Geol. Soc. South Africa Spec. Publ. 2, p. 245-274.

Walker, R.R., Matulich, A., Amos, A.C., Watkins, J.J., and Mannard, G.W., 1975, The geology of the Kidd Creek Mine: Econ. Geol., v. 70, no. 1, p. 80-89.

Windley, B.F., 1977, Archean greenstone belts: *in* The evolving continents: John Wiley and Sons Chapter 2, p. 23-44.

The Continental Crust and Its Mineral Deposits, edited by D.W. Strangway,
Geological Association of Canada Special Paper 20

Cu-PYRITE MINERALIZATION AND SEAWATER CONVECTION IN OCEANIC CRUST – THE OPHIOLITIC ORE DEPOSITS OF CYPRUS

E.T.C. Spooner
Department of Geology, University of Toronto, Toronto, Ontario M5S 1A1

ABSTRACT

One of the major discoveries of recent years about the physical and chemical behaviour of the solid Earth and the oceans is the scale of the phenomenon of convective seawater circulation within the upper 3 to 5 km of the oceanic crust at spreading ridges. Analysis of the discrepancy between observed conductive heat flow and that predicted on the basis of a purely conductive cooling model suggests that the total ocean mass may circulate through basaltic oceanic crust at ridges once every 3 to 10 Ma.

An important point concerning hydrothermal circulation to arise from studies of ophiolitic rocks is that the formation of economically significant (on land) cupriferous pyrite ore deposits appears to be a natural side effect of seawater convection. This suggestion has recently received considerable support from the discovery of massive sulphide mounds (Francheteau et al., 1979) and turbulent, buoyant plumes of hot water at 380°C ± 30°C precipitating sulphides at 21°N on the East Pacific Rise (Spiess et al., 1980).

The geometry of circulatory flow in the ophiolitic sequence of Cyprus appears to have consisted of axially symmetric cells containing central plumes of hot ascending fluid which were positionally fixed through time with respect to enclosing rock. Parmentier and Spooner (1978) have modeled such hydrothermal circulation by finite difference approximations. Fluid inclusion studies and theoretical models suggest that the principal factors which caused localized ore deposition were surface and near-surface cooling due to mixing, and conductive heat loss.

The possible effects of subduction of compositionally modified oceanic crust on the generation of magmas and associated mineral deposits at convergent plate boundaries are clear. For example, large quantities of reduced seawater sulphate, in association with hydroxyl and chloride, could be added to oceanic crust by seawater/rock interaction at ridges. Release of water from the descending slab at subduction zones might then cause wet melting of mantle material and could produce volatile rich siliceous magmas enriched in chloride and sulphur. Such a simple model could explain the amounts of magmatically released water, chlorine and sulphur required for the formation of porphyry Cu ± Mo± Au deposits spatially associated with calc-alkaline intrusive rocks.

RÉSUMÉ

Une des principales découvertes au cours des dernières années sur le comportement physique et chimique de la terre solide et des océans a été l'échelle du phénomène de circulation par convection de l'eau de mer dans les 3 à 5 km supérieurs de la croûte océanique le long des crêtes en expansion. L'analyse de l'écart entre le flux thermique convectif observé et celui qu'on prédit sur la base d'un modèle de refroidissement par convection uniquement suggère que la masse totale de l'océan peut circuler dans la croûte basaltique de l'océan le long des crêtes une fois tous les 3 à 10 Ma.

Un point important concernant la circulation hydrothermale qui a ressorti des études sur les roches ophiolitiques est que la formation de dépôts de minerais de pyrite cuprifère d'importance économique (sur terre) semble être un effet secondaire naturel de la convection de l'eau de mer. Cette suggestion a récemment reçu un support considérable par la découverte de monticules de sulfures massifs dans le glacis continental de l'est du Pacifique (Francheteau et al., 1979).

La géométrie de l'écoulement circulatoire dans la séquence ophiolitique de Chypre semble avoir consisté en cellules axialement symétriques contenant en leur centre des panaches de fluide chaud ascendant dont la position s'est fixée au cours du temps par rapport à la roche encaissante. Parmentier et Spooner (1978) ont modélé une telle circulation hydrothermale à l'aide d'approximations par différences finies. Les études sur les inclusions fluides et les modèles théoriques suggèrent que les principaux facteurs qui sont responsables du dépôt localisé de minerai ont été le refroidissement en surface et près de la surface par mélange et la perte de chaleur par conduction.

On voit clairement les effets possibles de la subduction de la croûte océanique dont la composition a été modifiée sur la génération des magmas et des dépôts minéraux associés aux limites de plaques convergentes. Par exemple, de grandes quantités de sulfates d'eau de mer réduits, en association avec les chlorures et les groupes hydroxyles, pourraient s'ajouter à la croûte océanique par l'interaction eau de mer/roche le long des crêtes. La libération de l'eau provenant d'une plaque descendante dans les zones de subduction pourrait alors causer la fusion humide du matériel du manteau et ainsi produire des magmas siliceux riches en volatiles et enrichies en chlorures et en soufre. Un tel modèle simple pourrait expliquer les quantités d'eau, de chlore et de soufre libérées du magma qui sont requises pour la formation des dépôts porphyriques de $Cu \pm Mo \pm Au$ associés dans l'espace avec les roches intrusives calco-alcalines.

INTRODUCTION

One of the major discoveries of recent years about the physical and chemical behaviour of the solid earth and the oceans is the scale of the phenomenon of convective seawater circulation within the upper 3 to 5 km of the oceanic crust at spreading ridges (Lister, 1972; Spooner and Fyfe, 1973; Williams et al., 1974; Wolery and Sleep, 1976; Anderson et al., 1979). Analysis of the discrepancy between observed conductive heat flow and that predicted on the basis of a purely conductive cooling model (the "heat flow anomaly") suggests convective removal of about 50×10^{18} cal/yr (Sleep and Wolery, 1978). This is equivalent to about 16% of the Earth's total heat loss of 32×10^{19} cal/yr, as estimated by Williams and Von Herzen (1974). The estimate implies that the total ocean mass (1.41×10^{24}g) may circulate through basaltic oceanic crust at spreading ridges once every 3 to 10 Ma, for an average hot water discharge temperature between 100°C and 300°C (mass fluxes from Sleep and Wolery, 1978). Fluid inclusion data obtained from identified discharge zones in the Upper-Cretaceous ophiolitic complex of Troodos, Cyprus suggest that 300°C is a reasonable figure (Spooner and Bray, 1977) and that, therefore, an estimate for the mean recirculation time of 10 Ma is likewise reasonable.

In a geological conceptual framework 10 Ma is, of course, a relatively short length of time. It implies that the total present ocean mass may have circulated through oceanic crust at least 400 times in the last 4000 Ma. Assuming that there is some relationship between the amount of convective heat transfer and total heat production, the mean recirculation time may have been about 3 Ma in early Archean time (3500 Ma ago). As noted by Fryer et al. (1979), it appears that a significant amount of intrusion of plutonic granitoids occurred in the submarine environment in the Archean, rather than in sub-aerial continental crust as at present. This effect may have lowered the recirculation time further.

Marked complementary chemical changes occur during basalt/seawater interaction at elevated temperatures (e.g., Mottl and Holland, 1978). Hence, compositionally modified oceanic crust returns to the mantle in subduction zones. The nature of these various phenomena and their implications with respect to the chemical composition and evolution of the oceans, the oceanic crust, the mantle and the continental crust are gradually beginning to be detected and appreciated (e.g., Wolery and Sleep, 1976). For example, Spooner (1976) developed and tested a possible quantitative model to explain the isotopic composition of strontium dissolved in ocean water. It involved a balance between strontium delivered by continental runoff and strontium derived from the oceanic crust by exchange during hydrothermal convection.

A side effect of fluid mass transfer associated with convective heat transfer at oceanic ridges appears to be the formation of sulphide mineral deposits. The suggestion that the oceanic crust might contain sulphide deposits was first made simply on the basis of an empirical comparison between ophiolitic complexes and the oceanic crust (Sillitoe, 1972; Spooner and Fyfe, 1973). The speculative hypothesis that such ore deposits may have been formed during seawater convection (Spooner and Fyfe, 1973) has been tested geochemically in some detail, and has essentially been validated (e.g., Spooner, 1977; Heaton and Sheppard, 1977; Spooner and Bray, 1977; Chapman and Spooner, 1977; Spooner et al., 1977; Parmentier and Spooner, 1978). Similarly, the prediction that massive sulphide mineral deposits should be found actively forming on, and within, oceanic spreading ridges has recently been vindicated, firstly, by the discovery of fine grained sulphide mounds up to 10 m high on the East Pacific Rise about 240 km south of the entrance to the Gulf of California, with no associated active hot water discharge (Francheteau et al., 1979), and, secondly, by the discovery recently announced (May 4, 1979) by the U.S. National Geographic Society of discharge of very hot water (300°C - 400°C) associated with mineral deposition in the same general area.

Spiess et al., (1980) have now described the latter discovery in greater detail. Turbulent, conical plumes of very hot water at 380°C ± 30°C have been observed discharging at velocities of several metres/sec. from "chimneys" 1 - 5 m high and as much as 30 cm in diameter. The buoyant jets, referred to as "smokers", appear black because of entrained particles consisting mainly of aggregates of hexagonal pyrrhotite platelets, typically 20μ across, together with lesser amounts of pyrite, sphalerite and Cu-Fe sulphides. The chimneys occur on sulphide mounds having typical lateral dimensions of 15 x 30 m. Both consist mainly of pyrite-chalcopyrite-sphalerite, with minor amounts of such Fe-Cu-Zn sulphides as pyrrhotite, marcasite, wurtzite, bornite, cubanite and chalcocite. Ca, Ba sulphates (anhydrite, gypsum and barite), talc, amorphous silica and secondary iron hydroxyoxides (goethite, limonite) have also been identified.

SEAWATER INTERACTION DURING DOWN-FLOW

Much information supporting the general hypothesis of seawater circulation within oceanic crust has now been obtained from the examination of ophiolitic complexes. For example, clear evidence for water/rock interaction is found in the non-isochemical, hydrothermal metamorphism of ocean-floor origin which has been reported from ophiolitic assemblages in Cyprus (Gass and Smewing, 1973; Spooner *et al.*, 1977), Newfoundland (Williams and Malpas, 1972; Coish, 1977), E. Liguria, Italy (Spooner and Fyfe, 1973),S. Chile (Stern *et al.*, 1976) and Taiwan (Liou and Ernst, 1979). Metamorphic mineral assemblages of the zeolite to amphibolite facies, which pseudomorph original igneous textures and which also occur as open-space fillings, have been observed in metabasic pillow lavas, metadolerite sheeted dykes and the upper parts of layered plutonic sequences. The use of isotopes (H/D; $^{13}C/^{12}C$; $^{18}O/^{16}O$; $^{87}Sr/^{86}Sr$) as geochemical tracers has confirmed that the interacting fluid was of seawater origin (Spooner *et al.*, 1974; Heaton and Sheppard, 1977; Spooner *et al.*, 1977).

Example 1. Zeolite to amphibolite facies metabasic rocks from the Troodos Massif, Cyprus are variably enriched in ^{87}Sr relative to fresh analogues (Spooner *et al.*, 1977). Initial $^{87}Sr/^{86}Sr$ ratios as high as 0.70760 ± 0.00003, for an interstitial zeolite sample, and 0.7069, for a metabasic rock, have been recorded. These compare with low values obtained for fresh gabbroic rocks of 0.70338 ± 0.00010 to 0.70365 ± 0.00005. Upper Cretaceous seawater, with a ratio of about 0.7076, was the only reasonable contaminant (Spooner *et al.*, 1977).

Example 2. Relative to a $\delta^{18}O$ value of about 6‰ for fresh basaltic rocks, the hydrothermally metamorphosed rocks of Cyprus show strong oxygen isotopic modifications which change from relative ^{18}O enrichments in the upper part of the sequence (e.g., + 12.40‰) to relative ^{18}O depletions in the lower part of the sequence (e.g., + 3.31‰). These changes indicate interaction with large volumes of hot water and are consistent with interaction with heated seawater (Spooner *et al.*, 1974).

Example 3. Heaton and Sheppard (1977) have shown that the isotopic composition of hydrogen in water in equilibrium with chlorite and amphibole samples from metadolerite dykes and metagabbros from Cyprus was indistinguishable from that of seawater.

Calculations based on the degree of oxidation of the basaltic rocks suggest that the bulk integrated water/rock ratio may have been as much as 10^3:1 (Spooner *et al.*, 1977). In Cyprus, the degree of hydration decreases in the upper part of the plutonic sequence. This observation suggests that the depth of seawater penetration and circulation was about 2 to 3 km. Comparison of compressional wave velocities, shear wave velocities and Poisson's ratios measured at 1 kb for a variety of altered and fresh ophiolitic lithologies with model profiles for in situ oceanic crust (Christensen, 1978) suggest that the equivalent penetration depth in "average" oceanic crust may be about 3 to 5 km.

THE VOLCANOGENIC CUPRIFEROUS PYRITE ORE DEPOSITS

Whereas hydrothermal metamorphism occurred during in-flow and through-flow of seawater, it appears that, in Cyprus, sulphide ore deposits formed at the positions of discharge. In general terms these deposits consist of a lens of massive ore underlain by

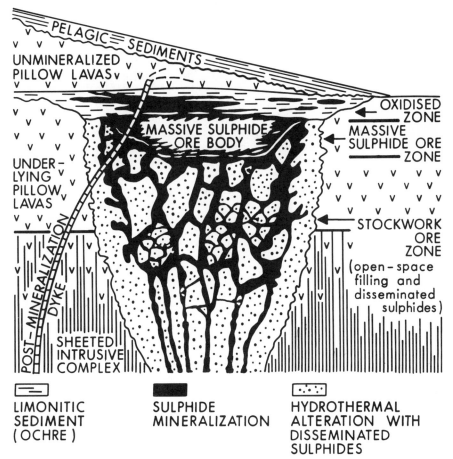

LIMONITIC SEDIMENT (OCHRE)

SULPHIDE MINERALIZATION

HYDROTHERMAL ALTERATION WITH DISSEMINATED SULPHIDES

Figure 1. Schematic diagram of an ophiolitic (Cyprus-type) volcanogenic sulphide ore deposit (modified after Hutchinson and Searle, 1971).

The section shows the surficially oxidized massive pyritic lens, which is intercalated within the metabasic pillow lavas, the underlying pipe-shaped stockwork of hydrothermally metamorphosed and mineralized material which extends down into the dyke complex, and the fact that the sulphide ore bodies are usually overlain by unmineralized basaltic pillow lavas, which are in turn overlain by pelagic marine sediments.

an approximately pipe-shaped stockwork (Fig. 1). The massive ore consists of a porous, crudely colloform-textured mass of fine-grained pyrite and chalcopyrite, with accessory quantities of marcasite, sphalerite and galena. The voids in the massive ore are frequently partially filled with fine-grained, sooty pyrite. The lenses occur interca-lated within the pillow lava sequence and their attitudes are conformable with the local stratigraphy. (Description summarized from Hutchinson and Searle, 1971; Constan-tinou and Govett, 1973.) A large lens, for example that at Skouriotissa, had a maximum thickness of about 50 m, and was approximately elliptical in plan (long axis about 500 m; short axis about 350 m) (Constantinou and Govett, 1973). It contained pre-production ore reserves of about 6 million tonnes at 2.3% Cu (Bear, 1963). An important point is that the evidence suggests that this massive ore actually formed *on* the ocean floor since it does not replace pre-existing rock, can show alteration near the top to a goethitic material (ochre), which has been interpreted as a submarine weather-ing product of ore (Constantinou and Govett, 1972), and is itself overlain either by unmineralized deep-water marine sediments, or by unmineralized pillow lavas which are, in turn, overlain by pelagic sediments (Robertson, 1975).

Stockworks, which occur beneath massive ore and which were clearly feeder zones, consist of highly altered and mineralized material of basaltic origin. Mineraliza-tion occurs as veins, veinlets and disseminated impregnations of sulphides. The alteration silicate mineralogy of the reconstituted pillow lava consists of quartz-chlorite-illite-sphene. Uneconomic alteration pipes characterized by pyritization penetrate 1 to 2 km down into the ophiolitic sequence. Stockworks are normally elliptical in plan. A good idea of dimensions is provided by the Limni deposit in western Cyprus, which has a long axis of about 800 m and a short axis of about 400 m (Trennery and Pocock, 1972; Adamides, 1975). This deposit contained pre-production reserves of around 4 million tonnes at 1.37% Cu (Bear, 1963) and has proved to be economically exploitable for about 200 m below the original ocean floor.

Approximately 20 significant sulphide ore deposits have been mined in Cyprus. Along the 80 km long continuous northern pillow lava outcrop, 15 deposits occur in an imperfectly regular distribution (Fig. 2) which is characterized by a half-spacing of 2.6

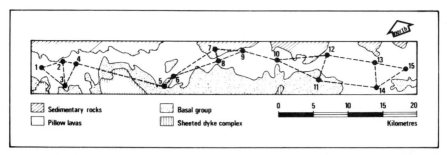

Figure 2. The imperfectly regular distribution of 15 cupriferous pyrite ore deposits along the northern pillow lava outcrop of the Troodos ophiolitic complex, Cyprus. The average of the 21 inter-ore deposit half-spacings is 2.6±1.4 km (1 standard deviation).

1, Mavrovouni; 2, Lefka; 3, Apliki; 4, Skouriotissa; 5, Alestos; 6, Memi; 7, Kokkinoyia; 8, Kokkinopezoula; 9, Agrokipia; 10, Klirou district; 11, Kapedhes; 12, Kambia; 13, Mathiati; 14, Lythrodondha; 15, Sha.

TABLE I
PRE-PRODUCTION RESERVES AND GRADES OF OPHIOLITIC AND OTHER VOLCANOGENIC MASSIVE SULPHIDE ORE DEPOSITS

Mine	Pre-production reserves (million tons)	Cu (%)	S (%)	Zn (%)	Pb (%)	Ag (ppm)	Au (ppm)	Source
Mavrovouni; Cyprus	15	4	47	0.5	–	39	0.3	Bear (1963)
Skouriotissa; Cyprus	6	2.3	48	0.06	—	—	—	Bear (1963)
Kalavasos; Cyprus	2.3	1.5	33	0.7	0.01	5.7	1.6	Bear (1963)
Limni; Cyprus	4	1.4	—	0.29*	—	2.7*	0.51*	Bear (1963) *Trennery and Pocock (1972)
Lasail, Aarja, Bayda; Oman	15	2.1	—	—	—	—	—	Mining Journal, 22 July 1977, p. 63.
York Harbour; Newfoundland (ophiolitic)	0.28	1.92	—	4.67	—	—	—	Duke and Hutchinson (1974)
Arithmetic average Precambrian volcanogenic massive sulphide deposit in the Canadian Shield (n = 110)	6.9	1.9	—	4.6	(90% <1%)	45	0.84	Boldy (1977) *Sangster (1977)
Japanese Kuroko deposits (ave. 1976 production)	2-35†	2.0	—	4.2	1.3	63	0.7	Hashimoto (1977)

— indicates no data available to author.

† Range of pre-production reserves for Kuroko deposits (Lambert and Sato, 1974).

km ± 1.4 km (1 standard deviation; 21 inter-ore deposit measurements; Spooner, 1977).

In Cyprus, ophiolitic sulphide deposits have been worked principally for copper and for high-quality non-arsenical pyrite suitable for the manufacture of sulphuric acid (e.g., Spooner, 1975). Minor amounts of gold and silver have been produced from veins of "Devil's Mud" which occur in the flamboyant gossans (Bear, 1963). Examples of pre-production commercial reserves and grades of some attractive ophiolitic sulphide deposits in Cyprus and Oman are given in Table I, together with comparative data for the small York Harbour ophiolitic deposit in Newfoundland, for Precambrian volcanogenic massive sulphide deposits in the Canadian Shield and for the Kuroko deposits of Japan.

Total pre-production reserve tonnages can be as high as 15 million tonnes (e.g., Mavrovouni). The average for eleven deposits in Cyprus listed by Bear (1963) and Constantinou and Govett (1973) is 3.6 ± 4.0 (1 standard deviation) million tonnes. This is similar to normal Kuroko type deposits (e.g., 2 to 35 million tonnes for those in Japan) and Precambrian volcanogenic massive sulphide deposits (e.g., an average of 6.9 million tonnes for 110 deposits in the Canadian Shield; Boldy, 1977). The copper contents of the ophiolitic deposits of Cyprus vary widely, ranging from low values (e.g., 0.24% Cu for Mathiati; Constantinou and Govett, 1973) to as much as 4% Cu (Mavrovouni). Zinc and lead contents in economic, not geochemical, terms are characteristically low for ophiolitic deposits (e.g., Table I, Zn = 0.06% to 0.7%; Pb = 0.01%), in comparison wtih Kuroko and Precambrian deposits. An exception is the York Harbour, Newfoundland, deposit which contains 4.67% Zn. However, an important, but generally unremarked, characteristic of ophiolitic deposits is that they can contain quite significant concentrations of silver and gold. For example, Mavrovouni ore contained 39 g/tonne silver, which is comparable to the average 1976 Kuroko production grade (63 g/tonne) and the average Canadian Precambrian volcanogenic massive sulphide deposit (45 g/tonne). Kalavasos (Cyprus) ore contained 1.6 g/tonne gold, which exceeds both the average 1976 Kuroko production grade (0.7 g/tonne) and the average Canadian Precambrian volcanogenic massive sulphide deposit grade (0.84 g/tonne).

In on-land economic terms, the massive sulphide deposits recently discovered on the East Pacific Rise, 240 km south of the mouth of the Gulf of California (Francheteau et al., 1979), are simply minor showings. They consist of partially hollow, sponge-like mounds up to 10 m high which are aligned along faults subparallel to the spreading axis. Active hot spring discharge has ceased, and the sulphide material is oxidizing to red, yellow, and brown amorphous iron hydroxyoxides (cf. ochre) and yellow native sulphur. The primary sulphide mineral assemblage consists mainly of sphalerite and pyrite with minor chalcopyrite and marcasite. Analysis of four small sulphide samples (26.5 g to 90.7 g) reveals both pyrite rich (14.9% to 29.6% Fe) and sphalerite rich (23 to 28.7% Zn) types with higher copper concentrations (0.2% to 6% Cu). Hence, this material shows chemical and mineralogical similarities to that from ophiolitic deposits.

SEAWATER DISCHARGE DURING MINERALIZATION

The hypothesis of metal leaching and transport in convecting hot seawater (see Fig. 3), which was speculatively proposed to account for the formation of ophiolitic massive sulphide ore deposits (Spooner and Fyfe, 1973), has been tested by geochemical methods. Examination of fluid inclusion properties and the hydrogen, oxygen and strontium isotopic composition of mineralized material has revealed no evidence for a component of the hydrothermal fluid which was *not* of seawater origin (Spooner and Bray, 1977; Heaton and Sheppard, 1977; Chapman and Spooner, 1977; Spooner, 1977).

Example 1. Microthermometric examination of fluid inclusions in ore material from Cyprus has shown that samples of the ore forming fluid trapped in small cavities in quartz intergrown with sulphides have a freezing point indistinguishable from that of seawater. A mean fluid inclusion freezing point of $-1.9°C \pm 0.3°C$ (1 standard deviation) was obtained from 273 measurements from four mineralized localities (Spooner, in prep.; updated from Spooner and Bray, 1977). This freezing point is statistically identical to that of ordinary seawater. Since the freezing point of a solution is a reflection of its salinity, this information indicates a close similarity in bulk composition between the hydrothermal fluid, which was at temperatures between 300°C and 350°C (Spooner and Bray, 1977), and normal seawater.

Example 2 Mineralized material from four deposits in Cyprus is significantly enriched in ^{87}Sr (Chapman and Spooner, 1977). Initial $^{87}Sr/^{86}Sr$ ratios as high as, but not higher than, upper Cretaceous seawater have been recorded (e.g., 0.7075 $\pm$ 0.0002).

Example 3 The hydrogen isotope composition of the hydrothermal fluid has been shown to have been essentially the same as that of seawater (Heaton and Sheppard, 1977).

THE NATURE OF THE CONVECTIVE PROCESS IN THE OPHIOLITIC ROCKS OF THE TROODOS MASSIF, CYPRUS

The evidence summarized above successfully traces the complete cycle of seawater circulation (Fig. 3). Several deductions can now be made about the nature of the convective process which occurred in the ophiolitic rocks of Cyprus. Although it is uncertain how representative of oceanic crust the Troodos complex is and, for that matter, how variable the oceanic crust itself is, some of these points may be relevant to an understanding of convective heat and mass transfer within spreading ridges.

1) As shown diagrammatically in Figure 3, the first-order geometry of circulatory flow appears to have consisted of axially symmetric cells containing central plumes of hot ascending fluid which were positionally fixed through time with respect to enclosing rock but not, therefore, with respect to the spreading axis (Spooner, 1977; Parmentier and Spooner, 1978). The explanation for this pattern is uncertain. It could either be a reflection of some periodicity in the distribution of discrete magma intrusions within the spreading ridge, or it might be a natural pattern of convection in a quasi-uniformly heated, unconfined, permeable layer. The latter explanation would appear less likely, however, since the geometry of flow would probably reflect the decrease in basal heat flow away from a ridge axis. It would then consist of linear two-dimensional rolls sub-parallel to the axis.

2) The distribution of major plumes, as deduced from the arrangement of the major

ore desposits in Cyprus, was imperfectly regular in a direction normal to the original spreading axis, and was characterized by a half-spacing (2.6 ± 1.4 km) comparable to the thickness of the permeable layer (2 to 3 km) (Spooner, 1977; Fig. 2).

3) In Cyprus, ore deposits are frequently cross-cut by unmineralized dykes and/or overlain by unmineralized pillow lavas (Fig. 1). This suggests that formation of the ore deposits occurred *within* the volcanically active zone at a spreading ridge. For a half-width of about 6 km and a spreading rate of about 5 cm/yr, the ore desposits may, therefore, have formed in a time on the order of 10^5 years (Spooner, 1977). The mean internal upward flow rate may have been about 10^{-6} g/cm^2/s (Spooner, 1977). Later finite difference calculations have shown this rate to be reasonable (Parmentier and Spooner, 1978).

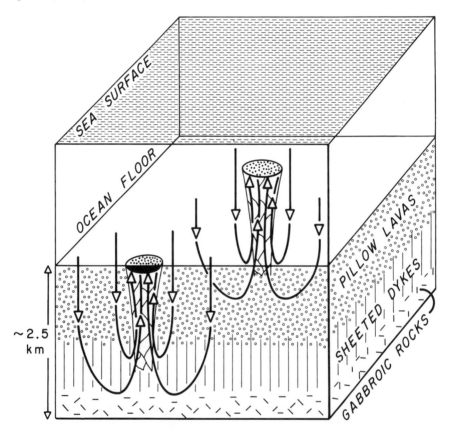

Figure 3. Schematic three-dimensional diagram of the geologically inferred mode of hydrothermal convection, metamorphism and mineralization within the upper Cretaceous ophiolitic sequence of the Troodos Massif, Cyprus.

General metamorphism is thought to have occurred in the zones of recharge flow, whereas the cupriferous pyrite ore deposits are thought to have formed at the positions of discharge of approximately axially symmetric plumes of rising hot fluid of sea water origin.

4) The depth of the original ocean above the permeable oceanic crust has been derived from the filling temperatures of fluid inclusions which homogenize into the liquid phase, and which co-exist with aqueous inclusions homogenizing into the gaseous phase. This evidence indicates that boiling occurred during mineral deposition (Spooner, in prep.). The inclusions were found in quartz from quartz-sulphide veins within the high-level Mathiati stockwork, and were located only about 5 m below the original ocean floor, as indicated by the presence of massive ore. The equilibrium pressure for boiling at a measured temperature for a measured composition has been determined as 250 ± 30 bars. This is equivalent to an original ocean depth of 2.5 ± 0.3 km. This deduction is very reasonable, since the elevations of spreading ridge axes are typically between −2 km and −4 km.

5) Using the ocean depth discussed above and estimates of stratigraphic depth, fluid inclusion filling temperatures may be pressure corrected to yield mineralization temperatures (Spooner, in prep.). Data from four mineralized localities in Cyprus are presented in Table II.

Except for information from the Mathiati mine, which applies to the near-surface cooling and mixing environment, the data suggest that the temperature of the rising plume, as defined by the range of mean ± standard deviation values for the three other stockworks, was about 300°C to 370°C. This range is in good agreement with a value of 300°C estimated for the temperature of last equilibrium with quartz of end-member hydrothermal fluid from the Galapagos Ridge axis (Edmond, 1978), is consistent with a maximum value of 285°C estimated by Crane and Normark (1977) for hydrothermal activity on the East Pacific Rise at 21°N and corresponds closely with the discharge temperatures of 380°C ± 30°C measured at 21°N on the East Pacific Rise (Spiess *et al.*, 1980).

TABLE II

Trapping temperatures (pressure corrected filling temperatures) for fluid inclusions from ophiolitic sulphide ore deposits in Cyprus (Spooner, in prep.)

Locality	Range (°C)	Arithmetic Mean (°C), standard deviation and number of measurements	Mode(s) (°C)
Mathiati Mine (very high level stockwork)	212 - 321; 109	286 ± 22 (93)	298
Limni Mine (high level stockwork)	283 - 372; 89	320 ± 18 (95)	325
Limni Mine; Mineralized Basal Group (deep stockwork)	271 - 368; 97	336 ± 19 (80)	341
Alestos Mine (deep stockwork)	(274) 323 - 394; 71	352 ± 21 (61)	335 and 370 (poorly defined)

6) The temperature immediately below massive ore was high during mineralization. Fluid inclusion studies of material from the Mathiati Mine, Cyprus and the Lasail
deposit, Oman give a temperature range of 264°C to 348°C and indicate mixing with
overlying cold seawater (Spooner, in prep.). These data confirm discharge of very hot
seawater, and also confirm that discharge of hot seawater may be an effective heat
removal mechanism (see Ribando *et al.*, 1976 for discussion).

NUMERICAL FLUID DYNAMIC MODELLING OF CONVECTION RELEVANT TO THE ORIGIN OF THE OPHIOLITIC SULPHIDE ORE DEPOSITS OF CYPRUS

Parmentier and Spooner (1978) have carried out numerical modelling by finite
difference approximations on a 21 X 21 grid of convection in a permeable layer relevant
to the origin of the ophiolitic massive sulphide deposits of Cyprus (see Fig. 4). The
model parameters chosen to be appropriate include a cylindrical geometry, an aspect
ratio (height:radius) of 1:1, a cell top open to inward and outward flow, constant
bottom heat flux and a constant upper boundary temperature. For mathematical
reasons the model did *not* allow discharge of *hot* water. This assumption is known to
be incorrect, but it should not affect the gross aspects of flow. It should only be
relevant to boundary layer phenomena in the discharge zone. (For further mathematical and computational aspects see Parmentier and Spooner, 1978.)

A solution for a Rayleigh number of 50 was dimensionalized by choosing a cell
height (radius) of 2 km, a maximum basal temperature of 450°C, derived from basal
metamorphic mineral assemblages, and an upper boundary temperature of 0°C (see
Fig. 4). The principal geological findings may be summarized as follows:

1) The flow geometry inferred to have existed naturally, which was characterized
by axially symmetric rising plumes, was experimentally found to be stable at Rayleigh
numbers of 50 and 100 for a cell aspect ratio of 1:1.

2) The overall zone of rising fluid contains a hot core which is dimensionally
comparable to mineralized stockworks observed in Cyprus.

3) The water in the hot core of the rising plume remains essentially isothermal at a
temperature of 350°C to 400°C for much of its upward transit. The temperature of the
rising plume is in reasonable agreement with that found independently from fluid
inclusion studies of deep-level stockworks (317°C to 373°C; see Table II).

4) The rising plume cools conductively only within the topmost 200 m of the
permeable layer. This depth corresponds to the economically exploitable depth of
mineralization within the Limni stockwork (about 200 m).

5) The upward fluid flux of the hot plume is on the order of 10^{-6} cm/sec. This
corresponds closely to the flux estimated crudely on the basis of geochemical arguments (Spooner, 1977).

6) The model has reproduced several factors necessary for the successful formation of an economically exploitable mineral concentration. In this case, three principal
factors may be identified: a dimensionally restricted hot core zone, a hot core zone
which also contains the maximum fluid flux and a dimensionally restricted surface
boundary layer within which cooling occurs. Thus, a theoretical volume of maximum
mineralization may be defined as that volume through which fluids flowed which had
been at temperatures greater than 350°C, and which cooled below a temperature of
300°C (volume 'a' in Fig. 4). The leaching and depositional threshold temperatures

were chosen in the light of experimental solubility work by Crerar and Barnes (1976). It can be seen that this volume ('a') is comparable in dimensions to the Limni mine stockwork (volume 'c' in Fig. 4).

7) A theoretical volume of maximum leaching may also be defined as that for which the temperature was greater than 350°C (volume 'b' in Fig. 4). This is comparable in size to a volume of basaltic material which would contain all the copper contained in the Limni deposit (volume 'd' in Fig. 4).

8) Absolute values of average permeability (0.1 millidarcies) and basal heat flux

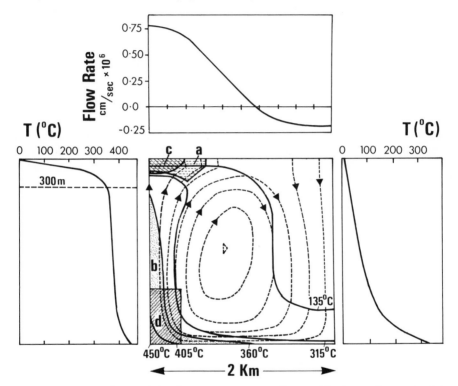

Figure 4. The central part of the diagram shows streamlines and isotherms for steady-state convection of aqueous fluid with temperature dependent viscosity and coefficient of thermal expansion in a permeable medium at a Rayleigh number of 50 in a cylindrical cell with an open top and a constant bottom heat flux (from Parmentier and Spooner, 1978). The solution has been dimensionalized by choosing realistic values of cell height/radius (2 km), maximum basal temperature (450°C) and upper boundary temperature (0°C).

Area a = a theoretical volume of most intense mineralization. Area b = a theoretical volume of maximum leaching. Area c = approximate dimensions of the Limni mine stockwork. Area d = approximate volume of basalt which would contain the copper contained in the Limni ore deposit. (For further discussion see text.)

Temperature profiles along the vertical boundaries of the cell are shown on the left- and right-hand sides of the central diagram, and the vertical flow rate through the open upper boundary is shown above the central diagram.

(50.4 HFU) obtained by dimensionalizing the model are reasonable (Parmentier and Spooner, 1978).

In conclusion, it can be stated on the basis of the above discussion that a seawater convection model for the origin of the ophiolitic sulphide deposits, which had already been shown to be qualitatively reasonable, has also been shown to be semi-quantitatively satisfactory.

THE MECHANISM OF SULPHIDE ORE PRECIPITATION

By combining information from the fluid dynamic modeling discussed above with results from microthermometric examination of fluid inclusions in material immediately below massive ore in Cyprus (Mathiati deposit) and Oman (Lasail deposit) (Spooner, in prep.), it is possible to begin to discuss the actual mechanism of mineral precipitation both below and on the original ocean floor.

Conductive Cooling

The fluid dynamic model (Fig. 4) indicates that strong conductive cooling may start within 200 m to 300 m of the seawater/rock interface. This is the approximate maximum depth of the economically exploitable Limni stockwork. Microthermometric examination of fluid inclusions in this stockwork shows that inferred trapping temperatures have a well-defined symmetrical statistical distribution with a low coefficient of variation of only 6% and an arithmetic mean of 320°C (see Table II). This type of distribution, which exhibits no negative skewness, is consistent with conductive cooling without significant mixing with variable quantities of cooler, overlying seawater. The information suggests, therefore, that the principal factor in mineral precipitation in the deeper parts of stockworks was simple conductive cooling.

Cooling by Mixing

If the effect of dilution on precipitation is outweighed by the effect of cooling by mixing on solubility, then it is possible for temperature decrease induced by mixing with a cooler fluid to be an effective mechanism for exceeding mineral solubility products. This can be demonstrated by using data on copper solubility in equilibrium with the sulphide assemblage chalcopyrite-pyrite-bornite at 350°C and 250°C, in near-neutral solutions of 1.0 molal chloride ion content (Crerar and Barnes, 1976). Between 350°C and 250°C, the copper concentration of the solution drops by two orders of magnitude, from about 1000 ppm to about 10 ppm. It is possible to lower the temperature of a fluid from 350°C to 250°C by mixing with 0°C seawater in the ratio of 1:0.4. While dilution alone would lower the copper concentration to 714 ppm, the temperature drop would lower it to 10 ppm. Hence, it can be seen that the effect of temperature decrease on solubility far outweighs the counteracting effect of dilution; therefore, cooling by mixing can be a very effective precipitation mechanism. It is worth noting that the temperatures discussed above are *exactly* those which fluid inclusion studies have shown to have occurred during formation of the ophiolitic sulphide ore deposits of Cyprus.

That this mechanism does occur is shown by the fact that volcaniclastic sediments in a bay of the island of Vulcano in the Tyrrhenian Sea are being actively cemented by pyrite and marcasite in shallow water submarine fumaroles (Honnorez et al., 1973). The recent observations on the chemistry and temperature of hot springs on the Galapagos spreading ridge (Edmond, 1978) provide similar evidence of mixing. Although measured discharge temperatures only rise as high as 17°C (15°C above ambient), it can be calculated from the point of intersection of the dissolved silica temperature dilution line with the quartz solubility curve that the end-member hydrothermal fluid was at a temperature of at least 300°C. Two other important points in this study are, firstly, that mixing with overlying seawater can occur *below* the ocean floor in hot spring regions and, secondly, that despite the extreme dilution with oxygenated ocean water, the warm water still contains dissolved reduced sulphur species. The latter are metabolized by sulphide oxidizing bacteria which form the base of a complex ocean-bottom food chain with essentially no connections to the photic zone (Corliss and Ballard, 1977).

Microthermometric determination of fluid inclusion filling temperatures and freezing points in material immediately below massive sulphide ore indicates that shallow sub-surface mixing also occurred during formation of ophiolitic sulphide deposits. In the Mathiati deposit (Cyprus), the distribution of fluid inclusion filling temperatures shows a pronounced negative skewness (Pearson measure of skewness derived from data in Table II is -0.55), with values as low as 212°C relative to a modal value of 298°C. A similar relationship, combined with a drop in salinity from high values produced by boiling, has also been detected immediately below massive sulphide ore in the Lasail deposit, Oman (Spooner, in prep.).

Hence, cooling caused by mixing of hot hydrothermal fluid with cold seawater combined with pH increase caused precipitation immediately beneath the original seawater/basalt interface, in the upper parts of stockworks and, especially, during discharge into overlying seawater.

SPECULATIONS ON THE IMPLICATIONS OF SUBDUCTION OF COMPOSITIONALLY MODIFIED OCEANIC CRUST WITH RESPECT TO THE ORIGIN OF CALC-ALKALINE IGNEOUS ROCKS AND SPATIALLY ASSOCIATED PORPHYRY CU ± Mo ± Au DEPOSITS

The geochemical effects of subduction of compositionally modified, in particular hydrated and [87]Sr-enriched, oceanic crust (Spooner, 1976; 1978) are just beginning to be detected and appreciated. It is possible that the phenomena of hydrothermal metamorphism and mineralization of oceanic crust could be quantitatively important aspects of the geochemical evolution of the crust and the mantle. The various possible connections and processes are shown diagrammatically in Figure 5.

The precise mechanism of generation of calc-alkaline igneous rocks and associated mineral deposits which occur spatially related to subduction zones has long been a matter of dispute. Recent geochemical work, in particular that involving combined analysis of the neodymium and strontium isotopic compositions of calc-alkaline lavas (Hawkesworth et al., 1977, 1979) may be pointing towards a resolution of the problem. For andesitic lavas associated with subduction zones in the Scotia Arc, Ecuador, Chile, the Marianas and the Lesser Antilles, $^{87}Sr/^{86}Sr$ ratios have been determined which are anomalously high relative to the well-defined $^{143}Nd/^{144}Nd$ −

^{87}Sr/^{86}Sr antipathetic correlation which has been observed for mid-ocean ridge basaltic rocks. On the basis of a variety of arguments, Hawkesworth *et al.* (1979) suggest that this effect, combined with the major and trace element chemistry of the lavas, may be best explained not by partial melting of the subducted oceanic crust, but by devolatilization which produced an alkali (K, Rb, Sr)-rich, high ^{87}Sr/^{86}Sr aqueous fluid phase which activated partial melting of the overlying mantle wedge (see Fig. 5). Devolatilization would occur because the upper 3 to 5 km of oceanic crust are hydrated by seawater/basalt interaction during hydrothermal circulation within spreading ridges. For example, the H$_2$O+ contents of nine ophiolitic metabasalts and metadolerites reported by Spooner *et al.* (1977) range from 0.6% to 4.5 wt.% and average 2.5

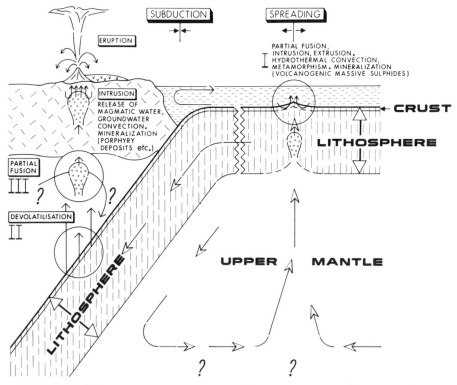

Figure 5. Highly schematic cartoon showing possible geochemical connections which may occur related to the processes of sea-floor spreading and subduction. The principal objective of the diagram is to portray a technically feasible series of linkages between seawater sulphate, massive sulphide deposits in hydrated oceanic crust produced by seawater convection within spreading ridges, devolatilization of hydrated oceanic crust in subduction zones, partial fusion of the overlying mantle wedge and formation of porphyry copper deposits spatially related to high level, sub-volcanic, calc-alkaline plutons of intermediate composition.

It is interesting to note how convective phenomena manifest themselves in a wide variety of media, forms and locations. The media include solid mantle material, silicate magma, seawater in oceanic crust, groundwater in continental crust, the oceans and the atmosphere.

wt.%. These data compare with a mean water content for fresh deep-sea basaltic material of about 0.30 wt.% (Moore, 1970), and suggest an increase in water content of about 2 wt.%.

If this figure is representative, then at the present rate of subduction of hydrated oceanic crust (about 4×10^{16} g/yr; Spooner, 1976), the total ocean mass could be recycled through the mantle in about 1.8×10^9 yr.

A closely related implication concerns the possible presence of sulphide deposits in oceanic crust, which could be comparable in size and density of occurrence to those in Cyprus. The isotopic composition of contained sulphur suggests that the ore sulphide is mainly of reduced seawater sulphate origin (Bachinski, 1977; Spooner, 1977; Heaton and Sheppard, 1977). Hence, large quantities of sulphide, in association with hydroxyl and chloride, could be added to the oceanic crust during seawater/rock interaction at ridges (Fig. 5). Release of water from the descending slab at subduction zones may cause wet melting of mantle material, as suggested by the arguments of Hawkesworth *et al.* (1979), and could produce volatile rich siliceous magmas enriched in chlorine and sulphur (Fig. 5). Such a simple model could begin to explain the amounts of magmatically released water, chlorine and sulphur required for the formation of porphyry Cu ± Mo ± Au deposits. Water is needed as the transport medium, chloride for metal complexing and sulphur for fixing the metals as solid phases (Spooner, 1978).

An oceanic crust depleted in volatiles and associated elements to an unknown extent is then returned to the active mantle where it may become involved in subsequent ocean ridge and other melting events in the slow, continuous cycle of mantle convection and partial fusion.

CONCLUSIONS

Further implications of the subduction of chemically and isotopically modified oceanic crust remain to be evaluated. Many difficult geochemical problems are encountered, particularly if semi-quantitative models are attempted. Some examples are given below:

1) What is the bulk average change in the chemical composition of the upper part of the oceanic crust affected by seawater circulation?

2) How much of the modified material returned to the mantle during subduction is returned irreversibly to the continental crust by the processes which occur in and above subduction zones?

3) How much of the modified material is, therefore, completely returned to the mantle?

4) How much of the mantle is, and has been, involved?

5) How is the mantle changing in composition now as a result of these processes?

6) What sort of elemental and isotopic heterogeneities may be produced in the mantle?

7) If the combination of processes is important now, then how much more important may it have been in the early stages of the chemical evolution of the Earth, when radiogenic heat production was approximately three times what it is today?

These geochemical questions arise from the realization by several people, including J. Tuzo Wilson, that the Earth is a rather complex dynamic system.

ACKNOWLEDGEMENTS

Receipt of NSERC grant no. A6114 is very gratefully acknowledged since it helped significantly in the preparation of this work. I would also like to express my thanks to Subhash Shanbhag (draftsman), Brian O'Donovan (photographer) and Sylvia Skinner (secretary) for invaluable assistance.

REFERENCES

Adamides, N.G., 1975, Geological history of the Limni concession, Cyprus, in the light of the plate tectonics hypothesis: Trans. Instit. Mining and Metallurgy (Section B: Applied Earth Science), v. 84, p. B17-B23.

Anderson, R.N., Hobart, M.A. and Langseth, M.G., 1979, Geothermal convection through oceanic crust and sediments in the Indian Ocean: Science, v. 204, p. 828-832.

Bachinski, D.J., 1977, Sulfur isotopic composition of ophiolitic cupriferous iron sulfide deposits, Notre Dame Bay, Newfoundland: Econ. Geol., v. 72, p. 243-277.

Bear, L.M., 1963, The mineral resources and mining industry of Cyprus: Bull. Geol. Survey Department of Cyprus, v. 1, 208 p.

Boldy, J., 1977, (Un)certain exploration facts from figures: Canadian Instit. Mining Bull., v. 70, no. 781, p. 86-95.

Chapman, H.J. and Spooner, E.T.C., 1977, [87]Sr enrichment of ophiolitic sulphide deposits in Cyprus confirms ore formation by circulating sea water: Earth and Planetary Sci. Letters, v. 35, p. 71-78.

Christensen, N.I., 1978, Ophiolites, seismic velocities and oceanic crustal structure: Tectonophysics, v. 47, p. 131-157.

Coish, R.A., 1977, Ocean floor metamorphism in the Betts Cove ophiolite, Newfoundland: Contrib. Mineralogy and Petrology, v. 60, p. 255-270.

Constantinou, G. and Govett, G.J.S., 1972, Genesis of sulphide deposits, ochre and umber of Cyprus: Trans. Instit. Mining and Metallurgy (Section B: Applied Earth Science), v. 81, p. B32-B46.

——————————, 1973, Geology, geochemistry and genesis of Cyprus sulfide deposits: Econ. Geol., v. 68, p. 843-858.

Corliss, J.B. and Ballard, R.D., 1977, Oases of life in the cold abyss: National Geographic Magazine, v. 152, p. 441-453.

Crane, K. and Normark, W.R., 1977, Hydrothermal activity and crustal structure of the East Pacific Rise at 21°N: Jour. Geophys. Research, v. 82, p. 5336-5348.

Crerar, D.A. and Barnes, H.L., 1976, Ore solution chemistry V. Solubilities of chalcopyrite and chalcocite assemblages in hydrothermal solution at 200°C and 350°C: Econ. Geol., v. 71, p. 772-794.

Duke, N.A. and Hutchinson, R.W., 1974, Geological relationships between massive sulfide bodies and ophiolitic volcanic rocks near York Harbour, Newfoundland: Canadian Jour. Earth Sci. v. 11, p. 53-69.

Edmond, J.M., 1978, Chemistry of the hot springs on the Galapagos ridge axis (abst.): The Second Maurice Ewing Memorial Symposium (Abstracts), p. 13.

Francheteau, J., Needham, H.D., Choukroune, P., Juteau, T., et al., 1979, Massive deep-sea sulphide ore deposits discovered on the East Pacific Rise: Nature, v. 277, p. 523-528.

Fryer, B.J., Fyfe, W.S. and Kerrich, R., 1979, Archean volcanogenic oceans: Chemical Geol., v. 24, p. 25-33.

Gass, I.G. and Smewing, J.D., 1973, Intrusion, extrusion and metamorphism at constructive margins: evidence from the Troodos Massif, Cyprus: Nature, v. 242, p. 26-29.

Hashimoto, K., 1977, The Kuroko deposits of Japan – geology and exploration strategies: Asociacion de Ingenieros de Minas Metalurgistas y geologos de Mexico Memoria Tecnica, v. 13, p. 25-88.

Hawkesworth, C.J., O'Nions, R.K., Pankhurst, R.J., Hamilton, P.J. and Evensen, N.M., 1977, A geochemical study of island-arc and back-arc tholeiites from the Scotia Sea: Earth and Planetary Sci. Letters, v. 36, p. 253-262.

Hawkesworth, C.J., Norry, M.J., Roddick, J.C. and Baker, P.E., 1979. $^{143}Nd/^{144}Nd$, $^{87}Sr/^{86}Sr$, and incompatible element variations in calc-alkaline andesites and plateau lavas from South America: Earth and Planetary Sci. Letters, v. 42, p. 45-57.

Heaton, T.H.E. and Sheppard, S.M.F., 1977, Hydrogen and oxygen isotope evidence for sea-water-hydrothermal alteration and ore deposition, Troodos complex, Cyprus: Geol. Soc. London Special Publ., no. 7, p. 42-57.

Honnorez, J., Honnorez-Guerstein, B., Valette, J. and Wauschkuhn, A., 1973, Present day formation of an exhalative sulphide deposit at Vulcano (Tyrrhenian Sea), part II: active crystallization of fumarolic sulphides in the volcanic sediments of the Baia di Levante: in Amstutz, G.C. and Bernard, A.J., eds., Ores in Sediments: Berlin – Heidelberg – New York, Springer-Verlag, p. 139-166.

Hutchinson, R.W. and Searle, D.L., 1971, Stratabound pyrite deposits in Cyprus and relations to other sulphide ores: Soc. Mining Geol. of Japan Special Issue, no. 3, p. 198-205.

Lambert, I.B. and Sato, T., 1974, The Kuroko and associated ore deposits of Japan: a review of their features and metallogenesis: Econ. Geol., v. 69, p. 1215-1236.

Liou, J.G. and Ernst, W.G., 1979, Oceanic ridge metamorphism of Taiwan ophiolite: Contrib. Mineralogy and Petrology, v. 68, p. 335-348.

Lister, C.R.B., 1972, On the thermal balance of a mid-ocean ridge: Geophys. Jour. Royal Astron. Soc., v. 26, p. 515-535.

Moore, J.G., 1970, Water content of basalt erupted on the ocean floor: Contrib. Mineralogy and Petrology, v. 28, p. 272-279.

Mottl, M.J. and Holland, H.D., 1978, Chemical exchange during hydrothermal alteration of basalt by sea water – I. Experimental results for major and minor components of sea water: Geochim. et Cosmochim. Acta, v. 42, p. 1103-1116.

Parmentier, E.M. and Spooner, E.T.C., 1978, A theoretical study of hydrothermal convection and the origin of the ophiolitic sulphide ore deposits of Cyprus: Earth and Planetary Sci. Letters, v. 40, p. 33-44.

Ribando, R.J., Torrance, K.E. and Turcotte, D.L., 1976, Numerical models of hydrothermal circulation in the oceanic crust: Jour. Geophys. Research, v. 81, p. 3007-3012.

Robertson, A.H.F., 1975, Cyprus umbers; basalt-sediment relationships on a Mesozoic ocean ridge: Jour. Geol. Soc. London, v. 131, p. 511-531.

Sangster, D.F., 1977, Some grade and tonnage relationships among Canadian volcanogenic massive sulphide deposits: Geol. Survey Canada Paper, v. 77-1A, p. 5-12.

Sillitoe, R.H., 1972, Formation of certain massive sulphide deposits at sites of sea-floor spreading: Transactions of the Institution of Mining and Metallurgy (Section B: Applied Earth Science), v. 81, p. B141-B148.

Sleep, N.H. and Wolery, T.J., 1978, Egress of hot water from midocean ridge hydrothermal systems: some thermal constraints: Jour. Geophys. Research, v. 83, p. 5913-5922.

Spiess, F.N., Macdonald, K.C. Atwater, T., Ballard, R., et al., 1980, East Pacific Rise: hot springs and geophysical experiments: Science, v. 207, 1421-1433.

Spooner, E.T.C., 1975, Cyprus pyrite today: Sulphur, no. 121, p. 23-27.

_____, 1976, The strontium isotopic composition of seawater and seawater-oceanic crust interaction: Earth and Planetary Sci. Letters, v. 31, p. 167-174.

_____, 1977, Hydrodynamic model for the origin of the ophiolitic cupriferous pyrite ore deposits of Cyprus: Geol. Soc. London Special Publ., no. 7, 58-71.

—————————, 1978, Ophiolitic rocks and evidence for hydrothermal convection of sea water within oceanic crust (abst): The Second Maurice Ewing Memorial Symposium (Abstracts), p. 36-37.

Spooner, E.T.C., Beckinsale, R.D., Fyfe, W.S. and Smewing, J.D., 1974, [18]O enriched ophiolitic metabasic rocks from E. Liguria (Italy), Pindos (Greece) and Troodos (Cyprus): Contrib. Mineralogy and Petrology, v. 47, p. 41-62.

Spooner, E.T.C. and Bray, C.J., 1977, Hydrothermal fluids of sea water salinity in ophiolitic sulphide ore deposits in Cyprus: Nature, v. 266, p. 808-812.

Spooner, E.T.C., Chapman, H.J. and Smewing, J.D., 1977, Strontium isotopic contamination and oxidation during ocean floor hydrothermal metamorphism of the ophiolitic rocks of the Troodos Massif, Cyprus: Geochim. et Cosmochim. Acta, v. 41, p. 873-890.

Spooner, E.T.C. and Fyfe, W.S., 1973, Sub-sea-floor metamorphism, heat and mass transfer: Contrib. Mineralogy and Petrology, v. 42, p. 287-304.

Stern, C., De Wit, M.J. and Lawrence, J.R., 1976, Igneous and metamorphic processes associated with the formation of Chilean ophiolites and their implication for ocean floor metamorphism, seismic layering and magnetism: Jour. Geophys. Research, v. 81, p. 4370-4380.

Trennery, T.O. and Pocock, B.G., 1972, Mining and milling operations at Limni Mine, Cyprus: Transactions of the Institution of Mining and Metallurgy (Section A: Mining Industry), v. 81, p. A1-A12.

Williams, D.L. and Von Herzen, R.P., 1974, Heat loss from the Earth: new estimate: Geology, v. 2, p. 327-328.

Williams, D.L., Von Herzen, R.P., Sclater, J.G. and Anderson, R.N., 1974, The Galapagos spreading centre: lithospheric cooling and hydrothermal circulation: Geophys. Jour. Royal Astronomical Soc., v. 38, p. 587-608.

Williams, H. and Malpas, J., 1972, Sheeted dikes and brecciated dyke rocks within transported igneous complexes, Bay of Islands, Western Newfoundland: Canadian Jour. Earth Science, v. 9, p. 1216-1229.

Wolery, T.J. and Sleep, N.H., 1976, Hydrothermal circulation and geochemical flux at mid-ocean ridges: Jour. Geol., v. 84, p. 249-275.

The Continental Crust and Its Mineral Deposits, edited by D.W. Strangway,
Geological Association of Canada Special Paper 20

GEOLOGY AND STRUCTURAL CONTROL
OF KUROKO-TYPE MASSIVE SULPHIDE DEPOSITS

S.D. Scott
Department of Geology,
University of Toronto,
Toronto, Ontario M5S 1A1

ABSTRACT

The Miocene Kuroko ores of Japan, particularly those of the Hokuroku district of Northern Honshu, are an end-member "type" of volcanogenic massive Cu-Zn-Pb-Ag sulphide deposit hosted by felsic lava domes and pyroclastic rocks. The deposits lie at the intersections of NNW and NE sets of linears which are discordant to the contemporaneous Miocene structures. The NNW set is parallel to major structural features and tectonic zones in the Paleozoic-Mesozoic basement. The NE set is parallel to weaker basement structures and to a group of lineaments of unknown origin but which are clearly visible on LANDSAT mosaics.

The association of lineaments and Kuroko-type massive sulphide ores as documented in the Miocene of Japan can be seen in much older rocks in Canada. The Buchans deposit of upper Ordovician to lower Silurian age in central Newfoundland is strikingly similar to Kuroko deposits. As in the Hokuroku, the ores at Buchans lie in sinuous troughs which are aligned subparallel to two major tectonic features: 1) major NE-striking tectonic units (analogous to the basement zones of northern Honshu) and 2) a major lineament of unknown origin which crosses the entire island in a NW direction and passes through Buchans (analogous to the major NE lineaments crossing Northern Honshu). In the Noranda area of Quebec, Archean massive sulphide deposits show a spatial relationship to NE and NW lineaments on both a regional and a local scale.

RÉSUMÉ

Les minerais du Miocène de Kuroko, au Japon, en particulier ceux du district de Hokuroku dans le nord du Honshu, représentent le "type" même des dépôts de sulfures massifs volcanogéniques de Cu-Zn-Pb-Ag qu'on retrouve dans les dômes de lave felsique et les roches pyroclastiques. Les dépôts se retrouvent à l'intersection de familles de lignes NNO et NE qui sont discordantes par rapport aux structures contemporaines du Miocène. La famille NNO est parallèle aux traits structuraux et aux zones tectoniques majeures du socle paléozoïque-mésozoïque. La famille NE est parallèle à des structures moins marquées du socle et à un groupe de linéaments d'origine inconnue mais qui sont clairement visibles sur les mosaïques Landsat.

L'association des linéaments et des minerais de sulfures massifs du type Kuroko telle qu'on l'a documentée au Japon peut s'observer dans les roches beaucoup plus anciennes au Canada. Le dépôt de Buchans, datant de l'Ordovicien supérieur au Silurien inférieur dans le centre de Terre-Neuve, ressemble de façon frappante aux dépôts de Kuroko. Comme dans le Hokuroku, les minerais de Buchans se trouvent dans des dépressions sinueuses alignées à peu près parallèlement à deux ensembles tectoniques majeurs: 1) des unités tectoniques de direction NE (analogues aux zones de socle du nord du Honshu) et 2) un linéament majeur d'origine inconnue qui coupe toute l'île dans une direction NO et passe par Buchans (analogue aux linéaments majeurs NE qui traversent le nord du Honshu). Dans la région de Noranda, au Québec, les dépôts de sulfures massifs de l'Archéen montrent une relation spatiale avec les linéaments NE et NO à échelle locale et régionale.

INTRODUCTION

The Kuroko ores of Japan, particularly those of the Hokuroku district in Northern Honshu, are generally regarded to be the type example of volcanogenic massive Cu-Zn-Pb-Ag sulphide ores in island arcs. Because the deposits are young (Miocene, about 13 Ma), relatively undeformed and only very weakly metamorphosed, they offer considerable insight into a particular ore-forming process. Decades of research by Japanese and a few foreign geologists and geochemists have revealed this process to be hydrothermal emanations onto the sea floor during a short hiatus in what was otherwise a turbulent submarine environment of predominantly explosive felsic volcanism (Horikoshi, 1969; Sato, 1974; Sato et al., 1974; Lambert and Sato, 1974; Hashimoto, 1977; Urabe and Sato, 1978).

The ores exhibit a remarkable stratigraphic control in their distribution. For example, within the Hokuroku district, all of the mineable deposits (representing eight clusters of ore) occur within the upper portion of the Nishikurosawa stage, which is represented here by relatively thin (50-600 m) formations of felsic volcanic rocks and mudstone. There is paleontological and paleomagnetic evidence that other Kuroko deposits as far as 300 km away were formed at the same time as those of the Hokuroku district (Ueno, 1975). Thus, the search for new deposits is aided immensely by the knowledge that the ores occur only within a particular stratigraphic unit, i.e., the upper Nishikurosawa. If the controls on the distribution of ores within the horizontal plane were known as well, the search would be narrowed even further.

This paper discusses some geological features of Kuroko deposits and compares them with the much older Canadian massive sulphide deposits of Buchans, Newfoundland (Paleozoic) and Noranada, Quebec (Archean). The distribution of ores in all three cases appears to have been controlled by major lineaments for which there is as yet no satisfactory (plate) tectonic explanation.

THE HOKUROKU DISTRICT

The geology of the Hokuroku district and its ores is described in detail in a number of recent compilations and reviews (e.g., Tatsumi, 1970; Ishihara, 1974; Lambert and Sato, 1974; Hashimoto 1977, 1979). The following brief description is based in part on personal observations during 1977 and 1978.

General Features

The Kuroko ores sit within a crudely circular depression or "basin" (Fig. 1) that is filled with a well known succession of Miocene submarine volcanic rocks, associated mudstones and intrusions belonging to the extensive "Green Tuff" volcanic belt of Japan. The belt is so named because of its colour, imparted by low-grade (sea-floor?) alteration. The volcanic and intrusive rocks of the basin are predominantly dacitic or rhyolitic in composition, although alteration has obscured their primary petrochemistry. Basalt occurs in significant amounts in the lower Nishikurosawa and also is found in the hanging wall of about three quarters of the ore deposits (Urabe, 1979). A thick andesite is at the base of the pile. The felsic volcanic rocks occur as small domes, but are much more abundant as volcaniclastic products of explosive activity. They typically consist of tuffs (some pumiceous), tuff breccias and volcanic breccias. A characteristic feature of the basin is the clay mineral and zeolitic alteration that all rocks have suffered to some degree, but which is most evident and intensely developed within the ore-forming upper Nishikurosawa horizon, particularly in the immediate vicinity of the ore deposits. The overall aspect of the basin is that of a fossil geothermal field with discharge points represented by the ore deposits.

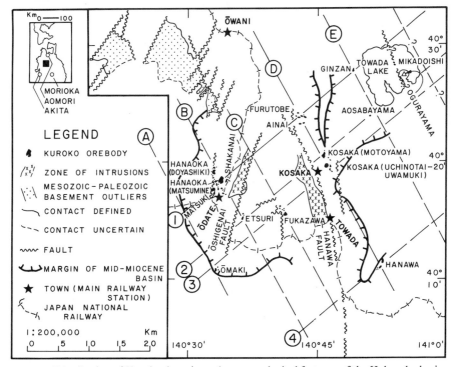

Figure 1. Distribution of Kuroko deposits and some geological features of the Hokuroku basin. All deposits lie within 500 m of a NNW or NE line defining a nearly orthogonal grid. Largest deposits are at the intersections of the grid (Revised from Scott, 1978).

The majority of Japanese geologists believe that the Hokuroku was a sedimentary basin that underwent rapid and continuous subsidence as it was filled with volcanic debris (Tanimura, 1973). Ohmoto (1978), on the other hand, has argued convincingly that a caldera more reasonably explains the geological features of the district. Whatever the correct tectonic explanation for the Hokuroku structure, the basin is probably underlain by a large felsic intrusion for which there is only indirect evidence. Ohmoto (1978), Cathles (1978), and Urabe and Sato (1978) have pointed out that the thermal energy required to drive the hydrothermal system responsible for the alteration and mineralization is far greater than can be accounted for by the known igneous rocks of the area. Furthermore, breccia dykes consisting of rounded and polished fragments within a matrix of rock flour are common, and are most reasonably explained by gas venting from a large magmatic reservoir. The dykes contain, in addition to local and basement rocks, fragments of quartz porphyry and quartz diorite that are similar to nearby Tertiary intrusions (Takenouchi, 1978) and may be actual samples of the postulated deep intrusion.

Ore Deposits

A typical Kuroko deposit, as illustrated in Figure 2, consists of one or more lenses of gypsum and of massive stratiform sulphide ore draped above or over the flank of a single "white rhyolite" dome or group of domes. The domes are characteristically only a few hundred metres in diameter and have been intensely altered to a chalky white appearance. The two types of massive stratiform ore are commonly in sharp contact, although in a few places (e.g., Kanayamazawa orebody of Fukazawa mine) there is a very thin layer of clay separating them. The footwall zone of yellow ore consists predominantly of pyrite and chalcopyrite, while the hanging-wall zone of black ore contains sphalerite, galena, barite, and much less pyrite. The black ore is typically fragmental on a scale of a few milimetres to tens of centimetres. The barite content of the black ore increases upwards, sometimes culminating in a barite bed as

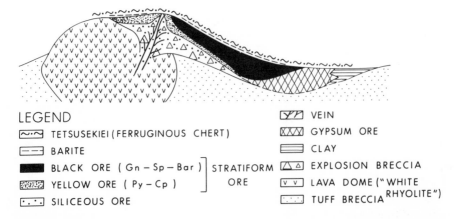

LEGEND

~·~	TETSUSEKIEI (FERRUGINOUS CHERT)
⊏−−⊐	BARITE
■■■	BLACK ORE (Gn − Sp − Bar)
▦▦	YELLOW ORE (Py − Cp)
⊏·.·⊐	SILICEOUS ORE

} STRATIFORM ORE

▨	VEIN
▧▧	GYPSUM ORE
☰	CLAY
△△	EXPLOSION BRECCIA
v v	LAVA DOME ("WHITE RHYOLITE")
⊏·.·⊐	TUFF BRECCIA

Figure 2. Schematic section of a typical Kuroko deposit. Hanging-wall rocks are omitted (Redrawn and modified from Sato, 1974).

much as one metre thick on top of the black ore. Massive stratiform ore that is in place is commonly underlain by a stockwork or disseminated zone of siliceous chalcopyrite-pyrite mineralization developed within the "white rhyolite" or in associated tuff breccias. However, it is not unusual for the stratiform portion to have slumped as much as 100 m from its original site of deposition (Horikoshi, 1969). Ferruginous chert ("tetsusekiei") sits on top of the massive stratiform ore and extends as a discontinuous layer some distance beyond it. The chert takes many forms, but is most commonly a silicified and ferruginous tuff with the iron in the form of pyrite and chlorite, and of hematite, which gives the rock its distinctive red colour. The hanging-wall rocks capping the deposits may be dacite tuff, basalt tuff and flows, or mudstone.

There are several local variations to the average Kuroko deposit described above and shown in Figure 2: not all of the gypsum is bedded, some is in veins; the proportions of black ore to yellow ore can be quite variable from deposit to deposit (Horikoshi and Shikazono, 1978) as can those of stratiform to siliceous ore; slumping has produced a fragmental mixed yellow and black ore ("hankuroko") in a few places; most siliceous ores are copper-rich, but there are examples of sphalerite-galena-barite siliceous ores (e.g., Uwamuki deposits); the shape and size of "white rhyolite" domes are not always obvious and some published sections are highly schematic; some ores are not even associated with rhyolite domes.

In addition to the Kuroko ores, there are polymetallic vein deposits within and surrounding the Hokuroku district. Most are mined for their copper, and some are very large (e.g., Osarizawa). Such veins occur in the Nishikurosawa and older rocks, but their genetic relationship, if any, to the Kuroko ores is not known.

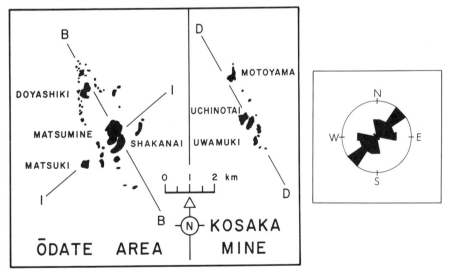

Figure 3. (Left) Alignment of ore clusters in the Ōdate and Kosaka areas. Labelled lines are from Figure 1. (Modified from Scott, 1978). (Right) Strike of Miocene veins within and to the southwest of the Hokuroku district in Akita and Yamagata Prefectures (From unpublished diagram by E. Horikoshi).

Structural Control of the Kuroko Ores

Examination of Figure 1 reveals that most of the Kuroko deposits are located close to the edge of the Hokuroku basin. Over the decades exploration geologists have assumed that fractures around the margins of the basin were responsible for the distribution of the ore. However, this theory was repudiated when, in 1969 and 1976, the Dowa Mining Company discovered the Fukazawa and Etsuri deposits in the *centre* of the basin. Scott (1978) subsequently proposed that the ores lay at or close to the intersection of a nearly orthogonal (actual 87°) set of linears that reflect basement structures and define a rectilinear grid within the basin (Fig. 1). This proposal is currently being tested by the Metal Mining Agency of Japan (T. Sato, pers. commun., 1978).

The grid is apparent on the scale of individual mines and of the entire Hokuroku district. On a local scale, Japanese geologists have long recognized that ore clusters within the basin were scattered along linear depressions or "troughs" on the Miocene sea floor (e.g., see Figure 2 of Horikoshi and Shikazono, 1978). This can be seen most clearly in the distribution of ores in the Odate and Kosaka areas (Fig. 3) and is the basis upon which Scott (1978) drew the set of lines intersecting the other ore clusters in Figure 1. All of the known ore occurrences lie within 500 m of one of the NNW or NE lines and the larger deposits are at their intersections.

The dominant Cenozoic structures in Northern Honshu strike N-S, parallel to the axis of the island arc. This is illustrated regionally in Figure 4 by most of the faults and by the distribution of Quaternary volcanoes. The N-S structural grain is also evident in the Hokuroku basin from many of its faults (e.g., Oshigenai fault east of Odate in Figure 1) and fold axes (see Sato *et al.*, 1974, Fig. 3). In contrast to the younger N-S trends, the basement rocks exhibit pronounced NNW and weaker NE structures. The NNW trend is most evident in the distribution of basement belts in Northern Honshu (see Fig. 9). Both trends are found in the faults within the Kitakami Mountainland (Fig. 4), a large massif of basement rocks 75 km southwest of the Hokuroku basin.

These NNW and weaker NE basement directions have had a profound influence on the geology of the Hokuroku district, resulting in a variety of features that cross-cut the dominant N-S grain. For example, the Hanawa fault, which is part of a 450-km long system of N-S faults, is deflected in strike to the NNW through the Hokuroku district (Fig. 1 and 4). Many small faults around the ore deposits, particularly near Kosaka and Fukazawa, have NNW and NE strikes (Hashimoto, 1979). Northwest of the Hokuroku basin small blocks of basement rocks are bounded by NNW and NNE to NE faults. Coincidentally, one of the NNW faults is on strike with line C (Fig. 1) through the Fukazawa deposit. The regional NE fracture pattern is clearly displayed by the preferred orientation of Miocene veins, shown in Figure 3 (right).

Perhaps the best evidence for the influence of basement on younger structures is seen around Lake Towada, a Quaternary Krakatoa-type caldera northeast of the Hokuroku basin (Fig. 1). Although resurgent activity has added two peninsulas to its southern margin, the distinctly rectilinear shape of the steep-walled caldera has been retained. Such rectilinear shapes are common in Krakatoan calderas and reflect the influence of deep-seated structures (Oide, 1968). In the case of Towada, these struc-

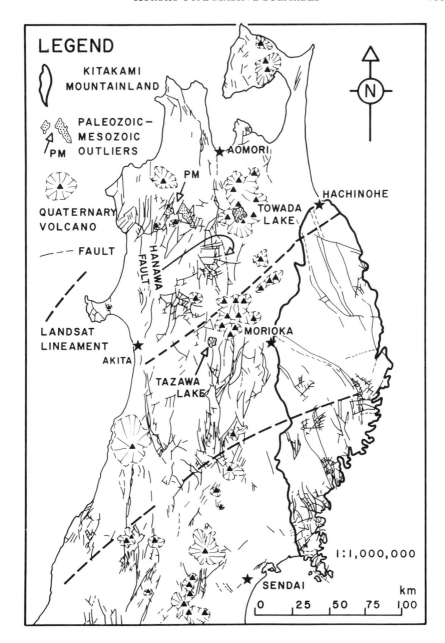

Figure 4. Structural features of Northern Honshu, redrawn and simplified from Kitamura (1976). LANDSAT lineaments are from Remote Sensing Research Group (1977) and Scott (1978). The Hokuroku district is immediately southwest of Towada Lake. (Revised from Scott, 1978).

bodies. The distribution of the breccias is controlled by deep elongate depressions or "troughs" not unlike those of the Hokuroku district. Most of the troughs extend in a NE or NW direction from the area of in situ orebodies, which presumably represents the source area of sulphides in the ore breccias. As discussed by Thurlow *et al.* (1975), the main features of a Kuroko deposit that are missing or are poorly represented at Buchans are gypsum ore, rhyolite domes, and yellow ore (although the latter two are not always well-developed in Kuroko deposits either).

The distribution of the different ore types at Buchans is shown in Figure 5. The footwall mineralization is more widespread than the in situ ore and underlies much of the area of the troughs containing the ore horizon breccias. The footwall mineralization, ore horizon breccias, and transported ores are preferentially aligned in two nearly orthogonal directions, NE and NW. One breccia channel extends NE from the Oriental orebody. An equally well-developed, ulthough somewhat more sinuous, channel extends NW from the Lucky Strike orebody and contains the Rothermere and MacLean transported ores. Preferred orientation of structural features can be seen on an even smaller scale (E.A. Swanson, pers. commun., 1979). For example, the Clementine prospects lie within small depressions that strike NE and NW; a zone of clay and pyrite alteration to the southwest of MacLean has an apparent NE strike.

As the NE trend is parallel to the axis of the deformed Ordovician-Silurian island-arc sequences, fractures and structural grains can be expected to develop along it (compare Figure 4). The reason for the NW strike is not as clear, but a major lineament plainly visible in a LANDSAT mosaic of Newfoundland (Fig. 6) crosses the entire island on a NW strike and passes through Buchans. Swanson and Brown (1962) pointed out that in the mine area airborne magnetic patterns and fold axes are strongly affected by the lineament. Although the origin of the lineament is unknown (H. Williams, pers. commun., 1979), its relation to the distribution of ores at Buchans is perhaps not coincidental. It may also not be coincidental that the lineament is nearly on strike with the linear southern margin of the continental shelf at the Grand Banks off the coast of Newfoundland and with the Newfoundland fracture zone on the deep ocean floor (see Auzende *et al.,* 1970; Le Pichon and Fox, 1971, Fig. 1; Sykes, 1978, Fig. 16). Certainly, if the lineament has influenced the emplacement of the Buchans orebodies, it must have been a very long-lived structure.

Noranda, Quebec

As summarized by Sangster (1972), Spence (1975), Spence and de Rosen-Spence (1975), and Sangster and Scott (1976), the massive sulphide ores of Noranda have many features in common with those of the Hokuroku district, both on the scale of individual deposits and regional geology. The Noranda ores are Archean in age and yet, despite their antiquity, their host rocks are only weakly metamorphosed (Jolly, 1978) and mildly deformed. Comparison of Figures 2 and 7 demonstrates the essential morphological indentity of deposits in the two districts. Significant differences between them, such as the lack of sulphates and lead and the preponderance of chloritic footwall alteration at Noranda, have been ascribed to the primitive geochemical evolution of the Earth in the Archean as compared with the Miocene (Hutchinson, 1973, 1980; Sangster and Scott, 1976). The regional similarity is underscored by the work of Spence and de Rosen-Spence (1975) and de Rosen-Spence (1976). These authors showed that the majority of the massive sulphide deposits in the central

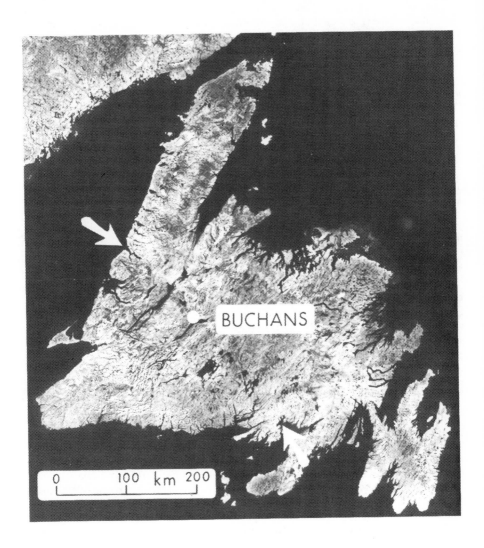

Figure 6. LANDSAT mosaic of Newfoundland. Original photograph supplied by the Surveys and Mapping Branch of the Department of Energy, Mines and Resources, Canada. The arrows locate a strong lineament extending across the island and through Buchans.

tures are in basement rocks and have constrained the alignment of all but the northern wall of the caldera in either a NNW or NE direction.

All of these features indicate that basement, rather than contemporaneous Miocene structures, have provided the main tectonic controls on the Hokuroku district. Fractures in underlying basement rocks are most likely responsible for the orientation of the ore-bearing "troughs" on the Miocene sea floor and for the distribution of ore clusters along the NNW and NE linears in Figure 1. Presumably, jostling of the basement during the middle Miocene under a tensional tectonic regime (Ishihara, 1974; Horikoshi, 1975, pers. commun., 1979; Uyeda and Nishiwaki, 1980) reactivated basement fractures. Highly permeable zones would have been developed along the fractures, particularly at their intersections. This would have eased the upward movement of rhyolite domes from the postulated main magmatic reservoir, and provided a suitable hydrothermal plumbing system for the formation of the ores. Since no major faults are known within the Miocene along lines A to E and 1 to 4 in Figure 1, it is unlikely that the basement fractures fully penetrated the Miocene cover. However, once fluids and magmas were focused at particular points at the base of the Miocene, it would be relatively easy for them to pass through the thin and unconsolidated section to the sea floor. Thus, although lineaments appear to have controlled the distribution of ores, they are in the basement rocks and are largely hidden from view.

What tectonic event reactivated the basement fractures? The required stress pattern could have been produced by caldera collapse, which is the mechanism that Ohmoto (1978) has proposed to explain the geological and geochemical features of the Hokuroku basin. In this regard, it is interesting that the margins of the basin (Fig. 1), some of which are fault-controlled, are not as irregular as they may seem at first glance, but have predominantly NNW and NE orientations. Basement structures may have played an important role in the development of the Hokuroku basin, as they did in the Towada caldera (Oide, 1968), which would lend additional support to Ohmoto's model.

One other feature deserves mention. LANDSAT mosaics reveal major NE-trending lineaments crossing Northern Honshu (Fig. 4). The origin of these lineaments is not known, but the coincidence of their orientation with Miocene veins (Fig. 3), with the many geological features described above ranging in age from Paleozoic to Quaternary, and with the NE alignment of ore clusters is striking. Clearly, major tectonic stresses having a NE trend helped to shape the geological fabric of Northern Honshu and were sufficiently long-lived to be visible today.

COMPARISON WITH CANADIAN MASSIVE SULPHIDE DISTRICTS

Volcanogenic massive sulphide deposits are found from coast to coast in Canada in rocks ranging in age from Archean to Mesozoic. The Phanerozoic deposits occur in what are probably island-arc environments. Many of the Precambrian deposits, which greatly predominate in total numbers, are in a tectonic regime and volcanic setting not unlike island arcs of the plate tectonic era (Hutchinson, 1973, 1980; Goodwin and Ridler, 1970; Sangster and Scott, 1976). As most of the Canadian deposits have volcanogenic, morphological, and mineralogical features in common with those of the Hokuroku district, it is not surprising that they exhibit a similar structural control on their emplacement as well. Two examples, one from the Paleozoic and one from the Archean, will illustrate.

Buchans, Newfoundland

The Buchans deposits in central Newfoundland have been described by Thurlow *et al.* (1975) and in a recent comprehensive compilation edited by Swanson *et al.*, (in press). Regionally, Buchans sits within a volcano-sedimentary island-arc complex of upper Ordovician-lower Silurian age (447 ± 18 Ma; Bell and Blenkinsop, in press). Metamorphic grade is low ("subgreenschist", Strong, 1977) and, in the immediate mine area, dips are on the order of 30°. The ores are within a sequence of dacitic tuffs, rhyolites, and siltstones that fills irregular depressions in a thick, dominantly basaltic volcanic terrane. Three types of mineralization are recognized. *In situ massive ore*, consisting of sphalerite, galena, chalcopyrite, barite, and lesser pyrite, occurs as conformable lenses. These overlie highly chloritized and silicified zones containing *disseminated and stockwork mineralization* of chalcopyrite + pyrite and of galena + sphalerite (compare Uwamuki deposits of the Hokuroku district). The most spectacular feature at Buchans, which is also seen in Japan but on a smaller scale, is the ore horizon breccias. The breccias are a chaotic mixture of all the rock types of the area, including fragments of massive sulphides. In places, the concentration of sulphide fragments is sufficiently high to be mined as *"transported"* ore-

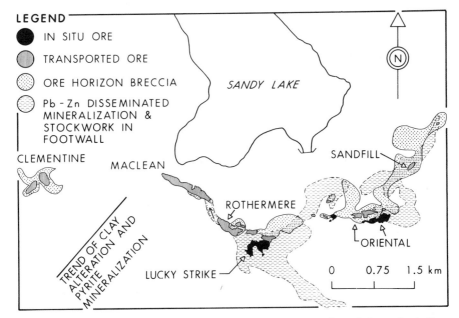

Figure 5. Distribution of ore types and ore horizon breccias at Buchans. Information is from ASARCO company maps provided by E.A. Swanson.

Noranda camp lie on or close to a particular stratigraphic horizon at the top of the Mine Series (compare Hokuroku's Nishikurosawa) within what they interpreted to be a caldera complex. The major exceptions to the stratigraphic control are the newly discovered Corbet deposit, located about 1000 m stratigraphically below the other deposits (Knuckey and Watkins, 1978), and the economically less important West Macdonald, Delbridge, and Mobrun deposits which are stratigraphically higher. The "favourable horizon" which hosts most of the Noranda deposits is marked by a cherty tuff that is pyritic in places and, except for the absence of hematite, has many features in common with the Japanese tetsusekiei.

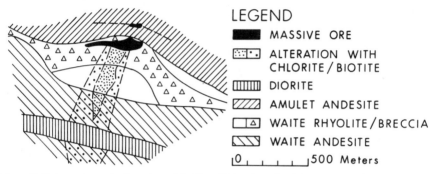

LEGEND

■■■■ MASSIVE ORE

▦ ALTERATION WITH CHLORITE/BIOTITE

▥ DIORITE

▨ AMULET ANDESITE

▥ WAITE RHYOLITE/BRECCIA

▧ WAITE ANDESITE

|0_____|500 Meters

Figure 7. Schematic vertical section of the East Waite deposit of the Noranda district. Note the association of the massive ore with rhyolite breccia in a dome. Stringer mineralization occurs within the alteration pipe beneath the massive ore. The diorite was intruded after mineralization. (Redrawn from Spence and de Rosen-Spence, 1975).

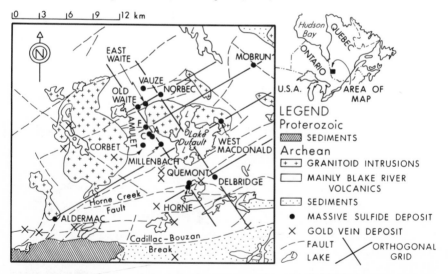

LEGEND

Proterozoic

▧ SEDIMENTS

Archean

[+ +] GRANITOID INTRUSIONS

☐ MAINLY BLAKE RIVER VOLCANICS

▦ SEDIMENTS

● MASSIVE SULFIDE DEPOSIT

✕ GOLD VEIN DEPOSIT

-- FAULT

⟋ LAKE

✕ ORTHOGONAL GRID

Figure 8. Simplified geological map of the Noranda district. A nearly orthogonal grid defines the distribution of the massive sulphide deposits. (Base map from Spence and de Rosen-Spence, 1975).

Figure 8 is a simplified geological map of the Noranda district showing the location of the massive sulphide and gold vein deposits. The distribution of the massive sulphides can be fitted into a nearly orthogonal grid. The NE set of lines is parallel to the major faults of the district (so-called "070 direction") which have long been suspected to have had an important influence on the distribution of the ores. The NW line drawn from Horne to Old Waite is defined by more than half of the deposits of the central Noranda district and is subparallel to a much weaker fault pattern.

The relationship of the Noranda ores to the dominant NE lineaments has been discussed by Sangster (1972; see also Sangster and Scott, 1976). Sangster pointed out that a line drawn parallel to the pronounced Old Waite-East Waite-Norbec trend passes through two widely separated orebodies, Mobrun and Amulet F. The same feature is observed for Horne and Delbridge and for Aldermac and West Macdonald. Since Old Waite, East Waite, and Norbec are relatively close together and lie on the same stratigraphic horizon, it is plausible that their distribution was controlled by a local feature within an environment of active synvolcanic faulting. Indeed, such cases have been recently documented at Millenbach and Amulet (Knuckey et al., 1978) and at Corbet (Knuckey and Watkins, 1978). However, the other three NE lines in Figure 8 connect deposits that are separated both in space and in time. If lineaments are responsible for the distribution of the Noranda massive sulphides, they must have been significant and long-lived features.

CONCLUSIONS

Massive Cu-Zn-Pb-Ag sulphide ores span almost the entire geological time scale and yet they exhibit a surprising number of common features. The minor differences among the three examples chosen here — Noranda (Archean), Buchans (Paleozoic), and Hokuroku (Miocene) — are probably due to the maturation of the Earth's crust with time rather than to fundamental differences in the ore-forming process. For example, the Buchans and Hokuroku districts are in island arcs whereas Noranda was formed within a pre-plate tectonic environment which has many geological and volcanological features in common with modern island arcs. Hutchinson (1973, 1980) has shown in detail how evolutionary trends in the geochemistry and tectonics of the Earth's crust have affected massive sulphide mineralization through time. One geological constraint that does not appear to have changed much with the passage of time is the role of regional structure in the emplacement of individual deposits and in the distribution of ore clusters within a district.

The similarity between the Hokuroku model of structural control and that proposed for Buchans is evident when a reoriented map of Newfoundland is compared with the basement map of Northern Honshu (Fig. 9). The major LANDSAT lineaments are superimposed. In both districts, an orthogonal set of fractures appears to have been involved in the emplacement of ore. The major differences between the districts is that the structural control was generated in the basement rocks of the Hokuroku and in the supracrustal rocks at Buchans. Common to both, however, is that one direction of preferred orientation is parallel to the strike of major tectonic belt and the other is parallel to a LANDSAT lineament. The LANDSAT lineaments are of unknown origin, but they appear to be manifestations of long-lived tectonic features that have influenced the structural fabric of each area.

Sangster (1972, p. 11), referring specifically to the alignment of ores parallel to major faults in the Noranda district, has commented that ". . . the possibility exists that the faults are merely the more obvious manifestations of a pervasive structural grain which began very early in the geological history of the area and continued to affect the rocks long after they were deposited. The structural stress may have been active during actual deposition of the volcanics as subtle zones of weakness along which the ore-bearing solutions rose to precipitate sulphides at or near the existing ocean floor . . .". And " . . . further investigations of the distribution of massive sulphide deposits adjacent to volcanic centres may reveal that the ore clusters are not as random as they might appear at first sight . . ." Sangster's suppositions seem most apt in the light of our new knowledge of the Hokuroku district and the analogous structural relationship at Buchans.

ACKNOWLEDGEMENTS

The ideas expressed in this paper arose from detailed study of the Hokuroku district in 1977 and in 1978 under the sponsorship of the scientist exchange programme of the Japan Society for Promotion of Science and the Natural Sciences and Engineering Research Council of Canada, and from visits to Buchans in 1979. My

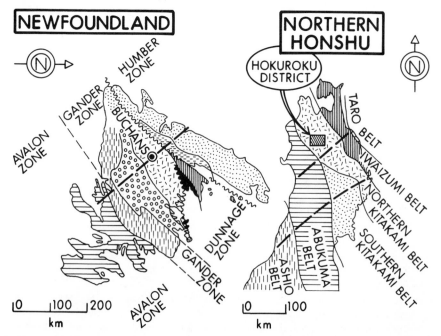

Figure 9. Simplified geological map of Newfoundland compared with the basement map of Northern Honshu. Note that the two maps are at different scales. Data for the Newfoundland map are from Williams (1978) and the LANDSAT lineament in Figure 6. The basement map of Northern Honshu is modified from Figure 1 of Ishihara and Terashima (1974). LANDSAT lineaments are superimposed from Figure 4.

second visit to Japan was as a member of the "U.S.-Japan-Canada Co-operative Research Project on the Genesis of Volcanogenic Massive Sulfide Deposits" coordinated by Professor Hiroshi Ohmoto of The Pennsylvania State University. I am indebted to many Japanese geologists who shared with me their knowledge of the Hokuroku district. They are properly acknowledged in Scott (1978). To those names I now add Professor Ei Horikoshi of Toyama University for his stimulating discussions on the genesis of Kuroko deposits and for the compilation of vein strikes used in Figure 3. J.G. Thurlow of the Price Co. Ltd. and E.A. Swanson of ASARCO spent a most exciting day with me discussing the application of my Hokuroku model to Buchans, and provided the information for Figure 5. My application of the model to Noranda was influenced by the writings of and discussions with D.F. Sangster of the Geological Survey of Canada. Any misrepresentations which may have arisen from my extension of his views are my responsibility. The manuscript has benefitted from incisive reviews by D.F. Strong of Memorial University, E.A. Swanson, and J.G. Thurlow. P. Ohlendorf, style editor for this volume, helped clarify the presentation. The editors of *Economic Geology* and *Mining Geology* are thanked for giving me permission to reproduce several figures from their journals.

My continuing study of massive sulphide deposits is supported by an NSERC Operating Grant A7069.

REFERENCES

Auzende, J.M., Olivet, J.L. and Bonnin, J., 1970, La marge du Grand Banc et la fracture de Terra-Neuve: C.R. Acad. Sci. Paris, v. 271, p. 1063-1066.

Bell, K. and Blenkinsop, J., in press, A geochronologic study of the Buchans area, Newfoundland: in Swanson, E.A., Strong, D.F. and Thurlow, J.G., eds., The Buchans Story: Geol. Assoc. Canada Spec. Paper.

Cathles, L.J., 1978, Hydrodynamic constraints on the formation of Kuroko deposits: Mining Geol., v. 28, p. 257-265.

Goodwin, A.M. and Ridler, R.H., 1970, The Abitibi orogenic belt: in Baer, A.J., ed., Symposium on Basins and Geosynclines of the Canadian Shield: Geol. Survey Canada Paper 70-40, p. 1-30.

Hashimoto, Koji, 1977, The Kuroko deposits of Japan – geology and exploration strategies: Asociacion de Ingenieros de Minas Metalurgistas y Geologos de Mexico, Memoria Tecnica XII, p. 25-88.

_____, 1979, The Kuroko deposits of the Hokuroku district – structural control and exploration implications: Unpublished Ph.D. Thesis, Waseda University, Japan.

Horikoshi, Ei, 1969, Volcanic activity related to the formation of the Kuroko-type deposits in the Kosaka district, Japan: Mineralium Deposita, v. 4, p. 321-345.

_____, 1975, Genesis of Kuroko-stage deposits from the tectonical point of view: Kazan, Series 2, v. 20, p. 341-353 (in Japanese).

Horikoshi, Ei and Shikazono, Naotatsu, 1978, Sub-types and their characteristics of Kuroko-type deposits: Mining Geol., v. 28, p. 267-276.

Hutchinson, R.W., 1973, Volcanogenic sulfide deposits and their metallogenic significance: Econ. Geol., v. 68, p. 1223-1246.

_____, 1980, Massive base metal sulphide deposits as guides to tectonic evolution: in Strangway, D.W., ed., The Continental Crust and Its Mineral Deposits: Geol. Assoc. Canada Spec. Paper 20, in press.

Ishihara, Shunso, ed., 1974, Geology of Kuroko Deposits: Soc. Mining Geologists of Japan, Mining Geol. Spec. Issue No. 6, 435 p.

Ishihara, Shunso and Terashima, Shigeru, 1974, Base metal contents of the basement rocks of Kuroko deposits: in Ishihara, Shunso, ed., Geology of Kuroko Deposits: Soc. Mining Geologists of Japan, Mining Geol. Spec. Issue No. 6, p. 421-428.

Jolly, W.T., 1978, Metamorphic history of the Archean Abitibi belt: in Fraser, J.A. and Heywood, W.W., eds., Metamorphism in the Canadian Shield: Geol. Survey Canada Paper 78-10, p. 63-78.

Kitamura, Nobu, 1976, Basement rocks from the Green Tuff region in Northeast Japan: in Minato, Masao, ed., Research Report on the Paleozoic Orogenic Movement in Northern Japan, no. 2, p. 9-20 (in Japanese).

Knuckey, M.J., Comba, C.D.A., Riverin, G., 1978, The Millenbach deposit, Noranda district, Quebec – an update on structure, metal zoning and wallrock alteration (abst.): Geol. Assoc. Canada and Mineral. Assoc. Canada, Abstracts with Programs, GAC-MAC- GSA Joint Annual Meeting v. 3, p. 436.

Knuckey, M.J. and Watkins, J.J., 1978, The Corbet mine and its environment of ore deposition, Noranda, Quebec (abst.): Geol. Assoc. Canada and Mineral Assoc. Canada, Abstracts with Programs, GAC-MAC-GSA 1978 Joint Annual Meeting, v. 3, p. 436.

Lambert, I.B. and Sato, Takeo, 1974, The Kuroko and associated ore deposits of Japan: a review of their features and metallogenesis: Econ. Geol., v. 69, p. 1215-1236.

Le Pichon, Xavier and Fox, P.J., 1971, Marginal offsets, fracture zones, and the early opening of the North Atlantic: J. Geophys. Res., v. 76, p. 6294-6308.

Ohmoto, Hiroshi, 1978, Submarine calderas: a key to the formation of volcanogenic massive sulfide deposits?: Mining Geol., v. 28, p. 219-231.

Oide, Keiji, 1968, Geotectonic conditions for the formation of the Krakatau-type calderas in Japan: Pacific Geol. v. 1, p. 119-135.

Remote Sensing Research Group, 1977, On the relationship between lineaments and ore deposits in the Tohoku district: Chishitsu News, no. 274, p. 1-19 (in Japanese).

de Rosen-Spence, A.F., 1976, Stratigraphy, Development and Petrogenesis of the Central Noranda Volcanic Pile, Noranda, Quebec: Unpublished Ph.D. Thesis, University of Toronto.

Sangster, D.F., 1972, Precambrian volcanogenic massive sulphide deposits in Canada: a review: Geol. Survey Canada Paper 72-22, 44 p.

Sangster, D.F. and Scott, S.D., 1976, Precambrian, strata-bound, massive Cu-Zn-Pb sulfide ores of North America: in Wolf, K.H., ed., Handbook of Strata-bound and Stratiform Ore Deposits: Amsterdam, Elsevier Scientific Publishing Company, v. 6, p. 129-222.

Sato, Takeo, 1974, Distribution and geological setting of the Kuroko deposits: in Ishihara, Shunso, ed., Geology of Kuroko Deposits: Soc. Mining Geologists of Japan, Mining Geol. Spec. Issue No. 6, p. 1-9.

Sato, Takeo, Tanimura, Shojiro and Ohtagaki, Tohru, 1974, Geology and ore deposits of the Hokuroku district, Akita Prefecture: in Ishihara, Shunso, ed., Geology of Kuroko Deposits: Soc. Mining Geologists of Japan, Mining Geol. Spec. Issue No. 6, p. 11-18.

Scott, S.D., 1978, Structural control of the Kuroko deposits of the Hokuroku district, Japan: Mining Geol., v. 28, p. 301-311.

Spence, C.D., 1975, Volcanogenic features of the Vauze sulfide deposit, Noranda, Quebec: Econ. Geol., v. 70, p. 102-114.

Spence, C.D. and de Rosen-Spence, A.F., 1975, The place of sulfide mineralization in the volcanic sequence at Noranda, Quebec: Econ. Geol., v. 70, p. 90-101.

Strong, D.F., 1977, Volcanic regimes of the Newfoundland Appalachians: in Baragar, W.R.A., Coleman, L.C. and Hall, J.M., eds., Volcanic Regimes in Canada: Geol. Assoc. Canada Spec. Paper 16, p. 61-90.

Swanson, E.A. and Brown, R.L., 1962, Geology of the Buchans Orebodies: Canadian Instit. Mining and Metallurgy Bull., v. 55, p. 618-626.

Swanson, E.A., Strong, D.F. and Thurlow, J.G., in press, The Buchans Story: The Geological Association of Canada, Special Paper.

Sykes, L.R., 1978, Intraplate seismicity, reactivation of preexisting zones of weakness, alkaline magmatism, and other tectonism postdating continental fragmentation: Rev. Geophys. Space Sci., v. 16, p. 621-688.

Takenouchi, Sukune, 1978, High salinity fluid inclusions in breccias of intrusive breccia dikes in the Hokuroku Kuroko area, Japan (abst.): Internat. Assoc. on the Genesis of Ore Deposits: Programs and Abstracts, 5th Symposium, p. 188.

Tanimura, Shojiro, 1973, Development of submerged sedimentary basin and environment of Kuroko mineralization in the Hokuroku district: Mining Geol., v. 23, p. 237-243 (in Japanese).

Tatsumi, Tatsuo, ed., 1970, Volcanism and Ore Genesis: Tokyo, University of Tokyo Press, 448 p.

Thurlow, J.G., Swanson, E.A. and Strong, D.F., 1975, Geology and lithogeochemistry of the Buchans polymetallic sulfide deposits, Newfoundland: Econ. Geol., v. 70, p. 130-144.

Ueno, Hirotomo, 1975, Duration of the Kuroko mineralization episode: Nature, v. 253, no. 5491, p. 428-429.

Urabe, Tetsuro, 1979, Rhyolitic and basaltic volcanisms associated with Kuroko (black ore) mineralization in Japan (abst.): Geol. Assoc. Canada and Mineral. Assoc. Canada, Program with Abstracts, v. 4, p. 83.

Urabe, Tetsuro and Sato, Takeo, 1978, Kuroko deposits of the Kosaka mine, Northeast Honshu, Japan – products of submarine hot springs on Miocene sea floor: Econ. Geol., v. 73, p. 161-179.

Uyeda, Seiya and Nishiwaki, Chikao, 1980, Stress field, metallogenesis and mode of subduction: in Strangway, D.W., ed., The Continental Crust and Its Mineral Deposits: Geol. Assoc. Canada Spec. Paper 20.

Williams, Harold, 1978, Tectonic Lithofacies Map of the Appalachian Orogen: Memorial University of Newfoundland.

The Continental Crust and Its Mineral Deposits, edited by D.W. Strangway,
Geological Association of Canada Special Paper 20

DISTRIBUTION AND ORIGIN OF PRECAMBRIAN MASSIVE SULPHIDE DEPOSITS OF NORTH AMERICA

D. F. Sangster
Geological Survey of Canada,
601 Booth Street, Ottawa, Ontario K1A 0E8

ABSTRACT

Conformable Precambrian massive sulphide deposits are widely distributed within the Canadian Shield as well as in isolated "windows" of Precambrian rocks in the remainder of North America. Within the Shield, massive sulphide deposits have been recorded in all structural provinces except Nutak and Bear.

The approximately 150 deposits can be grouped into four main periods coinciding with the formation of thick supracrustal accumulations: 1) 2750 to 2650 Ma, 2) 1900 to 1700 Ma, 3) 1400 to 1100 Ma, and 4) 800 to 600 Ma. Although the number of deposits are about evenly distributed between Archean and Proterozoic, a decisive majority of those in the Proterozoic were formed during the 1900 to 1700 Ma ore-forming period.

Alternatively, the deposits may be grouped in terms of the dominant lithology of locally associated host rocks: 1) a tholeiitic to calc-alkaline volcanic suite including a felsic component, 2) basalt-shale intercalations with little or no felsic volcanic rocks present, 3) clastic sediments of greywacke affinity with little or no direct volcanic component, and 4) shallow-water sediments such as lithic arenites, impure carbonates, and evaporites with minor associated volcanics. No ophiolite-associated deposits are known in the Precambrian of North America.

As a group, the deposits exhibit a range of Cu-Pb-Zn ratios, with the Cu-Zn association predominating. Many, but not necessarily most, of them display prominent alteration zones in the stratigraphic footwall. In the undeformed deposits these zones are markedly discordant to enclosing strata.

The deposits are considered to have originated as syn-volcanic or syn-sedimentary precipitates from hydrothermal solutions discharging directly into the submarine environment. Lead isotope evidence permits derivation of metals either directly from a magma or by leaching from the underlying rocks. Sulphur isotopes in the volcanic-associated deposits are close to meteoritic values, suggesting that the sulphur has been derived, perhaps indirectly, from an uncontaminated magmatic source; sulphur in the sediment-associated deposits may have been indirectly derived from seawater sulphate.

Although many deposits are enclosed in volcanic rocks having compositions comparable to island arc magmatic suites, conclusive evidence for plate margins, at least in terms of relatively modern analogues, is lacking in regions containing Precambrian massive suphide deposits. Rather, the dominant tectonic process leading to favourable environments for Precambrian massive sulphide formation appears to have been local rifting of a continental mass. Within these intracratonic basins massive sulphide deposits are enclosed in supracrustal rocks which rest on sialic basement.

RÉSUMÉ

Les dépôts de sulfures massifs précambriens concordants sont distribués un peu partout dans le Bouclier canadien de même que dans des "fenêtres" de roches précambriennes dans le reste de l'Amérique du Nord. Dans le bouclier, on a observé des dépôts de sulfures massifs dans toutes les provinces structurales à l'exception des provinces de Nutak et de l'Ours.

On peut regrouper les dépôts au nombre d'environ 150 en quatre périodes principales correspondant à la formation d'accumulations très épaisses sur la croûte: (1) de 2750 à 2650 Ma, (2) de 1900 à 1700 Ma, (3) de 1400 à 1100 Ma, et (4) de 800 à 600 Ma. Bien que le nombre de dépôts soit distribué de façon à peu près égale dans l'Archéen et le Protérozoïque, une claire majorité de ceux qui se sont formés au Protérozoïque datent de la période allant de 1900 à 1700 Ma.

D'une autre façon, on peut regrouper les dépôts en termes de lithologie dominante des roches encaissantes localement associées: (1) les suites tholéiitiques à volcaniques calco-alcalines comprenant une composante felsique, (2) des intercalations basalte-shale avec ou sans la présence de roches volcaniques felsiques, (3) des sédiments clastiques se rattachant aux grauwackes avec peu ou pas d'évidence directe de volcanisme, et (4) des sédiments d'eau peu profonde comme les arénites lithiques, les carbonates impurs et les évaporites associées à une certaine quantité de roches volcaniques. Il n'y a pas de dépôts associés aux ophiolites dans le Précambrien d'Amérique du Nord.

Comme groupe, les dépôts montrent des rapports variables de Cu-Pb-Zn quoique les associations Cu-Zn prédominent. Plusieurs, mais pas nécessairement tous, montrent des zones importantes d'altération dans le mur stratigraphique. Dans les dépôts non déformés, ces zones sont en discordance de façon marquée par rapport aux strates encaissantes.

On considère que ces dépôts ont leur origine sous forme de précipités syn-tectoniques ou syn-sédimentaires de solutions hydrothermales se déversant directement dans un milieu sous-marin. L'évidence tirée des isotopes de plomb permet d'expliquer l'origine des métaux soit directement à partir d'un magma soit par lixiviation des roches sous-jacentes. Les isotopes de soufre dans les dépôts associés aux roches volcaniques sont semblables aux valeurs établies pour les météorites, suggérant ainsi que le soufre provient, au moins indirectement, d'une source magmatique non contaminée; le soufre dans les dépôts associés aux sédiments a pu provenir indirectement des sulfates de l'eau de mer.

Bien que plusieurs dépôts soient compris dans des roches volcaniques ayant des compositions comparables aux suites magmatiques des arcs insulaires, il n'y a pas d'évidence concluante pour la présence de bordures de plaques, au moins en termes d'analoques relativement modernes, dans les régions contenant des dépôts de sulfures massifs précambriens. Au contraire, le processus tectonique dominant qui conduit à des milieux favorables pour la formation de dépôts de sulfures massifs au Précambrien semble avoir été l'effondrement local d'une masse continentale. A l'intérieur de ces bassins intracratoniques, les dépôts de sulfures massifs se retrouvent dans des roches superficielles qui reposent sur un socle sialique.

INTRODUCTION

Stratiform massive sulphide deposits are a distinct, economically important, and worldwide deposit type. They are characterized by their conformability with enclosing strata, internal layering, generally high iron sulphide content, and their composition. In addition to Cu-Pb-Zn, massive sulphide deposits may contain economically attractive amounts of silver and gold. They are found throughout the geological time scale and in association with a great variety of volcanic and sedimentary lithologies.

Most of the approximately 150 massive sulphide deposits known in the Precambrian of North America occur in Canada. As the geological features and distribution of the Canadian deposits have been reviewed by the author (Sangster, 1972) and those of

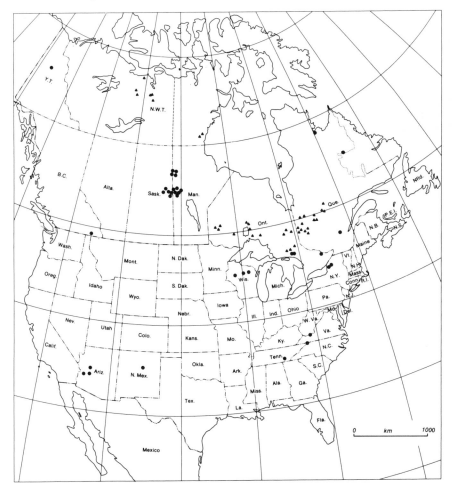

Figure 1. Distribution of Precambrian massive sulphide deposits in North America. Triangles signify Archean deposits; circles are Proterozoic.

North America by Sangster and Scott (1976), the present paper largely summarizes these earlier presentations and offers a few thoughts on the possible relation of plate tectonics to North American Precambrian massive sulphide deposits.

DISTRIBUTION

Distribution in Space

Precambrian massive sulphide deposits occur in all structural provinces of the Canadian Shield (Douglas, 1973) except Nutak and Bear, and also in isolated "windows" of Precambrian rocks elsewhere in the North American continent (Fig. 1).

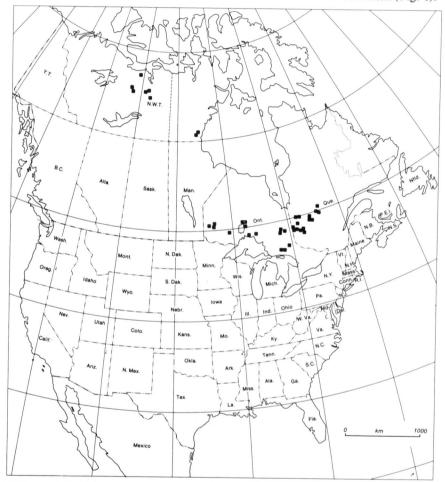

Figure 2. 2750 to 2650 Ma age group of massive sulphide deposits. Symbols in Figures 2, 3, 4, and 5 represent host rock lithologies as follows: squares = tholeiitic to calc-alkaline volcanic rocks and associated sediments; triangles = basalt-shale intercalations; circles = clastic sediments of greywacke affinity; diamonds = shallow water sediments (see text for details).

Major massive sulphide districts in the Shield include those of western Quebec and northern Ontario (Superior Province), northern Manitoba and Saskatchewan (southern Churchill Province), northwestern Northwest Territories (Slave Province), northern Wisconsin (Southern Province), and the Balmat-Edwards district of New York (Grenville Province). Elsewhere on the continent, the Ducktown-Gossan Lead-Ore Knob district (southern Appalachians) and the Prescott-Bagdad district (Arizona) contain significant massive sulphide deposits as well.

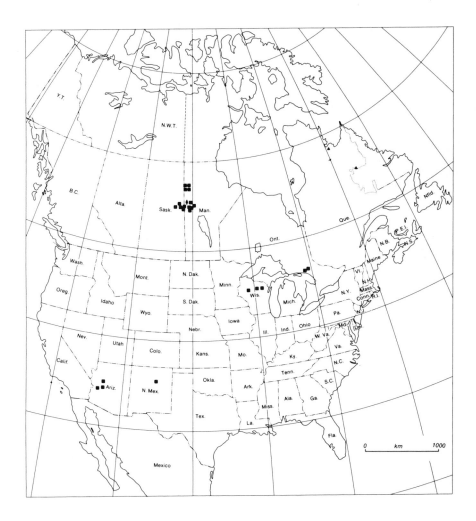

Figure 3. 1900 to 1700 Ma age group of massive sulphide deposits. Symbols as in Figure 2.

Distribution in Time

To the extent that the geochronolgy of Precambrian-terranes is known, the rocks enclosing the 150 or so North American deposits can be grouped into four main periods: 1) 2750 to 2650 Ma, 2) 1900 to 1700 Ma, 3) 1400 to 1100 Ma, and 4) 800 to 600 Ma (Fig. 2, 3, 4, and 5). Of these, the first two are by far the most important, for about 95% of the deposits and probably about 98% of the total ore tonnage are enclosed in rocks of these ages.

The four age groups should not be considered as marking unique metallogenic phenomena punctuating an otherwise orderly and continuous sequence of Precambrian rock deposition. Instead, they coincide with the deposition of four thick supra-

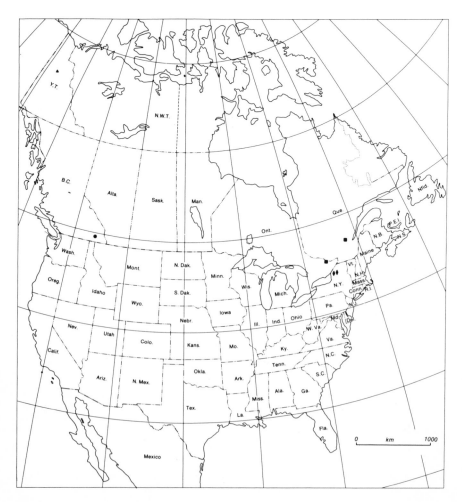

Figure 4. 1400 to 1100 Ma age group of massive sulphide deposits. Symbols as in Figure 2.

crustal accumulations which took place in the time-spans *between* major tectonic "events" affecting large portions of the North American continent, e.g., "early Archean", Kenoran, Hudsonian, Grenvillian and Taconic. The Hudsonian and Grenvillian events can probably be considered "orogenies" in the Phanerozoic sense of the word because they were accompanied by compressive-style tectonics and the accumulation of tholeiitic to calc-alkaline submarine volcanic rocks. The Kenoran and early Archean events may not be orogenies in the same sense because they appear to have been mainly plutonic in character. Thus, although the nature of the events bounding the periods of supracrustal accumulations is not entirely clear, it *is* apparent that these events have defined the time-spans in which such accumulations took place and hence have indirectly constrained the formation of stratiform massive sulphide deposits into four main periods.

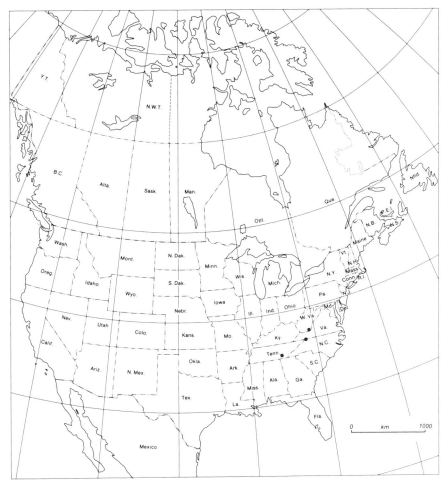

Figure 5. 800 to 600 Ma age group of massive sulphide deposits. Symbols as in Figure 2.

It is perhaps noteworthy that sulphide deposition did not take place uniformly *throughout* each of the four periods of supracrustal accumulation, but appears to have been most abundant in the younger half of each supracrustal "package".Thus, compare the four time spans of sedimentary and volcanic accumulations (3.2 to 2.6 Ga; 2.6 to 1.8 Ga; 1.8 to 1.0 Ga; and 1.0 to 0.45 Ga) with the four sulphide age groups noted previously (2.7 to 2.6 Ga; 1.9 to 1.7 Ga; 1.4 to 1.1 Ga; 0.8 to 0.6 Ga). In each case except the last, the massive sulphide age group occurs in the younger half of the supracrustal accumulation age group. The fourth supracrustal accumulation age group actually spans the Precambrian-Paleozoic boundary and the Precambrian massive sulphides are, by definition, in the older "half". If, however, the many dozens of North American Lower Paleozoic massive sulphides are taken into consideration, then the predominant sulphide age group would be approximately 0.55 to 0.47 Ga, i.e., in the younger half.

The tectonic and metallogenic implications of the age distribution outlined above are difficult to assess but it seems clear that, at the very least, formation of massive sulphides appears to precede or accompany (rather than follow) major tectonic episodes.

Distribution among Depositional Environments

The wide range in lithologies of host rocks enclosing massive sulphide deposits suggests a similarly wide range in depositional environments. In the Precambrian of North America, four main lithological associations are recognized (see Fig. 2, 3, 4 and 5):

1. *A tholeiitic to calc-alkaline volcanic suite including a felsic component.* This is the characteristic lithology enclosing the most common and widespread massive sulphide deposits in North America (refer to discussion in Sangster and Scott, 1976). Exact figures are difficult to obtain, but perhaps 80% or more of the deposits considered in this review occur in this lithological association. Certainly all the Archean ones do, as well as the many Proterozoic ones in northern Saskatchewan and Manitoba. Petrochemically the rocks resemble modern island arc suites but whether the Precambrian tectonic environment was analogous to island arcs is unclear. The abundance of pillows, cherts, and carbonates in these host rocks indicates that all deposits formed in a subaqueous (probably submarine) enviroment; depth of water was not a major controlling factor, however, because deposits have been found in rocks showing deep- as well as shallow-water characteristics.

2. *Basalt-shale intercalations with few, if any, felsic volcanic rocks present.* This lithological assemblage is comparable to that described by Kanehira and Tatsumi (1970) for certain Upper Paleozoic bedded cupriferous iron sulphide deposits in Japan. Precambrian examples in North America include certain deposits in northern Labrador Trough and the Hart River deposit in Yukon Territory. The association of mafic volcanics, quartzose (chert?) and pelitic schists suggest a relatively quiet volcano-sedimentary deep-water environment, possibly resulting from rifting.

3. *Clastic sediments of greywacke affinity with little or no direct volcanic component.* This lithological assemblage is characterized by conglomerates, greywackes (turbi-

dites), and pelites, with virtually no recognizable volcanic rocks present. Examples in North America are the Sullivan deposit in British Columbia and the Ducktown deposit in Tennessee. The depositional environment suggested by such lithologies is one of relatively deep water in a subsiding, and possibly fault-bounded, trough.

4. *Shallow-water sediments such as lithic arenites, impure carbonates, and evaporites with minor associated volcanics*. Examples include deposits in the Sherridon area of northern Manitoba (lithic arenites and associated carbonates) and the Balmat-Edwards, New York deposits (impure carbonates and evaporites).

Distribution of Metal Ratios

As a group, North American massive sulphide deposits display a range of Cu-Pb-Zn ratios, with the Cu-Zn association predominating. Of Archean volcanogenic massive sulphide deposits (Type 1 lithology) of the Canadian Shield (Fig. 6), the relatively anomalous lead-rich ones occur mainly in Slave Province, Northwest Territories and the Sturgeon Lake area (Superior Province), western Ontario. Proterozoic deposits in Type 1 lithologies are also dominantly Cu-Zn (Fig. 6), although relatively Pb-rich deposits occur in volcanics of the Grenville Central Metasedimentary Belt and in fine-grained sediments in the Sudbury Basin. Deposits in the three non-volcanic lithologic associations (Types 2, 3, and 4) are also of the Cu-Zn type (Fig. 7) with the notable exception of the Sullivan deposit in British Columbia, which is the most Pb-rich Precambrian massive sulphide deposit in North America. The zinc-rich Balmat-Edwards deposits also tend to be Zn(Pb) rather than Zn(Cu) deposits. It is perhaps instructive to note that both these Cu-poor deposits are contained in sedimentary sequences entirely lacking in volcanic rocks.

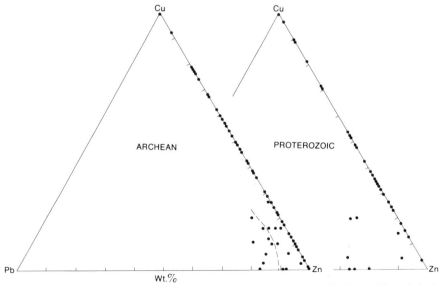

Figure 6. Metal ratios in Archean and Proterozoic deposits (Type 1 lithology). The relatively lead-rich Archean deposits to the left of the dotted line occur in Slave Province and the Sturgeon Lake area of Superior Province.

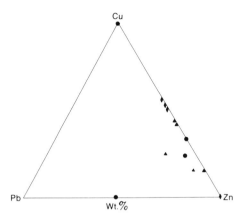

Figure 7. Metal ratios in Proterozoic deposits (type 2, 3, and 4 lithologies). Symbols as in Figure 2.

ORIGIN

Mode of Formation

Precambrian massive sulphide deposits in volcanic terranes, particularly those in North America, have been compared (e.g., Sangster and Scott, 1976) to the Kuroko deposits in the Miocene of Japan, for which a submarine exhalative origin has been proposed (see e.g., Ishihara, 1974). The concept of hot, metal-bearing solutions issuing into submarine basins through a network of fissures is now accepted by a majority of economic geologists. The most compelling geological evidence for such a mode of formation is most commonly present in the volcanic-associated variety of massive sulphide deposits (i.e., Type 1 lithology), namely the discordant stringer sulphide zone and its accompanying wall-rock alteration, which, if present, is always *underneath* the layered, concordant, massive sulphide zone. Massive sulphide deposits precipitated on and around the fumarolic vent are regarded as *proximal* deposits, that is to say they were deposited in close proximity to the exhalative orifice. Proximal deposits are the most common form of massive sulphide deposit in Type 1 lithology. Massive sulphide deposits lacking such obvious expressions of an exhalative origin as a discordant footwall stringer zone and wallrock alteration, are termed *distal* deposits, the implication being that they have been deposited at some unknown distance from their presumed exhalative vent. Distal massive sulphide deposits are much more common in the non-volcanic lithologies (Types 2, 3 and 4), although many examples are known in the volcanic environment as well. It is merely that the *proportion* of proximal deposits in the volcanic environment is higher that that in the non-volcanic environment. The best North American example of a proximal deposit in a *non*-volcanic environment would be the Sullivan mine, which is a relatively undeformed, large, well-layered massive sulphide deposit underlain by minor discordant sulphide veins and wall-rock alteration.

Whether a deposit is *proximal* or *distal* is a function, among other things, of the density of the ore-forming exhalations relative to seawater. This concept was proposed by Sato (1972) and its application to North American massive sulphide deposits has been discussed by Sangster and Scott (1976, p. 203-205). Briefly summarized, Sato considers four theoretical paths of temperature and density that would be followed by brine mixing with seawater (at 20°C and 0.5 molal NaC1, assuming hydrologic equilibrium is established). Type I brine, because of its high density, flows down-slope from the vent, collects in a sea-floor depression, and retains its density layering for a considerable time while precipitating metals. In Type IIa, the ore-forming brine is originally more dense than seawater. Upon mixing, the density rises to a maximum before decreasing and sulphides are rapidly precipitated close to, but downhill from, the exhalative vent. Type IIb brines are originally less dense than seawater but upon mixing become more dense, rising to a maximum before declining as in Type IIa. The resulting deposit would be similar to that produced in IIa except that sulphides could be deposited anywhere around the vent. Since Type III brines are less dense than seawater at all times, mixing results in a steady increase in density. Deposition of sulphides would occur as a thin layer over a large area remote from the vent.

Distal deposits would be produced by Sato's Type I brines, and proximal deposits by Types IIa and b. Type III brine rarely produces economic concentrations of base metals. Although most deposits in sedimentary or mixed volcanic-sedimentary environments are distal, deposits such as the Sullivan demonstrate that ore-forming exhalations *can* take place in a purely sedimentary environment. Consequently, massive sulphide deposits in *either* volcanic or non-volcanic environments may result from Type I brines (giving rise to distal deposits) or Type II brines (resulting in proximal deposits).

The hydrothermal system, of which the metalliferous brine is a part, is likely to have used seawater as the heat and metal transfer medium. Most current models propose that seawater is drawn down through a large volume of rock, heated, and then expelled through a restricted discharge zone such as a fracture, the entire system being driven by thermal gradients comparable to those present in modern subaerial hydrothermal systems (Spooner and Fyfe, 1973; Spooner *et al.*, 1977; Hodgson and Lydon, 1977). These principles have been used to model the hydrothermal systems which produced Precambrian massive sulphide deposits in the Canadian Shield (MacGeehan, 1978; Roberts and Reardon, 1978).

Origin of Metals

Lead isotope data probably provide the most direct method of tracing the source of base metals in a massive sulphide deposit. However, detailed studies of individual deposits are uncommon; published data span several generations of mass spectrometers and come from many different laboratories. The most abundant lead isotope data are from volcanogenic deposits of tholeiitic to calc-alkaline volcanic affinity (Sangster, 1976, 1978); fewer modern data are available for deposits in the sedimentary and mixed volcano-sedimentary host lithologies. Thus, comparisons and generalizations among deposits are difficult to make.

A summary of available Pb-isotope data is displayed in Figure 8, from which some general trends are discernible. Note, for example, how "tight" the Archean grouping is, indicating a homogeneity of compositions in spite of great differences in geographic

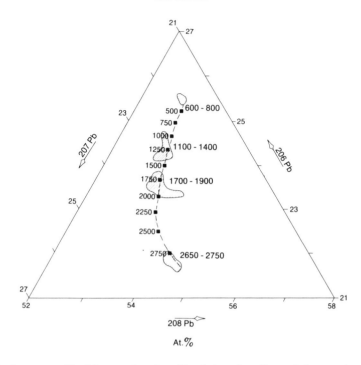

Figure 8. Summary of lead isotope data from North American Precambrian massive sulphide deposits. Age groupings (Ma) on right side of curve are those of host rocks to massive sulphide deposits. Growth curve and ages (Ma; left side) are from Stacey and Kramers (1975).

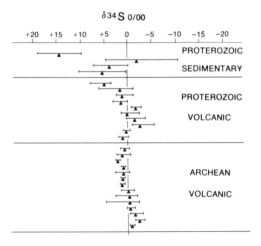

Figure 9. Summary of sulphur isotope data from North American Precambrian massive sulphide deposits. Deposits represented by sulphur isotope data are not necessarily the same as those shown in the lead isotope diagram (Fig. 8).

location, metamorphic grade, size of deposit, lead content, etc. On a 207/204 versus 206/204 diagram, these points also plot close to the normal growth curve and, as a rule, give model ages in relatively good agreement with radiometric ages of host rocks. Such lead is usually regarded as being of primitive or primordial composition and as having been derived from the mantle or, perhaps more likely, from volcanic rocks which themselves have been derived more or less uncontaminated from the primitive lithosphere.

The Proterozoic deposits enclosed in predominantly volcanic rocks also contain leads that are relatively isotopically homogeneous; these deposits plot close to the average growth curve and presumably have been derived in much the same manner as the Archean ones. Proterozoic deposits enclosed in sedimentary or mixed volcanic-sedimentary sequences, however, are heterogeneous by comparison, contain a higher proportion of radiogenic lead, and give younger model ages than the enclosing strata. These lead-isotope features could be explained by a model in which lead is leached from an assemblage containing a high proportion of sedimentary rocks.

Origin of Sulphur

The available sulphur isotope composition data for Precambrian massive sulphide deposits are summarized in Figure 9, where the narrow range in values for the Archean volcanic group of deposits is apparent. Moreover, the averages are close to zero per mil, suggesting a primitive or primordial origin for the sulphur. Whether the sulphur was leached from accessory pyrite in footwall rocks or whether it was initially present in hydrothermal ore fluids as a dissolved sulphur species cannot be resolved with the data available. Sulphur in the primitive Earth is likely to have had an isotopic composition close to zero per mil. A short "residence time" as accessory pyrite in Archean volcanic rocks would not be recorded in the sulphur isotopic composition; sulphur released into the ore-forming solutions by leaching of pyrite would have isotopic compositions indistinguishable from sulphur contributed directly from mantle sources.

In contrast to the Archean volcanic massive sulphide deposits, Proterozoic volcanic deposits have a somewhat wider range in δ^{34} S values. This trend is even more evident in Proterozoic sedimentary deposits (Fig. 9). The wider range in isotopic composition and the tendency to depart from zero per mil average composition could reflect heterogeneous sources of sulphur such as the mixing of sulphur from seawater sulphate with primitive sulphur in varying proportions. Sulphur derived from seawater is most likely to be a major factor in the sediment-hosted deposits, but may also be important in the mixed volcanic-sedimentary type. Sulphur may be contributed to the ore solutions either *directly*, by downward precolating seawater or by seawater entrapped in the sediments and/or volcanics, or *indirectly*, through leaching of previously precipitated sulphate or sulphide minerals both of which would contain sulphur from seawater.

Plate Tectonics and the Origin of North America Precambrian Massive Sulphide Deposits

Although it has been proposed that the origin and distribution of Phanerozoic massive sulphide deposits are related to both divergent and convergent movements of lithospheric plates (e.g., Mitchell and Garson, 1976), the relationship of Precambrian deposits to plate tectonic processes is very much open to question. Obviously a complete evaluation of plate tectonics in the Precambrian of North America is beyond the scope of this paper. There are, however, a few features which any such evaluation must take into account.

1) The only area in North America where a complete Precambrian Wilson Cycle *has* been proposed, i.e., the Coronation System (Hoffman, 1978), is in the Bear Province, one of two structural provinces in the Canadian Shield where massive sulphides have *not* been found.

2) Perhaps the most favourable evidence of widespread Precambrian plate tectonic activity in the regions where massive sulphide deposits do occur is the typical island-arc composition of the tholeiitic to calc-alkaline volcanic rocks enclosing a majority of these deposits. Goodwin (1977, p. 239), for example, concluded that "Archean volcanic rocks of Superior Province compare closely with those of modern island arcs in terms of class abundance, series and class successions, chemical compositions of class and series, and mean composition of entire volcanic belts. Also, the linear geometry of Archean volcanic superbelts is roughly comparable to some modern island arcs". Moore (1977) and Stauffer (1974) reached a similar conclusion for Proterozoic volcanic rocks of Northern Manitoba and Saskatchewan. Thus, although some authors (e.g., Condie, 1976) have pointed out certain differences in the chemistry between modern and Archean volcanic rocks, Precambrian greenstone belts appear to be broadly similar to modern island arcs. Volcanic rocks of tholeiitic to calc-alkaline composition, together with volcanogenic massive sulphide deposits, are normally attributed to subduction of oceanic crust at convergent plate margins (see, for example, Mitchell and Garson, 1976, pp. 142-156).

3) If plate convergence and subduction were as common in the Precambrian as the widespread distribution of island-arc-like volcanics suggests, then it seems reasonable that evidence for plate *divergence* in the Precambrian, such as ocean-floor volcanics, should be available. While some, or even most, oceanic crust would be destroyed by subduction, that which is obducted should be preserved just as it has been in the Phanerozoic. The fact is, however, that evidence of Precambrian ocean-floor volcanics and/or sediments is amazingly scarce. No ocean crust ophiolites, for example, have yet been identified in the Precambrian of North America. In fact, the evidence appears to favour a sialic basement to Precambrian greenstone belts (e.g., Windley and Bridgewater, 1971), including those containing massive sulphide deposits. This concept was proposed by Baragar and McGlynn (1976) to apply, in particular, to Canadian Archean greenstone belts. It was challenged by Glikson (1978) and redefended by Baragar and McGlynn, who concluded that "for the Canadian Shield at least there is good reason to believe that the greenstone belts were emplaced upon a sialic foundation" (Baragar and McGlynn, 1978, p. 15).

The arguments of Glikson (1978) and Baragar and McGlynn (1978) were directed mainly to Archean greenstone belts, but could apply equally well to certain Proterozoic belts. For example, the circum-Kisseynew volcanic belt of Manitoba and Saskatchewan (Sangster, 1978), containing at least 30 massive sulphide deposits, not only does not contain any identified ocean-floor volcanics, but contains dated Archean inliers at its southwestern end (see Bailes and McRitchie, 1978, Fig. 1). Similarly, in the Southern Province of the Shield in northern Wisconsin, a Proterozoic greenstone belt containing several volcanogenic massive sulphide deposits contains no recorded oceanic crust but is itself surrounded by Archean basement (Sims, 1976). Sims states, "it seems certain from the distribution of exposed lower Precambrian (Precambrian W) rocks . . . that these rocks underlie the Precambrain X sequences" (Sims, 1976, p. 1106). Massive sulphide deposits also occur in the Grenville Central Metasedimentary Belt, a subdivision of the Grenvillian Orogenic Belt which "was built upon a pre-existing continental plate. There is so far no evidence of an ophiolite or ensimatic belt in the Grenville Province . . ." (Wynne-Edwards, 1972, p. 325-326). Similarly, the Labrador trough, in which massive sulphides are enclosed in supracrustal rocks demonstrably resting on Archean basement, has been viewed as an intracratonic basin (Dimroth, 1972). The trough is somewhat unique in that it contains basaltic rocks which have been interpreted as ophiolites. Dimroth, however, citing the presence of shallow-water marine lithologies *beneath* the ophiolites, and the absence of sheeted gabbro complexes, has concluded that "it is evident that the ophiolites of the Labrador trough do not represent remnants of an oceanic plate. Features that would prove the presence of a consuming plate margin are also absent; the Labrador trough is not a Precambrian suture" (Dimroth, 1972, p. 496).

In summary, the lack of oceanic crustal material relative to the abundance of sialic basement suggests that volcanic rocks in both Archean structural provinces of the Shield (Superior and Slave), as well as in the Proterozoic circum-Kisseynew volcanic belt, the Southern Province in Wisconsin, the Grenville Central Metasedimentary Belt, and the Labrador Trough developed on a pre-existing craton.

In contrast, massive sulphide deposits in the Prescott-Bagdad district of Arizona are contained in typical tholeiitic to calc-alkaline volcanic rocks which, in turn, rest on dominantly pyroxenitic and peridotitic basement rocks (Anderson and Guilbert, 1978). No Archean rocks are present and no granitic basement is known beneath the volcanic belts. The authors conclude that the Prescott-Bagdad volcanic belts "originated as Precambrian island arcs built upon some form of oceanic crust" (Anderson and Guilbert, 1978). This interpretation, if substantiated by further studies, would appear to make the Arizona massive sulphide district unique in North America in that it is of Precambrian age yet did not develop on a sialic crust.

Since the areas mentioned above contain over 90% of the Precambrian massive sulphide deposits in North America, it seems clear that a majority of the ore-containing greenstone belts probably developed on a sialic basement (the Prescott-Bagdad area being a possible exception). This is difficult to reconcile with modern plate tectonic concepts. The author concludes, as have several other workers (e.g., Groves *et al.*, 1978), that ensialic or intracratonic rifting is a plausible alternative to subduction-related processes for the development of Precambrian greenstone belts. Groves *et al.* (1978, p. 461) concluded, after reviewing the evidence for rift zones or marginal basins as two possible models to explain the evolution of greenstone belts, that "all available evidence points to the development of the Yilgarn greenstone belts (of western Australia) on sialic crust, probably in rift zones . . ." and that "currently proposed plate tectonic models are not adequate to explain all aspects of Archean evolution" including, one might add, the origin and distribution of massive sulphide deposits.

The rifting process leading to environments favourable to the formation of a large majority of Precambrian massive sulphide deposits results in predominantly vertical, rather than lateral, movement of crustal material. The term "plate tectonics", with the attendant emphasis on lateral motion, has little or no meaning throughout most of the Precambrian. Terms such as "rift" or "block" tectonics may be preferable when referring to Precambrian tectonics, in order to emphasize the characteristic vertical motion involved.

All these lines of evidence (abundance of sialic basement, lack of oceanic crust, vertical tectonics) combine to present a general environmental picture of intracratonic basins in which massive sulphide deposits are enclosed in supracrustal rocks resting on sialic basement. From this concept, two important conclusions regarding Precambrian sulphide deposits of the volcanic variety (by far the most common type in North America) can be inferred: 1) Massive sulphide deposits and their enclosing volcanic host rocks, in spite of their similarity to relatively modern island-arc analogues, probably were *not* developed solely as a result of lateral plate movement and subduction in an oceanic or continental margin regime. Intracratonic rifting is capable of producing island-arc-like rocks and ores. 2) Consequently, modern plate tectonics will *not* aid the understanding of, and exploration for, Precambrian volcanogenic massive sulphide deposits. Because massive sulphide deposits appear to have formed in different tectonic environments at different times, a detailed understanding of intracratonic rifting processes might be more useful in locating deposits in Precambrian rocks.

ACKNOWLEDGEMENTS

Constructive advice by J.M. Duke, Geological Survey of Canada, improved an early draft of this manuscript and is herein respectfully acknowledged.

REFERENCES

Anderson, P. and Guilbert, J.M., 1978, The Precambrian massive sulphide deposits of Arizona: A distinct metallogenic epoch and province: in International Association on the Genesis of Ore Deposits (IAGOD), 5th Symposium, Snowbird, Utah, Programs and Abstracts, p. 39.

Bailes, A.H. and McRitchie, W.D., 1978, The transition from low to high grade metamorphism in the Kisseynew sedimentary gneiss belt, Manitoba: Fraser, J.A. and Heywood, W.W., eds., Metamorphism in the Canadian Shield: Geol. Survey Canada, Paper 78-10, p. 155-178.

Baragar, W.R.A. and McGlynn, J.C., 1976, Early Archean basement in the Canadian Shield: A review of the evidence: Geol. Survey Canada, Paper 76-14, 21p.

_____, 1978, On the basement of Canadian greenstone belts: Discussion: Geosci. Canada, v. 5, p. 13-15.

Condie, K.C., 1976, Trace-element geochemistry of Archean greenstone belts: Earth Science Reviews, v. 12, p. 393-417.

Dimroth, E., 1972, The Labrador geosyncline revisited; American Journal of Science, v. 272, p. 487-506.

Douglas, R.J.W., 1973, Geological provinces of Canada: Department of Energy, Mines and Resources, National Atlas of Canada.

Gilkson, A.Y., 1978, On the basement of Canadian greenstone belts: Geosci. Canada, vol. 5, p. 3-12.

Goodwin, A.M., 1977, Archean volcanism in Superior Province, Canadian Shield: Baragar, W.R.A., Coleman, L.C. and Hall, J.M., eds., in Volcanic Regimes in Canada: Geol. Assoc. Canada, Spec. Paper 16, p. 205-241.

Groves, D.I., Archibald, N.J., Bettenay, L.F., and Binns, R.A., 1978, Greenstone belts as ancient marginal basins or ensialic rift zones: Nature, v. 273, no. 5662, p. 460-461.

Hodgson, C.J. and Lydon, J.W., 1977, Geological setting of volcanogenic massive sulphide deposits and active hydrothermal systems: Some implications for exploration: Canadian Instit. Mining and Metallurgy, Bull., v. 70 (786), p. 95-106.

Hoffman, P.F., 1978, Geology of the Coronation Geosyncline (Aphebian), Hepburn Lake Sheet (86J), Bear Province, District of Mackenzie: Geol. Survey Canada, Current Research, Part-A, Paper 78-1A, p. 147-151.

Ishihara, S., ed., 1974, Geology of Kuroko deposits: Soc. Mining Geologists of Japan, Spec. Issue 6, 435p.

Kanehira, K. and Tatsumi, T., 1970, Bedded cupriferous iron sulphide deposits in Japan, a review: in Tatsumi, T., ed., Volcanism and Ore Genesis: Univ. of Tokyo Press, Tokyo, p. 51-76.

MacGeehan, P.J., 1978, The geochemistry of altered volcanic rocks at Matagami, Quebec: a geothermal model for massive sulphide genesis; Canadian Jour. Earth Sciences, v. 15, p. 551-570.

Mitchell, A.H.G. and Garson, M.J., 1976, Mineralization at plate boundaries: Minerals Science Engineering, v. 8, no. 2, p. 129-169.

Moore, J.M., 1977, Orogenic volcanism in the Proterozoic of Canada: in Baragar, W.R.A., Coleman, L.C. and Hall, J.M., eds., Geol. Assoc. Canada Spec. Paper 16, p. 127-148.

Roberts, R.G. and Reardon, E.J., 1978, Alteration and ore-forming processes at Mattagami Lake Mine, Quebec: Canadian Jour. Earth Sciences, v. 15, p. 1-21.

Sangster, D.F., 1972, Precambrian volcanogenic massive sulphide deposits in Canada: A review; Geol. Survey Canada, Paper 72-22, 44p.

_____, 1976. Sulphur and lead isotopes in stratabound deposits: in Wolfe, K.H., ed., Handbook of Stratabound and Stratiform Deposits, v. 2: Elsevier Scientific Publishing Company, Amsterdam, p. 219-266.

_____, 1978, Isotopic studies of ore-leads of the circum-Kisseynew volcanic belt of Manitoba and Saskatchewan: Canadian Jour. Earth Sciences, v. 15, p. 1112-1121.

Sangster, D.F. and Scott, S.D., 1976, Precambrian stratabound massive Cu-Zn-Pb sulfide ores of North America: in Wolfe, K.H., ed., Handbook of Stratabound and Stratiform Deposits, v. 6: Elsevier Scientific Publishing Company, Amsterdam, p. 129-222.

Sato, T., 1972, Behaviours of ore-forming solutions in sea water: Mining Geol. v. 22, p. 31-42.

Sims, P.K., 1976, Precambrian tectonics and mineral deposits, Lake Superior region: Econ. Geol. v. 71, p. 1092-1127.

Spooner, E.T.C., Chapman, H.J., and Smewing, J.D., 1977, Strontium isotopic contamination and oxidation during ocean floor hydrothermal metamorphism of the ophiolitic rocks of the Troodos Massif, Cyprus: Geochim. Cosmochim. Acta, v. 41, p. 873-890.

Spooner, E.T.C. and Fyfe, W.J., 1973, Sub-sea-floor metamorphism, heat and mass transfer: Contrib. Mineralogy and Petrology, v. 42, p. 287-304.

Stacey, J.S. and Kramers, J.D., 1975, Approximations of terrestrial lead isotope evolution by a two-stage model: Earth and Planet. Sci. Letters, v. 26, p. 207-221.

Stauffer, M.R., 1974, Geology of the Flin Flon area: A new look at the Sunless City; Geosci. Canada, v. 1, no. 3, p. 30-35.

Windley, B.F. and Bridgewater, D., 1971, The evolution of Archean low- and high-grade terrains: in Glover, J.E., ed., Symposium on Archean Rocks: Geol. Soc. Australia, Spec. Publ. 3, p. 33-46.

Wynne-Edwards, H.R., 1972, The Grenville Province: in Price, R.A. and Douglas, R.J.W., ed., Variations in Tectonic Styles in Canada: Geol. Assoc. Canada, Spec. Paper 11, p. 263-334.

The Continental Crust and Its Mineral Deposits, edited by D.W. Strangway,
Geological Association of Canada Special Paper 20

GRANITOID ROCKS AND ASSOCIATED MINERAL
DEPOSITS OF EASTERN CANADA AND
WESTERN EUROPE

D.F. Strong
Department of Geology,
Memorial University of Newfoundland,
St. John's, Newfoundland A1B 3X5

ABSTRACT

Paleozoic granitoid plutons in the Newfoundland sector of the Appalachian-Caledonian orogen comprise four main suites which appear to persist throughout the orogen. These are: 1) composite mafic-silicic hornblende-bearing suites; 2) microcline-megacrystic biotite granites; 3) biotite-muscovite ("two mica") leucogranites; and 4) alkaline-peralkaline granites. These suites are characteristic of three major tectonic zones, viz. 1 and 4 are found in the Dunnage zone, a region of Lower Paleozoic island arc sequences overlying oceanic crust, and in the Avalon zone, which is dominated by Proterozoic basalts and ignimbrites; suites 2 and 3 are found primarily within the high temperature-low pressure type Paleozoic metamorphic rocks of the Gander zone. Available geochemical and istopic data suggest a crustal origin for the granitoid rocks of these suites, which have characteristic chemical features for each tectonic zone. They also show distinct evolutionary patterns of initial $^{87}Sr/^{86}Sr$ ratios increasing in time, although there is no systematic correspondence between these initial ratios and the tectonic zones in which the plutons occur.

Granitoid suites comparable to those of Newfoundland have been well-documented in other regions as containing characteristic types of mineralization. Suite 1 is typically host to porphyry copper and molybdenum deposits, e.g., in the American Cordillera; suite 2 is generally barren; suite 3 contains the uranium, tin, tungsten, beryllium and associated deposits of Western Europe; and suite 4 is associated with Sn deposits in Nigeria, fluorspar in Newfoundland and uranium in Namibia. However, the formation of granitoid mineral deposits requires the concurrence of many processes ranging from tectonic to geochemical; without the full range of mineralization controls, such deposits would not have formed in the Appalachian-Caledonian orogen. Thus, the absence of subduction processes in the generation of suite 1 might account for the lack of any significant porphyry deposits in them. On the other hand, all indications are favourable for deposits associated with suites 3 and 4.

RÉSUMÉ

Les plutons granitoïdes du Paléozoïque dans le secteur terre-neuvien de l'orogène appalachien-calédonien comprend quatre suites principales qui semblent persister dans tout l'orogène: (1) des suites composites mafiques-felsiques riches en hornblende; (2) des granites à biotite et microcline mégacristalline; (3) les leucogranites à deux micas, biotite et muscovite; et (4) les granites alcalins-peralcalins. Ces suites sont caratéristiques de trois zones tectoniques majeures: les suites 1 et 4 se retrouvent dans la zone de Dunnage, une région de séquences d'arcs insulaires du Paléozoïque inférieur surmontant une croûte océanique et dans la zone d'Avalon qui est dominée par les basaltes et ignimbrites du Protérozoïque; les suites 2 et 3 se retrouvent principalement dans les roches métamorphiques du type haute température-basse pression du Paléozoïque de la zone de Gander. Les données géochimiques et isotopiques disponibles suggèrent une origine dans la croûte pour les roches granitoïdes de ces suites qui ont des caractéristiques chimiques bien définies pour chaque zone tectonique. Elles montrent aussi des patrons d'évolution distincts pour les rapports initiaux $^{87}Sr/^{86}Sr$ qui augmentent dans le temps, bien qu'il n'y ait pas de correspondance systématique entre ces rapports initiaux et les zones tectoniques dans lesquelles les plutons se rencontrent.

Des études bien documentées sur des suites granitoïdes comparables à celles de Terre-Neuve dans d'autres régions indiquent qu'elles contiennent des types caractéristiques de minéralisation. La suite 1 renferme typiquement des dépôts de cuivre et de molybdène porphyriques comme dans la Cordillère américaine; la suite 2 est généralement stérile; la suite 3 contient de l'uranium, de l'étain, du tungstène, du béryllium et les dépôts associés en Europe de l'Ouest; et la suite 4 est associée aux dépôts de Sn au Nigéria, de cryolite à Terre-Neuve et d'uranium en Namibie. Toutefois, la formation de dépôts minéraux granitoïdes exige la concurrence de plusieurs processus allant de la tectonique à la géochimie; sans l'éventail complet des contrôles de minéralisation, de tels dépôts n'auraient pu se former dans l'orogène appalachien-calédonien. Ainsi, l'absence de processus de subduction dans la génération de la suite 1 pourrait expliquer l'absence dans cette suite de dépôts porphyriques significatifs. D'autre part, toutes les indications sont favorables à la mise en place associés aux suites 3 et 4.

INTRODUCTION

All of Western Europe's tin, tungsten, beryllium and lithium, most of its uranium, and many associated metals are derived from Paleozoic granitoid rocks. Significant occurrences of a similar nature are now being investigated in eastern North America. An understanding of these deposits requires knowledge of a complex interplay of factors, ranging from the tectonic setting and evolution of the host rocks to the plethora of physical and chemical variables in the local environment of metal transport and deposition. The purpose of this paper is to examine some of these variables in the context of the granitoid rocks in Newfoundland.

TECTONIC SETTING

A new era in tectonic, and therefore metallogenic, studies of the Appalachian-Caledonian orogen began when J.T. Wilson (1966) suggested that its evolution related to the opening and closing of a "Proto-Atlantic Ocean" (Iapetus). Since then the orogen has become a focus for the application of plate tectonic concepts to ancient orogenic processes, with spectacular success in explaining the formation and evolution of the ancient continental margins, the ophiolites, the island arc volcanic-sedimentary sequences, and their associated mineral deposits. However, plate tectonic concepts have not yet succeeded in explaining the origin and metallogeny of the granitoid plutons of the orogen. For example, the hypotheses relating granitoid plutonism to subduction zones around the eastern Pacific margins are inapplicable, as it is now known that most Appalachian-Caledonian plutons were intruded well after Iapetus had closed and subduction ceased (Bell *et al.*, 1977). It appears more probable that granitoid plutonism following the closure of Iapetus was related to crustal melting caused by thickening due to continued plate compression, and to both shearing and extensional processes resulting from relative plate rotation, which produced a "megashear" environment throughout the mid-late Paleozoic.

Although Iapetus was closed by Caradocian times, compression continued well into the late Paleozoic, with thrust faulting and folding of post-closure redbeds resulting in significant crustal shortening and thickening. Folding and thrusting of the Upper Devonian and Carboniferous sedimentary rocks in some areas coincided with apparently extensional graben-type tectonics in other areas. The inconsistency in nature and orientation of these post-closure events probably reflects changing patterns of plate behaviour from collision and crustal shortening to rotation and oblique interaction. Such an interpretation is supported by paleomagnetic and structural data from different parts of the North Atlantic region.

Morel and Irving (1978) summarized available paleomagnetic data for the Phanerozoic and synthesized it into a number of possible interpretations of plate movements which offer clues to post-Iapetus tectonic processes. Several of their reconstructions are shown in Figure 1, spanning the 450-280 Ma period which encompasses most granitoid plutonism of the North Atlantic region. The main feature of these diagrams is the suggestion of relative rotation (i.e., with strike-slip displacement between the North American-European [Laurasia] and African [Gondwana] plates following Ordovician collision and closure of Iapetus). Such a process has likewise been indicated by geological evidence, at least for late Variscan structures (300-280Ma), which Arthaud and Matte (1977) explain as resulting from movement along many strike-slip faults. These, together with other structures, make up what can be called the "Hercynian Megashear" (Fig. 2). In such a large-scale dominantly shearing environment there would be local variations in tectonic processes, ranging from thrusting and folding to strike-slip faulting, to extension and normal faulting, as clearly illustrated by a modern analogue, the San Andreas fault system, which is the site of relative dextral rotation between the Pacific and North American plates (e.g., Koide and Bhattacharji, 1978). It is expected, therefore, that magmas generated in this post-Iapetus megashear environment would show local variations typical of all types from extensional to compressional tectonism.

PETROLOGY OF NEWFOUNDLAND GRANITOID ROCKS

The granitoid rocks of Newfoundland can be readily grouped into four dominant suites as follows (Fig. 3, Table 1):

1. *Composite mafic-silicic hornblende-bearing suites* can be described as calc-alkaline, and are generally intruded at shallow depths within the crust. These rocks form two main age-groups, viz. latest Precambrian to Cambrian ("Pan-African")

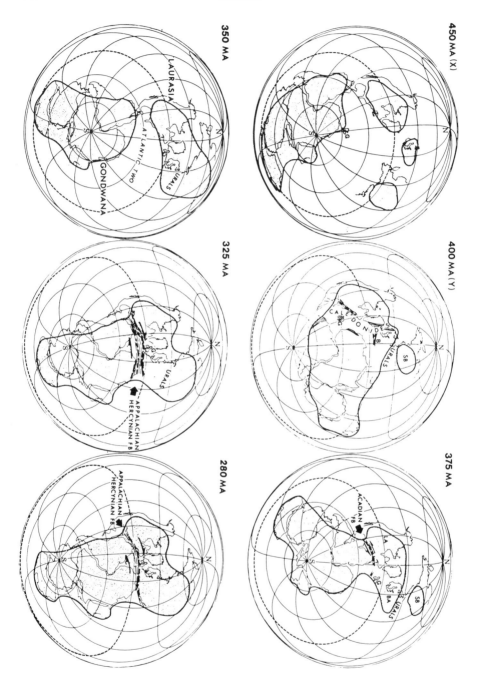

plutons intruding Precambrian rocks of the Avalon Zone of eastern Newfoundland, and post-Ordovician plutons intruding Paleozoic rocks of the Notre Dame, Exploits, and Botwood Zones of central Newfoundland.

2. *Biotite-bearing granites-granodiorites* are the dominant rock type, and range from aphanitic to the more typical variety which is rich in microcline megacrysts. They are characteristic of the amphibolite-facies metamorphic rocks of the Gander and Fleur de Lys Zones, but also intrude the lower-grade environments of the Avalon and Dunnage Zones. They range from pre- to post-tectonic and have Rb-Sr dates of 440 to 312 Ma.

3. Muscovite ± *biotite granites,* like the biotite granites, are dominantly found within metamorphic rocks of the Gander Zone, and are most abundant within the metasedimentary rocks of the zone. The few dates available indicate Silurian ages for these rocks, and they are typically syn-tectonic.

Figure 1. Possible arrangement of major continental plates between the Ordovician and Carboniferous suggested from paleomagnetic data summarized by Morel and Irving (1978). The displacement vectors at 325 and 280 Ma have been added for this paper, based on relative positions between 375 and 280 Ma. Geological evidence suggests that closure of the proto-Atlantic (Iapetus) ocean in Newfoundland was completed by the Caradocian (~ 445 Ma).

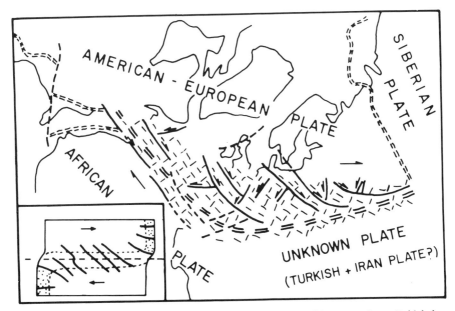

Figure 2. Late Hercynian structures of the North Atlantic region interpreted as a Reidel shear system (inset) in the context of a "Hercynian megashear" as outlined by Arthaud and Matte (1977). Such an environment appears to have existed from closure of Iapetus in the Caradocian to at least Triassic times.

4. *Alkaline-peralkaline granites* are found on both margins of the Appalachian orogen in Newfoundland. The Topsails batholith in the west intrudes both the upper Ordovician-Silurian volcanic rocks of the Buchans and Springdale Groups, and its own peralkaline ignimbritic cover, with radiometric dates straddling the Silurian-Devonian boundary (Bell and Blenkinsop, in press). Other peralkaline intrusives are

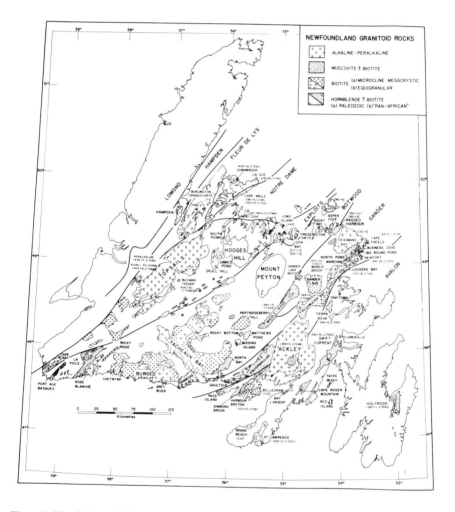

Figure 3. Distribution of Paleozoic plutonic rocks of Newfoundland, showing the main lithological variations, and tectonostratigraphic zones after Williams *et al.* (1974). Sources of radiometric dates are given in Table I. K/Ar dates are in square brackets and Rb/Sr dates in round brackets. Note that the "Dunnage Zone" includes all those zones between the Gander and Lomond zones.

represented in western Newfoundland by the La Scie Intrusive Complex (DeGrace *et al.*, 1976), dated by Bell and Blenkinsop (1977) at 328 ± 14 Ma. The St. Lawrence batholith intrudes Middle Cambrian sedimentary rocks of the Avalon Zone, as well as its own peralkaline ignimbritic cover, and is also dated at 328 ± 5 Ma (Bell and Blenkinsop, 1977).

GEOCHEMISTRY

A currently popular interpretation of the geochemistry of granitoid rocks is that of White and his co-workers (1974, 1977; Griffen *et al.*, 1978; Chappell and White, 1974; Chappel, 1978; Hine *et al.*, 1978), who have emphasized the "restite model" for granitoid magma genesis, whereby linear variation trends on Harker diagrams are interpreted as resulting from a mixture of granitic liquid and differing proportions of the residuum (or restite) of partial melting. According to these workers, the source of partial melting, and therby the restite, may be either igneous ("I-type") or sedimentary ("S-type"), which results in Harker diagram variation trends, and other chemical features, characteristic of each type.

Geochemical data are not available for all the Paleozoic granitoid plutons of Newfoundland, but all the suites described above are represented by abundant analyses. Each suite shows a reasonably coherent grouping of chemical data, but there are numerous differences in detail which indicate differences in origin, either through differing processes or differing source regions. There is little correspondence between the I- and S-type chemical patterns in Newfoundland suites (e.g., as illustrated in Figure 4), suggesting that the restite model cannot be simply applied to these suites.

Eastward increases in $^{87}Sr/^{86}Sr$ initial ratios and decreasing ages of granitoid rocks across the Sierra Nevada batholith (Hurley *et al.*, 1965; Kistler *et al.*, 1971) and the central Andes (McNutt *et al.*, 1975) have been interpreted in terms of mantle origins related to subduction processes; Brown and Hennessey (1978) have interpreted most of the British Calendonian granites in a similar way. However, as pointed out by Pidgeon and Aftalion (1978), such an interpretation for the latter is complicated by a number of features, such as the post-Iapetus age of most of these granites, and differing crustal sources indicated by relict zircons.

The limited isotopic data available for Newfoundland granites are summarized on Figures 3 and 5 along with two data points for post-Iapetus volcanic rocks. Figure 3 shows that there is no systematic variation of $^{87}Sr/^{86}Sr$ initial ratios as a function of geographic position across the orogen, or relative to specific tectonic zones. Most ratios cluster between 0.704 and 0.710 regardless of position or lithological suite, although the two lowest are calc-alkaline suites from the Fleur de Lys zone, and the highest is from the peralkaline St. Lawrence granite.

There appear to be two parallel trends of increasing $^{87}Sr/^{86}Sr$ with decreasing age (Fig. 5). Some workers (e.g., Moorbath, 1978) argue that such trends reflect the evolution of juvenile continental crust with time, i.e., that the younger granites are derived from progressively younger continental crust which would have evolved from the mantle along linear growth lines. As seen from Figure 5, many data points of the upper trend could be accommodated by derivation from sources younger than

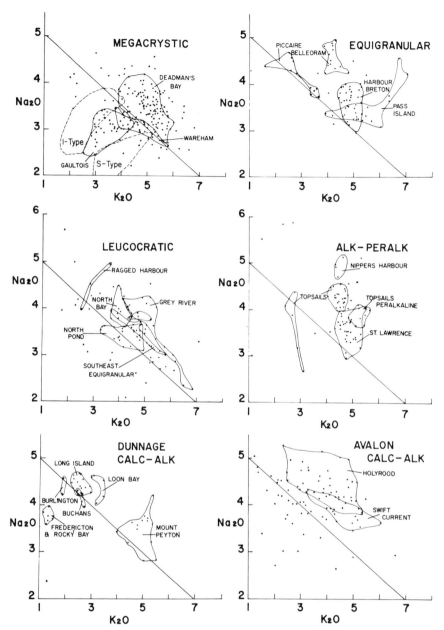

Figure 4. Variations in K₂O vs. Na₂O (wt%) of some Newfoundland granitoid rocks. The fields of "I-" and "S-type" granites shown on the "megacrystic" diagram are taken from Hine *et al.* (1978). The "megacrystic" and "equigranular" diagrams are both for biotite granites-granodiorites, with and without megacrysts of microcline, respectively. The compositional fields of only a few of the plutons are identified.

about 600 Ma and with Rb/Sr ratios between 0.4 and 1.0, while the lower line approximates a trend produced from a younger source (440 Ma) with Rb/Sr = 1.0. Although the relatively young sources are in accord with the tectonic models which have the crust of central Newfoundland formed by closure of Iapetus at about 440 Ma, these Rb/Sr ratios are rather high. For example, the Rb/Sr ratios of ophiolitic and island arc rocks of the Notre Dame Zone are about 0.50. Furthermore, the curvature of the trends shown on Figure 5 requires increasingly higher Rb/Sr ratios in sources of the younger rocks. If further investigations confirm these apparent trends, an interesting coincidence between the start of the lower trend at about 440 Ma and the Caradocian closure of Iapetus would also be confirmed. This coincidence could be most readily interpreted in terms of cause and effect: crustal thickening following the closure of Iapetus initiated a new input of low $^{87}Sr/^{86}Sr$ material from the lower crust or upper mantle. mantle.

Whatever the precise petrogenetic details, it is clear that granitoid suites with broadly similar major element chemistry must form from a variety of sources and processes. Since it is the local petrogenetic history that governs the formation of ore deposits, petrogenetic processes will be discussed in more detail.

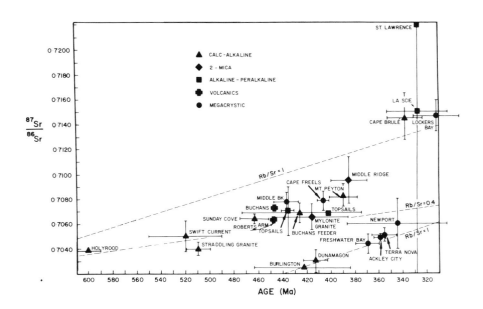

Figure 5. Variation of $^{87}Sr/^{86}Sr$ initial ratios in Newfoundland granitoid rocks with age. Data from Table I. Dashed lines show evolutionary trends of $^{87}Sr/^{86}Sr$ in source rocks with differing ages and Rb/Sr ratios.

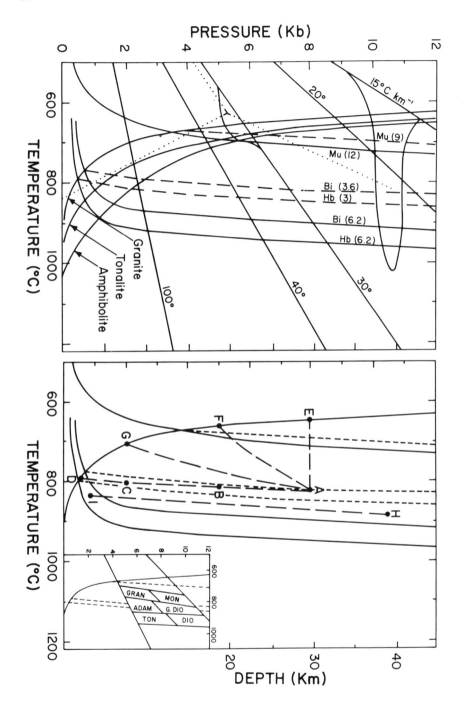

PETROGENESIS

Introduction

There are two aspects of the genesis of granitoid rocks which should be emphasized: the original melting processes that produced the magma, and the crystallization behaviour of the magma during cooling. In many cases, it is clear that the melts have moved from their original site of formation into areas of cooler rocks, as would be the case for most of suites 1 and 4 described above, so these two aspects of their origin might be easily separated. In other cases, such as the granites in migmatitic terrane like the Gander Zone, it might be difficult to separate the two.

The biotite-granites of Newfoundland (suite 2) may show evidence of significant movement from the site of generation (e.g., the Ackley batholith) or may be confined to metamorphic terrane and associated with migmatites from which they could have been generated in situ. The muscovite-bearing granites (suite 3) may show excellent evidence of in situ partial melting of pelitic sedimentary rocks (cf. Currie and Pajari, 1977), or they might form an integral part of larger biotite granite plutons (e.g., North Bay, Coleman-Sadd, 1976). The following discussion reviews the crucial role of water in the formation and behaviour of such melts, as this water is also important for the formation of ore deposits.

Magma Generation

Recent experimental studies have shown that granitoid rocks can be generated by partial melting of a variety of rocks under varying conditions, especially pressure, temperature, and water content (cf. Brown and Fyfe, 1970; Wyllie, 1977; White and Chappell, 1977; Burnham, 1979). Figure 6 summarizes some of these data, with reaction curves for relevant metamorphic minerals. The close similarity of water-saturated solidi for granite, tonalite, and amphibolite, especially above 5 kb, indicates that, with excess water, melts could form in a narrow temperature interval

Figure 6. (left side) Pressure-temperature projection of some melting and reaction relations relevant to the genesis of granitoid rocks. The solid lines labelled "granite", "tonalite", and "amphibolite" are the solidi of these compositions as shown by Wyllie (1977). The straight lines labelled 100°, etc., are geotherms showing these rates of temperature change per kilometre depth. The solid and dashed lines labelled Mu, Bi and Hb are liquidus and solidus curves respectively for each of the minerals muscovite, biotite and hornblende, as shown by Burnham (1979). The aluminosilicate triple point and reaction curves (dotted lines) are from Richardson et al. (1969). The perturbation of the 30°C km[1] geotherm is the 100°C temperature increase allowed by frictional heating as calculated by Reitan (1968), but note that any melting would result in lubrication and consequently limit the extent of such temperature increase. The numbers in brackets on solidi and liquidi are the approximate minimum amounts of water (wt%) present in any melt in equilibrium with these phases, over the interval where they are straight, as calculated by Burnham (1979). The water contents change significantly with pressure over the intervals where these lines are curved (cf. Wyllie, 1977).
(right side) Different crystallization paths (AE, etc.) which may be followed by granitoid melts depending upon their rate of ascent and level to which they rise (see text for discussion). Inset shows approximate melt compositions which might be produced at differing pressures and temperatures (after Brown and Fyfe, 1979). Melting curves, etc., from left side.

(between 600 and 700°C) over a wide range of source compositions and pressures (Wyllie, 1977).

In the absence of a hydrous fluid (water-undersaturation), melting would not take place until temperatures reached the solidi of hydrous phases which might be present, i.e., muscovite, biotite, or amphibole. Such undersaturated melts would have different compositions depending on the depth (pressure) of melting and the particular hydrous phases involved (Brown and Fyfe, 1970). The negative slope of the granite solidus requires such undersaturated melting in order to allow for ascent of melts to shallow crustal levels (Burnham, 1967; Harris *et al.*, 1970; Brown and Fyfe, 1970). This observation, of course, does not imply that all granites formed under water-undersaturated conditions; the muscovite-bearing leucogranites of northeastern Newfoundland, for example, did not. Currie and Pajari (1977) have demonstrated in situ partial melting of pelites in the Ragged Harbour area to form leucogranite under conditions close to the aluminosilicate triple point, while Jayasinghe (1979) has shown that in the Wesleyville area nearby, leucogranites which may have formed by fractionation from biotite granite have crystallized sub-solidus sillimanite, indicating a very narrow P-T range around the aluminosilicate triple point (Fig. 6, leftside). Thus, similar conditions of melting and crystallization indicate that no significant movement of the melt took place for leucogranites of the Ragged Harbour area, while removal from the source region is possible for those of the Wesleyville area.

The biotite granites, however, do not generally show evidence of in situ partial melting; they may rise into greenschist facies crustal levels, and are typically less silicic in composition than the muscovite granites. These features indicate higher temperatures and more advanced degrees of partial melting, with water content buffered by biotite, which allows for their rise into the higher crustal levels (see Fig. 6, left side). The same arguments can be used to suggest that the hornblende-bearing quartz-diorites were generated at even higher temperatures, and rose to even higher crustal levels than the biotite granites.

These differing degrees of melting could reflect progressive melting along a single "geotherm", but the complex tectonic situation outlined above, i.e., ocean closure and plate collision, crustal thickening, thrusting and strike-slip faulting, renders this unlikely. A number of "geotherms" are drawn on Figure 6 (left side), ranging from that of a stable shield at about 15°C/km^{-1} to that of mid-ocean ridges at about 100°C/km^{-1}. The first could represent the ambient condition immediately following collision, so that no granitoid melts would be generated except at local thermal highs, such as may have been caused by intrusion of mafic magmas. The Mount Peyton (Strong, 1979) and Fogo (Cawthorn, 1978) batholiths are two examples of this phenomenon. Melting under these conditions, presumably in a crust of low-H$_2$O intermediate composition dominated by lower Paleozoic island arc sequences, would have been responsible for the hornblende-biotite quartz-diorites and granites. The blanketing effect of sedimentary and volcanic rocks overlying the previous oceanic geotherm could cause a rapid increase in the geothermal gradient, with widespread crustal melting and granite plutonism. Local perturbations of relatively steep geotherms might have occurred, e.g., in the metamorphic Gander Zone, where water-saturated metasediments and continental crust would melt to give rise to the more silicic and potassic biotite and 2-mica granites, the former being able to rise to higher levels than the latter. If the melting were controlled by other processes, e.g., frictional heating along shear zones (cf. Nicolas *et al.*, 1977), such heating would be limited by the lubricating effect of the melt, so that only initial melts like the muscovite-bearing leucogranites would be produced. This may explain the typical close association of such granites with shear belts; the low degrees of melting might also serve to concentrate the economically important lithophile elements which typify such granites.

Magma Crystallization

Figure 6 (right side) shows paths which melts might follow during crystallization within the crust, ranging from in situ cooling at constant pressure (path AE) to rapid adiabatic ascent with little cooling (path ABCD). If a melt were generated at point A, with biotite melting at 8 kb, the resulting liquid would contain about 3% H_2O (Burnham, 1979), leaving orthopyroxene and plagioclase as residua (Naney, 1979). If this melt rose rapidly to a few kilometres depth along path ABCD, biotite would continue to melt until its liquidus were intersected, i.e., until complete breakdown, and amphibole would be the only stable hydrous phase on the water-saturated solidus at point D. The same would be true for melts generated at higher pressures and temperatures, e.g., rising along path HI. If melt A ascended only to point C and cooled along path CG, or if it ascended along path AG, biotite and amphibole would be stable on the water-saturated solidus at G. If the melt ascended only to point B and crystallized along path BF, or if it rose along path AF, muscovite would be a stable phase on the water-saturated solidus at F, as it would due to in situ cooling along path AE.

It is noteworthy that the water contents of melts buffered by muscovite (9 to 12%, Fig. 6,) are much higher than those buffered by biotite or amphibole (3 to 6%). This availability of mineralizing hydrous fluid probably explains why the former have economically important concentrations of lithophile elements. On the other hand, melts which rise to shallow crustal depths (such as at D, Fig. 6b) in the porphyry environment can thermally induce groundwater convection systems to leach and hydrothermally concentrate certain elements. Neither process would operate at the intermediate levels of biotite-granite crystallization (e.g., at G in Fig. 6b), and such rocks seldom contain economically important concentrations of elements.

The granitic rocks of suite 4 show some compositional similarities to those of the other suites, but they plot at lower pressure cotectics in "granite" phase diagrams, in accord with the fact that they typically reach shallow crustal levels and erupt to the surface. An important question in the petrogenesis of these rocks is how they attain their peralkaline character, a problem for which there are realtively few experimental data (see reviews by Bailey, 1974, 1976). Teng and Strong (1976) suggested that the St. Lawrence granite could have been derived from compositions like that of the Belleoram biotite-granite by fractionation of plagioclase and biotite. This process may equally apply to the Topsails suite, since the Topsails biotite-granite is somewhat similar to that of the Belleoram stock (Fig. 4); negative europium anomalies (Taylor *et al.*, 1979) support such an interpretation. The coexistence of mafic magmas with these peralkaline liquids may have genetic significance (cf. MacDonald, 1974), or the magmas may only provide a heat source for crustal melting. The high initial $^{87}Sr/^{86}Sr$ ratios support the latter intrepretation.

Aqueous fluids incongruently dissolve feldspar and other aluminous silicates (i.e., are peralkaline), at least up to 600°C and 3.5 Kb (Luth and Tuttle, 1969; Currie, 1968), and have decreasing SiO_2 and Al_2O_3 contents with increasing pressure (Burnham, 1967). If such peralkaline aqueous fluids were to develop during melting within the crust (cf. Bailey, 1964), and they initially mixed with or dissolved in a granitic melt, they could impart a peralkaline chemistry to it. Furthermore, it has been shown (Mustart, 1972) that peralkaline melts show continuous solubility with vapour; this allows them to exist at very low temperatures which would promote their extreme differentiation and concentration of a range of trace elements (Edgar and Parker, 1972). In this connection, there is good evidence for peralkaline or sodic fluids in both the Topsails (Taylor *et al.*, 1979) and St. Lawrence (Strong *et al.*, 1978) suites. The rapid ascent of peralkaline melts from depth would be most likely with a rapid release of pressure. Perhaps this explains why peralkaline suites are characteristic of extensional tectonic environments.

METALLOGENY

Introduction

Mineral deposits in granitoid rocks may be thought of as forming a spectrum ranging from the strictly *chalcophile* elements such as Cu to the strictly *lithophile* elements such as Nb, Ta, Li and Be. The former characteristically form covalent bonds, and the latter ionic bonds (cf. Taylor, 1965). Economic concentrations of the chalcophile group in granitoid rocks typically occur as "porphyry" deposits formed by hydrothermal processes during sub-solidus cooling of high-level sub-volcanic, dominantly calc-alkaline plutons (Sillitoe, 1972; Sutherland Brown, 1976). Granitoid economic deposits of the lithophile group typically occur in veins and pegmatites formed during the magmatic stage of cooling of leucogranite bodies at intermediate depths. A whole spectrum of "lithochalcophile" (Tischendorf *et al.*, 1978) elements (Pb, Zn, Ag, Sb, Bi, As) span these two extremes and typically form veins in country rocks marginal to a variety of plutonic rock types which crystallized under different conditions. It should be emphasized that these are just the typical patterns in which such elements are concentrated, and that reversals do occur, e.g., Sn in a sub-volcanic "porphyry" setting in Bolivia (Sillitoe *et al.*, 1975; Grant *et al.*, 1977) and in Mount Pleasant, New Brunswick (Dagger, 1972) and Be in a comparable setting in the Spor Mountain District, Utah (Shawe, 1968). A more valid generalization might be that there is a specific relation between granitoid composition and deposit type. Thus, porphyry Cu deposits are not found in high-silica leucogranites, and U, Nb, Sn, Be, etc., deposits are not typically associated with intermediate composition calc-alkaline plutons.

There is a voluminous literature on the geochemical behaviour of these elements in magmatic and hydrothermal solutions (e.g., Taylor, 1965; Wedepohl, 1969), and it is unnecessary to review it here. However, Figure 7, showing two of the main properties which control their behaviour, ionic radius and "ionic potential" (a rough measure of the tendency to form ionic bonds – Taylor, 1965), illustrates their natural grouping. Thus, the strongly lithophile group tend to form tetrahedral complexes ($(SnO^4)^{4-}$, $(MoO_4)^{2-}$, $(NbO_4)^{3-}$, $(TaO_4)^{3-}$, etc.) and be strongly concentrated in "dif-

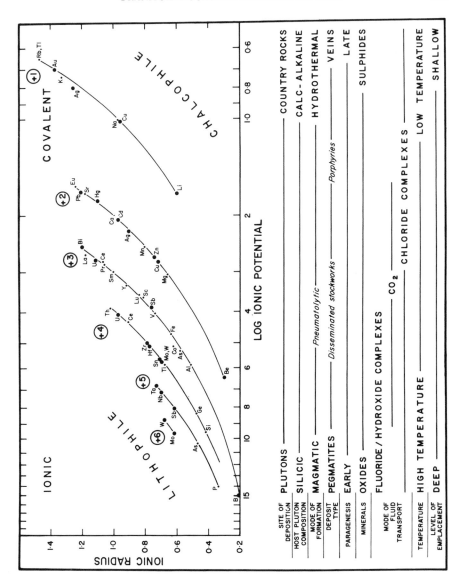

Figure 7. Schematic outline of the relation between ionic radius and ionic potential, two of the many variables which control the geochemical behaviour of elements forming different types of granitoid mineral deposits. Curves connect elements of similar ionic charges, as shown in circles. The terms "lithophile" and "chalcophile" refer to the geochemical behaviour of those economically important elements shown as large dots, except for Li, as discussed in the text. Some geological features corresponding to these variables are summarized in the lower half of the diagram, indicating the common patterns of occurrence from the more lithophile or ionically bonded elements on the left to the more chalcophile or covalently bonded elements on the right.

ferentiated" silicic magma formed either by crystal fractionation or as initial partial melts. Those elements with very small radii (Li, Be) may likewise form such complexes, but they might also be similarly concentrated due to their exclusion from common silicate structures because of their small radii. The strongly chalcophile Cu can either enter silicate lattices or form sulphides, and thus tends to be randomly distributed during cooling of a magma or partial melting. The other "lithochalcophile" elements, although comparable in ionic radii to silicate-forming ions such as Ca^{++}, have strongly covalent bonds with oxygen which exclude them from silicate lattices. They would thus be concentrated in differentiated liquids, *unless* sulphur activity is such that sulphides of certain elements such as Mo can form and cause the early or continuous removal of "lithochalcophile" elements from magmas during cooling.

When aqueous fluids coexist with silicate melts, e.g., at points E, F, G, or D in Figure 6 (right side), the behaviour of the elements depends critically upon their fluid-melt partition coefficients. Data for such coefficients and for the effects of these elements on granitic melts are sparse, but show consistent results (Stewart, 1978; Flynn and Burnham, 1978; Chorlton, 1973; Holland, 1972; Jahns and Burnham, 1969; Burnham, 1967; Wyllie and Tuttle, 1961). Thus, Wyllie and Tuttle (1961) showed that the volatiles HCl, CO_2 and NH_3 tend to raise the solidus temperatures of granitic liquids, SO_3 has very little effect (although H_2S is similar to H_2O (Burnham, 1979)) and H_2O, P_2O_5, HF and Li_2O lower the solidus temperatures. Figure 8 illustrates these effects, as well as the fluid-melt partitioning of these elements, which increases from Li_2O to HCL (Burnham, 1967). Stewart's (1978) experimental study gives an especially clear illustration of the effects of Li on lowering granite melting temperatures and allowing for extreme differentiation of Li-bearing melts; Chorlton (1973)

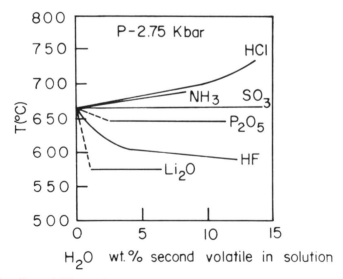

Figure 8. The effects of different elements or compounds on the melting temperature of granitic composition, as summarized by Luth (1976).

has shown similar effects for BO₃. Thus, although one might predict from Figure 7 that Li would be concentrated in the late-stage residual liquids, this does not mean that it would be partitioned into aqueous fluids to form hydrothermal deposits. The same can be said for F (Burnham, 1967), which accounts for the dominance of such elements in non-hydrothermal magmatic orebodies such as pegmatites, and also in veins with little evidence of hydrothermal alteration, such as the St. Lawrence fluorite deposits (Teng and Strong, 1976). Similar behaviour might be predicted for other non-volatile elements such as Be, Nb and Ta because of their similar geological occurrence.

Since Mo, W, or Sn may cover the full range from magmatic to hydrothermal deposits, often in the same pluton (cf. Whalen, 1976), it is clear that although these elements may first be concentrated during magmatic crystallization, they are also soluble in aqueous fluids. However, Mo also commonly forms a sulphide that might be removed from melts during the early stages of crystallization, to be available for later concentration as a porphyry deposit by circulation of hydrothermal solutions through the crystallized parts of the pluton.

Hildreth (1979) has presented a case study of element concentration during cooling of the large siliceous magma chamber which gave rise to the Bishop Tuff of southern California. He showed that the lithophile elements Li, Be, Nb, Mo, Sn, Ta, W, Th, U and the heavy rare earths, along with H₂O, Cl, and F were concentrated

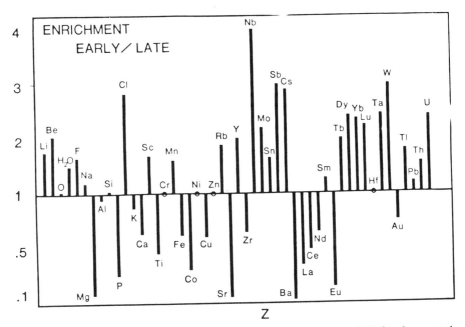

Figure 9. Relative enrichment of 48 elements estimated by Hildreth (1979) for the erupted portion of the Bishop Tuff magma chamber, calculated as a ratio of the earliest to the latest erupted samples, assumed to represent the top and deeper parts of the magma chamber respectively.

towards the top of the magma chamber, whereas the chalcophile elements Fe, Co, Cu and Au were concentrated in the lower portions, as illustrated in Figure 9. Although there is a close correlation between the volatiles and most lithophile elements of economic interest, Hildreth has interpreted this as indicative of a thermal diffusion process linked to gradients in temperature and corresponding structure of the melt which affects each element independently, rather than as a cause and effect relationship such as transportation by voltaile complexes. Nevertheless, the enrichment patterns are in striking agreement with predictions from the available experimental data.

Hypotheses to explain the formation of porphyry copper and molybdenum deposits have ranged from derivation from purely magmatic fluids (e.g., Burnham, 1967, 1979; Neilsen, 1976; Rose, 1970; Whitney, 1975) to varying degrees of interaction with groundwaters (e.g., Norton and Cathles, 1973), as summarized in Figure 10. Regardless of the source of fluid, all investigators seem to agree that the metals and probably most of the sulphur were derived from the calc-alkaline magmas, either through magmatic enrichment, sub-solidus leaching, or some combination of both.

Burnham (1979) has demonstrated that 20% partial melting of average oceanic basalt could produce up to 400 ppm Cu in a calc-alkaline melt, which is well above that of typical calc-alkaline rocks which host porphyry Cu deposits. This removes any need for the partial melting of anomalously enriched oceanic crust as required by the model of Sillitoe (1972), although the close association of porphyry deposits with subduction zones, e.g., around the Pacific margins or the Tethyan belts (Mitchell and Garson, 1972; Pereira and Dixon, 1971; Sillitoe, 1972), strongly suggests a genetic relationship. Nevertheless, the apparent concentration of the Cu-Au porphyries in island arc and Cu-Mo deposits in continental environments (Kesler, 1973) suggests a crustal input of the metal components of porphyry deposits. In general it appears that such deposits require a combination of factors for their formation. The *subduction-related magma* must be intruded into *shallow crustal levels* which allows resurgent boiling and separation of a *hydrothermal vapour plume*. This is typically in a *non-marine environment* where *groundwater* is available to interact with the vapour plume, leaching and redepositing metals, and producing the characteristic zoned alteration sequences. If either of these controls is absent, true porphyry-type deposition will be unlikely, although there may be some porphyry-like effects such as the stockwork alteration zones beneath massive sulphide deposits formed in submarine environments.

The transport and deposition of the "lithochalcophile" elements is a complex process which depends among other things upon the concentration of metals, Cl, and S in the aqueous fluid, as well as upon temperature and pressure. Consequently, "lithochalcophile" deposits show a range of characteristics: differing conditions of deposition, complex mineral assemblages, different forms of orebodies, etc. In very general terms, any systematic patterns of paragenesis and zonation can be related to differing metal solubilities, which in turn mainly reflect temperature variations temporally and spatially related to the cooling pluton. Thus Pb, Zn, and Sb veins are associated with both high-level porphyry deposits, as in the southwestern U.S.A. (e.g., Guild, 1978), Bolivia (Grant *et al.*, 1977) and New Brunswick (Dagger, 1972), and with the mesozonal leucogranites of western Europe, those of Cornwall being the classic example (e.g., Hosking and Shrimpton, 1964). In general, there seems to be a zonation pattern of progressively more chalcophile elements farther away from the pluton, reflecting their partitioning with Cl into the aqueous fluid and their differing solubilities with cooling of that fluid.

The lithophile elements not partitioned into the aqueous fluid may be retained by the silicate melt through extreme differentiation, to concentrate eventually in pegmatites and other intraplutonic deposits. Although there are no experimental data available, Bowden and Jones (1978) suggest that Sn would be partitioned towards the aqueous phase. Thus, retention of the aqueous phase by peralkaline melts may explain the disseminated "primary" tin mineralization of the Nigerian Younger Granites, while movement of aqeuous fluids exsolved from peraluminous melts would account for the vein-type deposits in Nigeria where they are associated with albitization, greisening, etc.

In summary, it would seem that the formation of granitoid mineral deposits depends upon whether the element of interest is dispersed or concentrated during crystallization or partial melting. With dispersal (e.g., Cu in general or Mo in S-bearing melts), a stage of hydrothermal leaching is necessary to form economically important concentrations. With magmatic concentration (e.g., the "lithochalcophile elements"), hydrothermal vein-type deposits may be formed during cooling of a water-saturated melt if the element is preferentially partitioned into the aqueous phase; this does not require particularly strong fractionation or, therefore, particularly silicic compositions. With fractionation of a water-undersaturated magma, or with preferential partitioning of the element into the silicate melt (as is the case with most lithophile elements), extreme differentiation (indicated by very high SiO_2, etc.) may be necessary to produce economically important concentrations. These patterns are schematically outlined in the lower half of Figure 7.

Appalachian Examples

Such a model provides a crude framework for the understanding of granitoid deposits in the North Atlantic region; it is somewhat different from the porphyry model which Hollister et al. (1974) applied to Appalachian granitoid deposits. These authors interpreted the characteristics of such deposits as reflecting different levels of erosion of zoned "porphyry" systems. However, it is a major contention of this paper that true porphyry deposits require shallow hydrothermal leaching and redeposition. Many other granitoid deposits may reflect entirely different processes of mineralization.

Hollister et al. (1974) interpreted the molybdenum mineralization along the southern margin of the Ackley batholith (Fig. 3) as a deeply eroded porphyry deposit. Whalen (1976) has shown, however, that the Ackley pluton was intruded and crystallized at relatively shallow depths (about 2 to 3 km), that the Mo was concentrated by extreme differentiation of granitic magma, and that the mineralization shows a range of types from pegmatitic to hydrothermal porphyry-like. According to N.C. Higgins (pers. commun., 1979), the tungsten mineralization at Grey River is intimately associated with a highly differentiated leucogranite phase of the Burgeo batholith (Fig. 3) occurring as hydrothermal veins and greisens with no porphyry-type alteration zones. The fluorspar deposits at St. Lawrence (Fig. 3) are a classic example of mineralization resulting from the separation of fluorine from an apparently water-deficient granitic melt during the final stages of crystallization, and they show no porphyry-type features (Teng and Strong, 1976).

Some of the other deposits discussed by Hollister *et al.* (1974), e.g., the Gaspé and Woodstock Cu and possibly the Mount Pleasant Mo-W-Bi (-Sn-Cu-Zn-Pb) deposits, would fit even the present narrow definition of porphyry deposits – products of hydrothermal solutions associated with hydrothermal alteration of a high-level pluton. However, the Mount Pleasant deposit is especially interesting because of its complex mineralogy and element assemblage and its occurrence in a relatively young (Mississippian) post-tectonic sub-volcanic pluton, which appear to represent the end stages of magmatism which began in Middle Devonian times.

According to Dagger (1972) the mineralization at Mount Pleasant is zoned; W-Mo-Bi are found within the pluton on joint coatings and disseminated replacements, and lower grades of Sn-Cu-Zn occur in the contact zones as irregular replacements. In general, the paragenetic sequence was one of early deposition of oxides and molybdenite followed by later deposition of sulphides. Greisen-type alteration accompanies the inner zone, with chloritization and silicification in the outer zones. Dagger suggested that the two porphyritic plugs that host the mineralization were enriched in W, Mo, Bi, and Sn as a result of differentiation of granitic magma, and he did not consider leaching of the host rocks to have been an important process. This suggests an orthomagmatic origin for the metal concentrations, with any Mo or W in excess of several ppm being partitioned into the aqueous fluid during a differentiation. Dagger did, however, allow that Cu and Zn may have required post-magmatic leaching for their concentration. It thus appears that the Mount Pleasant deposit is as much a primary magmatic as a porphyry type, as are the Ackley Mo deposits and St. Lawrence fluorspar veins, while the Grey River tungsten veins have no similarity to porphyry deposits.

Smith and Turek (1976) have pointed out a number of geochemical features of the New Ross pluton in the South Mountain batholith of Nova Scotia which suggest the presence of tin. Significant occurrences of Sn and U associated with Hercynian leucogranites of southern Nova Scotia are currently being investigated by mining companies (Northern Miner, October, 1978). Although little is known of the uranium occurrences, Smith (1974) and Smith and Turek (1976) suggest that the tin concentrations result from extreme differentiation of the South Mountain batholith, that is, *not* from porphyry-type hydrothermal leaching.

Similar environments to those discussed above can be recognized elsewhere in the Appalachians. For example, the Burnt Hill, New Brunswick tungsten deposits (Potter, 1968), and the U-enriched Concord granite of New Hampshire (Bothner, 1978) are probably analogous to the South Mountain batholith, while the U-enriched post-tectonic peralkaline Conway granite (Hoisington, 1977) compares with the St. Lawrence pluton. In general, most Appalachian mineral deposits associated with granitoid rocks in eastern Canada are better described as lithophile and magmatic-hydrothermal than as chalcophile and porphyry-type; they belong to the left rather than the right side of Figure 7.

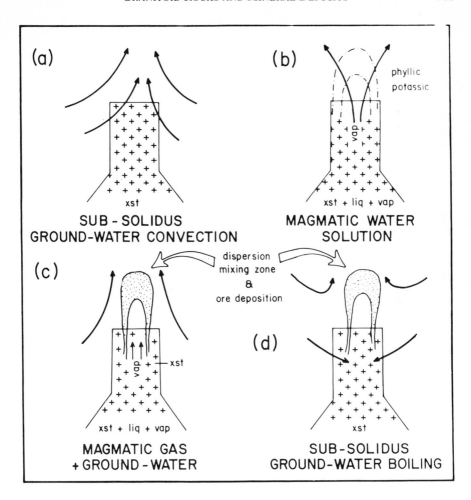

Figure 10. Different models of flow pattern and origin of aqueous fluids involved in formation of porphyry copper deposits as summarized by Henley and McNabb (1978).
(a) Groundwater convection through subsolidus pluton (cf. Norton and co-workers).
(b) Exsolution of magmatic fluid which subsequently "migrates" into country rock and develops characteristic alteration assemblages (cf. orthomagmatic model of Rose, 1970).
(c) A plume of low-salinity magmatic gas exsolved from the crystallizing pluton penetrates and interacts with a pre-existing groundwater convection system.
(d) Modification of the plume model (c) to allow for generation of "magmatic" vapour by boiling of groundwater in contact with the cooling subsolidus pluton. Little data support this model, according to Henley and McNabb (1978).

Western Europe

There have been a number of recent reviews of the granitoid rocks and associated mineral deposits of western Europe, e.g., Chauris (1977) and the three-volume proceedings of the "MAWAM" symposium (Stemprok *et al.*, 1978). As pointed out by Hall (1971, 1972), among others, the Paleozoic granitoid rocks of Western Europe can be readily grouped into Caledonian and Hercynian types, respectively intruded before and during the Devonian and during and after the Devonian. Hall (1971, 1972) has shown that in general Caledonian granitoid rocks are less silicic than those of the Hercynian. He interprets this as reflecting a lower geothermal gradient for the Caledonian orogen, requiring deeper levels of crustal melting. Leake (1978) has suggested that many Caledonian granites were produced and emplaced along deep crustal faults, due to crustal melting along these faults where the deep crust was juxtaposed against higher temperature upper mantle. As discussed above, Nicolas *et al.* (1977) have suggested that the Hercynian granites of Brittany were produced by frictional heating along fault zones, a process which might also have been active for any fault-controlled Caledonian granites. Bard *et al.* (1971) have demonstrated the close relationship between leucogranite production and Hercynian deformation in the Iberian massif. Although numerous models have been proposed which involve some form of subduction for both the Caledonian and Hercynian granitoid rocks (e.g., Brown and Hennessy, 1978; Badham and Halls, 1975; Hanmer, 1977; LeFort, 1978; Mitchell, 1974; – see review by Windley, 1978), these models appear to be improbable for the same reasons as those given for Appalachian granites (distribution on both sides of proposed subduction zones, younger ages than any subduction zones, etc.).

Whatever the precise tectonic controls of Caledonian and Hercynian granitoid magmatism, there is certainly a profound difference in their economic importance. Only the relatively small Carrock Fell tungsten mine is currently producing from any Caledonian granite. A significant porphyry Cu deposit is known in calc-alkaline diorites near Coed-y Brenin in North Wales (Rice and Sharp, 1976), but these diorites are sub-volcanic to the subaerial Ordovician volcanics around the Harlech Dome and not part of the Caledonian granitoid suite proper. Disseminated Cu mineralization has been recognized in early Devonian granodiorite-quartz diorite at Kilmelford in Argyllshire, Scotland (Ellis, 1977) and Cu-Mo in the Ballachulish granite, Argyllshire (Evans, 1977), but neither is of economic importance at present. Other areas of significant mineralization are known in Caledonian granites, e.g., the Li-pegmatites of the Leinster batholith (Brück and O'Connor, 1977; Steiger and von Knorring, 1974), but none are currently productive.

The Hercynian granites contrast dramatically with the Caledonian ones, having a long history of production of Sn, W, U, Mo, Cu, Pb, Be, Li, etc., as reviewed by numerous authors (e.g., Hosking and Shrimpton, 1964; Stemprok *et al.*, 1978; Rich *et al.*, 1977; Moreau and Ranchin, 1973; Tischendorf *et al.*, 1978; Chauris, 1977). The dominant characteristic of these deposits is their almost universal association with two-mica leucogranites. Although details of alteration, paragenesis, structural control, etc., may differ between deposits, they are just variations on the theme of metal concentration in water-saturated, highly silicic peraluminous granites which, as discussed above, were formed by either extreme differentiation of biotite granite melts or wet melting of aluminous metasediments.

DISCUSSION AND CONCLUSIONS

The factors which control economically important concentrations of elements in granitoid rocks are the tectonic setting, the petrogenetic history and source of the rocks, and local physical-chemical environments which determine the geochemical behaviour of the elements.

Most Paleozoic granitoid rocks of the northeastern Appalachians and western Europe were intruded well after the destruction of any pre-existing ocean such as Iapetus, and hence, unlike the Pacific batholiths, cannot have originated by subduction processes. They are more readily explained in terms of a "megashear" environment, which allows the simultaneous production of different granitoid types in locally different tectonic settings.

Using the Newfoundland sector of the Appalachians as an example, the granitoid rocks naturally group into four lithological suites. The first three are simply based on hydrous accessory minerals – muscovite, biotite or hornblende – and can be thought of as representing increasing degrees of partial melting, all in compressive or shearing tectonic environments. The fourth suite – alkaline to peralkaline granites – requires a more complex petrogenetic explanation, probably tied in with this group's typical occurrence in locally extensional environments. Isotopic data indicate a variety of source rocks, with no systematic relation to rock types.

Mineral deposits associated with granitoid rocks can be thought of in terms of two extremes: from purely chalcophile porphyry Cu-Mo to purely lithophile orthomagmatic/pegmatitic deposits, with a range of vein-type deposits in between. The chalcophile type requires the operation of groundwater convection systems, which in turn require very shallow intrusion of the pluton. This in turn requires the melt to have been undersaturated in water, and presumably explains why porphyry deposits are typically associated with calc-alkaline biotite-bearing or hornblende-bearing water-undersaturated melts (see Fig. 6). Subduction of oceanic crust seems to be essential to the formation of such deposits, whether it be as a source of metals or sulphur through sea-floor concentration processes, or simply as a source of chlorine, which acts as a leaching and transporting agent. This presumably accounts for the scarcity of porphyry Cu-Mo type deposists in the post-subduction Appalachian-Caledonian calc-alkaline granitoid rocks, despite their being otherwise similar to mineralized ones.

The lithophile element deposits are readily concentrated in melts during differentiation or initial meting, via processes greatly enhanced by both high water content and the effect of these elements on lowering melt temperatures. Hence, these deposits are typically associated with highly silicic, water-saturated, muscovite-bearing granites that cannot rise to shallow crustal depths (Fig. 6), and they occur within metamorphic terranes such as the Hercynian of Western Europe. Biotite granites are often neither sufficiently water-rich for such lithophile element concentrations to form, nor sufficiently water-deficient to rise to shallow enough depths for hydrothermal leaching and concentration, hence they are generally barren, e.g., the typical Caledonian granites of Britain. Although barren biotite granitoids are dominant in the Canadian Appalachians, there are abundant Acadian to Hercynian muscovite granites which should be a more fruitful ground for exploration.

TABLE I: WHOLEROCK Rb/Sr RADIOMETRIC DATES FOR NEWFOUNDLAND GRANITOID ROCKS ($\lambda Rb^{87} = 1.42 \times 10^{11}$)

Pluton	Radiometric Age (MSWD)	$^{87}Sr/^{86}Sr$ Initial Ratio	Reference
Dunnage Zone Hornblende Bearing			
1. Burlington Granodiorite	422 ± 40	0.7024 ± 0.0020	DeGrace *et al.* (1976)
3A. Sunday Cove	464 ± 13 (0.09)	0.7064 ± 0.0004	Bostock (1978)
6. Buchans Feeder	426 ± 50 (1.3)	0.7068 ± 0.0008	Bell & Belenkinsop (in press)
23. Mount Peyton	384 ± 9	0.7091 ± 0.0008	Bell *et al.* (1977)
Avalon Zone Hornblende Bearing			
9. Swift Current	520 ± 30 (12.8)	0.7050 ± 0.0012	Bell *et al.* (1977)
11. Holyrood	597 ± 11	0.704	McCartney *et al.* (1966)
13. Straddling	510 ± 10 (2.7)	0.7039 ± 0.0006	Blenkinsop *et al.* (1976)
Megacrystic Biotite			
16. Newport	345 ± 42 (0.7)	0.7059 ± 0.0020	Bell *et al.* (1979)
17. Freshwater Bay	369 ± 10 (0.5)	0.7043 ± 0.0008	Bell *et al.* (1977)
18. Middle Brook	437 ± 20 (4.6)	0.7077 ± 0.0008	Bell *et al.* (1977)
19. Lockers Bay	312 ± 18	0.7146 ± 0.0013	Bell *et al.* (1977)
20. Cape Freels	416 ± 5 (0.6)	0.7078 ± 0.0008	Bell *et al.* (1977)
22. Ackley	359 ± 5 (1.9	0.7048 ± 0.0005	Bell *et al.* (1977)
22A. Mylonitic Granite	416 ± 30 (1.4)	0.7039 ± 0.0006	Blenkinsop *et al.* (1976)
Equigranular Biotite			
24. Terra Nova	356 ± 10 (1.8)	0.7050 ± 0.0006	Bell *et al.* (1977)
27. Dunamagon	429 ± 10	0.7030 ± 0.0010	DeGrace *et al.* (1976)
28. Harbour Breton	343 ± 10	0.708 ± 0.001	Bell (1975)
Two-Mica Leucogranite			
38. Middle Ridge	385 ± 15	0.7094 ± 0.0018	Bell *et al.* (1977)
Alkaline − Peralkaline Granites			
41. La Scie Complex	328 ± 5	0.7150	Bell & Blenkinsop (1975)
42. St. Lawrence	328 ± 5 (2.6)	0.7220 ± 0.003	Bell & Blenkinsop (1975)
43. Topsails Peralkaline	436 ± 5 (2.3)	0.7070 ± 0.002	Bell & Blenkinsop (in press)
44. Topsails Alkali Feldspar	402 ± 9 (2.0)	0.7067 ± 0.0009	Bell & Blenkinsop (in press)

ACKNOWLEDGEMENTS

I am grateful to R.P. Taylor for discussion and criticism, and S.J.P. Burry for data compilation, processing and drafting, both of whom were essential to the preparation of this manuscript. I thank C.W. Burnham, W. Hildreth, and N. Jayasinghe for access to their unpublished manuscripts. For discussion, criticism of the manuscript and other help, I am grateful to J.P.N. Badham, K. Bell, R.F. Blackwood, T. Calon, L. Chorlton, D.B. Clarke, S.P. Colman-Sadd, W.L. Dickson, C. Dupuy, P. Elias, C. Halls, S. Hanmer, N.C. Higgins, T. Hudson, E.C. Irving, S.J. O'Brien, C.F. O'Driscoll, J. Platt, R.R. Potter, N. Rast, H.S. Swinden, B.F. Windley, P.F. Williams, and H. Williams. I thank W. Marsh for continuing photographic service and Cynthia Neary and Glenys Woodland for typing. This study was financed by the National Science and Engineering Research Council of Canada Operating Grant No. A-7975.

REFERENCES

Arthaud, F., and Matte, P., 1977, Late Paleozoic strike-slip faulting in southern Europe and northern Africa: result of right-lateral shear zone between the Appalachians and the Urals: Geol. Soc. Amer. Bull., v. 88, p. 1305-1320.

Badham, J.P.N., and Halls, C., 1975, Microplate tectonics, oblique collisions, and evolution of the Hercynian orogenic systems: Geology, v. 3, p. 373-376.

Bailey, D.K., 1964, Crustal warping – a possible tectonic control of alkaline magmatism: Jour. Geophys, Res., v. 69, p. 1103-1111.

Bailey, D.K., 1974, Continental rifting and alkaline magmatism: in Sorensen, H., ed., The Alkaline Rocks: Wiley and Sons, N.Y., p. 148-159.

Bailey, D.K., 1976, Application of experiments to alkaline rocks: in Bailey, D.K. and Mac-Donald, eds., The Evolution of Crystalline Rocks: Academic Press, N.Y., p. 419-469.

Bard, J.P., Capdevila, R., and Matte, P., 1971. La structure de la Chaine Hercyienne de la Meseta Iberique: Comparison avec les segments voisins. Publ. de L'Institut Français du Petrole, Collection Colloques et Seminaires 22, tome 1, p. I.4-1 to I.4-68.

Bell, K. and Blenkinsop, J., 1975, Geochronology of eastern Newfoundland: Nature, v. 254, p. 410-411.

Bell, K. and Blenkinsop, J., 1975. Preliminary report on radiometric (Rb/Sr) age-dating, Burlington Peninsula, Newfoundland. Newfoundland Department of Mines and Energy (unpublished report) 9 p.

Bell, K. and Blenkinsop, J., 1977, Geochronological evidence of Hercynian activity in Newfoundland: Nature, v. 265, No. 5595, p. 616-618.

Bell, K. and Blenkinsop, J., in press, A geochronological study of the Buchans Area, Newfoundland: Geol. Assoc. Canada Spec. Paper.

Bell, K., Blenkinsop, J. and Strong, D.F., 1977, The geochronology of some granitic bodies from eastern Newfoundland and its bearing on Appalachian Evolution: Canadian Jour. Earth Sci., v. 14, p. 456-476.

Bell, K., Blenkinsop, J., Berger, A.R. and Jayasinghe, N.R., 1979, The Newport granite: its age, geological setting and implications for the geology of northeastern Newfoundland: Canadian Jour. Earth Sci., v. 16, p. 264-269.

Blenkinsop, J., Cucman, P.F., and Bell, K., 1976, Age relationships along the Hermitage Bay-Dover fault system, Newfoundland: Nature, v. 262, 377-378.

Bostock, H.H., 1978, The Roberts Arm Group, Newfoundland: Geological notes on a Middle or Upper Ordovician island arc environment: Geol. Survey Canada, Paper 78-15, 21 p.

Bothner, W.A., 1978, Selected uranium and uranium – thorium occurrences in New Hampshire: U.S. Geol. Survey Open File Report 78-482, 42 p.

Bowden, P. and Jones, J.A., 1978, Mineralization in the younger granite province of northern Nigeria: in Stemprok, M., Burnol, L. and Tischendorf, G., eds., Metallization Associated with Acid Magmatism: Ustredni Ustav Geologicky, Prague, v. 3, p. 179-190.

Brown, G.C. and Fyfe, W.S., 1970, The production of granitic melts during ultrametamorphism: Contrib. Mineral. Petrol., v. 28, p. 310-318.

Brown, G.C. and Hennessy, J., 1978, The initiation and thermal diversity of granite magmatism: Phil. Trans. Royal Soc. London, Series A288, p. 631-643.

Brück, P.M. and O'Connor, P.J., 1977, The Leinster Batholith: geology and geochemistry of the northern units: Geol. Survey Ireland Bull. v. 2, p. 107-141.

Burnham, C.W., 1967, Hydrothermal fluids at the magmatic stage: in Barnes, H.L., ed., Geochemistry of Hydrothermal Ore Deposits: Holt, Rinehart and Winston, p. 34-76.

——————————, 1979, Magmas and hydrothermal fluids: in Barnes, H.L. eds., Geochemistry of Hydrothermal Ore Deposits: John Wiley and Sons, N.Y., Chapter 2, Second Edition.

Cawthorn, R.G., 1978, The petrology of the Tilting Harbour Igneous complex, Fogo Island, Newfoundland: Canadian Jour. Earth Sci., v. 15, no. 4, p. 526-539.

Chappell, B.W., 1978, Granitoids from the Moonbi District, New England Batholith, eastern Australia: Jour. Geol. Soc. Australia, v. 25, p. 267-283.

Chappell, B.W. and White, A.J.R., 1974, Two contrasting granite types: Pacific Geology, v. 8, p. 173-174.

Chauris, L., 1977, Les associations paragenetiques dans la metallogenic varisque du Massif amoricain: Mineral. Deposita, v. 12, p. 353-371.

Chorlton, L., 1973, Effect of Boron on Phase Relations in the Granite-Water System: Unpublished M.Sc. Thesis, McGill University, Montreal.

Colman-Sadd, S.P., 1976, Geology of the St. Albans Map Area (1M/13), Newfoundland: Newfoundland Mines and Energy Rept. 78-10, p. 43.

Currie, K.L., 1968, On the solubility of albite in supercritical water in the range 400 to 600°C and 750 to 3500 bars: Amer. Jour. Sci., v. 266, p. 321-341.

Currie, K.L. and Pajari, G.E., 1977, Igneous and metamorphic rocks between Rocky Bay and Ragged Harbour, northeastern Newfoundland: in Report of Activities, Part A, Geol. Survey Canada Paper 77-1A, p. 341-346.

Dagger, G.W., 1972, Genesis of the Mount Pleasant tungsten-molybdenum – bismuth deposit, N.B., Canada: Trans. Instit. Mining Metallurgy Section B, v. 81, p. B73-B102.

DeGrace, J.R., Kean, B.F., Hsu, E. and Green, T., 1976, Geology of the Nippers Harbour Map Area (2E/13), Newfoundland: Newfoundland Mining Devel. Div. Rept. 76-3, 73 p.

Edgar, A.D. and Parker, L.N., 1974, Comparison of some melting relations of some plutonic, and volcanic peralkaline undersaturated rocks: Lithos, v. 7, p. 263-273.

Ellis, R.A., 1977, Disseminated copper in a Caledonian calc-alkaline intrusion, Argyllshire, Scotland: Trans. Instit. Mining Metallurgy, v. 86, p. B52-54.

Evans, A.M., 1977. Copper-molybdenum mineralization in the Ballachulish granite, Argyllshire, Scotland: Trans. Instit. Mining Metallurgy, v. 86, p. B152-153.

Flynn, R.T., and Burnham, C.W., 1978, An experimental determination of rare earth partition coefficients between a chloride-containing vapor phase and silicate melts: Geochim. Cosmochim. Acta, v. 42, no. 6A, p. 679-684.

Grant, J.N., Halls, C., Avila, W., and Avila, G., 1977, Igneous geology and the evolution of hydrothermal systems in some sub-volcanic tin deposits of Bolivia: Geol. Soc. London Spec. Publ. No. 7, p. 117-126.

Griffin, T.J., White A.J.R. and Chappell, B.W., 1978, The Moruya Batholith and geochemical contrasts between the Moruya and Jindabyne Suites: Jour. Geol. Soc. Australia, v. 25, pt. 4, p. 235-247.

Guild, P.W., 1978, Metallogenesis in the western United States: Jour. Geol. Soc. London, v. 135, p. 355-376.

Hall, A., 1971, The relationship between geothermal gradient and the composition of granitic magmas in orogenic belts: Contrib. Mineral. Petrol., v. 32, p. 186-192.

_____, 1972, New data on the composition of Caledonian granites: Mineral. Mag., v. 38, p. 847-862.

Hanmer, S.K., 1977, Age and tectonic implications of the Baie d'Audierne basic-ultrabasic complex: Nature, v. 270, p. 336-338.

Harris, P.G., Kennedy, W.Q., and Scarfe, C.M., 1970, Volcanism versus plutonism – the effect of chemical composition: in Newall, G., and Rast, N., eds., Mechanism of Igneous Intrusion: Liverpool, Gallery Press, p. 187-200.

Henley, R.W. and McNabb, A., 1978, Magmatic vapour plumes and groundwater interaction in porphyry copper emplacement: Econ. Geol., v. 73, p. 1-20.

Hildreth, W., 1979, The Bishop Tuff: Evidence for the origin of compositional zonation in silicic magma chambers: Geol. Soc. Amer. Symp. Volume on Ash-Flow Tuffs (in press).

Hine, R., Williams, I.S., Chappell, B.W. and White, A.J.R., 1978, Contrasts between I- and S-type granitoids of the Kosciusko batholith: Jour. Geol. Soc. Australia, v. 25, p. 219-234.

Holland, H.D., 1972, Granites, solutions, and base metal deposits: Econ. Geol., v. 67, p. 281-301.

Hollister, V.F., Potter, R.R., Barker, A.L., 1974, Porphyry-type deposits of the Appalachian Orogen: Econ. Geol., v. 69, p. 618-630.

Hoisington, W.D., 1977, Uranium and Thorium Distribution in the Conway Granite of the White Mountain Batholith: Unpublished M.A. Thesis, Dartmouth College, Hanover, N.H., 107 p.

Hosking, K.F.G. and Shrimpton, G.J., 1964, Present views on some aspects of the geology of Cornwall and Devon: Royal Soc. Cornwall, Penzance, 330 p.

Hurley, P.M., Bateman, P.C., Fairbairn, H.W., and Pinson, W.H., Jr., 1965, Investigation of [87]Sr/[86]Sr ratios in the Sierra Nevada Plutonic Province: Geol. Soc. America Bull., v. 76, p. 165-174.

Jahns, R.H., and Burnham, C.W., 1969, Experimental studies of pegmatite genesis: a model for the derivation and crystallization of granitic pegmatites: Econ. Geol., v. 64, p. 843-864.

Jayasinghe, N., 1979, A Petrological, Geochemical and Structural Study of the Wesleyville Area, Newfoundland: Unpublished Ph.D. Thesis, Memorial University of Newfoundland.

Kesler, S.E., 1973, Copper, Molybdenum and gold abundances in porphyry copper deposits: Econ. Geol., v. 68, p. 106-111.

Kistler, R.W., Evernden, J.F. and Shaw, H.R., 1971, Sierra Nevada Plutonic Cycle: Part I. Origin of composite batholiths: Geol. Soc. America Bull., v. 82, p. 853-868.

Koide, H., and Bhattacharji, 1978, Geometric patterns of active strike-slip faults and their significance as indicators for areas of energy release: in Saxena, S.K. and Bhattacharji, S., eds., Energetics of Geological Processes: p. 47-66.

Leake, B.E., 1978, Granite emplacement: the granites of Ireland and their origin: in Bowes, D.R. and Leake, B.E., eds., Crustal Evolution in Northwestern Britain and Adjacent Regions: The Seel House Press, Liverpool, p. 221-248.

LeFort, Pm, 1978, Relations of two-mica granites with crustal lineaments – examples of the Haut-Dauphine (French Alps) and Himalaya: Sciences de la Terre, No. 11, p. 31-34, May, 1978.

Luth, W.C., 1976, Granitic Rocks: in Bailey, D.K. and MacDonald, I., eds., The Evolution of the Crystalline Rocks: Academic Press, N.Y., p. 335-417.

Luth, W.C. and Tuttle, O.F., 1969, the hydrous vapour phase in equilibrium with granite and granite magmas: in Larsen, L., Mason, V. and Prinz, M., eds., Geol. Soc. Amer. Mem. 115, Igneous and Metamorphic Geology: p. 513-548.

MacDonald R., 1974, Tectonic settings and magma associations: Bull. Volcanol., v. 38, p. 575-593.

McCartney, W.D., Poole, W.A., Wanless, R.Km, Williams, H., and Loveridge, W.D., 1966, Rb/Sr age and geological setting of the Holyrood granite, southeast Newfoundland: Canadian Jour. Earth Sci., v. 7, p. 1485-1498.

McNutt, R.H., Crocket, J.H., Clark, A.H. Caelles, J.C., Farrar, E., Hynes, S.J. and Zentilli, M., 1975, Initial $^{87}Sr/^{86}Sr$ ratios of plutonic and volcanic rocks of the central Andes between latitudes 26° and 29° South: Earth Planet. Sci. Letters, v. 27, p. 305-313.

Mitchell, A.H.G., 1974, Southwest England granites: magmatism and tin mineralization in a post-tectonic setting: Trans. Instit. Mining Metall., v. 83, p. B95-B97.

Mitchell, A.H.G. and Garson, M.S., 1972, Relationship of porphyry copper and circum-Pacific tin deposits to paleo-Benioff zones: Trans. Instit. Mining Metall., v. 81, p. B10-B25.

Moorbath, S., 1978, Age and isotope evidence for the evolution of continental crust: Phil. Trans. Royal Soc. London, A288, p. 401-414.

Moreau, M. and Ranchin, G., 1973, Alterations hydrothermales et contrôles tectoniques dans les gîtes filoniens d'uranium intragranitiques du Massif Central Français: in Morin, P., ed., Les Rôches Plutoniques dans Leurs Rapports avec les Gîtes Mineraux: Masson, Paris, p. 77-100.

Morel, P. and Irving, E., 1978, Tentative paleocontinental maps for the early Phanerozoic and Proterozoic: Jour. Geol., v. 86, p. 535-561.

Mustart, D.A., 1972, Phase Relations in the Peralkaline Portion of the System $Na_2O - Al_2O_3 - SiO_2 - H_2O$: Unpublished Ph.D. Thesis, Stanford University.

Naney, M.T., 1978, Stability and Crystallization of Ferromagnesian Silicates in Water-Vapour Undersaturated Melts at 2 and 8 kb Pressure: Unpublished Ph.D. Thesis, Stanford University.

Nicholas, A., Bouchez, J.L., Blaise, J., and Poirier, J.P., 1977, Geological aspects of deformation in continental shear zones: Tectonophysics, v. 42, p. 55-73.

Nielsen, R.L., 1976. Recent developments in the study of porphyry copper geology: A review: Canadian Instit. Mining Metall. Spec. v. 15, p. 487-500.

Norton, D.L., and Cathles, L.M., 1973, Breccia pipes-products of exsolved vapour from magmas: Econ. Geol., v. 68, p. 540-546.

Pereira, J. and Dixon, C.J., 1971, Mineralization and plate tectonics: Mineral. Deposita, v. 6, p. 404-405.

Pidgeon, R.T. and Aftalion, M., 1978, Cogenetic and inherited zircon U-Pb systems in granites: Paleozoic granites of Scotland and England: in Bowes, D.R. and Leake, B.E., eds., Crustal Evolution in Northwestern Britain and Adjacent Regions: The Seel House Press, Liverpool, p. 183-220.

Potter, R.R., 1968, Geology of the Burnt Hill Area With Special Reference to the Ore Controls in the Vicinity of the Burnt Hill Tungsten Mines: Ph.D. Thesis, Carleton University.

Reitan, P.H., 1968, Frictional heat during metamorphism. 1. Quantitative evolution of concentration of heat generation in time: Lithos 1, p. 151-163. 2. Quantitative evolution of concentration of heat generation in space: Lithos 1, p. 268-274.

Rice, R. and Sharp, G.J., 1976, Copper mineralization in the forest of Coed-y-Brenin, North Wales: Trans. Instit. Mining Metall., v. 85, p. B1-B13.

Rich, R.A., Holland, H.D. and Petersen, U., 1977, Hydrothermal Uranium Deposits: Elsevier, N.Y., 264 p.

Richardson, S.W., Gilbert, M.C. and Bell, P.M., 1969, Experimental determination of the kyanite-andalusite and andalusite-sillimanite equilibria; the aluminum silicate triple point: Amer. Jour. Sci., v. 267, p. 467-288.

Rose, A.W., 1970, Zonal relations of wallrock alteration and sulfide distribution in porphyry copper deposits: Econ. Geol., v. 65, p. 920-936.

Shawe, D.R., 1968, Geology of the Spor Mountain Beryllium District, Utah: in Ridge, J.D., ed., Ore Deposits of the United States: v. II. A.I.M.E., N.Y.

Smith, T.E. and Turek, A., 1976, Tin-bearing potential of some Devonian Granitic rocks in S.W. Nova Scotia: Mineral. Deposita (Berl.), v. 11, p. 234-245.

Sillitoe, R.H., 1972, A plate tectonic model for the origin of porphyry copper deposits: Econ. Geol., v. 67, p. 184-197.

Sillitoe, R.H., Grant, J.N., and Halls, C., 1975, Porphyry tin deposits in Bolivia: Econ. Geol., v. 70, p. 913-927.

Stemprok, M., Burnol, L. and Tischendorf, G., 1978, Metallization Associated with Acid Magmatism: Ustredni Ustav Geobgicky. Praha, v. I (1974), 410 p., II (1977) 166 p., and III (1978), 446 p.

Stewart, D.B., 1978, Petrogenesis of lithium-rich pegmatites: Amer. Miner., v. 63, p. 970-980.

Strong, D.F., 1979, The Mount Peyton batholith, central Newfoundland: A bimodal calc-alkaline suite: Jour. Petrol., v. 20, p. 119-138.

Strong, D.F., O'Brien, S.J., Taylor, S.W., and Wilton, D.H., 1978. Geology of the St. Lawrence (1L/14) and Marystown (1M/3) Map Areas, Newfoundland: Dept. Mines and Energy, Min. Dev. Div. Rept. 77-8, 81 p.

Sutherland Brown, A., 1976, Porphyry Deposits of the Canadian Cordillera: Canadian Instit. Mining Metall. Spec., Vol. 15, 510 p.

Taylor, R.P., Strong, D.F., and Kean, B.F., 1979, Devonian peralkaline volcanism and plutonism in western Newfoundland Abst.: Ann. Mtg. Geol. Assoc. Canada, Quebec City, May , 1979.

Taylor, S.R., 1965, The application of trace element data to problems in petrology: in Physics and Chemistry of the Earth, v. 6, p. 133-213.

Teng, H.C. and Strong, D.F., 1976, Geology and geochemistry of the St. Lawrence granite and associated fluorspar deposits, southeast Newfoundland: Canadian Jour. Earth Sci., v. 13, p. 1374-1385.

Tischendorf, G., Schust, F., and Lange, H., 1978, Relation between granites and tin deposits in the Erzgebirge, G.D.R.: in Stemprok, M., Burnol, L., and Tischendorf, G., eds., Metallization Associated with Acid Magmatism: Ustredni Ustav Geologicky, Praha, v. 3, p. 123-138.

Wedepohl, K.H., 1969, Handbook of Geochemistry: Springer-Verlag, Berlin-Heidelberg, New York.

Whalen, J.B., 1976, Geology and Geochemistry of the Molybdenite Showings of the Ackley City Batholith, Fortune Bay, Newfoundland: Unpublished M.Sc. Thesis, Memorial University, St. John's, Newfoundland, 267 p.

White, A.J.R. and Chappell, B.W., 1977, Ultrametamorphism and granitoid genesis: Tectonophysics, v. 43, p. 7-22.

White, A.J.R., Chappell, B.W. and Cleary, J.R., 1974, Geologic setting and emplacement of some Australian Paleozoic batholiths and implications for some intrusive mechanisms: Pacific Geol., v. 8, p. 159-171.

White A.J.R., Williams, I.S. and Chappell, B.W., 1977, Geology of the Berridale 1:100,000 Sheet, 8625: Geol. Survey New South Wales, Dept. Mines, 138 p.

Whitney, J.A., 1975, Vapour generation in a quartz monzonite magma: A synthetic model with application to porphyry copper deposits: Econ. Geol., v. 70, p. 346-358.

Williams, H., Kennedy, M.J. and Neale, E.R.W., 1974, The Northeastward Termination of the Appalachian Orogen: in Nairn, A.E.M. and Stelhi, F.G., eds., The Ocean Basins and Margins: Plenum, N.Y., v. 2, p. 79-123.

Wilson, J.T., 1966, Did the Atlantic close and then re-open?: Nature, v. 211, p. 676-681.

Windley, B.F., 1978, The Evolving Continents: John Wiley and Sons, 385 p.

Wyllie, P.J., 1977, Crustal anatexis: an experimental review: Tectonophysics, v. 43, p. 41-71.

Wyllie, P.J. and Tuttle, O.F., 1961, Experimental investigation of silicate systems containing two volatile components, Part II: Amer. Jour. Sci., v. 259, p. 128-143.

The Continental Crust and Its Mineral Deposits, edited by D.W. Strangway,
Geological Association of Canada Special Paper 20

ORE-LEAD ISOTOPES AND GRENVILLE PLATE TECTONICS

R. M. Farquhar and I. R. Fletcher
Geophysics Laboratory,
Department of Physics,
University of Toronto,
Toronto, Ontario M5S 1A7

ABSTRACT

Recent advances in the "whole-earth" modelling of evolutionary processes of Pb isotopes shed light upon the origin of the metals found in various types of ore deposits. On the bases of these models and several recently published data sets, we believe that the ore deposits formed in various plate tectonic environments may carry "isotopic fingerprints" which, when used with other characteristics such as mineral assemblages, may identify the depositional environments of many ore bodies.

In the present study Pb-isotopic measurements have been made of a number of Precambrian mineralization types and localities throughout the Central Metasedimentary Belt of the Grenville Province. The data for individual deposits are at best ambiguous, but fall into two groups sufficiently distinctive to allow some degree of "fingerprint" identification.

Comparisons with data from other areas suggest that the major periods of sedimentation within the Central Metasedimentary Belt accompanied plate rifting and/or island arc tectonic activity, with most of the mineralized lead being derived from mantle sources.

Detailed comparisons between the Grenville and the other regions are uncertain, mainly because there are few detailed high-accuracy data sets from younger, tectonically unambiguous mineral occurrences. We suggest that once these data sets are available, isotopic fingerprinting may become diagnostic for deposits ranging well back into the Precambrian.

RÉSUMÉ

Les développements récents dans la modélisation "à l'échelle de la Terre" des processus d'évolution des isotopes Pb ont jeté de la lumière sur l'origine des métaux qu'on retrouve dans différents types de gîtes minéraux. En nous basant sur ces modèles et plusieurs groupes de données publiées récemment, nous croyons que les gîtes minéraux formés dans différents contextes de tectonique des plaques peuvent garder des "empreintes isotopiques" qui, lorsqu'on les utilise avec d'autres caractéristiques comme les assemblages de minéraux, peuvent identifier les milieux de dépôt de plusieurs gîtes minéraux.

Dans cette étude, on a fait des mesures des isotopes de Pb sur un certain nombre de types de minéralisation datant du Précambrien à plusieurs localités à travers la zone métasédimentaire centrale de la province de Grenville. Les données provenant de dépôt individuels sont au mieux ambiguës, mais elles tombent dans deux groupes suffisamment distincts pour permettre jusqu'à un certain degré l'identification des "empreintes".

Les comparaisons avec les données d'autres régions suggèrent que les périodes principales de sédimentation dans la zone métasédimentaire actuelle ont accompagné l'effondrement de la plaque et/ou l'activité tectonique d'arcs insulaires, et que la plus grande partie de la minéralisation de plomb proviendrait de sources dans le manteau.

Des comparaisons détaillées entre le Grenville et d'autres régions sont incertaines surtout parce qu'il y a peu de données détaillées de grande précision pour des gîtes minéraux plus récents et non ambigus au point de vue tectonique. Nous suggérons que lorsque ce type de données deviendra disponible, les empreintes isotopiques pourront devenir diagnostiques pour les dépôts datant du Précambrien.

INTRODUCTION

The past decade has seen considerable advances in both measuring and interpreting the lead isotopic compositions found in nature. Technical advances have improved analytical precision, allowing previously-unknown isotopic fine structure to be identified in the data from a number of geological settings, and much more sophisticated mathematical models have been developed to represent the observed lead isotopic patterns. Many of these models have plate tectonic implications, some applying to the data patterns in specific situations, such as mid-oceanic plate environments (e.g., Tatsumoto, 1978; Russell, 1972) and others reflecting improved understanding of "average-Earth" isotopic and elemental evolution (e.g., the differentiating-Earth "Plumbotectonics" model of Doe and Zartman, 1979).

In our study, we have searched for specific, possibly recurring and possibly plate-tectonically generated data patterns. We therefore make use of some of the more accurate data now available in the literature, but at this stage the interpretations do not rely on any form of mathematical modelling. However, if our approach here is successful, then the evolutionary and environmental significance of the data patterns can only be determined by applying such models.

THE GRENVILLE PROVINCE AND PLATE TECTONICS

The Grenville Province is one of the major segments of the Canadian Shield, extending for ~1500 km along its southeastern margin (Fig. 1). Although parts of the Grenville are considerably older than 1 Ga, its present characteristics result primarily from a major orogenic event at about that time. The geological and structural complexities resulting from this orogeny are so great that the pre-orogenic history of the region, and the nature of the orogeny itself, are very poorly recorded and understood. The geology and structure of the province have been well summarized by Wynne-Edwards (1972), and Baer (1974) has reviewed current opinions concerning its history. Since most of the models proposed recently for Grenville evolution have

TABLE I

Summary of some recent models of Grenville tectonism. All models are consistent with published geochronological data.

Model Features	Reference
Continental platform, deposition of volcanics and sediments ending prior to Grenville metalmorphism: based on regional structural, petrological analysis.	Wynne-Edwards, 1972
Island arc followed by ocean closure, sedimentation and continental collision: based on analysis of volcanism, sedimentation in Hastings basin.	Brown *et al*, 1975
Continental rifting, volcanism associated with aulacogen formation: based on regional structural, petrological, and geochronological analysis.	Baer, 1976
Continental platform, thickened, remobilized and deformed by lateral compression resulting from continental collision: based on regional structural, paleomagnetic analysis.	McWilliams and Dunlop, 1978
Wilson cycle: based on analysis of paleomagnetism, Rb-Sr ages of diabase dykes.	Patchett *et al*, 1978
Wilson cycle: based on analysis of paleomagnetism, continental reconstruction.	Seyfert, 1978

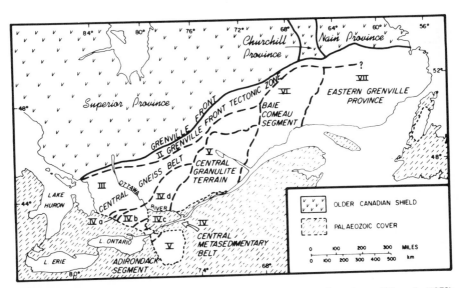

Figure 1. The Grenville province and its structural subdivisions, after Wynne-Edwards (1972) (Wynne-Edwards subdivision I, the Foreland Belt, is not shown).

analogies in modern plate tectonics, we have compared the lead isotopic patterns for the Grenville with those from other areas.

The Central Metasedimentary Belt (CMsB) is the only major segment of the Grenville which includes extensive exposures of metasedimentary rock units, more than half of the present surface being of this type (Fig. 2). Since these units almost certainly post-date previous orogenies, the CMsB provides the best opportunity for determining the details of the orogeny at ~ 1 Ga.

Despite the amount of work which has been directed towards this question, neither the time scale nor the style of the orogenic activity is yet well determined. Table I summarizes some of the differences of opinion found in the literature.

OUTLINE OF THIS STUDY

In this paper, we present new lead isotopic data fields which we believe to be representative of CMsB. We have attempted to interpret these data fields by identifying a number of similar fields from different regions, and then considering whether

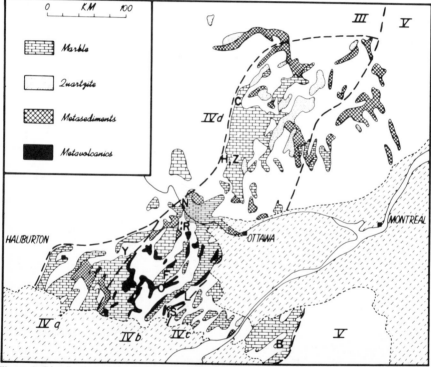

Figure 2. Distribution of the major rock types of the Grenville Supergroup, after Baer (1976), and the subsections of the Central Metasedimentary Belt, after Wynne-Edwards (1972). Sample sites within the Belt are identified by letters. B is Balmat-Edwards. Tétrault (T) lies in the Central Granulite Terrain 380 km east of N. T and N are classified as stratiform volcanogenic deposits. C,L,S,R,H, and Z are mineralized zones within metamorphosed carbonate rocks. Y occurs in a nepheline gneiss. O and F are associated with quartz veins.

the tectonic situations pertaining to these could also account for the formation of CMsB. We have, in effect, searched for "isotopic fingerprints" for plate tectonic environments which may apply to CMsB. We do not claim that our sets of comparative data from other regions are exhaustive. In the extensive literature on lead isotopes, there are other blocks of isotopic ratio data which we might have used (for example, the very recent work of Doe *et al.*, 1979, on the relationships among ore and rock leads in the San Juan volcanic field of southwestern Colorado). The examples we have selected, however, adequately illustrate certain features which may contain clues to common tectonic regimes.

This approach has been used previously, notably in attempts to apply lead isotopic data to problems in mineral exploration and mineral prospect evaluation (e.g., Cannon *et al.*, 1971; Gulson and Mizon, 1979). This interpretive method has several dangers, not the least of which is the prejudice of the investigators, so interpretations must be treated with caution. However, useful comparisons can usually be made, even when isotopic pattern matching results in convincing "negative proof", i.e., proof that certain proposed models are unacceptable.

The lack of suitable reference data also limits interpretations. It will be seen that the reference data sets for some tectonic environments are quite inadequate. With extension of these, this interpretive method may become much more useful.

Stanton and Russell (1959) were the first to recognize that the ratios of lead isotopes in certain sulphide deposits of the massive, stratiform class appeared to change through time in a rather simple manner. Using the data available at the time, they showed that the lead isotopic ratios for those deposits lay close to a growth curve representing isotopic evolution in a single closed reservoir and that the "model ages" computed for the leads were reasonably close to their geological ages defined in other ways. In the light of improved decay constant values, isotopic ratios of higher accuracy, and some additional data points, more complex models have been developed (Oversby, 1974; Stacey and Kramers, 1975; Cumming and Richards, 1975). As the growth curves deriving from these models do not fit all the data points exactly, these curves are now considered to be "averages" for orogenic leads. These recent models provide model ages which are generally within ± 50 Ma. of the times of initial mineralization, where these are known. More recently Doe and Zartman (1979) have developed a complex "plumbo-tectonic" model in which the isotopic compositions of continental leads result from the mixing of inputs from mantle, sub-crustal, and crustal reservoirs; the dynamic nature of the Earth is taken into account by allowing these reservoirs to exchange uranium, thorium, and lead throughout geological time.

In all of the models, growth curves are characterized by the parameter μ, the ratio $^{238}U/^{204}Pb$ (referred to the present, t = 0). Figure 3 shows two growth curves, the "average" curve of Stacey and Kramers (1975) ($\mu = 10.87$), which corresponds to the "orogene" curve of Doe and Zartman (1979), and the latter authors' mantle growth curve ($\mu = 8.92$). Most of the data we will discuss are bounded by these curves. Also shown on Figure 3 are a set of equiage lines similar to conventional isochrons; each line denotes the isotopic relations among leads extracted at a given time from the mantle and/or orogene. It is on the basis of these isochrons that the model ages for leads are generally determined. Since our data are for leads which

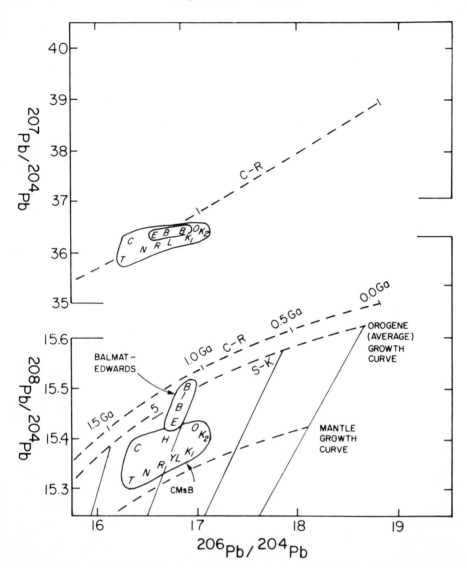

Figure 3. Lead isotopic data fields for samples from the Central Metasedimentary Belt and the Balmat-Edwards area. Letters within the CMsB field identify data points for locations shown on Figure 2. S and Z have low lead contents and are not plotted. K_1 and K_2 are data for feldspars from Westport, Ontario and Balmat, New York (Zartman and Wasserburg, 1969). C-R and S-K identify the "average" growth curves of Cumming and Richards (1975) and Stacey and Kramers (1975) respectively. The second-stage isochrons of Stacey and Kramers are shown. The mantle growth curve is that of Doe and Zartman (1979).

may have been derived from volcanic processes at continental margins, island arcs, or rift zones, the curve representing mantle lead is clearly relevant. The "average" curve is also relevant, since it is defined by data points for massive, stratiform sulphide deposits, most of which are classed as volcanogenic in origin.

The growth curve concept can also be applied to ^{208}Pb-^{206}Pb-^{204}Pb data as shown on Figure 3. Growth curves corresponding to average continental crust and mantle are defined by relationships among ^{232}Th, ^{238}U and Pb, a system which is generally less constrained than the ^{238}U- ^{235}U -Pb system. Since ^{208}Pb-^{206}Pb-^{204}Pb variations are less distinctive, this paper will focus on ^{207}Pb-^{206}Pb-^{204}Pb relationships.

DATA FIELDS AND ISOTOPIC CHARACTERISTICS

The new data used in the following sections came out of a survey of Pb isotope ratios in Pb-Zn mineralization associated with the CMsB (Fletcher, 1979). Although an exhaustive sampling has not been possible, all major geological classes of deposits and all of the best-known deposits are represented. These data, together with published data for some CMsB rocks and minerals are plotted in Figure 3. They clearly fall into two fields.

A) *Central Metasedimentary Belt.* The first Grenville data field in Figure 3 includes all published lead isotopic data for Precambrian mineral deposits and rocks within the CMsB except for those relating to Balmat-Edwards ore deposits. The data and a detailed discussion of the isotopic-mineralogical relationships will be published elsewhere. It is notable that the data field includes K-feldspars from the Balmat area.

As there is true fine structures in the data, with discernible differences among the mineral data as well as between minerals and rocks, it is not reasonable to represent these data by single lines. This fine structure is certainly important in defining the relationship between the lead isotopes and the detailed geology of the Grenville supracrustal rocks, but for the purpose of this paper, the detailed variations are less important than the features of the overall data field.

Taken as a group, and regarded qualitatively, the ^{207}Pb- ^{206}Pb- ^{204}Pb field has several characteristics which may be diagnostic. These are: i) *consistently low* μ: The entire data set falls well below the orogene curve; most of it in fact lies rather close to the mantle curve; ii) *broad linearity:* There is a spread of points along a low-angle, roughly linear trend. For comparison purposes, the slope of the trend can be regarded as roughly tangential to the orogene growth curve at about the oldest (or least radiogenic) end of the data field. The data distribution is bounded on the left by the 1.3 Ga isochron and on the right by the 0.8 Ga isochron. We do not wish to suggest that this age range has any direct geochronological significance, although the ^{207}Pb - ^{206}Pb isochron age (Cumming and Richards, 1975) of the least radiogenic lead (Tétrault ~ 1.29 Ga) is in reasonable accord with limiting age estimates (1.38 to 1.23 Ga) for the sedimentation which preceded the metamorphism in the CMsB (Barton and Doig, 1974). Our purpose in defining the data range is to point out that the low-angle distribution of these CMsB lead isotopic ratios is relatively short.

B) *Balmat-Edwards.* Our data set for this field includes 31 analyses of composite ore concentrates and 9 analyses of galenas from various orebodies, which fills out the distributions observed and commented on by earlier authors (Doe, 1962a, b;

Reynolds and Russell, 1968). Our data for this field will be published elsewhere. Several general characteristics of the Balmat-Edwards field can be clearly specified, and may serve as bases for qualitative comparisons with other fields. They are: 1) *extreme linearity of the isotopic data:* Within the limits of experimental uncertainty, all of our data, (galenas and composite ore samples) conform to single straight lines on the ^{207}Pb-^{206}Pb-^{204}Pb and ^{208}Pb-^{206}Pb-^{204}Pb graphs. The slopes of the lines on these graphs are $R' = 0.36 \pm .04$ (^{207}Pb-^{206}Pb-^{204}Pb) and $R'' = 0.44 \pm .07$ (^{208}Pb-^{206}Pb-^{204}Pb). The slope of the ^{207}Pb-^{206}Pb-^{204}Pb best-fit line is too high to permit interpretation of the trend as a secondary isochron, but for comparison purposes, it can be taken as roughly parallel to an isochron through the primordial lead isotopic ratios (a "geochron"). Although the slope of the ^{208}Pb-^{206}Pb-^{204}Pb line is notably lower than that of the adjacent growth curves, no useful comparisons have yet been found; 2) *variable apparent μ:* This is a corollary to i). It may or may not be significant that the upper end of the trend falls on the orogene curve; 3) *spatial extent:* The samples represented by this field cover a considerable stratigraphic range ($\sim$ 700 m) but quite limited areal range (maximum separation $\sim$ 15 km).

ISOTOPIC DATA FIELDS FOR OTHER AREAS

The differences among isotopic ratios that delineate the two groups of CMsB leads are relatively small. Although there are a great many lead isotopic analyses in the literature, most are early measurements having comparatively large analytical uncertainties, or are for epigenetic fissure-vein deposits. When we consider only highly accurate isotopic ratios for sulphides believed to be of volcanogenic or of volcanic-sedimentary origin, the data available for comparison with the CMsB results are remarkably few. Since the nature and time scales of tectonic processes may have been different in the Archean, we have also confined study to data sets for regions in which extensive orogenic events took place more recently than $\sim$ 2 Ga. For these data sets, we will identify isotopic distributions which seem to be similar to the two Grenville data fields.

DISTRIBUTIONS SIMILAR TO CMsB DATA

1) *Montana/Idaho, U.S.A.* In an extensive survey of lead isotopic relationships among sulphide deposits in northwestern Montana and northern Idaho, Zartman and Stacey (1971) identified the trend shown on Figure 4. The mineralization (which they term "Coeur d'Alene" type) occurs mainly in veins in Belt Supergroup rocks, although a few stratabound sulphide deposits exist. Zartman and Stacey believe that the lead whose isotopic ratios define this trend was introduced into the host rocks at or soon after sedimentation. The isotopic ratios lie between the 1.5 Ga and 0.8 Ga isochrons. Like the more highly metamorphosed CMsB rocks, the age of the Belt Supergroup sedimentation is not well known; Zartman and Stacey suggest a range of 1.7 to 1.1 Ga, based on geochronological measurements on rocks and minerals which predate or intrude the supracrustals.

The distribution of isotopic data on Figure 4 is broadly reminiscent of that observed in CMsB stratiform zinc-lead sulphides. It is a relatively short, low angle trend, which lies below the "orogene" growth curve, but above the CMsB data field.

2) *Namaqualand, South Africa*. Koeppel (1978) has described the results of a study of lead isotopic variations in stratiform ore deposits occurring in rocks of the Namaqualand Metamorphic Complex. Within the Bushmanland Supergroup of this complex, the ores at Prieska lie in metavolcanics which were originally andesitic to rhyolitic (Middleton, 1976). The lead isotopic ratios in this deposit are homogeneous; lying on a broad band to the right of the data for Prieska are isotopic ratios for other deposits in the area (Fig. 4). On Figure 4 these data points lie between the 1.3 Ga and 1.0 Ga isochrons. Thus the outline and scale of the Namaqualand data is similar to that for the CMsB, but is shifted to a region above the "orogene" growth curve, suggesting a derivation from a higher μ source.

3) *Southwestern Japan*. Sato and Sasaki (1976, 1978) have published and discussed the results of a number of isotopic analyses of lead in Paleozoic and Mesozoic stratiform sulphide deposits of the Besshi-type (bedded cupriferous iron sulphides).

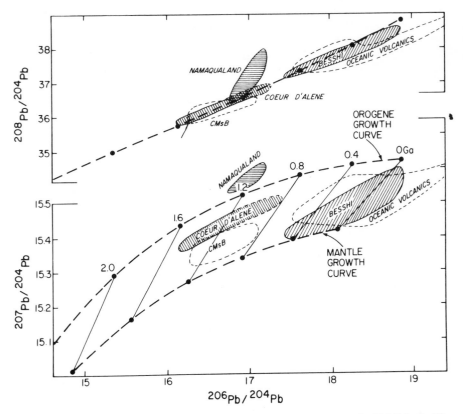

Figure 4. Data field outlines for isotopic ratio distributions similar to those for CMsB leads. The "orogene" growth curve of Doe and Zartman (1979) is similar to the "average" curve of Stacey and Kramers (1975). Lines joining points of equal time on "orogene" and mantle growth curves are equivalent to isochrons.

These deposits are associated with basic schists derived from basaltic rocks and are believed to be volcanogenic. There are significant isotopic variations among these sulphides, as shown on Figure 4. The data set forms a low-angle trend which lies to the right of the .5 Ga isochron, and extends significantly to the right of the 0 Ga (present day) isochron. This time range is much greater than that required for the sulphide deposition, as Sato and Sasaki point out. The trend lies below the "orogene" growth curve, close to that suggested for the mantle.

DISTRIBUTIONS SIMILAR TO BALMAT-EDWARDS DATA

The deposits at Balmat, New York, lie within a metasedimentary sequence which is locally devoid of metavolcanics (Lea and Dill, 1968). The genesis of the ores and, in particular, the role of volcanic processes in that genesis remain unresolved problems. With two possible exceptions, there seem to be no other stratiform ore deposits in similar host rocks for which extensive studies of lead isotopes have been made. We have, therefore, searched among published lead isotopic data on stratiform sulphides for ^{207}Pb-^{206}Pb-^{204}Pb distributions like that observed at Balmat, without regard to the type of rocks which enclose the mineral deposits.

1) *Flin Flon-Snow Lake, Canada.* Sulphide mineralization in these two areas is stratiform within metavolcanic host sequences. The two areas are approximately 100 km apart, and it is not clear whether they are separate volcanic complexes or parts of the same complex (Sangster, 1972). Slawson and Russell (1973) have analyzed lead from the Flin Flon mine and from the Chisel Lake mine and Snake Lake deposits in the Snow Lake region. The results are plotted on Figure 5; they define a steep distribution which roughly parallels the 1.6 Ga isochron. Almost the full range of isotopic variation is recorded in each of the areas, with the greatest ^{207}Pb-^{206}Pb-^{204}Pb isotopic ratio differences occurring between the samples from the Flin Flon mine. The observed differences have roughly the same magnitude as the Balmat-Edwards distribution, and the data carry a markedly low-μ imprint.

2) *Sullivan, Canada.* The sulphide ores of the Sullivan mine are stratiform and occur within a sequence of metamorphic rocks consisting mainly of quartzites, silt-stones, and argillites, among which there are very few volcanic components. Sangster (1972) has suggested that the Sullivan mineralization is the result of an exhalative process, and that its genesis differs from the normal volcanic style only in the depositional environment. Only two lead isotopic data points of high precision are available, and their $^{206}Pb/^{204}Pb$ and $^{207}Pb/^{204}Pb$ ratios differ by amounts only slightly greater than the experimental uncertainties quoted for the data (Stacey *et al.*, 1977). It is probably unwise to regard the short, steep trend which these two points define as significant until more analyses have been undertaken, but the average clearly lies above the data distribution for the Coeur d'Alene type leads, just as the Balmat-Edwards data lies above the main CMsB field.

3) *Namaqualand, South Africa.* The isotopic studies of Koeppel (1978) include measurements on two stratiform sulphide deposits which occur in schists and quartzites within the Bushmanland Supergroup. Like the metasediments which host the Balmat-Edwards zinc ores, metavolcanics are not "conspicuous" (Koeppel) among the formations in which these deposits lie. Figure 5 indicates the lead isotopic ratios vary significantly within these sulphide deposits; the distributions are short, lie

TABLE II

Pb isotopic analyses of galena samples from the Buchans Mine area, Newfoundland

Sample No.	Location	$^{206}Pb/^{204}Pb$	$^{207}Pb/^{204}Pb$	$^{208}Pb/^{204}Pb$
1033	20′ Level	$17.826 \pm .014$	$15.494 \pm .014$	$37.607 \pm .041$
1036	200′ Level	$17.841 \pm .014$	$15.503 \pm .012$	$37.633 \pm .041$
1035	1100′ Level	$17.821 \pm .012$	$15.502 \pm .012$	$37.637 \pm .049$
1032	2600′ Level	$17.820 \pm .014$	$15.492 \pm .017$	$37.605 \pm .053$
203	exact location unknown	$17.823 \pm .027$	$15.481 \pm .023$	$37.57 \pm .06$

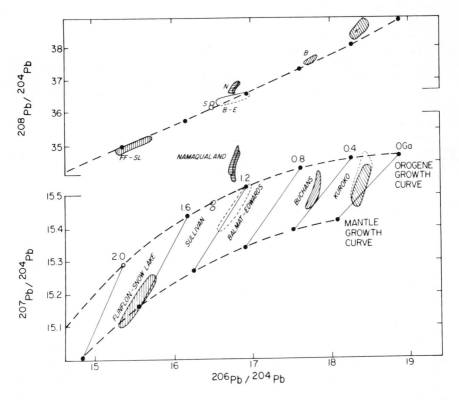

Figure 5. Data field outlines for isotopic ratio distributions similar to those for Balmat-Edwards leads. The growth curves and isochrons are defined as in Figure 4.

roughly parallel to the 1.25 Ga isochron, and extend above the data point for the Prieska mine. The isotopic distributions are in fact remarkably like those observed for Balmat-Edwards, except that they are shifted to higher $^{207}Pb/^{204}Pb$ ratios (higher apparent μ values), and they are not strictly linear.

4) *Buchans, Canada.* The sulphides in this mining area occur in a volcanic sequence of andesite flows, pyroclastics, silicic tuff, and agglomerates. Bell and Blenkinsop (1980) present lead isotopic analyses of five samples of galena from one of the mines (Fig. 5). We have measured lead isotopic ratios in five lead-rich sulphides from drill cores extending from near-surface to 800 m depth in the Buchans area, and as shown in Table II, we observe no differences in isotopic composition greater than analytical uncertainty. Our results lie on the trend defined by Bell and Blenkinsop's data, which is a short distribution approximately parallel to the 0.5 Ga isochron. As Bell and Blenkinsop have observed, the data points lie below the "orogene" growth curve.

5) *Northeastern Japan.* Associated with the Miocene volcanism in the northeastern tectonic province of Japan are the stratiform Kuroko-type deposits, which lie within a 40-km wide belt extending from Hokkaido to the southwestern boundary of the province in Honshu. Sato (1975) and Sato and Sasaki (1978) have analyzed lead for a number of the deposits in this belt, and their data are plotted in Figure 5. The results for a group of deposits in a small segment of the belt (Akita Prefecture) do not define a substantially different distribution than the entire data set. This is a rather broad but steeply inclined band of points on the ^{207}Pb-^{206}Pb-^{204}Pb graph; some of the breadth may be the result of the analytical uncertainties quoted by Sato and Sasaki, which are somewhat larger than those of other measurements used in the present paper. The distribution appears to be substantially different from that for the Cenozoic and Mesozoic ores of Japan described in the previous section.

GENERAL DISCUSSION   CMsB-TYPE ISOTOPIC DISTRIBUTIONS

In addition to the common feature of small but significant isotopic heterogeneity, the examples we include in this class (with the exception of the Namaqualand data) have μ values lower than that which identifies the "average" growth curve. Doe and Zartman (1979) have suggested that this "low μ" characteristic can be used as a measure of the amount of mantle-derived lead in the ores. As noted by Sato and Sasaki (1978, $^{207}Pb/^{206}Pb/^{204}Pb$ and $^{208}Pb/^{206}Pb/^{204}Pb$ ratios for lead from Besshi-type sulphide deposits lie on the trend of modern ocean ridge basalts (MORB); this is consistent with the rift zone and basic volcanism with which many of the ores are associated. The isotopic relationships for MORB (Tatsumoto, 1969, 1978) are shown on Figure 4. The low-angle distribution of the Besshi isotopic ratios would thus be a reflection of the heterogeneities of lead isotopes in the oceanic mantle beneath the Japanese Paleozoic arc.

It is questionable whether this relationship explains the isotopic distribution for any of the sulphide deposits of greater age. The Coeur d'Alene type deposits are believed to be the same age as the sediments in which they lie, but there is no evidence that the sulphides are volcanogenic. The apparent μ values of the Coeur d'Alene leads lie between the growth curves for the mantle and the orogene. If the model proposed by Doe and Zartman (1979) applies to these deposits, then they contain lead derived from the crust as well as the mantle. The mantle component might possibly have reached the crust via the continental rift, the remnants of which Kanasewich (1968) has suggested underlie the crust beneath part of the Belt Supergroup. If substantial mixing of lead from continental and mantle sources took place, it is difficult to see how the original mantle isotopic distribution could have been preserved. It is more likely that the linear trend observed on the ^{206}Pb-^{207}Pb-^{204}Pb plot is due to purely crustal processes, as Zartman and Stacey (1971) have suggested. Certainly, the chronological limits which the slope of the ^{207}Pb-^{206}Pb-^{204}Pb distribution requires can be fitted without difficulty into the rather wide limits on the time of the Belt Supergroup deposition.

The influence of crustal sources of lead is even more apparent in the isotopic data from the sulphides in the Namaqualand Metamorphic Complex. The isotopic ratios all lie above the "orogene" growth curve, suggesting that the older basement rocks underlying the Bushmanland Supergroup may have played an important role in determining the lead isotopic composition of the sulphides (Koeppel, 1978). These sulphides are all stratiform, and at least in some cases (Prieska, Rosh Pinah) are within metavolcanics or volcaniclastics (originally of calc-alkaline composition) (Middleton, 1976; Page and Watson, 1976). Radiometric ages for the host rocks of these sulphides (quoted by Koeppel, 1978) range from 1.3 Ga to 1.0 Ga, so that the ores may be the products of a protracted series of volcanic events rather than the result of a relatively short-lived pulse of mineralization from a source having variable isotopic composition. If this is so, the lead isotopic ratios in the sulphides seem to be registering the chronology of the events in a remarkably accurate manner, since the model ages range from 1.25 Ga to 1.1 Ga.

The isotopic ratios in the CMsB sulphides may reflect a similar set of events. These deposits have evidently derived much if not all of their lead from the mantle, which is in keeping with the proposition that this section of the Grenville was generated as new continental crust at $\sim$ 1.0 Ga (McCulloch and Wasserburg, 1978). Two of the deposits for which we have data (Tétrault, New Calumet) are classed as volcanogenic (Sangster, 1972), and the differences in isotopic composition may be due to real age differences. The model ages for these two deposits are $\sim$ 1.3 Ga and $\sim$ 1.2 Ga; Silver and Lumbers (1966) give uranium-lead dates on zircons of $\sim$ 1.3 Ga for the lower Tudor metavolcanics of the Hastings Basin, and $\sim$ 1.2 Ga for volcanics in the Adirondack Lowlands. Brown et al. (1975) interpret the succession of metavolcanic rocks in one section of the CMsB as products of the island arc phase of a tectonic cycle that also included subduction, the generation of calc-alkaline volcanism, and granodioritic intrusions. It is possible that two periods of sulphide deposition differing in age by $\sim$ .1 Ga could have taken place during the island arc stage; however, the ^{207}Pb-^{206}Pb-^{204}Pb distribution of the total data set of CMsB leads is too wide to be explained entirely by island arc-related volcanogenic processes. Given that the lead was probably derived fairly directly from the mantle, the spread of values may simply reflect the heterogeneity of the mantle at $\sim$ 1 Ga.

BALMAT-EDWARDS-TYPE ISOTOPIC DISTRIBUTIONS

The youngest deposits which record the short, steep isotopic distributions of the Balmat-Edwards type are the Kuroko ores of Japan. These sulphides were deposited in association with calc-alkaline volcanism, and are richer in lead than the Besshi-type ores. Doe and Zartman (1979) and Sato (1975) note that the steep distribution is unlikely to be the result of a second stage of lead isotopic generation, because this would require far too old a source for the leads. Doe and Zartman favour the mixing of various proportions of lead of either continental crustal or mantle isotopic composition with lead from pelagic sediments. Lead model ages could be calculated only if the composition of the undiluted mantle/crustal isotopic component were known, since the lead in pelagic sediments is derived mainly from the erosion of continental rocks, and is variable in isotopic composition.

Again, it is not clear to what extent the Kuroko deposits may be considered a modern equivalent of the older deposits for which we have isotopic data. The mineralization at Flin Flon and Snow Lake is associated with metavolcanic rocks, but older rocks may underlie the area. Slawson and Russell (1973) in fact interpreted the steep trend of the ^{206}Pb-^{207}Pb-^{204}Pb data as indicating a second stage of isotopic evolution within the crust which commenced sometime between 3.7 and 2.45 Ga.

The data for the Sullivan mine are difficult to interpret because of the limited number of analyses and the enigmatic nature of the genesis of the Sullivan ores (cf. Sangster, 1972). As mentioned earlier, Kanasewich (1968) proposed that mineral deposits at Sullivan (and possibly also those of the Coeur d'Alene district) resulted from Precambrian continental rifting accompanied by processes of metal extraction and concentration similar to those now producing the hot metal-rich brines in the Red Sea. The lead in these brine pool sediments varies in isotopic composition, so that several sources must be contributing to the metals being precipitated. If indeed this process produced the Sullivan and Coeur d'Alene sulphides, then the ore at Sullivan evidently has a higher fraction of continentally derived lead than Coeur d'Alene.

The high-angle isotopic trends for the deposits in the Namaqualand Metamorphic complex have also been attributed to mixing processes (Koeppel, 1978). One of the components of the mixture is assumed to be lead having a composition similar to that at the Prieska mine. At least two other components are necessary to explain the scatter of the data; Koeppel has proposed that the sources of these components are in the older basement rocks of the supracrustals that host the ores.

The mixing of two lead components might help to explain the Balmat-Edwards lead isotopic distribution. One of the components would have a low-μ character. Volcanism adjacent to the Balmat-Edwards site could contribute lead of this kind, but its isotopic composition (in particular its $^{208}Pb/^{204}Pb$ ratio) would have to be somewhat different from any of the CMsB volcanogenic sulphides for which we have data. The other (high-μ) component would presumably be derived from older continental crust, but we have no record of its composition. Doe and Zartman (1979) point out that low-angle data distributions on the ^{208}Pb-^{206}Pb-^{204}Pb graph also mark mantle-derived leads. Balmat-Edwards sulphides and marbles exhibit this characteristic (Doe, 1962b) as, in fact, do the other CMsB leads. However, low-angle distributions do not appear to be as strong a characteristic as the low-μ nature of the leads, when the other data sets are examined. The lack of extensive local occurrences of metavolcanic rocks at Balmat and the mineralogical character of the deposits (predominantly sphalerite, with subordinate pyrite, minor galena and only trace element levels of copper) argue against a genesis like that of the Kuroko ores.

The zinc-lead-copper deposits in the Buchans area occur in basic metavolcanic rocks of Ordovician age. The isotopic data distribution is reminiscent of the trends for the Kuroko and Balmat ores, and is also centred between the mantle and orogene growth curves. The distribution supports the impression that steep angled linear ^{207}Pb-^{206}Pb-^{204}Pb trends are a lead isotopic phenomenon reflecting a process that has recurred throughout the genesis of stratiform sulphide ores during the past 2 Ga.

GENERAL CONCLUSIONS

The lead isotopic distributions observed for Grenville sulphides appear to have counterparts in other geological terranes and at other times in geological history. Trends like that of the Balmat-Edwards suite, the Namaqualand deposits, the Kuroko ores, and possibly Buchans are most easily explained by mixing on a scale quite different from that envisaged by Doe and Zartman in their "plumbotectonic" model of lead isotopic evolution. The low-angle isotopic distributions of the Coeur d'Alene, CMsB, and Namaqualand sulphides are subject to a much wider range of interpretations, and may bear only a superficial resemblance to the lead isotopic variations that exist among the Paleozoic ores of Japan. This is unfortunate, because the ability to define a specific model of sulphide genesis based on the distribution of lead isotopes would have important consequences. Baer (1976) has suggested that the metavolcanic rocks to which the sulphides are related were associated with an aulacogen produced during continental rifting. Brown *et al.* (1975) have interpreted the same suite of volcanic rocks as evidence for an island arc complex and hence for a different tectonic history than that proposed by Baer. If the lead isotopic distributions in sulphides associated with volcanism were characteristic of a specific tectonic process (rifting or island arc), then we would have a useful tool for identifying those processes where the normal petrological and structural evidence is ambiguous. However, the isotopic consequences of mature island-arc tectonism, crustal remobilization of lead, repeated volcanogenic sulphide deposition, and ancient mantle heterogeneities are not sufficiently well catalogued or understood. We expect that as geochronological and isotopic studies become more extensive, high-quality lead isotopic data sets for various recent deposits will enable us to distinguish some of these tectonic processes, and that isotopic "fingerprinting" of volcanogenic leads will ultimately assist in both defining tectonic models and applying them to otherwise ambiguous areas such as the Central Metasedimentary Belt.

ACKNOWLEDGEMENTS

We wish to thank P. R. Kuybida for preparing samples for analysis. Discussions with J. R. Richards and the comments of B. R. Doe and W. F. Slawson were most helpful. Mineral specimens from Balmat were obtained through the courtesy of the St. Joe Zinc Co., with the assistance of Mr. D. B. Dill; the Buchans galenas were provided by R. H. Relly. One of the authors (I.R.F.) was financially assisted by scholarships from the National Research Council of Canada and the University of Toronto. Continuing support from the National Research Council in the form of operating grants to R.M.F. is gratefully acknowledged.

REFERENCES

Baer, A.J., 1974, Grenville geology and plate tectonics: Geosci. Canada, v. 1, no. 3, p. 54-61.

——————, 1976, The Grenville province in Helikian times: a possible model of evolution: Phil. Trans. Royal Soc. London A, v. 280, p. 499-515.

Barton, J.M. and Doig, R.D., 1974, Temporal relationships of rock units in the Shawinigan area, Grenville province, Quebec: Canadian Jour. Earth Sci., v. 11, p. 686-690.

Bell, K., and Blenkinsop, J., 1980, A geochronological study of the Buchans area, Newfoundland: in Swanson, E.A., Strong, D.F. and Thurlow, J.G., eds., The Buchans Story: Geological Association of Canada Special Paper, in press.

Brown, R.L., Chapell, J.F., Moore, J.M. Jr., and Thompson, P.H., 1975, An ensimatic island arc and ocean closure in the Grenville province of southeastern Ontario, Canada: Geosci. Canada, v. 2, p. 141-144.

Cannon, R.S., Pierce, A.P. and Antweiler, J.C., 1971, Suggested uses of lead isotopes in exploration: in Boyle, R.W. and McGerrigle, J.I., eds., Geochemical Exploration: Canadian Instit. Mining Metallurgy Spec. Vol. 11, p. 457-463.

Cumming, G.L. and Richards, J.R., 1975, Ore lead isotope ratios in a continuously changing earth: Earth and Planet. Sci. Letters, v. 28, p. 155-171.

Doe, B.R., 1962a, Distribution and composition of sulfide minerals at Balmat, New York: Geol. Soc. America Bull., v. 73, p. 833-854.

——————, 1962b, Relationships of lead isotopes among granites, pegmatites, and sulfide ores near Balmat, New York: Jour. Geophys. Research, v. 67, p. 2895-2906.

Doe, B.R., Stevens, T.A. Delevaux, M.H., Stacey, J.S., Lipman, P.W. and Fisher, F.S., 1979, Genesis of ore deposits in the San Juan volcanic field, Southwestern Colorado lead isotope evidence: Econ. Geol., v. 74, p. 1-26.

Doe, B.R. and Zartman, R.E., 1979, Plumbotectonics, the Phanerozoic: in Barnes, H., ed., Geochemistry of Hydrothermal ore deposits, p. 22-70.

Fletcher, I.R., 1979, A lead isotope study of lead-zinc mineralization associated with the Central Metasedimentary Belt of the Grenville province: Ph.D. Thesis, Department of Physics, University of Toronto, 165 p.

Gulson, B.L. and Mizon, K., 1979, Lead isotopes as a tool for gossan assessment in base metal exploration: Jour. Geochem. Exploration, in press.

Kanasewich, E.R., 1968, Precambrian rift: genesis of stratabound ore deposits: Science, v. 161, p. 1002-1005.

Koeppel, V., 1978, Lead isotope studies of stratiform ore deposits of Namaqualand, northwest Cape Province, South Africa, and their implications on the age of the Bushmanland Supergroup: in Zartman, R.E., ed., Short Papers of the Fourth International Conference, Geochronology, Cosmochronology, Isotope Geology: United States Geol. Survey Open-file Report 78-701, p. 223-226.

Lea, E.R. and Dill, D.B., 1968, Zinc deposits of the Balmat-Edwards district, New York: in Ridge, J.D., ed., Ore deposits of the United States, 1933-1967: American Instit. Mining, Metallurgical and Petroleum Engin., Graton-Sales Volume, p. 20-48.

McCulloch, M.T. and Wasserburg, G.J., 1978, Sm-Nd and Rb-Sr chronology of continental crust formation: Science, v. 200, p. 1003-1011.

McWilliams, M.O. and Dunlop, D.J., 1978, Grenville paleomagnetism and tectonics: Canadian Jour. Earth Sci., v. 15, p. 687-695.

Middleton, R.C., 1976, The geology of Prieska Copper Mines Limited: Economic Geology, v. 71, p. 328-350.

Oversby, V.M., 1974, A new look at the lead isotope growth curve: Nature, v. 248, p. 132.

Page, D.C. and Watson, M.D., 1976, The Pb-Zn deposit of Rosh Pinah Mine, South West Africa: Econ. Geol., v. 71, p. 306-327.

Patchett, P.J., Bylund, C. and Upton, B.G.J., 1978, Paleomagnetism and the Grenville orogeny: new Rb-Sr ages from dolerites in Canada and Greenland: Earth and Planetary Science Letters, v. 40, p. 349-364.

Reynolds, P.H. and Russell, R.D., 1968, Isotopic composition of lead from Balmat, New York: Canadian Journal of Earth Sciences, v. 5, p. 1239-1245.

Russell, R.D., 1972, Evolutionary model for lead isotopes in conformable ores and in ocean volcanics: Geophy. Space Phys. Reviews, v. 10, p. 529-549.

Sangster, D.F., 1972, Precambrian massive sulphide deposits in Canada: Geol. Survey Canada, Paper 72-22, 42 p.

Sato, K., 1975, Unilateral Isotopic variation of Miocene ore leads from Japan: Econ. Geol., v. 70, p. 800-805.

Sato, K. and Sasaki, A., 1976, Lead isotope evidence on the Genesis of pre-Cenozoic stratiform sulfide deposits in Japan: Geochem. Jour. v. 10, p. 197-203.

_____, 1978, Two major evolutionary systems for stratiform ore leads as exemplified by Japanese samples: in Zartman, R.E., ed., Short papers of the Fourth International Conference, Geochronology, Cosmochronology, Isotope Geology: United States Geological Survey Open-file report 78-701, p. 378-379.

Seyfert, C.K., 1978, Paleomagnetic evidence for three Wilson cycles during the Precambrian (Abst.): Program with Abst. Geol. Assoc. Canada, v. 3, p. 490.

Silver, L.T. and Lumbers, S.B., 1966, Geochronological studies in the Bancroft-Madoc area of the Grenville province, Ontario, Canada (Abst.): Geol. Soc. America Spec. Paper, v. 87, p. 156.

Slawson, W.F. and Russell, R.D., 1973, A multistage history for Flin Flon lead: Canadian Jour. Earth Sci., v. 10, p. 582-583.

Stacey, J.S. and Kramers, J.D., 1975, Approximation of terrestrial lead isotope evolution by a two-stage model: Earth and Planet. Sci. Letters, v. 26, p. 207-221.

Stacey, J.C., Doe, B.R., Silver, L.T. and Zartman, R.E., 1977, Plumbotectonics IIA, Precambrian massive sulfide deposits: in Karpenko, S.F., ed., Geochronology and the problems of ore formations: also United States Geological Survey open-file report 76-476, 26 p.

Stanton, R.L. and Russell, R.D., 1959, Anomalous leads and the emplacement of lead sulfide ores: Econ. Geol., v. 54, p. 588-607.

Tatsumoto, M., 1969, Lead isotopes in volcanic rocks and possible ocean-floor thrusting beneath island arcs: Earth and Planet. Sci. Letters, v. 6, p. 369-376.

_____, 1978, Isotopic composition of lead in oceanic basalt and its implication to mantle evolution: Earth and Planet. Sci. Letters, v. 38, p. 63-87.

Wynne-Edwards, H.R., 1972, The Grenville province: in Price, R.A. and Douglas, R.J.W., eds., Variations in Tectonic Styles in Canada: Geological Association of Canada Special Paper 11, p. 263-334.

Zartman, R.E. and Stacey, J.S., 1971, Lead isotopes and mineralization ages in Belt Super-group rocks, northwest Montana and northern Idaho: Econ. Geol., v. 66, p. 849-860.

Zartman, R.E. and Wasserburg, G.J., 1969, The isotopic composition of lead in potassium feldspars from some 1.0-b.y. old North American igneous rocks: Geochim. et Cosmochim. Acta, v. 33, p. 901-942.

The Continental Crust and Its Mineral Deposits, edited by D.W.Strangway, Geological Association of Canada Special Paper 20

PALEOMAGNETISM APPLIED TO THE STUDY OF TIMING IN STRATIGRAPHY WITH SPECIAL REFERENCE TO ORE AND PETROLEUM PROBLEMS

F.W. Beales, K.C. Jackson, E.C. Jowett, G.W. Pearce and Y. Wu
Department of Geology,
University of Toronto,
Toronto, Ontario M5S 1A1

ABSTRACT

With the development of cyrogenic magnetometers, a whole series of investigations of more weakly magnetized specimens has become possible, which will increasingly assist the unravelling of the complex evolutionary histories of many sedimentary basins. This is illustrated by a selection of studies that were difficult or impossible earlier. Measurements on recent carbonate muds in Florida as well as laboratory experiments have demonstrated that even very pure limestones can acquire sufficient impurities to develop a measurable detrital remanence. The timing of mineralization for Mississippi Valley-type ore deposits has been an enigma, but magnetic remanence imprinted in the ore from the Viburnum Trend, southeastern Missouri, suggests that it is Upper Pennsylvanian (± 300 Ma). The adjacent host rock of Upper Cambrian age also has an Upper Pennsylvanian direction, which suggests that it has been remagnetized, presumably by the movement of fluids through then-existing porosity trends. Very close links are considered to exist between the sediments and the interstitial fluids which transport both ores and hydrocarbons in sedimentary basins.

RÉSUMÉ

Avec le développement de magnétomètres cryogéniques, il devient maintenant possible de procéder à toute une série de mesures sur des roches plus faiblement aimantées, ce qui aidera de plus en plus à éclaircir l'histoire complexe de l'évolution de plusieurs bassins sédimentaires. On illustre cette technique par un choix d'études qui étaient difficiles ou impossibles auparavant. Les mesures sur des boues calcaires récentes en Floride en plus d'expériences en laboratoire ont démontré que même des calcaires très purs peuvent incorporer suffisamment d'impuretés pour développer une rémanence détritique mesurable. La chronologie de la minéralisation dans les gisements métallifères du type "Vallée du Mississippi" a toujours été une énigme, mais la rémanence magnétique imprimée dans le minerai de Viburnum Trend, dans le sud du Missouri, suggère une minéralisation datant du Pennsylvanien supérieur (± 300 Ma). Les roches encaissantes du Cambrien supérieur montrent aussi une direction typique du Pennsylvanien supérieur, probablement à cause du mouvement des fluides selon des zones poreuses préexistantes. Il semble exister des liens très étroits entre les sédiments et les fluides interstitiels qui transportent les minerais et les hydrocarbures dans les bassins sédimentaires.

INTRODUCTION

Paleomagnetic studies have contributed more to the development of plate tectonic theory than any other branch of geology or geophysics. However, the use of this powerful research tool in many stratigraphic problems has been hampered by the low intensities of the natural remanent magnetization (NRM) in most common sediments. The recent development of cryogenic magnetometers, in which the total noise level of the magnetometer has reduced to 10^{-7} emu, allows us to measure NRM intensities of 10^{-8} to 10^{-9} emu/g using large volume samples. This improved capability has given a new impetus to many investigations, and the mechanisms of sediment magnetization are only now being studied effectively (Barton and McElhinny, 1979; Cain *et al.*, 1979; Ellwood, 1979; Giovanoli, 1979; Verosub, 1979).

The NRM of a sediment can be acquired either at the time of deposition by alignment of detrital particles – detrital remanent magnetization (DRM) – or at the time of subsequent diagenesis by chemical alterations – chemical remanent magnetization (CRM). Furthermore, the rock may be subjected to other magnetizing events at any time in its history by chemical or thermal (TRM) influences (Verosub, 1977). Thus, the complexity of the magnetization process means that a variety of genetic and diagenetic evidence is potentially decipherable, a development that would increase the usefulness of paleomagnetism as a stratigraphic tool.

With widespread acceptance of plate tectonic theories, stratigraphic problems can be looked at as a logical sequence of cause-and-effect changes. Sedimentary basin evolution, recorded in the stratigraphy, embodies not only the sediment types and structures, which reveal local and regional environments at the time of deposition, but also many post-depositional changes such as compaction, thermal history, cementation, and interstitial fluid evolution and migration. The maturation and migration phenomena are of vast economic importance relative to the oil industry (Cordell, 1972; Tissot and Welte, 1978; McAuliffe, 1979; Hunt, 1979) and to the genesis of certain types of metallic mineral deposits (Anderson, 1975, 1978; Beales, 1975, 1976; Hitchon, 1977; Dunsmore and Shearman, 1977; Macqueen, 1979). The timing of maturation and migration is one of the least understood processes of sedimentary basin evolution.

In particular, both petroleum reservoirs and Mississippi Valley-type ore deposits are clearly related to sedimentary basin evolutionary processes. The former require a *physical* trap to ensure accumulation of an economically important deposit, while the latter require a *chemical* trap, i.e., a precipitation mechanism (Anderson, 1975; Beales, 1975, 1976). Both are epigenetic deposits that occur within pre-existing porosity in lithified sedimentary rocks and their economic importance is dependent, up to a point, upon the extent, interconnection, and degree (percentage of voids) of that porosity. They are characteristically unrelated to either metamorphism or igneous activity, but have important links with burial diagenesis (Philippi, 1965; Cordell, 1972; Connan, 1974; Abelson, 1978; McAuliffe, 1979, for oil; Macqueen, 1979, for the MVT Ores).

There are also many obvious differences of chemistry and sedimentary milieu. For example, Mississippi Valley-type ores (Ohle, 1959) occur preferentially in dolostones on the flanks of sedimentary basins; another example is the antipathetic occurrence of economic concentrations of the two. However, to discuss these examples in detail would be beyond the scope of this paper, which focuses on the similarities. Petroleum reservoirs and Mississippi Valley-type ore deposits may share similar host rocks, interstitial fluids, and above all, comparable histories. Therefore, comparable paleomagnetic remanence acquisition is most probably decipherable.

DETRITAL REMANENT MAGNETISM OF LIMESTONES

A number of studies have been made relating to the magnetic imprinting of unconsolidated siliciclastic sediments of recent and geologically young ages. There have been no previous reports of remanance in modern limestones because of the low magnetic intensities involved. In order to better understand and interpret the results from orebody host rocks, an investigation was made of some Florida limemuds (Jowett, 1977; Jowett and Pearce, 1977).

In the present study, sub-recent, oriented cubes of pure carbonate mud were collected from two localities in the Florida Keys. The Swamp on Sunshine Key is an isolated, backwater marsh separated from the Straits of Florida by a low ridge of rubble piled up by storms and stabilized by grasses and bushes. The sandy mud here accumulates in small depressions in the irregular Key Largo Limestone surface. The mud is bound in place by blue-green algal mats. At the other locality sampled on Sugarloaf Key, the mud is accumulating on the Florida Bay side and is stabilized by red mangroves and by blue-green algal mats in an intertidal environment. The thickness of the recent mud is less than one metre and it overlies the Pleistocene Miami Oolite Formation. At both locations the top leathery algal mat layer was removed and a grid, oriented with respect to magnetic north, was marked on the level surface. Samples were taken by pushing a five centimetre cubic aluminum box into the mud, allowing air to escape through a hole in the top, and carefully excavating the sample. The mud block was then placed, still oriented, onto a non-magnetic screen raised off the ground. To see whether any realignment could occur upon drying, two Sunshine Key samples were turned 90° to the west and two Sugarloaf Key cubes were turned 90° to the east. The samples were protected from possible rain and left to dry oriented as close to in situ as possible. Several days later an extra cube was broken open to reveal that the samples were dry except for a moist central portion. The formerly plastic mud was then competent and hard. Next, samples were placed into cubic plastic sample holders and taken to the Toronto laboratory in a magnetically-shielded box to prevent possible realignment. Laboratory storage was in a magnetically shielded room (field in order of 50 gammas except during measurements and treatment).

The NRMs of the dried mud samples were low in intensity (5.6×10^{-8} to 1.6×10^{-7} emu/g), with those from Sugarloaf Key being somewhat higher than those from

Sunshine Key. The directions of the NRM of those samples dried in the original orientation closely reproduced the direction of the in situ magnetic field (Fig. 1, Table I). The samples that had been misoriented for drying all showed a shift towards the direction in which they had been rotated (e.g., for Sunshine Key samples the resultant declination of the properly oriented sample was 356°, whereas that of the two samples rotated 90°W was 14°, representing an 18° swing). It thus appears that during the two days of drying, the NRM was able to partly realign toward the new direction. It is possible that the realignment might have continued to completion if the samples had dried more slowly.

Under alternating field demagnetization the samples retained their NRM directions to 300 to 500 Oersteds, whereupon the directions became erratic (Fig. 2). The

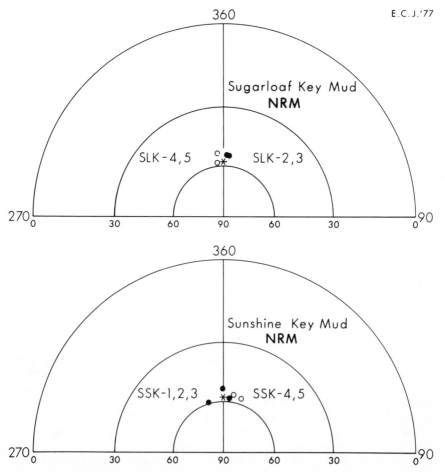

Figure 1. Declination and inclination plot of NRM for samples from Sugarloaf Key and Sunshine Key, Florida. Present magnetic field for Florida localities shown by *. All inclinations are positive (downward).

TABLE 1
NRM OF FLORIDA MUD

	Intensity (emu/g)	Declination D	Inclination I
Sunshine Key			
SSK-1	9.16×10^{-8}	358°	+53°
2	6.76×10^{-8}	342°	+59°
3	5.58×10^{-8}	006°	+58°
4*	1.00×10^{-7}	010°	+56°
5*	8.77×10^{-8}	018°	+57°
Sugarleaf Key			
2	1.34×10^{-7}	003°	+54°
3	1.24×10^{-7}	004°	+54°
4**	1.60×10^{-7}	004°	+53°
5**	1.42×10^{-7}	352°	+58°

*these samples were turned 90° to the west to dry
**these samples were turned 90° to the east to dry

behaviour of the intensity of the NRM under AF demagnetization is roughly a logarithmic decay with the field strength required to reduce the intensity to half being about 150 Oe.

Thermal demagnetization of the NRM of several samples yielded behaviour such as that illustrated in Figure 3. Direction of the NRM remained stable until the sample was heated to 375° to 425°C. The intensity decreased logarithmically with temperature, reaching half of the original at 200°C. Above about 450°C, the generation of additional magnetic material (identified from its Curie point as magnetite) caused complicated behaviour in the thermal demagnetizations.

Although the sample location suggested a detrital origin for the NRM, additional evidence was provided by an experiment in which mud samples from each site were slurried with water and redeposited in the Earth's field in the laboratory. The remanent moments possessed by these samples after drying again reproduced the ambient magnetic field accurately, and in particular showed no significant inclination error. The intensities and AF demagnetization behaviour of these artificial remanent moments were also similar to those of the NRM. The NRMs of these lime muds were acquired and fixed to accurately reproduce the Earth field direction by the drying process. When drying in nature, some early vertical compaction of the mud would be expected. Thus, some inclination error might develop from this source, but no other source of error appears important for these materials. However, as the bulk of limestone compaction is achieved by a process of pressure solution (stylolitization) with the remaining limestone fabric undisturbed, this potential source of error will probably be less with limestones than with siliciclastic rocks.

These experiments reveal detrital remanent moments in modern muds of similar magnetic characteristics to those encountered in ancient carbonate rocks, and are taken to indicate that such rocks can and probably do acquire a detrital or penecontemporaneous diagenetic remanence at the time of sedimentation. This remanence is presumably carried by cosmic, aeolian, and mud detritus and by chemical and possibly biochemical precipitates.

STUDIES OF ANCIENT LIMESTONES AND ASSOCIATED ORES

We have investigated the magnetic remanence of the ores and host rocks of Mississippi Valley-type ore deposits, to see if the host rocks contain a measureable magnetic imprint, and, if so, if the apparent pole position corresponds with the polar wander path for the age of the rock. This would indicate that the fossil remanence had been neither reset nor tectonically translated. The ores themselves are clearly epigenetic, that is, they were deposited in void space in the host carbonate rock, but exactly when the ore precipitation took place is anyone's guess. As D.F. Sangster (pers. commun.) has expressed it for Pine Point, N.W.T., Canada, "The ore is post-Mid-Devonian, the age of the host rock, and pre-1850 when the native peoples

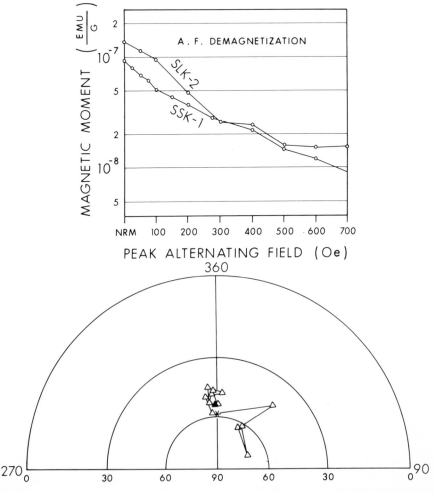

Figure 2. A.F. Demagnetization curves for Sugarloaf and Sunshine Key samples. Directional changes of magnetic moments which retained the NRM direction to 500 Oe for Sunshine sample.

first showed it to the white man." This is typical of most Mississippi Valley-type ore deposits and, because the leads in such deposits are normally of complex origin, they cannot be readily dated isotopically (Doe and Delevaux, 1972; Sangster, 1976a, 1976b; and Fletcher and Farquhar, 1977). Therefore, if the ores contain entrapped or co-precipitated magnetic mineral impurities and if an apparent pole position can be calculated, this might be related to the apparent polar wander curve for the plate in question.

Early work on two mines (the Newfoundland Zinc Mines and the St. Joseph #8 Mine in "the Old Lead Belt" in southeastern Missouri) indicated that mineralization occurred shortly after deposition of the enclosing host rocks (Table II). Unfortu-

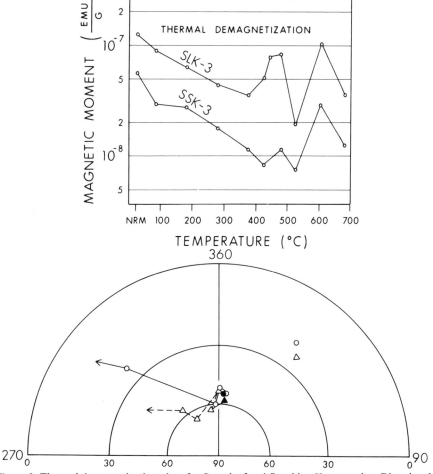

Figure 3. Thermal demagnetization plots for Sugarloaf and Sunshine Key samples. Directional changes in NRM begin in the Sunshine Key sample (circles) above 180°C and for the Sugarloaf sample (triangles) above 280°C; above these temperatures new minerals are probably forming and the changes become more apparent.

TABLE II

PALEOMAGNETISM OF SELECTED MISSISSIPPI VALLEY-TYPE ORE DEPOSITS

Site	Apparent Pole Posn	# of Samples	Fisher Statist. k	α_{95}	Remarks	References
Newfoundland Zinc Mines (W. Newfoundland) Combined Host and Ore	126°E 26°N	80	202.6	6.9	Host rock is Lr. Ordovician Ore is probably M. Ord. with nearby host rock 'Reset' to M. Ord.	Beales *et al.*, 1974 Discussion of result in WU & Beales 1980.
St Joe #8 Mine Bonneterre, Mo. Combined Host and Ore	170°W 35°S	21	468.6	4.5	Probably complex, Lower Paleozoic	
Viburnum Trend, Mo., Ore and Host Dolostone Near Ore	124.3°E 45°N	16	176	5.8	Host rock is U. Cambrian, is probably U. Pennsylvanian, about 300 Ma (Fig. 5.), with adjacent host rock 'Reset' to same	WU & Beales 1980.
Bonneterre Dolostone 10 Km from Ore (S.Q.)	132.7°E 12.7°N	7	$Q = 4.6 \times 10^{-4}$ (Q = quality factor, Halls, 1976, 1978)		Probably Upper Cambrian	

nately, the apparent polar wander curve for the North American continent for early Paleozoic times is not well defined owing to a lack of reliable data points (Irving, 1979), so that there is some doubt attached to the ages assigned. The late Early to early Medial Ordovician age of mineralization for the Newfoundland Zinc Mines ores suggested by Beales *et al.* (1974) is probably more reliable than the age given for the #8 Mine in the "Lead Belt" of southeastern Missouri (see Table II). In the latter case, the Lamotte Sandstone, which was studied by Al Khafaji and Vincenz (1971) and was assigned a Late Cambrian age, is suspected to have suffered considerable post-depositional resetting and/or overprinting (Wu and Beales, 1980).

At the time the "Old Lead Belt" was studied, only spinner magnetometers were available to us. When a cryogenic magnetometer became available, we revisited southeastern Missouri to resample the "New Lead Belt", or "Viburnum Trend", which is now the only producing area.

VIBURNUM TREND MINERALIZATION

The southeastern Missouri area, owing to its long history and presently pre-eminent production, has come to be regarded by many geologists as the type area for the "Mississippi Valley type" class of ore deposits that are now recognized worldwide (Ohle, 1959). A series of modern mines along the "New Lead Belt", or Viburnum Trend, currently exploit one of the major Pb/Zn mineral deposits of the world and supply about 85% of the lead mined in the U.S.A. with reserves that will see them through the millenium (Vinegard, 1977).

The predominantly galena ores, with lesser sphalerite and minor chalcopyrite, occur in the Upper Cambrian Bonneterre dolostone along a north-south trend that lies along the westerly flank of the St. Francois Mountains. These "mountains" are Precambrian basement highs exposed by erosion of high points of the much more extensive Ozark Dome. The host rocks at the time of mineralization were dolomitic oolitic grainstones, algal boundstones, and collapse breccias, with minor dolomite and pyrite gangue. Approximately one kilometre west of the trend, the dolostones give way to tight, fossiliferous limestone. To the east a dolomitized limestone platform facies, referred to locally as Whiterock facies, is the more restricted, probably hypersaline facies onlapping the Upper Cambrian St. Francois islands (Fig. 4). The Viburnum ore trend approximately marks the ancient facies front, an inferred shoal and probably barrier island coastline with elongate salinas, that lay between the Saint Francois Platform and the deeper water, open marine shelf to the west (Gerdemann and Myers, 1972; Rogers and Davis, 1977; Larsen, 1977; Beales and Hardy, 1977).

Along the trend, successive mines are associated with the feather edge of the underlying Lamotte Sandstone and draping structures over hummocks of the Pre-cambrian basement. No mappable or geophysically detectable faults have yet been discovered to account for the sublinear trend of the facies front. The regional structure is simple, apart from minor faulting and draping (Kisvarsanyi, 1977). The lack of structural complications was, in fact, one of the chief attractions of the study area. However, paleomagnetically, plenty of problems can occur. Since none of the major minerals are magnetic, completion of the reconnaissance study depended on the chance occurrence of very small amounts of entrapped or co-precipitated magnetic impurities, as has been the case in all our studies to date. A majority of samples were either discarded at once for showing undetectable magnetic moment or for 'washing-out' during demagnetization without providing a stable end point. However, 16 samples gave what appears to be a reliable Late Pennsylvanian pole position (Table II and Fig. 5).

There were some surprises in these results: 1) The indicated age conflicted with earlier paleomagnetic work that had suggested a Late Cambrian age for the "Old Lead Belt" (Beales *et al.*, 1974); 2) Upper Cambrian host rocks from the mines, i.e., in close proximity to ore, had been remagnetized to an Upper Pennsylvanian pole position, while host rock samples obtained well away from mineralization preserved a presumed Upper Cambrian or much earlier remanence (Table II). 3) Since both reversed and normal field orientations were encountered, magnetization, and presumably mineralization, must have been spread over a considerable period of time.

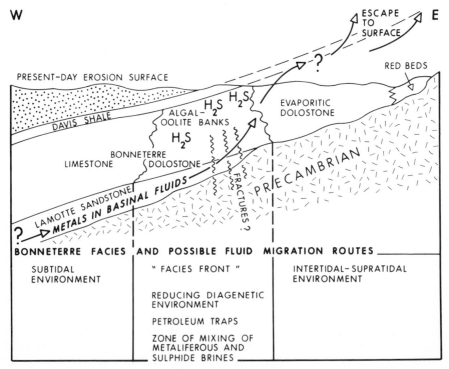

Figure 4. The Bonneterre facies pattern and possible fluid migration routes. Ore precipitation probably took place where basinal brines carrying the metals mixed with formational fluids carrying reduced sulphur. H_2S is considered to have formed from the reduction of sulphate by petroleum which was trapped in the porosity generated in the vicinity of the 'Facies Front', i.e., the complex zone marking the edge of the former shallow marine platform to the east where it abuts on the deeper water area to the west. The relatively narrow, sub-linear, now heavily mineralized, north-trending zone is about 70 km long and is referred to as the "Viburnum Trend".

Is the result geologically reasonable, i.e., could mineralizing fluids have been moving at the time indicated? There is no doubt that they could have, but this is not to say that they were. An alternative hypothesis would be that the mineralization was emplaced earlier and that both ore and host were remagnetized in Late Pennsylvanian times. However, many related events were required as prerequisites to the actual precipitation of the ore, and these appear to have culminated in Late Pennsylvanian times. This would have been a feasible time for ore emplacement because the rapid sedimentation and the onset of orogeny, which were taking place in the

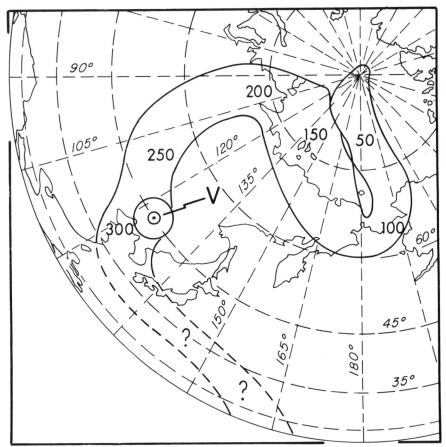

Figure 5. Polar wander curve (slightly modified after Irving, 1979), for the North American craton during the Phanerozoic; the ages assigned are approximate, as there is considerable spread in latitutde and longitude at any time period or from any sampling locality. However, the last 300 Ma is much more securely defined than the earlier Phanerozoic, because in the later part of the wander path, cross checks with ocean-floor spreading permit recognition of spurious results. The Viburnum Trend (V) geomagnetic pole falls just within the better defined tract, which lends confidence to the age assignment of about 300 Ma to the mineralization; i.e., it is highly unlikely that the age of mineralization could be early Paleozoic.

Ouachita geosyncline and the Arkansas/Arbuckle basins to the south and southwest in Late Pennsylvanian times, could have provided a compactional drive for escaping interstitial fluids. The Lamotte Sandstone, underlying the Bonneterre Dolostone host rock, is a widespread basal blanket sandstone resting uncomfortably on the underlying rocks (Fig. 4; and Wu and Beales, 1980). The Lamotte Sandstone, therefore, could well have been a regional aquifer that was pressurized by these tectonic events occurring 150 to 200 miles to the south. Sediment thicknesses indicate that the Lower Paleozoic source rocks there were sufficiently depressed to have passed through the so-called "oil window", and burial metamorphism must have chemically matured the sediments and their interstitial fluids. Southeastern Missouri, one of the areas with the least Upper Pennsylvanian superincumbent sedimentary load over the Bonneterre host rock, would, therefore, have been a likely break-out point for pressurized fluids. Routes of fluid flow would undoubtedly have been complex and varying through time, but our currently favoured hypothesis is that petroleum migrated first and became trapped in the Bonneterre Dolostone, where it reacted with gypsum to form sour gas (Anderson, 1975; Beales, 1975). Basinal dewatering continued and the continuing escape of fluids altered the regional pressure gradients. Reservoirs pressurized to regional grade started leaking to the escape route when pressure in the reservoir exceeded that in the conduit. The mixing of fluids from basinal sources (carrying metals as chloride complexes) with formational fluids from the Bonneterre Dolostone (carrying H_2S or reducants) is the most probable precipitation mechanism (Dunsmore and Shearman, 1977; Anderson, 1975; Carpenter et al., 1974). This mixing took place in the vicinity of the facies front where porosity relationships were favourable for petroleum entrapment, sulphate reduction, and ore precipitation (Fig. 4; and Beales and Hardy, 1977).

As a possible model, one might suggest that the stress that developed during plate interactions caused folding in the Ouachita belt and also much more widespread, minor, mostly lateral movements in the basement. These could have rejuvenated even more ancient planes of weakness in the Precambrian basement that are reflected in the linearity of the sedimentary and evolutionary features of Viburnum Trend area. Basement structural influences could have caused a gentle hinge line to develop along this trend, thereby controlling the sub-linear N-S facies front at the host rock depositional stage. Associated elongate salinas were generated in the zone of frontal, higher-energy sedimentary facies. Later, preferential solution of the more susceptible rocks along the same trends took place in response to either subsurface fluid escape or, possibly, to surface karstification. This enhanced the permeability of the host rocks and removed almost all trace of the evaporites (Beales and Hardy, 1977). Solution-generated collapse formed the porous breccias that provided a highly permeable plumbing system for the escape and mixing of fluids. Precipitation took place in the interstices of the breccias that host the higher grade ore sections of the Viburnum Trend.

During a visit to the Ozark Mine near the south end of the Viburnum Trend, the common occurrence of traces of bitumiuous residue in the host rock adjacent to Mississippi Valley-type ores was confirmed. A sample of sticky residual oil that we collected there was kindly analyzed by the Tulsa Research Center of Amoco Production Company. The analysis confirmed that the oil was a normal mature "Paleozoic type" oil, so heavily altered by bio-degradation that is was impossible to correlate it with a specific source area like the Illinois or Anadarko basin. Such biodegradation can take place very rapidly (i.e., in a few days of exposure in the aerobic setting of the mine), and is quite normal. It could also have occurred by exposure to oxygenated groundwater.

However, the presence of fully mature oil even in trace amounts at that location is interesting, because, apart from being consistent with a theory of ore precipitation, it also constitutes the first report of oil possibly derived from the Arkansas basin, where burial metamorphism has been considered to have matured all the hydrocarbons to gas. The Ozark Mine oil traces could represent liquid hydrocarbons that migrated into cooler rocks before the source areas were buried to super-mature depths. If this is the case, a large area of southern Missouri between the south end of the Viburnum Trend and the Arkansas basin is potential ground for oil exploration. The oil industry has contributed many ideas related to basin evolution to base metals exploration philosophy. This is one of the few examples of the mining industry uncovering information of direct interest to the petroleum industry.

The uniform 'resetting' of the Viburnum Trend host rock close to ore was an unexpected finding of the original research, fully as important as our aim of dating mineralization. Much more work will be required, but the safest present course is to be very cautious when establishing apparent polar wander curves, as Irving (1979) has re-emphasized. Although many sediments may have acceptable detrital or early diagenetic remanences, until their whole histories are understood — particularly for more porous sediments — caution should be used.

DATING OF DIAGENETIC EVENTS
ASSOCIATED WITH OIL MIGRATION AND ENTRAPMENT

We noted in the introduction that a characteristic feature common to hydrocarbon accumulation and Mississippi Valley type ore deposits is that both can be regarded as products of the normal evolution of sedimentary basins. Thus, they can be hosted (or reservoired) in the same types of rocks, generally carbonate rocks, which were formed in comparable genetic environments both sedimentologically and tectonically. These similarities are complemented to a remarkable degree by similarities in associated formational fluids (Collins, 1975; Carpenter et al., 1974). Fluid inclusion studies on Mississippi Valley-type ores (Roedder, 1967, 1977a, 1977b) have strongly suggested that the ore-transporting brines were highly saline, Na-, Ca-, chloride-brines, and that ore precipitation took place in the 50° to 200°C range. Hitchon (1977) made a detailed comparison between oil field brines, mainly from the western Canada sedimentary basin, and proposed sedimentary ore-forming brines. He noted that, with the exception of the higher temperatures of the so-called geothermal brines, overlap occurred in all characteristics, and he concluded that there is essentially one source for both: expressed "formation water". This means that the paleomagnetic signatures of Mississippi Valley type ore host rocks and hydrocarbon reservoir rocks should have much in common. We may hypothesize that a reservoir rock well away from hydrocarbon migration and accumulation may preserve a DRM or a very early CRM related to the time of sedimentation and "locking" (Verosub,

1977). Magnetization connected with the migration of formational fluids could be associated with redox changes when oil enters a reservoir, with temperature changes, or with the introduction of magnetic impurities. For example, the magnetic anomalies over the Cement oilfield in Oklahoma have been interpreted as due to the formation of diagenetic magnetite over a leaky hydrocarbon reservoir (Donovan *et al.*, 1979). To test the possibility that diagenetic CRMs might be associated with fluid migration, we have made measurements on the few standard oil industry oriented cores that we had available for a pilot study. Successful interpretation depends upon being able to decipher the causes for diagenetic resetting in each individual case. However, knowing *when* an event occurred should assist us in determining *why*. Research effort should, therefore, first concentrate on searching for the time when the 'anomalous' remanence was acquired. Hopefully, in some cases these results may be related to the time of migration of oil. A pilot study has been made of four Cretaceous Muddy Sandstone cores from Colorado. Two contain stable magnetizations, but these do not coincide with either a Cretaceous pole or the modern pole. A single sample of Devonian Swan Hills limestone was unstable. Additional laboratory measurements on many more oriented samples will have to be made to resolve the remanence acquisition histories of these rocks.

DISCUSSION

It is becoming increasingly clear that plate tectonic processes are directly involved with fluid/sediment interactions and therefore with ore (and oil) emplacement mechanisms:

1) by generating pregnant sediments in an orderly facies juxtaposition;

2) by generating thick sedimentary piles and compressing them to drive out fluids;

3) by providing a heat source;

4) by pressuring aquifers;

5) by generating structures that both impede and enhance fluid flow; and

6) by determining the sites of emplacement and preservation of economically important deposits.

Paleomagnetic measurements on related suites of sediments have the potential to reveal the timing of differing events of sedimentary basin evolution that are difficult to measure in any other way.

ACKNOWLEDGEMENTS

We are deeply indebted to the Natural Sciences and Engineering Research Council of Canada and to the Department of Energy, Mines and Resources of Canada for financial support; to Mining and Oil Company geologists too numerous to mention by name but without whose aid the studies would have been impossible; and to David Dunlop, Henry Halls, David Redman, John Rylaarsdam, and David Strangway who contributed to numerous discussions.

Abelson, P.H., 1978, Organic matter in the Earths' Crust: Annual Rev. Earth Planet. Sci., v. 6, p. 325-351.

Al Khafaji, S.A. and Vincenz, S.A., 1971, Magnetization of the Cambrian Lamotte Formation in Missouri: Geophys. Jour. Royal Astron. Soc., v. 24, p. 175-205

Anderson, G.M., 1975, Precipitation of Mississippi Valley-type ores: Econ. Geol., v. 70, p. 937-942.

_____, 1978, Basinal brines and Mississippi Valley-type ore deposits: Episodes, Geological Newsletter, Internat. Union Geol. Sci., v. 1978, p. 15-19.

Barton, C.E., and McElhinny, M.W., 1979, Detrital remanent magnetization in five slowly redeposited long cores of sediment: Geophys. Res. Letters, v. 16, p. 229-232.

Beales, F.W., 1975, Precipitation Mechanisms for Mississippi Valley-type Ore Deposits: Econ. Geol., v. 70, p. 943-948.

_____, 1976, Precipitation mechanisms for Mississippi Valley-type Ore Deposits- A Reply: Econ. Geol., v. 71, p. 1062-1064.

Beales, F.W., and Hardy, J.L., 1977, The problem of recognition of occult evaporites with special reference to southeast Missouri: Econ. Geol., v. 72, p. 487-490.

Beales, F.W., Carracedo, J.C. and Strangway, D.W., 1974, Paleomagnetism and the origin of Mississippi Valley-type ore deposits: Canadian Jour. Earth Sci., v. 11, p. 211-223.

Cain, B., Payne, M.A., Shulik, S., Donahue, J., Rollins, H.B., and Schmidt, V.A., 1979, The recovery of paleomagnetic polarities from cyclothemic sediments in the Carboniferous Appalachian Basin, U.S.A.: Geophys. Res. Letters, v. 6, p. 261-264.

Carpenter, A.B., Trout, M.L., and Pickett, M.E., 1974, Preliminary report on the origin and chemical evolution of lead-and zinc-rich oil field brines in central Mississippi: Econ. Geol., v. 69, p. 1191-1206.

Collins, A.G., 1975, Geochemistry of oilfield waters: Developments in Petroleum Science #1: Amsterdam, Elsevier Co., 496 p.

Connan, J., 1974, Time-temperature relation in oil genesis: American Assoc. Petrol. Geol. Bull., v. 58, p. 2516-2521.

Cordell, R.J., 1972, Depths of oil origin and primary migration: a review and critique: American Assoc. Petrol. Geol. Bull., v. 56, p. 2029-2067.

Doe, B.R., and Delevaux, M.H., 1972, Source of lead in southeast Missouri galena ores: Econ. Geol., v. 67, p. 409-425.

Donovan, T.J., Forgey, R.L. and Roberts, A.A., 1979, Aeromagnetic detection of diagenetic magnetite over oil fields: American Assoc. Petrol. Geol. Bull., v. 63, p. 245-248.

Dunsmore, H.E. and Shearman, D.J., 1977, Mississippi Valley-type lead zinc ore bodies: a sedimentary and diagenetic origin: in Garrard, P., ed., Proceedings Forum Oil and Ore in Sediments: p. 189-205.

Ellwood, B.B., 1979, Particle flocculation: one possible control on the magnetization of deep sea sediments: Geophys. Res. Letters, v. 6, p. 237-240.

Fletcher, I.R., and Farquhar, R.M., 1977, Lead isotopes in the Grenville and adjacent Palaeozoic formations: Canadian Jour. Earth Sci., v. 14, p. 56-66.

Gerdemann, P.E. and Myers, H.E., 1972, Relationships of carbonate facies patterns to ore deposition and to ore genesis in the southeast Missouri lead district: Econ. Geol., v. 67, p. 426-433.

Giovanoli, F., 1979, A comparison of the magnetization of detrital and chemical sediments from Lake Zurich: Geophys. Res. Letters, v. 6, p. 233-235.

Halls, H.C., 1976, A least squares method to find a remanence direction from converging remagnetization circles: Geophys. Jour. Royal Astron. Soc., v. 45, p. 297-304.

Halls, H.C., 1978, The use of converging remagnetization circles in paleomagnetism: Phys. of the Earth and Planetary Interiors, v. 16, p. 1-11.

Hitchon, B., 1977, Geochemical links between oil fields and ore deposits in sedimentary rocks: Garrard, P., ed., Proceedings Forum on Oil and Ore in Sediments, p. 1-37.

Hunt, J.M., 1979, Petroleum geochemistry and geology: San Francisco, W.H. Freeman, 617 p.

Irving, E., 1979, Paleopoles and paleolatitudes of North America and speculations about displaced terrains: Canadian Jour. Earth Sci., v. 16, p. 669-695.

Jowett, E.C., 1977, Acquisition and retention of magnetization in modern lime-mud, ancient limestones and dolostones, and carbonate hosted sulphides: Univ. of Toronto, Unpubl. M.A.Sc. Thesis.

Jowett, E.C. and Pearce, G.W., 1977, Detrital remanent magnetization of modern lime muds from the Florida Keys: Abstract, Joint I.A.G.A./I.A.M.A.P.Meeting, Seattle.

Kisvarsanyi, G., 1977, The role of the Precambrian igneous basement in the formation of the stratabound lead-zinc-copper deposits of southeast Missouri: Econ. Geol., v. 72, p. 435-442.

Larsen, K.G., 1977, Sedimentology of the Bonneterre Formation, southeast Missouri: Econ. Geol., v. 72, p. 408-419.

Macqueen, R.W., 1979, Base metal deposits in sedimentary rocks: Some approaches: Geosci. Canada, v. 6, p. 3-9.

McAuliffe, C.D., 1979, Mechanics of secondary hydrocarbon migration-chemical and physical constraints: American Assoc. Petrol. Geol. Bull., v. 63, p. 761-781.

Ohle, E.L., 1959, Some considerations in determining the origin of ore deposits of the Mississippi Valley type: Econ. Geol., v. 54, p. 769-789.

Philippi, G.T., 1965, On the depth, time and mechanism of petroleum generation: Geochim. et Cosmochim. Acta., v. 29, p. 1021-1049.

Roedder, E., 1967, Environment of deposition of stratiform (Mississippi Valley-type) ore deposits, from studies of fluid inclusions: in Brown, J.S., ed., Genesis of Stratiform Lead-Zinc-Barite-Fluorite, Deposits in Carbonate Rocks: Econ. Geol. Mon. 3, p. 326-332, p. 349-361.

————————————, 1977a, Fluid inclusion studies of ore deposits in the Viburnum Trend, southeast Missouri: Econ. Geol., v. 72, p. 474-479.

————————————, 1977b, Fluid inclusions as tools in mineral exploration: Econ. Geol., v. 72, p. 503-525.

Rogers, R.K. and Davis, J.H., 1977, Geology of the Buick Mine, Viburnum Trend, southeast Missouri: Econ. Geol., v. 72, p. 372-380.

Sangster, D.F., 1976a, Sulphur and lead isotopes in stratabound deposits: in Wolf, K.H., ed., Handbook of Stratabound and Stratiform Ore Deposits: Elsevier, New York, v. 2, p. 219-266.

————————————, 1976b, Carbonate-hosted lead-zinc deposits: in Wolf, K.H., ed., Handbook of stratabound and stratiform ore deposits: Elsevier, New York, v. 6, p. 447-456.

Tissot, B.P. and Welte, D.H., 1978, Petroleum Formation and Occurrence: Springer-Verlag, Berlin Heidelberg, 538 p.

Verosub, K.L., 1977, Depositional and post-depositional processes in the magnetization of sediments: Rev. Geophys. and Space Phys., v. 15, p. 129-143.

————————————, 1979, Paleomagnetism of varied sediments from western New England: Variability of the paleomagnetic recorder: Geophys. Res. Letters, v. 6, p. 241-244.

Vinegard, J.D., 1977, Preface, to An Issue devoted to the Viburnum Trend, southeast Missouri: Econ. Geol., v. 72, p. 337-338.

Wu, Y. and Beales, F.W., 1980, A reconnaissance study of the age of mineralization along Viburnum Trend, southeast Missouri, by paleomagnetic methods: Econ. Geol., submitted ms.

(The "Viburnum Trend Issue" of Economic Geology, v, 72, no. 3, May, 1977 contains a recent description of regional and mine geology and an extensive bibliography.)

The Geological Association of Canada Special Papers

Orders and requests for information should be sent to:
Geological Association of Canada Publications,
Business and Economic Service, Ltd.,
111 Peter Street, Suite 509,
Toronto, Ontario M5V 2H1.